The
FUNGAL COMMUNITY

MYCOLOGY SERIES

Edited by

Paul A. Lemke

Department of Botany,
Plant Pathology, and Microbiology
Auburn University
Auburn, Alabama

The
FUNGAL COMMUNITY

ITS ORGANIZATION AND ROLE IN THE ECOSYSTEM

EDITED BY

Donald T. Wicklow
Northern Regional Research Center
Agricultural Research
Science and Education Administration
U.S. Department of Agriculture
Peoria, Illinois

George C. Carroll
Department of Biology
University of Oregon
Eugene, Oregon

MARCEL DEKKER, INC.   New York and Basel

Library of Congress Cataloging in Publication Data

Main entry under title:

The Fungal community.

    (Mycology series ; v. 2)
    Includes indexes.
    1. Fungi--Ecology. I. Wicklow, Donald T. [date].
II. Carroll, George C. [date]. III. Series.
QK604.F86   589.2'045   81-5433
ISBN 0-8247-6956-2   AACR2

MARCEL DEKKER, INC.
270 Madison Avenue, New York, New York 10016

Current printing (last digit):
10 9 8 7 6 5 4 3 2 1

PRINTED IN THE UNITED STATES OF AMERICA

Mycology is the study of fungi, that vast assemblage of microorganisms which includes such things as molds, yeasts, and mushrooms. All of us in one way or another are influenced by fungi. Think of it for a moment--the good life without penicillin or a fine wine. Consider further the importance of fungi in the decomposition of wastes and the potential hazards of fungi as pathogens to plants and to humans. Yes, fungi are ubiquitous and important.

Mycologists study fungi either in nature or in the laboratory and at different experimental levels ranging from descriptive to molecular and from basic to applied. Since there are so many fungi and so many ways to study them, mycologists often find it difficult to communicate their results even to other mycologists, much less to other scientists or to society in general.

This Series establishes a niche for publication of works dealing with all aspects of mycology. It is not intended to set the fungi apart, but rather to emphasize the study of fungi and of fungal processes as they relate to mankind and to science in general. Such a series of books is long overdue. It is broadly conceived as to scope, and should include textbooks and manuals as well as original and scholarly research works and monographs.

The scope of the Series will be defined by, and hopefully will help define, progress in mycology.

Paul A. Lemke

CONTRIBUTORS

Felix Bärlocher, Dipl. Sc. Nat., Ph.D., Privatdozent, Botanisches Institut, Universität Basel, Basel, Switzerland

Ralph Baker, Ph.D., Professor of Botany and Plant Pathology, Department of Botany and Plant Pathology, Colorado State University, Fort Collins, Colorado

R. J. Bandoni, Ph.D., Professor of Botany, Department of Botany, The University of British Columbia, Vancouver, British Columbia, Canada

George C. Carroll, Ph.D., Associate Professor of Biology, Department of Biology, University of Oregon, Eugene, Oregon

Martha Christensen, Ph.D., Professor of Botany, Department of Botany, University of Wyoming, Laramie, Wyoming

Kermit Cromack, Jr., Ph.D., Assistant Professor of Forest Science, Department of Forest Science, School of Forestry, Oregon State University, Corvallis, Oregon

David C. Culver, Ph.D., Associate Professor of Biological Sciences, Department of Biological Sciences, Northwestern University, Evanston, Illinois

Michael W. Dick, Ph.D., Senior Lecturer in Botany, Department of Botany, Plant Science Laboratories, University of Reading, Whiteknights, Reading, England

Paul Dowding, Ph.D., School of Botany, Trinity College, Dublin University, Dublin, Ireland

Ralph Emerson (deceased), *Ph.D.*, Professor of Botany, Department of Botany, University of California, Berkeley, California

Jack W. Fell, Ph.D., Professor of Biology and Living Resources, Department of Biology and Living Resources, Rosenstiel School of Marine and Atmospheric Science, University of Miami, Miami, Florida

Peter Jack Fisher, Ph.D., Department of Biological Sciences, University of Exeter, Exeter, Devon, England

Patrick W. Flanagan, Ph.D., Honorary Research Fellow, Institute of Arctic Biology, University of Alaska, Fairbanks, Alaska

Robert Fogel, Ph.D.,[*] Research Assistant, Department of Forest Science, School of Forestry, Oregon State University, Corvallis, Oregon

Juliet C. Frankland, Ph.D., Merlewood Research Station, Institute of Terrestrial Ecology, Grange-over-Sands, Cumbria, England

S. Gianinazzi-Pearson, D.Sc.,[†] Station de Physiopathologie Végétale, Institut National de la Recherche Agronomique, Dijon, France

V. Gianinazzi-Pearson, Ph.D.,[†] Station de Physiopathologie Végétale, Institut National de la Recherche Agronomique, Dijon, France

S. E. Gochenaur, Ph.D., Associate Professor of Biology, Department of Biology, Adelphi University, Garden City, New York

E. B. Gareth Jones, Ph.D., D.Sc., Reader in Biological Sciences, Department of Biological Sciences, Portsmouth Polytechnic, Portsmouth, England

Bryce Kendrick, Ph.D., D.Sc., Professor of Biology, Department of Biology, University of Waterloo, Waterloo, Ontario, Canada

K. A. Kershaw, Ph.D., D.Sc., F.R.S.C., Professor of Biology, Department of Biology, McMaster University, Hamilton, Ontario, Canada

M. J. Klug, Ph.D., Associate Professor, Kellogg Biological Station, Hickory Corners, and Department of Microbiology and Public Health, Michigan State University, East Lansing, Michigan

Y. Koltin, Ph.D., Professor of Microbiology, Department of Microbiology, Faculty of Life Sciences, Tel-Aviv University, Ramat-Aviv, Israel

Martin J. Lechowicz, Ph.D., Assistant Professor of Biology, Department of Biology, McGill University, Montreal, Quebec, Canada

Ching Yan Li, Ph.D., Forestry Sciences Laboratory, Pacific Northwest Forest and Range Experimental Station, U.S. Forest Service, Corvallis, Oregon

Gösta Lindeberg, Pn.D., Professor of Forest Mycology and Pathology, Department of Forest Mycology and Pathology, Swedish University of Agricultural Sciences, Uppsala, Sweden

John L. Lockwood, Ph.D., Professor of Botany and Plant Pathology, Department of Botany and Plant Pathology, Michigan State University, East Lansing, Michigan

---

Present affiliation:

[*]Assistant Curator and Assistant Professor, Herbarium and Division of Biological Sciences, University of Michigan, Ann Arbor, Michigan

[†]Station d'Amélioration des Plantes, Institut National de la Recherche Agronomique, Dijon, France

Robert D. Lumsden, Ph.D., Research Plant Pathologist, Soilborne Diseases Laboratory, Plant Protection Institute, Agricultural Research Science and Education Administration, Agricultural Research Station, U.S. Department of Agriculture, Beltsville, Maryland

John Lussenhop, Ph.D., Associate Professor of Biological Sciences, Department of Biological Sciences, University of Illinois at Chicago Circle, Chicago, Illinois

Barry J. Macauley, Ph.D., Department of Microbiology, La Trobe University, Bundoora, Victoria, Australia

Samuel J. McNaughton, Ph.D., Professor of Biology, Department of Biology, Syracuse University, Syracuse, New York

Dieter Mueller-Dombois, Ph.D., Professor of Botany, Department of Botany, University of Hawaii at Manoa, Honolulu, Hawaii

T. H. Nash III, Ph.D., Associate Professor of Botany, Department of Botany and Microbiology, Arizona State University, Tempe, Arizona

Donald O. Natvig, M.A., Department of Botany, University of California, Berkeley, California

Steven Y. Newell, Ph.D.,[*] Research Associate, Department of Biology and Living Resources, Rosenstiel School of Marine and Atmospheric Science, University of Miami, Miami, Florida

Lawrence H. Pike, Ph.D., Adjunct Assistant Professor, Department of Biology, University of Oregon, Eugene, Oregon

J. Remacle, Ph.D., Maître des Conferénces, Département de Botanique, Université de Liège, Sart-Tilman, Liège, Belgium

Richard P. Seifert, Ph.D., Assistant Professor of Biological Sciences, Department of Biological Sciences, The George Washington University, Washington, D.C.

L. Sigal, Ph.D.,[†] Research Assistant, Department of Botany and Microbiology, Arizona State University, Tempe, Arizona

R. N. Sinha, Ph.D., Principal Research Scientist, Research Station, Agriculture Canada, and Honorary Professor, Faculty of Graduate Studies, University of Manitoba, Winnipeg, Manitoba, Canada

Phillip Sollins, Ph.D., Research Associate, Forest Science, Oregon State University, Corvallis, Oregon

---

Present affiliation:
[*] Research Associate, Marine Institute, University of Georgia, Sapelo Island, Georgia
[†] Research Associate, Environmental Sciences Division, Oak Ridge National Laboratory, Oak Ridge, Tennessee

Judith Stamberg, Ph.D.,[*] Senior Lecturer in Genetics, Microbiology Department, Faculty of Life Sciences, Tel-Aviv University, Ramat-Aviv, Israel

William T. Starmer, Ph.D., Assistant Professor of Biology, Department of Biology, Syracuse University, Syracuse, New York

Jack S. States, Ph.D., Associate Professor of Biological Science, Department of Biological Science, Northern Arizona University, Flagstaff, Arizona

Keller Suberkropp, Ph.D.,[†] Assistant Professor of Biological Sciences, Department of Biological Sciences, Indiana University-Purdue University at Fort Wayne, Fort Wayne, Indiana

John S. Waid, Ph.D., Professor of Microbiology, Department of Microbiology, La Trobe University, Bundoora, Victoria, Australia

Henry A. H. Wallace, M.S., Adjunct Professor of Agricultural Engineering, Department of Agricultural Engineering, University of Manitoba, and Research Scientist Emeritus, Research Station, Agriculture Canada, Winnipeg, Manitoba, Canada

Roy Watling, Ph.D., F.R.S.E., Department of Agriculture and Fisheries for Scotland, Royal Botanic Garden, Edinburgh, Scotland

John Webster, Ph.D., D.Sc., Professor of Biological Sciences, Department of Biological Sciences, University of Exeter, Exeter, Devon, England

Donald T. Wicklow, Ph.D., Northern Regional Research Center, U.S. Department of Agriculture, Peoria, Illinois

Paul Widden, Ph.D., Associate Professor of Biology, Department of Biology, Loyola Campus, Concordia University, Montreal, Quebec, Canada

Chandrani Wijetunga, M.S., Research Assistant, Department of Botany and Plant Pathology, Colorado State University, Fort Collins, Colorado

----

Present affiliation:

[*]Department of Pediatrics, Johns Hopkins School of Medicine, and Genetics Department, John F. Kennedy Institute, Baltimore, Maryland

[†]Department of Biology, New Mexico State University, Las Cruces, New Mexico

What justification can there be for producing a book on fungal ecology? Learned volumes abound, written by specialists, for specialists, and probably read by none other than specialists. Is this to be the fate of this volume? It is certainly not the aim of the editors or the contributors. The aim indeed is nothing less than evangelical. Too long, say the editors, have the ecologists working with fungi and the ecologists working with other organisms labored in isolation and ignorance of one another. It is time they joined forces. By this fusion they enrich both mycology and mainstream ecology. The need for this fusion and the means whereby it might be achieved are clearly spelled out in the Preface and reiterated through the text. Quite simply, it could be put like this: Fungal ecology has a wealth of observation and a great potential for experimentation but a dearth of hypotheses; mainstream ecology has a superabundance of theory, much of which has proved difficult to test. From this book mycologists can learn to interpret their data, ecologists to substantiate their theory.

This is a bold claim and perhaps the editors may not thank me for so overstating it. But I believe there should be no apology for such boldness, for I see little danger of failure. The stimulus is bound to work. Those of us privileged to work in the International Biological Programme know how persuasive the atmosphere of integrated ecological work is. For better or worse we (and our subject) will never be the same again. This book, I suggest, is the next step in that educational process, a further juxtaposition of differing concepts and approaches which are surely bound to mix and interpenetrate. But this is not an easy process nonetheless. "Uncertainty" said Bertrand Russell (1946), "in the presence of vivid hopes and fears, is painful, but must be endured if we wish to live without the support of comforting fairy tales."[*] To venture beyond the relatively safe confines of one's own data, terminology, and conceptual framework is certainly to tempt the gods, not to mention the reviewers; but the attempt must be made. It is only by being bold enough to ask fundamental questions that we earn the rights to some of the answers.

Is this merger something of merely academic importance? No, for the main ecological significance of fungi lies in their quantitative importance in ecosystems as

---

[*] Russell, B. (1946). *A History of Western Philosophy*. Allen and Unwin, London.

decomposers and plant pathogens. All estimates show that in most terrestrial ecosystems fungi form a crucial bridge between the plants and the rest of the biotic
community, a bridge which is the pathway for the main flow of nutrients to and from
the plant. This in itself is surely justification enough for looking more closely
at the ecology of this group. But this has not till now been the case. Only compare the number of references to birds and to fungi in any standard ecological text
and then compare these two groups on the basis of their significance to almost any
aspect of the functioning and structure of communities and ecosystems; this will give
a measure of the present dire state of integration.

But even when the awaited marriage takes place, the story is only just begun.
While fungi offer tremendous potential for certain types of ecological investigation
their study is also beset by many practical problems. Any survey of the literature
of fungal ecology shows an obsession with methods. Many fundamental parameters of
fungal activity can still not be estimated with any precision. To paraphrase the
words of one of the pioneers of modern microbial ecology, S. D. Garrett (1952):
"What you can measure you can't identify and what you can identify you can't measure!"[*] But it is equally true that, as Burgess (1965) has said, "micro-organisms
are simpler than human beings"[†] and therefore lend themselves to investigation under
circumstances where other organisms are totally unsuitable.

This has proved of tremendous value in other aspects of science. If an up-to-
date microbiologist had been asked in the 1950s what the state of the art in microbial genetics would be in the 1980s his answer would have undoubtedly been horrendous-
ly wrong. Dare we hope that in the year 2000 the perspective of microbial ecology
will have changed equally dramatically and with equally far-reaching effects?

M. J. Swift

Department of Plant Biology and Microbiology
Queen Mary College
University of London
London, England

---

[*]Garrett, S. D. (1952). The soil fungi as a microcosm for ecologists. *Sci.
Progr. 40*: 436–450.

[†]Burgess, A. (1965). The soil microflora--its nature and biology. In *Ecology
of Soil-Borne Plant Pathogens*, K. F. Baker and W. C. Snyder (Eds.). Murray, London,
pp. 21–32.

We have organized this volume in response to a need for a single, adequately integrated, up-to-date book covering research pertaining to the ecology of fungi within the current conceptual framework of the ecological sciences. This book is designed for the mycologist who is being asked to become more directly involved in problems of ecological relevance and for the ecologist who finds himself examining fungal communities for the first time. Our aim is to present mycological ecology as a rational, organized body of knowledge and to promote the interfacing of mycological data with ecological theory. We considered that the most effective way of achieving this aim was to invite contributions from ecologists as well as mycologists. The ecologists were asked to introduce generalizations, definitions, and ecological ideas, while the mycologists were encouraged to provide relevant examples extracted from their own research or from the voluminous mycological literature. Both groups of researchers faced difficult tasks; the ecologist in assessing the potential value of fungi as tools for examining ecological ideas and the mycologist in identifying those aspects of mycological research that may be of importance to ecology. A synthesis of the two disciplines constitutes a challenge for the 1980s and beyond.

The editors acknowledge with gratitude the invaluable help of Gordon L. Adams, who superbly handled various editorial tasks, and Elsie Mumbach, who produced the camera-ready copy in its final form.

Donald T. Wicklow
George C. Carroll

CONTENTS

The
FUNGAL COMMUNITY

INTRODUCTION

Donald T. Wicklow

Northern Regional Research Center
U.S. Department of Agriculture
Peoria, Illinois

George C. Carroll

University of Oregon
Eugene, Oregon

In the past century, the realized niche of mycological activity has changed continually as mycologists have identified new research problems; funding agencies have established new priorities; and industry, government agencies, and universities have reevaluated programs. This niche has expanded greatly through the recognition of the importance of fungi as pathogens of plants and animals, agents in the production of fermented foods, and experimental tools for molecular and developmental biology. At the same time, we have seen in an increasing number of institutions a realignment of teaching/research groups to coincide with current operational definitions of the biological sciences (e.g., molecular biology; cell and developmental biology; ecology, systematics, and evolutionary biology). Organism group affiliation is becoming less important as a determining factor in staff recruitment. Instead, the contribution of the research to formulation of general principles and concepts in a given area of biology often determines whether bacteria, mushrooms, goldenrods, moths, parrots, or caribou are used as experimental systems.

Ecological research is directed toward a better understanding of the natural world and its function as a system. Fortunately for mycologists, fungi are important biotic components of ecosystems. The collaborative efforts of biologists in various biome studies (International Biological Program) have gone far in introducing mycologists to ecosystem process studies (e.g., energy flow; nutrient transfer), and ecosystem level process modeling is where fungal ecologists now appear to be making the greatest contribution to ecology. If, however, we examine the role that fungi have had in the development of principles and concepts of population and community ecology, we find few examples. Papers by "fungal ecologists" are rarely cited in general ecology texts or review articles concerned with central concepts in ecology. This is due in part to the zoological or botanical backgrounds of many ecologists and in part to the reluctance of mycologists to step outside their own literature and attempt generalizations on the significance of their observations and experimental results. In this country, for example, research proposals involving ecological studies with fungi are reviewed by ecologists. If the research objective of a proposal were primarily the description of fungal succession on an as yet uninvestigated type of litter, it would most probably generate little interest or support. However, if the observations were directed toward questions of community organization, colonization-extinction phenomena associated with the theory of island biogeography, or the mechanisms determining patterns of species replacement in successional seres, the proposal would have great appeal. Mycologists might argue that ecologists are welcome to examine published work on the fungi and use those results which might have ecological relevance. However, even if ecologists were so inclined, chance observations are usually of little value; such investigations ought to be designed to test particular hypotheses. At a time when ecologists working with large organisms are actively developing and testing general principles and concepts, it would be short-sighted of mycologists to avoid examining the fungi with similar objectives.

We would be remiss in suggesting that the problems described above are exclusive to mycology. Until recent years, there have been very few ecologists who have attempted to discuss their results in terms of cross-organism group phenomena. How often has the aquatic ecologist attempted to find common ground with results from studies of terrestrial systems? How many higher plant or animal ecologists discuss their results in connection with data generated by microbial ecologists? In developing this text, we have emphasized bridges between mycological ecology and general ecology theory (Table 1). The book has been organized to follow the format of a general ecology textbook, with emphasis on examples from the mycological literature. Several ecologists were invited as contributors to introduce generalizations and definitions pertinent to population and community ecology. Some provide their own assessment of the current status of fungal ecology with suggestions designed to promote a greater involvement of mycologists in examining ecological ideas.

The text begins (Part I) by examining the relative contribution of fungal ecology in generating ecological theory. In the first of these chapters, Seifert attempts to assess why this contribution has been minimal with regard to population and community ecology. He then identifies several theoretical concepts that should be amenable to study using fungal populations. Drawing primarily upon his experiences with the coniferous forest biome (USIBP), Carroll (Chap. 2) describes how mycological input has evolved from a peripheral to a central role in determining program direction, defining research objectives, and establishing priorities. Meaningful descriptions of how fungal communities are organized depend ultimately on the method(s) of analysis used to determine what is potentially an interacting group of fungi. In Chap. 3, Lussenhop emphasizes the importance of sampling at the scale of fungal interaction and resource use before attempting to provide biological explanations for patterns of species abundance. Wicklow (Chap. 4) expounds the use of coprophilous fungal communities in ecological research by illustrating some of the unique features of the dung microcosm that make it a preferred system for experimental manipulation and hypothesis testing.

The local abundance of fungal species and the organization of communities is determined by how individual populations subdivide a space and its contained resources. Part II of this volume focuses on the theory of the niche and the competitive

Table 1 Mycological Data and Ecological Organization

		Levels of organization for ecological research
		Biome--biogeography
		Ecosystem--structure; function (processes; evolution)
Original contributions to ecological theory	Results from experiments designed to test principles and concepts of ecology based upon data from studies of organism groups other than fungi	Community--composition; interactions (within or among communities)
		Population--characterization; growth and regulation; natural selection; evolution; realized vs. fundamental niches
		Organism--adaptation to the environment

Mycological data base

Herbarium and culture collection records

Natural history observations

Phenological data: seasonal patterns of fungal importance/abundance and regulation of life history stages

Physiological studies: resources and regulators; environmental cues; fungal response

Reports on the distribution and dynamics of fungal populations in native and man-managed ecosystems: terrestrial and aquatic fungi in decomposition processes; plant pathogenic fungi; mycorrhizal fungi; entomophagous fungi; mycotoxigenic fungi; domesticated fungi

Fungal population genetics: adaptation to environment

Biodeterioration studies: energy flow; detritus pathway; secondary productivity

Biogeochemical cycling: element/nutrient transfer

Fungal biochemistry: products of decomposition; secondary metabolites

exclusion principle. McNaughton's introduction (Chap. 5) traces the development of the niche concept, examines different approaches used in niche characterization, and identifies several current hypotheses that could be effectively tested using fungi. The next three chapters examine the physiological/biochemical adaptations of selected fungal populations which determine their fundamental and realized niches. Lechowicz (Chap. 6) thoroughly characterizes the niche of several lichens in an effort to answer important questions about the adaptive significance of a species fundamental niche. The adaptations of zoosporic fungi to aquatic environments that are low or lacking in oxygen and high in carbon dioxide, extreme environments for most fungi, are reviewed by Emerson and Natvig (Chap. 7), who emphasize the importance of rigorous laboratory studies of fungal physiology in order to accurately characterize the

fundamental niche of an organism. Starmer's contribution (Chap. 8) compares the
realized niche with the fundamental niche of a group of yeasts isolated from the
decaying stems of giant columnar cacti. The metabolic capabilities of yeasts, the
chemical environment of their habitats, and various dispersal agents (*Drosophila*
spp.) are shown to have a role in determining yeast distribution. The ability of a
fungal population to rapidly adapt to environmental change can determine the dimensions of both its fundamental and its realized niche. In the concluding chapter of
Part II, Stamburg and Koltin (Chap. 9) theorize as to how different mechanisms controlling the amount of genetic heterogeneity in a population convey selective advantage to fungi inhabiting either a stable or a changing environment.

Part III deals primarily with quantitative methodology used in measuring and
describing how fungal populations and communities are organized. The introductory
chapter by Mueller-Dombois (Chap. 10) reviews different approaches to community analysis with specific reference to problems in analysis created by the distinctive properties of soil microfungi. In Chap. 11, States considers some of the morphological
and physiological attributes of soil microfungi that account for the prevalence of
certain groups in different ecosystems. He also describes a simple graphic technique
that has been used to compare the soil fungi of different plant communities by illustrating the proportional representation of species in several major taxonomic units.
Next, Christensen (Chap. 12) provides an important synthesis of the literature pertaining to the global distribution of soil microfungal populations by quantitatively
comparing patterns of fungal dominance and diversity at the ecosystem level. Principal component analysis (PCA) is examined by Wallace and Sinha (Chap. 13) as a means
of screening ecological data to identify the important biotic and abiotic variables
determining fungal distribution and abundance in cereal grain agricultural ecosystems. The subject of population growth and regulation in fungi is examined by Baker
and Wijetunga (Chap. 14), who focus primarily on the mathematical description of
fungal growth, propagule formation, and survival of soil-borne plant-pathogenic
fungi. Fluctuations in population numbers through time and mechanisms regulating
population size are important to an understanding of how coexisting populations of
fungi compete for and process shared resources. The final contribution in Part III
(Chap. 15 by Dick) is also concerned with the resources and regulators that influence
dynamics of fungal populations. However, Dick examines factors controlling the distribution and abundance of Oomycetes in a lentic environment.

Important to the organization of any community are the biotic interactions
among individual species populations. Part IV examines the various types of fungal
interactions that structure fungal communities. An introduction to the theory of
species interactions is presented by Culver (Chap. 16). Because fungi have evolved
numerous important mutualistic associations among themselves and with other organism
groups, Culver recognizes their potential utility in field and laboratory studies
designed to investigate the validity of the theory. In Chap. 17, by Lumsden, attention is directed to studies dealing with the ecology of mycoparasitism (fungal pre-

dation), with particular reference to the potential for mycoparasite regulation of populations of soil-borne plant-pathogenic fungi. The next two chapters in this section focus on types of competitive interactions that structure fungal communities (e.g., exploitation competition and interference competition). Lockwood (Chap. 18) reviews studies on mechanisms of exploitation competition where one or more populations of soil microorganisms deplete the resource pool of a fungal competitor. Mechanisms of interference competition are reviewed by Wicklow in Chap. 19. The conceptual view is presented that fungal antagonism is an important mechanism of species replacement in fungal succession. Wicklow contends that secondary invaders capable of producing antifungal antibiotics might free nutrients concentrated in the cells of initial colonists. Mycologists have only recently begun to assess the importance of protozoan and microarthropod predators in limiting the population size and distribution of fungal competitors, and there is not yet a sufficient body of literature to justify a review. By identifying this as an important deficiency in our understanding of how fungal communities are structured, we hope to encourage multidisciplinary research in this area.

Part V examines processes of fungal community development (succession) associated with the colonization and degradation of individual substrates and the changes in fungal community structure that occur in response to ecosystem maturation or disturbance. Kershaw's introduction (Chap. 20) emphasizes the importance of studying community patterns and processes within a multivariate ecological framework, recognizing both the physical aspect of pattern and the temporal component. It is important that mycologists not confuse phenological changes with patterns of increasing population densities that occur with successional age. In Chap. 21, Widden examines the response of saprophytic fungal populations to changes in seasonal climatic factors (phenology). The various mechanisms by which fungal populations replace one another on individual substrates or during ecosystem maturation are explored by another on individual substrates or during ecosystem maturation are explored by Frankland (Chap. 22). Frankland's contribution represents a major step forward in integrating ideas about mechanisms regulating fungal successions with successional theory generated by ecologists studying large organisms. Numerous mycorrhizal fungi in Britain were apparently able to switch to alternative plant hosts (*Salix* spp.) following catastrophic ecosystem disturbance centuries earlier. Watling (Chap. 23) illustrates that one can document the former vegetational history of a disturbed area by its present assemblage of macrofungi. The final two chapters of this section consider the response of fungal communities to perturbation. In Chap. 24, Gochenaur contrasts patterns of soil microfungal community organization in disturbed and undisturbed communities. The various morphological and physiological attributes of individual species populations, which convey selective advantage in a disturbed habitat or enable a fungus to survive a catastrophic disturbance, are considered. Nash and Sigal (Chap. 25) provide a summary of research documenting changes in lichen communities following short- or long-term exposure to specific air pollutants.

Parts VI and VII deal with the role of fungi in ecosystems, with so-called process studies. The estimation of fungal biomass, conversion efficiencies, and turnover rates have posed persistent, intractable problems for the microbial ecologist. Any quantitative assessment of fungal activities in ecosystems requires that these variables be known rather accurately, and such accuracy is achieved only with great labor. Part VI is devoted to the question of fungal biomass and productivity estimation for several very different groups of fungi in several different habitats. The initial chapter by Macauley and Waid (Chap. 26), although directed specifically toward leaf-inhabiting fungi, provides an admirable overview of much of the recent literature on methods and their application. Chapter 27 by Pike should also prove of interest to a broad audience; in his discussion of the estimation of lichen biomass and production, Pike reviews and makes accessible a body of information on the statistical treatment of ratios, a subject ignored not only by microbial ecologists but by classical ecologists as well. Since data from both the field and laboratory are frequently taken and/or expressed as ratios (e.g. kilograms/hectare, units activity/gram of soil, or grams/milliliter), this chapter may well be counted among the most significant in the book. In the next contribution, Fogel (Chap. 28) deals with problems in sampling populations of large fungal fruiting bodies which are, however, hidden from view: those of hypogeous Ascomycetes and Basidiomycetes. These results are considered from a phenological point of view and should be compared with those of Widden presented in Chap. 21. Flanagan (Chap. 29) treats the estimation of fungal biomass and description of community structure in several arctic and temperate soils and attempts to relate this information to the process of decomposition. This contribution, then, serves as a bridge to Part VII.

In Part VII, processes mediated by fungal populations in ecosystems are discussed. Remacle (Chap. 30) reviews all the various roles fungi may play in elemental cycling and suggests approaches whereby some of these activities may be quantified. In Chap. 31, Sollins, Cromack, Fogel, and Li present recent, attractive hypotheses on the role of organic acids in elemental uptake (notably phosphorus) by fungi. These ideas have not received wide circulation among mycologists and deserve serious consideration. Because of their close association with roots of vascular plants, mycorrhizal fungi have long been recognized as important factors in plant mineral nutrition. A report by Dowding follows (Chap. 32) in which nitrogen and phosphorus uptake and allocation were determined for *Chaetomium globosum* in a series of ingenious microcosm studies. The next two chapters in this section consider the activities of mycorrhizal fungi in vegetated ecosystems. Pearson-Gianinazzi and Gianinazzi (Chap. 33) provide current insights on the role of endomycorrhizae in phosphorus uptake. Lindeberg (Chap. 34) reviews and updates a large and illustrious literature on nitrogen cycling by litter-decomposing and ectomycorrhizal fungi in temperate coniferous forests. The final contribution in Part VII by Fell and Newell (Chap. 35) examines the role of fungi in affecting both energy and nutrient transfers during the decomposition of coastal marine litter. Efforts to develop realistic models of the decomposition process depend on an understanding of the interaction between carbon flow, nutrient immobilization, and the population dynamics of heterotrophic microorganisms.

Finally, Part VIII was intended as an example--a demonstration of what might be accomplished when investigations on both community organization and nutrient- and energy-cycling processes were conducted with a single ecologically circumscribed group of fungi, the aquatic hyphomycetes. In retrospect it has become apparent that process studies with these fungi are far advanced whereas focused studies on community organization such as those described by Wicklow in Chap. 4 have barely begun. What emerges from the early chapters of this section are elegant descriptions of fungal natural history--species lists with habitat preferences and details of conditions for growth and sporulation. With so much of the basic taxonomic and physiological information on hand, this should prove an attractive group for community ecologists in search of an experimental system.

In soliciting contributions for this volume, we have found some lacunae, i.e., gaps in the proposed outline for which information and/or contributors were unavailable. Prime among these are studies on the genetic variability actually present in natural populations of fungi. Although isozyme analysis has been widely used to assess such variability in higher plants and animals, similar work on fungi has, to our knowledge, not yet been done. A second broad area which has been slighted is that of interactions among fungi and insects. While mycorrhizal and lichen symbioses have received their due, fungus-insect symbioses have been virtually ignored (although see Seifert, Chap. 1, for discussion of a theoretical ecologist's view of these associations). The question of consumption of fungal biomass by insects and other animals has been considered here only for the aquatic hyphomycetes, although few would doubt the importance of this process in soils and other habitats where fungi occur.

We trust that this text will serve as an important reference for contemporary mycologists in that they will be able to determine, through the organizational framework and examples provided, where their current research might have direct relevance to ecological theory. On the other side, ecologists can turn to the text for an assessment of the potential of fungi as tools for examining ecological ideas, for examples of the types of ecological studies being conducted by mycologists, and for an understanding of the unique properties of fungal organism groups in different ecosystems and microcosms. The late Professor Sparrow, an eminent mycologist, once observed "considerable thought was given to attempting here the application of the terminology of classical ecology as has recently been done for the microorganisms by Brock." He continues: "The importance to science of precision in usage of terms and the danger here of misapplying them discouraged for the most part their use in this account." We believe that we are not alone in thinking that the time has come when fungal ecologists should attempt to develop the conceptual content of their subject and to do so within the current framework of the ecological sciences.

The eventual contribution of "fungal ecology" to the development of ecological theory will depend in large part on whether (1) "fungal ecologists" are willing to ask more penetrating questions of their data while discussing their results within the broader context of cross-organism group phenomena; (2) fungi as an organism group provide a number of satisfactory experimental tools with which the "fungal ecologist" can test and clarify specific problem areas of ecological theory or principle.

REFERENCES

Sparrow, F. K. (1968). Ecology of freshwater fungi. In *The Fungi*, Vol. III: *The Fungal Population*, G. C. Ainsworth and A. S. Sussman (Eds.). Academic Press, New York, pp. 41-93.

Brock, T. D. (1966). *Principles of Microbial Ecology*. Prentice-Hall, Englewood Cliffs, N.J.

Part I

FUNGAL ECOLOGY AND ECOLOGICAL THEORY

Chapter 1

APPLICATIONS OF A MYCOLOGICAL DATA BASE
TO PRINCIPLES AND CONCEPTS
OF POPULATION AND COMMUNITY ECOLOGY

Richard P. Seifert

The George Washington University
Washington, D.C.

I. INTRODUCTION

The impact of studies of fungi on ecological theory has been minimal. Mycological studies have not generated theories in population and community ecology, and mycological data have seldom been used to test existing ecological hypotheses derived from mathematics or empirical field studies on other, more well-studied groups such as vertebrates (particularly birds and mammals) and flowering plants. Studies on the ecology of fungi have typically been relegated to mycological journals and have not received wide attention by ecologists. Similarly mycologists studying population and community processes have been interested primarily in descriptive studies, such as descriptions of the succession of fungi on leaves, and have not focused their experiments on more recently generated ecological ideas. Quantitative methodology has also been lacking in most ecologically based mycological studies.

We can see how little mycology has influenced ecology by examining the amount of mycological literature published in major ecological journals. Over a 10-year period, from 1966 through 1975, *The American Naturalist, Ecology, Ecological Monographs,* and *The Journal of Ecology* have all had noticeably few mycological papers. *The Journal of Ecology* devoted the most proportional space to mycological studies. Mycological papers made up 1.31% of the articles in *The Journal of Ecology* over the 10-year period. (Papers which considered several aspects of soil biology were included when a considerable amount of the paper examined fungi. Papers which had only brief references to fungi and were primarily devoted to other aspects of soil biology were not included.) *Ecological Monographs* had the poorest record and did not publish a single mycological paper over this time. *Ecology* and *The American Naturalist* had intermediate records: 0.85% of the articles published in *Ecology* were on fungi, while only one article (0.16%) in *The American Naturalist* dealt with some aspect of fungi. A similar situation occurs for the *Annual Review of Ecology and Systematics* which, from 1970 through 1976, published only one (0.87%) paper on fungi. The *American Naturalist* article (Levin, 1971) was not primarily a study of fungi but was a general discussion of plant phenolics and included some of the mechanisms by which plants reduce fungal damage. The record in *The American Naturalist* is perhaps the most telling. This journal states that it "publishes research papers and reviews in which theoretical interpretation and synthesis are predominant...." (See the inside front cover of any issue in 1976.) To the extent that mycological papers with general import to theoretical population and community ecology exist, they have not been published in the leading journals devoted to theoretical ecology and evolution.

The situation is not better if we look at some recent ecology textbooks and determine what percent of the references cited are mycological papers. Of recent general ecology texts by Odum (1971), Krebs (1972), Colinvaux (1973), McNaughton and Wolf (1973), and Ricklefs (1973) only 0.76%, 0.46%, 0.31%, 0.78%, and 0.38% of the references, respectively, are devoted to fungi. It appears that fungal ecology has made little impact on either writers of ecological textbooks or on students of ecology.

There are problems here in determining cause and effect. Are papers on fungi rarely cited in general ecology textbooks because they are seldom published in general ecological journals or are they rare because mycology has little to offer ecology? These are not mutually exclusive propositions. References to mycology are rare because mycologists appear to have expressed little interest in the growth of populations and the integration of ecological communities, but this need not be a continuing phenomenon. In texts, fungi are almost always discussed in relationship to decomposition (and these discussions are typically ecosystem oriented rather than population or community oriented) or in relationship to symbiosis. The symbiotic discussions revolve around mycorrhizae (often discussed with reference to ecosystem processes) or around lichens. It is remarkable that ecological texts do not discuss fungi when discussing interference competition. The antibiotic ability of fungi to stop growth of other organisms even in situations where resources are not limiting has long been known (Waksman, 1941; Waksman and Horning, 1943; Brian, 1951; Broadbent, 1966). The mechanisms of competition among fungi are quite varied. For example, Dennis and Webster (1971) have shown how fungal competition may include the physical coiling of *Trichoderma* hyphae around foreign hyphae as well as use of antibiotics and extracellular enzymes. A prime example of interference competition is given by the fungistatic action of heather (*Calluna vulgaris*) which kills the mycorrhizae of birch (*Betula pendula*) and spruce (*Picea abies*) (Robinson, 1972). Under these conditions, birch and spruce plants are stunted and eventually die, while heather remains dominant. This example clearly demonstrates both interference competition (here produced by heather rather than fungus) and the importance of the symbiotic mycorrhizal associations. It is unfortunate that studies such as this are not presented in ecological texts more frequently.

II. PROBLEMS

Fungi as a group show several characteristics which make them difficult to use for the purpose of studying population and community ecology. First, the nature of an

individual is never clear and it may be difficult to determine the number of individuals in a given location. Both in practice and in principle the individuals of
only a few fungal taxa can be recognized (e.g., marine yeast, holocarpic chytrids
on pollen; zoospores of Saprolegniaceae; perennial mycelia of certain individual
basidiomycetes which can be recognized by incompatibility reactions in culture;
Phellinus wierii, Schizophyllum commune). Since the number of individuals cannot
often be determined, the term *population* also has little meaning. Similarly, the
term *generation* has a somewhat special meaning, indicating the time from the occurrence of sporulation to the next occurrence of sporulation (spores can be either
sexual or asexual). Generation time may be exceedingly long, although the size of
the organism (area covered) may increase dramatically within this time period. These
problems all make it difficult to estimate certain parameters traditionally considered important in population ecology, including the intrinsic rate of natural increase
(r), the carrying capacity of the environment (K), and the interaction coefficients
(α). Estimates of exponential population growth rates, for example, which take into
account only asexual growth actually overestimate the parameter r. Nonetheless, such
estimates may have use, particularly if estimations of asexual growth rates are made
over a variety of species. Estimations of the parameter K encounter even more severe
problems since the number of individuals at maximum densities is also very difficult
to determine for most fungal taxa. What can be determined is the maximum amount of
cover in the absence of competing species; this measure is directly analogous to the
K of population growth theory. Estimations of α are beset by the same problem of
expressing growth rates in terms of individuals. Nevertheless, a close analogy is
some measurement of the cover lost by species i when in the presence of species j
relative to that lost by species i in conflict with another colony of species i.
Under these conditions r, K, and α may have the same meanings that they do in the
Lotka-Volterra equations, and the dynamics of a two-species fungal system could be
studied by the usual procedures of population ecology.

The inability to distinguish the number of individuals of each species in a
community leads to problems in the study of community ecology. The standard methods
of estimating species diversity in a community (such as the Shannon-Wiener or Bril-
louin information indices or the Simpson diversity index) all depend on not only the
number of species in a community (or sample) but also their relative abundances.
Since the bases for these indices are not biological but rather come from information
or probability theory (Pielou, 1969), one could use area covered per species or proportion of plates on which a given species germinated as equivalences of species
abundance. The diversity indices so computed would not be species diversity indices
in the strict sense but rather represent some sort of cover or germination diversity.
Such measures could be used in comparative community studies, just as they are used
in nonfungal community ecology (Lloyd et al., 1968; Sanders, 1968; Poulson and Culver, 1969; Lussenhop, 1973; Abele, 1974). For example, one could compare the diversity of fungal communities in pre- or postburn prairie soil (Wicklow, 1973) or see if

there exists a consistent relationship between the ascomycete and basidiomycete diversity in several types of communities.

One of the most frequently expressed complaints in studying fungal communities
is that all the species in a community are never known. Studies of soil fungal communities in particular are dependent on plating techniques which may leave a large
number of species unobserved. For this reason some workers have attacked fungal
communities where there is a high probability of obtaining all the species, such as
fungal communities on dung (Wicklow and Moore, 1974; Angel and Wicklow, 1975). This
problem is not unique to mycology and faces ecologists studying plant, insect, or
vertebrate communities. The term *community* has been used in a variety of ways (Wilbur, 1972), but seldom are all species involved in a given study. For example, tropical insect communities living in floral parts of *Heliconia* may have as many as 50
species (Seifert, 1975), although interesting aspects of the community interactions
are found when as few as six species are studied (Seifert and Seifert, 1976). Experimental studies cannot handle large numbers of species, and typically they don't.
If the organisms involved are animals, the vagility of certain species, the times of
diapause of others, and the rarity of still others may limit the number of species
found in a series of samples to less than the total number which actually exists in
the community. Repeated sampling over periods of years may be necessary to obtain
most of the species (Preston, 1948, 1962a,b). Both spatial and temporal separation
of species render complete collections exceedingly difficult and may contribute to
the inability to record all the species in a community.

Although the proportion of fungal species collected at a given site may be
small compared with the total number of species present, so long as collecting methods
are standardized, species lists could be compared over a variety of locations and
experiments. I see no a priori reason why one should not expect to collect similar
proportions of the total number of species in all locations or at least in all locations with certain similar characteristics. Unless one can generate reasons for sus-
pecting that one should collect a higher proportion of the species which exist in,
for example, oak forests, species lists, even though incomplete, can be used to compare communities at least in a general way. Such lists could be used to analyze
species richness in tropical versus temperate communities, or on islands versus mainland locations, or to relate area to species richness [as Strong and Levin (1975)
have done; their work is discussed in the next section].

III. PROSPECTS

There are several areas in community and population ecology and evolution in which
the study of fungi can be important. An area of current interest in population and
community ecology which could be fruitfully approached through fungi is the relationship of colonizing ability to competitive ability. A variety of workers have described ways by which competitively inferior species can survive if the environment

contains constantly renewable patches of satisfactory habitats (Hutchinson, 1951; Levins and Culver, 1971; Horn and MacArthur, 1972; Slatkin, 1974; Levin, 1974). Fungi must be among the most appropriate organisms for experimental demonstration of the relationship between competition and colonizing ability since many fungal species are dependent on patchy, renewable environments. Armstrong (1976) has published a stimulating paper on the relationship of *Aspergillis* and *Penicillium* in a laboratory system. He found that *Aspergillus* was a better competitor and would exclude *Penicillium* when both were present in a limited environment. Similarly, if occasional extinction occurred on all plates at the same time and the plates were inoculated with spores from the previous generation, *Aspergillus* would exclude *Penicillium*. However, when there was a repeated extinction of a few plates and inoculation occurred from differently aged plates, *Penicillium* could be maintained in the system. The continued existence of *Penicillium* depended on the colonizing ability of *Penicillium* and the temporal nature of new habitats rather than strictly the competitive ability of the two species.

In natural communities similar situations may exist. Swift (1976) has suggested that *Coriolus versicolor* and *Stereum hirsutum* occupy similar niches on branches and that their coexistence in the field probably depends on the patterns of branch colonization by the two species. Field work necessary to support this hypothesis could be undertaken. Mycologists have studied a variety of successional processes in a descriptive manner (Webster, 1956; Hering, 1965; Frankland, 1966; Hogg and Hudson, 1966). These descriptions could be followed by studies on the relationship of colonizing ability and competitive ability. Such studies could include: (1) determining if species coexistence or extinction will occur when a constant supply of resource is available; (2) examining species coexistence when constant eradication of some of the resource occurs; (3) determining the factors which influence the rapidity of colonization, including the relationships of spore size and spore number.

Fungi are appropriate organisms for study of other biogeographic problems. Strong and Levin (1975) have compared the species richness of fungi on 23 species of British trees and found that fungal species richness increased as the distributional range of the tree species increased. They did not find that introduced tree species had a depauperate flora: tree species accumulate fungal species rapidly after their introduction. The work of Strong and Levin bears on two important aspects of evolutionary and ecological theory. First, these results imply that evolutionary time has little to do with the number of species in a community since trees which have been in England for only a short time have a similar number of fungal species as those trees with similar distributions which have been in England for long periods of time. This is contrary to models which have proposed that differences in species richness in areas are dependent on long-term evolutionary processes rather than short-term ecological processes (Richards, 1952; Fischer, 1960; Southwood, 1961). Second, some aspects of general island biogeographic theory (MacArthur and Wilson, 1967) were shown to be applicable to ecological islands. Insect-host plant

relationships had already been shown to follow portions of island biogeographic
theory (Opler, 1974; Strong, 1974a,b; Seifert, 1975), but the work of Strong and
Levin has extended ecological island analyses into an additional kingdom.

Fungi should be amenable to study of biogeographical problems involving area
relations or latitudinal variations. Diversity of fungal flora may be related to
differences in latitude. Tropical forests may have more fungal species than do
temperate forests, and fallen leaves in tropical regions may have more species than
do fallen leaves in temperate regions. This would be consistent with results from
other organisms (Pianka, 1966; MacArthur, 1969, 1972; Slobodkin and Sanders, 1969),
but the quick cycling of leaves (acceleration of leaf utilization by fungi) in the
tropics may mean that there are actually few species of fungi per species of leaf.
Thus, not only may the quantitative aspects of species interactions change, but the
qualitative nature of community structure may change as a consequence.

One can also ask if there are changes in reproductive strategies with latitudes.
Some reports suggest that temperate organisms should show a higher reproductive rate
than tropical organisms (r selected) while tropical organisms should be less fecund
but more able to survive at high densities (K selected) or be able to respond to
competitive pressures (α selected) (Dobzhansky, 1950; MacArthur and Wilson, 1967;
Slobodkin and Sanders, 1969; Gill, 1974). Comparative studies on reproductive rates
of fungi (again obscured by the problem of generation time) in different latitudes
might well lead to profitable insights into the processes of selection and competition among fungi.

Symbiotic relationships are among the most important interactions of fungi
with other organisms. However, symbioses are seldom discussed in connection with
the general theory of population and community ecology because these relationships
cannot regulate population densities, i.e., symbioses are destabilizing (May, 1973;
Seifert and Seifert, 1976). For example, in an obligate association of algae and
fungi, increases in algae should allow the fungal population to increase while an
increase in the fungal population should allow the algae to increase. These increases cannot go on continuously but will be limited by some other factor such as
the lack of space, poor humidity, or lichen destruction by herbivores. Population
regulation in communities is thought to occur through competition (Levins, 1968;
Vandermeer, 1970), predation (Hairston et al., 1960; Paine, 1966; Slobodkin et al.,
1967), or climate factors (Andrewartha and Birch, 1954).

Even though symbiosis is not thought to regulate ecological communities, one
must be careful to reserve judgment on the amount of symbiosis which occurs in
nature. Clearly, the mycorrhizal associations of trees and fungi seem to be important in forest structuring (Harley, 1968; Marks and Kozlowski, 1973), and one competitive mechanism which trees use to reduce interactions with other tree species
is based on the inhibition of mycorrhizal development (Robinson, 1972). The extent
to which symbiosis (including mutualism and commensalism) is important in the interactions of organisms is probably underestimated. Symbiosis may be important in

structuring tropical communities and allowing them to be initially resistant to
perturbations (Futuyma, 1973). How fungal symbiotic relationships are limited may
reflect on the limiting processes of other symbiotic relationships.

Symbiotic relationships of fungi and insects could profit from further study,
both in an ecological and an evolutionary sense. These relationships are found in
several insect groups and primitive insects and fungi probably have shown symbiosis
since the Devonian period (Kevan et al., 1975). Ambrosia beetles (families Scolytidae, Paltypodidae, and Lymexylomidae) construct galleries in the wood of trees,
carry fungal spores in a specialized pouch, and inoculate the galleries with these
spores. The fungus becomes the food for developing larval beetles. In some cases
fungi and beetles are obligate mutualistic symbionts (Batra, 1963; Batra and Batra,
1967), whereas in others reproduction is earlier among those adult beetles which
feed on fungi than among those without fungi (Kingsolver and Norris, 1977).

Another apparently symbiotic relationship is shown by beetles in the genus
Gymnopholus (Curculionidae; Leptopiine) which live in cloud forests in New Guinea
and have fungi growing on their elytra (Gressitt, 1969). The beetles are peculiarly
adapted to retain these fungi: their elytra are fused and the beetles do not fly;
the bodies of the beetles have grooves and pits where fungi (as well as vascular
cryptogams) can grow; and projecting pegs and scales on the bodies decrease the
probability that the fungi will be scraped off when the insect moves. Orbatid mites
often live in and feed on the fungal mass. While the mites are detrimental to the
fungus, they may actually be beneficial to the beetles since the feeding patches
give the beetle a more mottled appearance which confers additional camouflage on the
beetle. The dispersal abilities and patterns of these species are unknown, as is
the extent to which the species can live separately.

Perhaps the best-known mutualistic relationships of insects and fungi involve
the fungus-growing or leaf-cutter ants (Attini) and also the fungus-growing termites.
Leaf-cutter ants collect leaves and flowers, deposit them in underground cavities,
and then raise fungi on the plant material. The ants then feed on this fungus.
This ant-fungus relationship seems to be obligate: the fungus can seldom be cultured in the lab and is not found in nature outside ant colonies, while ant colonies
deprived of the fungus die (Weber, 1966; Wilson, 1971).

Although the leaf-cutter ants are probably better known as tenders of fungal
gardens, in some respects the fungus-growing termites (tribe Macrotermitinae) represent a more interesting evolutionary situation. Fungus-growing termites are restricted to tropical Africa and Asia and are most commonly associated with fungi in
the genera *Termitomyces* and *Xylaria* (Wilson, 1971). Within the Macrotermitinae
there is variable dependence on the fungal associates. The most primitive members
of this tribe maintain fungi but can be cultured in the lab in the absence of any
fungus. These primitive species (genera *Pseudacanthotermes* and *Ancistrotermes*)
appear to grow the fungus in association with fecal cartons (Sands, 1960; Wilson,
1971), and the fungal gardens may play an important role in the ventilation system

of large nests. More recently derived members of the tribe (*Macrotermes*, *Odonto-termes*, and some *Microtermes*) cannot survive without the fungus, whereas the most advanced genera either feed facultatively on fungi (some *Microtermes*) or do not feed on fungi at all (*Sphaerotermes*). In all of these genera symbiotic protozoans, char-acteristic of the majority of termites, do not exist. Fungi and bacteria are used to break down cellulose. Notice that this system represents a set of evolutionary sequences of feeding behavior. The initial association of the fungus and termites, though not obligate, appears to have been a preadaptation to subsequent obligate feeding associations.

A final category of mutualism between insects and fungi involves some species of bugs in the family Coccidae (particularly *Aspidiotus osborni*) and fungi in the genus *Septobasidium*. Here, the interactive system is the reverse of those previous-ly discussed: the insect provides food rather than receiving it, and the fungus provides shelter (Couch, 1931; Batra and Batra, 1967). Young coccids which invade new host plants carry with them innoculum of the fungus. This grows as a protective covering over the insect and over any other coccids which might join the colony. The fungal covering protects the insects from desiccation as well as from parasit-oids. At least some of the coccids in a colony are invaded by fungal haustoria, and the fungus derives its nutrition through the coccids. The invaded coccids die with-out reproducing. Most of these colonies persist over more than one season, and some of the individuals in the colonies seem to be mothers and their daughters. This remarkable situation presents an interesting and unusual series of evolutionary and ecological processes for study. For such a system to have evolved, at least some of the individual coccids which are protected by the fungus must be close relatives of those coccids which do not reproduce. Here some coccids must be invaded by the haustoria before fungal growth can occur, so new colonies should be started primar-ily by groups of related coccids. If this were not the case, there would be no selective advantage to the coccids in being invaded by haustoria and an individual coccid would have a higher probability of passing genes by being exposed to insect parasitoids and climatic fluctuations than by being protected but parasitized by the fungus. However, if fungal parasitism insures that the relatives of the parasitized individual will survive, selection can favor such an individually disadvantageous behavior. To my knowledge, the importance of kin selection in this coccid and fun-gus interaction has never before been suggested. If kin selection is operating in this case, it would be just about the only such instance among the nonsocial insects. The importance of kin selection versus natural selection in field situations is not well understood (Williams, 1966; Wilson, 1975) and could profit from further exami-nation.

All of the preceding insect-fungus interactions involve coevolutionary pro-cesses which, with few exceptions, have largely been neglected until recently (Gil-bert and Raven, 1975). A mycologist studying insect-fungus mutualism would have a variety of potential problems for study including (1) the evolution of such an association; (2) the rates of evolution of the fungi with respect to the rates of evolution of the insects; (3) the relative rates of speciation; (4) the particular use of the fungus by the insect (food versus shelter); (5) the transportation of the fungus; (6) the biogeography of the association; (7) the obligate nature of the association in relationship to environmental predictability and insect density; and (8) where the association is facultative, the increase in fitness derived by individuals when mutualism occurs. The results of studies on these subjects might lead to a better understanding of the general role and evolution of symbionts in natural systems.

In a more general system of community ecology, fungal associations support associations of collembola and orbatid mites. How are these herbivorous arthropods related to the fungal community? Does an increase in fungal species diversity increase the arthropod diversity, or is the arthropod diversity based strictly on quantity of fungus irrespective of species composition? These questions involve the trophic relationships of the organisms rather than a simple analysis of one or two populations. Here again, the colonization rates of the organism may be important. When the substrate for the fungi is quite variable temporally and specially, one might find that the arthropod diversity is strictly related to the area covered by the fungi; on the other hand, if the substrate for the fungi is large and constant, arthropod diversity might be directly related to the number of species of fungi that are available to the arthropods.

A final aspect of the biology of fungi relates to the evolutionary and ecological importance of asexuality. It has been proposed that sexuality should occur when the environment is subject to irregular perturbations whereas asexuality should be most common in stable environments (Williams, 1975). The reassortment of genetic material that occurs in sexual reproduction should result in a variety of genetic combinations at least some of which will be able to survive and reproduce in future unpredictable environments. If an environment is constant and predictable, an asexually reproducing organism with the most fit genome should eventually replace any sexually reproducing organism which has similar ecological requirements. Here, the predictability and constancy of the environment are defined by the physiological tolerances and ecological perception of the organism. The timing of sexual versus asexual reproduction in fungi should relate to environmental fluctuations. The fungi are especially useful for the analysis of the relationship of sexuality and environmental predictability since many species have both sexual and asexual stages. In this respect the Fungi Imperfecti represent an especially interesting group. Answers to questions revolving around the fitness of the Fungi Imperfecti and the reasons for the loss, or at least reduction (to heterokaryosis and mitotic recombination), of sexuality in this group will be important in furthering knowledge on the evolution of the loss of sexuality.

The prospects for future joining of mycological and ecological research are good. Mycological goals should include a quantitative methodology and a focus on

more general ecological problems as well as on problems which involve the interactions of fungi with organisms from other kingdoms. The evolutionary ecologist can
draw a great deal from certain kinds of mycological studies and should be able to
use fungi as organisms with which to answer questions about sexuality, symbiosis,
species diversity, and biogeography.

ACKNOWLEDGMENTS

This paper has profited from discussions with George C. Carroll, David C. Culver,
Douglas E. Gill, Daniel H. Janzen, Florence Hammett Seifert, and Donald T. Wicklow.
Financial expenses associated with the presentation of this paper at the Second
International Mycological Congress were borne by Columbian College and the Graduate
School of Arts and Sciences, The George Washington University.

REFERENCES

Abele, L. G. (1974). Species diversity of decapod crustaceans in marine habits.
Ecology 55: 156-161.

Andrewartha, H. G., and Birch, L. C. (1954). *The Distribution and Abundance of
Animals*. Univ. of Chicago Press, Chicago.

Angel, K., and Wicklow, D. T. (1975). Relationships between coprophilous fungi
and fecal substrates in a Colorado grassland. *Mycologia 67*: 63-74.

Armstrong, R. A. (1976). Fugitive species: experiments with fungi and some theoretical considerations. *Ecology 57*: 953-963.

Batra, L. R. (1963). Habitat and nutrition of *Dipodascus* and *Cephaloascus*.
Mycologia 55: 508-520.

Batra, S. W. T., and Batra, L. R. (1967). The fungus gardens of insects. *Sci.
Amer. 217*(Nov.): 112-120.

Brian, P. W. (1951). Antibiotics produced by fungi. *Bot. Rev. 17*: 357-430.

Broadbent, D. (1966). Antibiotics produced by fungi. *Bot. Rev. 32*: 219-242.

Colinvaux, P. A. (1973). *Introduction to Ecology*. Wiley, New York.

Couch, J. N. (1931). The biological relationship between *Septobasidium retiforme*
(B. and C.) Pat. and *Aspediotus osborni* New. and Ckll. *Quart. J. Microsc. Sci.,
Ser. II 74*: 383-437.

Dennis, C., and Webster, J. (1971). Antagonistic properties of species-groups of
Trichoderma. III. Hyphal interaction. *Trans. Brit. Mycol. Soc. 57*: 363-369.

Dobzhansky, T. (1950). Evolution in the tropics. *Amer. Sci. 38*: 209-221.

Fischer, A. G. (1960). Latitudinal variations in organic diversity. *Evolution
14*: 64-81.

Frankland, J. C. (1966). Succession of fungi on decaying petioles of *Pteridium
aquilinum*. *J. Ecol. 54*: 41-63.

Futuyma, D. J. (1973). Community structure and stability in constant environments.
Amer. Natur. 107: 443-445.

Gilbert, L. E., and Raven, P. H. (1975). *Coevolution of Animals and Plants*. Univ.
of Texas Press, Austin, Texas.

Gill, D. E. (1974). Intrinsic rate of increase, saturation density, and competitive ability. II. The evolution of competitive ability. *Amer. Natur. 108*: 103-116.

Gressitt, J. L. (1969). Epizoic symbiosis. *Entomol. News 80*: 1-5.

Hairston, N. G., Smith, F. E., and Slobodkin, L. B. (1960). Regulation in terrestrial ecosystems and the implied balance of nature. *Amer. Natur. 101*: 109-124.

Harley, J. L. (1968). Mycorrhiza. In *The Fungi,* Vol. III: *The Fungal Population,* G. C. Ainsworth and A. S. Sussman (Eds.). Academic Press, New York, pp. 137-178.

Hering, T. F. (1965). Succession of fungi in the litter of a Lake District oakwood. *Trans. Brit. Mycol. Soc. 48*: 391-408.

Hogg, B. M., and Hudson, H. J. (1966). Micro-fungi on leaves of *Fagus sylvatica.* I. The micro-fungal success. *Trans. Brit. Mycol. Soc. 49*: 185-192.

Horn, H. S., and MacArthur, R. H. (1972). Competition among fugitive species in a harlequin environment. *Ecology 53*: 749-752.

Hutchinson, G. E. (1951) Copepodology for the ornithologist. *Ecology 32*: 571-577.

Kevan, P. G., Chaloner, W. G., and Savile, D. B. O. (1975). Interrelationships of early terrestrial arthropods and plants. *Palaeontology 18*: 391-417.

Kingsolver, J. G., and Norris, D. M. (1977). The interaction of *Xyleborus ferrugineus* (Coleoptera: Scolytidae) behavior and initial reproduction in relation to its symbiotic fungi. *Ann. Entomol. Soc. Amer. 70*: 1-4.

Krebs, C. J. (1972). *Ecology: The Experimental Analysis of Distribution and Abundance.* Harper & Row, New York.

Levin, D. A. (1971). Plant phenolics: an ecological perspective. *Amer. Natur. 105*: 157-181.

Levin, S. A. (1974). Dispersion and population interaction. *Amer. Natur. 108*: 207-228.

Levins, R. (1968). *Evolution in Changing Environments.* Princeton Univ. Press, Princeton, N.J.

Levins, R., and Culver, D. C. (1971). Regional coexistence of species and competition between rare species. *Proc. Nat. Acad. Sci. U.S. 68*: 1246-1248.

Lloyd, M., Inger, R. F., and King, F. W. (1968). On the diversity of reptile and amphibian species in a Bornean rain forest. *Amer. Natur. 102*: 497-515.

Lussenhop, J. (1973). The soil arthropod community of a Chicago expressway margin. *Ecology 54*: 1124-1137.

MacArthur, R. H. (1969). Patterns of communities in the tropics. *Biol. J. Linn. Soc. 1*: 19-30.

MacArthur, R. H. (1972). *Geographical Ecology.* Harper & Row, New York.

MacArthur, R. H., and Wilson, E. O. (1967). *The Theory of Island Biogeography.* Princeton Univ. Press, Princeton, N.J.

McNaughton, S. J., and Wolf, L. L. (1973). *General Ecology.* Holt, Rinehart and Winston, New York.

Marks, G. C., and Kozlowski, T. T., Eds. (1973). *Ectomycorrhizae: Their Ecology and Physiology.* Academic Press, New York.

May, R. (1973). *Stability and Complexity in Model Ecosystems.* Princeton Univ. Press, Princeton, N.J.

Moore-Landecker, E. (1972). *Fundamentals of the Fungi.* Prentice-Hall, Englewood Cliffs, N.J.

Odum, E. P. (1971). *Fundamentals of Ecology,* 3rd ed. Saunders, Philadelphia.

Opler, P. A. (1974). Oaks as evolutionary islands for leaf-mining insects. *Amer. Sci. 62*: 67-73.

Paine, R. T. (1966). Food web complexity and species diversity. *Amer. Natur. 100*: 65-75.

Pianka, E. R. (1966). Latitudinal gradients in species diversity: a review of concepts. *Amer. Natur. 100*: 33-46.

Pielou, E. C. (1969). *An Introduction to Mathematical Ecology*. Wiley (Interscience), New York.

Poulson, T. L., and Culver, D. C. (1969). Diversity in terrestrial cave communities. *Ecology 50*: 153-158.

Preston, F. W. (1948). The commonness, and rarity of species. *Ecology 29*: 254-283.

Preston, F. W. (1962a). The canonical distribution of commonness and rarity. *Ecology 43*: 185-215.

Preston, F. W. (1962b). The canonical distribution of commonness and rarity. Part II. *Ecology 43*: 410-432.

Richards, P. W. (1952). *The Tropical Rain Forest*. Cambridge Univ. Press, New York.

Ricklefs, R. E. (1973). *Ecology*. Chiron Press, Newton, Mass.

Robinson, R. K. (1972). The production of a factor inhibitory to growth of some mycorrhizal fungi. *J. Ecol. 60*: 219-224.

Sanders, H. L. (1968). Marine benthic diversity: a comparative study. *Amer. Natur. 102*: 243-282.

Sands, W. A. (1960). The initiation of fungus comb construction in laboratory colonies of *Ancistrotermes guineensis* (Silverstri). *Insectes Sociaux 12*: 49-58.

Seifert, R. P. (1975). Clumps of *Heliconia* inflorescences as ecological islands. *Ecology 56*: 1416-1422.

Seifert, R. P., and Seifert, F. H. (1976). A community matrix analysis of *Heliconia* insect communities. *Amer. Natur. 110*: 461-483.

Slatkin, M. (1974). Competition and regional coexistence. *Ecology 55*: 128-134.

Slobodkin, L. B., and Sanders, H. L. (1969). On the contribution of environmental predictability to species diversity. *Brookhaven Symp. Biol. 22*: 13-24.

Slobodkin, L. B., Smith, F. E., and Hairston, N. G. (1967). Regulation in terrestrial ecosystems and the implied balance of nature. *Amer. Natur. 101*: 109-124.

Southwood, T. R. E. (1961). The number of species of insects associated with various trees. *J. Anim. Ecol. 30*: 1-8.

Strong, D. R., Jr. (1974a). Nonasymptotic species richness models and the insects of British trees. *Proc. Nat. Acad. Sci. U.S. 11*: 2766-2769.

Strong, D. R., Jr. (1974b). Rapid asymptotic species accumulation in phytophagous insect communities: the pests of cacao. *Science 185*: 1064-1066.

Strong, D. R., Jr., and Levin, D. A. (1975). Species richness of parasitic fungi of British trees. *Proc. Nat. Acad. Sci. U.S. 72*: 2116-2119.

Swift, M. J. (1976). Species diversity and the structure of microbial communities in terrestrial habitats. In *The Role of Terrestrial and Aquatic Organisms in Decomposition Processes*. J. M. Anderson and A. Macfadyen, Eds. Blackwell Sci. Publns., Oxford, England, pp. 185-222.

Vandermeer, J. H. (1970). The community matrix and the number of species in a community. *Amer. Natur. 104*: 73-83.

Waksman, S. A. (1941). Antagonistic relations of microorganisms. *Bacteriol. Rev. 5*: 231-291.

Waksman, S. A., and Horning, E. S. (1943). Distribution of antagonistic fungi in nature and their antibiotic action. *Mycologia 35*: 47-65.

Weber, N. A. (1966). Fungus-growing ants. *Science 153*: 587-604.

Webster, J. (1956). Succession of fungi on decaying cocksfoot culms. I. *J. Ecol. 44*: 517-544.

Wicklow, D. T. (1973). Microfungal populations in surface soils of manipulated prairie stands. *Ecology 54*: 1302–1310.

Wicklow, D. T., and Moore, V. (1974). Effect of incubation temperature on the coprophilous fungal succession. *Trans. Brit. Mycol. Soc. 62*: 411–415.

Wilbur, H. M. (1972). Competition, predation, and the structure of the *Ambystoma-Rana sylvatica* community. *Ecology 53*: 3–21.

Williams, G. C. (1966). *Adaptation and Natural Selection*. Princeton Univ. Press, Princeton, N.J.

Williams, G. C. (1975). *Sex and Evolution*. Princeton Univ. Press, Princeton, N.J.

Wilson, E. O. (1971). *The Insect Societies*. Harvard Univ. Press (Belknap), Cambridge, Mass.

Wilson, E. O. (1975). *Sociobiology*. Harvard Univ. Press (Belknap), Cambridge, Mass.

Chapter 2

MYCOLOGICAL INPUTS TO ECOSYSTEMS ANALYSIS

George C. Carroll
University of Oregon
Eugene, Oregon

I. INTRODUCTION

In considering the ecology of fungi, various contributors to the present volume have noted the dearth of studies in which fungi have been employed to develop or test some fundamental ecological concept. In short, as a discipline, mycology has made little contribution to the development of ecological theory, and general ecologists have adopted a take-it-or-leave-it attitude toward the fungi when considering possible organisms for testing such theory. In contrast, studies initiated under the International Biological Program (IBP) have repeatedly demonstrated the centrality of fungi in the functioning of ecosystems, and as a consequence an attitude of indifference toward fungi is less prominent among practitioners of ecosystems analysis than among professional ecologists as a whole. The question for large, integrated studies of ecosystems has become not whether microorganisms in general and fungi in particular will be considered but whether essentially mycological investigations will be undertaken by workers with backgrounds in classical mycology or by investigators from other disciplines forced by necessity to confront the fungi and their activities.

In the following pages I hope to demonstrate that the mycologist can make unique contributions to programs in ecosystems analysis and that such collaborative efforts can provide information about the biology of fungi unobtainable in any other fashion. For this discussion I have relied heavily on my own experiences with the Coniferous Forest Biome (US/IBP) and on consultation with Drs. James Trappe and William C. Denison, who were involved with the Coniferous Forest Biome research effort from its inception. This work has also benefited from useful conversations with several colleagues at Oregon State University and the University of Oregon who have also been heavily involved with research in coniferous forests. (See the Acknowledgments at the end of this chapter.)

In summary, then, I wish to address two questions within the historical context of IBP ecosystems research in western coniferous forests: (1) Which lines of productive inquiry have been initiated on the advice of mycologists, and how might the program have faltered without such mycological expertise? (2) What have mycologists learned about fungi in the process of collaborating in ecosystems research?

II. FUNGAL TAXONOMY AND ECOSYSTEMS ANALYSIS

Since its inception the organizers of biome projects under the IBP have demonstrated justifiable concern that research not become bogged down with interminable census studies and species lists. Rather, it was hoped that from the beginning efforts could be focused on process studies and computer modeling. It was recognized, however, that vastly different amounts of floristic and faunistic information existed for various groups of organisms, with rather complete species lists available for flowering plants and mammals, and almost nothing known about the composition of microbial and microfaunal communities. Since process studies frequently involve laboratory research with organisms prevalent in the field or organisms chosen to represent functional groupings, preliminary census studies often proved a necessity.

Within the Coniferous Forest Biome, mycological opinion dictated that preliminary species lists be prepared for a number of groups of fungi, including hypogeous fungi (Fogel, 1976), mycorrhizal and litter-decomposing fleshy fungi (Rhoades, 1972), microfungi (Sherwood, 1973; Sherwood and Carroll, 1974; Bernstein and Carroll, 1977a, b), and lichens (Pike, 1972; Pike et al., 1975). As a result of such studies, the dominance of just a few taxa in each of the habitats surveyed has been recognized, and more focused investigations using these species have been initiated. Examples are cited below.

The degree to which temperate zone ectomycorrhizal fungi degrade litter has long been debated in the mycological literature, with recent papers suggesting that such fungi have negligible capacities for litter decomposition (Hacskaylo, 1973). In view of the enormous standing crops and rather short turnover times reported for mycorrhizal fungi in the forest floor of a Douglas fir stand in Oregon (Fogel and Hunt, 1979), this question appears of considerable importance for modeling carbon fluxes in coniferous forests. Current work with *Laccaria laccata* (Fr.) Berk. and Br. and *Cenococcum geophilis* Fr., species found to be prevalent in early census studies, has shown these fungi to degrade coniferous litter to a significant extent in two-membered cultures with Douglas fir seedlings (A. Todd, personal communication).

For the lichens the foliose cyanophycophilous species *Lobaria oregana* (Tuck.) Müll. Arg. was early identified as predominant in Douglas fir canopies in low-elevation stands (Pike et al., 1975). Because of its capacity for nitrogen fixation, investigations on growth rates (Rhoades, 1977, 1978), litterfall (Rhoades, 1978; Pike and Carroll, unpublished data) and rates of nitrogen fixation (W. Denison, M. Roose, and L. Pike, personal communication) have been carried out (discussed later).

Estimation of microbial biomass from visual measurements of cell volume requires that dry weight/volume ratios be calculated for the cells which are measured. In attempting to estimate standing crops of needle microepiphytes in a Douglas fir canopy, Carroll (1979) determined these ratios for cultures of *Atichia glomerulosa* (Ach. ex Mann.) Stein, *Aureobasidium pullulans* (De Bary) Arn., and *Cladosporium* spp., fungi which had previously been identified as prevalent on needles and twigs in the Douglas

fir canopy (Sherwood and Carroll, 1974). In all of these cases the recognition of
the preeminence of just a few characteristic fungi in any given habitat was first
revealed through taxonomic surveys which required mycological expertise and which
necessarily preceded field and laboratory process studies.

In several cases, the mycologist's familiarity with fungal taxa has led to the
recognition of certain fungi in unexpected places, revealing clues about habitats
and processes. While carrying out a preliminary census of microepiphytes in the
canopy of Douglas fir, I attempted to culture several of the more common species.
One of these, *A. glomerulosa*, typically forms a nonmycelial pseudoparenchymatous
thallus on needles and twigs (Sherwood and Carroll, 1974; Meeker, 1975). When grown
on agar media, however, the fungus produces pigmented yeastlike cells in the film
of water on the agar surface. *A. pullulans*, another prevalent canopy fungus, produces similar yeastlike cells in liquid films. These observations suggested an
aquatic dispersal of vegetative fungal cells and prompted me to look for additional
evidence of aquatic spore dispersal in canopy fungi. Samples of throughfall from
an old-growth Douglas fir stand were collected within several hours of rainstorms
intermittently over a 2-year period. The water samples were brought back to the laboratory and filtered through Nuclepore filters (0.2 μm pore size). Microscopic examination of the filters revealed numerous triradiate and tetraradiate spores including
representatives of *Tripospermum*, *Tridentaria*, *Ceratosporium*, and many other genera
(see also Bandoni, Chap. 37). Conidia from needle endophytes from such forests typically occur in gloeoid masses and are either known or presumed to be also water
dispersed (Carroll and Carroll, 1978). Low elevation Douglas fir forests in the
Pacific Northwest normally receive 200-250 cm of rain a year, 70-80% of which may
fall from November through March. Canopy fungi clearly respond to such seasonal
inundation as if they existed in an aquatic habitat; many, like *Seuratia* (= perfect
state of *Atichia*), probably produce aerially dispersed ascospores during the dry
season and aquatically dispersed conidia or vegetative cells during the wet season.

Such fungi, then, constitute a guild of arboreal aquatic hyphomycetes which may
function in canopies much as classical Ingoldian aquatic hyphomycetes function in
streams (see Part VIII). Investigators should expect to find such fungi implicated
in the decomposition of litter lodged in the canopy and in the processing of such
litter for consumption by canopy arthropods. The litter may be smaller than the
allochthonous materials degraded by stream fungi, as with pollen attacked by *Retiarius* in the phyllosphere of evergreen forests in South Africa (Olivier, 1978). The
arthropods will also prove to be smaller than those in streams; D. Voegtlin in our
laboratory has found fungivorous mites and collembolans to be the most abundant grazers in the Douglas fir canopy. Further, the absorption and concentration of dilute
organic substrates from canopy leachates by fungi, while of little significance in
streams, appears to be an important process in the canopy subsystem (discussed later).
Although set apart by such superficial distinctions, the fundamental role of the arboreal aquatic fungi appears similar to that of aquatic hyphomycetes in streams; they

serve as ecological transducers, converting and concentrating recalcitrant and/or dilute substrates to available food sources which serve as the base for arthropod food chains.

This conclusion is further supported by another set of mycological clues. During the warmer months of the year microfrass (<1 mm diameter) becomes very abundant in the Douglas fir canopy. Examination of such frass under the microscope reveals the presence of abundant cell wall fragments from algal and fungal cells, largely *Protococcus viridis* Agh., *Atichia* sp., and various metacapnodiaceous taxa; apparently the resident microarthropods graze almost exclusively on canopy microorganisms.

The recent recognition of spores from hypogeous fungi in the gut contents of small forest mammals constitutes a second situation in which expertise on fungal taxonomy has yielded important information about processes in ecosystems. Maser et al. (1978) have shown that such fungi constitute a major food source for many rodents in the Pacific Northwest and that certain species (northern flying squirrel and red-backed vole) may subsist almost exclusively on hypogeous fungi. They have further demonstrated that habitat preferences of *Clethrionomys californicus,* the California red-backed vole, appear to be tied to the availability of sporocarps from obligate mycorrhizal hypogeous fungi; when a tract of forest is clear-cut, populations of this species vanish within a year, presumably because the fungi upon which they are known to feed have disappeared with the trees.

Aside from the obvious applicability of such information to models of carbon cycling and secondary productivity in forest ecosystems, these discoveries have further implications for an understanding of forest succession and for practical aspects of forest management. Animal species such as the Oregon vole (*Microtus oregoni*), which feed on fungal sporocarps in the forest and on grasses in adjacent clear-cuts are strongly suspected to serve as agents of dispersal for the fungi on which they feed, depositing viable spores in fecal pellets over the entire area where they range. Such pellets may well serve as the inoculum for mycorrhizal fungi during the natural afforestation of denuded areas. Since mycorrhizal associations are an essential requirement for the survival and growth of coniferous stands, the mycophagous activities of small mammals assume considerable importance for forest regeneration. These findings also cast doubt on the wisdom of scattering poison baits in clear-cuts as a tactic for protecting young seedlings from rodent grazing pressures (Maser et al., 1978).

III. THE MYCOLOGIST AS COMMENTATOR ON FUNGAL PROCESSES
      IN ECOSYSTEMS

While few classically trained mycologists would consider themselves fungal physiologists, most have grown fungi in laboratory culture and have an intuitive appreciation for fundamental aspects of fungal metabolism. In culture fungi are seen to be capable of rapid growth in even dilute liquid medium, a process which may bring impres-

sive absorptive and translocational capacities into play. On solid medium fungi may
also grow rapidly, decomposing recalcitrant substrates such as filter paper or crys-
talline cellulose. Repeated pure culture work from a variety of natural substrates
reveals that the fungi are truly ubiquitous. Thus, in viewing processes in ecosystems, the mycologist looks for the insidious but consequential fungi at every turn.

Rapid growth responses and short turnover times are frequently stated attributes
of microbial populations. The implications of these characteristics for ecosystems
analysis have often been explicitly set forth (e.g., by Odum, 1971): comparisons
of standing crops in an ecosystem overemphasizes the importance of larger organisms;
energy flow or biomass production provide the only really suitable index for comparing all components in an ecosystem. The above notwithstanding, general forest ecologists have frequently been reluctant to give fungi serious consideration in ecosystems analysis because their standing crops are dwarfed by those of the trees themselves. Nowhere is this situation more pronounced than in the coniferous forests
in the Pacific Northwest where the aboveground biomass of trees in old-growth stands
may range from 500-1000 t/ha* (Grier and Logan, 1977).

In spite of this disparity in standing crops, mycologists associated with the
Coniferous Forest Biome and subsequent research programs have recognized the possibility for the rapid microbial turnover times alluded to earlier. Standing crops
and turnover times have been investigated for three different guilds of fungi:
mycorrhizal and saprophytic fungi in the forest floor (Fogel and Hunt, 1979); *L.
oregana*, an arboreal epiphytic lichen (Rhoades, 1978); and epiphytic fungi associated with needles and twigs in the Douglas fir canopy (Carroll, 1979; Carroll et
al., 1980). Perhaps the most impressive data emerge from a study by Fogel and Hunt
(1978) on fungal biomass and turnover in the forest floor in a young second-growth
Douglas fir stand in the Oregon Coast Range. In this stand total tree biomass accounted for approximately 320 t/ha while total fungal biomass in the forest floor
was estimated at 20 t/ha, half of which was localized in mycorrhizal mantles. However, when turnover times and annual throughput were considered, the fungi accounted
for 50% of the total stand throughput of 30 t/ha. Clearly fungi in the forest floor
must be considered for any models of carbon cycling in forest ecosystems.

Rhoades (1978) has carried out similar investigations for one of the most abundant lichens in the Douglas fir canopy, *L. oregana*. Sampling of *Lobaria* on a number of individual old-growth Douglas fir trees has led to estimates of 500-600 kg/ha
for standing crops. Photographic measurements of thallus growth rates have yielded
estimates for annual production of 150-200 kg/ha. Since *Lobaria* and similar cyanophycophilous canopy lichens fix nitrogen and, indeed, appear to be the principal
source of newly fixed nitrogen in old-growth Douglas fir forests, this production

---

*Tonnes (i.e., metric tons) per hectare.

assumes an importance larger than that suggested by simple comparison with annual
primary productivity of the trees themselves. *Lobaria* thalli typically contain
about 2% nitrogen; thus, annual *Lobaria* production should account for 3-4 kg of
fixed nitrogen per hectare, a substantial contribution to the nitrogen economy of
the forest. Epiphytic lichens may also play a significant role in the cycling of
other minerals such as phosphorus and potassium (Pike, 1978).

Needle and twig surfaces in mature Douglas fir trees represent a third habitat
in the Pacific Northwest for which the estimation of fungal standing crops has been
attempted (Carroll, 1979; Carroll et al., 1980). Here standing crops of algal and
fungal microepiphytes appear to be on the order of 40-50 kg/ha when sampled at the
end of the summer dry season. Data on annual production are completely lacking.
However, annual turnover rates are unlikely to be less than 100%, as with fungi in
the forest floor (Fogel and Hunt, 1979), or more than 1000%, as with microorganisms
in several other phyllosphere habitats (see references in Carroll, 1979). Thus,
annual secondary production of microbial cells in this habitat probably amounts to
50-500 kg/ha, a total small by comparison with total primary production in the forest, but large by comparison with the substrates available for growth of heterotrophic microepiphytes in the canopy. Clearly such organisms must be important in influencing throughfall chemistry.

Recognition of the abilities of microorganisms to absorb and utilize organic
substances from dilute solutions has led to investigations of interactions between
epiphytic microorganisms and nutrients leached from the coniferous canopy in rainstorms (G. C. Carroll, unpublished data). Various canopy components, notably thalli
from *L. oregana*, moss bolsters, chlorophycophilous lichens such as *Alectoria* spp.,
Douglas fir needles of several age classes, and living twigs have been brought into
the laboratory, picked clean of extraneous materials, and misted with rainwater previously collected for the purpose. Leachates collected in bottles beneath the funnels containing the samples were then analyzed routinely for total nitrogen, total
suspended particulates, and total dissolved solids; occasionally samples were analyzed for total phosphorus, total polyols, total protein, and various cations (Na^+,
K^+, Ca^{2+}, Mg^{2+}).

When concentrations of these substances in the leachates are compared with concentrations in the incident rainwater, net fluxes of substances during the misting
episode can be easily computed. Consideration of lumped data from 20 one-hour leaching episodes on samples collected over a 17-month period (Sept. 20, 1976, through
Feb. 12, 1978) reveals the following trends: (1) All components show initial losses
of dissolved solids. (2) Nitrogen is taken up by mosses and chlorophycophilous
lichens but is lost by cyanophycophilous lichens and the various tree components;
such losses are particularly pronounced after periods of dry weather. (3) Release
of suspended solids is high, particularly for the epiphytes and particularly when
the samples have been exposed to prolonged periods of wet weather in the field prior
to collection and misting in the laboratory. These trends become more explicable

when filters bearing the released particulates are examined under the microscope. Filterable solids from *Lobaria* leachates consist almost exclusively of bacterial cells from populations resident on the surface of the lichen; those from needles and twigs are composed largely of fungal and algal cells from populations resident on those surfaces. When prolonged leaching experiments are carried out in the laboratory, it is found that all components eventually begin to take up dissolved nitrogen from the incident rainfall while continuing to export particulates in the leachates. Analysis of the nitrogen content of such particulates reveals that they largely account for this nitrogen uptake. The point at which uptake begins appears to be correlated with the growth rates of the responsible microorganisms. Thus, for *Lobaria*, nitrogen uptake (by bacteria) begins after several hours; microbial responses to moisture and nutrients on foliage (by algae and fungi) require several days.

These results have implications not only for models of nutrient cycling within canopies. They also relate to studies on inputs to the forest floor in rain and throughfall, suggesting that the standard technology for throughfall collection may be inadequate. When recently collected throughfall is subjected to microbiological filtration and the filters are examined under a microscope, they are found to be coated with assorted microbial cells; in view of the microbiological uptake of organic materials from canopy leachates already described, it should come as no surprise to find that throughfall enters ground-level collectors laden with actively metabolizing microbial cells which are probably adapted to growth over a wide range of temperatures. Such microbial cells persist and grow in slimy layers on the inner surfaces of the containers in which tne throughfall is collected, sequestering nutrients from the throughfall during the time the cumulative sample is taken and altering their concentrations prior to analysis. These slimy layers cannot be dislodged by simple mechanical agitation; their buildup can be restricted only partially through the use of metabolic poisons in the collectors. Impregnation of plastic containers with iodine (Heron, 1962) prevents the buildup of slimes but may alter the water chemistry (e.g., pH) of samples to an unacceptable degree. Frequent collection of samples and the replacement of dirty containers or liners with each collection may ameliorate the situation, the integrity of the samples improving with the frequency of collection (Carlisle et al., 1966; G. C. Carroll and J. Perkins, unpublished data). While metabolic poisons are widely used in collectors for throughfall studies, other precautions to minimize microbial growth during the sampling interval are less frequently seen. Consequently, results from published throughfall studies should be evaluated with respect to methods used to preserve the integrity of the samples; they should be regarded with caution if they fail to address the possibility of microbial growth in the samplers.

Although several studies have been carried out on decomposition in the coniferous forest biome, all of them have relied chiefly on weight loss data from litter bags and thus have not required any special mycological expertise (e.g., see Fogel and Cromack, 1977). However, results from at least one such study can be more eas-

ily interpreted if the remarkable absorptive and translocational abilities of fungi
are kept in mind. Pike and colleagues (unpubl.) have shown an actual increase in the
total Ca^{2+} capital during decomposition of several classes of litter (*Lobaria,
Alectoria*, and Douglas fir needles) over a 14-month period. Temporary increases
in the total K^+ capital were also demonstrated for these litter types following bud
burst of the trees the second season the litter bags were in place. The probable
absorption and mobilization of Ca^{2+} and K^+ by fungi into the rotting substrates provides a ready explanation for these otherwise inexplicable effects.

IV. ECOSYSTEMS ANALYSIS AS A CATALYST FOR MYCOLOGICAL DISCOVERIES

If integrated studies of ecosystem function have gained from the involvement of professional mycologists, what has mycology, the study of fungi for their own sake,
gained from entanglement in ecosystems analysis? Several obvious benefits have ac-
crued to the discipline. Because of the possible direct applicability of results
from ecosystems research to problems in the management of natural resources, programs
in ecosystems analysis have been liberally funded in recent years. In a less merce-
nary, more scholarly vein, mycologists have been forced to deal with the fungi beyond the confines of the laboratory, in habitats where they really grow and function.
In the process, mycologists have acquired skills and adopted techniques previously
little in evidence in the mycological literature. Most notably we have seen an increased competence for experimental design and the statistical analysis of data on
the part of mycoecologists (for instance, see Pike, Chap. 27, and Fogel, Chap. 28).
In many instances sampling of fungal populations in the field has proved so laborious that mycologists have sought more sophisticated and efficient sampling schemes
than those normally used (Pike et al., 1977; Carroll et al., 1980; Pike, Chap. 27).
Because mycological data sets seldom meet the assumptions for standard parametric
statistical procedures, mycologists can be expected to rely increasingly on nonparametric tests (e.g., see Carroll, 1979).

Beyond these generalities, programs in ecosystems analysis have yielded discoveries about the fungi that deserve widespread recognition among professional mycologists and mention in any introductory mycology textbook. Among these I would
count the detection of extensive calcium oxalate production by soil fungi and the
elucidation of its role in the inorganic nutrition of higher plants and fungi (cf.
Sollins et al., Chap. 31; also Cromack et al., 1979). A second, more diffuse discovery has involved recognition of animals as a significant selective force in fungal habitats. Thus, standing crops of arboreal microepiphytes appear to be grazed
extensively by canopy microarthropods, and the intensity of the grazing pressure
may be a major factor in regulating microbial standing crops (Carroll, 1979). Much
of the structure of the fungal thallus becomes explicable in light of such pressures.
A coenocytic organization in which individual compartments can be sealed off quickly

by an elaborate septal pore apparatus has obviously evolved in an environment in which injury to the thallus is a common event; grazing activities of small arthropods would appear to be the most frequent source of such injury. Many other attributes of phyllosphere fungi such as thick, heavily melanized cell walls, gelatinous or slimy coverings, and very large cells (as in the Metacapnodiaceae) may also constitute highly evolved responses to grazing pressures. The discovery of hypogeous sporocarps as a major constituent in the diets of a number of small mammals (Maser et al., 1978) has explained the rank or fragrant odors associated with these subterranean fruiting bodies as well as the widespread distribution of the fungi that produce them. Further observations of animals gorging on hypogeous sporocarps plainly reveal aerial dispersal of spores from certain species (*Elaphomyces* spp.) and explain the anatomical features which ensure such dispersal: a powdery spore mass and resilient capillitium (J. Trappe, personal communication).

The aforementioned examples are striking and known to me. Mycologists working in biomes other than the coniferous forest biome could surely cite numerous other equally relevant case histories. In summary, I wish to urge the professional mycologist toward involvement with ecosystems studies. Individuals with formal training in the taxonomy and biology of fungi can frequently posit the crucial questions and interpret the important clues with regard to the roles of fungi in ecosystems as no one else can. The collaborating mycologist may be rewarded with financial support for his research and with the opportunity, even necessity, for scholarly growth. With any imagination he or she will certainly be rewarded with new insights about the behavior of fungi in real habitats.

ACKNOWLEDGMENTS

Various people have provided valuable suggestions for this paper. I particularly wish to acknowledge the input of Dr. William C. Denison (Dept. of Botany and Plant Pathology, Oregon State University, Corvallis, Oregon) and Dr. James Trappe (U.S. Dept. of Agriculture Forest Service, Corvallis, Oregon), whose early efforts in the Coniferous Forest Biome served to catalyze much of the research described herein. I also wish to acknowledge useful conversations with Drs. Kermit Cromack, Kenneth Cummins, and Phillip Sollins (Oregon State University) and Dr. Lawrence Pike (University of Oregon). Much of this work was initiated with funding from the National Science Foundation to the Coniferous Forest Biome (IBP). Subsequent work has been funded with separate research grants from the National Science Foundation to Drs. George C. Carroll, Kermit Cromack, William Denison, Robert Fogel, Charles Grier, Lawrence Pike, Phillip Sollins, and James Trappe. This paper is contribution number 303 from the Coniferous Forest Biome (NSF Grant GB 20963 to the Coniferous Forest Biome, Ecosystems Analysis Studies, US/IBP).

REFERENCES

Bernstein, M. E., and Carroll, G. C. (1977a). Internal fungi in old-growth Douglas fir foliage. *Can. J. Bot. 55*: 644-653.

Bernstein, M. E. , and Carroll, G. C. (1977b). Microbial populations of Douglas fir needle surfaces. *Microbial Ecol. 4*: 41-52.

Carlisle, A., Brown, A. H. F., and White, E. J. (1966). The organic matter of nutrient elements in the precipitation beneath a sessile oak canopy. *J. Ecol. 54*: 87-98.

Carroll, G. C. (1979). Needle microepiphytes in a Douglas fir canopy: biomass and distribution patterns. *Can. J. Bot. 57*: 1000-1007.

Carroll, G. C., and Carroll, F. E. (1978). Studies on the incidence of coniferous needle endophytes in the Pacific Northwest. *Can. J. Bot. 56*: 3034-3043.

Carroll, G. C., Pike, L., Perkins, J., and Sherwood, M. (1980). Biomass and distribution patterns of conifer twig microepiphytes in a Douglas fir forest. *Can. J. Bot. 58*: 624-630.

Cromack, K., Sollins, P., Graustein, W. C., Speidel, K., Todd, A., Spycher, G., Li, C., and Todd, R. L. (1979). Calcium oxalate accumulation and soil weathering in mats of the hypogeous fungus *Hysterangium crassum*. *Soil Biol. Biochem. 11*: 463-468.

Fogel, R. (1976). Ecological studies of hypogeous fungi. II. Sporocarp phenology in a western Oregon Douglas fir stand. *Can. J. Bot. 54*: 1152-1162.

Fogel, R., and Cromack, K. (1977). Effect of habitat and substrate quality on Douglas fir litter decomposition in western Oregon. *Can. J. Bot. 55*: 1632-1640.

Fogel, R., and Hunt, G. (1979). Fungal and arboreal biomass in a western Oregon Douglas-fir ecosystem: distribution patterns and turnover. *Can. J. Forest Res. 9*: 245-256.

Grier, C. C., and Logan, R. S. (1977). Organic matter distribution and net production in plant communities of a 450-year-old Douglas-fir ecosystem. *Ecol. Monogr. 47*: 213-300.

Hacskaylo, E. (1973). Carbohydrate physiology of ectomycorrhizae. In *Ectomycorrhizae*, G. C. Marks and T. T. Kozlowski (Eds.). Academic Press, New York, pp. 207-230.

Heron, J. (1962). Determination of PO_4^{3-} in water after storage in polyethylene. *Limnol. Oceanogr. 7*: 316-321.

Maser, C., Trappe, J. M., and Nussbaum, R. A. (1978). Fungal-small mammal interrelationships with emphasis on Oregon coniferous forests. *Ecol. 59*: 799-809.

Meeker, J. A. (1975). Revision of the Seuratiaceae. I. Morphology of *Seuratia*. *Can. J. Bot. 53*: 2462-2482.

Odum, E. P. (1971). *Fundamentals of Ecology,* 3rd ed. Saunders, Philadelphia.

Olivier, D. L. (1978). *Retiarius* gen. nov.: phyllosphere fungi which capture wind-borne pollen grains. *Trans. Brit. Mycol. Soc. 71*: 193-201.

Pike, L. H. (1972). *Lichens of the H. J. Andrews Experimental Forest: Preliminary Checklist*. Coniferous Forest Biome Internal Report No. 46. Biome Central Office, Univ. of Washington, Seattle, 13 pp.

Pike, L. H. (1978). The importance of epiphytic lichens in mineral cycling. *Bryologist 81*: 247-257.

Pike, L. H., Denison, W. C., Tracy, D. M., Sherwood, M. A., and Rhoades, F. M. (1975). Floristic survey of epiphytic lichens and bryophytes growing on old-growth conifers in western Oregon. *Bryologist 78*: 389-402.

Pike, L. H., Rydell, R. A., and Denison, W. C. (1977). A 400-year-old Douglas fir tree and its epiphytes: biomass, surface area, and their distributions. *Can. J. Forest Res.* 7: 680-699.

Rhoades, F. M. (1972). *Fleshy Fungi Fruiting in the H. J. Andrews Experimental Forest: A Partial List of Collections from Fall 1970 to Spring 1972.* Coniferous Forest Biome Internal Report No. 45. Biome Central Office, Univ. of Washington, Seattle, 16 pp.

Rhoades, F. M. (1977). Growth-rates of the lichen *Lobaria oregana* as determined from sequential photographs. *Can. J. Bot. 55:* 2226-2233.

Rhoades, F. M. (1978). *Growth, Production, Litterfall, and Structure in Populations of the Lichen Lobaria oregana (Tuck.) Müll. Arg. in Canopies of Old-growth Douglas Fir.* Ph.D. dissertation, Univ. of Oregon, Eugene, 139 pp.

Sherwood, M. A. (1973). *Microfungi of the H. J. Andrews Experimental Forest: A Preliminary Checklist.* Coniferous Forest Biome Internal Report No. 58. Biome Central Office, Univ. of Washington, Seattle, 12 pp.

Sherwood, M. A., and Carroll, G. C. (1974). Fungal succession on needles and young twigs of old-growth Douglas fir. *Mycologia 66:* 499-506.

Chapter 3

ANALYSIS OF MICROFUNGAL COMPONENT COMMUNITIES

John Lussenhop
University of Illinois at Chicago Circle
Chicago, Illinois

I. INTRODUCTION

The purpose of this paper is to review methods which might aid in describing the organization of fungal communities, especially those of soil and litter. The viewpoint
is that of a nonmycologist seeking to recommend those methods best suited to the kind
of data gathered by mycologists--well-replicated frequency of occurrence data.

Fungal communities have been studied with conservative methods because the relation between collected samples and natural units of fungal activity is not clear.
This is due to the small scale of fungal activity and the complex sequence of sampling and culturing. The mycological literature reflects appropriate concern for the
way sample collection, storage, culturing, and sampling of colonies for identification all affect the picture of fungal assemblages in nature. In addition, size of
sample units and the selectivity with which they are collected affect our picture
of fungal communities in nature.

A detailed picture of the basidiomycete community of Skokholm Island was developed by Parker-Rhodes (1951), who adapted sampling and statistical methods to the
special problems of sampling basidiomycetes. On the basis of four censuses of sporocarps over the complete island, Parker-Rhodes estimated the total number of species
and their densities. He showed that species-relative abundance was very uneven and
that it fit the logseries distribution (Parker-Rhodes, 1952). In explanation, he
hypothesized that this was due to prevalent species which could always exclude other
species from suitable sites in the soil if they colonized them. A similar analysis
would be much more difficult for microfungi: How can one define and sample from a
potentially interacting group of microfungi? What insights into community organization might relative abundance provide?

II. COMPOUND AND COMPONENT COMMUNITIES

Selective isolation of fungi from small samples of specific substrates makes biological and statistical interpretation easier than for fungi isolated from large heterogeneous samples. The biological value of selective isolation is that it matches
sample units with natural units of fungal activity (fungi on leaf blades, petioles,

root surfaces, for example). Such substrate groups have been called component communities by Root (1973), who defines them as "the species associated with some microenvironment or resource." Swift (1976) has used the term *unit community* for a similar concept.

The statistical value of selective isolation is twofold. First the sample is
likely to include active, possibly interacting colonies. Second, replicates of such
species groups will be collected if a single substrate is indeed sampled. The importance of a sample unit sized to include the normal sphere of fungal interactions is
not included in discussions of the statistical basis of plate sampling (Eisenhart
and Wilson, 1943) where it is assumed that sample volume taken from nature is much
greater than the size of naturally occurring units of fungal activity which represent component communities. When sample units are this large, estimates of species
interactions are not meaningful. The usual solution to such a problem is to choose
the most efficient sample size for each species by collecting replicate samples of
each of a number of different sizes and plotting the change in the variance of individual numbers against the mean (Greig-Smith, 1964). However, the wide distribution
of fungal propagules makes this a less useful method than isolating fungi from portions of single substrates which support component communities. Second, the difficulty of sampling from a component community supported by a single substrate is illustrated by a study of Mignolet and Gerard (1973). They found that 1-g samples of forest litter are large relative to the size of aggregations of four fungi by demonstrating that for no species did number of occurrences fit the binomial distribution.
Mignolet and Gerard (1973) collected 100 samples and from each prepared five agar
plates at two dilutions. The number of uninfected plates/sample was best approxi-
mated by a two-parameter contagious distribution (binomial log-gamma) for which the
parameter reflecting aggregation was independent of dilution and the parameter reflecting density increased with dilution. If a smaller, more homogeous sample such
as the blade or petiole of one leaf species were used, the number of uninfected plates
might have fit the binomial distribution.

III. INDICES OF SIMILARITY AND ASSOCIATION

If species are sampled from compound communities, hypotheses about relations between
samples might be best tested by using indices of similarity as one might do for regional floras of higher plants. However, if component communities are sampled, tests
of hypotheses might be based on association indices, as well as the classifications
based on them. The many useful measures of association and similarity are reviewed
by Goodall (1973), whose terminology I use here. Pielou (1977) discusses hypothesis
testing using such measures.

Sorensen's index of similarity does not utilize information from either pairwise comparisons or species frequency; both may be valuable. For example, Hubalek
(1978) prepared dendrograms expressing the affinities between fungal species assoc-

iated with birds by using Jacard's index of association calculated for all species pairs. Aberdeen (1956) used Cole's index to test the hypothesis that there are no interactions between fungal species on washed roots of the grasses *Chloris gayana* and *Tagetes minuta*. He found that *Trichoderma* sp. decreased where *Cladosporium* sp. or two unidentified species were present. *Aspergillus* sp. decreased in presence of a *Fusarium* species. Aberdeen (1956) also quantified the effect of plant species on fungal interactions by transforming frequency data and performing an analysis of variance.

As an alternative to Aberdeen's use of data transformation and parametric statistics, Kent (1972) examined the value of seven nonparametric tests in distinguishing responses of sand dune fungi to fertilization, soil, and incubation temperature. Nonparametric methods could not test interactions, and both parametric and nonparametric tests were affected by low frequencies of fungi. These limitations may be overcome by use of discrete multivariate methods because zero observed frequencies need not affect the power of the tests, and models including interactions can be fitted to data in order to test specific hypotheses (Bishop et al., 1975; Fienberg, 1977; see also Fienberg, 1970, for an ecological example).

IV. RELATIVE ABUNDANCE PATTERNS

Frequency of isolation data, used directly or to estimate relative abundance, might be utilized to test whether species (alike or not) in two communities have the same relative abundance (Pielou, 1975). Relative abundance data might also be used to decide between alternative controls of community structure: either stochastic events or many independent factors control relative abundance; alternatively, either incomplete sampling or abundance by a single, dominant factor controls relative abundance.

Relative abundance of many organisms has been fitted to frequency distributions derived from models of niche structure, but this approach does not provide an unambiguous understanding of specific causes of relative abundance distributions (Pielou, 1975). Nevertheless, fit of relative abundance data to the lognormal distribution does appear to offer basic insight into control of species abundance. Anderson (1974) argues that fit of data to the lognormal distribution reflects stochasticity rather than deterministic processes. Anderson (1974) found no evidence that evolutionary laws control relative abundance of taxa; he found that numbers of species per genus fit the lognormal distribution for all the mammalian and insect taxa examined. May (1975) came to a similar conclusion for the distribution of individuals among species. May argues that fit of relative abundance data to the lognormal distribution may indicate whether few or many independent factors control relative abundance. When organisms are collected in ways which mix species associated with different resources or microhabitats, that is when species are collected from many component communities, relative abundance is expected to fit the lognormal distribution. It is not surprising that many factors should control the relative abundance of species in heterogeneous collections. Examples of compound communities for which relative abundance fits the lognormal distribution are insects caught in light traps (Williams, 1964), trees, shrubs, and herbs of Great Smoky Mountains cove forests (Whittaker, 1965), or birds of large geographic areas (Preston, 1962; Williams, 1964).

The fungi isolated from soils of two dry northern Wisconsin forests by Christensen (1969) are a compound community because her 15-cm^3 collections must have contained roots and detritus carrying different component communities. I will estimate relative abundance of species in Christensen's collections by assuming that the sample volume was large enough to randomly sample substrates supporting each type of fungal component community. To the extent that the fungi were randomly sampled, relative abundance may be estimated by setting each frequency equal to the number of samples containing at least one individual expected under the Poisson distribution, and then solving for the mean [frequency = 1 − exp(mean) (Greig-Smith, 1964)]. If individuals are aggregated at the scale of the collection, this will overestimate relative abundance by a small percent at low frequencies and by over 100% at frequencies of 0.8 or more.

The relative abundance of fungi in Christensen's collections fit the lognormal distribution more closely than do trees in the dry forest (Fig. 1A), suggesting that relative abundance of dry forest trees is controlled by fewer factors than in the dry-mesic forest. Fit to the lognormal distribution is easiest to judge when the cumulative number of species is plotted on a probability scale ordinate so that points fitting a lognormal distribution fall on a straight line (Harding, 1948; Williams, 1964; see Fig. 1B). The poor fit of trees in the dry forest to the lognormal distribution is emphasized by the points lying above or below values for the dry-mesic forest, whereas both sets of points for fungi in the forests are intermingled.

Relative abundance of organisms in many component communities does not fit the lognormal distribution. This may be due to insufficient sampling, or it may be due to the effect of a dominating factor controlling relative abundance. For example, rigorous environments may minimize biological interactions and be the primary cause of less even relative abundance than the lognormal among plants of alpine or heath bald communities (Whittaker, 1965). Parker-Rhodes (1952) found that relative abundance of basidiomycetes was also less even than the lognormal distribution. In apparently benign environments in which species appear to be evolutionarily adjusted to one another, there are móre species of intermediate abundance. Examples are bird communities of homogeneous plant communities (MacArthur, 1965).

In spite of many examples of component communities in the ecological literature in which relatie abundance is either more or less even than the lognormal, the relative abundance of fungal component communities I have examined all fit the lognormal distribution. My literataure sample consists of 16 data sets of rhizosphere fungi for which I estimated fungal abundance from frequency: *Fagus sylvatica* (Harley and

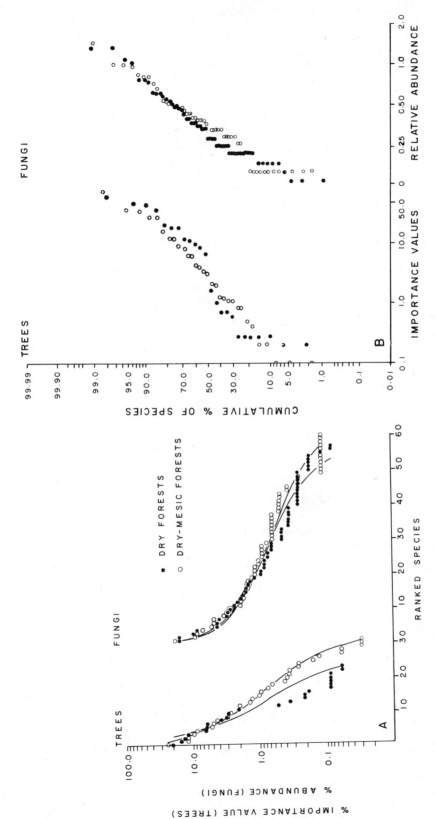

Fig. 1 (A) Species abundance and fitted lognormal curves for trees and fungi of northern Wisconsin dry and dry-mesic forests. Forest data are based on the average importance value given in Curtis (1959): fungal abundance is derived from Christensen's (1969) frequency data using the relationship abundance = ln(1 – frequency/100). (B) A probability scale plot of the same data.

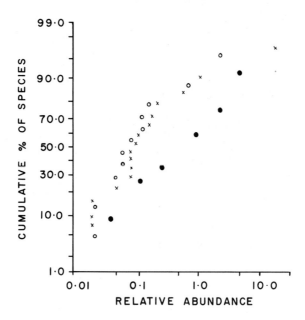

Fig. 2 A probability scale plot of fungal relative abundance on bean (*Phaseolus vulgaris*) root surfaces (Dix, 1964): mature roots (crosses), 3 days of decay (solid circles), and 14 days of decay (hollow circles).

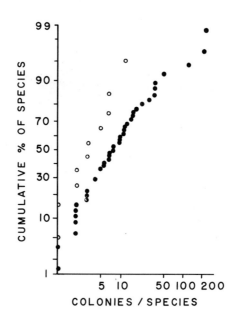

Fig. 3 A probability scale plot of fungal relative abundance in wheat-field soil. The data are from Warcup (1955). Hollow points are colonies from hyphae; solid points are all colonies.

Waid, 1955), *Phaseolus vulgaris* (Dix, 1964), *Vicia faba* (Mahiques, cited in Waid, 1974), and *Hordeum vulgare* (Parkinson and Pearson, 1965). The actively growing mycelia which Warcup (1955) isolated from wheat field soil also fit the lognormal distribution (Fig. 2), although when compared with the entire propagule population the range of abundance is less, as indicated by the steeper slope of the hollow points in Fig. 3.

As roots decompose, fungal species composition changes but relative abundance remains lognormal. The community on decomposing roots simply has a greater range of abundance than species on living roots. Figure 2 is an example of the pattern common to 13 fungal data sets for decomposing roots (Dix, 1964; Mahiques, cited in Waid, 1974; Parkinson and Pearson, 1965).

Why do relative abundance patterns of microfungal component communities show so little variety? It is possible that there are too few species in some data sets to adequately judge whether the points fit any line. But it is more likely, given the consistency of the 31 data sets examined, that fungal relative abundance patterns simply reflect the large number of independent environmental factors controlling fungal abundance. Rather than one factor, such as antibiotics, controlling fungal abundance, it is not hard to imagine many. For example, Garrett (1970), in defining competitive saprophytic ability, lists four factors: germination and growth rates, cellulytic and lignolytic enzymes, tolerance of fungistatic compounds, and ability to produce antibiotics.

V. CONCLUSIONS

Descriptive understanding of microfungal community structure has been slowed by the difficulty of sampling at the scale of fungal interaction and resource use. With large fungi it is easier to collect basic data and relate fungal component communities to higher plant communities as Parker-Rhodes (1955) did for the Skokholm Island basidiomycetes. For the microfungi of litter and soil, the development of selective isolation methods allows collection from component communities, and this makes it reasonable to quantify species associations and to compare relative abundance patterns. The fit of relative abundance distributions to the lognormal distribution has descriptive value and is at least suggestive of whether one or many independent factors control relative abundance. Further, if patterns observed among higher organisms are followed among fungi, relative abundance more even than the lognormal might indicate reciprocally adapted species in benign environments, while relative abundance less even might indicate the effect of rigorous environmental factors.

ACKNOWLEDGMENTS

I am grateful to T. Poulson, R. Seifert, and D. Wicklow for discussion and encouragement, and to the National Science Foundation for support under grants DEB75-09443 and DEB77-09443.

REFERENCES

Aberdeen, J. E. C. (1956). Factors influencing the distribution of fungi on plant roots. Part I. Different host species and fungal interactions. *Pap. Dept. Bot., Univ. Queensland 3*: 113-124.

Anderson, S. (1974). Patterns of faunal evolution. *Quart. Rev. Biol. 49*: 311-332.

Bishop, Y. M. M., Fienberg, S. F., and Holland, P. W. (1975). *Discrete Multivariate Analysis*. MIT Press, Cambridge, Mass.

Christensen, M. (1969). Soil microfungi of dry to mesic conifer-hardwood forests in northern Wisconsin. *Ecology 50*: 9-27.

Curtis, J. T. (1959). *The Vegetation of Wisconsin*. Univ. of Wisconsin Press, Madison.

Dix, N. J. (1964). Colonization and decay of bean roots. *Trans. Brit. Mycol. Soc. 47*: 285-292.

Eisenhart, C., and Wilson, P. W. (1943). Statistical methods and control in bacteriology. *Bacteriol. Rev. 7*: 57-137.

Fienberg, S. E. (1970). The analysis of multidimensional contingency tables. *Ecology 51*: 419-433.

Fienberg, S. E. (1977) *The Analysis of Cross-Classified Categorical Data*. MIT Press, Cambridge, Mass.

Garrett, S. D. (1970). *Pathogenic Root-Infecting Fungi*. Cambridge Univ. Press, New York.

Goodall, D. W. (1973). Sample similarity and species correlation. In *Handbook of Vegetation Science*, Pt. V, R. H. Whittaker (Ed.). Junk, The Hague, pp. 107-156.

Greig-Smith, P. (1964). *Quantitative Plant Ecology*. Butterworth, London.

Harding, J. P. (1948). The use of probability paper for the graphical analysis of polymodal frequency distributions. *J. Mar. Biol. Ass.* (U.K.) *28*: 141-153.

Harley, J. L., and Waid, J. S. (1955). A method of studying active mycelia on living roots and other surfaces in the soil. *Trans. Brit. Mycol. Soc. 38*: 104-118.

Hubalek, Z. (1978). Coincidence of fungal species associated with birds. *Ecology 59*: 438-442.

Kent, J. W. (1972). Application of statistical techniques to the analysis of fungal populations. *Trans. Brit. Mycol. Soc. 58*: 253-268.

MacArthur, R. H. (1965). Patterns of species diversity. *Biol. Rev. 40*: 410-533.

May, R. M. (1975). Patterns of species abundance and diversity. In *Ecology and Evolution of Communities*, M. L. Cody and J. M. Diamond (Eds.). Harvard Univ. Press, Cambridge, Mass., pp. 81-121.

Mignolet, R., and Gerard, G. (1973). Répartition spatiale de la mycoflore dans la litière d'une chênaie. *Oikos 24*: 123-127.

Parker-Rhodes, A. F. (1951). The basidiomycetes of Skokholm Island. VII. Some floristic and ecological calculations. *Ann. Bot.* (London) *50*: 227-243.

Parker-Rhodes, A. F. (1952). The basidiomycetes of Skokholm Island. VIII. Taxonomic distributions. *Ann. Bot.* (London) *51*: 216-228.

Parker-Rhodes, A. F. (1955). The Basidiomycetes of Skokholm Island. XII. Correlation with the chief plant associations. *Ann. Bot.* (London) *54*: 259-276.

Parkinson, D., and Pearson, R. (1965) Factors affecting the stimulation of fungal development in the root region. *Nature 205*: 205-206.

Pielou, E. C. (1975). *Ecological Diversity*. Wiley, New York.

Pielou, E. C. (1977). *Mathematical Ecology*. Wiley, New York.

Preston, F. W. (1962). The canonical distribution of commonness are rarity. Part II. *Ecology 43*: 410-432.

Root, R. B. (1973). Organization of a plant-arthropod association in simple and diverse habitats: the fauna of collards (*Brassica oleracea*). *Ecol. Monogr. 43*: 95-124.

Swift, M. J. (1976). Species diversity and the structure of microbial communities in terrestrial habitats. In *Tne Role of Terrestrial and Aquatic Organisms in Decomposition Processes*, J. M. Anderson and A. MacFadyen (Eds.). Blackwell Sci. Publns., Oxford, England, pp. 185-222.

Waid, J. S. (1974). Decomposition of roots. In *Biology of Plant Litter Decomposition*, C. H. Dickinson and G. J. P. Pugh (Eds.). Academic Press, New York, pp. 175-211.

Warcup, J. H. (1955). On the origin of colonies of fungi developing on soil dilution plates. *Trans. Brit. Mycol. Soc. 38*: 298-301.

Whittaker, R. H. (1965). Dominance and diversity in land plant communities. *Science 147*: 250-260.

Williams, C. B. (1964). *Patterns in the Balance of Nature*. Academic Press, New York.

Chapter 4

THE COPROPHILOUS FUNGAL COMMUNITY:
A MYCOLOGICAL SYSTEM FOR EXAMINING
ECOLOGICAL IDEAS

Donald T. Wicklow

Northern Regional Research Center
U.S. Department of Agriculture
Peoria, Illinois

I. INTRODUCTION

The characterization and experimental manipulation of various microfloral and microfaunal groups involved in decomposition processes comprise major obstacles in attempting to examine factors that determine the organization of heterotrophic communities
in ecosystems. The mycologist must deal with the formidable task of first uncover-
ing (isolating) and then attempting to identify and quantify the entire spectrum of
fungi occupying the "composite fungal niche" of a particular ecosystem. Requirements
for isolation, growth, and sporulation may vary considerably among individual taxa.
In addition, thorough identification and quantification of the entire mycoflora of
a forest or grassland ecosystem would be an unusually cumbersome process and has yet
to be accomplished. The development of fruiting bodies of many fleshy fungi may not
always coincide with the investigator's field schedule; further, many fungi cannot
easily be isolated into culture or do not sporulate following isolation. As a consequence, it is difficult to obtain a complete taxonomic picture of the fungus flora.
Seasonal patterns of vegetative activity are not known for most species of fungi,
and even less is known about the environmental cues triggering events in their life
histories (see Widden, Chap. 21).

Over the past three decades a number of mycologists from the United Kingdom have
focused their efforts on intensive studies involving the fate of individual substrates in ecosystems (Hudson, 1968; Swift, 1976). However, many problems remain
both in recognizing and in characterizing the role (niche) of each fungal colonist.
For example, many fungal colonists of leaf litter in the organic horizons of a forest soil do not develop fruiting bodies on the individual litter components. Instead, their mycelium is part of a larger network which supports the development of
a few larger fungal sporocarps. Nutrient resources from one zone of the extended
mycelial network may be allocated to the hyphae involved in colonization and biodegradation of an individual leaf or twig. Physicochemical or biotic factors affecting this mycelial network in different soil horizons or stages of litter decomposi-

tion or in different phases of ecosystem maturation will possibly modify the activities of hyphal cells in contact with specific litter components. In other words,
factors external to the immediate microenvironment of a litter component have an influence on its microfloral decomposer community.

How does one evaluate how a few hyphae on a single litter fragment relate to
the success (fitness) of the entire mycelium to which those hyphae are connected?
If one removes individual litter components from the organic horizon in order to
work with them as experimental units in the laboratory, a number of variables are
introduced: (1) disconnected mycelial networks; (2) difficulty in reproducing micro-
environments identical to those of the microhabitat from which the fragment was collected; (3) inability to authentically reproduce patterns of microbial colonization;
(4) inability to identify all colonists; and (5) inability to determine origin of
inoculum, specific vectors, and inoculum potential. Moreover, experimental manipulation of populations of individual species is not readily attainable, because it is
difficult to add species to or remove them from natural substrates without substantially altering substrate composition, microenvironment, species composition, and
community organization of microfloral and microfaunal colonists--in sum, decomposition processes.

The purpose of this chapter is to illustrate the potential utility of coprophilous fungal communities associated with herbivore dung as a microcosmic or model system for examining the organization and function of a naturally occurring fungal community and for studying interactions among decomposer functional groups. The dung
microcosm has unique features which make it a suitable system for experimental manipulation and hypothesis testing.

II. FACTORS DETERMINING THE COPROPHILOUS FUNGAL NICHE

A. Dispersal Mechanisms

Coprophilous fungi are the most easily recognized component of dung microflora. They
include numerous representatives in the Zygomycetes, Ascomycetes, Basidiomycetes, and
Deuteromycetes (Fungi Imperfecti). In addition, both plasmodial and cellular slime
molds occur commonly on dung. Because fecal substrates are the equivalent of sinking
islands on which the fungal colonists (vegetative cells) are doomed as the dung decomposes and becomes incorporated into the soil, it is critical that coprophilous
fungi have efficient dispersal mechanisms enabling them to colonize fresh substrate.
Many of these fungi are believed to rely almost entirely on a cyclic process involving herbivore ingestion of spores with foliage; germination of spores following passage through the gut; mycelial growth within dung and eventual sporulation thereon;
and dispersal of spores to herbage utilizing phototropic spore discharge mechanisms,[*]
whereby some spores are violently discharged (shot) great distances (Fig. 1). Since

---

[*]Phototropism enables spores to be fired clear of substrate and avoid hitting
surrounding objects.

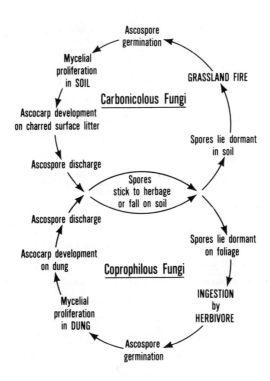

Fig. 1 Schematic representation of the relationships between the life cycles of coprophilous and carbonicolous ascomycetes in a burned prairie. (From Wicklow, 1975.)

it is essential that discharged spores remain affixed to herbage until ingested by a warm-blooded herbivore, one finds that many of these spores show gelatinous outer membranes or appendages and some spores are surrounded by mucilaginous substances. In addition, the spore walls often contain a dark pigment(s) that probably protects them from ionizing radiation and also prevents enzymatic degradation of spore walls during or immediately following gut passage.

The spores of many coprophilous fungi require passage through the gut of a warm-blooded herbivore to trigger their germination. According to Webster (1970) the warm-blooded condition appears to be important in the evolution of the coprophilous habit, because large numbers of fungal species have been recorded from the dung of birds and mammals but very few have been reported from the dung of reptiles and amphibia.

It is not known whether coprophilous fungi are capable of increasing their propagule density (growth) during gut passage. The potential for such growth would provide a distinct competitive advantage, because the fungus, by virtue of increasing its inoculum density, could (1) become a more competitive saprophyte (sensu Garrett, 1956) in individual fecal droppings or (2) distribute inoculum to greater numbers of fecal droppings. *Sporomiella minima*, a coprophilous ascomycete, was repeatedly isolated from the rumen and found to grow under near anaerobic conditions (Brewer et al., 1972).

Many, if not the majority, of coprophilous fungi are homothallic, an adaptation that has increased survival value when an individual spore becomes isolated during dispersal or gut passage in a single fecal pellet or pat. Were the isolated spore from a heterothallic fungus, the developing mycelium, in the absence of an appropriate mating type, would remain sterile—and therefore be unable to escape, in the form of spores, from an otherwise doomed substrate. *Ascobolus furfuraceus* is heterothallic and produces oidia (spermatia) which are dispersed by wind or carried by flies and mites (Dowding, 1931). Oidia of one mating type promote the development of sex organs on a vegetative colony of the opposite mating type (Bistis and Raper, 1963). The absence of a particular insect vector at a time when spore inoculum is ready for dispersal to other dung pats may serve to restrict the distribution of some coprophilous fungi, although such examples have never been documented.

Cleistothecial ascomycetes such as *Kernia nitida*, which do not forcibly discharge spores into the air, produce hooked appendages on their spore-bearing ascocarps which enable the spore mass to be carried on animal hair from one fecal substrate to another. One could expect that such fungi would be particularly suited to colonizing dung of animals which regularly return to the locations at which they deposit feces. This would be important to fungi hitching rides on animal fur because after cleistothecia had matured, the animal producing these feces would return to serve as a vector. According to D. Malloch (personal communication, 1971), cleistothecial ascomycetous fungi are particularly prevalent on the dung of herbivores and carnivores with clearly demarked areas of fecal deposition ("toilets") or showing predictable patterns of movement throughout a territory. In either instance the chances improve that the animal providing a source of substrate will also represent the most reliable vector for fungi colonizing that substrate. Many cleistothecial fungi on carnivore dung represent the perfect stages of fungal pathogens of vertebrates. As such they are often able to attack keratin (carnivore dung often contains animal fur; vertebrates also shed hair, skin, or feathers in their burrows and nests).

B. Importance of Dung Type

On a global scale the niche of any coprophilous fungus includes all of the dung types on which it will sporulate. Likewise, the dimensions of the coprophilous fungal niche in a given ecosystem is determined, in part, by the species composition and relative abundances of the mammals living there. In his presidential address to the British Mycological Society, Webster (1970) observed that coprophilous fungi were generalists in regard to their ability to colonize different types of feces. Citing studies of Massee and Salmon (1901, 1902) based on dung collected at the London Zoo, he notes that there was little evidence of specialization. This might be expected if many of the animals were fed from hay or grain harvested in the south of England and therefore contaminated with a local fungal inoculum. Webster (1970) observed, however, that in a field where cows, horses, sheep, and rabbits graze together, *Coprobia granulata* is usually found only on ruminant dung; therefore, further systematic observations on animals which share a common feeding territory might yield useful information. Since then, there have been several attempts to quantify the relationships between individual species of coprophilous fungi and the feces of various herbivores. Mitchell (1970) recorded a predominance of discomycetes on ostrich feces and an abundance of pyrenomycetes on goat feces colected from the same grassland enclosure. Most coprophilous ascomyetes from Sweden were found to be restricted to one or a few categories of feces (Lundqvist, 1972). Examination of 137 collections of ruminant and lagomorph feces, collected primarily from differ- ent habitats in England and Scotland at different times of the year, revealed that certain fungi were associated with ruminant feces, others with lagomorph feces, and several were common to both fecal types (Richardson, 1972). Angel and Wicklow (1975) calculated similarity indices for communities of coprophilous fungi colonizing cattle, pronghorn, rabbit and small-mammal feces collected from the Pawnee National Grass- land in northeastern Colorado. The communities on ruminant feces (cattle and prong- horn) were most similar in species composition, whereas those on pronghorn and small- mammal feces showed the least similarity. Wicklow and his colleagues (1980) have demonstrated experimentally that when hay from the same field and containing the same fungal inoculum was fed to a rabbit and a sheep housed in separate locations, the coprophilous fungal communities appearing on feces of equivalent sample weight from each animal differed substantially. Pyrenomycetes and loculoascomycetes were found more commony on rabbit feces, whereas discomycetes represented a greater por- tion of the colonists on sheep dung. I wonder whether at the turn of the century, i.e., the time of Massee and Salmon's studies (1901, 1902), the systematics of the coprophilous fungi were as well known as they are today; if quantitative methods were applied to sampling and analysis of fungal populations, a greater number of differences in fungal community composition might have been observed in their studies of fungi on dung from the London Zoo.

In Table 1, I have summarized the distribution patterns of 16 species of *Sporor- miella* selected alphabetically (a-d), on dung types found in Ontario, Canada, and/or western regions of the United States. Professor Roy F. Cain and his former students have assembled at the Cryptogamic Herbarium, University of Toronto, the largest collection of coprophilous fungi in the world. In suggesting *Sporormiella* as a good candidate with which to contrast species distribution, Prof. Cain pointed out that its ascocarps survive for relatively long periods of time on dung whether in the field or in moist chamber cultures. Therefore, it is more likely that members of this genus would be regularly encountered and easily recognized by both graduate students and laboratory assistants who, for over 30 years, assisted Prof. Cain in the screening of dung collections for fungi. The records I examined show plainly that some species were recorded for Ontario but not in the western United States and vice versa. There appear to be greater numbers of species of *Sporormiella* in the

Table 1 Distribution of *Sporormiella* spp. on Dung Types Collected from Ontario and Western Areas of the United States

	Dung type													
Taxa	Prairie dog	Marmot	Prong-horn	Elk	Sheep/ Goat	Cow	Moose	Horse/ Burro	Deer	Undeter-mined	Rabbit	Porcupine	Wolf	Partridge
Sporormiella affinis														
Ontario											6			
Western states		1	1					3		1	6			
Sporormiella alloiomera														
Western states			1	1	3	2	1	2						
Sporormiella ambigua														
Western states						1					1			
Sporormiella anisomera														
Western states						3		1						
Sporormiella americana														
Western states	1		3	2	7	2	6	6	2	2	25			
Sporormiella antarctica														
Ontario											2			
Sporormiella australis														
Ontario					11	12		1	11		15			
Western states			2	5	15			13	10	1	41	15	1	

Species	Region											
Sporormiella bipartis	Ontario	2					2		1	30	1	1
Sporormiella capybare	Western states	1	3	3	7	7		10	4	5		
Sporormiella chaetomioides	Ontario					2			1			
	Western states									3		
Sporormiella commutata	Western states									2		
Sporormiella corynespora	Western states	2			1		2		3	5		
Sporormiella confusa	Ontario						1	3	1	4		
	Western states		1	1	2	10		11	3	1		
Sporormiella cylindrospora	Western states			5	4	4	3	1	1	1		
Sporormiella cymatomera	Western states		1	1	5	19	1		1	4		
Sporormiella dakotensis	Ontario						1					
	Western states					2	1		1	6		

Source: Cryptogamic Herbarium, University of Toronto, Ontario, Canada; specimens recorded prior to December 1974.

western states than in Ontario. Moreover, the data also indicate that some species of *Sporormiella* occurring on dung types common to each region are using different sections of their fundamental niche by switching to the dung of locally occurring host species.

What factors contribute to such differences in substrate preference among coprophilous fungi? First, the nature of the germination stimulus may be related to the specific properties of a digestive system such as (1) duration of interval between ingestion of spores and their being voided with feces and (2) the different physicochemical or microbiological factors and biochemical processes associated with specific segments of an animal's gut (Lodha, 1974). It is also possible that some coprophilous fungi do not survive passage through the digestive system of a particular herbivore, although this has never been experimentally documented. Secondly, the volume, the texture, and the chemical composition of different dung types provide potentially unique environments for fungi surviving gut passage and capable of germinating in feces. Nature and abundance of utilizable energy resources is an important niche dimension. For coprophilous fungi, one segment of that resource gradient might be the amount of available carbohydrate, while another segment might consist of total cellulose, hemicellulose, or lignocellulosic residues. Animals that consume the stems and leaves of mature herbage produce feces with a greater proportion of lignocellulosic residues and undigested plant tissues. Fungi capable of enzymatically attacking lignocellulosic residues such as coprophilous Agaricaceae in pastures may be analogous to white rotting basidiomycetes in forests and are likely to have greater importance and an expanded niche in dung of this type.

Within individual samples of herbivore dung, the composite fungal niche has well-defined spatial limitations. Fungal growth in dung is affected by the degree of aeration or anaerobiosis (Dickinson and Underhay, 1977), by the moisture content and moisture retaining capacity of individual fecal droppings, and by temperatures generated in decomposing feces. Some of the larger dung insects indirectly modify the physical environment for microbial growth by fragmenting the dung or creating tunnels in dung pats.

C. Conditions of Incubation (Habitat and Exposure Interval)

The niche is further defined by geographic or seasonal climatic factors associated with individual ecosystems in which these dung types are deposited. In some temperate climate ecosystems, coprophilous fungi may depend upon fecal pats produced in the late fall or winter when animals are grazing on senescent or dead herbage. These fecal droppings remain intact (frozen) throughout the winter and serve as a source of inoculum for subsequent spring herbage. By fall, herbage has accumulated fungal spores discharged throughout the entire growing season. Therefore, fungal species richness in feces from winter pastures might expectedly be greater, although there is no evidence that such is the case.

To compare, under different environmental regimes, the response(s) of a coprophilous fungal community associated with mesic habitats in Pennsylvania with that
from a semiarid grassland in eastern Colorado, I fed baled alfalfa hay harvested
from each region to separately caged laboratory rabbits. By feeding alfalfa hay
from both sites, rather than the native vegetation, I had hoped to avoid differences
in composition of feces (structural/chemical) which could make the results more difficult to interpret. The effort proved futile because during the period in which
the particular cutting of alfalfa hay was taken from irrigated fields in eastern
Colorado, there was no precipitation. Only four fungal species appeared on the dung
as contrasted with more than 20 species on rabbit dung collected from surrounding
grasslands. It would appear that there is little opportunity for the alfalfa to
become contaminated with spore inoculum discharged from dung in surrounding range
lands if there is not sufficient moisture during this interval to promote spore discharge.

Climatic factors, vegetational cover, and soil properties will determine temperature and moisture relations in fecal substrates and therefore indirectly regulate metabolic activities and competitive abilities of individual colonists. Temperature influences the composition of coprophilous fungal communities on rabbit
dung (Wicklow and Moore, 1974); *Thelebolus* spp. and *Sporormiella intermedia* appeared
only on cohorts of dung incubated at 10°C and not 24°C or 37.5°C. Furthermore, the
order of initial sporocarp appearance of certain individual species pairs at 10°C
was reversed at 24°C. Daily fluctuation of temperatures during the growing season
in temperate ecosystems may explain why psychrophilic species such as *Thelebolus*
(Wicklow and Malloch, 1971) can be found fruiting alongside species with higher temperature optima on the same dung collection.

Microfloral activity can be equated directly with rate of substrate decomposition and therefore with the interval a fecal substrate will be available for fungal colonization. Furthermore, the activities of arthropods, birds, small mammals,
and earthworms in the fragmentation and/or removal of portions of a fecal pat also
affects the niche dimensions of individual fungal colonists. For example, those
fungi whose vegetative development is delayed until after the activity of other microbes has ceased, or those fungi requiring a prolonged period of vegetative development before they can sporulate, would become extinct in habitats where dung rapidly disappears before such sporocarps can be formed. Lodha (1974) provides an interesting tabular summary of the interval (days) between first and last initial appearance of coprophilous fungal species from different taxonomic groupings appearing on
laboratory-incubated dung collected from temperate (incubation temperature = 20-24°C)
and tropical (incubation temperature = 23-25°C) ecosystems. My interpretation of
his data is that communities of coprophilous fungi in tropical ecosystems comprise
species which are capable of initiating sporocarp development within a shorter time
frame than those on dung collected from temperate ecosystems. Rapid maturation of

the fungal community colonizing dung in tropical habitats may be important, because rate of disappearance (decomposition/fragmentation) of dung is undoubtedly more rapid in the tropics.

D. Competitive Interactions

Since coprophilous fungi depend entirely on spores as a means of escaping from spent fecal substrates and reaching fresh ones, sporocarp density can be used as a direct measure of ecological success (fitness). The greater the number of fungal species in a fecal pellet or pat, the greater the likelihood that niches will overlap along certain critical parameters and competition will occur. The outcome of interference or exploitation competition should be reflected in the reduced sporocarp density of each competitor relative to sporocarp densities produced in the absence of competitors. Evidence has been presented that interference competition results in a reduction in sporocarp numbers among coprophilous fungi (Harper and Webster, 1964; Ikediugwu and Webster, 1970).

Park (1954) has shown that the outcomes of competition among *Tribolium* populations depended on environmental conditions in the laboratory. Because competition cannot occur for physical factors of the environment (pH; temperature), reduction in realized niche hypervolume will result from active interactions along gradients other ˌthan temperature and pH. Organisms compete for resources, not regulators. Thus a fungus might be excluded from one habitat not because it cannot grow there but because other coprophilous fungi can take up a limiting nutrient more efficiently under those habitat parameters. An expectation for coexisting species in natural systems is that competition is either not so important or the spatial structure of the environment allows competitively inferior species to be maintained.

E. Predation

Noncompetitive limiting factors may keep populations below the size at which competition plays an important role. The predation hypothesis suggests that predators, by keeping the abundance of prey in check, prevent competitive exclusion and thus permit or maintain a higher species richness than would occur in their absence (Paine, 1966, 1971; Janzen, 1971). Some studies of the effect of grazing on grassland composition (A. J. Nicholson et al., 1970) indicate that herbivores tend to increase the proportion of rare species and decrease the abundance of dominant species in grassland. Studies showing the importance of predator-mediated coexistence in generating species diversity, as well as evidence that predation actually decreases the number of coexisting species, have been cited by Caswell (1978).

I wanted to examine the possibility that predation by dung-inhabiting arthropods may reduce the mycelial biomass of certain dominant fungi and thus enable rarer species to develop a mycelium capable of supporting the development of fungal sporocarp(s). In ecological terms, the niche of the dominant species would be reduced, and this reduction would enable a less competitive fungal species to expand its niche

sufficiently to sporulate and potentially reach another fecal substrate(s). An experiment was designed to study the effects of grazing by larvae of the sciarid fly
Lycoriella mali Fitch. on species diversity of fungal communities in rabbit dung
collected in the laboratory from an animal fed alfalfa hay (D. T. Wicklow and D. H.
Yocom, unpublished data). Fertile eggs that were deposited on the surface of dung
extract agar were transferred to the surface of approximately 2.0-g samples of rabbit
feces in individual moist chambers in petri dishes. Five replicates were inoculated
for each treatment, which consisted of inoculum densities of from 5 to 125 eggs. The
species producing sporocarps on fecal surfaces were recorded following a 30-day incubation period, an interval allowing for both the emergence of all first generation
adults of *L. mali* (Helsel and Wicklow, 1979) and the completion of the life cycle
(sporulation) of nearly all members of the fungal community in these rabbit feces
(Wicklow and Moore, 1974). The effect of increased larval density was to reduce both
the total number of species per treatment (all replicates) and the mean number of
species per replicate. Treatment replicates receiving 75 eggs or more averaged only
three species while the control (no eggs), as well as the treatment receiving only
five eggs per replicate, averaged six species. *Preussia* sp. appeared solely on dung
inoculated at low larval densities. It should be noted that in this study only the
frequency of occurrence of a species was recorded and not its relative abundance
(number of sporocarps) per treatment replicate. Substantial differences in sporocarp
abundance were observed among treatments in which a species was recorded at the same
frequency. These results, obtained from studies of the impact of a dipteran larval
predator (*L. mali*) on populations of coprophilous fungi (prey) in rabbit dung, do not
support theoretical arguments that predation promotes diversity in prey communities.
Instead, we find a decrease in the number of coexisting species under the impact of
predation. Negative effects or zero effects of predation on coexistence have also
been reported by Harper (1969), Paine and Vadas (1969), Hurlbert et al. (1972), Adicott (1974), and Janzen (1976). Lubchenco (1978) examined the effects of the marine
snail *Littorina littorea* on the abundance and type of algal species in tide pools.
She presents evidence that predators or herbivores do not simply increase or decrease
the species diversity of their food but can potentially do both. Lubchenco envis-
ions a more complex field of interrelationships in which the precise effect of a
consumer (herbivore or predator) on prey community oreganization depends on the consumer's feeding preferences, on the food's (prey) competitive abilities, and on the
intensity of the grazing or predation pressure.

In an earlier study designed to examine the effects of larval colonization by
the sciarid fly *L. mali* on the decomposition of laboratory-collected rabbit feces
(Helsel and Wicklow, 1979), it was observed that larvae consume both the mycelium
and sporocarps of coprophilous fungi. The feces used in these experiments were
known to be colonized by approximately 20 species of coprophilous fungi; however,
I noticed that *Chaetomium bostrychodes* Zopf was nearly the sole survivor on samples
colonized by larvae at levels approaching or exceeding the carrying capacity of the

substrate. Ornamented perithecial hairs of *Chaetomium* and related genera (*Achaeto-mella* Arx; *Thielavia* Zopf) may have evolved as a mechanical deterrent against pre-dation by larger fungivorous arthropod detritivores such as dipteran larvae; these hairs enable such species to successfully sporulate on the surfaces of nutrient rich, relatively short-lived substrates (i.e., dung, plant detritus) (Wicklow, 1979). The importance of coprophilous fungi-detritivore interactions in generating diversity among dung-inhabiting organisms should be examined in much the same way that coevo-lutionary theory has been applied to other organism groups (Janzen, 1971, 1977). *Chaetomium* species may also have evolved chemical modes of defense against arthro-pods. For example, *Chaetomium affine* Corda produces chemical defenses (anthraqui-nones) analogous to those of higher plants, and *C. cochliodes* produces chetomin (or chaetomin), long recognized as an antimicrobial antibiotic (Gieger, 1949). Its act-ual function in nature may be that of an insect-feeding retardant (Janzen, 1977).

Mycoparasitism, a form of predation, is a phenomenon often identified with the Mucorales (e.g., *Syncephalis*, *Piptocephalis*, *Chaetocladium*) on other mucorine host species found growing on dung (Zycha and Siepmann, 1969). However, very little can be said about causal factors affecting the distribution and abundance of these mycoparasites or their hosts; therefore, it is not yet possible to assess their role in the organization of fungal communities on fecal substrates.

F. Grassland Fires

Wicklow (1975) observed that a number of ascomycetes recorded from dung collected in a Midwestern prairie also developed ascocarps on burned soil surfaces in the weeks following a prairie fire. Grassland fires provide an alternative to gut passage for initiating fungal vegetative growth and ascocarp development in coprophilous ascomy-cetes. The timing of the development of sporocarps and subsequent spore discharge on burned soil surfaces cooincides with the emergence of prairie vegetation follow-ing spring burning, and therefore discharged spores can become affixed to newly emer-gent prairie plants and upon herbivore ingestion, reenter the coprophilous cycle (Fig. 1). Heat-activated ascospores of coprophilous fungi that were applied to un-burned and unheated prairie soil were unable to complete their life cycle in that sporocarps never developed in such soils during subsequent laboratory incubation (Wicklow and Zak, 1979). The restriction of carbonicolous ascomycetes to burned sites results from their inability to compete successfully in unburned soils (El-Abyad and Webster, 1968). In mesic prairies of the Midwest, fire occurred frequent-ly enough so that a number of coprophilous fungi were periodically able to expand their realized niche (Wicklow, 1975).

III. CRITERIA USED IN IDENTIFYING COPROPHILOUS FUNGI
AS AN EXPERIMENTAL SYSTEM FOR POPULATION AND
COMMUNITY ECOLOGISTS

In order to conduct exacting experiments with fungal populations in their natural condition, it would be helpful if one could manipulate individual species populat-

ions. This means that some control of the inoculum of each fungus composing the natural community would be important and would allow an investigator the opportunity to
affect the inoculum potential (Garrett, 1970) of these species populations and to add
or to delete species from the community. The Cambridge method of studying the competitive saprophytic ability of root-infecting fungi (Garrett, 1970) represents such
an attempt to manage the various fungal components in an artificial laboratory system. The method incorporates a contrived substratum, fungal populations that don't
necessarily interact in nature (see Lussenhop, Chap. 3), and the absence of other
microfloral components generally associated with cultivated soils (i.e., bacteria,
actinomycetes, non-root-infecting saprophytic fungi). Attempts to manipulate fungal populations in natural terrestrial ecosystems with the idea of studying fungal
community organization have been limited to assessing the impact of a major system
perturbation on a target organism (i.e., plant pathogen) or to correlating differences in the propagule or sporocarp density of fungal populations with same.

Experiments with coprophilous fungi can be conducted both in the field or under
controlled laboratory conditions, and the response of the microfungi can be compared
directly. One is not faced with attempting to interpret results based on artificially constructed communities versus naturally occurring ones. Estimates of reproductive success can be conveniently.quantified by examining dung surfaces for sporing structures. Fungi which fail to produce spores cannot escape from decomposing
fecal substrates and therefore may become locally extinct.

Using direct hyphal counting procedures one can relate reproductive potential
of single or mixed species to the total fungal biomass in the fecal substrate on
which the sporocarps are produced. Microbial decomposition of dung represents a
naturally occurring solid substrate fermentation. Time-course studies involving gas
chromatographic headspace analysis of incubating solid substrates provide a quick
and accurate measure of carbon dioxide production and oxygen consumption or the respiration of microbial growth. Such methods can easily be applied to the dung microcosm in measuring respiratory activity at various stages in development of a fungal
community and during decomposition of the substrate.

Fecal substrates represent a complete resource base for all members of the fungal community. Variation in the resource quality of any habitat or ecosystem will
not directly affect the ability of individual fungal populations to grow and produce
spores. Substrate units consisting of nearly identical propagule numbers and kinds
of fungal species can be easily prepared and handled as experimental units without
any disturbance to the resident fungal community: (1) Vegetation from a given field
can be harvested and fed to caged animals, and the collected feces can be shown to be
populated with coprophilous fungi determined by natural field inoculum. (2) Wild
or domesticated herbivores can be placed in predator-safe enclosures in the field,
and the feces, containing inoculum from ingested herbage, can be collected periodically. (3) Different combinations of coprophilous fungal spores can be added to commercial, pelletized rabbit ration that is otherwise free of coprophilous fungi.

Spores of potential fungal contaminants are for the most part unable to survive gut passage (Harper and Webster, 1964). Lodha (1974) implies that several species of imperfect fungi (e.g., *Aspergillus* spp., *Alternaria alternata*, *Chlamydomyces palmarum*, *Cladosporium herbarum*, and *Trichoderma lignorum*) found on the dung samples he examined from India may have survived gut passage, because Bonner and Fergus (1959) recorded them from cattle food. These very species, however, are also commonly isolated from soil and from air exposure plates. In our efforts to collect rabbit feces aseptically from caged animals fed sterilized rabbit chow, such contaminants have been a formidable problem. Presumably the dung samples examined by Lodha were not collected aseptically.

Complete expression (sporulation) of the resident fungi in smaller pellet-like feces (of rabbit, sheep, deer, etc.) usually takes less than 40 days in the laboratory. Laboratory incubation of cattle feces representing different age classes, from habitats in which fecal decomposition rates are restricted by low levels of available moisture, produces fungal communities characteristic of the particular age class examined (Angel, 1977). Individual developmental responses to experimental manipulation of feces can be recognized readily, since the interval between spore germination (in fresh-moistened feces) and sporocarp appearance of individual species growing alone on these feces is consistent at a particular incubation temperature (Harper and Webster, 1964; Wicklow and Moore, 1974). Lodha (1974) points out that the earliest date of appearance for fungal species colonizing more than one dung type varies according to dung type. For example, *Lophotrichus ampulus* sporulated on rat dung after 14 days of laboratory incubation, whereas it appeared on monkey dung after 22 days. Moreover, *Saccobolus versicolor* produced sporocarps on cow dung from India after 14 days of laboratory incubation, but in Denmark it first appeared on deer dung after 40 days. It is important to recognize that before Harper and Webster (1964) demonstrated experimentally that patterns of initial fungal sporocarp appearance on rabbit dung are not affected by the presence or absence of other fungal populations, it was generally believed that the activities of earlier colonists, in depleting substrate nutrient reserves, influenced patterns of growth and sporulation among the later forms (Burges, 1958; Garrett, 1956). Others have now shown that lignin-degrading basidiomycetes do not necessarily occur at the end of the sequence, but rather they fruit together with the discomycetes, pyrenomycetes, and loculoascomycetes (Larsen, 1971; Lodha, 1974; Wicklow and Moore, 1974).

My enthusiasm for the dung microcosm as a model system for ecological studies with fungi can be traced to a paper by Harper and Webster (1964) in which they introduced the technique of combining various species of coprophilous fungi onto sterile rabbit food and then feeding this inoculated diet to 6-month-old rabbits that were starved for 48 hr prior to feeding. The rabbits were housed in sterile metal cages with sterile metal collecting funnels and were given only sterile distilled water. Rabbit feces were collected on the three successive mornings following feeding with inoculated diet and incubated on moist filter paper in petri dishes under fluorescent light. Using this technique, Harper and Webster (1964) were able to study the effects of competitive interactions among coprophilous fungal populations on the time required for the production of mature fruiting bodies of individual species, either growing alone or in combination with other fungi. What is uniquely important about this approach is that the normal gut microflora and the structural integrity of the fecal samples are preserved in the natural condition, and the only difference is the inoculum density of individual spore populations associated with an experimental sample cohort. Consider for a moment the difficulty a plant or animal ecologist would have in effectively duplicating such experiments using other communities from aquatic or terrestrial ecosystems without substantial destructive or artificial manipulation of the ecosystem involved.

Ikediugwu and Webster (1970) used this procedure to evaluate the effect of *Coprinus heptemerus* M. Lange and A. H. Smith on the sporulation of *Ascobolus crenulatus* Karst. *Pilaira anomala* (Ces.) Schroet., which Harper and Webster (1964) found not to affect sporulation of *A. crenulatus* in mixed culture, was employed in combination with *A. crenulatus* to serve as a control for the combination involving *A. crenulatus* and *C. heptemerus*. Mixed spore suspensions were prepared by combining 3 ml portions of a suspension of each spore type at a concentration of about 5000 per milliliter. Estimates of sporocarp abundance and sporulation included the following: (1) examination of 40-50 individual pellets incubating on moist filter paper in petri dishes at 3-day intervals (28 days) for the presence of fungal contaminants (*Mucor*, *Actinomucor*, *Circinella*, *Chaetomium*, *Cladosporium*, and *Penicillium* appeared on a number of rabbit pellets and were probably airborne contaminants from the animal room); (2) examination of petri dish lids (replaced every 3 days) for discharged fungal spores; and (3) examination of fecal pellets and filter paper for mature apothecia of *A. crenulatus* and quantification of these ascocarps on the tenth day of incubation. Harper and Webster (1964) observed that fewer apothecia of *A. crenulatus* were produced in the presence of *C. heptemerus*, that these apothecia were smaller in size and ill developed, and that they discharged spores for a shorter interval, as contrasted with pellets in which *C. heptemerus* was not present. In addition, Harper and Webster (1964) present evidence that *C. heptemerus* was antagonistic to and limited the fruiting of *P. anomala* and *Pilobolus crystallinus* under the same experimental conditions. Ikediugwu and Webster (1970) experimentally determined that "hyphal interference," a mode of antagonism which operates on contact or at close proximity to the hyphae of the antagonist (*C. heptemerus*), brings about a drastic alteration of the permeability of the cell membrane leading to cell death in *A. crenulatus* and *P. crystallinus*. They found no evidence that a stable antifungal antibiotic was produced by *Coprinus*, nor could they demonstrate that nutrient depletion was responsible for the antagonistic effect. What is particularly significant about these results from an ecological point of view is the fact that Harper and Webster (1964) and later Ikediugwu and Webster (1970) were able to experimentally alter the realized niche of certain coprophilous fungi by incorporating an aggressive competitor into the fungal community.

Such a technique could also be used to measure the stepwise, cumulative effects of species additions (species packing) on the species composition and relative abundance (a partial measure of fitness) of sporocarp populations. Species additions presumably lead to overlap along critical niche parameters and thus competition occurs. It would be interesting to determine whether one might demonstrate "higher order" competitive interactions (McNaughton and Wolf, 1973) by using abundance of sporulation to learn whether the competitive effects of species A and B on species C are additive or whether there is a competitive coalition between A and B resulting in a more serious depression of C. Some effort must be made to determine how important it is that each population of coprophilous fungal spores have equal spore densities at the beginning of any experiment. In other systems, the outcome of competition may depend on which of two species achieves numerical predominance first (Slobodkin, 1961).

It should be pointed out that in both plant and animal populations, intraspecific competition related to high initial colonization densities could be expected to have an adverse effect on the population whereas individual germlings of the same fungus are often capable of hyphal anastamosis. The latter enables fungi to avoid the potentially disastrous effects of intraspecific competition for substrate resources. It is easy to envision situations in which there is enough substrate to support the germination of numerous fungal spores but not enough to allow all of the germlings in turn to produce sporocarps. In similar cases of intraspecific competition involving plant or animal populations, some genotypes would become lost from the population while the better competitors would survive and reproduce. However, hyphal anastomosis would enable less competitive genotypes (haploid nuclei) to survive in the composite mycelial network. Webster (1970) reports that in an experiment conducted in his laboratory in which wild-type black ascospores of *Sordaria fimicola* and ascospores from a "melanin-free" white-spored mutant were mixed together on pelletized chow fed to a caged laboratory rabbit, hybrid perithecia containing both white and black ascospores were observed on the rabbit's feces. If different physiological or biochemical characteristics (e.g., the ability to enzymatically attack specific organic compounds) were linked to a given ascospore color mutant, there would be a unique opportunity for research that applies the methodology of biochemical genetics to ecological studies designed to evaluate the relative importance of individual physiological or biochemical attributes of coprophilous fungi in defining niche dimensions. In the process of free cell formation in the ascus, ascospore wall pigmentation is regulated by the haploid (n) nucleus it envelopes. Therefore, the number of surviving mutant versus wild-type nuclei in any heterokaryon will be expressed as a ratio of black to mutant (colored) ascospores produced on fecal surfaces. An approximate measure of the fitness attributed to individual biochemical attributes might be estimated by calculating the ratio of black to white ascospores following sporocarp maturation on feces containing mostly black-spored inoculum, mostly white-spored inoculum, or equal numbers of each spore type. One

would have to work with species in which vegetative incompatibility is not a problem. I believe that this approach has enormous potential as an experimental system
for workers in a number of related disciplines (e.g., population genetics, population and community ecology, fungal/microbial ecology, and evolutionary biology) in
that performance of mutant genotypes can be measured in situ against a background
of naturally occurring dung microbes and potential competitors.

IV. FUNGAL COMMUNITY DIFFERENTIATION AND STAGE OF ECOSYSTEM MATURATION (DUNE SUCCESSION)

Explanations for variations in fungal community composition within different habitats or ecosystems have traditionally been linked to differences in the diversity of
the vegetation among the study areas and the diversity of the primary resources available for fungal colonization (Christensen et al., 1962; Hering, 1965; Swift, 1976).
The importance of microclimate in determining the structure of fungal communities
has not been documented experimentally.

An experimental plan was devised in which laboratory-collected rabbit feces,
from animals fed baled alfalfa hay, were exposed in different habitats (sand dune
exposed site, sand dune beneath *Ammophila*, drift beach, *Myrica* thicket, and oak-
maple forest) associated with the dune succession at Presque Isle for either 2-, 4-,
or 10-week intervals (Fig. 2) and then returned to the laboratory for examination
and subsequent moist chamber incubation.

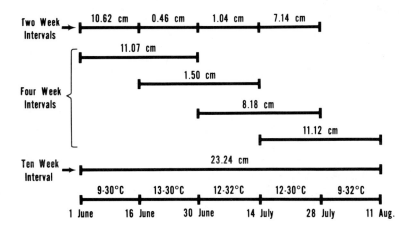

Fig. 2 Maximum-minimum temperatures (°C) and total precipitation for field exposure
intervals (time line) at Presque Isle State Park, Pennsylvania, 1975.

Individual microclimates of the plant communities associated with the dune succession at Presque Isle (northwestern Pennsylvania)[*] form a type of stress gradient, at one extreme of which is the relatively stable microclimate within a deciduous forest, and at the other extreme is the comparatively harsh (unstable) microclimate of an exposed dune. Therefore, it was possible to examine the effects of such microclimatic regimes on species diversity of coprophilous fungal communities developing on samples of rabbit dung exposed in different successional habitats (Yocom and Wicklow, 1980). It is important to remember that fecal substrates provide a complete resource base for a self-contained fungal community that is independent of microhabitat.

Fungal community structure and composition differed substantially among experimental cohorts, responding to both the habitat in which samples were exposed and the climatic patterns unique to each exposure interval. A community ordination of all sample cohorts (all habitat/4-week exposure interval combinations) appears to position the communities along the first axis (Fig. 3) according to a moisture gradient combining the effects of soil moisture content, relative humidity, and rate of moisture loss from dung due to habitat factors and climatological events during particular exposure intervals.

Individual fungal distribution patterns suggested that microclimate of the habitat affects fungal community composition. For example, four species of *Podospora* occurred most frequently on dung exposed in the *Myrica* thicket (i.e., *P. curvicolla*, *P. glutinans*, *P. pleiospora*, *P. tetraspora*), while *P. miniglutinans* attained its highest frequency on samples placed beneath dense culms of *Amnophila*, and *P. fimicola* appeared most frequently on samples exposed in the oak-maple forest. Climatological patterns associated with each exposure interval also affected the development of individual fungal populations. *Iodophanus carneus*, *Podospora curvicola*, *P. pleiospora*, and *Thielavia* sp. appeared at higher frequencies on samples exposed during the first and final 4-week intervals, which received 11.1 cm of rainfall, than during the second and third 4-week intervals, which received less rain. Another group of species (i.e., *Ascobolus furfuraceus*, *Coprotus marginatus*, *C. lacteus*, *C. sexdecimsporus*, *Lasiobolus* sp. 1, *P. miniglutinans*, *P. tetraspora*, *Saccobolus depauperatus*, *Sporormiella intermedia*) attained their highest frequency among samples exposed during the first 4-week interval. During this interval there was a regular cloud cover and slightly lower temperatures. Substrate moisture at the soil surface was therefore higher on sites that at other times might have been dry and less supportive of fungal growth. In contrast, *Sporormiella minima* was recorded from every sample, whereas *Chaetomium bostrychodes*, *Sordaria fimicola*, and *S. humana* developed on more than 80% of the samples irrespective of habitat or exposure interval.

Sporocarp densities were estimated by counting individual sporocarps on visible surfaces of rabbit pellets (Yocom, 1977). *Sordaria* spp. and *Sporormiella* were

---

[*]A peninsula in Lake Erie forming Presque Isle Bay.

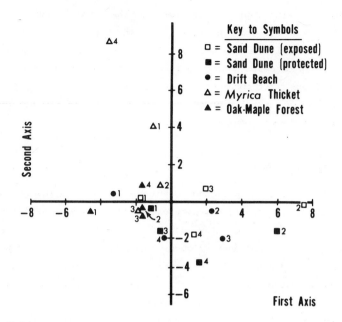

Fig. 3 Ordination of coprophilous fungal communities. Similarity coefficients
based upon differences in the frequencies (%) of occurrence of individual species
following 4 weeks of field exposure and 40 days of moist chamber incubation (Yocom
and Wicklow, 1980).

the principal genera responsible for high densities of sporocarps on certain samples.
Mean numbers of *Sordaria* and *Sporormiella* ascocarps were always considerably lower
on fecal samples exposed in the *Myrica* thicket and the oak-maple forest. This is
possibly correlated with the increased incidence of *Podospora* spp. recorded in the
former habitat. Yocom (1977) further observed that increased sporocarp density
of *Sordaria* could be correlated with a decrease in species richness of the fungal
community. Initial conditions of field exposure in each site favor the early development of a few species (e.g., *S. fimicola* and *S. humana*). These fungi might be
compared to the initial colonists on an island, which fill the available niche space
and make later colonization by other potential colonists more difficult (MacArthur,
1972). Therefore, at the time samples are returned to the laboratory and incubated
under uniform conditions, those fungi that initially become established (vegetative
growth) during field exposure strongly influence patterns of fungal growth and sporocarp abundance.

Our research at Presque Isle demonstrated clearly that local microclimate will
determine, in part, the final structure and composition of a fungal community as determined by the reproductive success of individual populations. Initially, each sample of laboratory-collected rabbit feces had the same potential in terms of the upper
limits of fungal species diversity. Therefore, microclimate was the principal factor
regulating fungal community sporulation patterns. The mean number of species per
habitat and exposure interval combination showed considerable variation, between-

habitat differences being significantly lower on the exposed dune site as contrasted with differences among all other habitats in which dung samples were exposed. In this instance, it is believed that widely fluctuating physical environments on the dune impose stresses on certain components of the microflora in dung. This study represents one of the first documented examples in which it can be shown that fungal species diversity of a naturally occurring community is affected by abiotic, micro-climatic factors of the environment.

V. FUNGAL COMMUNITY ORGANIZATION IN SHORTGRASS, MIXED-GRASS,
    AND TALLGRASS PRAIRIES

The study at Presque Isle was designed to examine the near-term effects of habitat and exposure interval on fungal community composition. Try to envision a hypothet-ical set of circumstances whereby the individual habitats at Presque Isle were geo-graphically isolated from one another such that spore inocula from coprophilous fun-gi in one habitat could not easily be dispersed to vegetation in another habitat. Furthermore, individual populations of the eastern cottontail rabbit would also be restricted to consuming only the local vegetation. Within a relatively short time, one might expect to find rather unique communities of coprophilous fungi on the rab-bit dung collected from each habitat. Because cattle dung and rabbit dung can easily be collected throughout the prairie regions of the Midwest as well as the Great Plains, it is possible to examine the long-term effects of local environment on the organization of coprophilous fungal communities.

     This problem was taken up by Angel and Wicklow (1975) and by Angel (1977) in a study of coprophilous fungal communities in three grassland ecosystems (U.S. Inter-national Biological Program study sites) including a semiarid, shortgrass prairie site (Pawnee National Grassland, Nunn, Colorado), a mixed-grass prairie site (Cotton-wood, South Dakota), and a tallgrass prairie site (Osage Site, Shidler, Oklahoma). A comparison is presented in Table 2 of some selected site parameters at each loca-tion.

     Moisture availability appears to be the most important factor in determining fungal community composition at different sites. Communities on dung from drier sites (Pawnee and Cottonwood) were more similar to one another (Sorensen's similar-ity coefficient (SC) = 53) than to communities from the more mesic Osage site (Pawnee versus Osage, SC = 33; Cottonwood versus Osage, SC = 37). Of the 70 fungi recognized as being important species in Angel's study (i.e., occurring at a frequency of 33% or greater on any dung collection within a given site), 14 species were found only on dung collected from the Pawnee site, 11 were limited to fecal samples collected from the Osage site, and 2 were found only on dung from the Cottonwood site. Fifteen coprophilous fungi were common on dung from the three sites.

     Community similarity coefficients were also calculated for all age classes of cattle dung within individual sites. Angel (1977) determined that as dung ages on the Pawnee site, the resident fungal community shows progressively less similarity

Table 2 Comparison of Some Selected Parameters Associated with Three Grassland Ecosystems

Site	Elevation (m)	Grassland type	Dominant grasses	Average annual max. temp.	Average annual min. temp.	Average annual precipitation(cm)	Average date of lilac bloom
Pawnee	1430	Shortgrass	*Bouteloua gracilis* *Buchloe dactyloides*	18°C	2.0°C	30	May 10
Cottonwood	850	Mixed-grass	*Agropyron smithii* *Stipa viridula*	17°C	1.0°C	38	May 20
Osage	380	Tallgrass	*Andropogon gerardi* *Panicum virgatum*	22°C	8.3°C	94	April 20

Source: From Angel (1977).

to the initial community in fresh dried feces. Fungal communities associated with
a particular age class of feces were most similar to those communities on dung samples nearest in age. Because the number of important species on each age class of
cattle feces remained relatively consistent (20-28, $\bar{X}$ = 23) throughout this 54-month
period of field incubation, changes in species composition (species replacement) and
in the relative importance of individual species (frequency of occurrence) accounted
for observed differences in fungal community similarity.

In semiarid grasslands, the slow rate of decomposition and fragmentation of cattle feces allows for the accumulation of different age classes of dung, each offer-
ing a somewhat unique assemblage of physicochemical environment(s) and therefore extending the niche axes (options) available to potential fungal colonists. The theory
that spatial heterogeneity leads to greater within-habitat diversity (Simpson, 1964;
MacArthur, 1965, 1972; Pearson, 1971)--and therefore more potential niches for species to occupy--may best explain the greater diversity of coprophilous fungi on the
Pawnee site (Angel, 1977).

Coleman et al. (1973) observed that moisture variability in semiarid grasslands
limits the numbers and kinds of consumers that can occupy such a system. Reduced
moisture levels might interrupt the growth and fruiting body development of dung
fungi but would not eliminate them from the site. It also seems apparent that the
longer feces survive fragmentation and eventual incorporation into the mineral soil,
the greater chance they have of becoming colonized by additional fungi. Earthworms,
which are very important in determining the lifetime of individual dung pats in mesic
grasslands, appear to be absent from the Pawnee site. In tallgrass prairies, earthworms rapidly consume fecal substrates, particularly during moist periods in the
spring (personal observation). Destruction of fecal habitats by earthworms and/or
losses of potential spore inoculum through grassland fires could be considered catastrophic events in the life histories of many coprophilous fungi and may serve to exclude them from mesic sites. To a coprophilous fungal detritivore, the Pawnee site
could represent the more stable environment in that the fecal resource base (e.g.,
the dung pat) remains intact for up to 10 years.

Except for phenomena associated with spore discharge and dispersal (Ingold,
1965), the mycological literature contains little information as to the potential
adaptive significance of individual morphological or physiological characteristcs
of coprophilous fungi. The information is descriptive in nature and there has been
only one attempt to link a particular character with the natural distribution of the
fungus, namely, the observation by Wicklow and Malloch (1971) that members of the
genus *Thelebolus* are psychrophilic and come from regions experiencing seasonal low
temperatures (at high altitudes and/or high latitudes). Wicklow and Angel (unpublished data) were able to identify habitat-related distributional patterns for certain morphological characters of coprophilous ascomycetes. Bristles (setae) and
tufts of hair surrounding an ascocarp or ostiolar opening may serve as a defense

against mycophagous invertebrates such as slugs (Buller, 1922). Setae were encountered more frequently among species of ascomycetes recorded from the wettest grassland site (Osage = 21%) as compared to species from drier sites (Cottonwood = 10%; Pawnee = 13%) where slugs are unknown. Another interpretation may be that the setae, through capillary action, can break a water film which might cover an ostiole and thus prevent optimal firing of spore guns. Were this the case, such appendages would be useless in a semiarid environment.

Ingold (1965) presents evidence that larger ascospore projectiles can be discharged further than smaller ones. There was a greater proportion of species with multispored asci (>8 spores/ascus) on the Osage site (11%) as compared with drier sites (Cottonwood = 0%; Pawnee = 2%). This adaptation would be expected to have greater significance in mesic sites, where taller grasses surrounding a dung pellet or pat might interfere with dispersal.

A third morphological character, that of the enclosed ascocarp or cleistothecial habit, represents an evolutionary shift away from the condition in which a fungus relies on violent spore discharge as its principal mode of dispersal (Cain, 1956; Malloch and Cain, 1971). Five species of cleistothecium-producing ascomycetes were recorded from the Pawnee dung collections, two from Cottonwood, and none from Osage. Each of these cleistothecium-forming species develops elaborate hairs or appendages on the ascocarp that aid in dispersal by passively attaching themselves to the fur of mammals or the bodies of insects which come in contact with dung surfaces. If the cleistocarpous habit is favored in semiarid ecosystems, it may be connected with the longevity of dung pats and the increased probability that a cleistothecium containing spores will be around long enough for a suitable animal vector to function in spore dispersal. In mesic grasslands, the cleistothecium would face a comparatively shorter life span on the dung and, since it has no other means of escape other than by animal vectors, might soon face extinction. What is particularly important about these observations is that they represent the first time that specific morphological attributes of microscopic fungi colonizing a given substrate have been linked through plausible biological explanations to ecological success in different ecosystems.

VI. DECOMPOSITION PROCESSES, SECONDARY PRODUCTIVITY, AND THE HETEROTROPHIC FOOD WEB

Although there are numerous estimates available for the decomposition rates of dead plant materials (Witkamp, 1966; Thomas, 1968; Reiners and Reiners, 1970; Witkamp and Frank, 1970; Hughes, 1971; Kaushik and Hynes, 1971; Van Cleve, 1971; and Mason, 1976), ecologists know comparatively little about rates of decay of animal remains or their feces (invertebrate feces: Nicholson et al., 1966; Mason and Odum, 1969; McDiffett, 1970; vertebrate feces: Angel and Wicklow, 1974; Booth, 1977). While the decomposition of invertebrate remains is largely attributed to soil microflora, feces of

large herbivores support a diversified community of fungal detritivores in addition
to populations of intestinal or soil bacteria.

Because fungal and bacterial growth takes place within the substratum and the
production and biodegradation of microbial biomass is occurring simultaneously with
the degradation of digested plant remains in feces, it is difficult to distinguish
between the amounts of caloric loss from either of these substrates. Moreover, the
adult arthropods and larvae colonizing the feces of vertebrates can be expected to
leave enriched fecal pellets as well as pupal cases and dead larvae within the matrix
of the fecal substrate they inhabit. It is much easier for ecologists to trace energy flow through detritivore systems in which a macroconsumer can be separated from
food and feces at the end of an experiment and the amount (standing crop) and rate
of assimilation (productivity) can be calculated.

Angel and Wicklow (1974) measured decomposition losses (weight and caloric) in
rabbit feces incubated 40 days at 10°, 24°, or 37.5°C in order to evaluate the influence of incubation temperature on decomposition rate. There was a general relationship between the rate of weight loss and the time required for certain coprophilous
fungal poulations to produce sporocarps on fecal surfaces as determined by Wicklow
and Moore (1974). This suggests that there is a connection between environmental
regulation of successional patterns involving sporocarp production by the fungal
community and the amount of biomass conversion and heat loss brought about by the
activities of the fecal microflora. Fecal weight losses of laboratory-collected
rabbit feces after 40 days incubation in moist chambers were as follows: 10°C =
11.4%; 24°C = 20.1%; 37.5°C = 26.4%. These relatively high rates of decomposition
for rabbit feces, as compared with other types of plant litter, pointed not only to
the nutrient-rich status of herbivore feces but also to the ecological specialization of the resident microflora. Indirect evidence is offered by Angel (1977) that
the microfloral community in aged and partially decomposed feces can efficiently
attack the more recalcitrant lignocellulosic structures in digested and decomposed
plant remains. Using uniform conditions of laboratory incubation, she compared decomposition rates for cohorts of cattle dung representing various ages (<1, 6, 18,
30, and 54 months) collected from the Pawnee National Grassland. While the composition of the fungal community changed considerably during field exposure, fecal weight
losses during laboratory incubation (80 days), with one exception, did not vary significantly (p > 0.05) among cohorts.

To measure the effects of various fungal species combinations on rate of fecal
decomposition, various coprophilous fungal species were combined in rabbit feces according to the method devised by Harper and Webster (1964). My initial expectation
was that the greater the number of species occupying a fecal substrate, the more rapid would be its rate of decomposition. This was predicated on the assumption that
the greater the diversity of the fungal decomposer community, the more enzymatic-bio-
chemical diversity it would be likely to muster in attacking fecal substrates.
The results (Wicklow and Yocom, 1981) that these authors summarized in Fig. 4 sug-

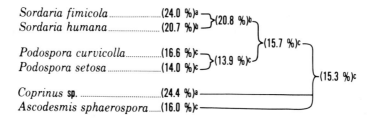

Fig. 4 Diagrammatic representation of effect(s) of single species of coprophilous fungi and various species combinations on decomposition (%) of rabbit feces (Wicklow and Yocom, 1981). Values (a, b, or c) differ significantly (p<0.05).

gest that, when two or more species are combined in the same fecal sample, some form of interference competition serves to limit the decomposition rate of each treatment to approximately that obtained by the fungal partner showing the slowest rate when occurring singly in the dung. In this example, interference competition would result in a decreased ability of competing microbes to utilize available substrate resources as has been theorized for other groups of organisms (Elton and Miller, 1954; Park, 1954; Miller, 1967) where interference involves some form of behavioral or chemical interction between individuals prior to the use of the resource (McNaughton and Wolf, 1973). Perhaps the efficiency of the multiple-species treatment as compared to one species alone or two-species combinations is only observed in a fluctuating environment where each memer of the fungal community is able to operate most efficiently at different points along temperature/moisture gradients.

Apart from seasonal field observations that describe rates of fragmentation or removal of feces by arthropods (Laurence, 1954; Anderson and Coe, 1974), only one attempt has been made to examine the total metabolic activity of dung colonized by a given population of arthropods (Helsel and Wicklow, 1979). This study examined the potential of the sciarid fly Lycoriella mali Fitch to affect total fecal metabolism by measuring gross decomposition (weight/caloric) losses of rabbit feces. Under conditions of high larval density and intense grazing pressure, fecal weight and caloric losses (30 days) were double those from fecal samples incubated without flies. The authors also observed directly that larvae consumed both the mycelium and sporocarps of coprophilous fungi. While the study represents a "black box" approach to the problem of arthropod detritivore-microfloral interactions in feces, and although it leaves many unanswered questions related to even the most basic properties of the system, these observations and those of earlier studies (Angel and Wicklow, 1974) do illustrate the potential usefulness of the dung microcosm for decomposer-consumer studies involving ecosystem function analysis and component interaction analysis.

A working model of this decomposer subsystem based on data from direct microscopic counts of fungal hyphae and bacterial cells at various stages in the decomposition of fresh feces and in the presence and absence of insect colonists is presently

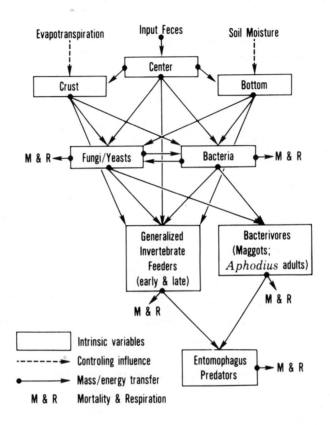

Fig. 5 Trophic structure and energy flow pathways in the dung microcosm.

being developed. Differences in microclimate, nutrient composition of feces, fecal decomposition rates, microbial biomass, and biomass of macroinvertebrate detritivores (e.g., dipteran larvae, beetle adults, and larvae) are examined and their interactions related (Fig. 5).

VII. SUMMARY

Coprophilous fungi allow us to overcome a fundamental problem in describing the niche of a species. At the microcosm level, both the fundamental niche and the realized niche can readily be determined and experimentally manipulated. The response of individual fungal poulations to these fecal environments, as well as to the other populations of microbes composing the microfloral community, largely determines the breadth of its niche and therefore the success (fitness) of a particular fungus as measured by sporocarp/spore production. By measuring changes in the relative numbers of sporocarps, it is possible to obtain a direct measure of the response of different members of the fungal community to phenomena such as interspecific competition as affected by changes in environmental parameters during field or laboratory incubation. Moreover, all of these fungal interactions can be measured against the background of decomposition losses, an ecosystem process. The dung microcosm provides experimental options for the ecologist that cannot be found in other experimental systems.

ACKNOWLEDGMENTS

Some of the research on which this paper is based was supported by National Science Foundation Grant DEB 75-21995. I am especially grateful to Drs. Roy Cain, David Coleman, Mel Dyer, Jack Lloyd, John Lussenhop, David Malloch, Charles Ralph, Richard Weigert, and particularly Richard Seifert for their encouragement and helpful suggestions at various stages in the development of the research program.

REFERENCES

Adicott, J. F. (1974). Predation and prey community structure: an experimental study of the effect of mosquito larvae on the protozoan communities of pitcher plants. *Ecology 55*: 475-492.

Anderson, J. M., and Coe, M. J. (1974). Decomposition of elephant dung in an arid, tropical environment. *Oecologia 14*: 111-125.

Angel, K. (1977). Structure, composition, and development of coprophilous fungal communities in three grassland ecosystems. Ph.D. thesis, University of Pittsburgh, Pittsburgh, Pa.

Angel, K., and Wicklow, D. T. (1974). Decomposition of rabbit feces: an indication of the significance of the coprophilous microflora in energy flow schemes. *J. Ecology 62*: 429-437.

Angel, K. and Wicklow, D. T. (1975). Relationships between coprophilous fungi and fecal substrates in a Colorado grassland. *Mycologia 67*: 63-74.

Bistis, G. N., and Raper, J. R. (1963). Heterothallism and sexuality in *Ascobolus stercorarius*. *Amer. J. Bot. 50*: 880-891.

Booth, T. (1977). Muskox dung: its turnover rate and possible role on Truelove Lowland. In *Truelove Lowland, Devon Island, Canada: A High Arctic Ecosystem*, L. C. Bliss (Ed.). Univ. of Alberta Press, Edmonton, Alberta, Canada.

Bonner, R. D., and Fergus, C. L. (1959). The fungus flora of cattle feeds. *Mycologia 51*: 855-863.

Brewer, D., Duncan, J. M., Safe, S., and Taylor, A. (1972). Ovine ill-thrift in Nova Scotia. IV. The survival at low oxygen partial pressure of fungi isolated from the contents of the ovine rumen. *Can. J. Microbiol. 18*: 119-128.

Buller, A. H. (1922). Researches on fungi. II. Further investigations upon the production and liberation of spores in hymenomycetes. Longmans, Green, London.

Burges, A. (1958). *Micro-organisms in the soil*. Hutchinson, London.

Cain, R. F. (1956). Studies of coprophilous ascomycetes. II. *Phaeotrichum*, a new cleistocarpous genus in a new family, and its relationships. *Can. J. Bot. 34*: 675-687.

Caswell, H. (1978). Predator-mediated coexistence: a nonequilibrium model. *Amer. Natur. 112*: 127-154.

Christensen, M., Whittingham, W. F., and Novak, R. O. (1962). The soil microfungi of wet-mesic forests in southern Wisconsin. *Mycologia 54*: 374-388.

Coleman, D. C., Dyer, M. I., Mills, J. E., French, N. R., Gibson, J. H., Innis, G. S., Marshall, J. K., Smith, F. M., and Van Dyne, G. M. (1973). Grassland ecosystem evaluation. *Bull. Ecol. Soc. Amer. 54(2):* 8-10.

Dickinson, C. H., and Underhay, V. H. S. (1977). Growth of fungi in cattle dung. *Trans. Brit. Mycol. Soc. 69:* 473-477.

Dowding, E. S. (1931). The sexuality of *Ascobolus stercorarius* and the transportation of the oidia by mites and flies. *Ann. Bot. 45:* 621-637.

El-Abyad, M. S. H., and Webster, J. (1968). Studies on pyrophilous discomycetes. II. Competition. *Trans. Brit. Mycol. Soc. 51:* 369-375.

Elton, C., and Miller, R. S. (1954). The ecological survey of animal communities with a practical system of classifying habitats by structural characters. *J. Ecol. 42:* 460-496.

Garrett, S. D. (1956). *Biology of Root-Infecting Fungi.* Cambridge Univ. Press, New York.

Garrett, S. D. (1970). *Pathogenic Root-Infecting Fungi.* Cambridge Univ. Press, New York.

Gieger, W. B. (1949). Chaetomin: an antibiotic substance from *Chaetomium cochliodes.* III. Composition and functional groups. *Arch. Biochem. 21:* 125-131.

Harper, J. L. (1969). The role of predation in vegetational diversity. *Brookhaven Symp. Biol. 22:* 48-62.

Harper, J. E., and Webster, J. (1964). An experimental analysis of the coprophilous fungal succession. *Trans. Brit. Mycol. Soc. 47:* 511-530.

Helsel, E. D., and Wicklow, D. T. (1979). Decomposition of rabbit feces: Role of the sciarid fly *Lycoriella mali* Fitch in energy transformations. *Can. Entomol. 111:* 213-217.

Hering, T. F. (1965). Succession of fungi in the litter of a Lake District oakwood. *Trans. Brit. Mycol. Soc. 48:* 391-408.

Hudson, H. J. (1968). The ecology of fungi on plant remains above the soil. *New Phytol. 67:* 837-874.

Hughes, M. K. (1971). Seasonal caloric values from a deciduous woodland in England. *Ecology 52:* 923-926.

Hurlbert, S. H., Zedler, J., and Fairbanks, D. (1972). Ecosystem alteration by mosquito fish (*Gambusia affinis*) predation. *Science 175:* 639-641.

Ikediugwu, F. E. O., and Webster, J. (1970). Antagonism between *Coprinus heptemerus* and other coprophilous fungi. *Trans. Brit. Mycol. Soc. 54:* 181-204.

Ingold, C. (1965). *Spore Liberation.* Oxford Univ. Press (Clarendon), New York, 210 pp.

Janzen, D. H. (1971). Seed predation by animals. *Ann. Rev. Ecol. Syst. 2:* 465-492.

Janzen, D. H. (1976). The depression of reptile biomass by large herbivores. *Amer. Natur. 110:* 371-400.

Janzen, D. H. (1977). Why fruits rot, seeds mold, and meat spoils. *Amer. Natur. 111:* 691-713.

Kaushik, N. K., and Hynes, H. B. N. (1971). The fate of the dead leaves that fall into streams. *Arch. Hydrobiol. 68:* 465-515.

Larsen, K. (1971). Danish endocoprophilous fungi and their sequence of occurrence. *Bot. Tidsskr. 66:* 1-32.

Laurence, B. R. (1954). The larval inhabitants of cow pats. *J. Anim. Ecol. 23:* 234-260.

Lodha, B. C. (1974). Decomposition of digested litter. In *Biology of Plant Litter Decomposition,* I. C. H. Dickinson and G. J. F. Pugh (Eds.). Academic Press, New York, pp. 213-241.

Lubchenco, J. (1978). Plant species diversity in a marine intertidal community: importance of herbivore food preference and algal competitive abilities. *Amer. Natur. 112:* 23-39.

Lundqvist, N. (1972). Nordic Sordariaceae s. lato. *Symb. Bot. Upsal. 20:* 1-374.

MacArthur, R. (1965). Patterns of species diversity. *Biol. Rev. 40:* 510-533.

MacArthur, R. (1972). *Geographical Ecology.* Harper & Row, New York.

McDiffett, W. F. (1970). The transformation of energy by a stream detritivore, *Pteronarcys scotti* (Plecoptera). *Ecology 51:* 975-988.

McNaughton, S. J., and Wolf, L. L. (1973). *General Ecology.* Holt, Rinehart and Winston, New York.

Malloch, D. and Cain, R. F. (1971). New cleistothecial Sordariaceae and a new family, Coniochaetaceae. *Can. J. Bot. 49:* 869-880.

Mason, C. F. (1976). *Decomposition* (Studies in Biology No. 74). Arnold, London.

Mason, W. H., and Odum, E. P. (1969). The effect of coprophagy on retention and bioelimination of radionuclides by detritus-feeding animals. *Proc. 2nd Nat. Symp. Radioecology,* D. J. Nelson and F. C. Evans (Eds.). Clearinghouse Fed. Sci. Tech. Inf., U.S. Dept. of Commerce, Springfield, Va., pp. 721-724.

Massee, G., and Salmon, E. S. (1901). Researches on coprophilous fungi. *Ann. Bot.* (London) *15:* 313-357.

Massee, G., and Salmon, E. S. (1902). Researches on coprophilous fungi. II. *Ann. Bot. 16:* 57-93.

Miller, R. S. (1967). Pattern and process in competition. *Advan. Ecol. Res. 4:* 1-74.

Mitchell, D. T. (1970). Fungus succession on dung of South African ostrich and Angora goat. *J. S. Afr. Bot. 36:* 191-198.

Nicholson, A. J., Patterson, I. S., and Currie, A. (1970). A study of vegetational dynamics: selection by sheep and cattle in *Nardus* pasture. *Brit. Ecol. Soc. Symp. 10:* 129-144.

Nicholson, P. B., Bocock, K. L., and Heal, O. W. (1966). Studies on the decomposition of the faecal pellets of a millipede (*Glomeris marginata* Villers). *J. Ecol. 54:* 755-766.

Paine, R. T. (1966). Food web complexity and species diversity. *Amer. Natur. 100:* 65-75.

Paine, R. T. (1971). A short-term experimental investigation of resource partitioning in a New Zealand rocky intertidal habitat. *Ecology 52:* 1096-1106.

Paine, R. T. and Vadas, R. L. (1969). The effects of grazing by sea urchins, *Strongylocentrotus* spp., on benthic algal populations. *Limnol. Oceanogr. 14:* 710-719.

Park, T. (1954). Experimental studies of interspecies competition. II. Temperature, humidity, and competition in two species of *Tribolium. Physiol. Zool. 27:* 177-238.

Pearson, D. (1971). Vertical stratification of birds in a tropical dry forest. *Condor 73:* 46-55.

Reiners, W. A., and Reiners, N. M. (1970). Energy and nutrient dynamics of forest floors in three Minnesota forests. *J. Ecol. 58:* 497-519.

Richardson, M. J. (1972). Coprophilous ascomycetes on different dung types. *Trans. Brit. Mycol. Soc. 58:* 37-48.

Simpson, G. (1964). Species density of North American recent mammals. *Syst. Zool. 13:* 47-73.

Slobodkin, L. B. (1961). *Growth and Regulation of Animal Populations.* Holt, Rinehart and Winston, New York.

Swift, M. J. (1976). Species diversity and the structure of microbial communities in terrestrial habitats. In *The Role of Terrestrial and Aquatic Organisms in Decomposition Processes*, J. M. Anderson and A. Macfadyen (Eds.). Blackwell Scientific Publns., Oxford, England, pp. 185-222.

Thomas, W. A. (1968). Decomposition of loblolly pine needles with and without additions of dogwood leaves. *Ecology 49*: 568-571.

Van Cleve, K. (1971). Energy- and weight-loss functions for decomposing foliage in birch and aspen forests in interior Alaska. *Ecology 52*: 720-723.

Webster, J. (1970). Coprophilous fungi. *Trans. Brit. Mycol. Soc. 54*: 161-180.

Wicklow, D. T. (1975). Fire as an environmental cue initiating ascocarp development in a tall grass prairie. *Mycologia 67*: 852-862.

Wicklow, D. T. (1979). Hair ornamentation and predator defence in *Chaetomium*. *Trans. Brit. Mycol. Soc. 72*: 107-110.

Wicklow, D. T., and Malloch, D. (1971). Studies in the genus *Thelebolus*. Temperature optima for growth and ascocarp development. *Mycologia 63*: 118-131.

Wicklow, D. T., and Moore, V. (1974). Effect of incubation temperature on the coprophilous fungal succession. *Trans. Brit. Mycol. Soc. 62*: 411-415.

Wicklow, D. T., and Yocom, D. H. (1981). Fungal species numbers and decomposition of rabbit faeces. *Trans. Brit. Mycol. Soc. 76*: (in press).

Wicklow, D. T., and Zak, J. C. (1979). Ascospore germination of carbonicolous ascomycetes in fungistatic soils: an ecological interpretation. *Mycologia 71*: 238-242.

Wicklow, D. T., Angel, K., and Lussenhop, J. (1980). Fungal community expression in lagomorph versus ruminant feces. *Mycologia 72*: 1015-1021.

Witkamp, M. (1966). Decomposition of leaf litter in relation to environment, microflora and microbial respiration. *Ecology 47*: 194-201.

Witkamp, M., and Frank, M. L. (1970). Effects of temperature, rainfall, and fauna on transfer of ^{137}Cs, K, Mg, and mass in consumer-decomposer microcosms. *Ecology 51*: 465-474.

Yocom, D. H. (1977). Community differentiation along a dune succession: an experimental approach with coprophilous fungi. M.A. thesis, University of Pittsburgh, Pittsburgh, Pa.

Yocom, D. H., and Wicklow, D. T. (1980). Community differentiation along a dune succession: an experimental approach with coprophilous fungi. *Ecology 61*: 868-880.

Zycha, H., and Siepmann, R. (1969). *Mucorales*. Cramer, Lehre, West Germany.

Part II

FUNGAL POPULATIONS,
ENVIRONMENTAL PARAMETERS,
AND THE NICHE CONCEPT

Chapter 5

NICHE: DEFINITION AND GENERALIZATIONS

Samuel J. McNaughton
Syracuse University
Syracuse, New York

I. INTRODUCTION

Ecologists have a way of taking words from common parlance and endowing them with special meaning. This practice creates hazards of interpretation manifested in a blizzard of papers purporting to provide more precise or more meaningful definitions of the term at hand. One such word is *niche*, which is taken directly from the French word for a cavity or hole in a wall that holds statuary or some other ornament. The word is ultimately derived from the Indo-European root *nizdo*, meaning a bird's nest but literally translated as "place where the bird sits down" (Morris, 1969). Perhaps ecologists would have been prudent to merely leave it at that.

Although usage of niche in an ecological context can be traced at least to Johnson in 1910 (see Gaffney, 1975), its entry into ecology is commonly traced to Grinnell's paper on the California thrasher (*Toxostoma redivivum*) in 1917. It is apparent from reading Grinnell that he probably was using a term that was a commonplace among field biologists, as he never really defines it but implies a meaning by example. Many concepts in ecology evolve in a like manner: a word in common usage is introduced in an ecological context, and its meaning grows by accretion of a bewildering variety of nuances usually inferred from the context in which the term is used. Someone eventually proposes an explicit definition of the term, and varying degrees of controversy thereupon arise--depending upon whose conceptual ox is being gored. Eventually, of course, after considerable intellectual and emotional blood has been spilled in the professional literature, someone proposes the ultimate solution: let's abandon the term. But this solution just begs the question. Much of ecology's accessibility to nonecologists lies in the use of terms that convey a part of their meaning to laypeople due to their familiarity; however, the professionals must restrict themselves to a more limited interpretation. It is this more limited interpretation which concerns us here.

Much of modern ecology can be traced to two small digressions by one of ecology's major intellects, G. Evelyn Hutchinson (1958, 1959). In 1959, Hutchinson proposed that an organism's niche be defined as a region in n-dimensional space, where n is the number of environmental factors affecting fitness. Any organism will be able to survive and reproduce over a certain range of a given environmental factor, and Hutchinson emphasized the multifactorial nature of the niche by stressing the combination of factors which influence organisms' survival and reproductive capabilities. He also pointed out that we can discriminate between a *fundamental niche*, determined by an organism's genetic properties, and a *realized niche*, defined by the expression of those genetic properties in the presence of competitors. That is, where the fundamental niches of two organisms overlap and efficiency of resource exploitation is unequal in the overlap zone, the less efficient competitor will be excluded. The realized niche of the poorer competitor will be only a portion of its fundamental niche. Since most studies of niche structure have been concerned with distribution patterns in nature, they have dealt with realized niches. However, what we observe in nature is the outcome of niche overlap rather than its extent and effects. Although there are numerous studies in physiological ecology that relate implicitly to fundamental niches, few of these have been related explicitly to niche theory.

Fungi would seem to be nearly ideal tools for examining both fundamental and realized niches. Realized niches can be characterized by the physical and biological conditions within which a population occurs. Fundamental niches can be defined in the laboratory by careful control of climate, substrate chemistry, and interacting organisms. Few other organisms would seem to lend themselves so nicely to an integrated approach to the niche.

In his 1959 paper, Hutchinson inquired into the factors that might be responsible for differences in species diversity in different communities. He identified at least seven hypotheses which could explain different species diversities: (1) Environmental rigor, and particularly its effect on productivity at the base of the food chain, may limit the number of species which can coexist. (2) Older areas may have allowed the evolution of more diverse life forms just because they are older. (3) Periodic catastrophes may eliminate some species, thereby lowering diversity. (4) Small areas, like islands, may lack refuges for some species during unfavorable seasons or from special kinds of competition. (5) Low-diversity aras may lack the type of environmental diversity necessary for species to co-occur in different niches at the same food chain level. (6) The divisibility of niches may differ in different ecosystems. (7) An environmental mosaic may promote species diversity by allowing some organisms to occur in some areas and similar organisms of other species to occur in different areas of the same community.

Hutchinson's two aforementioned papers and subsequent papers amplifying, modifying, and testing the ideas they contained have dominated ecological thinking in North America, as indicated by citation frequency, as have few others over the last 15 years. Much of ecology today deals with problems of species co-occurrence, niche breadth, niche overlap, species diversity, and their consequences for how ecosystems are organized. The emphasis, perhaps unfortunate, that Hutchinson placed on competition as a community organizer is also reflected in this literature, although there are refreshing exceptions (Connell, 1961; Paine, 1966; Connell and Slatyer, 1977).

The advantages of Hutchinson's definition, and particularly of applying it to hypothesis formulation, are many: It is abstract in the sense that it separates the inherent qualities governing species distributions and abundances from any actual organism or group of organisms. It therefore cuts across disciplines and infuses a variety of ecologists with a common intellectual framework. It is operational in both the sense of being applicable under real, specified conditions and in the philosophical sense of being derived from and defined by specified rules of procedure. At present, of course, it is impossible to completely characterize the niche of any organism, just as it is impossible to completely characterize a phenotype or genotype. And just as geneticists concentrate on a few characters in studies of inheritance, ecologists concentrate on a few dimensions in studies of the niche. But, like the chemist's ideal gas, it provides a model against which to test reality. It focuses our attention on factors influencing population processes, i.e., the survival and reproductive success of individuals. By contrasting the fundamental and realized niches, it focuses our attention on the contrast between physiological tolerance and distribution in nature. Although Hutchinson restricted population interactions to competition, it seems appropriate to recognize that predators and symbionts are both important in defining the extent to which the fundamental niche is translated into a realized niche.

As an example of predators as niche limiters, the interaction between Klamath weed (*Hypericum perforatum*) and a leaf-eating beetle (*Chrysolina quadrigemina*) is classic (Huffaker, 1957). The plant is native to Eurasia and North Africa but was introduced into California in 1900. It proved to be a vigorous competitor in rangeland and had invaded more than 2 million acres, to the extent of dominance, by the 1940s. Introduction of the predator resulted in a rapid deline in many areas, reducing abundance of the plant precipitously throughout much of its range. The beetle eats leaves in the spring, which prevents the buildup of root reserves prior to the dry summer season in California. Progressive depletion of root reserves results in death of the plant after 2 or 3 years. But in the Pacific Northwest, where summers are more mesic, the beetle has been a less effective control agent. In dry summer climates, the Klamath weed is presently restricted to shady sites. Its restriction to a suboptimal portion of its fundamental niche arises because the adult beetles avoid shady sites when laying eggs. This example is particularly instructive because we know the past dynamics leading to the present distribution of Klamath weed. If we examined its present distribution, we would conclude that the main factors controlling that distribution were light and moisture, that predation had little effect, and that the plant's optimum habitat was moist, shady conditions. But, in fact, the plant actually grows better and competes very effectively in bright sunny locations. We also see from this example that the effect of predation depended upon climate, highlighting the interactive nature of niche dimensions.

Many examples of symbionts as niche expanders also come to mind, but lichens are one of the classic cases (Ahmadjian, 1967). Many lichen algae have light-sensitive

photosynthetic pigment systems that tend to bleach out at light intensities far below full sunlight. The fungi in lichen associations often require various organic
compounds for growth. The shading effect of the fungus expands the range of light
intensities wherein the algae are capable of living, and transfer of organic molecules from alga to fungus allows the symbiont to grow on inorganic substrates. Thus,
the symbiotic association has a larger niche than either participant in isolation.

II. CHARACTERIZING THE NICHE

Along any single niche dimension, an individual will have some niche breadth over
which it is capable of existing (Levins, 1968). To the extent that a population is
genetically and developmentally heterogeneous, the range of the population will be
greater than that of any one member. And, to the extent that different populations
are ecotypic variants, the range that can be occupied by any one population will be
less than the total range of the species.

     Although it was not explicitly related to niche theory, Loucks's (1962) study of
the distribution of forest plants along environmental gradients is an informative
examination of the realized niches of vascular plants. He characterized vascular
plant niches in three dimensions: local climate, soil nutrient regime, and soil moisture regime. This probably represents a fairly complete evaluation of the most important complex dimensions of a vascular plant niche and the addition of another dimension, herbivore regime, would seem to allow a nearly complete description. Each
of these four dimensions, of course, can be factored and refactored. And, in fact,
Loucks combined many separate environmental measurement into each of the three dimensions he examined. It is probably most useful, however, to follow Loucks's lead and
characterize niches in terms of such complex dimensions rather than trying to factor
out each individual parameter. Although we may, for instance, speak of photoperiod-
ic and thermoperiodic adaptations, it is more realistic to recognize that adaptations
to local climate may be achieved by the evolutionary modification of physiological
response to photoperiod and thermoperiod (Heslop-Harrison, 1964). The selective factor is not photoperiod or thermoperiod per se but the combination of these two that
defines a local climate within which a genotype must express a phenotype with positive fitness if it is to persist.

     Niche breadth is a consequence of the interaction between genotypes and environ-
ments, but little is known directly about the genetic properties defining different
niche breadths. The ecologist's concept of niche breadth is similar to the geneti-
cist's concept of developmental homeostasis (Lerner, 1954), which may be increased
in individuals by increasing heterozygosity (Dobzhansky and Wallace, 1953) and in
populations by the additional mechanism of genetic heterogeneity (Dobzhansky and
Levene, 1955). A genotype with a broad fundamental niche is a genotype capable of
maintaining a similar phenotype over a broad range of environment; that is, it has a
high degree of developmental stability. We might expect this type of stability to
be selected in an environment that fluctuates considerably over the time range of an

individual's lifetime. Such physiological generalists, however, often tend to be poor competitors (McNaughton and Wolf, 1970; Morse, 1974; Colwell and Fuentes, 1975). Thus, some balance between physiological flexibility and canalization is expected, depending upon the environmental variance and the intensity of competition. Because of the ease and rapidity with which genetic and growth experiments can be performed with fungi, they are likely to be particularly suitable for examining the consequences of genetic foundation for niche breadth and competitive ability.

We expect fitness to show variation over the range of environments occupied by an individual or a population. Although species distributions often show bell-shaped abundance curves over a continuously varying environment (Whittaker, 1965), the frequency of threshold phenomena in physiological responses suggests that niches may have strikingly assymetrical fitness distributions. The rate at which fish eggs hatch, for instance, is often maximal just below the lethal temperatures. To the extent that such threshold responses determine fitness, we may expect both fitness and abundance to change abruptly with minor changes in environment. Most theoretical explorations of niche width and overlap, however, have been based on assuming a normal curve, or some variant thereof.

For species distributed normally along a single niche dimension, we can define a mean position of occupance, a distance between adjacent species, a range of occupance, and an overlap between adjacent species. Fungi, for instance, might be distributed according to substrate moisture content, with species segregating in space according to substrate moisture. Or, for species exploiting substrates with similar moisture contents, the species might segregate according to habitat temperature. May and MacArthur (1972) examined the relationship between mean spacing distance (d) and niche width (w) necessary for species composition to be stable in a varying environment. In one of the more robust mathematical results of theoretical ecology, they found that d/w must not be less than 1 for species composition to be stable. This result suggests that there is a limit to the amount of niche overlap possible in a fluctuating environment.

Let's assume that we have some range of resource exploited by potential competitors and that the amount of resource is constant across its range. For example, suppose growth is carbohydrate limited, species differ in the efficiency with which they can use carbohydrates of different sizes, and smaller carbohydrates are more common than larger carbohydrates in a way such that total energy available from carbohydrates is constant across molecular size. Different species might become distributed according to the efficiency with which they exploit different molecules, with overlap in areas where species are more or less equally efficient. Additional species might be added to the community in various ways. If we increase the range by adding, say, carbohydrates larger than any presently existing in the ecosystem, an additional colonist might be able to invade. If we keep the total amount constant when adding to the range, we expect each species to become slightly less abundant than it was before. Alternatively, we might just add to the total pool as we add to the range, so that species abundances are unaffected. Another way we could add species to the community would be to increase environmental complexity by adding variation in another environmental factor for which species had different tolerance ranges. If, for instance, two species were equally efficient at exploiting carbohydrates of a given size, they could coexist in a community if their temperature responses were different and temperature varied. Species could also be added if predation kept population densities low enough to avoid competition. In this case, niches can overlap to a degree, since no species could eliminate another by competition and abundances would be determined by predation intensity. If species are more specialized, additional species can also be added to the community since "holes" are left between species that can be invaded by other species. Ecologists have only limited and contradictory data on the importance of these various factors in determining community organization. Most data are from distributions in nature, which, as I pointed out earlier, show the outcome of species interactions but not their nature, intensity, or direction.

III. A PROPOSED RESEARCH AGENDA

Fungi should be valuable tools for examining many hypotheses about niche properties and their consequences for community organization. The ease and rapidity with which physiological and genetic experiments can be done allows direct, straightforward testing of many hypotheses current in the ecological literature. There is also considerable information of growth requirements and environmental tolerances of fungi.

Margalef (1968) concluded that "Species that interact feebly with others do so with a great number of other species [but] species with strong interactions are often part of a system with a small number of species." MacArthur (1972) called the former type of competition "diffuse competition" and concluded that species sandwiched between others on a resource gradient are in a particularly precarious position. Pianka's (1974) studies of desert lizard communities showed that mean overlap of realized niches decreased with an increase in the number of co-occuring species. Total overlap, in contrast, increased with the number of species. He concluded that stronger diffuse competition requires greater separation of exploitation centers. But we should be cautious about generalizing from studies of realized niches to fundamental niches, and Brown (1975) got exactly the opposite result in desert rodent communities. Here we have one of ecology's common perplexing problems: Is the theory faulty, or are these two communities organized according to different rules? Fungal ecologists are able to make an important contribution to our resolution of such seeming paradoxes.

We do know that communities are organized according to rules because several studies indicate that only a subset of the total species potentially capable of invading a community ever does so (Neill, 1975; Diamond, 1975). In a fascinating study of the role of immigration in colonization and succession, Maguire (1963) found

that a wide variety of stable species combinations with different species but similar diversities existed in simple environments. Many ecologists have been arguing
recently that there are a variety of permissable community "solutions" to any given
habitat and that which particular species actually coexist is influenced by both stochastic and deterministic factors (Lewontin, 1969; Gilpin and Case, 1976).

Consider also the relation between community diversity and community stability,
which have long been held to be positively associated (MacArthur, 1955; Margalef,
1963; Odum, 1969). May (1972), one of ecology's most gifted theoreticians, drew an
analogy between large cybernetic networks and ecosystems. These matrices have three
properties (Gardner and Ashby, 1970): (1) the number of elements in the interaction
matrix, which May characterized as the number of species, s, in the ecosystem; (2)
an average interaction term, i, which is the average effect of each species on each
other species--overall, i is bounded at +1 and -1; and (3) a connectance, c, which
is the proportion of nonzero i's in the matrix, that is, the proportion of all possible species interactions which actually occur. Each element in the matrix, in isolation, will return to its equilibrium value if perturbed because of negative self-
damping terms. May showed that systems are almost certainly stable if

$$i < (sc)^{-1/2}$$

and unstable if

$$i > (sc)^{-1/2}$$

This work contradicted the conventional ecological wisdom because these matrices are
increasingly liable to instability as the number of interacting elements--and connectance among them--increases. But it also laid ground rules that may explain what
properties large diverse ecosystems must have if they are to be stable. First, species which interact strongly should do so with few species, a conclusion essentially
the same as Margalef's. And, second, for a given average interaction strength, stability may be maintained as diversity increases if connectance also diminishes. This
organization of communities into guilds (Root, 1967) may be a general feature of ecosystem organization, so that diversity is increased by adding blocks of species that
interact among themselves but interact not at all or only feebly with members of
other guilds.

Lewontin (1969), in a paper that warrants and repays careful reading, distinguished between local and global stability. A community is locally stable if its
properties are restored after small perturbations; but it is globally stable only
if it recovers from all perturbations. Clements (1936) argued that one of the essential features of a true climax community was that it resulted from many different
starting communities. Thus, although early and intermediate successional stands
might differ from one another according to both chance and deterministic factors,
they all must eventually converge on the true climax. So, for Clements, a climax
community was globally stable. But there is abundant evidence from studies of suc-

cession in sessile animal communities of the intertidal zone that the initial colonization pattern may change the ultimate climax community drastically (Dayton, 1975;
Menge, 1976; Osman, 1977).

Following Lewontin, we can recognize several properties of the ecosystem we must
know if we are to describe it: First, what species are present and what are their
abundances? Second, what are the environmental factors and their values? (These
are two traditional concerns of community ecology.) Third, what species are changing, in what direction and at what rates? Fourth, what environmental factors are
changing, in what direction, and at what rates? (These two are the traditional concern of students of succession.) And, finally, what are the inertias of the environment and community, their tendency to remain the same, if constant, and to change
in the same way, if changing? Applying these ideas to the seasonal succession of
phytoplankton and associated water physicochemical properties , Allen et al. (1977)
found that both community and environment were characterized by domains of attraction separated by regions of rapid change. Some environmental and community tran-
sitions were simultaneous, but many were not, and Allen and associates argued that
many environmental changes merely signaled the beginning of the end of some community patterns, rather than forcing changes immediately. May (1977) suggested that
natural communities are likely to have several different equilibrium points, that
their patterns of change commonly possess thresholds beyond which breakpoints move
the community rapidly to a new state, even when environmental change is continuous
and gradual.

From the foregoing, I think it is apparent where fungal ecology can make important contributions to testing and aiding the formulation of ecological theory. Ecology has many detailed studies of community organization, on the one hand, and population dynamics, on the other. But only recently have these two fields begun to converge, and this convergence has mainly been brought about by theoreticians or experi-
mentalists; many of the latter have worked with microorganisms other than fungi. It
is clear that several levels of understanding are required if we are to understand
the mechanisms organizing ecosystem composition and dynamics: (1) an understanding
of the properties of fundamental niches of species that co-occur within and between
trophic levels; (2) similar knowledge of potentially co-occurring species that are
somehow excluded from these ecosystems; (3) information on the factors leading to
exclusion in one case and co-occurrence in the other, and their manifestation in
realized niches; (4) experimental data showing how groups of species fit together to
make up a natural community; (5) an understanding of the consequences of different
species compositions to ecosystem functional properties; and (6) a translation of
this understanding into a theory of ecology unifying the traditional fields of auto-
ecology, population ecology, community ecology, and ecosystem ecology. Fungi can
be useful tools at each level of this agenda, but they cannot be viewed in isolation.
Fungi compete, but they are also grazed on by herbivores which, in turn, are themselves prey. Concentrating on one level may distort our understanding of the role

that other levels play in determining the dynamics at that level. Clearly this agenda will be a long time in execution, but the recent activity in both theoretical and empirical ecology suggests that the stage may now be set for the resolution of many important ecological problems.

REFERENCES

Ahmadjian, V. (1967). *The Lichen Symbiosis*. Blaisdell, Waltham, Mass.

Allen, T. F. H., Bartell, S. M., and Koonce, J. F. (1977). Multiple stable configurations in ordination of phytoplankton community change rates. *Ecology 58*: 1076-1084.

Brown, J. H. (1975). Geographical ecology of desert rodents. In *Ecology and Evolution of Communities,* M. L. Cody and J. M. Diamond (Eds.). Harvard Univ. Press (Belknap), Cambridge, Mass., pp. 315-341.

Clements, F. E. (1936). Nature and structure of the climax. *J. Ecol. 24*: 252-284.

Colwell, R. K., and Fuentes, E. R. (1975). Experimental studies of the niche. *Ann. Rev. Ecol. Syst. 6*: 281-310.

Connell, J. H. (1961). The influence of interspecific competition and other factors on the distribution of the barnacle *Chthamalus stellatus*. *Ecology 42*: 710-723.

Connell, J. H., and Slatyer, R. O. (1977). Mechanisms of succession in natural communities and their role in community stability and organization. *Amer. Natur. 111*: 1119-1144.

Dayton, P. K. (1975). Experimental evaluation of ecological dominance in a rocky intertidal algal community. *Ecol. Monogr. 45*: 137-159.

Diamond, J. M. (1975). Assembly of species communities. In *Ecology and Evolution of Communities,* M. L. Cody and J. M. Diamond (Eds.). Harvard Univ. Press (Belknap), Cambridge, Mass., pp. 342-444.

Dobzhansky, Th., and Levene, H. (1955). Genetics of natural populations. XXIV. Developmental homeostasis in natural populations of *D. pseudoobscura*. *Genetics 40*: 797-808.

Dobzhansky, Th., and Wallace, B. (1953). The genetics of homeostasis in *Drosophila*. *Proc. Nat. Acad. Sci. US 39*: 162-171.

Gaffney, P. M. (1975). Roots of the niche concept. *Amer. Natur. 109*: 490.

Gardner, M. R., and Ashby, W. R. (1970). Connectance of large dynamic (cybernetic) systems: critical values for stability. *Nature 228*: 784.

Gilpin, M. E., and Case, T. J. (1976). Multiple domains of attraction in competition communities. *Nature 261*: 40-42.

Grinell, J. (1917). The niche-relationships of the California thrasher. *Auk 34*: 427-433.

Heslop-Harrison, J. (1964). Forty years of genecology. *Advan. Ecol. Res. 2*: 159-247.

Huffaker, C. B. (1957). Fundamentals of biological control of weeds. *Hilgardia 27*: 101-157.

Hutchinson, G. E. (1958). Concluding remarks. *Cold Spring Harbor Symp. Quant. Biol. 22*: 415-427.

Hutchinson, G. E. (1959). Homage to Santa Rosalia, or why are there so many kinds of animals? *Amer. Natur. 93*: 145-159.

Lerner, I. M. (1954). *Genetic Homeostasis*. Oliver & Boyd, Edinburgh.

Levins, R. (1968). *Evolution in Changing Environments*. Princeton Univ. Press, Princeton, N.J.

Lewontin, R. (1969). The meaning of stability. In *Diversity and Stability of Ecological Systems*, G. M. Woodwell and H. Smith (Eds.), Brookhaven Symp. Biol. No. 22. U.S. Dept. of Commerce, Springfield, Va., pp. 13-24.

Loucks, O. L. (1962). Ordinating forest communities by means of environmental scalers and phytosociological indices. *Ecol. Monogr. 32*: 137-166.

MacArthur, R. H. (1955). Fluctuations of animal populations and a measure of community stability. *Ecology 35*: 533-536.

MacArthur, R. H. (1972). *Geographical Ecology*. Harper & Row, New York.

McNaughton, S. J., and Wolf, L. L. (1970). Dominance and the niche in ecological systems. *Science 167*: 131-139.

Maguire, B., Jr. (1963). Passive dispersal of small aquatic organisms and their colonization of small bodies of water. *Ecol. Monogr. 33*: 161-185.

Margalef, R. (1963). On certain unifying principles in ecology. *Amer. Natur. 97*: 357-374.

Margalef, R. (1968). *Perspectives in Ecological Theory*. Univ. of Chicago Press, Chicago, p. 7.

May, R. M. (1972). Will a large complex system be stable? *Nature 238*: 413-414.

May, R. M. (1977). Thresholds and breakpoints in ecosystems with a multiplicity of stable states. *Nature 269*: 471-477.

May, R. M., and MacArthur, R. H. (1972). Niche overlap as a function of environmental variability. *Proc. Nat. Acad. Sci. US 69*: 1109-1113.

Menge, B. A. (1976). Organization of the New England rocky intertidal community: role of predation, competition, and environmental heterogeneity. *Ecol. Monogr. 46*: 355-393.

Morris, W. (ed.) (1969). *The American Heritage Dictionary of the English Language*. American Heritage and Houghton Mifflin, New York.

Morse, D. H. (1974). Niche breadth as a function of social dominance. *Amer. Natur. 108*: 818-830.

Neill, W. E. (1975). Experimental studies of microcrustacean competition, community composition, and efficiency of resource utilization. *Ecology 56*: 809-826.

Odum, E. P. (1969). The strategy of ecosystem development. *Science 164*: 262-270.

Osman, R. W. (1977). The establishment and development of a marine epifaunal community. *Ecol. Monogr. 47*: 37-63.

Paine, R. T. (1966). Food web complexity and species diversity. *Amer. Natur. 100*: 65-75.

Pianka, E. R. (1974). Niche overlap and diffuse competition. *Proc. Nat. Acad. Sci. US 71*: 2141-2145.

Root, R. B. (1967). The niche exploitation pattern of the blue-gray gnatcatcher. *Ecol. Monogr. 37*: 317-350.

Whittaker, R. H. (1965). Dominance and diversity in land plant communities. *Science 147*: 250-260.

Chapter 6

ADAPTATION AND THE FUNDAMENTAL NICHE:
EVIDENCE FROM LICHENS

Martin J. Lechowicz

McGill University
Montreal, Quebec, Canada

I. INTRODUCTION

When Hutchinson (1958) formalized the concept of the niche, he emphasized the species' limits of tolerance to environmental conditions. On an environmental gradient there is a range of tolerance over which a species can survive and reproduce. All the environmental variables affecting a species' survival and reproduction can be taken together as the axes of a multidimensional space. The hypervolume bounded by the species' tolerance limits on each environmental axis Hutchinson called the *fundamental niche*. This fundamental niche describes the breadth of environmental conditions over which species can exist when removed from competition. In natural communities of competing species an organism is often excluded by competitors from some part of the environment it could potentially utilize. Hutchinson called this truncated fundamental niche which arose under competition with other species the *realized niche*. These concepts of the niche, by explicitly relating the survival and reproduction of a species to environmental variables, are particularly appropriate in the analysis of possibly adaptive relationships between organisms and their environment.

The empirical and theoretical work on the niche that followed Hutchinson's seminal paper has primarily drawn on the concept of the realized niche to analyze competition and community structure (Colwell and Fuentes, 1975), while the evolution and the adaptive significance of a species' fundamental niche has received less attention. The observation that no organisms seem able to survive and reproduce equally well under all environmental conditions suggests two interesting questions about the fundamental niche of a species. First, can we predict the niche breadth that will be favored by natural selection in a given environmental regime? Some theoretical analyses of optimal niche breadths have been made (Roughgarden, 1972; Slatkin and Lande, 1976) but remain largely untested by empirical studies. Second, can we predict how a species' survival and reproductive capacities will vary within the boundaries of its fundamental niche? This second question involves an extension of Hutchinson's concept of the fundamental niche and has not yet been analyzed theoretically. The net photosynthetic responses of lichens discussed in this chapter bear on both questions. It is hoped that this review will encourage additional mycologi-

cal work toward a general theory predicting the optimal shape of the fundamental
niche for a population in a given environment. As McNaughton (Chap. 5) points out,
an analysis of the properties of fundamental niches will contribute to our understanding of the mechanisms organizing community composition and dynamics.

II. FITNESS AND THE FUNDAMENTAL NICHE

A. Definitions

The concept of fitness, rigorously defined by Sewall Wright in early work on population genetics, describes "the average contribution which the carriers of a genotype,
or a class of genotypes, makes to the gene pool of the following generation relative
to the contribution of other genotypes" (Dobzhansky, 1968a). Another only slightly
different view, stemming from the work of Fisher (1930), places less emphasis on the
relative contribution of genotypes to subsequent generations and instead measures
fitness as the intrinsic rate of increase for individuals of a particular genotype
in the population. Both measures of fitness depend on the survival characteristics
and lifetime reproduction of individuals in the population. Hutchinson's concept
of the fundamental niche, based on the environmental limits of survival and reproduction, provides an appropriate conceptual framework for an examination of the interrelationships between an organism's characteristics and its environment that together ultimately determine fitness.

   Since environmental conditions are variable and not wholly predictable, to successfully colonize a habitat an organism must survive both the normal environmental
conditions and occasional extreme variations from these norms. If there is a genetic basis for rates of growth and reproduction, natural selection will lead to the
evolution of responses maximizing successful reproduction in the particular environmental regime of the habitat. Elements of the growth and reproductive responses to
environmental variables will then reflect adaptations to environmental pattern. In
deciding how well-adapted an organism is to its environment, not only the boundaries
of the fundamental niche but also organismal responses within those boundaries are
of evolutionary significance (Maguire, 1973; Emlen, 1973, pp. 210-214). Figure 1
gives an example of variation in a lichen net photosynthetic response within the
boundaries of a fundamental niche defined for simplicity by only two environmental
axes. To the degree that net photosynthesis determines survival and reproduction
in this lichen, its fitness will clearly depend on the pattern of tissue temperatures
and water contents prevailing in its habitat.

B. Describing the Lichen Niche

In describing the response of an organism within its fundamental niche, two questions
have to be resolved: (1) What response best measures an organism's adaptation to
an environmental condition? (2) What environmental axes affect this response and
determine fitness?

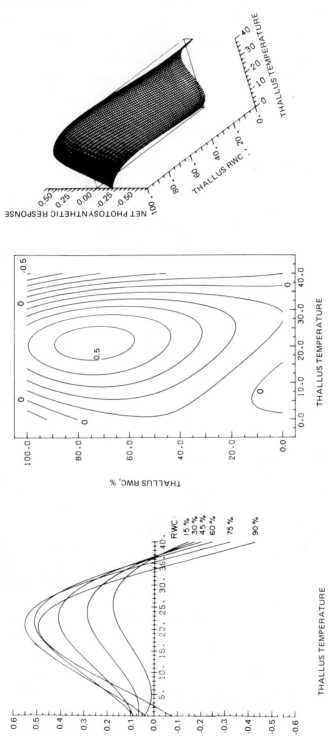

Fig. 1 The net photosynthetic response of the lichen *Cladina stellaris* on two axes of its fundamental niche, tissue temperature and relative water content. The same data are presented in the three graphs, each illustrating a common method of depicting multidimensional response surfaces. Net photosynthesis is measured as milligrams of CO_2 g^{-1} h^{-1} at an irradiance of 600 μEinsteins m^{-2} s^{-1}.

The response chosen should both be practical to measure and contribute significantly to fitness. Rates of growth and reproduction are perhaps most meaningful but not always easily determined over an extensive range of environmental conditions. Related variables such as rates of assimilation or net photosynthesis are feasible alternatives.

Most modern studies of the adaptation of higher plants have assumed that net photosynthetic responses to environmental variables are a critical determinant of fitness. For example, Björkman and Holmgren (1963) showed that, under high irradiance, populations of *Solidago virgaurea* from sunny habitats took up CO_2 at greater rates than did populations from shaded habitats. Conversely, in low light the shaded populations had the higher rates of net photosynthesis. These responses were shown to be in part genetically determined and were taken as evidence that natural selection favored genotypes leading to greater carbon gains in each population's habitat. Similar inferential evidence from numerous studies in physiological plant ecology has been recently reviewed (Strain and Billings, 1974; Cooper, 1975). In addition, Mooney (1972, 1975, 1976) has emphasized the central role of carbon metabolism in plant growth and reproduction and argued that, generally, the greater a plant's net carbon gain over time, the greater its fitness.

Comparable arguments apply in lichens, perhaps more strongly since complications from differing strategies of carbon allocation to specialized organs such as roots, leaves, and flowers do not arise. Compared to higher plants, lichen thalli are relatively undifferentiated (Jahns, 1973) and vegetative propagules are, in many taxa, the primary or only reproductive mode (Bowler and Rundel, 1975; Bailey, 1976). The greater an individual lichen's net carbon balance, the greater its available resources for growth and reproduction. In taxa reproducing largely by vegetative propagules, including simply the dispersal of thallus fragments, increased growth in itself represents an increased reproductive potential. Although information on the competitive ability of lichens is scanty (see Topham, 1977), it appears likely that faster-growing individuals have a competitive advantage. With both the survival and reproduction of lichens enhanced by increasing photosynthetic carbon gains, natural selection will favor individuals with net photosynthetic responses that maximize carbon gain in the lichen's habitat. Thus net photosynthetic rates provide a good response variable to define the fundamental niche of lichens.

The axes of the fundamental niche should ideally include all the environmental variables that alter the chosen response and thereby affect fitness. For simplicity it is useful to combine the interacting effects of many environmental factors by considering only the operational environment (Spomer, 1973), the environmental variables that directly affect the observed response. For example, tissue temperature is determined by the complex interaction of organismal characteristics and such environmental factors as air temperature, humidity, windspeed, solar radiation, and infrared radiation from the surroundings (Monteith, 1973). All these environmental variables can be subsumed on a single niche axis, tissue temperature, which directly con-

trols most biological processes. It may often be very demanding to determine which
environmental variables adequately describe the fundamental niche, but any analysis
of the adaptive significance of a niche rests on it complete description.

Without undue sacrifice of analytic precision, the environmental axes defining
the fundamental niche of many lichens can be reduced to three: tissue temperature,
tissue water status, and the incident radiation in the 400-700 μm waveband used in
photosynthesis (Farrar, 1973; Richardson, 1973; Kallio and Kärenlampi, 1975). Farrar (1973) suggested that tissue nutrient status did not limit lichen growth, and
recent experimental evidence supports this view (Farrar, 1976a; Carstairs and Oechel,
1978). The normal wind regimes of most lichen habitats allow for sufficient convec-
tive exchange to maintain fairly uniform ambient CO_2 concentrations; limited evidence (Larson and Kershaw, 1975a) also indicates that CO_2 concentration may not limit
net photosynthesis in lichens as it does in higher plants. The generally minor effects of vertebrate (Richardson and Young, 1977) and invertebrate (Gerson and Seaward, 1977) herbivores can be avoided by judicious choice of the species or populations studied. Pollutants like SO_2 which have a marked effect on lichen photosynthesis (Nieboer et al., 1976) are unlikely to represent an equilibrium adaptation
open to evolutionary interpretation. Populations from polluted areas can best be
neglected unless the actual process of natural selection is to be studied in a
population under pollution stress. It is unlikely that all lichen species have so
advantageously few niche dimensions; further research is required to determine additional significant environmental variables and their effects on the measured responses. For clarity the discussion in this chapter emphasizes the adaptive significance
of net photosynthetic responses to only light, temperature, and water in lichens from
diverse habitats.

III. ADAPTEDNESS OF LICHEN NICHES

A. Assessing Adaptedness

A well-chosen response and full definition of its environmental control are the foun-
dations for an analysis of what Dobzhansky (1968a,b) has called *adaptedness*, a measure of how a trait like the response characteristics of the fundamental niche contributes to the probability of successful survival and reproduction in a given environmental regime. Assuming that lichen fitness does increase with increasing carbon
gains, we can reason that the greater the net photosynthetic response to a particular
combination of light, temperature, and water content, the better adapted the lichen
is to that environmental condition. As the environment varies over time in the habitat, the adaptedness of the lichen's fundamental niche then increases as the integral of net photosynthetic response to the sequence of environmental conditions. We
lack the necessary long-term data on environmental regimes to quantitatively integrate
a lichen's net photosynthetic response over time and must remain limited to a qualitative analysis for the present.

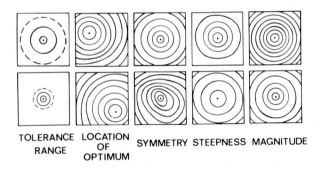

TOLERANCE   LOCATION   SYMMETRY STEEPNESS MAGNITUDE
RANGE   OF
                OPTIMUM

Fig. 2 Qualitative characteristics used to assess the adaptedness of the net photosynthetic responses to environmental variables. A response surface on only two environmental axes is drawn for simplicity. The upper and lower graphs above each
heading illustrate a qualitative contrast in that characteristic of the net photosynthetic response. For further discussion, see the text.

Five qualitative characteristics of a net photosynthetic response are useful in
assessing its adaptedness (Fig. 2); (1) the range of environmental conditions which
can be tolerated; (2) the environmental conditions giving maximal rates of carbon
gain; (3) the symmetry of the response surface away from this set of environmental
conditions; (4) the steepness of the response surface, that is, the rate of change
in CO_2 flux with change in environmental conditions; and (5) the actual magnitude of
the maximal CO_2 flux. These five traits qualitatively characterize the net photosynthetic response surface at a single point in time. Any temporal changes in the
response surface may also have adaptive significance, particularly regular seasonal
changes in response that contribute toward maximizing carbon gain in the lichen's
habitat.

Various authors have reviewed aspects of the environmental control of net photosynthesis in lichens, but most have not focused primarily on the adaptive significance of observed responses. Smith (1962) provides a good summary of early work in
this area; this is supplemented in recent reviews by Farrar (1973) and by Kallio and
Kärenlampi (1975). Kappen (1973) thoroughly considers the effects of extreme environmental conditions on net photosynthetic capacity. Unfortunately, much of the published work in the physiological ecology of lichens is difficult to interpret from
an evolutionary perspective. Usually either the net photosynthetic response to all
significant environmental variables has not been reported or insufficient information is available on the natural environmental regime in the lichen's habitat to permit judgment as to the adaptedness of the observed responses. The relatively complete studies that will be reviewed here illustrate the use of the niche concept to
assess physiological adaptations in lichens.

B. Tolerance Range

The range of tolerance to environmental extremes determines the boundaries of a li-

chen's fundamental niche and should generally exceed the range of environmental variation in the lichen's habitat. Most species studied seem well able to withstand the
periods of environmental stress likely to occur in their habitats. *Ramalina maci-*
formis from the Negev Desert in southern Israel, for example, can withstand a year
at a water content as low as 1% of dry weight without reduction of net photosynthesis under subsequently more favorably conditions (Lange, 1969a). Dry *R. maciformis*
can withstand brief exposures to 65°C without subsequent reduction of its net photosynthetic capacity (Lange, 1969a), but when the lichen is wetted temperatures above
35°C are detrimental (Lange, 1965). Temperatures as low as -196°C do not, however,
reduce the subsequent photosynthetic performance of wetted *R. maciformis* (Kappen and
Lange, 1972). The intense irradiance regime of the desert, with values from darkness
up to 886 W m^{-2} (Lange et al., 1970a,b), does not appear to impair net photosynthesis. Even the net photosynthesis of wetted *R. maciformis* is not inhibited by irradiance equal to about 50% of the maximal solar radiation measured in the Negev Desert (Lange, 1969a). Comparable data, very thoroughly reviewed by Kappen (1973), indicate a remarkably large tolerance range for most lichens, sometimes in excess of
any environmental conditions likely to occur in most natural habitats. While the
range limits of some lichens may be set by their intolerance of environmental extremes (Rogers, 1971; Kappen and Lange, 1972), the net photosynthetic responses within the niche boundaries probably more often control distribution and certainly control abundance.

C. Maximal Response

What is the optimal relationship between the environmental conditions giving maximum
net photosynthesis and the environmental regime that will be favored by natural selection? Early studies, usually based on response to a single environmental variable, lead to a number of generalizations founded on the premise that the most frequent environmental conditions should determine the optimal response. Thus lichens
from cold regions should have maximum net photosynthesis at lower temperatures than
lichens from warmer regions (Larcher, 1975; Rogers, 1977). Lichens from dry habitats should have maximum net photosynthesis at lower water contents than lichens
from more moist habitats (Kershaw, 1971). By analogy to higher plants, lichens from
shaded habitats would be expected to reach maximum net photosynthesis at lower irradiance than those from exposed habitats (Lechowicz and Adams, 1973). There are, unfortunately, a number of problems with these intuitively appealing predictions relating the maximal net photosynthetic response to environmental conditions of high frequency.

Kershaw (1977a) has discussed the danger of drawing conclusions from net photosynthetic responses that are not defined in relation to all significant environmental axes. The optimum response to a single environmental variable invariably depends
on the levels of other variables controlling net photosynthesis. In *Cladina stel-*
laris, for example, the optimum temperature for net photosynthesis at 60% relative

water content is 22°C at 600 µEinsteins m^{-2} s^{-1}, 20°C at 300 µEinsteins m^{-2} s^{-1}, and only 15°C at 50 µEinsteins m^{-2} s^{-1} (Lechowicz, 1978). A fully defined response surface has a net photosynthetic maximum at a single combination of light, temperature, and water content levels. It is this global maximum that characterizes the lichen's net photosynthetic response surface, not the innumerable local maxima recognized when response to only a single variable at a time is considered.

Recent, relatively complete studies of lichen net photosynthetic responses (Kallio and Kärenlampi, 1975; Kershaw, 1977a; Lange, 1969a; Lange et al., 1975, 1977; Larson and Kershaw, 1975b,c; Lechowicz, 1976, 1978) only partially support the early generalizations that the net photosynthetic optimum is determined by the most frequent environmental conditions. Lechowicz (1978) compared a *Cladina stellaris* population from northern Quebec with *Cladina evansii* from northern Florida. An analysis of hourly weather data for the collection sites (Lechowicz, 1976; Lechowicz and Adams, 1978) reveals that the two lichens had distinctly different environmental regimes. In the early fall when the response surfaces were measured, the active periods of *C. stellaris* were characterized by lower irradiance, lower tissue temperatures, and higher tissue water content than those of *C. evansii*. The maximum net photosynthesis of *C. stellaris* occurred at about 600 µEinsteins m^{-2} s^{-1}, 20°C, and 70% relative water content; the *C. evansii* maximum occurred at about 1000 µEinsteins m^{-2} s^{-1}, 28°C, and 75% relative water content. These responses to irradiance and temperature follow the early generalizations, but their nearly equal relative water contents at maximum net photosynthesis do not since *C. evansii* grows under considerably more arid conditions than *C. stellaris*. Similarly unexpected net photosynthetic responses to water content are reported for *Cetraria nivalis* and *Alectoria ochroleuca* growing on raised-beach ridges in northern Ontario (Larson and Kershaw, 1975b, c). The ridge tops are a more arid microhabitat than the ridge troughs (Kershaw and Larson, 1974), yet *C. nivalis* from either microhabitat had maximal net photosynthesis at water contents of about 150% dry weight, as did *A. ochroceuca* at about 100% dry weight. The apparent failure of either intra- or interspecific correspondence of the maximal net photosynthetic response and habitat moisture regime suggests that the frequency of environmental conditions alone cannot be the selection pressure explaining observed net photosynthetic responses.

Lechowicz and Adams (1979) have proposed an alternative explanation of the effects of environmental patterns on the evolution of net photosynthetic response. Suppose that irradiance dominates the evolution of the net photosynthetic response surface and that responses to water content and temperature are controlled in part by the light regime. Water contents or temperatures that tend to occur at low light levels offer less potential for carbon gain simply because fewer photons are available for capture. Natural selection may lead to maximal net photosynthesis at an infrequent temperature or water content associated with high irradiance. The frequency of temperatures and water contents in an environmental regime weighted by co-occurring photon flux densities perhaps determines the maximal net photosynthetic response favored by natural selection. The higher the photon flux densities occurring with a particular temperature or water content, the more its impact on natural selection for the maximal net photosynthetic response. This hypothesis assumes that natural selection operates to maximize carbon gain under all irradiances and applies only within the environmental limits of active photosynthesis. For most lichens high irradiance, high temperatures, and very low water contents are a frequent condition in their natural habitats, but photosynthetic activity is then limited by water content. Within these limits, however, the maximal net photosynthetic response of photosynthetically active lichens is hypothesized to be controlled by patterns of temperature and water content weighted by irradiance.

This hypothesis might explain the similar optimal water contents for net photosynthesis observed in *C. stellaris* and *C. evansii* (Lechowicz, 1978). During summer and fall, rains in Florida are usually intense late-afternoon thunderstorms, whereas in northern Quebec rains can come at any hour and are more frequent (Byers and Rodebush, 1948; Lechowicz, 1976; Lechowicz and Adams, 1978). Since neither lichen photosynthesizes when dry, the active periods of the two species are qualitatively distinct as a result of these different rainfall and drying regimes: *C. stellaris* will maintain high water contents for extended periods, whereas *C. evansii* will actively photosynthesize most often in the morning hours following a thunderstorm the previous day. The 70% optimal relative water content of *C. stellaris* corresponds to its water content during extended periods of overcast with low light and temperature levels; the 75% optimal water content of *C. evansii* corresponds to its water content during the high-light and moderate-temperature conditions that often prevail the morning after a storm. The relatively low optimal temperature for net photosynthesis in the desert lichen *Ramalina maciformis* gives additional support to this hypothesis, since active photosynthesis is limited to early mornings after dewfall (Lange, 1969a; Lange et al., 1970a). More work is necessary to establish whether or not the evolution of the optimal environmental conditions for lichen net photosynthesis depends on patterns of temperature and water content weighted by irradiance.

D. Symmetry of Response

The environmental conditions affecting net photosynthesis are strongly interrelated in most natural habitats, and net photosynthetic responses are likely to reflect adaptation to these interrelationshps. For example, air temperature generally increases with increasing irradiance, which suggests that even allowing for the role of evaporative cooling in the lichen energy balance (Hoffman and Gates, 1970; Lechowicz, 1976), tissue temperature will also increase with increasing irradiance. Figure 3 shows an example of this expected relation taken from hourly readings of tissue temperatures and irradiance in *Cetraria cucullata* from the Alaskan Arctic tundra. The *C. cucullata* was often dry during the monitoring period, but when the lichen is wet the relationship shown is maintained. Because of such correlations among environmental variables, the net photosynthetic response of lichens is unlikely to be symmetrical around its optimum.

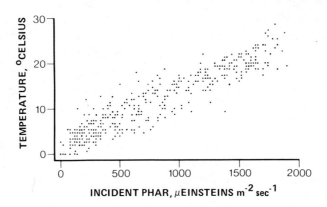

Fig. 3 The relationship between tissue temperature and irradiance for *Cetraria cu-cullata* growing in dry ridge tundra at Atkasook, Alaska (70°N, 157°W). Readings were taken hourly from June 28 through July 17, 1977.

Considering the relationship between tissue temperature and irradiance, we might predict that maximal net photosynthesis at a particular water content will occur at greater temperatures as the light level increases. This prediction is borne out by *Neuropogon acromelanus*, *Buellia frigida*, and *Xanthoria mawsoni* from Antarctica (Lange and Kappen, 1972), by *Ramalina maciformis* from the Negev Desert (Lange, 1969a), by *Cetraria nivalis* from subarctic Finland (Kallio and Kärenlampi, 1975), by *Cladina stellaris* from subarctic Quebec, and by *Cladina evansii* from temperate Florida (Lech-owicz, 1978). Comparable increases in the optimal temperature for net photosynthe-sis with decreasing water content have also been described (Lechowicz, 1978). Since carbon gains will be increased by responses coordinated with combinations of envir-onmental conditions that occur at disproportionately high frequency, such asymmetries in the response surface of lichen net photosynthesis are of adaptive significance.

E. Steepness of Response

If a narrow range of environmental conditions occurs predictably in a habitat, it may be advantageous for a lichen to specialize in fixing carbon rapidly under these conditions. This strategy could result in a low photosynthetic rate over a broad range of tolerated, relatively infrequent environmental conditions and a steep in-crease in net photosynthesis as the predictable and frequent environmental condit-ions are approached. In habitats of either low predictability or a predictably broad range of environmental conditions, more generalist responses may be favored. This aspect of net photosynthetic response to environmental variables requires investiga-tion but is beset by technical difficulties. The data needed to rigorously test eco-logical theories predicting the occurrence of generalist versus specialist responses (Levins, 1968a,b) are not easily obtained with lichens. Monitoring or estimating the environmental regime is difficult over a long enough period to establish the predict-

ability of environmental patterns, and the predictability of a multidimensional environmental condition cannot be quantitatively described by presently available methods (Colwell, 1974). Despite its possible adaptive significance, variation in the
steepness of net photosynthetic responses is unlikely to be carefully analyzed with
present limitations in environmental monitoring.

F. Magnitude of Response

If the magnitude of net photosynthesis could be increased without limit, discussion
of the response to particular environmental conditions would become largely irrelevant. Although this evolutionary option is impossible, the absolute rates of net
photosynthesis in lichens from different habitats can be of adaptive significance.
Data available from completely defined net photosynthetic responses (Fig. 4) show
that maximal rates of net photosynthesis decrease with increasing latitude. This
trend, also supported by earlier data tabulated in Kallio and Kärenlampi (1975), may
represent a trade-off of photosynthetic capacity against some other physiological
trait necessary for surviving stress associated with environments in the more north-

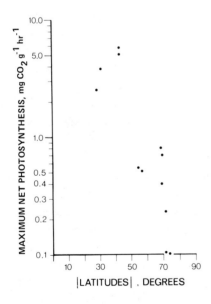

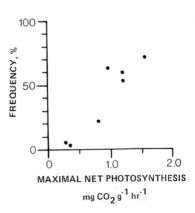

Fig. 4 Latitudinal trend in maximum
rates of lichen net photosynthesis.
[Data are from Kallio and Kärenlampi
(1975), Kershaw (1977a,b), Lange
(1969a), Lange and Kappen (1972),
Larson and Kershaw (1975a,b,c), and
Lechowicz (1978; also unpublished
data).]

Fig. 5 Relationship between the maximum rates of net photosynthesis and
the frequency of species occurrence
in an arboreal lichen community.
[Data are from Nowak (1973).]

erly or southerly latitudes (Berry, 1975). Data from a single site in Germany (Nowak, 1973) provide evidence that greater net photosynthetic rates enhance survival and reproduction in an arboreal lichen community (Fig. 5). Nowak examined both the dominance relations and CO_2 exchange responses of species in a lichen community growing on oak and beech in a forest unaffected by air pollution. If the frequency of the species is plotted against the maximal net photosynthetic rates observed for each species, it is apparent that dominance in the community increases with photosynthetic capacity. For lichens in general, increasing absolute rates of net photosynthesis probably increases fitness.

G. Seasonal Changes in Response

The extensive studies of Kershaw and Larson (see Larson and Kershaw, 1975b,c,d; Kershaw, 1975, 1977a,b) have confirmed seasonal changes in lichen net photosynthetic responses first noted as early as 1939 by Stålfelt. A central question is to what degree these observed seasonal changes in net photosynthetic responses are adaptive in either lichens or other plants (Oechel, 1976). Kershaw (1977a) has shown that the seasonal changes in the temperature responses of net photosynthesis in both *Peltigera polydactyla* and *Peltigera canina* var. *rufescens* are an acclimation to the thermal regime of their temperate forest habitat. Maximal net photosynthesis is attained throughout the year at about 500 μEinsteins $m^{-2} s^{-1}$ and at water contents about 250% of dry weight, but the temperature optimum in both species increases from spring through summer and then decreases to winter. Unfortunately, sufficient data on seasonal patterns of tissue water contents, temperatures, and irradiance in the habitat are unavailable and the adaptedness of these seasonal patterns in the net photosynthetic responses cannot be ascertained. Similarly, more environmental data are necessary to decide whether changes in net photosynthetic responses observed over a few days (Kershaw, 1977b) represent rapid acclimation of adaptive significance in the natural habitat or only transient responses to stress (Lange, 1969a; Kallio and Heinonen, 1971). Although it seems likely that the adaptedness of lichens can be enhanced by seasonal changes in net photosynthetic response, a comparison of changes observed in lichens from habitats with well-defined seasonal patterns of environmental conditions will be necessary to elucidate the full import of such acclimation.

IV. THE CARBON BALANCE OF LICHENS

Net photosynthetic responses contribute to increasing carbon gain, but the impact of these gains on fitness must be weighed against associated losses of carbon. It is growth, not rates of net photosynthesis, that is more closely tied to fitness. Growth may be defined as the change in lichen biomass over time and described by a simple mass balance equation:

$$B_{t+1} = B_t + kF_{CO_2} \tag{1}$$

The biomass B_{t+1} depends on the biomass, B_t, in the previous time interval and the carbon dioxide flux during the interval, F_{CO_2}. The constant, k, is a conversion of milligrams of CO_2 to milligrams of biomass derived by consideration of the chemical composition of lichen tissue. The flux of CO_2 has two components:

$$F_{CO_2} = {}_aF_{CO_2} + {}_{rs}F_{CO_2} \tag{2}$$

The term ${}_aF_{CO_2}$ represents this flux of CO_2 from the fully active lichen. During periods of activity CO_2 may be gained or lost at a rate determined by irradiance, temperature, and water content; this is the net photosynthetic response emphasized in the previous discussion. This response at zero irradiance is often distinguished as dark respiration. The term ${}_{rs}F_{CO_2}$ represents a negative flux of CO_2 that occurs for some time after a dry and dormant lichen is rewetted; this loss of carbon associated with rewetting has been called *resaturation respiration* (Smith and Molesworth, 1973) and may represent the carbon cost of reactivating the dormant lichen's metabolism (Farrar, 1976b; Farrar and Smith, 1976). These carbon fluxes integrated over time determine the lichen's net carbon balance and its consequent growth. The greater the coordination of flux responses to the environmental regime of a lichen's habitat, the greater the potential growth.

The contribution of these components of the lichen carbon balance to survival and reproduction has been most thoroughly studied in *Ramalina maciformis* (Lange and Bertsch, 1965; Lange, 1965, 1969a,b; Lange et al., 1969, 1970a,b, 1977; Kappen and Lange, 1972). These results are summarized in Lange et al. (1975). *Ramalina maciformis* growing in the Negev Desert was shown to be tolerant of the extremes of temperature and water content in this habitat. Water contents sufficient for photosynthetic activity could be attained by vapor uptake or dewfall as well as rainfall. At water contents as low as 20% of dry weight, *R. maciformis* could maintain positive net photosynthesis under favorable light and temperature conditions; maximal net photosynthetic responses were achieved at a water content of only 60% of dry weight, 20°C, and 48,500 lux (about 900 μEinsteins m^{-2} s^{-1}). Estimation of the lichen's annual carbon balance was simplified because, like another desert lichen, *Chondropsis semiviridis* (Rogers, 1971), *R. maciformis* did not have significant resaturation respiration. Field studies in the Negev Desert showed that net photosynthetic responses give sufficient carbon gain under the natural environmental regime to counter losses during darkness and allow an annual growth rate of some 5–10% dry weight.

In *Cetraria cucullata* from the Alaskan Arctic tundra, resaturation respiration is a significant component of the carbon balance (Fig. 6) and complicates assessment of the lichen's adaptedness. Fully active *C. cucullata* has maximal net photosynthesis at over 1200 μEinsteins m^{-2} s^{-1}, 10°C, and about 90% relative water content (Fig. 7). Although few environmental data are available (Conover, 1960; Haugen et al., 1976; Clebsch and Shanks, 1968), the net photosynthetic response of *C. cucullata* seems reasonably well adapted to the arctic summer. Yet, during the period July 1

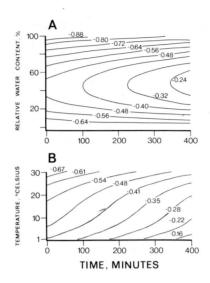

Fig. 6 The time course after wetting of CO_2 flux in the dark for *Cetraria cucullata* as a function of relative water content (A) and tissue temperature (B). The *C. cucullata* was collected in June-July from a dry ridge tundra community near Atkasook, Alaska (70°28'N, 157°23'W). The CO_2 flux is given as milligrams of CO_2 g^{-1} h^{-1}. In part A the tissue temperature is 10°C; in part B the relative water content is 60%.

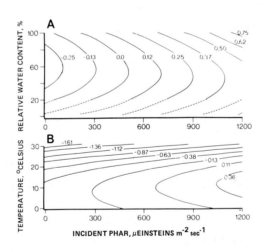

Fig. 7 The net photosynthesis of fully active *Cetraria cucullata* as a function of incident photosynthetically active radiation, relative water content (A), and tissue temperature (B). The *C. cucullata* was collected in June-July from a dry ridge tundra near Atkasook, Alaska (70°28'N, 157°23'W). The lichen was wetted at least 12 hr before the experiments began and was stored at 5°C. The CO_2 flux is given as milligrams of CO_2 g^{-1} h^{-1}. In part A the tissue temperature is 10°C; in part B the relative water content is 60%.

to 17, 1977, *C. cucullata* had a measured relative growth rate of -7.2 mg g^{-1} day^{-1}. During this unusually dry period, the lichen was frequently wetted incompletely by an early morning dew or fog and then rapidly dried by midmorning. The carbon losses sustained in resaturation respiration were thus not completely regained in net photosynthesis before the lichen reentered dormancy. These data support the early hypothesis of Ried (1960) that cycles of wetting and drying can determine a lichen's ability to survive in a habitat. Presumably *C. cucullata*, like the *Cetraria nivalis* studied by Kallio and Kärenlampi (1975), has compensating positive relative growth rates in another season that allow its survival on the dry ridge tundra.

These data also emphasize the importance of considering all components of the carbon balance in judging adaptedness in a particular habitat. The adaptedness of net photosynthetic responses alone can be considered since rates of carbon gain are probably maximized under natural selection, but in actuality lichen fitness is more directly tied to growth rates which are set by the balance of photosynthetic gains and respiratory losses in the habitat. This complicates analysis of the adaptive significance of lichen niches since lichen growth rates are difficult to assess over a broad range of controlled conditions (Armstrong, 1976).

V. NONLICHENIZED FUNGI AND THE NICHE

The growth or reproductive rates of many nonlichenized fungi are easier to measure than those of lichens and provide good fitness-related response variables. There has been considerable research on the environmental control of fungal growth and reproduction (Ainsworth and Sussman, 1968; Griffin, 1972), but studies of responses to single variables have predominated. The concept of the fundamental niche, by emphasizing the multivariate nature of environmental control over organismal response, has a useful heuristic role in the interpretation of fungal response data. Moreover, the qualitative response characteristics illustrated here with lichen net photosynthesis may also be applied to analysis of the adaptive significance of other fungal responses. Such analyses may be easier and more instructive using fungi in soil or aquatic habitats where long-term environmental patterns are more easily estimated than for plants exposed to the atmospheric environment. Fungal ecologists, with thoughtful choice of experimental taxa, should be able to thoroughly describe the relationships between fungal responses to multiple environmental variables and the pattern of those variables in the natural habitats of the taxa. A complete description of these relationships does not exist in the ecological literature and would be an important contribution toward measuring the adaptedness of a fundamental niche.

The limited evidence from lichens reviewed here does not conclusively establish that the fundamental niche has evolved under natural selection as an adaptation to environmental patterns. Adaptive characteristics of the fundamental niche can only evolve if there are heritable differences among the fundamental niches of individuals in a population that lead to differential lifetime reproductive success. Because of

the relative ease of isolating and culturing pure strains of many fungi, fungal ecologists can test these critical requirements for evolution of niche characteristics.
Clonal isolates from natural populations of fungi can provide valuable data on genetically based variation in the fundamental niche of a species. Competition experiments between strains can establish the adaptedness of responses within the boundaries of the fundamental niche under diverse environmental regimes. By studying relationships between variation in niche characteristics and fitness, fungal ecologists could advance our understanding of the evolution of the niche.

As a final caution, the likely importance of competition in the evolution of
fundamental niche characteristics must be noted. The physical environment alone may
affect differential survival and reproduction among individuals in populations occurring at low densities like some of the lichens or similar colonizing species. In
most communities, however, both intra- and interspecific competition exert selection
pressure on growth and reproductive responses. The dynamic interactions of species
in a community require that to understand the evolution of the fundamental niche we
must consider the effect of biotic as well as physical environmental factors.

ACKNOWLEDGMENTS

Work reported in this chapter was partly supported by grants from the Faculty of
Graduate Studies and Research, McGill University, and by a subcontract of a U.S.
National Science Foundation grant DPP-76-80646 to Dr. Philip C. Miller at San Diego
State University. This NSF grant was part of a coordinate research program called
Research on Arctic Tundra Environments (RATE). I thank Graham Bell, Don Kramer,
and Marcia Waterway for helpful comments on early drafts of this chapter and Susan
Brunette for typing the manuscript.

REFERENCES

Ainsworth, G. C., and Sussman, A. S. (Eds.). (1968). *The Fungi*, Vol. III: *The
Fungal Population*. Academic Press, New York.

Armstrong, R. A. (1976). Studies on the growth rates of lichens. In *Lichenology:
Progress and Problems*, D. H. Brown, D. L. Hawksworth, and R. H. Bailey, (Eds.).
Academic Press, New York, pp. 309-322.

Bailey, R. H. (1976). Ecological aspects of dispersal and establishment in lichens.
In *Lichenology: Progress and Problems*, D. H. Brown, D. L. Hawksworth, and R. H.
Bailey (Eds.). Academic Press, New York, pp. 215-247.

Berry, J. A. (1975). Adaptation of photosynthetic processes to stress. *Science
188*: 644-650.

Björkman, O., and Holmgren, P. (1963). Adaptability of the photosynthetic apparatus to light intensity in ecotypes from exposed and shaded habitats. *Physiol.
Plant. 16*: 889-914.

Bowler, P. A., and Rundel, P. W. (1975). Reproductive strategies in lichens. *Bot.
J. Linn. Soc. 70*: 325-340.

Byers, H. B., and Rodebush, H. R. (1948). Causes of thunderstorms of the Florida
peninsula. *J. Meteorol. 5*: 275-280.

Carstairs, A. G., and Oechel, W. C. (1978). Effects of several microclimatic factors and nutrients on net carbon dioxide exchange in *Cladonia alpestris* (L.) Rabh. in the subarctic. *Arctic and Alpine Res. 10*: 81-94.

Clebsch, E. E. C., and Shanks, R. E. (1968). Summer climatic gradients and vegetation near Barrow, Alaska. *Arctic 21*: 162-171.

Colwell, R. K. (1974). Predictability, constancy, and contingency of periodic phenomena. *Ecology 55*: 1148-1153.

Colwell, R. K., and Fuentes, E. R. (1975). Experimental studies of the niche. *Ann. Rev. Ecol. Syst. 6*: 281-310.

Conover, J. H. (1960). Macro- and microclimatology of the Arctic Slope of Alaska. Tech. Rept. EP-139, Quartermaster Research and Engineering Command, U.S. Army.

Cooper, J. P., Ed. (1975). *Photosynthesis and Productivity in Different Environments*. Cambridge Univ. Press, New York.

Dobzhansky, Th. (1968a). On some fundamental concepts of Darwinian biology. *Evol. Biol. 2*: 1-34.

Dobzhansky, Th. (1968b). Adaptedness and fitness. In *Population Biology and Evolution*, R. C. Lewontin (Ed.). Syracuse Univ. Press, Syracuse, pp. 109-121.

Emlen, J. M. (1973). *Ecology: An Evolutionary Approach*. Addison-Wesley, Reading, Mass.

Farrar, J. F. (1973). Lichen physiology: progress and pitfalls. In *Air Pollution and Lichens*, B. W. Ferry, M. S. Baddeley, and D. L. Hawksworth (Eds.). Univ. of Toronto Press, Toronto, pp. 238-282.

Farrar, J. F. (1976a). The uptake and metabolism of phosphate by the lichen *Hypogymnia physodes*. *New Phytol. 77*: 127-134.

Farrar, J. F. (1976b). Ecological physiology of the lichen *Hypogymnia physodes*. II. Effects of wetting and drying cycles and the concept of "physiological buffering." *New Phytol. 77*: 105-113.

Farrar, J. F., and Smith, D. C. (1976). Ecological physiology of the lichen *Hypogymnia physodes*. III. The importance of the rewetting phase. *New Phytol. 77*: 115-125.

Fisher, R. A. (1930). *The Genetical Theory of Natural Selection*. Oxford Univ. Press (Clarendon), New York.

Gerson, V., and Seaward, M. R. D. (1977). Lichen-invertebrate associations. In *Lichen Ecology*, M. R. D. Seaward (Ed.). Academic Press, New York, pp. 69-119.

Griffin, D. M. (1972). *Ecology of Soil Fungi*. Chapman & Hall, London.

Haugen, R. K., Brown, J., and May, T. A. (1976). Climatic and soil temperature observations at Atkasook on the Meade River, Alaska, Summer 1975. CRREL Special Report 76-1, CRREL, Hanover, N.H.

Hoffman, G. R., and Gates, D. M. (1970). An energy budget approach in the study of water loss in cryptogams. *Bull. Torrey Bot. Club 97*: 361-366.

Hutchinson, G. E. (1958). Concluding remarks. *Cold Spring Harbor Symp. Quant. Biol. 22*: 415-427.

Jahns, H. M. (1973). Anatomy, morphology, and development. In *The Lichens*, V. Ahmadjian and M. E. Hale (Eds.). Academic Press, New York, pp. 1-58.

Kallio, P., and Heinonen, S. (1971). Influence of short-term low temperature on net photosynthesis in some subarctic lichens. *Rep. Kevo Subarctic Res. Sta. 8*: 63-72.

Kallio, P., and Kärenlampi, L. (1975). Photosynthesis in mosses and lichens. In *Photosynthesis and Productivity in Different Environments*, J. P. Cooper (Ed.). Cambridge Univ. Press, New York, pp. 383-423.

Kappen, L. (1973). Response to extreme environment. In *The Lichens*, V. Ahmadjian and M. E. Hale (Eds.). Academic Press, New York, pp. 311-380.

Kappen, L., and Lange, O. L. (1972). Die Kälteresistenz einiger Makrolichenen. *Flora 161*: 1-29.

Kershaw, K. A. (1971). The relationship between moisture content and net assimilation rate of lichen thalli and its ecological significance. *Can. J. Bot. 50*: 543-555.

Kershaw, K. A. (1975). Studies on lichen-dominated systems. XIV. The comparative ecology of *Alectoria nitidula* and *Cladina alpestris*. *Can. J. Bot. 53*: 2608-2613.

Kershaw, K. A. (1977a). Physiological-environmental interactions in lichens. II. The pattern of net photosynthetic acclimation in *Peltigera canina* (L.) Willd var. *praetextata* (Floerke in Somm.) Hue, and *P. polydactyla* (Neck.) Hoffm. *New Phytol. 79*: 377-390.

Kershaw, K. A. (1977b). Physiological-environmental interactions in lichens. III. The rate of net photosynthetic acclimation in *Peltigera canina* (L.) Willd var. *praetextata* (Floerke in Somm.) Hue, and *P. polydactyla* (Neck.) Hoffm. *New Phytol. 79*: 391-402.

Kershaw, K. A., and Larson, D. W. (1974). Studies on lichen-dominated systems. IX. Topographic influences on microclimate and species distribution. *Can. J. Bot. 52*: 1935-1945.

Lange, O. L. (1965). Der CO_2-Gaswechsel von Flechten nach Erwärmung in feuchten Zustand. *Ber. Deut. Bot. Ges. 78*: 441-454.

Lange, O. L. (1969a). Experimentell-ökologische Untersuchungen an Flechten der Negev-Wüste. I. CO_2-Gaswechsel von *Ramalina maciformis* (Del.) Bory unter kontrollierten Bedingungen im Laboratorium. *Flora Abt. B 158*: 324-359. [Tech. transl. No. 1654, National Research Council of Canada: Ecophysiological investigations on lichens of the Negev Desert. I. CO_2 gas exchange of *Ramalina maciformis* (Del.) Bory under controlled conditions in the laboratory.]

Lange, O. L. (1969b). Die funktionellen Anpassungen der Flechten an die ökologischen Bedingungen arider Gebiete. *Ber. Deut. Bot. Ges. 82*: 3-22.

Lange, O. L., and Bertsch, A. (1965). Photosynthese der Wüstenflechte *Ramalina maciformis* nach Wasserdampfaufnahme aus dem Luftraum. *Naturwissenschaften 9*: 215-216.

Lange, O. L., and Kappen, L. (1972). Photosynthesis of lichens from Antarctica. In *Antarctic Terrestrial Biology*, G. A. Llano (Ed.). American Geophysical Union, Washington, D.C., pp. 83-95.

Lange, O. L., Koch, W., and Schulze, E.-D. (1969). CO_2-Gaswechsel und Wasserhaushalt von Pflanzen in der Negev-Wüste am Ende der Trockenzeit. *Ber. Deut. Bot. Ges. 82*: 39-61.

Lange, O. L., Schulze, E.-D., and Koch, W. (1970a). Experimentell-ökologische Untersuchungen an Flechten der Negev-Wüste. II. CO_2-Gaswechsel und Wasserhaushalt von *Ramalina maciformis* (Del.) Bory am natürlichen Standort während der sommerlichen Trockenperiode. *Flora 159*: 38-62. [Tech. transl. No. 1655, National Research Council of Canada: Ecophysiological investigations on lichens of the Negev Desert. II. CO_2 gas exchange and water relations of *Ramalina maciformis* (Del.) Bory in its natural habitat during the summer dry period.]

Lange, O. L., Schulze, E.-D., and Koch, W. (1970b). Experimentell-ökologische Untersuchungen an Flechten der Negev-Wüste. III. CO_2-Gaswechsel und Wasserhaushalt von Krusten- und Blattflechten am natürlichen Standort während der sommerlichen Trockenperiode. *Flora 159*: 525-538. [Tech. transl. No. 1656, National Research Council of Canada: Ecophysiological investigations on lichens of the Negev Desert. III. CO_2 gas exchange and water relations of crustose and foliose lichens in their natural habitat during the summer dry period.]

Lange, O. L., Schulze, E.-D., Kappen, L., Buschbom, U., and Evenari, M. (1975). Adaptations of desert lichens to drought and extreme temperatures. In *Environmental Physiology of Desert Organisms*, N. F. Hadley (Ed.). Dowden, Hutchinson, and Ross, Stroudsburg, Pa., pp. 20-37.

Lange, O. L., Geiger, I. L., and Schulze, E.-D. (1977) Ecophysiological investigations on lichens of the Negev Desert. V. A model to simulate net photosynthesis and respiration of *Ramalina maciformis*. *Oecologia 28*: 247-259.

Larcher, W. (1975). *Physiological Plant Ecology* (transl. by M. A. Biederman-Thorson). Springer-Verlag, New York.

Larson, D. W., and Kershaw, K. A. (1975a). Measurement of CO_2 exchange in lichens: a new method. *Can. J. Bot. 53*: 1535-1541.

Larson, D. W., and Kershaw, K. A. (1975b). Studies on lichen-dominanted systems. XIII. Seasonal and geographical variation of net CO_2 exchange of *Alectoria ochroleuca*. *Can. J. Bot. 53*: 2598-2607.

Larson, D. W., and Kershaw, K. A. (1975c). Studies on lichen-dominated systems. XVI. Comparative patterns of net CO_2 exchange in *Cetraria nivalis* and *Alectoria ochroleuca* collected from a raised-beach ridge. *Can. J. Bot. 53*: 2884-2892.

Larson, D. W., and Kershaw, K. A. (1975d). Acclimation in arctic lichens. *Nature 254*: 421-423.

Lechowicz, M. J. (1976). Environmental response structure of *Cladonia* lichens from contrasting climates. Ph.D. thesis, University of Wisconsin, Madison.

Lechowicz, M. J. (1978). Carbon dioxide exchange in *Cladina* lichens from subarctic and temperate habitats. *Oecologia 32*: 225-237.

Lechowicz, M. J., and Adams, M. S. (1973). Net photosynthesis of *Cladonia mitis* (Sand.) from sun and shade sites on the Wisconsin Pine Barrens. *Ecology 54*: 413-419.

Lechowicz, M. J., and Adams, M. S. (1978). Diurnal and seasonal structure of the climate at Schefferville, Quebec. *Arctic and Alpine Res. 10*: 95-104.

Lechowicz, M. J., and Adams, M. S. (1979). Net CO_2 exchange in *Cladonia* lichen species endemic to southeastern North America. *Photosynthetica 13*: 155-162.

Levins, R. (1968a). *Evolution in Changing Environments*. Princeton Univ. Press, Princeton, N.J.

Levins, R. (1968b). Toward an evolutionary theory of the niche. In *Evolution and Environment*, E. T. Drake (Ed.). Yale Univ. Press, New Haven, Conn., pp. 325-340.

Maguire, B., Jr. (1973). Niche response structure and the analytical potentials of its relationship to the habitat. *Amer. Natur. 107*: 213-246.

Monteith, J. L. (1973). *Principles of Environmental Physics*. Elsevier, New York.

Mooney, H. A. (1972). The carbon balance of plants. *Ann. Rev. Ecol. Syst. 3*: 315-346.

Mooney, H. A. (1975). Plant physiological ecology: a synthetic view. In *Physiological Adaptation to the Environment*, F. J. Vernberg (Ed.). Educational Publ., New York, pp. 19-36.

Mooney, H. A. (1976). Some contributions of physiological ecology to plant population biology. *Syst. Bot. 1*: 269-283.

Nieboer, E., Richardson, D. H. S., Puckett, K. J., and Tomassini, F. D. (1976). The phytotoxicity of sulphur dioxide in relation to measurable responses in lichens. In *Effects of Air Pollutants on Plants*, T. A. Mansfield (Ed.). Cambridge Univ. Press, New York, pp. 61-85.

Nowak, R. (1973). *Vegetationsanalytische und experimentell-ökologische Untersuchungen über den Einfluss der Luft verunreinigung auf rinden bewohnende Flechten*. Polyfoto--Dr. Vogt KG, Stuttgart, West Germany.

Oechel, W. C. (1976). Seasonal patterns of temperature response of CO_2 flux and acclimation in Arctic mosses growing *in situ*. *Photosynthetica 10*: 447–456.

Richardson, D. H. S. (1973). Photosynthesis and carbohydrate movement. In *The Lichens*, V. Ahmadjian and M. E. Hale (Eds.). Academic Press, New York, pp. 249-288.

Richardson, D. H. S., and Young, C. M. (1977). Lichens and vertebrates. In *Lichen Ecology*, M. R. D. Seaward (Ed.). Academic Press, New York, pp. 121-144.

Ried, A. (1960). Stoffwechsel und Verbreitungsgrenzen von Flechten. II. Wasser- und Assimilationshaushalt, Entquellungs- und Submersionresistenz von Krustenflechten Genachbarter Standorte. *Flora 149*: 345-385. [Tech. transl. No. 1774, National Research Council of Canada: Metabolism and frequency distribution limits of lichens. II. Water utilization and assimilation patterns of crustaceous lichens in adjacent locations and their resistance to drought and submersion.]

Rogers, R. W. (1971). Distribution of the lichen *Chondropsis semiviridis* in relation to its heat and drought tolerance. *New Phytol. 70*: 1069-1077.

Roughgarden, J. (1972). Evolution of niche width. *Amer. Natur. 106*: 683-718.

Slatkin, M., and Lande, R. (1976). Niche width in a fluctuating environment-density independent model. *Amer. Natur. 110*: 31-55.

Smith, D. C. (1962). The biology of lichen thalli. *Biol. Rev. 37*: 537-570.

Smith, D. C., and Molesworth, S. (1973). Lichen physiology. XIII. Effects of rewetting dry lichens. *New Phytol. 72*: 525-533.

Spomer, G. G. (1973). The concepts of "interaction" and "operational environment" in environmental analyses. *Ecology 54*: 200-204.

Stålfelt, M. G. (1939). Der Gasaustauch der Flechten. *Planta 29*: 11-31.

Strain, B. R., and Billings, W. D., Eds. (1974). *Vegetation and Environment*. Junk, The Hague.

Topham, P. B. (1977). Colonization, growth, succession, and competition. In *Lichen Ecology*, M. R. D. Seaward (Ed.). Academic Press, New York, pp. 31-68.

Chapter 7

ADAPTATION OF FUNGI TO STAGNANT WATERS

Ralph Emerson[*] and Donald O. Natvig
University of California
Berkeley, California

> [La] fermentation est la conséquence de la vie sans air... .
> --Louis Pasteur (1876, p. 271)

I. INTRODUCTION

Perhaps our concepts of extreme environments are too closely linked to our own human physiological requirements. At all events, atmospheres low or lacking in oxygen and high in carbon dioxide are traditionally considered extreme for the fungi. Thus the subconscious thoughts of ecological mycologists about where the various groups of fungi occur and what they are doing in the biosphere tend to be strongly influenced by this point of view. It is our contention that anaerobiosis is more widely distributed among the molds than has been previously recognized and that an appreciation of the activities of fungi in microaerobic and anoxic environments is essential if we are to develop a sound understanding of the overall ecology of these universally distributed and highly significant microorganisms.

In this chapter, after briefly considering the aerobiosis dogma and the evidence that is accumulating against it, we will focus particular attention upon the striking anaerobic capacities of certain zoosporic fungi in the Chytridiomycetes and Oomycetes. For it is here in aqueous environments that eukaryotes are being found that have evolved such adaptations to lack of oxygen that they are obligately fermentative and, in some most recently explored situations, probably even obligately anaerobic. Finally, we will see that anaerobiosis is sometimes coupled with carboxyphilism, an actual requirement for high carbon dioxide levels, and very possibly always involves auxotrophy for essential metabolic substances that cannot be fabricated without oxidative energy metabolism. Our rudimentary knowledge of the field ecology of water molds in microaerobic habitats will be considered in the final section.

---

[*]Deceased.

II. FUNGAL AEROBIOSIS, A WORN-OUT DOGMA

Since the time of Anton de Bary, more than a century ago, one of the general tenets
of fungal physiology has been that fungi require oxygen for growth. Even though the
facultative fermentation of yeasts and species of Mucorales was established by Pasteur (1876), the requirement of aerobic conditions for continued fungal growth has
been reiterated right up to recent times. A few quotations from distinguished sources will serve to make the point. For example, some three decades ago: "One of the
major metabolic differences between molds and bacteria is that there are no anaerobic molds, either obligate or facultative" (Foster, 1949, p. 162). Or "All filamentous fungi are obligate aerobes, unable to grow in the complete absence of oxygen"
(Hawker, 1950), p. 198). Also some two decades ago: "the filamentous fungi require
oxygen for growth" (Cochrane, 1958, p. 212); and "While all, or nearly all, fungi
have an absolute requirement for oxygen in the atmosphere, the actual concentration
needed is usually rather low...." (Hawker, 1966, p. 454). In the present decade:
"The growth of molds is limited by their aerobic nature. Molds will not thrive in
the absence of oxygen and are, therefore, limited to surface growth. In liquids,
they grow as a floating mat or pellicle" (Wyss and Eklund, 1971, p. 168). Or, most
recently: "The ability of filamentous fungi to grow under strict anaerobic conditions is limited to a very few species and these are mostly restricted to *Mucor* and
Fusarium" (Watson, 1976, p. 107). Gleason (1976, p. 545), in discussing the physiology of the lower freshwater fungi, starts his section headed "Requirement for Oxygen" with the words "Fungi generally are thought to be highly aerobic."

Unquestionably very many, probably most, fungi are obligate aerobes. Nevertheless, it is becoming increasingly apparent that many--how many we cannot yet tell--
are true facultative anaerobes. Until recently the information to support such a
statement was not available because few if any tests had been made under sufficiently rigorously defined conditions to permit positive conclusions to be drawn. Earlier studies were sometimes performed under the naive assumption that essentially
oxygen-free atmospheres could be obtained with commercial tank nitrogen without special precautions to remove the last traces of oxygen. Only when such procedures are
followed, and the elimination of free oxygen is signaled by conversion of a standard
methylene blue indicator to the leuco state, can anaerobic conditions be claimed in
accordance with long-standing practice in bacteriology. Since 1968 a number of studies meeting this definition of anaerobiosis have been made with miscellaneous common laboratory molds. The results of two of these (Tabak and Cooke, 1968a; Curtis,
1969), particularly significant here, are summarized in Table 1. Tabak and Cooke
(1968a) found that 13 species of molds, mostly isolated from sewage sludge, produced
20-58% of their dry weight in 21% oxygen when they were grown on the same medium in
a 100% nitrogen atmosphere where methylene blue was in the leuco state. These same
investigators (Tabak and Coke, 1968b) presented a particularly helpful review of the
various methods that have been used over the years in investigations of anaerobiosis.

Table 1 A comparison of the growth (dry weight) of various molds under strictly anaerobic and aerobic conditions

Fungus	Anaerobic/aerobic x 100 (= %)
Data from Tabak and Cooke (1968a)--no reductant	
Fusarium oxysporum	57.6
Mucor hiemalis	55.3
Aspergillus fumigatus	51.4
Geotrichum candidum	51.4
Phialophora jeanselmei	48.3
Aureobasidium pullulans	42.2
Penicillium sp.	39.5
Fusarium solani	38.7
Aspergillus niger	28.8
Phoma herbarum	28.5
Trichoderma viride	26.7
Rhodotorula mucilaginosa	24.4
Candida parapsilosis	20.2
Data from Curtis (1969)	
Mucor abundans	33.1
Mucor fragilis	27.5
Fusarium oxysporum	16.8-19.0
Zygorhynchus vuilleminii	14.9
Gibberella fujikuroi	8.7-11.6
Zygorhynchus moelleri	9.9
Mucor erectus	8.6
Lilliputia rufula	4.6
Geotrichum candidum	4.6
Absidia spinosa	2.0

We recognize that the strictest sort of anaerobic growth, say, for obligately anaerobic bacteria or rumen organisms (see Sec. III.D), may require control of the redox level of the environment with reducing agents. Future experiments with anaerobic fungi may require such ultimate refinements, but they are not of immediate concern here. Moreover, it is noteworthy that Tabak and Cooke (1968a) themselves employed sodium formaldehyde sulfoxylate in their series of tests and found little or no significant difference in the growth obtained with or without the reductant.

Seven molds from a random collection studied by Curtis (1969) showed a range of 8.6-33.3% of their growth in 21% oxygen when grown in a nitrogen-carbon dioxide atmosphere, again with methylene blue decolorized. These two studies represent only

a minute sampling of the thousands of species of fungi, but they are important because they suggest the possibility that wide search may reveal many more bona fide
facultatively anaerobic molds. Moreover, taken with other scattered observations on
Fusarium spp. (e.g., Gunner and Alexander, 1964) and dimorphic Mucorales (e.g., Bartnicki-Garcia and Nickerson, 1962) that show undoubted anaerobic development, these
data cast real doubt upon the validity of the aerobiosis dogma in mycology. In fact,
at least 20 common mold species exhibit anaerobic growth. That the dry weight they
produce fermentatively is sometimes less than one-tenth that of their aerobic growth
should not surprise us. Pasteur (1876) found comparable ratios between his microaerobic and aerobic cultures of *Saccharomyces* more than a century ago. It seems likely
that phenomena comparable to the Pasteur effect will be routinely found among facultatively anaerobic mold fungi. What is more interesting, as we shall see, is that
some fungi, e.g., *Fusarium oxysporum*, *Aspergillus fumigatus*, or *Mucor hiemalis* (cf.
Table 1) should achieve more than 50% of aerobic growth by fermentative metabolism.
The higher the ratio of anaerobic to aerobic growth, the more fully adapted the species may be to life in microaerobic or anaerobic habitats. As will be apparent
shortly, the ratio can become unity when fermentation has become a way of life--
indeed, the only way of life in certain instances. The dogma of strict and virtually universal fungal aerobiosis must be laid to rest.

III. ANOXYPHILES: WATERMOLDS IN THE LABORATORY

A. Widespread Lactic Acid Formation

In this section we will focus attention upon the demonstration of fermentative metabolism and facultative anaerobiosis in the zoosporic fungi, members of the subdivision
Mastigomycotina. The first work on fermentative metabolism in the water molds was
done some 30 years ago by E. C. Cantino (see Emerson and Cantino, 1948; Cantino, 1949)
on the chytridiomycete genus *Blastocladia*. Surprisingly, the species *B. pringsheimii*
grew almost equally well in 21% oxygen and in commercial tank nitrogen. This water
mold was even more robust in an atmosphere of tank carbon dioxide. Under all these
conditions, its major metabolic waste product was lactic acid (Table 2). More important, whether aerobic or nearly anoxic, the amount of carbon recovered as lactic
acid (more than four-fifths except in high CO_2) was virtually identical. The great
significance of this point escaped us at the time.

In the following decade lactic acid production was demonstrated in five other
chytridiomycete genera in three orders, viz., *Allomyces* (Ingraham and Emerson, 1954),
Blastocladiella (Cantino, 1951a), *Chytridium*, *Macrochytrium* and *Monoblepharella*
(Crasemann, 1954), and in the oomycetous genera *Pythiogeton* (Cantino, 1951b) of the
Peronosporales and *Sapromyces*, *Rhipidium*, and *Apodachlya* in the Leptomitales (Golueke, 1957). *Mindeniella*, a fourth genus of that order, was known to be a strong
acid producer (Emerson, 1958), and the acidic metabolite was again found to be lactic
acid by Gleason et al. (1966). These and other reports of lactic acid formation by

Table 2 Energy metabolism of *Blastocladia pringsheimii* under various conditions of aeration[a]

Aeration conditions	Carbon recovered (%)	
	As lactic acid	As succinic acid
Standing culture in air	85	11
Bubbled with air	86	11
Bubbled with CO_2-free air	86	10
Bubbled with commercial N_2	88	10
Bubbled with commercial CO_2	76	18

[a]Note particularly the complete absence of any oxygen effect upon the energy metabolism as measured by fermentation products formed.

Source: Data from Cantino (1949).

the zoosporic fungi are well outlined in Gleason's (1976) review. As these data were gathered and compared, a coherent picture developed of widespread lactic acid production, particularly in the Leptomitales and Blastocladiales. In both orders, lactic acid fermentation ranged (Fig. 1) from weakly developed in such obligate aerobes as *Blastocladiella* and *Apodachlya*, to more pronounced in apparently still obligate aerobes like *Allomyces*, and to strongly and uniformly developed in genera then believed to be facultative fermenters such as *Blastocladia*, *Sapromyces*, and *Pythiogeton*. In no instance, however, had strict anaerobiosis yet been documented for the water molds with methylene blue.

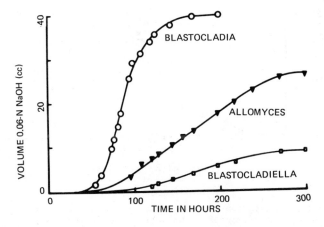

Fig. 1 Lactic acid production by three members of the family Blastocladiaceae. Acid formation will sometimes depend on the particular environment; however, some species are much stronger acid producers than others, as is shown here. (From Cantino, 1948a, p. 55a.)

B. Obligately Fermentative Energy Metabolism in
    Aqualinderella and *Blastocladia*

In 1957 we found the remarkable leptomitalean genus *Aqualinderella* in Costa Rica and
isolated *A. fermentans* in axenic culture for the first time in candle jars (Emerson
and Weston, 1967). Like other members of the family Rhipidiaceae, it proved to be
a strong acid producer with an apparent carbon dioxide requirement for full growth.
Intensive study of its physiology (Emerson and Held, 1969; Held, 1970) established
the following points: (1) Energy metabolism in *A. fermentans*, as measured by glucose consumption, lactic acid production, or dry weight accumulation is oxygen independent (Table 3 and Fig. 2). (2) Molar growth yields (mg dry weight/mM of glucose

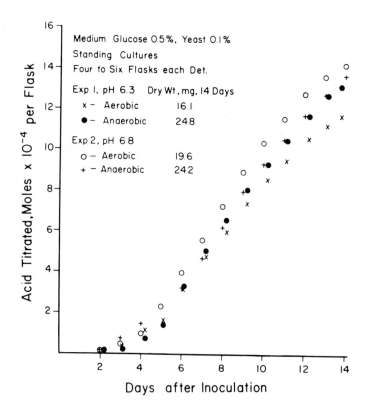

Fig. 2 A classic demonstration of obligately fermentative energy metabolism as it
occurs in *Aqualinderella fermentans*. No matter what oxygen level is provided, growth
and lactic acid formation occur with remarkable uniformity. The organism is oxygen
independent in its energy metabolism, and there is not the slightest indication of a
real Pasteur effect. *A. fermentans* is obligately fermentative but facultatively anaerobic. (From Emerson and Held, 1969, p. 1112, by permission of the copyright holder, *The American Journal of Botany*.)

Table 3 Comparison of aerobic and anaerobic[a] growth of four obligate or near-obligate fermenters

Fungus	Titratable acid (mM)		Dry weight (mg)		Molar growth yield[b]	
	O_2	No O_2	O_2	No O_2	O_2	No O_2
Blastocladia ramosa (54-8)[c]	--	--	33.1	28.0	37.9	30.0
Blastocladia ramosa (54-14)[c]	--	--	36.0	31.1	34.9	30.1
Aqualinderella fermentans[d]	2.05	1.99	50	55	34.6	39.4
Pythiogeton sp.[e]	0.88	1.09	26.2	25.5	42.3	36.5

[a]Leuco methylene blue.

[b]Molar growth yield = mg dry weight/mM glucose consumed.

[c]Data from Held et al. (1969); medium = 150 ml GY5 (0.3% glucose); harvest at 54 hr; titratable acid data not available.

[d]Data from Held (1967); medium = 75 ml IIW (0.72% glucose); harvest at 10 days.

[e]Data from Winans (1977); medium = 50 ml GYP (0.3% glucose); harvest at 6 days.

consumed) with and without oxygen (leuco methylene blue) are the same (Table 3).
(3) This fungus is a facultative anaerobe that is obligately fermentative, i.e.,
it has no capacity for oxidative energy metabolism. These conclusions were further
borne out (Held et al., 1969) when it was found that *A. fermentans* lacks recognizable cristate mitochondria and detectable cytochromes.

These remarkable findings in *Aqualinderella* drew attention again to *Blastocladia*. Cantino's (1949) carbon balances strongly suggested that *B. pringsheimii* is obligately fermentative. Studies reported by Held et al. (1969) showed that *B. ramosa*
is indeed in most respects similar to *A. fermentans*: its energy metabolism is essentially oxygen independent; it too is a facultatively anaerobic, obligately fermentative, homolactic fermenter; and it has only trace amounts of cytochrome and only rudimentary, noncristate mitochondria. Although closely similar to *B. ramosa* in all
other respects, *B. pringsheimii* required trace amounts of oxygen for growth, possibly,
as we shall see (in Sec. IV.B), because of what may be termed *anoxic auxotrophy*.

Here, then, are two clear instances of facultative anaerobiosis coupled with
obligately fermentative energy metabolism in two very different classes of aquatic
fungi, the Chytridiomycetes and the Oomycetes. To our knowledge, this particular
kind of energy metabolism has not been reported hitherto in any naturally occurring
fungi. The ecological implications are most interesting.

C. Anaerobiosis in *Pythiogeton* sp., CRB11

By 1970 the likelihood of facultative anaerobes occurring in widely different orders
of aquatic fungi was sufficiently clear that a search in anaerobic or near-anaerobic
environments appeared well justified. Five months' exploration in tropical stagnant
waters in Costa Rica (1972-1973) showed that certain fungi could be repeatedly collected from environments where readings with a field oxygen meter (Yellow Springs
Instrument Co., Model 51A) consistently showed less than 1 ppm dissolved oxygen. One
such fungus, found on numerous occasions as an inhabitant of fruit and seed baits at
50-100 cm depths in the water-lily pond on the grounds of the Centro Agronomico Tropical de Investigación y Enseñanza at Turrialba, proved to be an undescribed species
of *Pythiogeton* that will be designated here simply as CRB11. After considerable
difficulty in finding a medium that would support vigorous growth (see Sec. IV.B),
axenic cultures were established and it soon became apparent that this isolate, like
Cantino's (1949), was a strong acid producer. Our recent intensive studies, done
with S. C. Winans (not yet published), have revealed that this member of the economically important Pythiaceae in the order Peronosporales (1) is a true facultative
anaerobe with a fermentative, homolactic energy metabolism and (2) is very possibly,
like *A. fermentans* and *B. ramosa*, an obligate fermenter. The point at issue in this
regard is that there does appear to be a consistently higher (ca. 10%) molar growth
yield (Table 3) when CRB11 is grown in air with 2-3% carbon dioxide than when it is
grown at the same level of carbon dioxide but with no oxygen present. It is worth
noting here that Held et al. (1969) reported a correspondingly higher (by 20% in that

case) molar growth yield in *Blastocladia* spp. grown in carbon dioxide-supplemented
air even though cristate mitochondria and functional levels of cytochrome were lacking. The condition of mitochondria and cytochromes in *Pythiogeton* is not yet known,
but anoxic auxotrophy may provide an explanation for somewhat higher growth yields
of obligate fermenters growing in air (cf. Sec. IV.B). At all events, here is a
third genus of water molds, in a third order, and a second member of the Oomycetes
with undoubted facultative capacity for strictly anaerobic energy metabolism.

D. Rumen "Chytrids": The First Obligately Anaerobic Fungi

To any one as much involved as we have been during the past 30 years in tracking down
aspects of anaerobiosis in the lower fungi, C. G. Orpin's series of penetrating papers (1975, 1976, 1977a,b,c,d; see also Orpin and Bountiff, 1978) on the fungal microbiota of the sheep rumen was a revelation and a delight. The organisms involved,
Neocallimastix frontalis, *Sphaeromonas communis*, and *Piromonas communis*, were first
described, going back as far as the early 1900s, as flagellate protozoa. Orpin's
particular attention was centered upon these flagellated cells because they exhibited
an unexplained 25-fold increase in numbers an hour after feeding of the ruminant. He
discovered that the increase was in fact due to chemically coordinated mass release
of zoospores developed in eucarpic, mainly monocentric, chitin-walled, chytrid-like
thalli (Fig. 3) attached to and penetrating the food particles in the rumen by their
well-developed rhizoidal systems. Many fascinating aspects of their biology are
brought out in Orpin's work. In the present context we need only recall that the
rumen is certainly one of the most anaerobic natural environments to be found in the

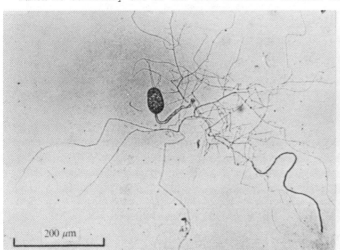
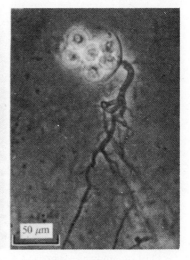

Fig. 3 Neocallimastix frontalis, a representative of the recently discovered chytrid-like fungi in the sheep rumen. The two thalli shown were raised in vitro.
Both show the typical, eucarpic, monocentric thallus with a conspicuously developed
rhizoidal system. These organisms are not capable of surviving as zoospores or vegetative thalli in the presence of oxygen, and they appear to represent the first known
obligately anaerobic fungi. (From Orpin, 1975, p. 225, by permission of the copyright holder, *The Journal of General Microbiology*.)

biosphere, and microorganisms adapted to grow therein occupy a very special niche indeed. The response of the rumen "chytrids" was investigated in the laboratory in antibiotic culture systems, free of living bacteria and protozoa. "Conditions for maximum [zoospore] production are similar to those occurring in the rumen: pH 6.5, 39°C, absence of O_2, presence of CO_2" (Orpin, 1975, p. 249). No release of zoospores occurred in air or in the absence of CO_2. In a recent statement C. G. Orpin (personal communication, 1978), has written "other experiments...confirm that *Neocallimastix* is obligately anaerobic...[and] both zoospores and vegetative growth are not viable after even short exposure to oxygen at low concentrations." It looks very much as though these are going to become famous as the first obligately anaerobic true fungi to be discovered. Whereas both *Sphaeromonas communis* and *Piromonas communis* zoospores are posteriorly uniflagellate and clearly chytrid-like, those of *Neocallimastix frontalis* have a complex sort of band-like locomotory organelle with up to 14 flagella. It would be surprising if some fairly extreme morphological evolution had not accompanied adaptation to such an extreme environment. The precise relationships of the rumen "chytrids" to known groups of other fungi will have to be worked out. Whatever their final disposition, they are fungi, and their significance to our discussion of the fungi from "stagnant waters" (sic!) is obvious. They must be obligate fermenters: will they be homolactic fermenters?

IV. CARBOXYPHILES AND ANOXIC AUXOTROPHY

A. Carbon Dioxide Relations

Increased levels of carbon dioxide often accompany reduced levels of oxygen in natural environments, and microorganisms often respond favorably to concentrations of carbon dioxide many times greater than that in air. Some three decades ago we reported (Emerson and Cantino, 1948) the more robust and normal thallus form of *Blastocladia pringsheimii* in broth cultures grown under nearly pure commercial carbon dioxide. The significance of carbon dioxide was again very apparent when it became clear (Emerson and Held, 1969) that it was the increased level of carbon dioxide in the candle chambers that had permitted the first successful gross and axenic laboratory cultures of *Aquilinderella fermentans*. On glucose-yeast media this fungus has an obligate requirement for high levels (5–20% or more) of carbon dioxide, and even in succinate-amended media it grows relatively poorly in air without supplemental carbon dioxide (Held, 1970). Hence *A. fermentans* can be correctly referred to as a carboxyphile.

We have no evidence that increased carbon dioxide is beneficial to *Pythiogeton* CRB11. Indeed, there is some evidence that 5% or more is toxic. Levels of 3% were used in the anaerobic experiments (see Sec. III.C) simply to be sure that a reasonable supply was available even in the presence of strong alkali being provided for neutralization (S. C. Winans and R. Emerson, unpublished data). The essentiality of carbon dioxide for the rumen fungi has been emphasized in Sec. III.D.

We do not yet have a thorough understanding of the biochemical significance of carbon dioxide for the growth of anaerobic aquatic fungi. It is likely (see Lynch and Calvin, 1952) that most or all fungi, whether anaerobic or not, have the capacity to fix carbon dioxide, but why this should be essential for some anaerobes, e.g., *A. fermentans*, and not for others, e.g., *B. ramosa* or *Pythiogeton* CRB11, is not known. Held (1967, 1970; see also Emerson and Held, 1969) went into the question in some detail and showed that the carbon dioxide requirement in *A. fermentans* could be partially replaced by succinic and several other dicarboxylic acids. Labeling experiments to follow the entry of CO_2 into the cellular metabolism of *A. fermentans* have long been needed. It is also noteworthy here that carbon dioxide may not be produced in a strict homolactic fermentation. Cantino (1949) found no carbon recovered as carbon dioxide in his carbon balance studies of the metabolism of *Blastocladia pringsheimii*, and Golueke (1957) detected no carbon dioxide in similar studies of the fermentative metabolism of *Rhipidium* and *Sapromyces* (Leptomitales) under anaerobic or microaerobic conditions. The availability of carbon dioxide from the environment for synthetic reactions may be of cardinal importance to a homolactic fermenter incapable of respiratory activity either because it is obligately fermentative or because it is growing in a strictly anoxic situation. Reviews of the role of carbon dioxide fixation in metabolism (e.g., Wood and Utter, 1965) are available, and essential biochemical pathways involving CO_2 are being extended and clarified each year. Anyone investigating the anaerobic capacities of the fungi should be clearly aware of the likely significance of externally provided carbon dioxide.

B. Anoxic Auxotrophy

Consideration of the possible role of carbon dioxide in the growth of anaerobic fungi also brings us face to face with the following general question: What special nutritional requirements may be exhibited by organisms adapted to life without oxygen? The answer is that quite a number of different ones are likely. More than 15 years ago Goldfine and Bloch (1963) had already compiled a list of essential metabolites requiring oxygen for their synthesis. Among others, it included sterols, unsaturated fatty acids, various vitamins, and nucleic acids, and the total has been enlarged since then by the march of cellular biochemistry. We are suggesting the term *anoxic auxotrophy* for this general phenomenon, viz., the requirement of specific nutrients in the environment as a result of oxygen lack or at least the incapacity to carry out the oxidative reactions necessary for their synthesis (e.g., in an obligate fermenter lacking all capacity for oxidative energy metabolism). Our attention was first drawn to this sort of auxotrophy by the work of Held (1970) on the nutrition of *A. fermentans*. He demonstrated the lipid requirement of that species and showed that it could be satisfied (fully) by wheat germ oil and partially by oleic acid plus cholesterol. He also noted that the nutritional requirements of *A. fermentans* were likely to be more complex than aerobic members of the order Leptomitales. Because of this complexity we still do not have a completely defined medium for *Aqua-linderella*.

Looking back now upon the early work on nutrition in the Blastocladiaceae, we can see a close parallel to *Aqualinderella*. The vitamin requirements of *Blastocladia* are noticeably more complex than those of other (aerobic) genera in the family. Cantino (1948b) demonstrated the need for nicotinamide and biotin as well as thiamin but did not succeed in working out a fully defined medium for his isolate (St. Helena, California) of *B. pringsheimii*. Not until a decade later did Crasemann (1957) establish a defined minimal medium for *B. pringsheimii* (Bakersfield, California, isolate) and *B. ramosa* (Cambridge, Massachusetts, isolate). She found that the vitamin requirements for *B. ramosa* were thiamin and nicotinamide; those for *B. prings-heimii* were thiamin, nicotinamide, and p-aminobenzoic acid. Thiamin is well known as the sole vitamin requirement for the other two intensively studied aerobic genera in the family, namely, *Allomyces* and *Blastocladiella*.

It is not clear to us why *B. ramosa* has not revealed any lipid requirements, but possibly this is because the lipids in yeast extract (used by Held et al., 1969) were sufficient to meet the need. On the other hand, a sterol and/or fatty acid requirement may explain why *B. pringsheimii* would not grow (Held et al., 1969) without a trace (0.2%) of oxygen. In a recent trial in our laboratory (unpublished data) *B. pringsheimii* (isolate CR62) grew as much in anaerobic culture (leuco methylene blue) as it did in aerobic culture when the medium in both instances was supplemented with wheat germ oil. This observation requires careful verification, especially in view of the comment (Held et al., 1969) that "considerable growth" of *B. prings-heimii* "took place if exposure to air was prolonged even slightly above the minimum required for inoculation." Nevertheless, there is a considerable likelihood that the small oxygen requirement in this species is somehow related to anoxic auxotrophy and has no direct bearing upon the fermentative energy metabolism per se.

Our interest in special nutrient requirements related to anoxia or loss of oxidative capacity in the fungi was further strengthened when *Pythiogeton* CRB11 proved to have an obligate need for wheat germ oil much like that displayed by *Aqualinder-ella*. Our great difficulty (unpublished data) in getting CRB11 into axenic culture and finding a medium on which it would produce more than the most meager growth certainly relates to this obligate lipid requirement. Even though zoospore production is copious and the individual spores are large, all efforts to obtain single-spore cultures ended in failure because no detectable growth occurred on a wide variety of standard media. Several years later axenic cultures were first obtained by plating individual hyphae on antibiotic agar, but the growth was still so delicate and slow that the isolates could barely be maintained. Wide but random search for a better growth medium was entirely unsuccessful until we recalled the wheat-germ-oil requirement in *Aqualinderella*. *Pythiogeton* CRB11 had now shown itself to be a fermenter and acid producer, and this suggested that it too might have special nutrient requirements. The addition of a trace of wheat germ oil (4.0 mg/100 ml of medium be-

came standard) caused an immediate and spectacular change. From then on growth of
CRB11 has been dense, vigorous, and fully comparable to what can be expected from
the usual members of the Pythiaceae. Satisfying the lipid requirement of this anaerobic *Pythiogeton* made possible the metabolism studies reviewed earlier in this chapter (Sec. III.C). The matter has been developed at some length here to underscore
the likely importance of auxotrophy in the isolation and study of anaerobic filamentous fungi that may have lost the oxidative capacity to synthesize certain essential
growth factors.

Orpin's rumen fungi (see Sec. III.D) provide a fitting close to this section on
auxotrophy. Rumen organisms are notoriously finicky (Hungate, 1966) in their growth
requirements, and it already appears likely (Orpin, 1975, 1976, 1977a,b,c,d; Orpin
and Bountiff, 1978) that the obligate anaerobes *Sphaeromonas communis* and *Neocallimastix frontalis* will prove to have special auxotrophic characteristics. We must
be aware of the likelihood of such requirements if we are going to locate, isolate,
and culture other groups of facultatively or obligately anaerobic fungi.

V. ANOXYPHILES AND CARBOXYPHILES: WATER MOLDS IN NATURE

In his pioneering work on zoosporic aquatic fungi inhabiting submerged fruits and
other soft plant parts, M. von Minden (1916) suggested more than half a century ago
that several genera, including *Gonapodya*, *Macrochytrium*, *Blastocladia*, *Rhipidium*,
and *Pythiogeton* (*Aqualinderella* and *Mindeniella* were unknown at the time) were capable of growth under conditions of low oxygen availability. Nearly 20 years later,
in what remains as the only really comprehensive ecological study of the water molds,
Lund (1934, p. 59) again named these same genera as fungi he believed to be "able to
thrive with a small supply of oxygen." Both investigators reached their conclusions
on the basis of the types of habitats in which these fungi occurred and also on growth
occurring in aging gross cultures in the laboratory. Neither reported measurements
of oxygen concentrations in the laboratory or the field.

The previous discussion in the present chapter shows that some of these fungi,
long believed to occur in microaerobic or anaerobic environments (e.g., *Blastocladia*,
Pythiogeton), have now been intensively studied in the laboratory along with such
more recently discovered genera as *Aqualinderella*. But it is only very recently
that the suggestions of von Minden and Lund have actually begun to be put to the test
in the field, and ideas about the ecology of the organisms involved are still primarily inferential. Field records of the occurrence of zoosporic fungi in various parts
of the world are available from many sources (see, for example, citations in Sparrow,
1960). Unfortunately, few of these reports include useful information about the precise environment in which a given species occurs. Furthermore, systematic sampling
methods that provide the kind of data necessary for valid comparative analyses have
seldom been employed. What follows is a summary of the meager field information that
is available on the fungi emphasized in this chapter, along with a consideration of
some of the interesting questions raised by the studies involved.

Aqualinderella fermentans is the most thoroughly studied species in both laboratory and field. Its unique physiology has already been discussed. Field data in
Costa Rica (ours, unpublished) and Nigeria (Alabi, 1972) confirm that it is indeed
an inhabitant of waters generally (but not always) low in oxygen and always much enriched in carbon dioxide. A most careful and comprehensive study has been published
by Alabi (1972). Using tomato baits, he sampled 145 aquatic habitats, monitoring
carbon dioxide and oxygen concentrations, organic matter content, pH, and temperature. He found *A. fermentans* in 41 locations. The percentage frequency of occurrence by habitat type was as follows:

Stagnant ponds	60%
Swampy ditches	20%
Sluggish streams	17%
Marshes	13%
Moist soils	0%

Temperature and pH did not differ significantly for these sites. However, *A. fermentans* was favored by habitats with relatively low oxygen (average = 5.1 ppm), high
carbon dioxide (average = 31.8 ppm), and high organic matter. Furthermore, only 25%
of the negative sites had overhanging vegetation, whereas 90% of the positive sites
did. Lack of light, hence algal activity, is also a marked component of the overgrown bog in southern Illinois where Bandoni and Parsons (1966) found *A. fermentans*
regularly in the late summer along with other species of Rhipidiaceae. Year-around
monitoring of dissolved gases in this bog showed (cf. Emerson and Held, 1969) less
than 1 ppm O_2 and 132 ppm CO_2 during the time each year when the fungi appeared.

We are currently studying a species of *Rhipidium* from Lake Anza, near Berkeley,
California (Natvig, 1977). For most of the year, this mesotrophic lake is thermally
stratified and the hypolimnion contains greatly reduced levels of dissolved O_2.
Apple baits suspended at depths where O_2 content is low are colonized almost exclusively by *Rhipidium*; but apples placed in well-oxygenated surface waters are colonized by filamentous aquatic phycomycetes and not by *Rhipidium* (Fig. 4). Laboratory
studies show that this *Rhipidium* is highly fermentative, and cultures bubbled with
air and those placed in an oxygen-free atmosphere show little or no difference in
growth. This alone cannot account for the observed distribution of *Rhipidium* in
the lake. It is possible that *Rhipidium* cannot compete under conditions favorable
to the growth of filamentous species. Another possibility is that *Rhipidium*, like
Aqualinderella and *Blastocladia*, is adapted to waters with high CO_2 concentrations.
Preliminary laboratory experiments indicate that the growth of our isolate is stimulated by high levels (ca. 5%) of atmospheric CO_2.

Little enough is known about the distribution of fungi adapted to environments
low in dissolved O_2, but almost nothing can be said about the ecological niches
these fungi occupy. This becomes clear when ecological distinctions are sought a-
mong any of the several facultative anaerobes common on submerged fruit baits. Just

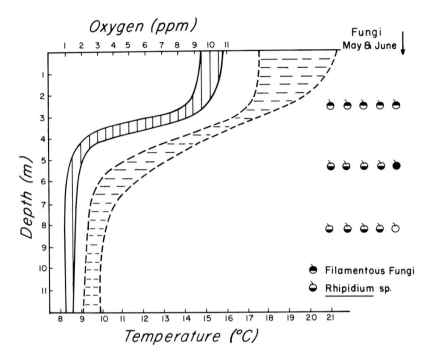

Fig. 4 The vertical distribution of *Rhipidium* on apples in Lake Anza, California. The apples were suspended three per nylon line, and the lines remained anchored at least 20 m apart for 2 weeks. The figure indicates which apples were colonized by filamentous oomycetes and which by *Rhipidium* sp. Note the restriction of *Rhipidium* to hypolimnetic waters. (From Natvig, 1977.)

as there are remarkable morphological and physiological similarities between such taxonomically unrelated fungi as *Blastocladia* and members of the Rhipidiaceae, there are also ecological parallels. But how far these parallels extend is not known. It is not at present possible to distinguish between the preferred habitats and sub- strates of different species. Geographical distributions, so far as they are known, show considerable overlap, and it is also common for two genera to occur on the same bait, even in the same pustule (von Minden, 1916; also, our personal observations). In the latter case, the way in which the natural habitat is apportioned is surely obscured by the sampling methods employed.

Another complicating and interesting fact is that species presumed to be adapt- ed to environments low in O_2 are also found in waters not having low O_2 levels. Rooney and McKnight (1972) listed *Rhipidium americanum* Thaxter as the most common species found during a two-summer study of aquatic phycomycetes in Lily Lake, a sub- alpine lake in Utah. Dissolved O_2 levels in their study averaged 7.2 ppm, far from anaerobic conditions. We have frequently isolated *Mindeniella spinospora* from waters with dissolved O_2 levels greater than 7 ppm; *M. spinospora* is a facultative anaerobe that is highly fermentative even when O_2 is plentiful (Gleason, 1967; D. O. Natvig, unpublished data).

Although there is nothing about the physiology of *Rhipidium* and *Mindeniella* that would preclude existence in well-oxygenated environments, the significance of microenvironmental levels of O_2 and CO_2 should not be overlooked. A common situation on field-collected material is to have an extensive rhizoidal system growing into the flesh of a fruit, with reproductive filaments penetrating through a lenticel or wound, and the whole covered with bacteria and other microorganisms (Thaxter, 1896, p. 328; Peterson, 1910, p. 526; our personal observations). The O_2 available to a water mold in this position is likely much less than the concentration in the surrounding environment.

Traditional approaches to aquatic mycology have been too broad to furnish more than superficial information about (1) where species occur, (2) what special adaptations allow them to occur where they do, and especially (3) what they do there. This situation will be greatly improved when answers are sought to specific questions about specific organisms. The way an organism is distributed through space and time can be partially determined by systematic sampling studies. Other questions will be more difficult to answer. The classical use of artificial baits, still the only satisfactory method to obtain many species, ensures that many mysteries about the roles of aquatic fungi in nature will remain for some time.

Whatever field ecologists may say about the direct inapplicability of laboratory studies to field situations, it is amply clear from the material covered in this chapter that rigorous laboratory studies of fungal physiology are going to provide an important baseline for ultimate clarification of fungal ecology. Certainly in the present instance, were it not for the detailed knowledge gained in the laboratory about zoosporic fungi from stagnant waters, we would still be in almost complete ignorance as to the niches these fungi occupy in nature. As it is, the stage is well set for a scientific characterization of these niches and correlation of this information with the physiology of the fungi concerned. The life of fungi in stagnant waters must relate, among other factors, to their fermentative abilities and their nutritional needs.

VI. SUMMARY AND CONCLUSIONS

Table 4 presents comparative molar growth yield data for several aquatic fungi grown under aerobic and strict anaerobic conditions. The range is from obligate aerobiosis to obligate anaerobiosis. *Allomyces*, an obligate aerobe that carries out lactic acid fermentation (Ingraham and Emerson, 1954), is appropriate at the head of the table. Orpin's anaerobic rumen fungi provide an appropriate end. With this table as a reference, we would like to conclude this chapter with the following points:

1. It is incorrect to say that all fungi other than yeasts are strictly aerobic.
2. Many fungi have the enzymes to produce, anaerobically, lactic acid, ethanol, or other products of true fermentation. Lactic acid production is rare outside the Phycomycetes but widespread therein.

Table 4 Phycomycetes: Molar growth yield vs. CO_2 and O_2 concentration

Source of data	Fungal species	Gas environment CO_2 (%)	O_2 (%)	Molar growth yield[a]
Presumed obligate aerobes				
Ingraham and Emerson (1954)	*Allomyces arbuscula*	0.03	21[b]	46
		0.03	0[c]	0
R. Emerson (unpubl.)	*Allomyces macrogynus*	0.03	21[d]	48W[e]
		0.0003	0[f]	0W
Facultative anaerobes				
Winans (1977)	*Phytophthora drechsleri*	4-5	0[f]	42W
Tabak and Cooke (1968a)	*Mucor hiemalis*	0.03	21[d]	46
		0	0[f]	26
Gleason (1967)	*Sapromyces elongatus*	0.03	21[g]	101
		5	0[f]	16
Gleason (1967)	*Mindeniella spinospora*	0.03	21[g]	52
		5	0	22
Presumed obligate fermenters				
Held et al. (1969)	*Blastocladia pringsheimii*	5	20	41
		5	0.2	38
		5	0[f]	0
R. Emerson (unpubl.)	*Blastocladia pringsheimii*	0.03	21[d]	14W
		0.0003	0[f]	19W
Held et al. (1969)	*Blastocladia ramosa*	5	20	36
		5	0.2	32
		5	0[f]	30
Held (1970)	*Aqualinderella fermentans*	5	20	35W
		5	0[f]	39W
Winans (1977)	*Pythiogeton* sp.	2-3	21[g]	42
		2-3	0[f]	37
Obligate anaerobes				
Orpin (1975, 1976, 1977a, b,c,d)	*Neocallimastix frontalis* *Piromonas communis* *Sphaeromonas communis*	Obligate rumen fungi Growth in 100% CO_2 No growth in air		

[a] Molar growth yield = mg dry weight/ mM glucose consumed.

[b] Slow bubbling.

[c] Anaerobic bottle.

[d] Shake culture.

[e] W = wheat germ oil.

[f] Leuco methylene blue.

[g] Standing culture.

3. The occurrence of strict facultative anaerobiosis (anoxic growth) is wide-spread and has been proven in Oomycetes, Chytridiomycetes, Zygomycetes, and higher fungi.

4. The occurrence of facultative anaerobiosis coupled with obligate fermenta-tion has been proven in *Blastocladia* (Blastocladiales) and *Aqualinderella* (Leptomitales) and almost certainly occurs in *Pythiogeton* (Peronosporales).

5. Apparent obligate anaerobiosis among the true (eucaryotic) fungi has just been discovered in a group of chytrid-like fungi inhabiting the sheep rumen.

6. Natural environments with anaerobic or near-anaerobic conditions have not been given just consideration as places satisfactory for the growth of fungi. Wide search is certain to reveal many more fungi with the ability to exist under anaerobic conditions. Laboratory proof of the capacity for anaerobic growth may require consideration of growth factors such as sterols, fatty acids, and vitamins.

ACKNOWLEDGMENTS

We are particularly grateful to Nancy Morin, whose patience and dexterity led to the first pure culture isolations of *Pythiogeton* CRB11. A special expression of apprec-iation goes here to the administration and staff of the Centro Agronomico Tropical de Investigación y Enseñanza at Turrialba, Costa Rica, for their generous hospitality to the senior author in 1972-1973. Various aspects of the research reported here were supported by grants from the committee on research of the Academic Senate, University of California, Berkeley.

REFERENCES

Alabi, R. O. (1972). Natural occurrence of *Aqualinderella fermentans* in Nigeria. *Trans. Brit. Mycol. Soc. 59*: 103-105.

Bandoni, R. J., and Parsons, J. D. (1966). Some aquatic phycomycetes from Pine Hills. *Trans. Illinois State Acad. Sci. 59*: 91-94.

Bartnicki-Garcia, S., and Nickerson, W. J. (1962). Induction of yeast-like devel-opment in *Mucor* by carbon dioxide. *J. Bacteriol. 84*: 829-840.

Cantino, E. C. (1948a). The nutrition and metabolism of the aquatic phycomycete *Blastocladia pringsheimii*. Ph.D. thesis, University of California, Berkeley.

Cantino, E. C. (1948b). The vitamin nutrition of an isolate of *Blastocladia pring-sheimii*. *Amer. J. Bot. 35*: 238-242.

Cantino, E. C. (1949). The physiology of the aquatic phycomycete, *Blastocladia pringsheimii*, with emphasis on its nutrition and metabolism. *Amer. J. Bot. 36*: 95-112.

Cantino, E. C. (1951a). Metabolism and morphogenesis in a new *Blastocladiella*. *Antonie van Leeuwenhoek J. Microbiol. Serol. 17*: 325-362.

Cantino, E. C. (1951b). Evidence for an accessory factor involved in fructose util-ization by the aquatic fungus, *Pythiogeton*. *Amer. J. Bot. 38*: 579-585.

Cochrane, V. W. (1958). *Physiology of Fungi*. Wiley, New York.

Crasemann, J. M. (1954). The nutrition of *Chytridium* and *Macrochytrium*. *Amer. J. Bot. 41*: 302-310.

Crasemann, J. M. (1957). Comparative nutrition of two species of *Blastocladia*. *Amer. J. Bot. 44*: 218-224.

Curtis, P. J. (1969). Anaerobic growth of fungi. *Trans. Brit. Mycol. Soc. 53*: 299-302.

Emerson, R. (1958). Mycological organization. *Mycologia 50*: 589-621.

Emerson, R., and Cantino, E. C. (1948). The isolation, growth, and metabolism of *Blastocladia* in pure culture. *Amer. J. Bot. 35*: 157-171.

Emerson, R., and Held, A. A. (1969). *Aqualinderella fermentans* gen. et sp. n., a phycomycete adapted to stagnant waters. II. Isolation, cultural characteristics, and gas relations. *Amer. J. Bot. 56*: 1103-1120.

Emerson, R., and Weston, W. H. (1967). *Aqualinderella fermentans* gen. et sp. nov., a phycomycete adapted to stagnant waters. I. Morphology and occurrence in nature. *Amer. J. Bot. 54*: 702-719.

Foster, J. W. (1949). *Chemical Activities of Fungi*. Academic Press, New York.

Gleason, F. (1967). Nutrition, fermentation, and terminal respiration in the Leptomitales. Ph.D. thesis, University of California, Berkeley.

Gleason, F. (1976). The physiology of the lower freshwater fungi. In *Recent Advances in Aquatic Mycology*, E. B. G. Jones (Ed.). Elek Science, London, pp. 543-572.

Gleason, F. H., Nolan, R. A., Wilson, A. C., and Emerson, R. (1966). D(-)-lactate dehydrogenase in lower fungi. *Science 152*: 1272-1273.

Goldfine, H., and Bloch, K. (1963). Oxygen and biosynthetic reactions. In *Control Mechanisms in Respiration and Fermentation*, 8th Symp. Soc. Gen. Physiol., B. Wright (Ed.). Ronald Press, New York, pp. 81-103.

Golueke, C. G. (1957). Comparative studies of the physiology of *Sapromyces* and related genera. *J. Bacteriol. 74*: 337-343.

Gunner, H. B., and Alexander, M. (1964). Anaerobic growth of *Fusarium oxysporum*. *J. Bacteriol. 87*: 1309-1316.

Hawker, L. E. (1950). *The Physiology of Fungi*. Univ. of London Press, London.

Hawker, L. E. (1966). Environmental influences on reproduction. In *The Fungi*, Vol. II: *The Fungal Organism*, G. C. Ainsworth and A. S. Sussman (Eds.). Academic Press, New York, pp. 435-469.

Held, A. A. (1967). A physiological study of *Aqualinderella fermentans*, a facultatively anaerobic, obligately fermentative water mold. Ph.D. thesis, University of California, Berkeley.

Held, A. A. (1970). Nutrition and fermentative energy metabolism of the water mold *Aqualinderella fermentans*. *Mycologia 62*: 339-358.

Held, A. A., Emerson, R., Fuller, M. S., and Gleason, F. H. (1969). *Blastocladia* and *Aqualinderella*: fermentative water molds with high carbon dioxide optima. *Science 165*: 706-709.

Hungate, R. E. (1966). *The Rumen and Its Microbes*. Academic Press, New York.

Ingraham, J. L., and Emerson, R. (1954). Studies of the nutrition and metabolism of the aquatic phycomycete, *Allomyces*. *Amer. J. Bot. 41*: 146-152.

Lund, A. (1934). Studies on Danish freshwater phycomycetes and notes on their occurrence particularly relative to the hydrogen ion concentration of the water. *D. Kgl. Danske Vidensk. Selsk. Skrifter, Naturv. Math., Ser. IV 6*(1): 1-97.

Lynch, V. H., and Calvin, M. (1952). Carbon dioxide fixation by microorganisms. *J. Bacteriol. 63*: 525-531.

Natvig, D. O. (1977). Observations on the distribution of aquatic fungi in Lake Anza. Student research paper for Forestry 178, Freshwater Ecology, University of California, Berkeley.

Orpin, C. G. (1975). Studies on the rumen flagellate *Neocallimastix frontalis*. *J. Gen. Microbiol. 91*: 249-262.

Orpin, C. G. (1976). Studies on the rumen flagellate *Sphaeromonas communis*. *J. Gen. Microbiol. 94*: 270-280.

Orpin, C. G. (1977a). Invasion of plant tissue in the rumen by the flagellate *Neocallimastix frontalis*. *J. Gen. Microbiol. 98*: 423-430.

Orpin, C. G. (1977b). The rumen flagellate *Piromonas communis*: its life-history and invasion of plant material in the rumen. *J. Gen. Microbiol. 99*: 107-117.

Orpin, C. G. (1977c). The occurrence of chitin in the cell walls of the rumen organisms *Neocallimastix frontalis*, *Piromonas communis* and *Sphaeromonas communis*. *J. Gen. .Iicrobiol. 99*: 215-218.

Orpin, C. G. (1977d). On the induction of zoosporogenesis in the rumen phycomycetes *Neocallimastix frontalis*, *Piromonas communis* and *Sphaeromonas communis*. *J. Gen. Microbiol. 101*: 181-189.

Orpin, C. G., and Bountiff, L. (1978). Zoospore chemotaxis in the rumen phycomycete *Neocallimastix frontalis*. *J. Gen. Microbiol. 104*: 113-122.

Pasteur, L. (1876). *Études sur la bière*. Gauthier-villars, Paris.

Peterson, H. E. (1910). An account of Danish freshwater Phycomycetes, with biological and systematical remarks. *Ann. Mycol. 8*: 494-560.

Rooney, H. M., and McKnight, K. H. (1972). Aquatic phycomycetes of Lily Lake, Utah. *Great Basin Naturalist 32*: 181-189.

Sparrow, F. K. (1960). *Aquatic Phycomycetes*, 2nd ed. Univ. of Michigan Press, Ann Arbor.

Tabak, H. H., and Cooke, W. B. (1968a). Growth and metabolism of fungi in an atmosphere of nitrogen. *Mycologia 60*: 115-140.

Tabak, H. H., and Cooke, W. B. (1968b). The effects of gaseous environments on the growth and metabolism of fungi. *Bot. Rev. 34*: 126-252.

Thaxter, R. (1896). New or peculiar aquatic fungi. 4. *Rhipidium*, *Sapromyces*, and *Araiospora*, nov. gen. *Bot. Gaz. 21*: 317-331.

von Minden, M. (1916). Beiträge zur Biologie und Systematik einheimischer submerser Phycomyceten. In *Mykologische Untersuchungen und Berichte*, R. Falck (Ed.), Vol. 1. Fischer, Jena, Germany, pp. 146-225.

Watson, K. (1976). The biochemistry and biogenesis of mitochondria. In *The Filamentous Fungi*, Vol. 2: *Biosynthesis and Metabolism*, J. E. Smith and D. R. Berry (Eds.). Wiley, New York, pp. 92-120.

Winans, S. C. (1977). Fermentative metabolism and anaerobic growth of the aquatic phycomycete *Pythiogeton*. Senior honors thesis, University of California, Berkeley.

Wood, H. G., and Utter, M. F. (1965). The role of CO_2 fixation in metabolism. In *Essays in Biochemistry*, P. N. Campbell and G. D Greville (Eds.), Vol. 1. Academic Press, New York, pp. 1-27.

Wyss, O., and Eklund, C. (1971). *Microorganisms and Man*. Wiley, New York.

Chapter 8

AN ANALYSIS OF THE FUNDAMENTAL AND REALIZED
NICHE OF CACTOPHILIC YEASTS

William T. Starmer
Syracuse University
Syracuse, New York

I. INTRODUCTION

In 1881 Emil C. Hansen made the basic distinction between yeasts used in the fermentation industry and "wild yeasts." He found the natural habitats of the wild yeasts
to be sweet, juicy fruits (cited in Lund, 1956). Since that time the majority of
work on the distribution of natural yeasts has been devoted to determining the species present in particular substrates. Consequently, during this century numerous
new taxa have been revealed from the study of yeast habitats.

In 1952 Lodder and Kreger-van Rij recognized 165 species of yeasts; by 1970
Lodder included 349 valid species in *The Yeasts: A Taxonomic Study*. A few years
later Barnett and Pankhurst (1974) listed 434 species of yeasts in their key to the
yeasts. Most recently Von Arx et al. (1977) listed almost 550 species of yeasts;
and the number is still growing, largely due to investigations into previously unexplored habitats. One such habitat, necrotic succulent tissue of cacti, is the subject of this chapter.

Laboratory investigations on the yeasts recovered from a survey of cactus habitats have uncovered a number of new species and revealed some interesting phenomena
related to the ecology of yeasts. In fact, the taxonomic study of the cactus-inhabiting yeasts provides a foundation for the study of the ecology of these yeasts. A
brief description of the methods of yeast taxonomy is therefore warranted. [See
Kreger-van Rij (1969) and Lodder (1970) for further discussion.]

The genera of sporogenous yeasts are largely determined by the appearance of the
vegetative cells and ascospores. However, the species of yeasts are defined and described mainly on the basis of patterns of aerobic and fermentative utilization of
simple compounds containing carbon and/or nitrogen (Van der Walt, 1970). Species
descriptions also contain information on the growth requirements and physiological
tolerance of the species being examined (e.g., salt tolerance, temperature ranges of
growth, and vitamin requirements). In addition, yeast taxonomists more recently have
employed serology (e.g., Kuroshima et al., 1976), DNA composition (e.g., Meyer and
Phaff, 1972), DNA-DNA homologies (Price et al., 1978), proton magnetic resonance
spectra (PMR) of the cell wall mannan (Gorin and Spencer, 1970), coenzyme Q (e.g.,

Yamada et al., 1976), and numerical methods (e.g., Campbell, 1973; Kocková-Kratochvílová et al., 1976) to the systematics of yeasts. Nevertheless, the physiological
properties of the yeasts still remain as the fundamental characteristics utilized
to key and identify the species (Lodder, 1970; Barnett and Pankhurst, 1974). In general, although there are some exceptions, the patterns of metabolism provide fairly
reliable means whereby yeast species can be described and differentiated.

The description of the basic metabolic capabilities and physiological tolerance
of a taxon is essentially a partial definition of its "fundamental niche" (see Mc-
Naughton, Chap. 5). On the other hand, knowing the distribution of yeasts in their
natural habitats is knowing the "realized niche" of the species. The realized niche
is determined by a number of ecological factors. These factors include: (1) habitat
chemistry and microclimate; (2) patterns of habitat distribution; (3) linkages among
habitats due to dispersal agents; and (4) relationships among the yeast species which
share habitats (e.g., facilitation and competition).

By comparing the realized niche with the fundamental niche we should be able to
develop an understanding of the ecological forces responsible for the distribution
and diversity of the yeasts observed in nature. The comparison of realized niche
and fundamental niche is thus a study of the natural distribution of isolates and
their taxonomic characteristics. The desirable comparison utilizes the methods of
several biological disciplines in a manner that yields working testable hypotheses.

In this chapter I will examine a group of yeasts found commonly in the decaying
stems of cacti. The study will be confined to the giant columnar cacti (cereoid) inhabiting the Sonoran Desert of North America[*] (Shreve and Wiggins, 1964). My primary
goal is to describe and analyze the ecological components that are important in maintaining the diversity and distribution of cactophilic yeasts. The ecological components considered include: (1) the hierarchical partition (or phylogenetics) of
the cactus habitats; (2) the effect due to the geographic distribution of the cacti;
(3) the effect due to insect dispersal among the habitats; (4) the patterns of resource utilization realized by the yeast species among the habitats; and (5) the
physiological capabilities of the yeasts as determined in the laboratory.

A. The Biological System

In general the process of interest to us here begins when the stem of a cactus undergoes some injury (say, freezing or windfall), rendering the succulent tissue vulnerable to infection and subsequent breakdown by pectinolytic bacteria (Lightle et al.,
1942). During succession, saprophytic yeasts become established in this soft rot,
producing vitamins, sterols, and concentrated protein necessary in the nutrition of
cactophilic *Drosophila*. The adult drosophilids are presumed to disperse the yeast
from soft rot to soft rot according to the particular host preferences of the flies,

---

[*] The Sonoran Desert is located in southern Arizona, southeastern California,
and northwestern Mexico.

Table 1 The cactus species and their geographic distribution within the Sonoran Desert

Cactus species	Common name	Baja California Peninsula	Mainland Sonoran Desert
Stenocereinae (Subtribe 1)			
Stenocereus gummosus (Engelm.) Gibs. and Horak	Agria	Throughout	Limited presence[a]
Stenocereus thurberi (Engelm.) Buxb.	Organ pipe	Central and southern	Southern Arizona
Stenocereus alamosensis (Coult.) Gibs. and Horak	Cina	Not present	Central and southern Sonora, Mexico
Pachycereinae (Subtribe 2)			
Pachycereus pringlei (S. Wats.) Britt. and Rose	Cardón	Throughout	Limited to Central Gulf Coast
Lophocereus schottii (Engelm.) Britt. and Rose	Senita	Throughout	Southern Arizona
Carnegiea gigantea (Engelm.) Britt. and Rose	Saguaro	Not present	Southwestern Arizona and Sonora, Mexico

[a]Agria cactus is found on the mainland along a 30-km strip of the Gulf of California from Desemboque del Rio San de Ignacio to Punta Chueca (Heed, 1978).

Source: Hastings et al. (1972).

whereas the larvae spread the yeasts within the soft rot for their own benefit. The rotting tissues may be localized and short-lived (weeks) in some of the smaller cactus species but long-lived (months) and involving large areas of tissue (several kilograms) in some of the larger cactus species. Although the main concern in this chapter is the microorganism, a brief account of the associations among the host plants and indigenous *Drosophila* will facilitate later discussion and analysis.

1. The cactus

In the Sonoran Desert the host plants for both the *Drosophila* and yeasts are the giant columnar cacti of the subtribes Pachycereinae and Stenocereinae of the tribe Pachycereeae (sensu Gibson and Horak, 1978). Six cactus species utilized in this study are listed in Table 1 along with their subtribe designation and geographic distribution within the Sonoran Desert (Hastings et al., 1972). The cacti will be referred to by their common names, as given in Table 1, throughout this chapter. The division of the two phylads is based partly on differences in the secondary plant products of the succulent tissue. In general the subtribe Stenocereinae is characterized by its content of triterpenes, whereas the subtribe Pachycereinae features alkaloids among its members (Gibson and Horak, 1978). There are specific chemical differences between species of the same subtribe, but the differences within a subtribe are less than those between species of the two phylads. With this in mind, we can view the two subtribes as two major chemical habitats available for colonization by the yeasts. In addition, the individual species of the subtribes can be taken as a secondary partition of the major chemical habitats. The phyletic changes experienced by the cacti will necessarily direct corresponding adaptive changes in the yeast species which reside in their decaying tissue. I have very little information on this aspect of the autecology of cactus yeast. However, when possible I have analyzed the genetic basis of metabolic differences for strains of the same species. The metabolic differences are correlated with switching of hosts plants and display control by single genetic loci.

2. *The Drosophila*

Four species of *Drosophila* are endemic in the Sonoran Desert (Heed, 1978). These phytophagous insects display varying degrees of host plant specificity, which is presumed to be determined by the toxic and nutritional nature of the host plants (Heed, 1978) as well as by competitive factors among the *Drosophila* in cacti where toxicity and nutrition are not factors (Fellows and Heed, 1972).

a. *Drosophila pachea* Patterson and Wheeler

Senita cactus is the only host plant capable of supporting *D. pachea*, and the cactus does not host any other desert *Drosophila*. The exclusion of other flies from the habitat is due to the presence of an alkaloid (pilocereine) in the tissues of senita cactus. The alkaloid is toxic to all cactophilic *Drosophila* except *D. pachea* (Kircher et al., 1967). Furthermore, unusual sterols [e.g., schottenol (Heed and Kircher, 1965)] occurring in senita cactus are required by *D. pachea* for development and fertility. The resulting monophagy of *D. pachea* may in part restrict the distribution of yeasts found in the host cactus, senita. In fact the yeast community diversity for the senita cactus habitat is the lowest among the cacti.

b. *Drosophila nigrospiracula* Patterson and Wheeler and *Drosophila mettleri* Heed

In Baja California *D. nigrospiracula* and *D. mettleri* utilize the decaying tissue of cardón and the soil wetted by the dripping, decaying tissue, respectively. On the mainland these flies inhabit both cardón and saguaro cacti where present (Heed, 1978). Most of the yeast data we have collected from the necrotic cacti come from the tissues, and few samples have been obtained from wetted soils. For this reason I will exclude *D. mettleri* from the forthcoming analysis. However, the novel niche separation of *D. nigrospiracula* and *D. mettleri* (Heed, 1977) is certainly worthy of further consideration at the level of the yeasts which may or may not foster the niche division.

c. *Drosophila mojavensis* Patterson and Crow

Drosophila mojavensis utilizes agria cactus where the cactus is present throughout Baja California and a small stand of agria on the mainland near Desemboque del

Rio de San Ignacio, Sonora, Mexico. The major host plant of *D. mojavensis* in Sonora, Mexico, and southern Arizona is organ-pipe cactus. Organ-pipe is present on the Baja California peninsula, but it is used as host infrequently there--possibly as a result of the rarity of decaying plants in this area. *Drosophila mojavensis* sometimes shares the saguaro and cardón cactus habitats with *D. nigrospiracula* (Heed, 1978; W. T. Starmer, unpublished data). However, *D. mojavensis* does not regularly use cardón or saguaro, presumably due to the superior competitive ability of *D. nigrospiracula* in these habitats (Fellows and Heed, 1972). *D. mojavensis* has also been bred from cina cactus, although this cactus is thought to be the primary host of *Drosophila arizonensis* Patterson and Wheeler (the sibling species of *D. mojavensis*).

II. COLLECTION LOCALITIES AND METHODS

A. The Collection

Samples were collected in the vicinity of Tucson, Arizona, in all seasons of the year between 1971 and 1974. Additional collections were made in the state of Sonora, Mexico, during various seasons from 1971 to 1974. Details on these collections can be found in Starmer et al. (1976). A total of 132 yeast isolates were recovered from 7 species of cereoid cacti. These isolates represent 13 species of yeast, 7 of which were isolated more than four times and 6 species of which were recovered once. During the late winter of 1976, samples were taken from numerous locations on the Baja California peninsula and in the vicinity of Guaymas, Sonora, Mexico. This collection resulted in an additional 302 yeast isolates from 7 species of cereoid cacti. A subsequent collection from cina cactus near Navojoa, Sonora, Mexico, was obtained from W. B. Heed and M. Wasserman in the fall of 1976. Overall more than 500 yeast isolates (representing 26 species of yeast) were obtained from 9 species of host plants. Only those host plants which were sampled frequently (>14 samples) will be studied here. Furthermore, all unique yeast isolates are excluded from the analysis, as well as isolates of *Cryptococcus albidus* (Saito) Skinner which were recovered in low frequency. The resulting collection is summarized in Table 2. This table shows 12 major yeast species recovered from 6 host cacti.

B. Methods of Collection

Decaying stems were located in the field and a sample was either directly streaked on AYM (Difco yeast-extract, malt-extract agar, acidified with HCl to a pH of 3.8) plates or diluted in sterile water before plating on AYM. The plates were incubated at room temperature (∿25-30°C) until colonies appeared. Each morphological colony type was brought into pure culture by two successive streakings. These isolates were studied by the methods described by Van der Walt (1970), as well as by additional methods to be discussed in the next subsection. When different morphological types from the same sample were identified as the same species they were treated as a single isolate of that species. Also, when an individual cactus was sampled more than once, the results of the multiple sample were pooled and treated as a single sample.

Table 2 Number of times a given yeast species was recovered from sampling of plants in the Sonoran Desert

Yeast species	Agria	Organ-pipe	Cina	Cardón	Senita	Saguaro	Total
Number of plants sampled	100	30	14	27	62	16	249
Pichia cactophila	72	21	14	16	7	8	138
Torulopsis sonorensis	40	16	4	5	3	1	69
Pichia heedii	1	0	0	1	39	9	50
Candida ingens	7	11	8	1	16	4	47
Cryptococcus cereanus	2	3	4	2	9	4	24
Pichia amethionina var. *amethionina*	19	1	1	0	1	0	22
P. amethionina var. *pachycereana*	1	0	0	2	2	4	9
Pichia species "K"	11	3	0	0	0	2	16
Candida tenuis	3	7	1	1	0	0	12
Torulopsis species "A"	10	1	0	0	0	0	11
Pichia opuntiae var. *thermotolerans*	0	0	0	5	2	2	9
Pichia species "A"	7	2	0	0	0	0	9
Total	173	65	32	33	79	34	416

After the collection was completed and the isolates sorted into species, we realized that two species (*Torulopsis sonorensis* and *Pichia amethionina*) which occur in the same habitat exhibited almost identical colony morphologies. This observation, in retrospect, indicates a possible underestimation of the frequency of the two yeast species in the habitats in which they occurred. Additionally, the collection method may have obscured yeast species which occurred in low frequency in the samples. However, the collection is still useful for a first approximation of yeast distribution. Only by selectively isolating and thus quantitatively enumerating the cactus yeasts can we better approximate the natural population.

C. Laboratory Taxonomy

Once isolates were in pure culture, they were studied with respect to 56 physiological characteristics listed in Tables 3 and 4. In addition the isolates were observed for morphological characteristics of their colonies, thallus, and ascospores when present. Those species which formed asci were dissected to obtain single spore cultures. These cultures were observed for self-sporulation (homothallic sexuality) or lack of self-sporulation (heterothallic sexuality). Single spore cultures that did not sporulate were mixed together in order to ascertain whether they were hetero-

Table 3 Compounds utilized in aerobic growth tests for classifying and identifying the collection

Hexoses	α-Glucosides	β-Glucosides
D-Glucose (12)	Sucrose (2)	Cellobiose (5)
D-Galactose (3)	Maltose (2)	Salicin (5)
L-Sorbose (3)	Trehalose (2)	
	Melezitose (2)	
	α-methyl-D-glucoside (1)	

α-Galactoside	β-Galactoside	Pentoses
Melibiose[b]	Lactose (1)	D-ribose (3)
		D-Xylose (5)
		L-Arabinose (3)
		D-Arabinose[b]

Polyols	Acids	Others
Glycerol (9)	Succinate (1)	Ethanol (12)
Erythritol (2)	Citrate (3)	L-Rhamnose (2)
Ribitol (4)	Lactate (10)	Methanol (1)
D-Glucitol (3)	Gluconate	D-Glucosamine (1)
D-Mannitol (5)	(glucono-δ-lactone) (2)	Inulin[b]
Inositol (1)		Starch[b]
Galactitol[b]		Ca-2-Ketogluconate[b]
		K-5-Ketogluconate[b]
		Raffinose[b]

[a]The number in parentheses indicates the number of species positive or variable for the resource state.

[b]Not utilized by any of the 12 major yeasts in the collection.

Source: Modified from Barnett (1968).

thallic and, if so, their mating types. Furthermore, original isolates of the same species that were asporogenous were mixed together to determine whether they existed as haploid heterothallic species. Once groups of isolates were recognized as potentially good species, each grouping was studied for homogeneity of their nuclear DNA composition. Employing the results of the physiological tests, morphological observations, sexuality, and DNA compositions, species were confirmed with existing taxa or described as new species. A number of the new species are still undescribed at this writing. Furthermore, species which were heterothallic and demonstrated variability for physiological characteristics were studied in order to determine the genetics of such differences.

1. The yeasts

a. *Pichia cactophila* Starmer et al.

By far the most common yeast encountered in the cactus habitat was *Pichia cactophila*. In previous publications on cactus-inhabiting yeasts (Heed et al., 1976; Starmer et al., 1976) we identified this species as *Pichia membranaefaciens* Hansen. This was largely due to the convergent physiologies of the two species and a lack of

Table 4 Additional physiological tests utilized in the classification and identification of the collection[a]

Fermentation:	Nitrogen assimilation:
D-Glucose (2)	Potassium nitrite[b]
	Potassium nitrate[b]
	Ethylamine (11)
	Ammonium sulfate[c]
Maximum temperature of growth:	Formation and synthesis of:
37°C[c]	Acid[c]
39°C (11)	Starch[c]
42°C (8)	
45°C[c]	
Growth on osmotic medium:	Growth in medium free of:
50% glucose[b]	Amino acids (10)
10% NaCl (5)	Vitamins[b]
Growth in presence of:	Liquefaction and hydrolysis of:
Cycloheximide (4)	Gelatin (1)
	Casein (3)
	Lipids (Tween-80) (2)
	Urea[c]

[a]The number in parentheses indicates the number of species positive or variable
for that resource state.

[b]Not observed for any of the 12 major yeasts in the collection.

[c]Not used in subsequent analysis due to insufficient data.

knowledge on the nuclear DNA content of the cactus isolates. The taxonomic information that allowed us to separate the two species is found in the description of *P.
cactophila* (Starmer et al., 1978a).

Pichia cactophila has been isolated from all cereoid cacti studied as well as
from *Opuntia* cacti from widespread geographic regions including the Sonoran Desert,
southern Mexico, the Hawaiian Islands, Australia, and Europe. It is undoubtedly
ubiquitous and may be found in necrotic cacti worldwide. It has not been recovered
from habitats other than cactus soft rots.

The physiological capabilities of *P. cactophila* are limited to aerobic metabolism of few compounds. Notable is the unique and strong utilization of the amino
sugar D-glucosamine as a carbon source. No other cactus-inhabiting yeast found in
this survey can utilize this compound.

   b. *Torulopsis sonorensis* Miller et al.

Another common cactus yeast, found both in cacti of the Sonoran Desert and
Opuntia elsewhere, is *Torulopsis sonorensis*. Among the cactus yeasts, this species
has the unique ability to assimilate methanol. Additionally, it is almost unique in
its fermentative ability. It shares this trait (fermentation) with *Candida tenuis*.

All other cactus yeasts are limited to aerobic respiration. During isolation *T. sonorensis* was never recovered from the same sample as *Pichia amethionina* var. *amethionina*. In retrospect, we found that the two species displayed the same colony morphology on the isolation plate. Since both species occur commonly in the same cactus habitats, we do not know if they co-occur. Thus, frequency of isolation for the two species is taken as a minimal estimate. *Torulopsis sonorensis* is cactus specific (Miller et al., 1976) and has not been recovered from any other source besides the *Drosophila* species which utilize and carry them.

c. *Pichia heedii* Phaff et al.

During the early phases of the survey, we recognized that some isolates recovered from senita cactus and then identified as *P. membranaefaciens* were unique among the cactus isolates in their ability to assimilate the pentose sugar D-xylose (Heed et al., 1976; Starmer et al., 1976). At that time we speculated that these isolates were representative of a new species. Subsequent collections throughout the Baja California peninsula confirmed this observation. On studying a large number of isolates of this form we found it to be distinct from *P. membranaefaciens* (even the xylose-positive strains) in its mol% G + C of the nuclear DNA, and so we described it as *Pichia heedii* (Phaff et al., 1978).

Pichia heedii is variable for its ability to utilize the organic acid citrate. The variability is accounted for by its distribution among host plants (senita cactus, citrate positive; other cacti, citrate negative). Furthermore, citrate utilization for this species was determined to be under the control of a single genetic locus. *Pichia heedii* is one of five cactus-inhabiting yeast species that are restricted to the Sonoran Desert. It has not been found in *Opuntia* cacti sampled elsewhere nor in cereoid cacti sampled in the Oaxaca region of southern Mexico (W. T. Starmer, unpublished data).

d. *Candida ingens* Van der Walt et van Kerken

The large cell size of *Candida ingens* is a distinctive feature of the species. It is not restricted to cacti and has been found in association with industrial effluent (Van Uden and Buckley, 1970). Recently *C. ingens* has been induced to sporulate by Rodrigues de Miranda and Török (1976), and the perfect form has been described as *Pichia humboldtii*. We have yet to find sporogenous strains from cacti; however, the techniques of Rodrigues de Miranda and Török have not been attempted on cactus strains. *Candida ingens* is almost unique among the cactus yeasts for its lipolytic activity. It shares this characteristic with some strains of *Candida tenuis*. Since some cacti do contain unusual amounts of short-chained fatty acids [e.g., organ-pipe (H. W. Kircher, personal communication, 1976)], the adaptation of *C. ingens* to this component of the habitat is apparent, especially when one considers the fungistatic property of short-chain fatty acids on other yeasts (Wyss et al., 1945). *Candida ingens* is also one of four yeasts resistant to the antibiotic cycloheximide. This resistance probably indicates a general adaptation of the species to toxins in

its environment. It is notable that a number of yeasts found associated with drosophilids and their habitats are also resistant to cycloheximide [e.g., *Saccharomyces montanus* (Phaff et al., 1956)].

  e. *Cryptococcus cereanus* Phaff et al.

  Cryptococcus cereanus is specific for decaying cactus tissue. Representatives
of the species have been obtained from all species of cacti sampled, including rotting cereoid cacti of the Oaxaca region of southern Mexico and the necrotic cladodes
of *Platyopuntia* in Australia. Among the common cactus yeasts, *C. cereanus* is the
only species which can assimilate the polyol, inositol. Unlike most other species
of *Cryptococcus* (Phaff and Fell, 1970), it lacks urease activity, does not produce
amyloid compounds, and has a relatively low mol% G + C of its nuclear genome (Phaff
et al., 1974).

  f. *Pichia amethionina* Starmer et al.

  The two varieties of *Pichia amethionina* are heterothallic, interfertile, and
have a number of unique characteristics. Both varieties have an absolute requirement for the sulfur-containing amino acid methionine or cysteine. This requirement
has been shown to be due to the lack of the ability to convert hydrogen sulfite into
hydrogen sulfide, a complex step in the metabolism of sulfate (Masselot and Surdin-
Kerjan, 1977; Breton and Surdin-Kerjan, 1977). Table 2 shows that *P. amethionina*
var. *amethionina* has been recovered mainly from the necroses of agria cactus, which
is in the subtribe Stenocereinae, whereas the variety *pachycereana* has been isolated
from saguaro, cardón, and senita cactus (subtribe: Pachycereinae). The two varieties of *P. amethionina* also differ in their ability to utilize D-mannitol as a carbon
source. *Pichia amethionina* var. *amethionina* cannot assimilate D-mannitol, whereas
the variety *pachycereana* is positive in this respect. The difference for mannitol
utilization is controlled by a single genetic locus (Starmer et al., 1978b).

  Pichia amethionina is indigenous to the Sonoran Desert and adjacent regions.
The frequency of isolation from the cactus habitats (Table 2) is probably an underestimate due to the confounding of colony morphologies of the species with *T. sonorensis* during isolation. In earlier publications (Starmer et al., 1976; Heed et al.,
1976) we identified *P. amethionina* var. *amethionina* as *P. membranaefaciens*, and *P.
amethionina* var. *pachycereana* as closely resembling *Pichia delftensis* Beech. Only
after extensive field collections, laboratory studies, and some serendipity were we
able to recognize *P. amethionina* as distinct from *P. membranaefaciens* and *P. delftensis*. Throughout the remainder of this chapter I will treat the varieties of *P. amethionina* as different species in order to facilitate the analytical procedures. In
fact postmating breakdown observed for some hybrids indicates that the varieties are
possibly incipient species.

  g. *Pichia* species "K"

  In earlier publications we identified *Pichia* species "K" as imperfect forms
(i.e., *Candida valida*) of *P. membranaefaciens* (Starmer et al., 1976; Heed et al.,

1976). Since a number of *C. valida*-like organisms did deviate from the salient characteristics of *P. cactophila*, *P. heedii*, and *P. amethionina* var. *amethionina*, we
examined the mol% G + C of their nuclear DNAs. These determinations established a
group of isolates which exhibited extremely low DNA base compositions [mol% G + C =
28% (H. J. Phaff, unpublished data)]. This low value clearly separated the isolates
from the aforementioned *Pichia* species. The isolates are similar to *Pichia kluyveri*
Bedford; however, they do not ferment D-glucose, whereas *P. kluyveri* does. We are
still studying this group and presently will refer to them as *Pichia* species "K."
The metabolic capabilities of *Pichia* species "K" are similar to those of *P. membranaefaciens* and are limited to aerobic utilization of very few compounds (D-glucose,
ethanol, glycerol, lactate, and succinate). In addition preliminary tests on media
containing food preservatives indicate that *Pichia* species "K" may be unique in its
resistance to some fungistatic compounds (M. Miranda, personal communication, 1978).

h. *Candida tenuis* Diddens et Lodder

Candida tenuis has broad metabolic capabilities as compared to most of the other
cactus yeasts. Of the 38 resource states tested and utilized in this study, *C.
tenuis* is positive for 30. It shows unique ability among cactus yeasts for assimilation of lactose and α-methyl-D-glucoside and liquefaction of gelatin, and shares
13 resource states with one or two other species (Table 5). However, *C. tenuis* is
not widespread among the cacti nor frequently isolated (Table 2). Most of the previous isolates of *C. tenuis* from outside the Sonoran Desert have come from bark
beetles infesting coniferous trees (Van Uden and Buckley, 1970).

i. *Torulopsis* species "A"

We encountered *Torulopsis* species "A" only in the late winter collections of
1976. It is recovered in low frequency from agria cactus, and we have isolated it
once from organ-pipe cactus. It is an unusual species among the cactus yeasts in
that it forms a pronounced capsule around its cells and has a highly mucoid and slimy
colony morphology. It has no unique resource states but shares most of the α-glucosides (listed in Table 3) with *C. tenuis*. *Torulopsis* species "A" is cactus specific; it has also been found in rotting *Opuntia* cladodes in Australia.

j. *Pichia opuntiae* Starmer et al.

Within the subtribe Pachycereinae we isolated *Pichia opuntiae* var. *thermotolerans* nine times. It is a heterothallic species which occurs primarily in the haploid
state. It is unusual in that crosses of opposite mating types found within the Sonoran Desert rarely form zygotes and therefore asci are infrequent. However, isolates
of *P. opuntiae* var. *opuntiae* have also been obtained from *Opuntia* cacti in Australia.
The Australian isolates mate well among eath other (resulting in heavy sporulation)
and only moderately with the Sonoran Desert strains, with sparse ascospore formation.
Pichia opuntiae may be limited in the cactus habitat due to its inability to grow at
or above 39°C. All other cactus-specific yeasts can grow at 39°C, and most can grow
above this temperature (Starmer et al., 1979).

Table 5 Fundamental resource utilization for the common cactus yeasts[a]

	Pichia cactophila	Torulopsis sonorensis	Pichia heedii	Candida ingens	Cryptococcus cereanus	Pichia amethionina[b]	Pichia species "K"	Candida tenuis	Torulopsis species "A"	Pichia opuntiae	Pichia species "A"	Number of species utilizing resource state
D-Glucosamine	+	-	-	-	-	-	-	-	-	-	-	1
Methanol	-	+	-	-	-	-	-	-	-	-	-	
Inositol	-	-	-	-	+	-	-	-	-	-	-	
Methionine less, gluconate	-	-	-	-	-	+	-	-	-	-	-	
Gelatin liquefaction, lactose, α-methyl-D-glucoside	-	-	-	-	-	-	-	+	-	-	-	
Melezitose, maltose, sucrose, L-rhamnose	-	-	-	-	-	-	-	+	v	-	-	2
Trehalose, erythritol	-	-	-	-	+	-	-	+	-	-	-	
Lipolytic activity	-	-	-	+	-	-	-	v	-	-	-	
Glucose fermentation	-	+	-	-	-	-	-	+	-	-	-	
D-Ribose	-	+	-	-	+	-	-	+	-	-	-	3
D-Galactose	-	-	-	+	-	-	-	+	+	-	-	
L-Sorbose	-	-	-	+	+	-	-	-	+	-	-	
L-Arabinose	-	+	-	-	-	-	-	+	+	-	-	
D-Sorbitol (D-glucitol)	-	+	-	-	+	-	-	+	-	-	-	
Citrate	+	-	v	-	-	-	-	+	-	-	-	
Casein hydrolysis	v	-	-	-	-	-	-	v	-	-	v	
Adonitol (ribitol)	-	+	-	-	+	-	-	+	+	-	-	4
Cycloheximide resistance	-	+	-	+	+	-	-	-	+	-	-	
Cellobiose, salacin	-	+	-	-	v	-	-	+	+	+	-	5
D-Xylose	-	+	+	-	+	-	-	+	+	-	-	
D-Mannitol	-	+	-	-	+	v	-	+	-	+	-	
10% NaCl	v	+	v	-	-	-	-	+	-	-	+	
Growth at 42°C	+	v	+	+	+	v	v	+	-	-	-	8
Glycerol	v	-	-	+	+	+	+	+	+	+	+	9
Lactate	+	+	+	+	+	+	+	-	-	+	+	
Succinate	+	-	+	+	+	+	+	+	-	+	+	
Growth at 39°C	+	+	+	+	+	+	+	v	v	-	+	10
Ethylamine	+	+	+	+	+	+	+	+	-	+	+	
D-Glucose, ethanol	+	+	+	+	+	+	+	+	+	+	+	11
Number of positive or variable resource states[c]	12	18	10	12	20	11	8	30	16	9	9	

[a] Key: + = positive response; v = variable response; and - = negative response.

[b] The two varieties of *P. amethionina* are combined in this table.

[c] These numbers exceed the actual number of + or v entries in the column, since some rows contain two or more resource states.

k. *Pichia* species "A"

Pichia species "A" is similar to the other *Pichia* species encountered in the cacti in that it is limited to very few resource states. The species was recovered in low frequency from agria and organ-pipe cactus. In addition it was found primarily in one geographic locality, the Cape region of the Baja California peninsula.

D. Analytical Methods

Resource utilization of the yeasts will be analyzed in two ways. The first is at the level of the cactus habitat. Considering each cactus as a different resource state, the yeast species resource utilization will be evaluated for the following partitions: (1) pooled overall cacti; (2) subtribes of cacti; (3) cactus species within each subtribe; and (4) subsets created by dispersal potential in conjunction with patterns of cactus geographic distribution. Second, the yeasts will be compared for use of fundamental resource states (e.g., assimilation of simple sugars) as determined in the laboratory. In effect the comparison at this level is a study of the fundamental niche. The fundamental resource utilization will be studied according to the frequencies observed for the yeasts in the collection. The analysis will be carried out on the same four partitions just listed for the cactus habitat resource states.

The various partitions into which the collection will be divided correspond to several factors presumed to operate in determining the distribution of the yeast species. The yeast species distribution pooled (summed) over all cacti corresponds to the null distribution to which all other partitions will be compared. The division into cactus subtribes represents the two major chemical habitats, as discussed earlier and described by Gibson and Horak (1978). The cacti within subtribes corresponds to the secondary chemical division of the major chemical habitats. The fourth partition represents the effect of *Drosophila* dispersal and geographic sympatry of the cacti. This effect is partially confounded with the partitions, cactus subtribes, and species within subtribes. This is true since one species of *Drosophila* (*D. pachea*) only visits one cactus (senita) and *D. nigrospiracula* is primarily restricted to two cacti in the subtribe Pachycereinae.

A multitude of niche metrics have been devised for the purpose of analyzing community structure, competition for common resources, components of species diversity, and other ecologically important factors (Pielou, 1972; Colwell and Futuyma, 1971; Hurlbert, 1971; Simpson, 1949; Levins, 1968; Alatalo and Alatalo, 1977). The choice of which niche metric to use depends on the data base, sampling techniques, and types of questions being posed. The basic question asked in this chapter is the following: What appear to be the major determinants of the yeast distribution in the cacti? The resulting analyses should not be viewed as tests of hypotheses but rather as methods for revealing working hypotheses. With this in mind, the results of the analyses will be taken as a first approximation in determining the causal factors important to the distribution of cactus yeasts.

Table 6 The estimates of interspecific encounter (PIE) and effective number of species (ENS) for the hierarchial partition of the cactus habitat and fundamental resource states

Level		Partition		
		Pooled cacti	Subtribes	Cacti
Cactus resource states	PIE	0.822	0.796	0.747
	±S.E.	--	0.026	0.021
	ENS	5.63	4.90	3.95
Fundamental resource states	PIE	0.470	0.402	0.369
	±S.E.	--	0.026	0.006
	ENS	1.89	1.67	1.58

I shall employ Simpson's (1949) "measure of concentration" as an index of the probabilistic nature of species interactions. In addition, I will utilize notions of Hurlbert (1971) for interpreting the results.

The mathematical procedures are found in the Appendix. This treatment results in estimates of yeast species concentrations for each partition of the two levels (i.e., cactus habitat and fundamental resource states).

Following the species diversity discussion of Hurlbert (1971), the complement of the total species concentration represents the probability of interspecific encounters (PIE). Furthermore, the reciprocal of the total species concentration represents the number of equally common species that would result in the same probability of interspecific encounters [i.e., the "effective number of species" (ENS)]. This notion is similar to that of effective population size utilized in population genetics (see Crow and Kimura, 1970). The PIE and ENS are estimated in Table 6 for both the fundamental and cactus resource states.

From the known geographic distribution of the cactus species (Hastings et al., 1972) and the actual collection records, the geographic sympatry of the collections can be determined. There are two minor divisions and one major division. First, all collections of yeasts from saguaro cactus were made in the vicinity of Tucson, Arizona, and are consequently allopatric with the other cactus collections. Second, cina cactus is geographically separated from agria cactus and was collected in only two localities. The major geographic sympatric collection is thus the collection from Baja California and parts of Sonora, Mexico, for agria, organ-pipe, cardón, and senita cactus. These four cactus habitats are used in conjunction with the estimation of dispersal probability among the cacti by the *Drosophila*.

The probability (V_{mj}) that a *Drosophila* species (m) will visit a cactus species (j) is estimated from the data of Fellows and Heed (1972; see their Table 3). They determined the relative attractiveness of various necrotic cactus tissues to natural

Table 7 The probabilities for the *Drosophila* dispersing yeasts to the four sympatric cactus species

Drosophila species	Agria	Organ-pipe	Cardón	Senita
D. *nigrospiracula*	0.08	0.06	0.84	0.02
D. *mojavensis*	0.66	0.17	0.15	0.02
D. *pachea*	0.01	0	0	0.99

populations of cactus-specific *Drosophila*. They did not determine the attractiveness of cardón in their study; however, the value should be close to that of saguaro. With this in mind, the potential dispersal probabilites of the three *Drosophila* species among the four sympatric cactus species are given in Table 7. Utilizing these dispersal probabilites (V_{mj}) as weights for the yeast species distributions in the cacti, we obtain a weighted yeast species distribution as a result of the dispersal patterns of the flies. Accordingly the values for the PIE and ENS are calculated from the data created by the dispersal patterns. The mathematical notation and calculations are identical with those of cactus resource states and the fundamental resource states (cf. the Appendix) except

$$P_{im} = \sum_{j=1}^{4} V_{mj} P_{ij}$$

for the level of cactus habitats (j), and

$$P_{ijm} = \sum_{k=1}^{4} V_{mk} P_{ijk}$$

for the level of fundamental resource states (j) and cactus (k). The estimates for the probability of interspecific encounter and effective number of species are then derived from these adjusted proportions and averaged over the three subsets (m) created by the dispersal patterns. These values are subject to a comparison only with the data for cactus species of the sympatric collections (agria, organ-pipe, cardón, and senita).

III. RESULTS

Utilizing the analytical tools described in the Appendix, we can reduce the data found in Tables 2 and 5 for comparison of the effect of cactus phylogeny and *Drosophila* dispersal on the yeast distribution. The values of PIE and ENS are given in Tables 6 and 8 for the cactus partition and dispersal partition, respectively. The decreasing values of PIE and ENS at both levels (cactus and fundamental resource) indicate an effect of cactus chemistry. However, the effect of *Drosophila* dispersal for the sympatric collection reduced the values of PIE and ENS to that of the cactus

Table 8 The estimates of interspecific encounter (PIE) and effective number of species (ENS) for the sympatric collection and the partition due to *Drosophila* dispersal[a]

Level		Partition			
		Pooled cacti	Subtribes	Cacti	Dispersal
Cactus resource states	PIE	0.832	0.796	0.735	0.737
	±S.E.[b]	--	0.013	0.025	0.021
	ENS	5.96	4.90	3.78	3.81
Fundamental resource states	PIE	0.474	0.414	0.376	0.410
	±S.E.[b]	--	0.012	0.004	0.026
	ENS	1.90	1.71	1.60	1.69

[a]Only agria, organ-pipe, cardón, and senita are included.

[b]S.E.: standard error.

partition for the level of cactus resource states and to that of the subtribe partition for the level of fundamental resource states. The difference in the reduction of PIE and ENS due to *Drosophila* dispersal for the two levels of resource states can be interpreted in the following manner. The PIE of the yeasts in the cactus habitats includes encounters of species which utilize the same fundamental resource states. The *Drosophila* dispersal weights these species equally in the two partitions, resulting in less reduction at the fundamental level. This means *Drosophila* acting as passive dispersal agents do not distinguish the fundamental dimensions of the yeasts but do distinguish the habitat dimensions of the yeasts. The result is that the selective force maintaining the yeast distribution primarily resides in the chemical nature of the cacti and is supported at the division of cactus subtribes by the dispersal of the flies.

The overall pattern of resource utilization for the yeasts is detailed in Table 5. Inspection of the table reveals that there are eight unique resource states, eight states shared by two species, etc. To a degree the distribution of resource states among the yeasts indicates the specialization of the yeast species on the 38 dimensions. If the yeasts which occurred in low frequency (<5%, i.e., *Pichia* species "K" through *Pichia* species "A" in Table 5) are excluded, then the specialization of the six most common cactus yeasts is apparent. Exclusion of low-frequency species results in 11 unique resource states, with all species except *P. heedii* having exclusive use of one or more resources. It should be noted that some of the resources listed may not be independent in that the same enzyme could be utilized to split one or more sugars. This is possible for the α-glucosides (i.e., sucrose and maltose) listed in Table 3 for *C. tenuis* and *Torulopsis* species "A" as well as the β-glucosides listed for a number of the species (see Barnett, 1976, for a full discussion). However, there was sufficient variability among the strains within each compound class to keep them separated for the analysis.

One other pattern which is apparent from Table 5 is the similarity among those organisms which utilize 12 or fewer resource states (i.e., all of the *Pichia* species; *C. ingens* = *P. humboldtii*). It is interesting that on considering the *Pichia* species as a group distinct from the other species, each species displays one or more unique resource states (except *Pichia* species "K") but shares most other resource states with the other species. In fact the distribution of resource sharing is as follows:

Number of resource states utilized by ... 10 3 1 0 1 2 5
Number of species 1 2 3 4 5 6 7

The distribution of resource states among the *Pichia* species is nonrandom, and the U-shaped pattern is another indication of specialization. On considering those four species which are positive or variable for 16 or more resource states, the distribution of resource sharing is much more even:

Number of resource states utilized by ... 8 12 8 7
Number of species 1   2 3 4

Although each of these species has exclusive use of one or more of the resource states, the general pattern indicates a relative increase in generalization for resource utilization.

The species can be classified as to (1) those endemic to the Sonoran Desert (*P. heedii*, *P. amethionina*, *P.* species "K," and *P.* species "A"); (2) those restricted to cacti (*P. cactophila*, *T. sonorensis*, *C. cereanus*, *Torulopsis* species "A," and *P. opuntiae*); and (3) those found in substrates other than cacti (*C. ingens* and *C. tenuis*). Comparing the average number of resource states utilized by the three groups reveals a trend of decreasing utilization going from the group using substrates other than cacti ($\bar{x} = 21 \pm 9.0$), to the group restricted to cacti ($\bar{x} = 15.0 \pm 2.0$), to the endemic group ($\bar{x} = 9.5 \pm 0.65$). Moreover the unique or infrequent isolates of yeasts (14 species) not reported here utilize a relatively large number of resource states (22.9 ± 1.6) and are not restricted to the cactus habitat. The trend, a loss of metabolic capabilities, may be interpreted as an evolutionary response of a yeast entering the cactus habitat and subsequently becoming established there. This trend is reflected in the correlation between the frequency of isolation and the number of positive resource states. On considering all yeasts isolated from cacti (26 species), the correlation between number of positive resource states and frequency of isolation is negative ($r = -0.353$, D.F.[*] $= 25$, $\alpha = 0.08$). Furthermore the correlation is decreased to $r = -0.572$ (D.F. $= 25$, $\alpha = 0.002$) on transforming the frequency of isolation to $\log_e$ (frequency). The reason for this additional decrease in the estimate

---

[*]Degrees of freedom.

of the correlation is that most yeast found frequently in the cacti use few resource states (<12) whereas those yeasts which occur infrequently utilize a number of resource states (>20), and there are few examples of yeasts in between these two groups. The significance of this correlation serves to indicate that the realized niche width is negatively correlated with the fundamental niche width for cactus-inhabiting yeast. The realized niche dimensions are the cactus habitats within the Sonoran Desert, while the fundamental niche dimensions are the physiological tests conducted in the laboratory. Some of the yeasts not restricted to the cactus habitat may have a broader realized niche outside the Sonoran Desert; however, within the realm of the cactus habitats, they generally display narrow realized niches and broad fundamental niches.

IV. DISCUSSION

The analytical results on the resource utilization of the common cactus yeasts lead to the following conclusions: (1) The decrease in the PIE and ENS according to the phylogenetic hierarchy indicates that cactus chemistry is functioning to dimension the niche (Table 6). (2) The large decrease in PIE and ENS going from cactus resource states to fundamental resource states indicates that the species can divide the niche more effectively at the level of their basic metabolism. In addition, if one takes the values of PIE as an indicator of competition, then relatively little competition is occurring among the yeasts at the level of their fundamental metabolism. This notion is of course predicated on the assumption that the conditions studied in the laboratory are representative of what the species encounters in the decaying cactus. We do know that some of the laboratory compounds are present in the necrotic tissue (e.g., glucose, ethanol, and methanol) but do not know which are absent. Furthermore, there are undoubtedly a number of compounds in the cacti that are utilizable by the yeasts and that we did not test in the laboratory. Thus the laboratory tests constitute a subset of the actual fundamental niche dimensions; this subset partially overlaps with the fundamental dimensions encountered in the cacti.

The analysis of the effect of the *Drosophila* dispersal shows two basic patterns. First, the overall average dispersal can only account for the reduction in PIE to the division of cactus subtribes for the fundamental resource states (Table 8). Therefore, cactus species chemistry is viewed as the principal factor responsible for partitioning the niche. Second, the *Drosophila* dispersal reduces the value of PIE and ENS to that of individual cacti at the level of the cactus habitats. Thus the dispersal agents distinguish the habitat dimensions of the yeasts but not the fundamental dimensions of the yeasts. The primary ecological force thus resides in the chemical nature of the cacti, which is confounded with and supported by the dispersal of the flies at the division of the cactus subtribes.

The metabolic potentials of a yeast dictate whether it will survive in any given environment. However, the ability to utilize a wide spectrum of compounds may in fact be maladaptive. This appears to be the case for some cactus-inhabiting yeasts and may apply to other yeasts as well [e.g., Wickerham (1969) describes *Hansenula* species dependent on coniferous trees]. The utilization of sugars in yeasts is well reviewed by Barnett (1976). In this review he summarizes the basic enzymatic steps necessary for conversion of sugars into energy and points out interrelationships among the various systems employed in metabolizing a resource. An important distinction is made between the ability to transport a compound across the plasmalemma and the ability to catabolize a compound once inside the plasmalemma. The entry of sugars is mediated by enzyme-like carriers which are under genetic control. The carriers may be inducible and can in some cases exhibit wide overlapping substrate specificities. The lack of certain carriers ("permeases") can cause cryptic loss of a yeast's ability to catabolize a substrate. Moreover, the lack of a carrier can result in impermeability for a number of similar compounds. With these factors in mind, the realized niche of a species could be determined and/or changed due to the following constraints on the fundamental niche dimensions: (1) The species may not have the genes necessary for the basic metabolism of a resource. (2) The yeast does have the genes necessary for metabolism of a resource but may be deficient (mutant) for one or more of the genes. (3) The genes necessary for the metabolism of a resource are present, but genes which provide transport across the plasmalemma may be either lacking or nonfunctional.

The realized niche could be expanded through loss of genetic activity when toxins are present in the habitat. This is especially true when a toxin is analogous in structure to normally harmless substances utilizable by the yeast. In fact, a laboratory technique for selecting permease mutants is to select strains resistant to a toxic analog of the substrate (Slayman, 1973). An enlightening example is the study of Breton and Surdin-Kerjan (1977) on *Saccharomyces cerevisiae*. Utilizing two toxic analogs of sulfate (selenate and chromate) in selecting for sulfate permease mutants, these investigators obtained a strain which could not concentrate sulfate and demonstrated the strain was mutant for two independent loci.

It is interesting that one of the common cactus yeasts, *P. amethionina*, is unable to synthesize methionine but is equipped with the enzymes for methionine biosynthesis past the sulfate assimilation pathway (Starmer et al., 1978b). The exact location of the deficiency for sulfate assimilation is not known. However, it could involve one or more enzymes, including sulfate permease (Masselot and Surdin-Kerjan, 1977; Breton and Surdin-Kerjan, 1977). The loss of ability to assimilate sulfate can be seen to be adaptive when this loss is accompanied by exclusion of toxins. I do not know if this is indeed the case, but recent experiments do support the notion that the cactus substrate is toxic to some yeasts. The experiments involved incorporating dried cactus powder (agria) into a complete yeast medium (YM: Difco yeast-extract, malt-extract agar) and observing growth on the medium of various cactus-specific and related noncactus yeasts. The results were clear, although preliminary. First, yeasts which are commonly found in decaying agria (see Table 2: *P. cactophila*, *T. sonorensis*, and *P. amethionina* var. *amethionina*) grow well on medium supp-

lemented with agria powder. However, other cactus yeasts not commonly found in agria
(e.g., *P. heedii*, *P. amethionina* var. *pachycereana*, and some strains of *C. cereanus*)
are completely inhibited on the agria-supplemented medium. In addition noncactus
yeasts (*P. delftensis*, *P. fluxuum*, and *P. scutulata*) with similar phenotypic properties as *P. amethionina* and *P. heedii* cannot grow on the cactus medium.

By studying the tetrads of one hybrid of the two varieties of *P. amethionina*,
I was able to demonstrate for this species that the ability to grow on the test medium is controlled by a single genetic locus. Furthermore all strains of *Cryptococcus cereanus*, *P. heedii* and *P. amethionina* that could not grow in the presence of
agria powder at 27°C could grow when the temperature was raised to 39°C. These
results are preliminary but indicate the following factors: (1) The cacti contain
compounds which inhibit yeast growth. (2) The ability to grow in a particular cactus
can be controlled by a single gene. (3) Growth can be temperature dependent. Since
the restrictive temperature is low and the permissive temperature high, it is possible that heat-sensitive permeases responsible for the transport of toxins may be involved.

Another example of genetic adaptation which indicates the possible loss of enzymes involved in transport is that of citrate metabolism in *P. heedii*. Strains of
P. heedii recovered from senita cactus were all capable of assimilating citrate,
whereas those from other cacti could not grow when given citrate as the sole carbon
source. Since citrate is an integral component in the tricarboxylic acid cycle, it
is probable that the variability for citrate utilization is due to differences in
the ability to transport the compound (see Barnett, 1968). A single genetic locus
is responsible for the differences in citrate metabolism seen in *P. heedii* (Phaff
et al., 1978). We have also determined that the ability to assimilate D-mannitol
for the two varieties of *P. amethionina* is under monogenetic control (Starmer et al.,
1978b). The gene involved in D-mannitol utilization is independent of the gene which
determines the growth on agria-supplemented medium, mentioned earlier (Starmer, unpublished data).

In general, it is probable that the ancestral yeasts of the present cactus-inhabiting yeasts had more diverse metabolic capabilities. In adapting to the cactus
habitat the yeasts would have had to contend with the unique secondary compounds
found in the cactus (see Gibson and Horak, 1978, for a review of some of these compounds). When these compounds are toxic, the mode of adaptation could be either the
acquisition of new genetic systems to cope with the toxins or degeneration of existing systems which allow the toxin to assert its effect (Waid, 1968). I think this
latter mode is the more probable since single genes are involved in the changes of
metabolic patterns observed for yeast species which switch host plants. In addition
the general adaptive mode for inhabiting the decaying tissue is loss of metabolic
activity. This is supported by two related results. First, the trend of decreasing
utilization of resource states--going from the yeasts which use substrates other than
cacti, to the yeasts restricted to cacti, to yeasts endemic to the cacti of the Son-

oran Desert--indicates loss of metabolic activity on becoming established in the cactus habitat. Second, the negatiave correlation between the numbers of positive resource states and frequency of isolation also indicates loss of metabolic activity on expansion of the realized niche (i.e., expanded cactus utilization).

In summary, loss of the ability to utilize fundamental resources may be an adaptive response of the yeasts to the toxins they encounter in the environment. The toxins in the habitats are presumably present as a chemical defense against herbivores and microorganisms responsible for the primary destruction of the living plants (Janzen, 1973). The yeasts existing as saprophytes would be responding secondarily to the toxins residing in the decaying tissue.

Another resource dimension which limits the species found in necrotic cacti is temperature. All common cactus yeasts grow well at a temperature of 37°C or above. Some species can grow at temperatures of 45-46°C, and most can grow at 42°C. When one considers the total known species of yeasts, fewer than 37% can grow at temperatures of 37°C or above (Barnett and Pankhurst, 1974). The only common cactus yeast found in the Sonoran Desert that cannot grow at 39°C or above is *P. opuntiae*. It is notable that *P. opuntiae* var. *thermotolerans* has only been recovered from cacti in the tribe Pachycereinae within the Sonoran Desert (Table 2). Two of these cacti, saguaro and cardón, have the largest stems among the cacti studied and possibly would buffer the temperature extremes experienced in the hot desert. In addition, *P. opuntiae* is not inhibited by media supplemented with agria cactus, yet it was not recovered from 100 agria cacti sampled. Agria is a relatively thin-stemmed plant, providing less buffer against temperature than the thicker-stemmed cacti of the tribe Pachycereinae. The maximum growth temperature of *P. opuntiae* may thus restrict its distribution.

The yeasts found in the Sonoran Desert are thermotolerant and may have adjusted to the temperature extremes they experience by adaptations in their membranes (Arthur and Watson, 1976). It is interesting that the basis of thermophily in fungi may also involve metabolism that is *slower* than that predicted from mesophilic fungi (Emerson, 1968). This view is compatible with the notion that a wider niche may be realized by degeneration of an attribute normally positive in the fundamental niche dimension.

V. SUMMARY

Most yeasts found in necrotic cacti are (in general) specialized for this habitat. The primary factors determining the distribution of the yeasts are the chemical environment of the habitats and the metabolic capabilities of the yeasts. The analysis indicates that the fundamental niche dimensions are sufficient to preclude competition as an ecological factor responsible for the distribution of the yeasts among the cacti. Furthermore, the dispersal agents (*Drosophila*) play a supportive role in determining the yeast distribution. The adaptative response of the yeasts to the cactus habitat is possibly a loss of metabolic activity as a result of toxins present in the tissues. Such a loss represents a decrease in the fundamental niche dimensions which can lead to an increase in the realized niche width. This response may involve basic transport across the plasmalemma of the yeast cell and can take the form of change at a single genetic locus. The concept of increasing the realized niche at the expense of the fundamental niche is on initial consideration contrary to basic ecological principles. However, adaptation to natural environments can certainly involve loss of fundamental niche dimensions as suggested in this chapter. The ideas and notions proposed as factors in the ecology of the cactophilic yeasts are viewed as working hypotheses which lend themselves to future investigations.

APPENDIX

A. Cactus Resource States

   1. Cactus within subtribes

   Let n_{ij} be the number of times yeast species i was recovered from N_j individual samples of cactus species j (Table 2). Since cactus species were not sampled in equal number ($N_1 \neq N_2$, etc.) the proportion of yeast species i recovered from cactus j is defined as

$$p_{ij} = \frac{n_{ij}}{N_j}$$

The total frequency of all yeast(s) recovered from cactus j is given as

$$p._{j} = \sum_{i=1}^{s} p_{ij}$$

The proportion of yeast species i, isolated from cactus j, relative to all other yeasts of cactus j is then

$$r_{ij} = \frac{p_{ij}}{p._{j}}$$

The squared value of r_{ij} will represent the probability of intraspecific encounter for yeast species i in cactus j. This is Simpson's (1949) measure of concentration. It follows that the average values for cacti within each subtribe [Stenocereinae (1); j = 1, 2, 3; Pachycereinae (2), j = 4, 5, 6] are

$$a_{i1} = \frac{1}{3} \sum_{j=1}^{3} r^2_{ij} \qquad \text{and} \qquad a_{i2} = \frac{1}{3} \sum_{j=4}^{6} r^2_{ij}$$

The summed values of a_{i1} and a_{i2} for all yeast species recovered in the respective subtribes are

$$a._{1} = \sum_{i=1}^{s} a_{i1} \qquad \text{and} \qquad a._{2} = \sum_{i=1}^{s} a_{i2}$$

It follows that the average value over both subtribes is

$$\bar{a}.. = \frac{1}{2}[a._1 + a._2]$$

2. *Cactus subtribes*

The pooled yeast frequency for each subtribe is subjected to a similar treatment. The total proportion of yeast species i recovered from subtribe m = 1, 2 is defined as

$$q_{i1} = \sum_{j=1}^{3} p_{ij} \quad \text{and} \quad q_{i2} = \sum_{j=4}^{6} p_{ij}$$

The total proportion of all yeast species from each subtribe (m) is given as

$$q._m = \sum_{i=1}^{s} q_{im}$$

The proportion of yeast species i recovered from each subtribe (m) relative to all other yeast species is then

$$r_{im} = \frac{q_{im}}{q._m}$$

The total measure of concentration (r^2_{im}) for all yeasts in each subtribe is

$$b._m = \sum_{i=1}^{s} r^2_{im}$$

and the average over subtribes is given as

$$b.. = \frac{1}{2}[b._1 + b._2]$$

3. *Pooled cacti*

The total proportion of yeast species i recovered from the collection is

$$u_i = (q_{i1} + q_{i2})$$

The total over yeast species is then

$$u. = \sum_{i=1}^{s} u_i$$

and the proportion of yeast species i in the collection is

$$r_i = \frac{u_i}{u.}$$

The measure of concentration totaled over all yeasts is derived as

$$c = \sum_{i=1}^{s} r^2_i$$

B. Fundamental Resource States

  1. *Cactus within subtribes*

     Considering the resource states (38) as the basic metabolic capabilities of the
cactus yeast (Tables 3 and 4), the measure of yeast species concentration is derived
for each level of the cactus hierarchy. The notation and derivation is identical
with the preceding description except for the following adaptation. Yet n_{ijk} = the
number of times yeast i, capable of utilizing resource j (j = 1, 38; Tables 3 and 4),
was recovered from N_k individual samples of cactus k. It follows that the three
levels of the cactus partition yields:

$$p_{ijk} = \frac{n_{ijk}}{N_k}, \qquad p \cdot_{jk} = \sum_{i=1}^{s} p_{ijk}$$

$$\rho_{ijk} = \frac{p_{ijk}}{p \cdot_{jk}}, \qquad \alpha_{ik} = \frac{1}{38} \sum_{j=1}^{38} \rho^2_{ijk}$$

where α is the species concentrations averaged over resource states (j) for cactus
k.

$$\alpha_{i1} = \frac{1}{3} \sum_{k=1}^{3} \alpha_{ik} \qquad \text{and} \qquad \alpha_{i2} = \frac{1}{3} \sum_{k=4}^{6} \alpha_{ik}$$

and

$$\alpha_{\cdot 1} = \sum_{i=1}^{s} \alpha_{i1} \qquad \text{and} \qquad \alpha_{\cdot 2} = \sum_{i=1}^{s} \alpha_{i2}$$

and

$$\overline{\alpha}_{\cdot \cdot} = \frac{1}{2} [\alpha_{\cdot 1} + \alpha_{\cdot 2}]$$

where ρ corresponds to r, and α corresponds to a in the cactus habitat analysis.

  2. *Cactus subtribes*

     The metrics for subtribes are defined as

$$q_{ij1} = \sum_{k=1}^{3} p_{ijk} \qquad \text{and} \qquad q_{ij2} = \sum_{k=4}^{6} p_{ijk}$$

$$\rho_{ijm} = \frac{q_{ijm}}{q \cdot_{jm}}, \qquad \beta_{im} = \frac{1}{38} \sum_{j=1}^{38} \rho^2_{ijm}$$

where β_{im} is the species concentration averaged over resource states (j) for subtribe (m).

$$\beta_{\cdot m} = \sum_{i=1}^{s} \beta_{im}, \qquad \overline{\beta}_{\cdot \cdot} = \frac{1}{2}[\beta_{\cdot 1} + \beta_{\cdot 2}]$$

where ρ corresponds to r and β corresponds to b in the subtribe analysis of the cactus habitats.

3. Pooled cacti

The mathematical notation for the fundamental resource states as realized for pooled cacti is

$$u_{ij} = (q_{ij1} + q_{ij2}), \qquad \rho_{ij} = \frac{u_{ij}}{u_{.j}}, \qquad \gamma_i = \frac{1}{38} \sum_{j=1}^{38} \rho^2_{ij}$$

where γ_i is the species concentrations averaged over resource states. Finally

$$\gamma_. = \sum_{i=1}^{s} \gamma_i$$

is the total yeast species concentrations, where ρ corresponds to r and γ corresponds to c in the pooled cactus notation of the cactus habitat resource states.

In summary the total yeast species concentrations for both cactus and fundamental resource states are listed in the tabulation.

	Cactus resource states	Fundamental resource states
Pooled cacti	c	γ
Subtribes	$\bar{b}_{..}$	$\bar{\beta}_{..}$
Cacti	$\bar{a}_{..}$	$\bar{\alpha}_{..}$

ACKNOWLEDGMENTS

This chapter was only possible through the efforts of a number of individuals. The initial stimulus and direction was given by Dr. William B. Heed. I thank him for introducing me to the fascinating problems which have led to the present understanding of the subject. The guidance of Dr. Herman J. Phaff is gratefully acknowledged. His exacting wisdom of yeast taxonomy facilitated the realization of the species encountered in this study. The methodical laboratory investigation of the isolates was mainly due to the efforts of Ms. Mary Miranda. I thank her for this indispensable information. The collections were assisted by Drs. Martin Miller, Don Vacek, Robert Mangan, William Johnson, and Alexander and Jean Russell. The cactus phylogeny which is rudimentary to the present study was due to Dr. Arthur Gibson. I thank him for his contribution and his keen interest in the subject. I am grateful to Drs. W. B. Heed, L. L. Wolf, and H. J. Phaff for their useful criticisms and comments on an earlier version of the manuscript. The collections and laboratory investigations were supported by grants from the National Science Foundation to W. B. Heed and H. W. Kircher (DEB 74-1918 A03) and from the National Institute of General Medical Sciences to H. J. Phaff (GM-16307-07).

REFERENCES

Alatalo, R. V., and Alatalo, R. H. (1977). Components of diversity: Multivariate analysis with interaction. *Ecology 58*: 900-906.

Arthur, H., and Watson, K. (1976). Thermal adaptation in yesat: growth temperatures, membrane lipid, and cytochrome composition of psychrophilic, mesophilic, and thermophilic yeasts. *J. Bacteriol. 128*: 56-68.

Barnett, J. A. (1968). Biochemical differentiation of taxa with special reference to the yeasts. In *The Yeasts*, Vol. I: *The Biology of Yeasts*, A. H. Rose and J. S. Harrison (Eds.). Academic Press, New York, pp. 557-595.

Barnett, J. A. (1976). The utilization of sugars by yeasts. *Adv. Carbohyd. Chem. Biochem. 32*: 125-234.

Barnett, J. A., and Pankhurst, R. J. (1974). *A New Key to the Yeasts*. North-Holland Publ., Amsterdam.

Breton, A., and Surdin-Kerjan, Y. (1977). Sulfate uptake in *Saccharomyces cerevisiae*: biochemical and genetic studies. *J. Bacteriol. 132*: 224-232.

Campbell, I. (1973). Numerical analysis of *Hansenula*, *Pichia*, and related yeast genera. *J. Gen. Microbiol. 77*: 427-441.

Colwell, R. K., and Futuyma, D. J. (1971). On the measurement of niche breadth and overlap. *Ecology 42*: 567-576.

Crow, J. F., and Kimura, M. (1970). *An Introduction to Population Genetics Theory*. Harper & Row, New York.

Emerson, R. (1968). Thermophiles. In *The Fungi: An Advanced Treatise*, Vol. III: *The Fungal Population*, G. C. Ainsworth and A. S. Sussman (Eds.). Academic Press, New York, pp. 105-128.

Fellows, D. P., and Heed, W. B. (1972). Factors affecting host plant selection in desert-adapted cactiphilic *Drosophila*. *Ecology 53*: 850-858.

Gibson, A. C., and Horak, K. E. (1978). Systematic anatomy and phylogeny of Mexican columnar cacti. *Ann. Missouri Bot. Gard. 65*: 999-1057.

Gorin, P. A. J., and Spencer, J. F. T. (1970). Proton magnetic resonance spectroscopy: an aid in identification and chemotaxonomy of yeasts. *Advan. Appl. Microbiol. 13*: 25-89.

Hastings, J. R., Turner, R. M., and Warren, D. K. (1972). An atlas of some plant distributions in the Sonoran Desert. *Univ. Arizona Inst. Atmos. Phys. Tech. Repts. 21*: 1-255.

Heed, W. B. (1977). A new cactus-feeding but soil-breeding species of *Drosophila Diptera*: (Drosophilidae). *Proc. Entomol. Soc. Wash. 79*: 649-654.

Heed, W. B. (1978). Ecology and genetics of Sonoran Desert *Drosophila*. In *Ecological Genetics: The Interface*. P. F. Brussard (Ed.). Springer-Verlag, New York, pp. 109-126.

Heed, W. B., and Kircher, H. W. (1965). Unique sterol in the ecology and nutrition of *Drosophila pachea*. *Science 149*: 758-761.

Heed, W. B., Starmer, W. T., Miranda, M., Miller, M. W., and Phaff, H. J. (1976). An analysis of the yeast flora associated with cactiphilic *Drosophila* and their host plants in the Sonoran Desert and its relation to temperate and tropical associations. *Ecology 57*: 151-160.

Hurlbert, S. H. (1971). The nonconcept of species diversity: a critique and alternative parameters. *Ecology 52*: 577-586.

Janzen, D. H. (1973). Community structure of secondary compounds in plants. *Pure Appl. Chem. 34*: 529-538.

Kircher, H. W., Heed, W. B., Russell, J. S., and Grove, J. (1967). Senita cactus alkaloids: their significance to Sonoran Desert *Drosophila* ecology. *J. Insect Physiol. 13*: 1869-1874.

Kocková-Kratochvílová, A., Wegner, K. A., and Slávíková, E. (1976). Die Beziehungen innerhalb der Gattung *Cryptococcus* (Sanfelice) Vuillemin. *Zbl. Bakteriol.* [*Abt. II*] *131*: 610-631.

Kreger-van Rij, N. J. W. (1969). Taxonomy and systematics of yeasts. In *The Yeasts,* Vol. 1: *The Biology of Yeasts.* A. H. Rose and J. S. Harrison (Eds.). Academic Press, New York, pp. 5-78.

Kuroshima, T., Kodaira, S., and Tsuchiya, T. (1976). Serological relationships among *Pichia* species. *Japan J. Microbiol. 20*: 485-492.

Levins, R. (1968). *Evolution in Changing Environments* (Monographs in Population Biology, No. 2). Princeton Univ. Press, Princeton, N.J.

Lightle, P. C., Standring, E. T., and Brown, J. G. (1942). A bacterial necrosis of the giant cactus. *Phytopathology 32*: 303-313.

Lodder, J., Ed. (1970). *The Yeasts: A Taxonomic Study,* 2nd ed. North-Holland Publ., Amsterdam.

Lodder, J., and Kreger-van Rij, N. J. W. (1952). *The Yeasts: A Taxonomic Study.* North-Holland Publ., Amsterdam.

Lund, A. (1956). Yeasts in nature. *Wallerstein Lab. Commun. 19(66)*: 221-236.

Masselot, M., and Surdin-Kerjan, Y. (1977). Methionine biosynthesis in *Saccharomyces cerevisiae.* II. Gene-enzyme relationships in the sulfate assimilation pathway. *Mol. Gen. Genet. 154*: 23-30.

Meyer, S. A., and Phaff, H. J. (1972). DNA base composition and DNA-DNA homology studies as tools in yeast systematics. In *Yeasts: Models in Science and Technics,* Kratochvílová and E. Minarik (Eds.). Publ. House Slov. Acad. Sci., Bratislava, Czechoslovakia, pp. 375-387.

Miller, M. W., Phaff, H. J., Miranda, M., Heed, W. B., and Starmer, W. T. (1976). *Torulopsis sonorensis*, a new species of the genus *Torulopsis. Inter. J. Syst. Bact. 26*: 88-91.

Phaff, H. G., and Fell, J. W. (1970). *Cryptococcus* Kützing emend. Phaff et Spencer. In *The Yeasts: A Taxonomic Study,* J. Lodder (Ed.). North-Holland Publ., Amsterdam, pp. 1088-1145.

Phaff, H. J., Miller, M. W., Recca, J. A., Shifrine, J., and Mrak, E. M. (1956). Studies on the ecology of *Drosophila* in the Yosemite regions of California. *Ecology 37*: 533-538.

Phaff, H. J., Miller, M. W., Miranda, M., Heed, W. B., and Starmer, W. T. (1974). *Cryptococcus cereanus*, a new species of the genus *Cryptococcus. Int. J. Syst. Bacteriol. 24*: 486-490.

Phaff, H. J., Starmer, W. T., Miranda, M., and Miller, M. W. (1978). *Pichia heedii*, a new species of yeast indigenous to necrotic cacti in the North American Sonoran Desert. *Int. J. Syst. Bacteriol. 28*: 326-331.

Pielou, E. C. (1972). Niche width and niche overlap: a method for measuring them. *Ecology 53*: 687-692.

Price, C. W., Fuson, G. B., and Phaff, H. J. (1978). Genome comparison in yeast systematics: delimitation of species within the genera *Schwanniomyces, Saccharomyces, Debaryomyces* and *Pichia. Microbiol. Rev. 42*: 161-193.

Rodrigues de Miranda, L., and Török, T. (1976). *Pichia humboldtii* sp. nov., the perfect state of *Candida ingens. Antonie van Leeuwenhoek J. Microbiol. Serol. 42*: 343-348.

Shreve, F., and Wiggins, I. L. (1964). *Vegetation and Flora of the Sonoran Desert,* Vol. 1. Stanford Univ. Press, Stanford, Calif.

Simpson, E. H. (1949). Measurement of diversity. *Nature 163*: 688.

Slayman, C. W. (1973). Genetic control of membrane transport. In *Current Topics in Membranes and Transport,* F. Bernner and A. Kleinzeller (Eds.), Vol. 4. Academic Press, New York, pp. 1-174.

Starmer, W. T., Heed, W. B., Miranda, M., Miller, M. W., and Phaff, H. F. (1976). The ecology of yeast flora associated with cactiphilic *Drosophila* and their host plants in the Sonoran Desert. *Microbial Ecology 3*: 11-30.

Starmer, W. T., Phaff, H. J., Miranda, M., and Miller, M. W. (1978a). *Pichia cactophila*, a new species of yeast found in the decaying tissue of cacti. *Int. J. Syst. Bacteriol. 28*: 318-325.

Starmer, W. T., Phaff, H. J., Miranda, M., and Miller, M. W. (1978b). *Pichia amethionina*, a new heterothallic yeast associated with the decaying stems of cereoid cacti. *Int. J. Syst. Bacteriol. 28*: 433-441.

Starmer, W. T., Phaff, H. J., Miranda, M., Miller, M. W., and Barker, J. S. F. (1979). *Pichia opuntiae*, a new heterothallic species of yeast found in decaying cladodes of *Opuntia inermis* and in necrotic tissue of cereoid cacti. *Int. J. Syst. Bacteriol. 29*: 159-167.

Van der Walt, J. P. (1970). Criteria and methods used in classification. In *The Yeasts: A Taxonomic Study*, J. Lodder, (Ed.). North-Holland Publ., Amsterdam, pp. 34-113.

Van Uden, N., and Buckley, H. (1970). *Candida* Berkhout. In *The Yeasts: A Taxonomic Study*, J. Lodder (Ed.). North-Holland Publ., Amsterdam, pp. 891-1087.

Von Arx, J. A., Rodriques de Miranda, L., Smith, M. Th., and Yarrow, D. (1977). The genera of the yeasts and the yeast-like fungi. *Studies in Mycology*, No. 14, pp. 1-42.

Waid, J. S. (1968). Physiological and biochemical adjustment of fungi to their environment. In *The Fungi: An Advanced Treatise*, Vol. III: *The Fungal Population*, G. C. Ainsworth and A. S. Sussman (Eds.). Academic Press, New York, pp. 289-323.

Wickerham, L. J. (1969). Yeast taxonomy in relation to ecology, genetics, and phylogeny. *Antonie van Leeuwenhoek J. Microbiol. Serol. 35*(Suppl.): 31-58.

Wyss, O., Ludwig, B. J., and Joiner, R. R. (1945). The fungistatic and fungicidal action of fatty acids and related compounds. *Arch. Biochem. 7*: 415-425.

Yamada, Y., Nojiri, M., Matsuyama, M., and Kondo, K. (1976). Coenzyme Q system in the classification of the ascoporogenous yeast genera *Debaryomyces*, *Saccharomyces*, *Kluyveromyces*, and *Endomycopsis*. *J. Gen. Appl. Microbiol. 22*: 325-337.

Chapter 9

*THE GENETICS OF MATING SYSTEMS: FUNGAL
STRATEGIES FOR SURVIVAL*

Judith Stamberg* and Y. Koltin

*Tel-Aviv University
Ramat-Aviv, Israel*

All mating systems have the function of deciding which, or whose, gene pools can mix,
hence regulating the amount of genetic variation or heterogeneity that will be present in the next generation. But how much heterogeneity is desirable in a population?
It is axiomatic that, for an established resident population in a stable environment,
it is desirable to minimize the amount of genetic variation in order to maintain those
genotypes which have enabled the population to be successful. On the other hand, a
new or colonizing population, or one in a constantly fluctuating environment, has need
of a large pool of heterogeneity from which to "choose" new phenotypes suitable to the
environmental conditions. The short-term success of a species depends on its suitability to prevailing conditions, whereas its long-term success depends on its ability
to adapt to new conditions. How well the balance between these needs is achieved is
crucial to the survival of a species (Fisher, 1930; Muller, 1932; Felsenstein, 1974).

I. WHAT GENERATES GENETIC HETEROGENEITY?

Gene mutations are, of course, the ultimate source of heterogeneity; these spread
through a population and combine with different genetic backgrounds via chromosomal
reassortment, rearrangement, and recombination at meiosis and occasionally at mitosis. It has only recently become apparent that the relationship between mutation and
recombination is even more intimate; these two fundamental genetic processes are related not just from an evolutionary point of view but also causally. A number of
fungi have been shown to have a higher frequency of spontaneous mutation after meiosis than after mitosis (Magni and von Borstel, 1962; Kiritani, 1962; Paszewski and
Surzycki, 1964; Bausum and Wagner, 1965; Friis et al., 1971; Koltin et al., 1975).
This phenomenon, first described in yeast, is known as the "meiotic effect" (Magni,
1964). It is frequently associated with recombination of markers bracketing the mu-
tated site, hence recombination is implicated as the cause of the mutation. In bacteria, also, recombination seems to be an error-prone, i.e., mutagenic, process. In
Bacillus subtilis, for example, the process of transformation, which includes a re-
combinational step, increases the spontaneous mutation frequency (Yoshikawa, 1966).

Present affiliation: Johns Hopkins School of Medicine, and Genetics Department,
John F. Kennedy Institute, Baltimore, Maryland

Thus, meiosis and recombination not only spread but are also a prime source for the generation of mutations.

II. HOW DO FUNGI CONTROL THE AMOUNT OF GENETIC HETEROGENEITY IN A POPULATION?

Any mechanism which permits individuals to inbreed or to reproduce asexually works in the direction of decreasing heterogeneity; conversely, outbreeding promotes heterogeneity.

A. Incompatibility Genes

Choice of a mating partner can be controlled in such a way that outbreeding will be favored over inbreeding. Morphological heterothallism (two morphologically distinct sexes) and physiological heterothallism (genes ensuring that morphologically similar mates will be of different genotypes) have this function. The number of such "incompatibility genes," their structure, and the number of alternate alleles of these genes found in natural populations together determine the inbreeding and outbreeding potentials of a species.

It should be noted at the outset that in the majority of fungi breeding is not regulated at all. Homothallism is the most common type of system among all groups of fungi except the Basidiomycetes (Raper, 1966a). Obviously, in homothallics inbreeding is strongly favored over outbreeding.

1. One gene--two alleles

Among the remainder of the ascomycete and basidiomycete fungi, the simplest kind of sexual incompatibility system consists of one gene with two alternate alleles (+/-, A/a, or a/α). Two individuals will be compatible only if they carry different alleles. Assuming that the two alleles are equally frequent in the population, any individual has a 50% chance of being compatible with a random member of the population. And since the meiotic products of one mating will be 50% of each allelic type, any two siblings have a 50% chance of being compatible. In other words, in this system the inbreeding and outbreeding potentials are each 50%. Such a system is found in all classes of fungi, and it is formally analogous to sexual dimorphism where the two sexes are separate insofar as its restrictions on inbreeding and outbreeding are concerned. The heterothallic ascomycetes are all of this type.

2. One gene--many alleles

When we turn to the Basidiomycetes, we see that incompatibility systems can become quite complex. Only about 10% of the higher basidiomycete species are homothallic (Whitehouse, 1949a). More than 30% have a one-gene incompatibility system (called "unifactorial"--or, confusingly, "bipolar" because progeny of a cross are of two types). However, in the Basidiomycetes there are numerous naturally occurring incompatibility alleles for the one gene. Any two individuals are compatible so long as they have different alleles; thus, the greater the number of randomly distributed al-

Table 1 Effect of incompatibility factor structure on breeding potential

Type of system	Component loci	Allelic number and distribution				Inbreeding potential (%)	Outbreeding potential (%)	Example
		Aα	Aβ	Bα	Bβ			
Unifactorial	A	2	--	--	--	50	50	Neurospora crassa, Yeasts
		4	--	--	--	50	75	--
		10	--	--	--	50	90	Polyporus palustris
		20	--	--	--	50	95	--
		30	--	--	--	50	97	Fomes cajanderi
Bifactorial	A,B	2	--	2	--	25	25	--
		10	--	10	--	25	81	--
		20	--	20	--	25	90	--
		30	--	30	--	25	93	--
	Aα,Aβ,Bα,Bβ	2	2	2	2	25–56[a]	56	--
		5	5	5	5	25–56[a]	95	--
		9	9	9	9	25–56[a]	97.5	--
		9	32	9	9	25–56[a]	98.4	Schizophyllum commune

[a]The exact value depends on the recombination frequencies between the component loci of each factor.

leles in nature, the greater the outbreeding potential (Mather, 1942; Whitehouse, 1949b). The exact value can be calculated from the formula $(n_A - 1)/n_A$, where n_A = number of alleles (Raper, 1966b). But since only two alleles are involved in any mating, the meiotic products will always include 50% of each type, and the inbreeding potential remains fixed at 50% regardless of how many alternate alleles are found in a population (Table 1). In the few unifactorial species analyzed, the number of alleles is always at least 20 (Burnett, 1965; Flexer, 1965; Neuhauser and Gilbertson, 1971; Ullrich and Raper, 1974); thus the outbreeding potential is at least 95% while the inbreeding potential stays at 50%.

3. *Two genes--many alleles*

Almost 60% of the higher Basidiomycetes are characterized by a still more complex incompatibility system, the "bifactorial" or "tetrapolar" system. Two factors, called A and B, together determine mating competence, and two individuals will be compatible only if they have different specificities for each factor. In all species studied, these factors are on separate linkage groups; their independent segregation at meiosis leads to production of four main types of progeny (hence the term "tetrapolar"). Since any one progeny is compatible with only 25% of its siblings, this system represents a drastic drop in inbreeding potential compared to the unifactorial system. It is reasonable to suspect that this drop in the inbreeding potential is desirable; otherwise there is no need for incompatibility factors.

The outbreeding potential of the bifactorial system depends on the number of alternate specificities for A and B in the population. A 95% outbreeding potential, comparable to that of the unifactorial species, would require the accumulation of more than 60 mutational events equally distributed in the two factors (Stamberg and Koltin, 1973a). However, a very great increase in outbreeding efficiency has been achieved in the bifactorials by the evolution of a two-locus structure for the incompatibility factors. It was first shown in *Schizophyllum commune* that the A and B factors each consist of two loci, designated Aα, Aβ, Bα, and Bβ (Papazian, 1951; Raper et al., 1958; Koltin et al., 1967). This complex structure is now known to be characteristic of the bifactorial fungi (Day, 1960; Terakawa, 1960; Takemaru, 1961). The two component loci of a factor jointly determine factor specificity, and an allelic difference at either the α or β locus is enough to confer a new specificity on the α-β combination. Four Aα and five Aβ alleles, for example, would provide 20 A incompatibility factor specificities from only nine mutational events. Thus, the two-locus structure confers a marked advantage on the outbreeding potential, because the number of factor specificities is the multiple of the number of alleles at each locus rather than their sum. The exact value of the outbreeding potential is calculated from the formula $(nA\alpha nA\beta nB\alpha nB\beta - nA\alpha nA\beta - nB\alpha nB\beta + 1)/nA\alpha nA\beta nB\alpha nB\beta$, where $nA\alpha$, $nA\beta$, $nB\alpha$, $nB\beta$ are the number of alleles at the four loci (Stamberg and Koltin, 1973a). Twenty alleles, five in each of the four loci, provide an outbreeding potential of 95%, which is attained by 20 alleles at the single locus of the unifactorials (Table 1).

Although the outbreeding potential is maximal when a given number of alleles is distributed equally among the four loci (Stamberg and Koltin, 1973a), there is a marked inequality in the actual number of alleles in the four loci of *Schizophyllum commune* (see Sec. III.B) and of other basidiomycete species as well. This may reflect the evolution of the complex incompatibility factors: the locus with the most alleles would be the ancestral locus, whereas the remaining three loci would have come into operation within a short time of each other, at a later stage (Stamberg and Koltin, 1973a).

The problem with the two-locus structure is that it makes possible a rise in the inbreeding potential because of recombination between the loci of a factor. Progeny carrying recombinant A or B factors are compatible with more than 25% of their siblings, and the inbreeding is therefore a function of the level of recombination between the α and β loci of each factor, according to the formula

$$\tfrac{1}{4}(1 + 2P_A - 2P_A^2)(1 + 2P_B - 2P_B^2)$$

where P_A and P_B are the recombination frequencies of the A and B factors (Simchen, 1967). If the loci of each factor were unlinked, the inbreeding potential would be 56%, which is higher than the 50% inbreeding level of the unifactorials and which would therefore negate the advantage of a bifactorial system. However, in all fungi studied to date the component loci of each factor are closely linked, thus lowering the actual inbreeding considerably.

B. Sporulation; Heterokaryosis

In addition to the inbreeding-outbreeding balance there are other ways to control the amount of heterogeneity in a population. One way to increase heterogeneity is to limit the production of asexual spores so that the major means of dispersal is through spores that result from a mating, followed by nuclear fusion and meiosis. Among the higher Basidiomycetes, sexual spores are the predominant means of propagation (Raper, 1966b); in the Ascomycetes asexual reproduction is very highly developed and seems to be the main means of propagation even for those species which can reproduce sexually (Bessey, 1950; Ingold, 1973).

Unrestricted heterokaryosis provides an additional source of heterogeneity in a population. In the Basidiomycetes, heterokaryon formation is a prerequisite to the sexual process which culminates in meiosis and formation of sexual spores. In the Ascomycetes the sexual process is not dependent on prior formation of a heterokaryon, but unrestricted heterokaryosis between any two neighboring individuals would allow an additional minor source of genetic variation via the parasexual cycle. That is, in a mycelium where free entry and establishment of foreign nuclei were possible, new nuclear types could be found due to rare occurrences of nuclear fusion followed by chromosomal segregation, loss, and occasionally even intrachromosomal recombination. These diverse nuclei can form mycelial sectors and asexual spores which are genetically unlike either of the original mycelial types. The parasexual cycle is a well-documented, if rare, occurrence in Ascomycetes and Basidiomycetes (Pontecorvo, 1959; Raper, 1966b; Mills and Ellingboe, 1971; Shalev et al., 1972; Burnett, 1975).

The heterokaryon can also serve as a storage mechanism for heterogeneity (Raper and Flexer, 1970). For most organisms with a dominant haplophase, the heterogeneity generated through sexual reproduction is exposed to immediate selective pressure. For a fungal species with no restrictions on heterokaryosis, however, this is not the case because heterokaryons will readily form between any adjoining mycelia. The phenotype of heterokaryons generally reflects the dominant alleles present in the nuclei. Thus genetic variation can accumulate as recessive alleles without being immediately exposed to selection.

In a number of fungi, however, it has been demonstrated that heterokaryosis is far from being unrestricted. Complex systems known as *vegetative* (or heterogenic) *incompatibility systems* regulate heterokaryosis; they work on a principle opposite to that of the *mating* (or homogenic) *incompatibility systems* (Davis, 1966; Esser, 1966). In the latter, sexual stages will not occur unless the two participating individuals are heterozygous at certain loci. In the former, a heterokaryon will not be formed unless the two individuals are homozygous at certain loci. In other words, vegetative incompatibility works to prevent heterokaryosis of genetically unlike individuals. The parasexual cycle, leading to reassortment of genetic information, thus can accomplish very little in the way of creating new genotypes since the participating nuclei, a priori, must be very similar (Caten and Jinks, 1966). For Basidiomycetes, restriction of heterokaryosis has the direct effect of limiting the sexual cycle, as well as the parasexual cycle, to individuals with at least a certain degree of homozygosity (Burnett, 1975). For Ascomycetes, too, genes restricting heterokaryosis may affect fertility or viability of sexually produced progeny (see Sec. III). Thus, a vegetative incompatibility system superimposed on the mating system acts antagonistically to it so as to restrict gene flow and heterogeneity. Caten (1972) suggested that the main function of vegetative incompatibility systems is to prevent the spread of viruses and other deleterious cytoplasmic agents from mycelium to mycelium in natural populations—the restriction of genetic variability being a side effect only, the consequence of a more immediate adaptive need.

III. FUNGAL STRATEGIES

With the aforementioned mechanisms in mind, we can now consider some of the overall strategies that fungal populations are exploiting.

A. Inbreeder: *Aspergillus nidulans*

Certain fungi are clearly inbreeders. An example is *Aspergillus nidulans*. This species produces very large quantities of asexual spores or conidia, and this is probably its major dispersal mechanism in nature. Sexual reproduction occurs, but the mating system is homothallic, i.e., a single individual is self-fertile. Thus, although crosses involving two individuals can occur as well, there is no restriction on inbreeding of the most intense kind. A stringent vegetative incompatibility system prevents formation of heterokaryons except when the participating mycelia are genetically identical for at least five specific loci (Jinks and Grindle, 1963); homozygosity for at least eight loci is required for heterokaryosis to occur more than occasionally (Jinks et al., 1966). Although two vegetatively incompatible individuals can undergo sexual crossing, their progeny are less vigorous than the progeny of a vegetatively compatible cross (Jinks et al., 1966). Since most wild isolates are heterokaryon incompatible (Grindle, 1963a,b), there is little opportunity in nature for the free flow and maintenance of heterogeneity between or even within populations. Even more extreme examples of the tendency toward increasing homogeneity are the large number of species of imperfect fungi, which, as far as is known, have dispensed with sex altogether and rely on asexual means of propagation and dispersal.

B. Outbreeder: *Schizophyllum commune*

An example at the other end of the spectrum is the basidiomycete *Schizophyllum commune*, which is clearly an outbreeder (Fig. 1). It produces no asexual spores and

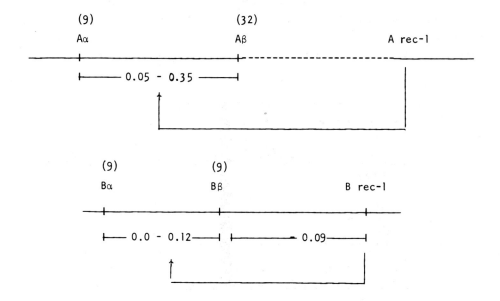

Outbreeding = f(nAα, nAβ, nBα, nBβ) = 0.98

Inbreeding = f(rec A, rec B) = 0.27 - 0.44

Fig. 1 Structure of the incompatibility factors in *Schizophyllum commune*. Recombination frequencies are indicated for the regions between the component loci of the A and B factors. A rec-1 and B rec-1 are fine-control genes controlling recombination frequencies for factors A and B, respectively. Above α and β are indicated the number (n) of alternate alleles for each locus.

thus has to rely on the products of sexual crosses for its dispersal. Its mating system is heterothallic and bifactorial, with both the A and B factors consisting of two linked loci. Each of these loci has a large number of alternate allelic forms which are randomly distributed in populations throughout the world: the estimated number of alleles is 9 Aα, 32 Aβ, and 9 each for Bα and Bβ (Raper et al., 1960; Parag and Koltin, 1971; Stamberg and Koltin, 1973a). Thus 288 different A factors and 81 B factors can theoretically exist, and the outbreeding potential would be 98.4%. The actual value is somewhat smaller, around 98.2%, as certain Bα-Bβ combinations do not occur in nature and cannot be synthesized in the laboratory (Stamberg and Koltin, 1971). It is thought that this is due to the association of these alleles with deletions (Stamberg and Koltin, 1973b).

Frequencies of recombination between the component loci of each factor are specifically regulated by genes that are part of a "fine control" system for recombination throughout the genome, as illustrated in Fig. 2 (Simchen and Stamberg, 1969a,b). Recombination frequencies within the A and B factors can vary independently within a wide range, depending on the genetic background and on environmental conditions such as temperature (Simchen, 1967; Simchen and Connolly, 1968; Stamberg, 1968; Stamberg and Simchen, 1970). One fine-control gene for A-factor recombination, A rec-1, is thought to be loosely linked to the A factor itself (Simchen, 1967). For the B factor one such gene, called B rec-1, has been located at 9 map units from Bβ, and is thought to act at a number of sites located between Bα and Bβ (Koltin and Stamberg, 1973; Stamberg and Koltin, 1973c). In addition, the presence of specific Bα or Bβ

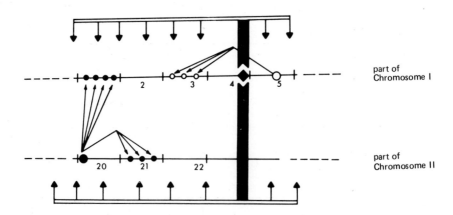

part of
Chromosome I

part of
Chromosome II

Fig. 2 Schematic representation of the systems controlling recombination in *Schizophyllum*. Numbers indicate short chromosomal regions. A gene in region 4 is part of the coarse control, a mutation in which would prevent recombination throughout the genome (e.g., the product of gene 4 could be DNA polymerase or a component of the synaptinemal complex). A gene in region 5 is part of the fine control and regulates the recombination frequency in segment 3, which has three recognition sites specific for the gene product of 5. Similarly, a fine-control gene in region 20 regulates recombination frequencies in regions 1 and 21. (Modified from Simchen and Stamberg, 1969a).

alleles can markedly inhibit or suppress B-factor recombination (Koltin and Raper, 1967; Koltin, 1969; Stamberg and Koltin, 1971).

The higher the recombination frequencies, the higher the inbreeding potential; so, as a result of recombination control and variation superimposed on the two-locus structure of the incompatibility factors, the inbreeding potential can vary from a low of 27% (if A- and B-factor recombination are at their minimal values of 5% and 0%, respectively) to a high of 44% (if the recombination frequencies are at the maximal values observed in the laboratory, 35% and 12%).But in nature it is unlikely that the inbreeding ever reaches this maximal value. The average recombination frequencies for the A and B factors, in a random sample of crosses made with strains from diverse worldwide locations, are 9.5% and 2.3%, respectively; recombination at these frequencies would allow 30.6% inbreeding (Koltin et al., 1972). A similar estimate is obtained from the data of Schaap and Simchen (1971), who studied recombination in a natural population of *S. commune*; the frequencies of recombination observed fixed the inbreeding potential at 25% to 29%.

In addition, the fine-control alleles for low recombination in both incompatibility factors are dominant to the alleles for high recombination and are found more frequently in nature than are the "high" alleles (Simchen, 1967; Stamberg, 1969). All of these data indicate that low recombination, hence a low inbreeding potential, close to the 25% minimum, is of selective advantage to the species. However, under special circumstances such as colonization of a new area by a small number of individuals, it would be advantageous to increase the inbreeding level, and this possibility exists due to the flexibility of the incompatibility structure and control (Papazian, 1951; Simchen, 1967; Stamberg and Koltin, 1973a).

C. No Clear Direction: *Neurospora crassa*, *Podospora anserina*, Yeasts

Many fungi show conflicting directions in their strategies for survival. One example is *Neurospora crassa*, which has a heterothallic mating system (one factor with two alleles) preventing self-fertility but allowing both inbreeding and outbreeding to occur with 50% efficiency. Notoriously large numbers of asexual spores are produced, however, and these are likely to be the major means of dispersal of the species. *Neurospora* is frequently found in areas of recently burned vegetation (Perkins and Barry, 1977), and this is due to the remarkable resistance of the sexual spores (ascospores) to extreme environmental conditions. Heat shock, while killing any vegetative growth, serves as a signal to initiate ascospore germination. Thus the sexual cycle is important in maintenance of a *Neurospora* population in situ (Spieth, 1975). A complex vegetative incompatibility system consisting of 10 or more genes (one of which is the mating factor) restricts stable heterokaryon formation to those instances in which the individuals have common alleles for all of the loci (Gross, 1952; Garnjobst and Wilson, 1956; Newmeyer et al., 1973; Mylyk, 1975; Perkins and Barry, 1977). In nature, it is rare for two neighboring mycelia to be compatible

and able to form a heterokaryon (Mylyk, 1976). The dual function of the mating factor in the life cycle underscores the basic ambiguity that *Neurospora* exhibits toward
the desirability of heterogeneity: two individuals are sexually compatible only if
one is of the A mating type and the other is a, but such individuals are vegetatively
incompatible and will not form a stable heterokaryon. Although *Neurospora* has retained its heterothallic mating system, the maintenance of heterogeneity may no longer be an important aim of its life-style, so to speak.

    Podospora anserina is another example of a species with ambivalent "attitudes"
toward gene flow. Its mating system is secondarily homothallic, i.e., although there
is one incompatibility gene with two alleles, the haploid + or - type nuclei are
usually oriented in the ascus after meiosis in such a way that two nuclei, one of
each mating type, enter each ascospore. The mycelium resulting from such a spore is
therefore heterokaryotic and self-fertile. The inbreeding potential is close to but
not quite 100% because of "leakage" in the homothallic mechanism, so that occasional
uninucleate (and therefore heterothallic + or - type) spores occur as errors during
ascospore formation (Esser, 1974). A complex vegetative incompatibility system affects both heterokaryon formation and sexual behavior between the uninucleate spores.
Esser (1971; see also Esser and Blaich, 1973) postulated the participation of four
unlinked genes with both allelic and nonallelic interactions to account for the observed variation in behavior between pairs of wild isolates. Normal, efficient heterokaryon formation, and perithecium production as well, require the two mycelia to
be homozygous at all four loci. The interactions between pairs of natural isolates
indicate that, because of inequalities in distribution among the alleles involved,
only 30% of random matings are fully fertile and only 17% are heterokaryon compatible.

    Heterothallic yeasts such as several of the *Saccharomyces* species and *Schizo-
saccharomyces pombe* have the classic one-factor, two-allele system which permits 50%
inbreeding and 50% outbreeding. Superimposed on this, however, is a constant low
background of mutations from one mating type to the other, making pure clones homothallic breeders. In *Saccharomyces* the mutations are under the control of specific
mutator or regulatory genes (Herman and Roman, 1966; Naumov and Tolstorukov, 1973;
Harashima et al., 1974; Hopper and Hall, 1975). It has been postulated that every
cell has the information for both mating-type alleles and that the function of the
dominant "HO" mutator gene is to activate one or the other of the alleles by inser-
tion of some essential site for its expression (Hicks and Herskowitz, 1976, 1977;
Hicks et al., 1977). In *Schizosaccharomyces* the incompatibility factor is composed
of two closely linked mutating loci which may constitute a "metastable" complex in
which duplications, excision, and transpositions frequently occur (Gutz and Doe,
1973, 1975; Egel, 1976a,b). In the case of the yeasts we seem to be observing the
evolution of secondary homothallism or a similar mechanism for flexible regulation
of the breeding system.

IV. CONCLUSION

It is difficult to draw conclusions about evolutionary trends among the fungi, partly because we have too few data on too few species. Certainly, however, in the Ascomycetes there is a trend toward decreasing heterogeneity within populations. *Neurospora*, *Podospora*, and the yeasts all exemplify basically heterothallic species with built-in restrictions on inbreeding; over many years and millions of generations evolution has superimposed, on this basic structure, shortcuts and ways around the restrictions so that today these species are, practically speaking, complete inbreeders or on the way to being so. The imperfect fungi represent the final stage in this evolution, having dispensed with sex altogether.

In the Basidiomycetes the emphasis is still, on the whole, on outbreeding and significant amounts of heterogeneity. According to evolutionary "dogma" this suggests that Ascomycetes, having speciated, diversified, and spread into every available niche, are now lazily resting on their conidia and gradually "ossifying" themselves out of prominence, while the Basidiomycetes may tend to become more prevalent. At the present time, however, one can only remark that the Ascomycetes, unaware of the dogma, are doing very well indeed.

REFERENCES

Bausum, H. T., and Wagner, R. P. (1965). "Selfing" and other forms of aberrant recombination in isoleucine-valine mutants of *Neurospora*. *Genetics 51*: 815-830.

Bessey, E. A. (1950). *Morphology and Taxonomy of Fungi*. Blakiston, Philadelphia.

Burnett, J. H. (1965). The natural history of recombination systems. In *Incompatibility in Fungi*. K. Esser and J. R. Raper (Eds.). Springer-Verlag, New York, pp. 98-113.

Burnett, J. H. (1975). *Mycogenetics*. Wiley, New York.

Caten, C. E. (1972). Vegetative incompatibility and cytoplasmic infection in fungi. *J. Gen. Microbiol. 72*: 221-229.

Caten, C. E., and Jinks, J. L. (1966). Heterokaryosis: its significance in wild homothallic ascomycetes and fungi imperfecti. *Trans. Brit. Mycol. Soc. 49*: 81-93.

Davis, R. H. (1966). Mechanisms of inheritance. 2. Heterokaryosis. In *The Fungi*, G. C. Ainsworth and A. S. Sussman (Eds.), Vol. II. Academic Press, New York, pp. 567-588.

Day, P. R. (1960). The structure of the *A* mating-type locus in *Coprinus lagopus*. *Genetics 45*: 641-651.

Egel, R. (1976a). Rearrangements at the mating type locus in fission yeast. *Mol. Gen. Genet. 148*: 149-158.

Egel, R. (1976b). The genetic instabilities of the mating type locus in fission yeast. *Mol. Gen. Genet. 145*: 281-286.

Esser, K. (1966). Incompatibility. In *The Fungi*, G. C. Ainsworth and A. S. Sussman (Eds.), Vol. II. Academic Press, New York, pp. 661-676.

Esser, K. (1971). Breeding systems in fungi and their significance for genetic recombination. *Mol. Gen. Genet. 110*: 86-100.

Esser, K. (1974). *Podospora anserina*. In *Handbook of Genetics*, R. C. King (Ed.), Vol. I. Plenum Press, New York, pp. 531-551.

Esser, K., and Blaich, R.   (1973).   Heterogenic incompatibility in plants and animals. *Adv. Genet. 17*:   107-152.

Felsenstein, J.   (1974).   The evolutionary advantage of recombination.   *Genetics 78*: 737-756.

Fisher, R. A.   (1930).   *The Genetical Theory of Natural Selection*.   Dover, New York. (1958 reprint ed.)

Flexer, A. S.   (1965).   Bipolar incompatibility in *Polyporus palustris*. *Amer. J. Bot. 52*: 634.

Friis, J., Flury, F., and Leupold, U.   (1971).   Characterization of spontaneous mutations of mitotic and meiotic origin in the *ade-1* locus of *Schizosaccharomyces pombe*. *Mutation Res. 11*: 373-390.

Garnjobst, L., and Wilson, J. F.   (1956).   Heterocaryosis and protoplasmic incompatibility in *Neurospora crassa*. *Proc. Nat. Acad. Sci. U.S. 42*: 613-618.

Grindle, M.   (1963a).   Heterokaryon compatibility of unrelated strains in the *Aspergillus nidulans* group.   *Heredity 18*: 191-204.

Grindle, M.   (1963b).   Heterokaryon compatibility of closely related wild isolates of *Aspergillus nidulans*. *Heredity 18*: 397-405.

Gross, S. R.   (1952).   Heterokaryosis between opposite mating types of *Neurospora crassa*. *Amer. J. Bot. 39*: 574-577.

Gutz, H., and Doe, F. J.   (1973).   Two different h- mating types in *Schizosaccharomyces pombe*. *Genetics 74*: 563-569.

Gutz, H., and Doe, F. J.   (1975).   On homo- and heterothallism in *Schizosaccharomyces pombe*. *Mycologia 67*: 748-759.

Harashima, S., Nogi, Y., and Oshima, Y.   (1974).   The genetic system controlling homothallism in *Saccharomyces* yeasts. *Genetics 77*: 639-650.

Herman, A., and Roman, H.   (1966).   Allele specific determinants of homothallism in *Saccharomyces lactis*. *Genetics 53*: 727-740.

Hicks, J. B., and Herskowitz, I.   (1976).   Interconversion of mating types in yeast. I.   Direct observations of the action of the homothallism (HO) gene.   *Genetics 83*: 245-258.

Hicks, J. B., and Herskowitz, I.   (1977).   Interconversion of yeast mating types. II.   Restoration of mating ability to sterile mutants in homothallic and heterothallic strains. *Genetics 85*: 373-393.

Hicks, J. B., Strathern, J. M., and Herskowitz, I.   (1977).   Interconversion of yeast mating types.   III.   Action of the homothallism (HO) gene in cells homozygous for the mating type locus. *Genetics 85*: 395-405.

Hopper, A. K., and Hall, B. D.   (1975).   Mutation of a heterothallic strain to homothallism. *Genetics 80*: 77-85.

Ingold, C. T.   (1973).   *The Biology of Fungi*.   Hutchinson Educ. Press, London.

Jinks, J. L., Caten, C. E., Simchen, G., and Croft, J. C.   (1966).   Heterokaryon incompatibility and variation in wild populations of *Aspergillus nidulans*. *Heredity 21*: 227-239.

Jinks, J. L., and Grindle, M.   (1963).   The genetical basis of heterokaryon incompatibility in *Aspergillus nidulans*. *Heredity 18*: 407-413.

Kiritani, K.   (1962).   Linkage relationships among a group of isoleucine and valine requiring mutants of *Neurospora crassa*. *Japan. J. Genet. 37*: 42-56.

Koltin, Y.   (1969).   The structure of the incompatibility factors of *Schizophyllum commune*: Class II factors. *Mol. Gen. Genet. 103*: 380-384.

Koltin, Y., and Raper, J. R.   (1967).   The genetic structure of the incompatibility factors of *Schizophyllum commune*: the resolution of class III *B* factors. *Mol. Gen. Genet. 100*: 275-282.

Koltin, Y., and Stamberg, J. (1973). Genetic control of recombination in *Schizophyllum commune*: location of a gene controlling *B*-factor recombination. *Genetics* 74: 55-62.

Koltin, Y., Raper, J. R., and Simchen, G. (1967). The genetic structure of the incompatibility factors of *Schizophyllum commune*: the *B* factor. *Proc. Nat. Acad. Sci. U.S.* 57: 55-62.

Koltin, Y., Stamberg, J., and Lemke, P. A. (1972). Genetic structure and evolution of the incompatibility factors in higher fungi. *Bacteriol. Rev. 36*: 156-171.

Koltin, Y., Stamberg, J., and Ronen, R. (1975). Meiosis as a source of spontaneous mutations in *Schizophyllum commune*. *Mutation Res. 27*: 319-325.

Magni, G. E. (1964). Origin and nature of spontaneous mutations in meiotic organisms. *J. Cell Comp. Physiol. 64*(Suppl. 1): 165-172.

Magni, G. E., and von Borstel, R. C. (1962). Different rates of spontaneous mutation during mitosis and meiosis in yeast. *Genetics 47*: 1097-1108.

Mather, K. (1942). Heterothally as an outbreeding mechanism in fungi. *Nature 149*: 54-56.

Mills, D. I., and Ellingboe, A. H. (1971). Somatic recombination in the common-AB diploid of *Schizophyllum commune*. *Mol. Gen. Genet. 110*: 67-76.

Muller, H. J. (1932). Some genetic aspects of sex. *Amer. Nat. 66*: 118-138.

Mylyk, O. M. (1975). Heterokaryon incompatibility genes in *Neurospora crassa* detected using duplication-producing chromosome rearrangements. *Genetics 80*: 107-124.

Mylyk, O. M. (1976). Heteromorphism for heterokaryon incompatibility genes in natural populations of *Neurospora crassa*. *Genetics 83*: 275-284.

Naumov, G. I., and Tolstorukov, I. I. (1973). Comparative genetics of yeast. X. Reidentification of mutators of mating types in *Saccharomyces*. *Genetika 9*: 82-91.

Neuhauser, K. S., and Gilbertson, R. L. (1971). Some aspects of bipolar heterothallism in *Fomes cajanderi*. *Mycologia 63*: 722-735.

Newmeyer, D. Howe, H. B., Jr., and Galeazzi, D. R. (1973). A search for complexity at the mating-type locus of *Neurospora crassa*. *Can. J. Genet. Cytol. 15*: 577-585.

Papazian, H. P. (1951). The incompatibility factors and a related gene in *Schizophyllum commune*. *Genetics 36*: 441-459.

Parag, Y., and Koltin, Y. (1971). The structure of the incompatibility factors of *Schizophyllum commune*: constitution of the three classes of *B* factors. *Mol. Gen. Genet. 112*: 43-48.

Paszewski, A., and Surzycki, S. (1964). "Selfers" and high mutation rate during meiosis in *Ascobolus immersus*. *Nature 204*: 809.

Perkins, D. D., and Barry, E. G. (1977). The cytogenetics of *Neurospora*. *Adv. Genet. 19*: 133-285.

Pontecorvo, G. C. (1959). *Trends in Genetic Analysis*. Columbia Univ. Press, New York.

Raper, J. R. (1966a). Life cycles, basic patterns of sexuality, and sexual mechanisms. In *The Fungi*, G. C. Ainsworth and A. S. Sussman (Eds.), Vol. II. Academic Press, New York, pp. 283-337.

Raper, J. R. (1966b). *Genetics of Sexuality in Higher Fungi*. Ronald Press, New York.

Raper, J. R., and Flexer, A. S. (1970). The road to diploidy with emphasis on a detour. *Symp. Soc. Gen. Microbiol. 20*: 401-432.

Raper, J. R., Baxter, M. G., and Middleton, R. B. (1958). The genetic structure of the incompatibility factors in *Schizophyllum commune*. *Proc. Nat. Acad. Sci. U.S. 44*: 889-900.

Raper, J. R., Baxter, M. G., and Ellingboe, A. H. (1960). The genetic structure of the incompatibility factors of *Schizophyllum commune*: the *A*-factor. *Proc. Nat. Acad. Sci. U.S. 46*: 833-842.

Schaap, T., and Simchen, G. (1971). Genetic control of recombination affecting mating factors in a population of *Schizophyllum commune*, and its relation to inbreeding. *Genetics 68*: 67-75.

Shalev, M., Stamberg, J., and Simchen, G. (1972). Sectoring and recombination in illegitimate di-mon matings of *Schizophyllum commune*. *Heredity 29*: 191-201.

Simchen, G. (1967). Genetic control of recombination and the incompatibility system in *Schizophyllum commune*. *Genet. Res. 9*: 195-210.

Simchen, G., and Connolly, V. (1968). Changes in recombination frequency following inbreeding in *Schizophyllum*. *Genetics 58*: 319-326.

Simchen, G., and Stamberg, J. (1969a). Fine and coarse controls of genetic recombination. *Nature 222*: 329-332.

Simchen, G., and Stamberg, J. (1969b). Genetic control of recombination in *Schizophyllum commune*: specific and independent regulation of adjacent and non-adjacent chromosomal regions. *Heredity 24*: 369-381.

Spieth, P. T. (1975). Population genetics of allozyme variation in *Neurospora intermedia*. *Genetics 80*: 785-805.

Stamberg, J. (1968). Two independent gene systems controlling recombination in *Schizophyllum commune*. *Mol. Gen. Genet. 102*: 221-228.

Stamberg, J. (1969). Genetic control of recombination in *Schizophyllum commune*: the occurrence and significance of natural variation. *Heredity 24*: 361-368.

Stamberg, J., and Koltin, Y. (1971). Selectively recombining *B* incompatibility factors of *Schizophyllum commune*. *Mol. Gen. Genet. 113*: 157-165.

Stamberg, J., and Koltin, Y. (1973a). The organisation of the incompatibility factors in higher fungi: the effect of structure and symmetry on breeding. *Heredity 30*: 15-26.

Stamberg, J., and Koltin, Y. (1973b). The origin of specific incompatibility alleles: a deletion hypothesis. *Amer. Nat. 107*: 35-45.

Stamberg, J., and Koltin, Y. (1973c). Genetic control of recombination in *Schizophyllum commune*: evidence for a new type of regulatory site. *Genet. Res. 22*: 101-111.

Stamberg, J., and Simchen, G. (1970). Specific effects of temperature on recombination in *Schizophyllum commune*. *Heredity 25*: 41-52.

Takemaru, T. (1961). Genetic studies on fungi. X. The mating system in Hymenomycetes and its genetical mechanism. *Biol. J. Okayama Univ. 7*: 133-211.

Terakawa, H. (1960). The incompatibility factors in *Pleurotus ostreatus*. *Sci. Papers Coll. Gen. Educ. Univ. Tokyo 10*: 65-71.

Ullrich, R. C., and Raper, J. R. (1974). Number and distribution of bipolar incompatability factors in *Sistotremma brinkmanni*. *Amer. Natur. 108*: 507-518.

Whitehouse, H. L. K. (1949a). Multiple-allelomorph heterothallism in the fungi. *New Phytol. 48*: 212-244.

Whitehouse, H. L. K. (1949b). Heterothallism and sex in fungi. *Biol. Rev. 24*: 411-447.

Yoshikawa, H. (1966). Mutations resulting from the transformation of *Bacillus subtilis*. *Genetics 54*: 1201-1214.

Part III

THE ORGANIZATION OF FUNGAL POPULATIONS AND COMMUNITIES

Chapter 10

ECOLOGICAL MEASUREMENTS AND MICROBIAL POPULATIONS[*]

Dieter Mueller-Dombois
University of Hawaii at Manoa
Honolulu, Hawaii

I. INTRODUCTION

For purposes of the theme of Part III, I interpret the title of my chapter more specifically as referring to ecological measurements for detecting *community organization* among microbial populations. With regard to this more closely defined topic,
two items need some explanation at the start: one relates to the idea of community
organization; the other to the ecological characteristics of microbial populations.
Only when these have received some clarification will it be possible to talk about
ecological measurements.

The idea of community organization brings to mind the more tangible concept of
integration in the community. I will attempt to explain this concept in some detail
next, after first giving attention to a few distinctive ecological properties inher-
ently associated with microbial populations, which I here relate in particular to
soil microfungi.

II. DISTINCTIVE ECOLOGICAL PROPERTIES OF SOIL MICROFUNGI

The underlying objective of this text is to review studies of fungi in light of general ecological principles. Thus far, most of our understanding of general ecological principles has been derived from studies of higher plants and animals. It was
Tansley (1935), a plant ecologist, who coined the now widely accepted term *ecosystem*
for the functional unity he saw in the biological community and its locality-specific
environment. It was Elton (1927), an animal ecologist, who realized that the "ground
plan" of every animal community is similar in spatially separated and otherwise different biomes. He thus pointed the way to the concept of *general niche*, which has
spawned many productive ecological studies.

As an organism group, soil microfungi are functionally as important in terrestrial ecosystems as are the higher plants and animals. However, as species assemblages or communities, they are hidden from view and as such they are much more dif-

---

[*]Invited conceptual paper, Second International Mycological Congress, Tampa,
Florida, August 27 to September 3, 1977.

ficult to analyze. Moreover, there are some distinctive ecological properties that
set this group apart from other organism groups in one way or another.

I will consider three basic properties of soil microfungi: (1) their wide dispersability; (2) their heterotrophy; and (3) their superior survival capacities under
adverse conditions. In my opinion, these properties will both impose limits and offer unique possibilities in ecological studies of these organisms.

Wide dispersability implies that soil microfungi are of very generalized, perhaps global, distribution. There may be restrictions in some species, but the dissem-
inules (spores) of most soil microfungi appear to have a chance to reach any natural
surface area on our planet. This is not so for higher plants and animals, which tend
to be rather provincial in their distribution. We speak of floristic provinces. For
example, the Karroo Desert in South Africa is occupied by succulents of several different plant families, while the American semideserts are famous for their cacti.
The Australian semideserts, in turn, lack typical succulents, because these deserts
do not have a floristic range of taxa with sufficient genetic and morphological flexibility for succulency. On the other hand, I would not be surprised should the soil
microflora of these three semideserts show a high degree of taxonomic similarity.

Heterotrophy, another basic ecological property of soil fungi, implies dependency. This property is shared by animals and to some extent also by higher plants.
The latter cannot grow unless atmospheric nitrogen is first converted by microorganisms into soil nitrogen. However, the degree of dependency of the soil microfungi
is greater by several magnitudes. They cannot survive unless there is some organic
matter in the soil. Soil organic matter provides the basic source of energy for soil
microfungi as does visible radiation for the higher plants. The heterotrophy of animals is usually much more specific than that of microfungi. Hence, the heterotrophy
of animals has aroused particular ecological interest, as in the study of predator-
prey relationships. The generalized heterotrophy of soil microfungi makes them more
similar to higher plants in terms of resource partitioning. Therefore, niche overlap
or competition for the same resource segments may be as common in soil microfungi as
it is in higher plants. Associated with heterotrophy, the soil microfungi exhibit a
characteristic "spatial confinement" within a terrestrial ecosystem.

A particularly important substrate realm is the surface soil (i.e., the overlying humus where present and the A_1 horizon in particular) and the rhizosphere of
the higher plants. Within the rhizosphere two major life-form types may be distinguished: parasitic and saprophytic. Alternation between these two ways of life can
exist in some microfungal species. Therefore, in fungal community ecology one may
encounter generalistic resource relationships (of saprophytes) as in higher plant
communities, as well as highly specific interactions (of parasites) as in predator-
prey relationships of some animal populations.

A third basic ecological property of the soil microfungi appears to be their
superior survival capacity under adverse conditions when compared to higher plants

and animals. Certain life forms among vascular plants also have developed interesting withdrawal strategies when conditions become unfavorable. For example, geophytes
can eliminate their entire shoot system and retain only subterranean organs (roots,
tubers, rhizomes, etc.) as a means of surviving an unfavorable season, and annuals
or therophytes have gone a step further in reduction by surviving only in the form
of seeds. Similar survival strategies can be cited for animals, and certain arthropods certainly rival the microfungi in metamorphic mechanisms for survival. But the
superior strategies that enable fungi to "disconnect from time" during periods of adverse conditions serve to make them particularly durable and stable members of the
ecosystem.

In view of these relationships, one can expect ecological studies of soil microfungi to yield many parallels to behavior of higher plant and animal populations and
communities, but one can also expect them to yield new ecological dimensions and theoretical insights because of their special characteristics. Having said something
on the ecological uniqueness of microfungi as well as their similarities with other
groups of organisms, I am now ready to consider the study of community organization.

III. COMMUNITY ORGANIZATION AND ITS MEASUREMENT

The term *organization* generally refers to a discernible level of order in contrast
to a haphazard arrangement; but opinions on the organization and nature of plant communities have varied from viewing them as tight organizations of components and functions comparable to organisms (Clements, 1916) to haphazard assemblages of plants
that happen to grow together primarily because of chance (Gleason, 1926). These extreme viewpoints are still held by some ecologists, but most seem to agree on a middle
ground, i.e., that some organization is detectable in many plant communities, although
chance factors also play a role in their total expression (structure and function).
Hence, when we speak of *community organization* we are usually implying some degree
of integration among populations of a community.

Integration among biological populations and thus community organization can be
detected in several ways. For example, we may detect species in a community that
compete directly for the same resource segment, or we may find species that favor the
presence of another species or simply favor the environmental modification provided
by the other species. Again, some species can be found that occur only because of
the presence of a certain species in the community. Many more forms of interdepen-
dencies exist in highly integrated communities. Their number and form, their complexities and state of equilibria may all give us an indication of the degree of integration or level of organization of a given community. This level may vary from
near absence of integration in vegetable gardens, where the natural community interaction is replaced by man's constant intervention, to very high integration as found,
for example, in complex tropical rain forests.

The fundamental question that should be asked for detecting the level of community organization seems to me: "Why do the species occur together in the assemblages as we find them in given locations?" There are several ways to approach this question. Here I would like to confine my discussion to two areas of community analysis:

1. Spatial distribution analysis of biological populations across several communities

2. Analysis of structure and niche differentiation within a community

A. Spatial Distribution Analysis

First I would like to state three principles referring to the spatial distribution of higher plants; these principles are encountered by anyone who has studied plant distribution over a large, coherent area:

1. Certain groups of species recur in similar combinations in different locations.

2. No two samples of vegetation (i.e., relevés or sample stands) are exactly alike.

3. In sampling similar combinations of species over a wide geographic range, one finds certain members of the species group replaced by new members in a more or less continuous pattern.

The first observation was emphasized by Braun-Blanquet (1928), who used the recombinations of the same species groups in different localities for establishing community types (plant associations), environmental correlations, and the plant indicator concept.

The second observation was emphasized by Gleason (1926), who said that every community or plant assemblage is unique and that its composition is largely due to chance. He questioned the validity of establishing community types.

When one examines vegetation samples or relevé records side by side, one finds that the contradictions expressed in the two approaches are more apparent than real. Among similar samples, one-third to two-thirds of the total species number can be expected to be alike. One of two indices is commonly used to express sample stand or relevé similarity. They are either Jaccard's (1928) index,[*] which relates the number of species common to two samples to all species in the two samples (i.e., IS_J = number of common species/all species), or Sørensen's (1948) index,[†] which relates the number of common species to the average number of species occurrences in the two samples (i.e., IS_S = number of common species/0.5 X all species occurrences).

---

[*] $IS_J = \dfrac{c}{a + b + c}$ X 100,

where c = No. of species common to two samples; a = No. of species unique to the first sample; and b = No. of species unique to the second sample.

[†] $IS_S = \dfrac{2c}{A + B}$ X 100,

where c = No. of species common to two samples; A = all species in the first sample; and B = all species in the second sample.

The apparent contradiction between Braun-Blanquet's and Gleason's approaches can be resolved with these mathematical indices, since they include both communality and uniqueness in species compositions and express community-sample similarities in degrees. However, within certain limits, it still requires judgment on the part of the investigator to determine what degree of similarity or dissimilarity between community samples may be accepted for the definition of community types. Ordination or other multivariate analysis techniques are useful tools in facilitating this decision.

It can be expected that the first two principles mentioned as referring to the distribution phenomena of higher plants will also hold for the soil microfungi. However, the third principle will probably not apply to the same degree to microbial populations as it does to higher plants. This principle says that locally well-defined plant species combinations change or lose their similarity when examined in similar recurring habitats over a wide geographic range such as a continent or, more obviously still, from one continent to another. This principle is related to the limited dispersability of higher plants. It was the cause of a certain frustration with the Braun-Blanquet system of community classification. The provinciality of higher plant species did not permit extrapolation of the system to a global scale as was possible, for example, with the taxonomic system of biological organisms or with the classification systems of rocks and soils. Therefore, for a worldwide classification of plant communities, the ecologist has to resort to vegetation parameters other than floristic composition, namely, the vegetation physiognomy, biomass structure, or life-form composition of the plants.

Because of the wide dispersability of soil microfungi, the aforementioned plant distribution principle may not apply to them. Instead, it appears possible to develop a worldwide classification system of soil microfungal communities. This would have important practical applications insofar as soil microfungal communities could then be used as biological indicators of similar environments or habitats on a comparative basis all over the globe. Their communities could serve as sensors of environmental change, such as that caused by industrial pollution. This idea poses a fundamental question: Are soil microfungi sufficiently sensitive to environmental change (in the spatial sense) to be good indicators? Their wide dispersability and superior survival capacities may mask the environmental gradients.

The basic data needed for any distributional analysis, including a contribution for such a worldwide analysis, are as follows: (a) species lists, based on the minimal area concept (Mueller-Dombois and Ellenberg, 1974); (b) information on the quantity of each species per sample site; and (c) samples or relevés of soil microfungi taken along a number of environmental gradients.

I am aware of some of the technical difficulties in obtaining a true sample of the microfungal community at a given soil site. But even with the bias introduced by the culture media, which always seem to be selective, the basic tests for community organization among microfungi can be made.

Two distributional phenomena would indicate, in my opinion, community organization among microfungi: (1) the existence of recurring combinations of soil microfungi in similar soil habitats; and (2) the presence of spatially associated species groups along environmental gradients. Both phenomena are testable by objective methods, and the degree to which they occur in nature can be expressed in numerical terms. The first relates to the Q-type analysis, in which community samples are tested for their degree of similarity by, for example, applying Jaccard's or Sørensen's indices as mentioned earlier. The second relates to the R-type analysis, in which the degree of spatial association is tested among species. For this, one can use the same indices (Jaccard's or Sørensen's), except that the meaning of the symbols then relates to the number of samples instead of species. Thus, by applying, for example, Jaccard's index for determining the degree of spatial association, one relates the number of samples in which the two species occur together to the total number of samples in which either one of the species occurs alone or in which both species occur together.[*] Further methodological detail is given in Mueller-Dombois and Ellenberg (1974).

We may now ask just how plant species have been found to behave in their spatial distribution along environmental gradients. Whittaker (1975) has answered this question by stating that distribution patterns are individual, i.e., nonassociated, with more or less random population ranges and modes. Others have shown that some species have similar ranges and approximately coinciding ecological optima along certain environmental gradients. For example, Ellenberg (1950) studied a large number of agricultural weeds in relation to gradients of soil pH, soil nitrogen, soil surface moisture, and soil surface temperature. He found that none of the species coincided in all respects, but for any specific gradient he was always able to classify a certain number of species into six spatial groups. These groups were as follows:

1. Species that are restricted to the high end of the gradient
2. Species whose distribution extends from the high end to more or less the midrange of the gradient
3. Species whose ecological optima are found across the midrange of the gradient
4. Species whose distribution extends from the low end to more or less the midrange of the gradient
5. Species that are restricted to the low end of the gradient
6. Species that respond indifferently to the gradient by being distributed over the entire gradient or by being absent in the middle but present at both ends of the gradient

---

[*]$$IA_J = \frac{c}{a + b + c} \times 100,$$

where IA = index of spatial association for two species; c = No. of samples in which the two species occur together; a = No. of samples in which one of the species occurs alone; and b = No. of samples in which the second species occurs alone.

Typically for plant communities, not all species can be classified into such distribution groups. There is always a remnant group of species which are too sporadic or rare to give any indication of a distribution trend.

It would be of considerable interest to find out whether soil microfungi can also be grouped into spatially associated species groups along such environmental gradients. If this were possible, the outcome could be interpreted as evidence for community organization among fungi. However, establishing correlation and integration tells us little about the underlying causes of such spatial patterns of distribution. The causes may be environmental adaptation, competition, or other factors of interaction. After the detection of spatial patterns and correlations, there is still much work to be done to establish specific causes for the observed distribution patterns. In the following subsection, we consider a further step in this direction.

B. Analysis of Community Structure and Niche Differentiation

This second area of community analysis aims at determining the ecological role of each species in a given community. If this seems to you to be a formidable task, you are right. It is! Detailed population analyses can usually be done only for the most important species of a community. Among higher plants and animals such population structure analyses involve determination of size and age structure, mortality and reproduction rates, maintenance patterns, resource use and partitioning, and many other behavioral traits that determine the status of a population in a community. On an experimental level, this may involve the comparative study of the "potential niche" (or "physiological optimum") and the "realized niche" (or "ecological optimum") of a species population. The term *niche* as used here means the space of a population in a community or ecosystem and the total functional behavior of that population in that space. The second part of the definition is also what I would define as the ecological role of a population.

There are several levels of detail with which one can address the analysis of community structure and niche differentiation. Here, I wish to emphasize only what I would consider the first level of information.

The basic data set for this analysis involves again (1) a complete species list of the organism group in question (e.g., the higher plants, birds, mammals, or soil microfungi) and (2) a quantification of all species on that list. From knowledge of the ecological behavior of each species (which may be supplemented by information in the literature), it will next be necessary to assign an ecological role to each species on that list.

For higher plants there are existing classifications relating to their ecological roles or "way of life" in the community. These are the so-called plant life-form classifications, of which the most well known is the system of Raunkiaer (1918).

Raunkiaer distinguishes five major plant life-forms:

1. Phanerophytes or woody perennials, which grow to heights of over 25 cm
2. Chamaephytes, usually low-growing woody perennials or dwarf shrubs
3. Hemicryptophytes, herbaceous plants that reduce their shoot system to one remnant, which survives close to the ground surface during the unfavorable season
4. Geophytes, which reduce their aboveground shoot system totally during the unfavorable season but survive by means of underground storage organs
5. Therophytes or annuals, which complete their whole life cycle within the favorable season and survive the unfavorable season only in the form of seeds

Other important subdivisions were recognized by Raunkiaer, but he conceived his system primarily with regard to summer and winter seasonality of the temperate zone. In later modifications of Raunkiaer's system, emphasis has also been given to the behavior of plants during the favorable season and in tropical and subtropical environments (e.g., see Walter, 1971). Another modified system (Ellenberg and Mueller-Dombois, 1967) takes account of not only seasonal plant behavior but also behavior in competition, response to herbivory, fire, and other mechanical damage.

Plants of the same life-form occurring together in the same community have been designated as *synusia* (Lippmaa, 1939). Several other terms have also been used (such as *union*, *consumer association*, or *eco-elements*) to designate the same phenomenon, namely, the structural and functional grouping of species in subcommunities below the level of the community. In animal ecology the terms *guild* and *functional group* have been used, which like the terms *synusia* and *plant life-form* lead naturally to the concept of *general niche*. The latter concept refers to the habitat segment as well as the function of a group of species occurring together in a community, which perform a closely similar ecological role. Thus, one may speak of the niche of a synusia or a functional group rather than simply of a species when referring to a general niche.

Once the various species of an organism group are identified as to their ecological roles or general niches in the community, the original quantified species list of the community (mentioned before as providing the first two necessary data sets) can be completed by adding the general niche designation to each species name. The population quantities of the species belonging to the same general niche or the same synusia or life-form group may then be summarized. The result can be presented diagrammatically in the form of a life-form spectrum or general niche spectrum of the community. This is simply a histogram, in which the life-form types or general niche types are identified side by side on the x axis and in which the histogram blocks show the population quantities found in each general niche.

As examples, I will show two such diagrams, one for the plant life-form spectrum of a tropical island rain forest (Fig. 1) and the other referring to two foliar arthropod communities occurring in the same forest (Fig. 2). (The data relate to the results of an interdisciplinary study carried out under the International Biological Program in Hawaii; these results will soon be available in the form of a synthesis

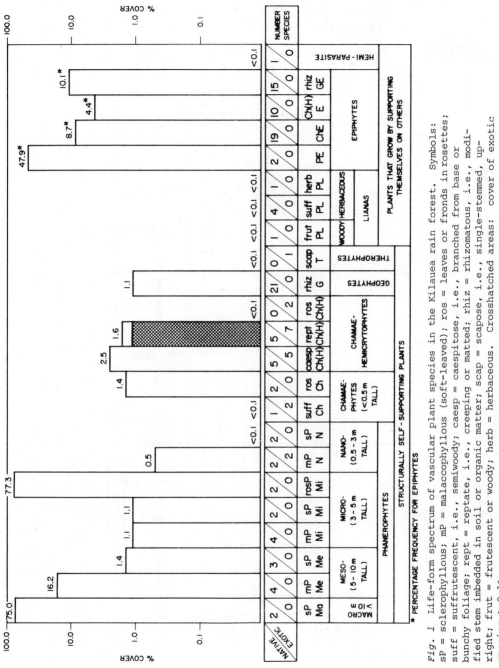

Fig. 1 Life-form spectrum of vascular plant species in the Kilauea rain forest. Symbols: sP = sclerophyllous; mP = malaccophyllous (soft-leaved); ros = leaves or fronds in rosettes; suff = suffrutescent, i.e., semiwoody; caesp = caespitose, i.e., branched from base or bunchy foliage; rept = reptate, i.e., creeping or matted; rhiz = rhizomatous, i.e., modified stem imbedded in soil or organic matter; scap = scapose, i.e., single-stemmed, upright; frut = frutescent or woody; herb = herbaceous. Crosshatched areas: cover of exotic species > 0.1%.

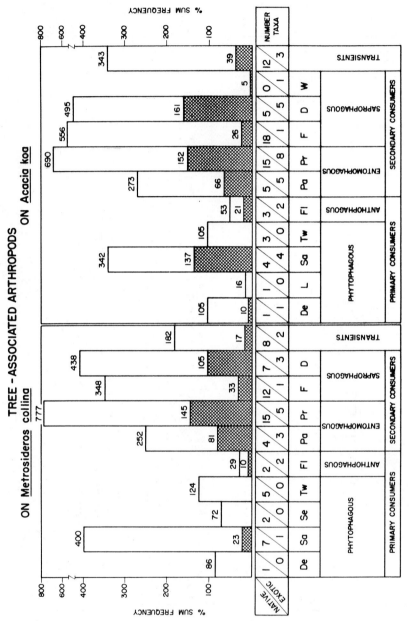

Fig. 2 Life-form spectra of two canopy arthropod communities in the Kilauea rain forest. Symbols: De = defoliators; Sa = sap suckers; Se = seed predators; Tw = twig borers; Fl = flower pollen and nectar feeders; Pa = parasites; Pr = predators; F = frugivores; D = detritivores of plant and animal matter on host; L = leaf miners; W = detritivores in dead wood. Crosshatched areas: combined frequency of exotics within respective life-form group. Transients include perching arthropods and those with undetermined ecological roles. (Data of W. C. Gagné and F. G. Howarth.)

volume published by Dowden, Hutchinson and Ross under the title *Island Ecosystems:*
Biological Organization of Selected Hawaiian Communities, US/IBP Synthesis Volume 15.)

The same kinds of diagrams can certainly be done for soil microfungi. However,
beyond the general life-form designations of saprophyte and parasite, it requires
special knowledge to differentiate the general niches of this organism group. Since
the categorization of general niches requires specialized knowledge, a considerable
subjective element is involved here. Two independent investigators may perhaps come
up with the same species list for the same community, but they will probably differ
somewhat in their general niche assignment of the species. For this reason it is
necessary to present the total species list together with the life-form spectrum
diagram. This then leaves no doubt as to how the investigator has differentiated
the general niches within his or her group of organisms.

IV. CONCLUSIONS

Such a general niche diagram of a major organism group in a community may be considered the third level of information beyond a species list and a quantification of the
species. It also provides a useful means of portraying the species diversity of a
community, because the diagram shows both the number of species considered to occupy
each general niche and the population quantities in each niche. This is certainly
a clearer statement than accomplished by any of the diversity indices, which usually
combine the two parameters merely in a single numerical expression.

A general niche spectrum established in this way is also a useful tool when we
are comparing different communities and ecosystems. It allows us to recognize quickly
important ecological differences and similarities. It can be used to compare spatially adjacent communities as well as communities on different continents. The latter
is of particular value for the comparison of organism groups with limited dispersibility, such as the higher plants and animals. In these organism groups, similar ecological functions are often performed by taxonomically totally different species in
spatially widely separated but otherwise similar ecosystems. However, this method of
data display might also be useful for comparing the niche functions of species with
wide dispersibility, such as the microfungi. It would be of interest, for example,
to know whether the soil microfungal communities of widely separated but similar ecosystems can also be portrayed in similar niche spectra (see Flanagan, Chap. 29).

It has now become apparent that the two approaches to community analysis which
I have discussed, namely, spatial distribution analysis and analysis of community
structure and niche differentiation, are quite complementary. The first works lat-
erally across different communities and habitats to derive and explain the community
and population distribution patterns. The second works by detailing the biological
content of a single community in terms of its ecological functioning.

Both approaches are based on methods of field analysis and data display that
have been developed and proven very productive in the area of vegetation ecology.
There appears to be no conceptual barrier to the usefulness of these approaches in
the area of fungal ecology.

REFERENCES

Braun-Blanquet, J. (1928). *Pflanzensoziologie*. Springer-Verlag, Berlin [2nd ed. (Vienna), 1951; 3rd ed. (Vienna and New York), 1964].

Clements, F. E. (1916). *Plant Succession: An Analysis of the Development of Vegetation*. Carnegie Inst. Washington, D.C.

Ellenberg, H. (1950). *Unkraut-Gemeinschaften als Zeiger für Klima und Boden*. Ulmer, Ludwigsburg, W. Germany.

Ellenberg, H., and Mueller-Dombois, D. (1967). A key to Raunkiaer plant life forms with revised subdivisions. *Ber. Geobot. Inst. ETH Stiftung Rübel* (Zürich) 37: 56-73.

Elton, C. (1927). *Animal Ecology*. Macmillan, New York. (2nd ed., 1935; 3rd ed., 1947).

Gleason, H. A. (1926). The individualistic concept of the plant association. *Bull. Torrey Bot. Club 53*: 7-26.

Jaccard, P. (1928). Die statistisch-floristiche Methode als Grundlage der Pflanzensoziologie. *Abderhalden, Handb. Biol. Arbeitsmeth. 11*: 165-202.

Lippmaa, T. (1939). The unistratal concept of plant communities (the unions). *Amer. Midland Natur. 21*: 111-145.

Mueller-Dombois, D., and Ellenberg, H. (1974). *Aims and Methods of Vegetation Ecology*. Wiley, New York.

Raunkiaer, C. (1918). Recherches statistiques sur les formations végétales. *Kgl. Danske Vidensk. Selsk. Biol. Medd.* (Copenhagen) *1*: 1-80.

Sørensen, T. (1948). A method of establishing groups of equal amplitude in plant sociology based on similarity of species content. *Kgl. Danske Vidensk. Selsk. Biol. Skr.* (Copenhagen) *5*(4): 1-34.

Tansley, A. G. (1935). The use and abuse of vegetational concepts and terms. *Ecology 16*: 284-307.

Walter, H. (1971). *Ecology of Tropical and Subtropical Vegetation*, transl. by D. Mueller-Dombois; J. H. Burnett, (Ed.). Oliver & Boyd, Edinburgh.

Whittaker, R. H. (1975). *Communities and Ecosystems*, 2nd ed. Macmillan, New York.

Chapter 11

USEFUL CRITERIA IN THE DESCRIPTION OF FUNGAL COMMUNITIES

Jack S. States
Northern Arizona University
Flagstaff, Arizona

I. INTRODUCTION

Terrestrial fungi, with their wide geographic distribution, considerable species diversity, and inconspicuous nature, provide an immense challenge to mycologists who search for recognizable and consistent patterns of community. The community concept as perceived here is predicated on the ability to recognize and measure differences among repeating assemblages of fungi that occur simultaneously in similar habitats. To some extent the existence of communities of fungi, in the ecological sense, is questionable. Based on the worldwide occurrence of large numbers of fungal species, particularly soil-inhabiting fungi, it has been concluded by some authors (e.g., Baker and Meeker, 1972) that essentially there are no geographic barriers to fungal distribution; that knowledge of their occurrence readily amplifies according to the location and interest of mycologists. Cooke (1975) in a discussion of the ubiquity of fungi concluded that, as our familiarity with mycofloras is broadened and techniques for isolation and identification are refined, it becomes increasingly evident that rather than being "cosmopolitan" in the broad sense of the word, fungi enjoy a wide distribution within particular areas widely separated geographically. Surely this must be true, for indications of habitat specificity have been reported in many studies where distinctive differences among groups of fungi within quite localized areas have been demonstrated (Pirozynski, 1968). Apinis (1972) presented substantial evidence for the association of fungal populations in more or less unique combinations with particular vascular plant communities and in soils of respective plant communities.

It is the task of today's mycoecologist to provide substantive evidence for fungal community patterns and ultimately to reach an understanding of community contributions to the structure and function of the ecosystem as a whole. Inadequate sampling procedures and inaccurate species identifications are in part responsible for the failure to recognize community patterns in the past. Also, the prolific nature of ubiquitous fungi tends to obscure the more discrete occurrence of fungal groups in particular habitats. Refinements in ecological methodology, some of which are discussed here, are needed to improve community description. There also exists a conceptual difficulty surrounding the definition of community. For some mycologists

the concept of community is founded on the premise that some form of interdependence
or mutual benefit among member species must be evident (Sparrow, 1968). However, it
is not beneficial for the purposes of community identification to restrict the definition of the concept to certain community attributes. Fungi coexist because of
many biotic and abiotic factors of environment favorable to their occupation of a
common habitat. These factors require careful consideration when attempting to account for the abundance and distributional amplitude of fungal taxa.

II. ECOLOGICAL CONCEPTS

The description of a vascular plant community is dependent upon an adequate sampling
procedure applied to a recognizable and preferably homogeneous vegetation segment.
By comparison, the delimitation of fungal community boundaries cannot be as easily
achieved through preliminary survey or "entitation" as is done by plant ecologists
(Mueller-Dombois and Ellenberg, 1974). The mycoflora consists of innumerable assemblages of species, separated spatially and occupying many microhabitats within the
general environment. However, fungi in an active state are rarely if ever randomly
distributed. Kershaw (Chap. 20) has demonstrated nonrandom distribution of lichens
at three different levels: (1) growth patterns; (2) position in relation to substrate; and (3) position of the plant on which the lichen is located in relation to
the entire plant community. These same relationships can be demonstrated for the
mycoflora.

Habitat boundaries for community analysis can be established indirectly through
a priori predictions of local fungal distributions based on a knowledge of habitat
factors that influence their occurrence and association. Factors considered to be
of major importance are presented in Fig. 1 and are discussed below. As will be indicated later on, an understanding of these factors within existing temporal and
spatial constraints is relevant and necessary for the selection of the methodology
to be used in community descriptions.

A. Nutrient Base

Most obvious to all who study fungi and their activities are their specific relationships with various kinds of substrates as a source of nutrition and energy. The substrate conditions of concentration, chemical and structural complexity, and availability are, in fact, conditions of habitat or substratum. Park (1968) has defined the
substratum to be the material medium on which a fungus occurs. The chemical composition of the substratum is variable and often selective for particular groups of
fungal species. Garrett (1951) has defined these favored associations of fungi as
"ecological groups." However, the resulting pattern of colonization and occupation
of the nutrient base is frequently characterized by the repeated occurrence of species
combinations, unique in structure and composition. Recognized in this way, they can
be considered to be fungal communities. It is then worthwhile to consider in some

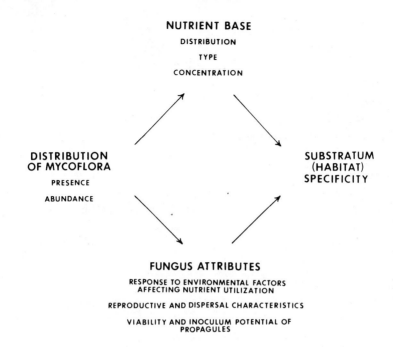

Fig. 1 Conceptual representation of the factors accounting for the distribution of fungi and their occurrence in particular habitats.

detail ecological classifications of fungi and their usefulness in the recognition of community structure.

Other than taxonomic classifications based principally on morphology, physiology, and reproduction, the most commonly recognized fungal groups are those delimited on the basis of biological function. In accordance with Raunkiaer's (1937) concept of plant "life forms" this classification regards fungi as either parasites, symbionts, or saprophytes; each group is usually associated with a single nutrient base of either plant, animal, or microbial origin. Where combinations of plant, animal, and microbial matter occur, as in dung and sewage, additional groups of fungi can be recognized. The resulting substrate-fungus associations, despite a tendency to merge one into another, are fundamentally discrete and can be observed to occur with regularity in nature (Cooke, 1979).

Further classifications have been constructed based on the position of the nutrient base in air, soil, and water. Soil fungi and aquatic fungi are broad classifications of this type, but even more specific groups have been identified. For example, *phyllosphere* fungi comprise a characteristic mycoflora located in the air-surface interface of plant leaves, whereas *phylloplane* fungi occupy the interior parts of the leaves. Similarly, fungi in the soil-root interface are classified as *rhizosphere* fungi and those located within the root tissues are termed *rhizoplane* fungi.

Ecological details of leaf- and root-inhabiting fungi are presented in greater detail
by Preece and Dickinson (1971) and Dickinson and Pugh (1974).

Nutritional groups of fungi have been described by Garrett (1951) and discussed
by Cooke (1958) and Hudson (1968). In this classification the differences in the
chemical composition of plant tissues are found to be selective for particular groups
of fungi. Three classes of nutrients--simple carbon compounds, cellulose, and lig-
nin--are present in the plant tissues and are decomposed, although not exclusively,
by three corresponding nutritional groups--sugar fungi, cellulose decomposers, and
lignin decomposers. Since the structural complexity and concentration of these nutrients varies throughout the plant body according to the type of plant (herbaceous
to woody) and the different plant organs, it is not surprising that fungal classifications have been based on these aspects of the substratum. Orlos (1966), in a complex classification of fungi on forest plants, combined fungus life forms with the
positional and compositional aspects of the nutrient base to establish five ecological groups. The groups were essentially described as: (a) *kormobionts*, parasites
on herbaceous plant tissues; (b) *xylobionts*, parasites or saprophytes on woody tissues; (c) *rhizobionts*, saphrophytes and symbionts in or on roots; (d) *pedobionts*,
saprophytes on litter and humus; and (e) *allobionts*, parasites or saprophytes on animal and microbial matter. In support of his classification scheme, Orlos has provided a substantial list of fungi characteristic of each group.

The process of substrate succession plays a major role in the interpretation of
community structure. As a consequence of colonization and consumption of the nutrient base, the capacity of the habitat to support any particular fungal community is
further diminished. The subsequent change in community structure is classically defined as heterotrophic succession. Changes in the substratum can be viewed as the
degradation of more complex substrates, like lignin, by specialist fungi to make
available the simpler organic fractions like sugar. Or conversely, when the more
soluble carbon sources are initially decomposed by generalist fungi, the more complex
organic debris remain. Unfortunately, the present concept of nutritional succession
places much emphasis on selective carbohydrate metabolism to the neglect of nitrogen
metabolism. Consideration of changes in carbon-nitrogen ratios may prove to better
illustrate substrate specificity of fungi. Regardless of how this process is viewed,
it has provided much information on important and rather predictable changes in the
fungal populations as a substratum is progressively decomposed (Frankland, 1966).
Sequential changes in fungal populations also occur on a higher level of habitat organization, namely, that of the plant community and its own successional processes.
Therefore two sources of variability may exist to influence the description of a fungal community when attempted at different points in time. Discussions of succession
in the ecology of terrestrial fungi have been presented in detail by Garrett (1963),
Park (1968), Hudson (1968), and Frankland (Chap. 22).

Growth patterns have also been categorized (Gray and Williams, 1975). They indicate spatial arrangements of the vegetative mycelium in soil and on various substrata. Detection of the pattern is restricted to localized sampling and observation. It nevertheless demonstrates differential distributions and arrangements of fungi in the environment, a knowledge of which is helpful in community descriptions.

B. Fungal Attributes

The survival and success of fungi as decomposers in nature is largely dependent upon their ability to adapt to the environment immediately surrounding the substratum and upon their ability to place, in a timely fashion, viable reproductive units in a favorable position for further activity. Although the substrate is the most important part of the fungal environment, and ever-changing factors like those of moisture, light, temperature, and oxygen and hydrogen ion concentration act singly and in concert to influence substrate colonization, decomposition, and reproduction. As perceptively stated by Austwick (1968), "the adjustment of fungi to environmental change is of great importance, for behind it operates environmental pressures of geological time, shaping the evolution of the organism and fitting it to its habitat." It may be concluded from this that the form of the fungus thallus, the physiological and biochemical attributes, and the mechanisms of sexual and asexual reproduction as governed by the genetic constitution are, in sum, manifestations of the fitness of a fungus to its substratum. Although there are many anomalies in the morphological and reproductive patterns of fungi as a whole, a respectable case can be made for the relationship between morphological and reproductive specialization in particular phylogenetic fungal groups and the substratum they are most apt to occupy.

In environments where the nutrient resources fluctuate, fungal populations, like those of animals, occupying the greatest variety of habitats and utilizing the largest range of nutrients will have the best chance for survival. A flexible or "generalist" life style is favored. When simple carbon compounds like hexose sugars become available, the fungi responding most rapidly and extensively will be able to obtain a larger share of this nutrient resource. Rapid germination and high growth rates, in addition to copious production and dispersal of asexual reproductive units, contribute to competitive saprophytic ability--the ability, according to Garrett (1963), of a fungus with these attributes to successfully colonize dead organic substrates. Such are the characteristics of the less specialized fungi in the Zygomycetes and some of the Ascomycetes including yeasts, and imperfect stages of Plectomycetes and Pyrenomycetes (Hyphomycetes and Coelomycetes). Characteristically these fungi have simple growth patterns ranging from unicellular forms to mycelial forms that are generally restricted to the immediate substratum. Less commonly they exhibit a diffuse, spreading pattern in soil which may or may not be associated with the substratum (Gray and Williams, 1975). Homothallism is the most common pattern of sexuality when it occurs (Raper, 1966).

In contrast to the above, the structurally and chemically complex nutrient bases like lignin are not conducive to rapid decomposition (Dickinson and Pugh, 1974). As a more energetically stable resource, lignin favors colonization by specialist fungi. Characterized by slow rates of spore germination and hyphal growth, lignin decomposers invest considerable energy in extensive mycelial development. The growth patterns most commonly found are in the form of well-differentiated strands and rhizomorphs that extend from one substratum to another and to reproductive structures. These fungi, called macromycetes because of the large and often elaborate fruiting bodies, are principally members of the Basidiomycetes. Although predominantly heterothallic, complex sexual mechanisms serve to prevent indiscriminate heterokaryosis, thus ensuring the constancy of physiological mechanisms specialized for substrate utilization. The intervals required for their sexual development and maturity are comparatively long, a situation permitted if not fostered by the temporal stability of the substratum. Longevity of the mycelium is promoted through its enclosure within the substratum and contact with a more constant nutrient supply.

How strongly correlated different phylogenetic groups of fungi may be with a particular nutrient base remains a subject for further ecological research. However, some relationships are apparent even in soil environments where description of fungal communities seems to be an almost impossible task. Soils, especially those not stabilized by continuous plant cover, as in deserts, represent an unstable mixture of organic and inorganic fractions where percentages change according to the contributions of the biota and climatic processes of soil genesis. Soil in the immediate vicinity of plants is more resistant to disruptive factors that tend to destroy spatial organization. Away from plants, soil mixing by physical and biotic factors results in the incorporation of diverse organic fractions in various concentrations. According to Margalef (1968), in environmental systems where spatial organization is poor, prolific species are those most capable of adjusting to disruptive factors and environmental stress. The ubiquitous occurrence of many soil fungi, especially those labeled as heavy sporulators, is not unexpected for they possess adaptive traits for survival under environmental stress and fluctuating trophic resources. In species lists from soil surveys, these characteristic pedobionts are most often found to be members of the Mucorales and asexual states of Ascomycetes. The proportional representation of these species in quantitative population measurements is highly variable. Therefore, the recognition of fungal communities in soil is complicated by two major factors: (1) the occurrence of prolific species favored by environmental conditions, and (2) a high diversity in kind and concentration of incorporated plant debris, much of which is already occupied by many different fungi.

Sexually reproducing Ascomycetes and Basidiomycetes are also present in soil. Most of these form elaborate reproductive structures on plants and plant debris in the soil. As kormobionts, rhizobionts, or xylobionts, their occurrence is indicative of a locally stable nutrient base. When isolated from fragments of plants, they often fail to reproduce (mycelia sterilia). Sterility often indicates reliance on a sexual process requiring the presence of a compatible mating partner. Where adaptations to both substrate and environment are highly specialized, both vegetative and reproductive responses may be poor or lacking under general cultural conditions.

Taxonomic groups of fungi are potentially good indicators of habitat. Together as communities, they reflect the conditions of the soil environment and the contributions of the herbaceous and woody components of the attendant plant community. Using the appropriate methodology, a strong relationship between soil fungal communities and the plant community can be demonstrated. How this can be successfully accomplished will be emphasized in subsequent sections of this chapter.

III. METHODS OF SAMPLING

When selecting or devising a method for detecting the abundance and associative patterns of fungi, it is instructive to consider the following criteria: (1) recognition of general patterns of distribution by means of a preliminary survey or "entitation;" (2) recognition of the uses and limitations of the methods in relation to the research objectives; and (3) acquisition of data in a form amenable to statistical analysis for the assessment of variability and repeatability.

A. Defining the Habitat

Establishment of the habitat boundaries may be accomplished through an initial quantitative sampling procedure applied to a defined area in which discrete fungal communities can be expected to occur. As discussed earlier, the various relationships of fungi to their nutrient base can be used in the entitation process. For example, in a grassland, a recognizable, homogeneous vegetation segment may be selected and sampled for fungi on the herbaceous substratum and/or in the soil. It is possible and appropriate to identify fungal communities at several levels through aspects of the vegetation. These aspects include general vegetation type, successional stage, composition and cover, seasonal condition, and particular parts of plants. When soils are the object of study, additional environmental factors unique to the soil system should be considered in the entitation process. Soil profile, topography, and physicochemical characteristics are among those of major importance. Although environmental factors within the vegetation segment (including those of soil) are to be evaluated, the determination of their causative relationship to the fungi present need not be a primary objective. Little specific mycological work has been done to appraise the process of community formation and development per se. (This topic is treated in the preceding chapter (by Mueller-Dombois.) It is difficult to account for all the possible temporal changes and environmental conditions pertinent to any habitat description. However, the research objectives should be carefully formulated within a temporal and spatial framework established through entitation. Once this is accomplished, valid comparisons of community descriptions can be made.

B. Sampling and Isolation Techniques

The fungi of different ecological classifications, by virtue of their special attributes, require different methods of sampling. No adequate methods exist for the determination of the in situ presence of all fungi within any given habitat. Even
though any one sampling technique cannot be assumed exhaustive, selective procedures
can be used effectively to identify community patterns. Care must be taken to obtain
an adequate, unbiased sample using an appropriate sampling procedure. The methods
must be explicit, closely followed, and the data tested for reproducibility. Frequently, the inaccuracies of community analysis are related to sample collection
rather than sample analysis. A grid or transect of regularly distributed quadrats
across a homogeneous vegetation segment will provide coverage of the range of variation of fungi present in most plant communities. If high variability obscures community heterogeneity, we may stratify the sample in a systematic fashion using either
random or systematic subsampling within subdivisions. This allows us to test for both
homogeneity of the system and the reproducibility of the sampling results. A very
useful discussion of principles of soil sampling, including sample number, sample
placement, and statistical analysis is given by Parkinson et al. (1971) and Williams
and Gray (1973). A lucid account of the mathematical theory applied to species-abun-
dance relations, relative to the sampling procedures employed, is given by Pielou
(1977).

    Few quantitative methods for higher fungi (macromycetes) are available (Parker-
Rhodes, 1951; Fogel, 1976). Studies of macromycetes are based primarily on the observed frequency of sporocarps, with the sometimes erroneous assumption that each
sporocarp recorded represents a single mycelial network. Erratic fruiting habits
and nonrandom distribution are additional, though not insurmountable, limitations
to sampling by observation (see Watling, Chap. 23; Fogel, Chap. 28). Similar and
perhaps more serious difficulties are encountered in studies of soil micromycetes.
Here, frequency and abundance records are gathered from representative soil samples
collected in the field and later processed in the laboratory under artificial culture conditions. Both vegetative and reproductive units contribute to the records
of species frequency and abundance, particularly to abundance in the case of spores.
Because records from spores do not indicate origin from a single mycelium, measurement of frequency rather than density is considered more indicative of a species'
contribution to community pattern.

    The dilution plate technique stands as one of the oldest and most used methods
for quantitative studies of fungi in soil environments. Despite severe but perhaps
undeserved criticism, its potential but qualified usefulness continues to be recognized. In many cases the dilution plate procedure as described by various authors
(Martin, 1950; Parkinson et al., 1971) has been inappropriately applied. It has been
shown to have limited usefulness when employed exclusively to study either rhizosphere fungi (Parkinson and Thomas, 1965) or kormobionts and xylobionts on and in

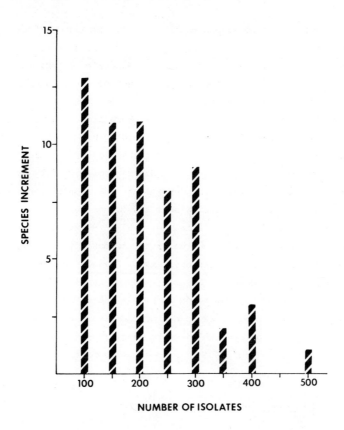

Fig. 2 Species increment diagram illustrating the number of additional fungus species encountered in the analysis of 10 soil samples. An initial 10 isolates per sample is increased by increments of 5. (After States, 1978.)

plants. The soil-washing technique and direct isolation more adequately sample these fungi, respectively (Parkinson et al., 1971).

As a selective technique the dilution plate method favors the isolation of pedobionts, species in soil with fast growth rates, low substrate specificity, and prolific production of asexual spores. The technique has been criticized because it does not allow the investigator to distinguish between fungi present as spores and those present as vegetative units. The importance of spores, however, cannot be ignored. When lowered metabolic states, due to nutrient depletion and low soil moisture, put severe stress on the vegetative mycelium, then it is the spore, whether chlamydospores or conidium, that provides advantage for survival and substrate colonization. Also, fungal spores have been shown to be metabolically active in transformation of chemical compounds in natural substrates (Vézina and Singh, 1975). Unadapted species in soil are not likely to persist long enough and in high enough numbers to be recorded as important members of the soil community (Christensen, 1969). It is not the purpose of the dilution plate method to evaluate the active nature of fungi in soil, but rather to sample populations for abundance, frequency, and patterns of community. Because the technique favors other generalist species like bacteria, actinomycetes, and yeasts, it may be necessary to discriminate against their growth and isolation, provided that the objectives prescribe such selectivity. The presence of heavy sporulators must be taken into account. They can adversely affect the isolation of prevalent species in samples where limited numbers of representative species are taken. Prevalent species, as defined by Parker-Rhodes (1951), are "those species whose enumerative habitats are larger than would be accountable by random distribution." Prevalent fungi are good indicators of community structure because of their constant presence over a long period of time. Sufficient representation of prevalent micromycetes in quantitative soil studies can be determined through species increment graphs (Gochenaur and Whittingham, 1967). Similar in purpose to species/ area curves, these graphs assist the researcher to determine sample size and provide, to some degree, a measure of species diversity. As illustrated in Fig. 2, the number of isolates required for adequate representation of fungal species in 10 soil samples from a desert shrub community was found to be some 45 to 50 isolates per site. The species isolated on this basis were found to demonstrate patterns of community structure different from the pattern exhibited by fungi from soils of neighboring plant communities. The treatment of data to demonstrate community structure is discussed below.

IV. EVALUATION OF COMMUNITY STRUCTURE

For several years European mycologists have been engaged in the evaluation of community composition and structure using phytosociological methods. Thus Peyronel (1955), who was interested in the comparative relationship between soil fungi of different plant communities, developed a graphic technique to illustrate the proportional representation of species in eight major taxonomic units. Each group was represented by an axis of the same origin whose length was proportional to the number of species in the group as a percentage of all species isolated. Lines connecting the ends of all axes resulted in a polygon which then represented the entire community, or *mycocenosis*. The shape of a polygon could then be compared to polygons representing fungal communities in soils of other plant communities. Upon comparison, the results indicated polygons of fungi isolated from similar plant communities were also similar. Further studies by Lupi Mosca (1964) and Badurowa and Badura (1967) confirmed the usefulness of the diagrams in making community comparisons. Particularly striking were the differences between fungal communities on the surfaces of leaves and those found on needles of various tree species (Badurowa and Badura, 1968). The differences were more distinctive than those found in soils. Quite clearly, the diagrams illustrate community structure owing to the specificity of the fungi for the plants serving as a nutrient base.

Without the support of quantitative data, the diagrams are susceptible to a low degree of accuracy because of improper species identifications, occurrence of sterile forms, presence of taxa not included on the diagram, and failure to employ uniform sampling techniques. Unfortunately assignment of taxonomic groups to different axes by some authors has disallowed valid community comparisons. Careful attempts to correct these points of weakness can result in rather precise comparisons. This has been done in desert soil communities, where it is especially difficult to recognize community patterns. Using species frequency data obtained with the dilution plate technique, the present author has identified fungal communities in soils of three major vegetation types in cool deserts (States, 1978). Quantitative soil sampling procedures as suggested in this chapter were followed. Species increment curves were prepared for a blackbrush shrub community, a grassland, and a pinyon-juniper woodland. Isolation procedures followed were essentially those of Christensen (1969). For community comparisons, species isolated from two sets of soil samples collected in different seasons were combined for each of three separate soils representing each vegetation type. Additional soil samples were collected from two of the three plant communities, analyzed as unknowns, and later compared with the knowns in the form of *mycographs*. Mycographs are essentially Peyronel's diagrams constructed on the basis of species whose relative frequency was 10% in both sets of data. All plant communities sampled in this study were within 20 km of each other. Physicochemical characteristics of the soils were similar.

The ability of the mycographs to provide community comparisons is illustrated in Fig. 3. Distinctive features of the grassland community are the abundance of dematiaceous species and the nearly equal representation of all other groups excluding Phycomycetes. Woodland communities are characterized by large numbers of species in the penicillia and low or no representation by Ascomycetes.

Woodland site No. 59 was somewhat similar to the grasslands, possibly because it had only a few widely spaced trees and high cover values for grasses and herbs. The shrub communities were distinctive, with high proportions of Coelomycetes and Moniliaceae. Aspergilli and Dematiaceae were also well represented. Mycographs constructed for the unknown communities, when matched with the others, were easily placed in the appropriate community type. The placement was verified by collection records for the soil samples.

It is clear that fungal communities, the mycocenoses of Peyronel, are recognizable within the soils of distinct vegetation types. The consistency of their occurrence in relationship to plants provides valuable predictive information. Vegetation very importantly influences the combination of species present. Mention should be made here of the danger of grouping species together in large taxonomic categories. For example, this procedure does not allow one to consider the biological significance of prevalent species. It is also important to know if a singly occurring species is locally rare or globally rare. Once recognized, the occurrence of a fungus

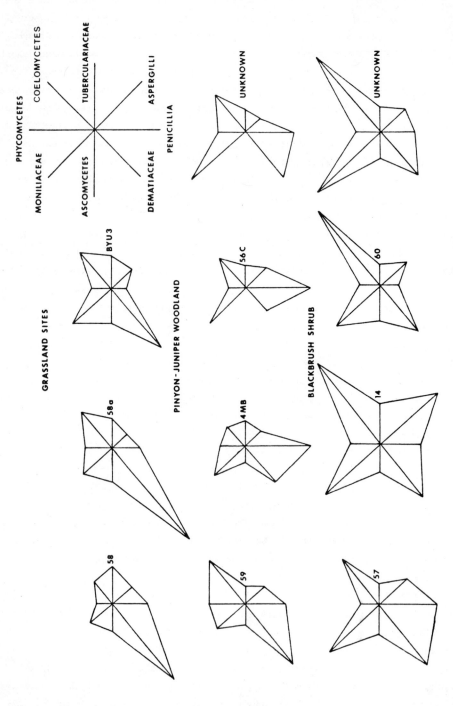

Fig. 3 Mycographs permitting comparison of fungal community structure based on the number of species found in eight major taxonomic groups as a percentage of all species isolated. Representative communities from soil of three cool-desert vegetation types are compared: grassland, blackbrush shrub, and coniferous woodland. (After States, 1978.)

because of a particular substrate or environmental factor(s) can provide additional and corroborative evidence for community pattern.

Several other synecological techniques borrowed from the plant ecologists have been successfully used to define and compare fungal communities. Species diversity indices and multidimensional ordinations were used by Christensen (1969) and Morall (1974) to compare soil microfungal communities with one another and with specific environmental factors. Studies of this nature deal with the more quantitative aspects of community analysis. (They are discussed in the following chapter by Christensen.)

V. CONCLUSION

Despite the useful criteria developed for describing fungal communities, as presented in this chapter, none of them can do justice to the natural variability of the mycoflora nor can they at present provide absolute comparisons. The present approach has been the recognition of structure and pattern that develops from the distribution of fungi and their association with a nutrient base. Quantitative techniques can advance the frontiers of this science and help us achieve a better understanding of factors accounting for community pattern and interactions. Community description is the first step in the dissection of the body of mycoecological information.

REFERENCES

Apinis, A. E. (1972). Facts and Problems. *Mycopathol. Mycol. Appl. 48*: 93-109.

Austwick, P. K. (1968). Effects of adjustment to the environment on fungus form. In *The Fungi*, Vol. 3: *The Fungal Population*, G. C. Ainsworth and A. S. Sussman (Eds.). Academic Press, New York, pp. 419-445.

Badurowa, M., and Badura, L. (1967). Further investigations on the relationship between soil fungi and the macroflora. *Acta Soc. Bot. Pol. 34*: 514-529.

Badurowa, M., and Badura, L. (1968). A comparative study on the occurrence of microscopic fungi on leaves and needles from different species of trees growing within the reserve "Kamień Slaski." *Ecol. Pollska (Ser. A) 14*: 253-260.

Baker, Gladys E., and Meeker, J. A. (1972). Ecosystems, mycologists, and the geographical distribution of fungi in the central Pacific. *Pacific Sci. 24*: 418-432.

Christensen, M. (1969). Soil microfungi of dry to mesic conifer-hardwood forests in northern Wisconsin. *Ecology 50*: 9-27.

Cooke, W. B. (1958). The ecology of the fungi. *Bot. Rev. 24*: 341-429.

Cooke, W. B. (1975). The ubiquity of fungi. *Rept. Tottori Mycol. Inst.* (Japan) *12*: 193-198.

Cooke, W. B. (1979). *The Ecology of Fungi*. CRC Press, Boca Raton, Florida.

Dickinson, C. H., and Pugh, G. J. F. (1974). *Biology of Plant Litter Decomposition*, Vols. 1 and 2. Academic Press, New York.

Fogel, R. (1976). Ecological studies of hypogeous fungi. II. Sporocarp phenology in a western Oregon Douglas fir stand. *Can. J. Bot. 54*: 1152-1162.

Frankland, J. C. (1966). Succession of fungi on decaying petioles of *Pteridium aquilinum*. *J. Ecol. 54*: 41-60.

Garrett, S. D. (1951). Ecological groups of soil fungi: A survey of substrate relationships. *New Phytol. 50*: 149-166.

Garrett, S. D. (1963). *Soil Fungi and Soil Fertility*. Pergamon Press, Elmsford, N.Y.

Gochenaur, S. E., and Whittingham, W. F. (1967). Mycoecology of willow and cottonwood lowland communities in southern Wisconsin. I. Soil microfungi in the willow-cottonwood forests. *Mycopathol. Mycol. Appl. 33*: 125-139.

Gray, T. R. G., and Williams, S. T. (1975). *Soil Microorganisms*. Longman, New York.

Hudson, J. (1968). The ecology of fungi on plant remains above the soil. *New Phytol. 67*: 837-874.

Lupi Mosca, A. M. (1964). Mycoflora of cultivated soil near Turin. (English summary.) *Allionia 10*: 7-16.

Margalef, R. (1968). *Perspectives in Ecological Theory*. Univ. of Chicago Press, Chicago.

Martin, J. P. (1950). The use of acid, rose bengal and streptomycin in the plate method for estimating soil fungi. *Soil Sci. 69*: 215-232.

Morrall, R. A. (1974). Soil microfungi associated with aspen in Saskatchewan: Synecology and quantitative analysis. *Can. J. Bot. 52*: 1803-1817.

Mueller-Dombois, D., and Ellenberg, H. (1974). *Aims and Methods of Vegetation Ecology*. Wiley, New York.

Orlos, H. (1966). *Forest Fungi Against the Background of Environment*. PWRIL, Warsaw. (Transl. from Polish, available from U.S. Dept. of Commerce, NTIS, Springfield, Va.)

Park, D. (1968). The ecology of terrestrial fungi. In *The Fungi*, Vol. 3: *The Fungal Population*, G. C. Ainsworth and A. S. Sussman (Eds.). Academic Press, New York, pp. 5-39.

Parker-Rhodes, A. F. (1951). The basidiomycetes of Skokholm Island. VII. Some floristic and ecological calculations. *New Phytol. 50*: 227-243.

Parkinson, D., and Thomas, A. (1965). A comparison of methods for the isolation of fungi from rhizospheres. *Can. J. Microbiol. 11*: 1001-1007.

Parkinson, D., Gray, T. R. G., and Williams, S. T. (1971). *Methods for Studying the Ecology of Soil Micro-organisms*. Blackwell, Oxford, England.

Peyronel, B. (1955). Proposta di un nuovo metodo di rappresentazione grafica della conposizione dei consorzi vegetali. *Nuovo Gion. Bot. Ital. [n.s.] 62*: 379-382.

Pielou, E. C. (1977). *Mathematical Ecology*. Wiley, New York.

Pirozynski, K. A. (1968). Geographical distribution of fungi. In *The Fungi*, Vol. 3: *The Fungal Population*, G. C. Ainsworth and A. S. Sussman (Eds.). Academic Press, New York, pp. 487-504.

Preece, T. F., and Dickinson, C. H. (1971). *Ecology of Leaf Surface Microorganisms*. Academic Press, New York.

Raper, J. R. (1966). Life cycles, sexuality and sexual mechanisms. In *The Fungi*, Vol. 2: *The Fungal Organism*, G. C. Ainsworth and A. S. Sussman (Eds.). Academic Press, New York.

Raunkiaer, C. (1937). *Plant Life Forms*. Oxford Univ. Press (Clarendon), New York.

Sparrow, F. K. (1968). Ecology of freshwater fungi. In *The Fungi*, Vol 3: *The Fungal Population*, G. C. Ainsworth and A. S. Sussman (Eds.). Academic Press, New York, pp. 41-93.

States, J. S. (1978). The soil fungi of cool-desert plant communities in northern Arizona and southern Utah. *Ariz. Acad. Sci. 13*: 13-17.

Vézina, C., and Singh, K. (1975). Transformation of organic compounds by fungal spores. In *The Filamentous Fungi*, Vol. 1: *Industrial Mycology*, J. E. Smith and D. R. Berry (Eds.). Wiley, New York, pp. 158-192.

Williams, S. T., and Gray, T. R. G. (1973). General principles and problems of soil sampling. In *Sampling--Microbiological Monitoring of Environments*, R. G. Board and D. W. Lovelock (Eds.). Academic Press, New York.

Chapter 12

SPECIES DIVERSITY AND DOMINANCE IN FUNGAL COMMUNITIES

Martha Christensen
University of Wyoming
Laramie, Wyoming

I. INTRODUCTION

In recent years, two very basic facts about fungal communities have been corroborated:
(1) it is apparent that there is an extremely high species diversity among the fungi
in any given ecosystem, and (2) habitat specificity for individual species and guilds
of species appears to be the rule. The microhabitats already noted by States (in
Chap. 11) support numerous species of fungi, operative in a vast number of ecosystem
functions, and the communities (synusiae) intergrade, both temporally and spatially.
One can begin to understand fungal community complexity in a single ecosystem by
thinking of a *Populus tremuloides* leaf successionally supporting phyllosphere, litter, fermentative layer, humus, and soil communities of microfungi, and, because of
proximity in the forest floor milieu, additionally sharing species with a cluster of
pine needles, a dead beetle, a plant rootlet, and an emerging basidiocarp. Despite
that complexity, however, the prominent microfungi obtainable from A_1 horizon soil
in any native association of plants constitute a distinctive assemblage for the given
association. Invariably the assemblages can be correlated to plant community composition, soil physicochemical characteristics, or both (Warcup, 1951a; Thornton, 1956;
Apinis, 1958; Brown, 1958; Gochenaur and Whittingham, 1967; Christensen, 1969; and
others).

II. DIVERSITY

Looking first at diversity, I'll follow Whittaker's (1975) definition of species diversity as simply "number of species, S, in a sample of standard size."

Several workers have reported a higher microfungal diversity in litter than in
the mineral soil beneath. Söderström (1975), in a Swedish *Picea* forest, obtained
about 26% more species from organic layers than from A_1 and B layers. D. T. Wicklow
and Whittingham (1974), Badurowa and Badura (1967), and Novak and Whittingham (1968)
also have reported species diversities in litter-humus composites that greatly exceed
those in the underlying soil. In a single *Acer-Ulmus-Fraxinus* forest in Wisconsin,
numbers of species represented among 2000 isolates from organic layers and from A_1
horizon soil, respectively, were 161 and 89 (Novak and Whittingham, 1968). The expla-

nation for litter versus soil diversity differences almost certainly has to do with
the fact that the litter-through-humus layer is a mixture of recent and declining materials, exposed to air and other spore vectors, and is itself diverse physically and
chemically.

In any relatively homogeneous northern temperate community, there is some indication that fungal species diversity may exceed vascular plant diversity by some 4-
20 times (Bisby, 1943; Novak and Whittingham, 1968; Apinis, 1972; Christensen, unpublished data). To the best of my knowledge, however, there are no serious estimates
even for total species of fungi in a single plant community.

Several workers have detected increases in numbers of species through the early
stages of fungal degradation of single substrates. Thus Lindsey and Pugh (1976), in
a study of microfungal succession on leaves of *Hippophaë rhamnoides*, found a steady
increase in numbers of species obtainable by five methods through an 8-month period,
May through December. In Wisconsin, G. A. Kuter (unpublished data) determined species
numbers among 100 microfungal isolates from standard samples of *Acer* leaves through
a span of about 15 months: green and senescent leaves yielded 17 species; diversity
per 100 isolates increased to a maximum of 46 by the following August; and there was
an apparent decline through the last 2 months to 38 and 25 species in September and
October, respectively. Apinis and associates at the University of Nottingham, England, in studies of colonization of *Phragmites* leavs by fungi (Apinis et al., 1972),
similarly reported a peak in species numbers at about 4-9 months and a decrease thereafter. Frankland's (1966) classical study of fungal succession on decaying petioles
of *Pteridium* revealed highest number of species in third-year litter, 177 species
from washed particles, and a subsequent decline in diversity in fourth- and fifth-
year litters. Maximum number of species in that study was coincident with termination of a period of relatively rapid dry weight and holocellulose loss (dry weight
remaining at 3 years was approximately 50% of the original weight; estimated time
for complete disintegration, 8-10 years). Garrett (1955) has suggested that microbial succession on a substrate is similar to higher plant succession except that in
the former instance the end point is not a persisting climax association, but zero.
In fungal succession as in plant succession (Loucks, 1970), it may be that highest
species diversity immediately precedes establishment of the climax community!

In soil communities, microfungal species diversity and vascular plant diversity
have been shown to be correlated in at least three regional studies (Apinis, 1958;
Christensen and Whittingham, 1965; M. C. Wicklow et al., 1974). In Wisconsin, in
three major vegetational units, this was the case: concomitant high species diversities for vascular plants and soil microfungi (Table 1) (Tresner et al., 1954; Orpurt and Curtis, 1957; Curtis, 1959; Christensen et al., 1962; Christensen and Whittingham, 1965; Christensen, 1969).

As one examines progressively drier grasslands across the United States, the
number of species per uniform number of isolates increases. Thus numbers of species

Table 1 Average numbers of species of soil microfungi among 180 or 200 isolates per community compared with numbers of vascular plants in the same vegetation subunits[a]

Southern hardwood continuum	Dry	D-M	Mesic	W-M	Wet	Average
Soil microfungi	37	53	54	*63*	--	52
Vascular plants, total species	289	275	230	*333*	175	282

Prairie continuum	Dry	D-M	Mesic	W-M	Wet	Average
Soil microfungi	35	35	*46*	39	34	38
Vascular plants, total species	132	245	*264*	252	186	216

Northern forest continuum	Dry	D-M	Mesic	W-M	Wet	Bogs	Average
Soil ficrofungi	32	32	34	*61*	37	31	38
Vascular plants, average species per stand	46	*56*	40	*56*	34	27	--

[a]The areas sampled are in Wisconsin. Numbers of communities sampled in the fungal surveys of southern forests, prairies, and northern forests and bogs are 18, 25, and 51, respectively. Vascular plant data have been taken from Curtis (1959). The highest figure in each rank is italicized for emphasis.

among 250 isolates were 46, 62, and 64 for grasslands in Wisconsin (D. T. Wicklow, 1973), South Dakota (Clarke, 1973) and Colorado (Scarborough, 1970), respectively. As will be pointed out later, however, those differences may be solely indicative of differences in component species equitabilities (Bazzaz, 1975), i.e., a few high-frequency species in the Wisconsin soils would dictate an *apparent* reduced richness. It is also possible, of course, that the western grassland soils contain a greater number and variety of microhabitats and support a longer survival of species than do the more mesic Wisconsin grassland soils.

Species diversity in soil fungal communities appears to be a multiple factor response. Some workers have related it to soil moisture (e.g., Mueller-Dombois and Perera, 1971). McLennan and Ducker (1954) related it directly to soil organic carbon content. Jensen (1963) in a study of Danish *Fagus* forests related it to existence of mull or mor humus. Apinis (1958) in Britain and M. C. Wicklow et al. (1974) in Oregon found positive correlations for soil microfungal species diversity and plant species diversity, as was reported in Wisconsin.

Diversity in the soil microfungal community *appears* to be reduced by cultivation, burning, clipping, grazing, slash-and-burn practices in tropical forests, irradiation, fertilization, and perhaps most other disturbance and manipulation activities (Martin, 1950; Meyer, 1963; Joffe, 1967; Scarborough, 1970; Apinis, 1972; D. T. Wicklow, 1973; Ciborska and Zadara, 1974; Gochenaur and Woodwell, 1974; Llanos and Kjøller, 1976). It also appears to be reduced by extremes in soil pH (Warcup, 1951a) and soil moisture (Orpurt and Curtis, 1957), and it probably is affected by occurrence of potent antifungal antibiotic producers. In his study on recolonization of soils sterilized by steam and formalin, Warcup (1951b) noted a consistently lower species diversity through 84 weeks in formalin-sterilized soils where *Trichoderma* was an early and persistent invader in contrast to steam-sterilized soils where *Trichoderma* was missing or very infrequent. Brown's (1958) report of a surface horizon of *Penicillium nigricans* "and relatively few associate species..." in Studland semifixed dune grass also suggests a possible depression in diversity as a consequence of dominance by a species with allelopathic chemicals (Jefferys et al., 1953; Bazzaz, 1975).

Brown's (1958) study in the Sandwich dune system revealed an increase in soil microfungal species numbers with vascular plant succession. She recorded just 13 species from open dunes, as compared with 38, 53, 45, and 47 species in later successional stages. Mallik and Rice (1966) later reported a similar finding for microfungi in three successional floodplain forests in Oklahoma: 73, 125, and 115 species in pioneer, "transitional," and climax communities, respectively. A rearrangement of data from the British study (see Brown, 1958, Table 2) additionally indicates a change in shape and position of microfungal dominance-diversity curves that may be comparable to the report by Bazzaz (1975) for vascular plants through secondary succession. In Fig. 1 it can be seen that as succession progresses homogeneity increases (position of the line moves to the right) and there is some indication of a change in slope

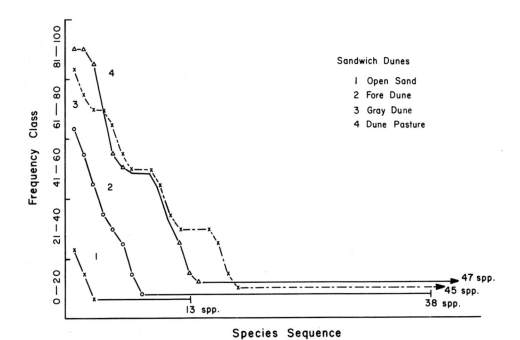

Fig. 1 Dominance-diversity curves derived from Brown (1958, Table 2). Species have been ranked on the abscissa in order of decreasing frequency (Bazzaz, 1975). Data for "semifixed yellow dunes," a stage following fore dunes and preceding gray dunes, have been omitted to simplify the figure. There were 53 microfungal species in that omitted stage; the dominance-diversity curve was very similar to the gray dune and dune pasture curves.

from geometric to lognormal. These findings, although admittedly tenuous, are in agreement with the concepts developed by Bazzaz.

A major difference for fungal communities as compared with vascular plant communities is that species richness (absolute number of species) in soil fungal communities is unknown--and perhaps undeterminable! Bazzaz (1975) and Whittaker (1972) have pointed out that diversity has two components: total species in the community and equitability. Equitability indicates the proportion of species occurring at high and intermediate levels of commonness; a dominance-diversity curve is a simple, graphic representation of equitability. In soil fungal communities, no one really knows where curves comparable to species-area curves for plants level off (Fig. 2). The more isolates you look at, the more species you find! If you sample long enough you are apt to encounter familiar forms, and recognition of infrequent species over and above quantitatively common species very likely led to the early, erroneous concept of a constant, characteristic, and cosmopolitan soil mycoflora. This situation, i.e., existence of an almost unfathomable reservoir of fungal species in soil, is very different from the situation for vascular plants in temperate communities, where species-

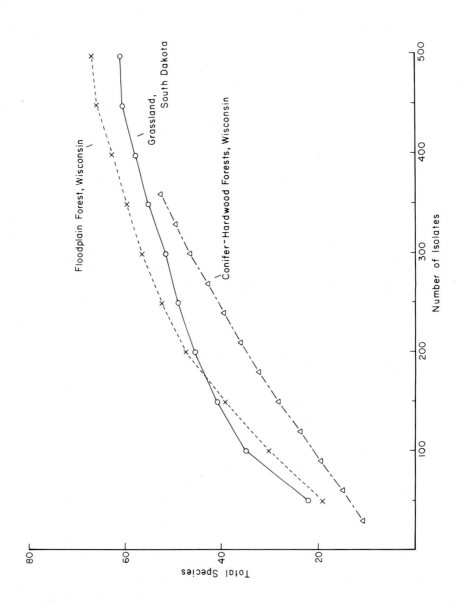

Fig. 2 Species-isolates curves plotted from data obtained in three soil microfungal community surveys. The Wisconsin floodplain forest and South Dakota grassland plots are for single communities, 10 soil samples (Gochenaur and Whittingham, 1967; Clarke, 1973). The conifer-hardwood data points are averages for 5 forests, 6 samples per forest (Christensen, 1969).

area curves level off at some reasonable point (Mueller-Dombois and Ellenberg, 1974). Apparent low species diversity, as shown in Fig. 3, example B, at 600 isolates, may be actual low total species (as probably is the case in the British open dunes) or it may result from incidence of comparatively many high- and intermediate-frequency forms, which then restricts rare form isolations (B versus A, in Fig. 3). Both cultivation and grazing have been shown to increase the number of species occurring at a frequency of 50% or more (England and Rice, 1957; Scarborough, 1970); diversity per standard number of isolates is decreased, but total species present may or may not be similarly affected. Synecologists looking at fungal communities unquestionably will find it illuminating to give more attention to community structure, including equitability (Fig. 1).

III. DOMINANCE

A. Introduction

R. H. Thornton in the mid-1950s came to the conclusion that, in relatively undisturbed soils, a small number of soil-inhabiting fungal species "assume dominance as a result of particular, favourable conditions" (Thornton, 1956). For mycologists, that comment signaled a turning point in synecological thought. It was based on Thornton's own detailed work with the soil microfungi of oak and heath communities in Britain, but actually had been or was soon to be corroborated by a number of other workers, including Warcup (1951a), Apinis (1958) and Brown (1958) in Britain, Peyronel (1955) and Sappa (1955) in Italy, Krzemieniewska and Badura (1954) in Poland, McLennan and Ducker (1954) in Australia, and Tresner et al. (1954) in the United States. Thornton and others of that period of the 1950s sensed fallacy in the notion of a constant, characteristic soil mycoflora and carried out the fundamental quantitative studies which, then and now, have allowed detection of "patterns" (Thornton's term) among and within fungal communities.

I will not review the early soil microfungal surveys and likewise will ignore here macrofungal synecology (see Wilkins and Patrick, 1940; Parker-Rhodes, 1955; Richardson, 1970; Lange, 1974; Lisiewska, 1974; and other references traceable through these) and phyllosphere, rhizosphere, and litter community ecology.

The remainder of this chapter describes my own attempt to synthesize an overview of soil microfungal synecology by examination of 33 surveys. Lists of the prevalent soil microfungi (about 30 quantitatively most common species) from 33 plant communities have been compared using a simple coefficient of similarity calculation. That calculation is $2w/(a + b)$, where a is the number of taxa in one list, b is the number in the other, and w is the number of taxa in common (Kershaw, 1973; Mueller-Dombois and Ellenberg, 1974). Franz (1963) strongly recommends use of this coefficient as a "first step for the delimitation of biocenoses."

Geographic distribution of the 33 surveys selected for comparison is shown in Fig. 4. Table 2 is a categorization of the studies by general vegetation type—grassland, desert, forest, tundra, heathland—and indicates also author, date, location,

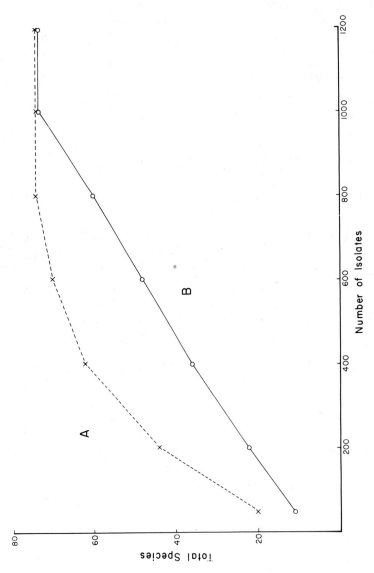

Fig. 3 Two hypothetical species-isolates curves. Communities A and B are equal in terms of total species present, but if fewer than 1000 isolates are examined, apparent diversity in B will be low because of species equitability differences in the two communities (see text).

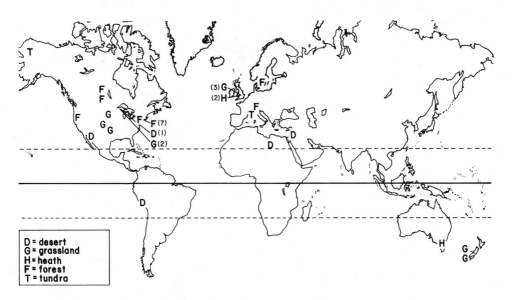

Fig. 4 Locations of the 33 soil microfungal surveys used in the comparison and ordinations. The selection includes 5 desert (D), 10 grassland (G), 13 forest (F), 3 heathland (H), and 2 tundra (T) studies.

and specific forest type. Each study accepted for use in the comparison met the following criteria: (1) it was a survey conducted in a native plant community, relatively undisturbed, in the Temperate Zone (except arctic tundra and Peruvian desert); (2) it involved isolation of microfungi by dilution plating, soil plating, soil washing (Söderström, 1975; Ruscoe, 1973), or screened immersion plates (Thornton, 1958); (3) it presented quantitative data (frequency or density), thereby making it possible to rank the taxa according to commonness; and (4) it carried identifications to the *species level*. However, 58 taxa, of 294 recorded and used in the comparison, were entered as genera (e.g., *Beauveria*, *Coniochaeta*, *Diplodia*, *Phoma*, *Trichoderma*, *Ulocladium*). Seventeen genera judged to be morphologically indistinct or too easily confused with other genera were omitted (e.g., *Acremonium*, *Cephalosporium*, *Monilia*, *Oospora*, *Tilachlidium*) and four genera usually requiring special isolation techniques (e.g., *Brevilegnia*, *Pythium*) also were not used in the calculations of similarity. Use of no more than some 23-32 forms per community in the comparison, the "top 30" in each study, ensured that my results would not be influenced by either the investigator's success in obtaining names for rare forms or by sample size.

As can be seen in Table 2 and Fig. 4, more than 50% of the studies are recent (1968 or later) and the sampled communities are widely distributed geographically. The insecurity of an earlier era over methods (Warcup, 1960) finally has given way to solid progress in synecology. Contemporary workers, in general, recognize the need for carefully drawn soil and vegetation descriptions and provide quantitative data for principal members of the fungal community.

Table 2 Lists of prevalent species for use in the comparison[a]

Grasslands:

1. Scarborough (1970), Colorado

2. D. T. Wicklow (1973), Wisconsin

3. Clarke (1973), S. Dakota

4. England and Rice (1957), Oklahoma

5. Brown (1958), Britain, alkaline dunes

6. Ruscoe (1973), New Zealand

7. Apinis (1958), Britain, alluvial grasslands

8. Thornton (1958), New Zealand

9. Warcup (1951a), Britain

10. Orpurt and Curtis (1957), Wisconsin, dry and dry-mesic sites

Deserts:

11. Gochenaur and Backus (1967), Wisconsin, sandbars

12. Ranzoni (1968), Arizona-California, Sonoran Desert

13. Borut (1960), Israel

14. Gochenaur (1970), Peru

15. Youssef (1974), Libya

Forests:

16. Gochenaur and Whittingham (1967), Wisconsin, *Populus-Salix*

17. Christensen et al. (1962), Wisconsin, *Acer-Ulmus-Fraxinus*

18. Tresner et al. (1954), Wisconsin, upland hardwood

19. Gochenaur and Woodwell (1974), New York, *Quercus-Pinus*

20. Morrall (1974), Canada, *Populus tremuloides*

21. Morrall and Vanterpool (1968), Canada, *Picea-Abies-Pinus*

22. Christensen (1969), Wisconsin, conifer-hardwood

23. Christensen and Whittingham (1965), Wisconsin, *Thuja-Abies*

24. Christensen and Whittingham (1965), Wisconsin, *Larix-Picea-Sphagnum*

25. Söderström (1975), Sweden, *Picea*

26. M. C. Wicklow et al. (1974), *Alnus-*conifer

32. Christensen (1960), Wisconsin, *Fagus-Tsuga*

33. M. Bauer and M. Christensen (unpublished data, 1973), Switzerland, *Fagus-Picea*

Tundra:

27a. Dowding and Widden (1974), Alaska

27b. Flanagan and Scarborough (1974), Alaska

28. Mosca (1960), Italy

Heathlands:

29. Brown (1958), Britain, two heath-dune units

30. Sewell (1959), Britain

31. McLennan and Ducker (1954), Australia

---

[a]Numbers shown are reference numbers and appear also in the ordination diagrams. The arctic tundra list is a combination from Refs. 27a and 27b.

B. Comparison and Ordination of the Microfungal Communities

Preliminary comparison of the 33 lists, each with each other, revealed an estimated 13-20 coefficients of zero (no "top-30" species in common) among the 528 matches. The estimate was based on comparison of all forests-heathlands-tundras with one another, deserts-grasslands-heathlands-tundras in all combinations, but just five forests and five grasslands in the combinations forest-grassland and forest-desert. The

principal combinations yielding zero coefficients were forest-desert (at least four),
heathland-desert (two), and tundra-desert (two). To avoid having to use low coefficients in the ordination, the communities were divided into two subgroups: a desert-
grassland subgroup (5 deserts, 10 grasslands) and a forest-heathland-tundra subgroup
(13 forests, 3 heathlands, and 2 tundras). Inclusion of four grasslands in the
forest-heathland-tundra comparison, however, provided a basis for confirming the
propriety of that division and, later, for linking the two single-axis ordinations.

Average numbers of prevalent soil microfungi shared between surveys are shown
in Table 3. The three heathlands (two British, one Australian) exhibited greatest
internal similarity in species composition--an average of 11 species shared, out of
27-30 prevalent species considered in each--whereas the deserts, on four different
continents, were least similar to one another. Between units, the forests and heathlands overall shared about 6 species; the two tundras, one arctic and one alpine,
were fairly dissimilar to one another and together were about equally forestlike and
grasslandlike; the deserts appeared to support soil mycofloras that differed considerably from those in the other vegetational units.

Considering the more than 400 coefficients calculated in this study, highest
similarities occurred between Wisconsin *Fagus* and Wisconsin conifer-hardwood communities (17 prevalent species shared), the two British heathlands (16 shared species), and between Wisconsin wet and wet-mesic conifer forests (16 species). Additionally, the Swedish *Picea* forest and each of two Wisconsin conifer forest units
(Table 2, items 22 and 23) shared 12 prevalent species, as did Canadian conifer forests and one Wisconsin conifer unit (Table 2, items 21 and 22). Southern Wisconsin
wet and wet-mesic hardwood forests shared 14 prevalent species. The New Zealand
grasslands and the arctic and alpine tundras shared an average of 7.8 species, whereas those tundras and North American and British grasslands shared an average of just
4.6 species. The arctic tundra shared 9 and 10 species with aspen stands in Canada
and a *Picea* forest in Sweden, respectively, and also shared 10 species with one New
Zealand grassland!

Tabulation of shared species proved to be an extremely useful technique for sum-
marizing soil mycofloral similarities in the Wisconsin studies. In Table 4 it can be
seen that in southern forests, where dominant vascular plants include both prairie
and northern forest species, soil microfungi are about equally prairie- and northern
forest-like and just 53% of all southern forest species are restricted to that unit.
The floristically dissimilar prairies and northern forests, however, have high percentages of restricted microfungal species and patterns for shared species are con-
sonant with floristic-edaphic similarity. A similar mycoflora-vegetation relationship is evident in Table 5. Thus dry northern forests (dominated by *Pinus banks-
iana*, *P. resinosa*, and three *Quercus* species) share many soil microfungal species
with dry-mesic communities (e.g., *Pinus strobus*, *Quercus rubra*), but share comparatively few species with *Larix-Picea* wet forests and the northern bog communities
dominated by *Sphagnum* and ericaceous shrubs. The Wisconsin conifer-hardwood unit

Table 3 Average soil microfungal species common to two component communities[a]

	Within unit	Between units				
		D	G	F	H	T
Deserts (5)	4.4	--	*4.4*	2.0[b]	2.8	3.8
Grasslands (10)	7.2	4.4	--	4.6[b]	4.2	*5.0*
Forests (13)	7.6	2.0[b]	4.6[b]	--	*5.9*	5.2
Heaths (3)	*11.0*	2.8	4.2	5.9	--	4.7
Tundras (2)	6.0	3.8	*5.0*	5.2	4.7	--

[a]Based on 23-32 (average 27.8) prevalent species in each study. The figure in parentheses is the number of deserts, grasslands, etc., in each vegetational unit; see Table 2. Highest figure in the column (within units) and in each rank of the matrix (between units) is italicized for emphasis.

[b]An estimate derived from use of just 5 of the 10 grasslands or 5 of the 13 forests or both.

Table 4 Soil microfungal similarity in three major vegetational areas of Wisconsin

Vegetational unit from which species were isolated	Total species	Percentage of species found only in unit of origin	Percentage of species found in the other units		
			Prairies	Southern upland forest	Northern upland forest
Prairies	96	70	--	25	14
Southern upland forest	107	53	22	--	24
Northern upland forest	87	71	15	29	--

Source: From Griffin (1972); after Christensen (1969).

in Table 2 includes the dry through mesic forests shown in Table 5. *Thuja-Abies* and *Larix-Picea-Sphagnum* communities in Table 2 are wet-mesic and combined wet forests and bogs in Table 5.

The lowest coefficient of similarity obtained in the comparison of 5 desert and 10 grassland lists of microfungi was a zero coefficient for Apinis's (1958) prevalents from an alluvial grassland once occupied by forest and Youssef's (1974) prevalents from a Libyan desert. All other desert-grassland combinations had 1-12 species in common, with coefficients ranging to 38.6 and 40.7 for various combinations of Colorado, South Dakota, and Wisconsin grasslands.

Table 5 Shared principal soil microfungal species in 51 northern Wisconsin forests and bogs[a]

	Dry	Dry-mesic	Mesic	Wet-mesic	Wet	Bog
D	*58*	41	35	21	18	12
DM	--	*60*	44	22	19	14
M	--	--	*59*	19	15	13
WM	--	--	--	*57*	29	21
W	--	--	--	--	*39*	22
B	--	--	--	--	--	*34*
Restricted	12(21%)	5(8%)	9(15%)	23(40%)	4(10%)	7(21%)

[a]The data were obtained by comparing lists of principal species in 36 dry through mesic forests (Christensen, 1969) and 15 open bogs and conifer swamps (Christensen and Whittingham, 1965). Principal species in each study included all microfungi present in three or more stands or with a frequency of 50% or more in one. Total principal species were 87 in the dry through mesic segment (D, DM, M) and 74 in the conifer swamp-bog segment (WM, W, B). Italic figures are the numbers of principal species in the indicated units; other inserted figures are shared principal species. The last rank shows absolute number and percent of subunit principal species that were restricted to the subunit.

Ordination of the 15 communities, following precisely the method described by Beals (1960) and Christensen (1969), resulted in the configurations shown in Fig. 5. Similar assemblages of prevalent microfungi in the grasslands dictated a tight constellation of grassland fungal communities in the first- versus second-axis plot. The third axis opposed British alkaline dune and Oklahoma tallgrass communities (points 5 and 4 in Fig. 5). End stands on the second axis were the Sonoran Desert and Negev Desert communities (points 12 and 13). Points representing the two New Zealand grasslands and the two Wisconsin grasslands were close on all three axes (6, 8, 2, 10). Deserts 11 and 14, although apparently close, were actually strikingly dissimilar (just one shared prevalent) and would have been at opposite ends of the fourth axis. Ordination in accordance with microfungal similarity has demonstrated that the desert and grassland assemblages of microfungi are different from one another but overlap via desert affinities in the guilds of prevalent species from Colorado and Oklahoma grasslands (points 1 and 4). There is substantial homogeneity among the prevalent species from grassland soils in contrast to a high diversity of principal species in the desert soils (see Table 3 and Fig. 5).

In the second table of coefficients, a 22 X 22 matrix constructed to show microfungal community similarities for 13 forests, 3 heathlands, 2 tundras, and 4 grasslands in all combinations, the 7 lowest coefficients were those for the 4 grasslands versus a Wisconsin conifer unit and for one grassland versus Canadian conifer and British heathland communities. The grasslands interjected in this comparison were

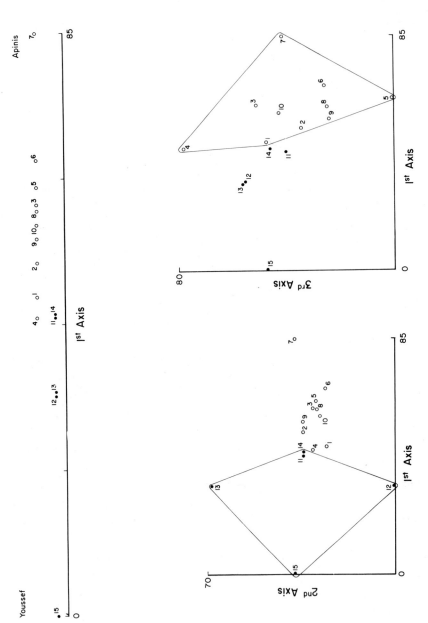

Fig. 5 Three-dimensional ordination of soil microfungal communities identified in 5 desert and 10 grassland surveys (see Table 2). The ordination was based on 2w/(a + b) comparisons, using presence, and involved prevalent ("top-30") microfungi only (see text). End stands on the first axis were the soil microfungal communities of Libyan deserts (Youssef, 1974) and alluvial grasslands once occupied by lowland forest in Britain (Apinis, 1958). The numbers are reference numbers (see Table 2).

those with mycofloras most different from the desert mycofloras: Apinis's alluvial
grassland, a New Zealand grassland, Brown's coastal grassland, and Clarke's South
Dakota grassland (points 7, 6, 5, and 3, respectively, in Fig. 5); they were included in the second comparison and ordination to test the appropriateness of having separated deserts-grasslands from forests-heaths-tundras. Comparison of coefficients
indicated that the grassland mycofloras were different from the forest mycofloras.
Each grassland had just one species in common with the *Larix-Picea-Sphagnum* unit and
an average of 4.7 species in common with each of the other forest units. In contrast,
the 13 forests shared an average of 7.6 species with one another. The average sum of
coefficients for the 4 grasslands versus the 13 forests was 218; sums of coefficients
for forests versus forests ranged from 232 to 461, average 332.

In contrast,Accordingly, Apinis's alluvial grasslands once occupied by forest was designated as one end stand in the first axis of the second ordination: that soil microfungal community, by comparison with those in the other 3 grasslands, most closely resembled the forest soil mycofloras. Apinis's community had 9 prevalent species in
common with several forests, 10 prevalent species in common with several grasslands,
and thus, compositionally, it appeared to truly bridge grasslands and forests. Since
among the forests-heaths-tundras the principal soil microfungi in the *Larix-Picea-*
Sphagnum unit least resembled prevalent species in the British alluvial grassland,
that unit became the other end stand on the first axis (Fig. 6).

In the second and third axis separations (Fig. 6): (1) the heath communities
(points 29, 30 and 31) remained together as a unit, opposed on the second axis by the
guild of microfungi from the Swiss *Fagus* forest (point 33); (2) the distinctness of
the microfungal community in the alpine tundra (point 28) was expressed; and (3) the
remarkable gradient of forest microfungal communities effected on the first axis persisted as the dominant configuration in the ordination. Wisconsin deciduous hardwood forests with rich, mull-type soils (points 17, 16, and 18) were found to harbor
microfungal communities compositionally similar to those in Apinis's grasslands.
That group of units is well-separated in the ordination from the fungal communities
characteristic in purely conifer and dominantly conifer forests in Sweden, Canada,
and Wisconsin (points 25, 21, 22-24) and the 5 assemblages of prevalents from mixed
forests occupy intermediate positions. The mixed forests are *Alnus*-conifer forests
in Oregon (point 26), *Populus tremuloides*-conifer stands in Canada (point 20), two
Fagus-conifer forests in Wisconsin and Switzerland (points 32, 33) and the *Quercus-*
Pinus community in New York (point 19).

In Fig. 7, first axes in the two ordinations have been combined by superimpos-
ing community "7" positions (in Figs. 5 and 6). The tundras and heathlands, with
microfungal assemblages generally distinct from those of the forests (Fig. 6), have
been omitted. The points on the baseline represent the 28 desert, grassland, and
forest soil microfungal communities examined in this study. Their positions relative
to one another have been dictated solely by compositional similarity in the microfun-

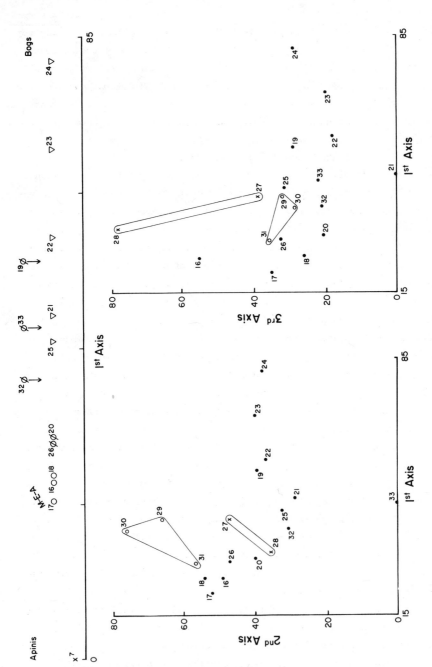

Fig. 6 Three-dimensional ordination of soil microfungal communities identified in 13 forest, 3 heathland, and 2 tundra surveys (Table 2). The ordination was based on $2w/(a+b)$ comparisons, using presence, and involved prevalent ("top-30") microfungi only (see text). End stands on the first axis were the soil microfungal communities of British alluvial grasslands (Apinis, 1958) and of *Larix-Picea-Sphagnum* communities ("bogs") in northern Wisconsin (Christensen and Whittingham, 1965). "M-E-A" refers to maple-elm-ash lowland hardwood forest communities in southern Wisconsin. The numbers are reference numbers (see Table 2).

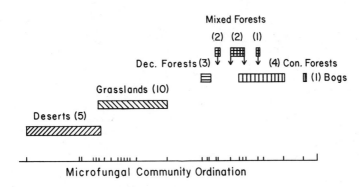

Fig. 7 Plant community types shown in relation to the combined first axes of the microfungal community ordinations. Indication of the nature of the vascular plant community has been superimposed on an ordination of the soil microfungal communities. The ordination was based on 2w/(a + b) comparisons, using presence, and involved prevalent ("top-30") microfungi only (see text).

gal communities, considering the approximately 30 prevalent species in each survey area.

Species composition in soil microfungal communities clearly is correlated with compositional variance in the cover vegetation. The complex of soil environmental factors directly involved in influencing fungal community composition and structure is only imprecisely known but presumably embraces physical and chemical features of the soil, the character of the litter, soil microclimate, and a vast array of specific biotic interactions.

C. Species Distributions

The ordinations indicated that prevalent microfungi in the 33 study areas were exhibiting habitat specificity. To confirm that relation and additionally to discover the guilds of species peculiar to deserts, deserts-grasslands, grasslands, etc., names of the 294 species used in the comparisons and occurrence as a prevalent for each were recorded on small cards. One hundred and forty-five of the 294 species used had occurred as a prevalent form in just one of the surveys. The remaining 149 species could be sorted, on the basis of occurrence, into 9 categories: exclusively or primarily desert, desert-grassland, grassland, grassland-forest, forest, forest-heathland, heathland, tundra-forest or tundra-forest-grassland, and unpatterned in respect to vegetation. Occurrences for 110 species with apparent distribution patterns in relation to vegetation and for 10 species prevalent in 5 or more of the 33 surveys but with unpatterned distributions are shown in Table 6. The sequence of the first 28 surveys in Table 6 is that shown in Fig. 7. Table 7 is an alphabetical arrangement of the 149 taxa present in 2 or more of the 33 lists of prevalent forms; apparent habitat specificity and occurrence in the 5 community types (desert, grassland,

Table 6 Occurrences for 120 taxa present in two or more communities as prevalent taxa[a]

	Deserts[b]	Grasslands	Forests	Heath	Tundra
DESERT					
Aspergillus ustus	X X				
Ulocladium	X X				
Stachybotrys	X X X				X
Fusarium semitectum	X X	X			
F. acuminatum	X	X			
Trimmatostroma	X				
Aspergillus niger	X X X	X			X
Chaetomium	X X	X	X		
Penicillium decumbens	X X		X	X	
Hormiscium	X	X			
Penicillium chrysogenum	X X X				
Aspergillus niveus	X	X			
Discosia	X	X			
Aspergillus alliaceus	X	X			
DESERT-GRASSLAND					
Alternaria	X X X X X	X X X	X	X	
Aspergillus fumigatus	X X X	X X X			
Torula	X	X X X X			
Cladosporium cladosporioides	X X X X	X X X	X		X X
Aspergillus terreus	X X X	X X X		X	X
Arthrinium	X	X X		X	
GRASSLAND					
Fusarium solani	X X	X X			X
F. nivale	X	X X			
F. moniliforme/oxysporum	X X X	X X X X X X	X	X	X
Penicillium lilacinum	X X X	X X X		X	
Aspergillus candidus	X	X X	X		
Pyrenochaeta	X	X X X X			
Stemphylium	X	X X X			
Coniothyrium	X X X	X X X	X	X	X
Papulospora	X	X X X	X		
Idriella lunata	X	X X			

Penicillium restrictum
P. stoloniferum
Gonytrichum macrocladium
Mortierella alpina
Humicola
Zygorhynchus moelleri
Myrothecium
Fusarium sambucinum
Doratomyces
Gliomastix (= Acremonium)
Aspergillus flavipes
Acrostalagmus
Fusarium avenaceum
Mortierella gracilis
Fusarium culmorum

GRASSLAND—FOREST
Penicillium funiculosum
P. janthinellum
Paecilomyces marquandii
Absidia glauca
Penicillium nigricans
Mucor hiemalis
Mortierella minutissima
Penicillium soppi
P. miczynskii

FOREST
Penicillium thomii
P. paxilli
P. palitans
P. herquei
Gliocladium roseum
Penicillium kapuscinskii
Paecilomyces carneus
Penicillium steckii
Absidia cylindrospora
Sporotrichum/Sporothrix
Penicillium multicolor
Oidiodendron citrinum
Penicillium variabile
Mortierella vinacea

Table 6 (cont.)

	Deserts[b]	Grasslands	Forests	Heath	Tundra
Penicillium ochro-chloron			X X		
Mortierella nana			X X X		
Penicillium pinetorum			X X X		
Trichoderma (white)		X	X X X X		X
Phialophora	X	X	X X		X
Mucor ambiguus			X X		
Penicillium nalgiovensis			X		
Torulomyces lagena			X X X X X		
Cryptococcus			X X X X X		
Thysanophora			X X		
Penicillium godlewskii			X X	X	
Oidiodendron echinulatum/cereale/chlamydosporum			X X X X X X X		
O. maius/tenuissimum/griseum			X X X X X		
O. flavum			X X X		
Chloridium chlamydosporum		X	X X X		
Oidiodendron periconioides			X X		
Mortierella macrocystis			X X		
Chalara			X X		
Tolypocladium/Pachybasium			X X		X
Penicillium terlikowskii	X		X X X	X	
P. montanense			X X X		
Mucor silvaticus			X X X		
Penicillium rolfsii			X X X		
P. radulatum			X X		
Torulopsis aeria			X X X		
Geotrichum candidum			X X X		
Histoplasma			X X		
Penicillium diversum var. aureum			X X X		
P. odoratum			X X X		
FOREST-HEATH					
Mortierella isabellina			X X X X	X X	
M. parvispora		X X X	X X	X X X	X
Penicillium brevi-compactum		X	X X	X X X X	
P. spinulosum		X	X X X X	X X X	

Mortierella ramanniana

HEATH
Penicillium adametzi
P. fellutanum
P. melinii
Absidia orchidis
Penicillium namyslowskii

TUNDRA FOREST/TUNDRA GRASSLAND
Chrysosporium
Trichocladium
Cylindrocarpon

UNPATTERNED, 5 OR MORE COMMUNITIES
Phoma
Penicillium canescens
P. cyclopium
P. frequentans
Aureobasidium pullulans
Penicillium simplicissimum
Cladosporium herbarum
Trichoderma (green)
Penicillium raistrickii
Beauveria

[a]Twenty-nine taxa present in 2–4 communities but unpatterned in respect to vegetation have been omitted. Community sequence for deserts–grasslands–forests was dictated by coefficients of similarity for lists of the prevalent microfungi obtained in each study area (see text). Community reference numbers, also in ordination sequence, are shown in Figs. 5 and 6; see Table 2. Heath and tundra sequences are "31", "30", "29", and "27", "28", respectively (Table 2).

[b]The fifth point is a grassland, "4" in Table 2 (see first axis of the ordination in Fig. 5).

Table 7 Alphabetical list of the 149 taxa present as prevalent forms in two or more of the 33 surveys[a]

Absidia cylindrospora   Forest   G-2   F-4   H-1
A. glauca   Grassland-forest   G-5   F-6
A. orchidis   Heath   H-2
A. spinosa   Unpatterned/infrequent   G-2   H-2
Acrostalagmus   Grassland   G-2
Alternaria   Desert-grassland   D-4   G-4   F-1   H-1
Arthrinium   Desert-grassland   D-1   G-2   H-1
Aspergillus alliaceus   Desert   D-1   G-1
A. candidus   Grassland   G-2
A. flavipes   Grassland   D-1   G-2
A. fumigatus   Desert-grassland   D-3   G-6
A. niger   Desert   D-3   G-2   T-1
A. niveus   Desert   D-1   G-1
A. sydowi   Unpatterned/infrequent   D-1   T-1
A. terreus   Desert-grassland   D-2   G-4   H-1
A. ustus   Desert   D-3
A. versicolor   Unpatterned/infrequent   D-1   G-1   T-1
Aureobasidium pullulans   Unpatterned   D-1   G-2   F-8   H-2   T-1

Beauveria   Unpatterned   G-2   F-2   H-2

Chaetomium   Desert   D-3   G-1   F-2
Chalara   Forest   F-2
Chloridium chlamydosporum   Forest   G-1   F-2
Chrysosporium   TF/TG   G-1   F-7   T-2
Cladosporium cladosporioides   Desert-grassland   D-4   G-3   F-2   T-2
C. herbarum   Unpatterned   D-1   G-5   F-7   H-3   T-1˙
Coniochaeta   Unpatterned/infrequent   D-1   F-1   T-1
Coniothyrium   Grassland   D-2   G-5   F-1   H-1   T-1
Cryptococcus   Forest   F-6
Cylindrocarpon   TF/TG   D-1   G-5   F-4   T-2

Discosia   Desert   D-1   G-1
Doratomyces   Grassland   G-2

Eupenicillium   Unpatterned/infrequent   G-1   T-1

Fusarium acuminatum   Desert   D-1   G-1
F. avenaceum   Grassland   G-2
F. culmorum   Grassland   G-3
F. episphaeria   Unpatterned/infrequent   F-1   T-1
F. moniliforme/oxysporum   Grassland   D-2   G-10   F-2   T-1
F. nivale   Grassland   D-1   G-3
F. roseum   Unpatterned/infrequent   D-1   F-1
F. sambucinum   Grassland   G-3
F. semitectum   Desert   D-1   G-1
F. solani   Grassland   D-2   G-6

Geotrichum candidum   Forest   F-2   T-1
Gliocladium roseum   Forest   D-1   G-2   F-3
Gliomastix (Acremonium)   Grassland   G-4   F-1   H-1   T-1
Gonytrichum   Grassland   G-1   F-1

Histoplasma   Forest   F-2
Hormiscium   Desert   D-1   G-1
Humicola   Grassland   G-4   F-2   H-1   T-1

Idriella lunata   Grassland   G-2

Margarinomyces   Unpatterned/infrequent   D-1   F-1
Mortierella alpina   Grassland   G-5   F-2   T-1
M. gracilis   Grassland   G-2

Table 7 (Continued)

M. isabellina   Forest-heath   F-9   H-2
M. macrocystis   Forest   F-2
M. marburgensis   Unpatterned/infrequent   F-1   H-1
M. minutissima   Grassland-forest   G-2   F-2   T-1
M. nana   Forest   F-6
M. parvispora   Forest-heath   F-3   H-2   T-1
M. pusilla   Unpatterned/infrequent   F-1   H-1   T-1
M. ramanniana   Forest-heath   F-7   H-3   T-1
M. spinosa   Unpatterned/infrequent   G-1   F-1
M. vinacea   Forest   F-9   H-1
Mucor ambiguus   Forest   F-2
M. hiemalis   Grassland-forest   D-1   G-4   F-5   H-1
M. silvaticus   Forest   F-2
Myrothecium   Grassland   D-1   G-3   F-3

Oidiodendron citrinum   Forest   F-3
O. echinulatum/cereale/chlamydosporum   Forest   F-7
O. flavum   Forest   F-4
O. maius/tennuissimum/griseum   Forest   F-8
O. periconioides   Forest   F-2

Paecilomyces carneus   Forest   G-2   F-9
P. marquandii   Grassland-forest   G-5   F-4
Papulospora   Grassland   D-2   G-3   F-1
Penicillium adametzi   Heath   D-1   G-1   F-1   H-3
P. brevi-compactum   Forest-heath   G-1   F-5   H-3   T-1
P. canescens   Unpatterned   D-2   G-1   F-2   H-1
P. chrysogenum   Desert   D-2   G-1
P. corylophilum   Unpatterned/infrequent   D-1   G-1   T-1
P. crustosum   Unpatterned/infrequent   D-1   F-1
P. cyclopium   Unpatterned   D-1   G-2   F-1   H-2
P. daleae   Unpatterned/infrequent   F-1   T-1
P. decumbens   Desert   D-2   H-1
P. diversum var. *aureum*   Forest   F-2
P. fellutanum   Heath   G-1   H-2
P. frequentans   Unpatterned   D-1   G-1   F-3   H-1   T-1
P. funiculosum   Grassland-forest   D-1   G-6   F-6   H-1
P. godlewskii   Forest   F-2   H-1
P. granulatum   Unpatterned/infrequent   G-1   F-1
P. herquei   Forest   F-2
P. janthinellum   Grassland-forest   D-2   G-5   F-6   H-1   T-1
P. kapuscinskii   Forest   F-2
P. lanosum   Unpatterned/infrequent   D-1   F-2
P. lapidosum   Unpatterned/infrequent   F-1   H-1
P. lilacinum   Grassland   D-2   G-5   F-1
P. lividum   Unpatterned/infrequent   G-1   F-1   H-1   T-1
P. martensii   Unpatterned/infrequent   F-1   H-1
P. meleagrinum   Unpatterned/infrequent   G-1   H-1
P. melinii   Heath   G-1   H-2
P. miczynskii   Grassland-forest   G-2   F-2
P. montanense   Forest   F-3
P. multicolor   Forest   D-1   F-3
P. nalgiovensis   Forest   F-2
P. namyslowskii   Heath   H-2
P. nigricans   Grassland-forest   D-1   G-6   F-6   H-3
P. notatum   Unpatterned/infrequent   G-2   F-1   T-1
P. ochro-chloron   Forest   F-3
P. odoratum   Forest   F-2
P. palitans   Forest   F-2
P. paxilli   Forest   D-1   F-2   H-1
P. pinetorum   Forest   F-4

Table 7   (Continued)

P. radulatum	Forest	F-2				
P. raistrickii	Unpatterned	G-2	F-3	H-1	T-1	
P. restrictum	Grassland	G-4	F-3	H-3		
P. rolfsii	Forest	F-2				
P. rubrum	Unpatterned/infrequent	G-1	T-1			
P. rugulosum	Unpatterned/infrequent	G-1	F-1	H-1		
P. simplicissimum	Unpatterned	D-1	G-1	F-2	H-1	
P. soppi	Grassland-forest	G-1	F-2			
P. spinulosum	Forest-heath	G-1	F-7	H-3		
P. steckii	Forest	D-1	F-5			
P. stoloniferum	Grassland	G-3	F-1	T-1		
P. terlikowskii	Forest	D-1	F-3	H-1		
P. thomii	Forest	G-2	F-6	H-2	T-1	
P. variabile	Forest	G-1	F-4			
P. waksmani	Unpatterned/infrequent	F-1	H-1			
Periconia	Unpatterned/infrequent	G-1	F-1			
Pestalotia	Unpatterned/infrequent	D-1	T-1			
Phialophora	Forest	D-1	G-1	F-4	T-1	
Phoma	Unpatterned	D-4	G-4	F-4	H-1	T-2
Pyrenochaeta	Grassland	D-1	G-5			
Rhinocladiella	Unpatterned/infrequent	G-1	F-1			
Rhizopus	Unpatterned/infrequent	D-1	H-1	T-1		
Scopulariopsis	Unpatterned/infrequent	G-1	H-1			
Sporotrichum/Sporothrix	Forest	G-2	F-6			
Stachybotrys	Desert	D-4	G-1	T-1		
Stemphylium	Grassland	D-1	G-4			
Thysanophora	Forest	F-2				
Tolypocladium/Pachybasium	Forest	F-2	T-1			
Torula	Desert-grassland	D-1	G-2			
Torulomyces lagena (=Monocillium humicola)	Forest	F-7				
Torulopsis aeria	Forest	F-3				
Trichocladium	TF/TG	G-2	F-2	T-2		
Trichoderma (green)	Unpatterned	D-2	G-10	F-12	H-3	T-2
Trichoderma (white)	Forest	G-1	F-7	T-1		
Trimmatostroma	Desert	D-1	G-1			
Ulocladium	Desert	D-2				
Zygorhynchus moelleri	Grassland	G-4	F-1	H-1	T-1	

[a]Apparent habitat specificity for each taxon is followed by a series of letters and figures which indicate number of deserts (D), grasslands (G), etc. wherein the given taxon was present as a prevalent. Total communities (maximum numbers) are: deserts 5, grasslands 10, forests 13, heathlands 3, tundras 2. The habitat specificity categories are: desert, desert-grassland, grassland, grassland-forest, forest, forest-heathland, heathland, tundra-forest/tundra-grassland, and unpatterned. One hundred forty-five taxa of the 294 used in the comparison occurred as prevalents in single surveys and are not listed here.

forest, heath, tundra) are indicated. Total survey areas examined in each vegetational unit were the following: 5 deserts, 10 grasslands, 13 forests, 3 heaths and 2 tundras.

About 74% of the taxa prevalent in 2 or more areas (110 of 149 taxa) exhibited discernible distribution patterns in relation to vegetation (Table 6). Principal broad-amplitude and restricted species in each of the 5 community types are noted below. Taxa apparently unpatterned with respect to vegetation (1) may be primary colonizers (e.g., *Phoma*, *Beauveria*, *Aureobasidium*, *Cladosporium herbarum*) less subject to selection through interaction than are secondary and tertiary colonizers, (2) may be controlled by factors unrelated to vegetation, or (3) may be mixtures of species or ecotypes (e.g., *Trichoderma*, *Penicillium nigricans*).

1. Deserts

The 5 desert communities supported a greater diversity of prevalent forms than did the grasslands and heaths, for example. No species was present as a principal form in all deserts examined (Table 6), but several are broad-amplitude, primarily desert taxa and a few may be restricted indicators of desert habitats, as shown in the accompanying list.

(a) Broad amplitude

Alternaria
Aspergillus fumigatus
Aspergillus niger
Chaetomium
Cladosporium cladosporioides
Stachybotrys

(b) Restricted

Aspergillus ustus
Fusarium semitectum
Ulocladium

Alternaria species, *Aspergillus fumigatus*, *Cladosporium cladosporioides*, *Aspergillus ustus*, and *Ulocladium* species also have been listed as common microfungi in Arizona and Utah desert soils (States, 1978). Apinis and Chesters (1964) in Britain obtained 12 species of *Chaetomium* from surface horizons of coastal sand dunes and salt marshes colonized by grasses. *Aspergillus nidulans* group members, especially *A. nidulans*, *A. quadrilineatus*, and *A. rugulosus*, also appear to be desert-adapted forms (Kuehn, 1960; Moubasher and Moustafa, 1970) but were present as prevalents in just one of the five deserts sampled here (Gochenaur, 1970). *Phoma* species are isolated frequently from desert soils but occur elsewhere as well, probably as primary colonizers (Frankland, 1966; Hudson, 1968). Dematiaceous microfungi, including dark "mycelia sterilia," and Sphaeropsidales appear to be characteristic groups in desert soils, contributing both a diversity of species and high proportions of isolates (Gochenaur, 1970; Moubasher and Moustafa, 1970). The desert vegetational unit was unique in being the only unit with *Trichoderma* (green) recorded as prevalent in fewer than 50% of the communities sampled. Kuehn (1960) also did not obtain *Trichoderma* from desert soils in New Mexico.

2. Grassland

Fusarium oxysporum or *Fusarium moniliforme* was present as a principal form in all of the 10 temperate grasslands, and 5 other species of *Fusarium* were prevalent in 2 to 6 grassland units but absent or infrequent in the other vegetational units (Table 6). The constancy of *Fusarium* as a genus characteristic in grassland soils now appears to be well established (Orpurt and Curtis, 1957; Morrall, 1974). In an inten-

sive literature search, Park (1963) found no record of isolation of *F. oxysporum*
from heath or woodland soils. Both of the forest units shown to contain *F. oxysporum*
in Table 7 are grassy forests, aspen "20" and willow-cottonwood "16". The occurrence
of several tropical grassland species in Oklahoma, Wisconsin, South Dakota, and Colorado grassland soils, but not in British and New Zealand grassland profiles, e.g.,
Aspergillus candidus, *A. flavipes*, *A. flavus*, *A. fumigatus*, *A. niveus*, *A. terreus*,
and *Syncephalastrum racemosum* (see Table 6; also Mueller-Dombois and Perera, 1971;
Dwivedi, 1966; and others), is consistent with the postulated great antiquity of New
World midcontinent grasslands and their derivation, in part, from the tropical and
subtropical Sonoran Floristic Province (Gleason and Cronquist, 1964). The broad-amplitude, primarily grassland taxa and the restricted grassland forms are listed here.

(a) Broad amplitude   (b) Restricted

    Aspergillus terreus   *Aspergillus candidus*
    Coniothyrium   *Doratomyces*
    Fusarium nivale   *Fusarium avenaceum*
    F. oxysporum/moniliforme   *F. culmorum*
    F. solani   *F. sambucinum*
    Gliomastix (Acremonium)   *Idriella lunata*
    Humicola   *Mortierella gracilis*
    Mortierella alpina
    Papulospora
    Penicillium lilacinum
    P. restrictum
    Pyrenochaeta
    Stemphylium
    Zygorhynchus moelleri

About 5 taxa are broad-amplitude species, i.e., in 7-15 communities as prevalents,
with highest frequencies of occurrence in grassland and deciduous forest soils. They
are *Absidia glauca*, *Mucor hiemalis*, *Myrothecium*, *Paecilomyces marquandii*, and *Penicil-*
lium janthinellum. *Penicillium funiculosum*, *P. nigricans*, and *P. brevi-compactum*
may be ecotypic, with grassland-deciduous forest adapted forms and coniferous forest-
heathland adapted forms. The "*Trichoderma* (green)" taxon was prevalent in all grasslands, all tundras, all heathlands, and all but one of the 13 forests examined.

    3. Forest

    The combinations of microfungi obtainable from forest soils readily distinguish
those soils from grassland soils. In general, *Mortierella isabellina* group members
(*M. isabellina*, *M. ramanniana*, *M. vinacea*, *M. nana*), species in *Oidiodendron*, and a
distinctive group of penicillia are the primary distinguishing species. Deciduous
forest soils, especially soils from wet to wet-mesic and mesic deciduous forest sites,
characteristically yield a great diversity of *Penicillium* species, and they account
for relatively high proportions of total isolates (Christensen et al., 1962; Gochenaur and Whittingham, 1967; Christensen, 1969). The common occurrence of species of
Oidiodendron in forest soils may be related to the claim of consumption and dispersal

of *Oidiodendron* conidia by certain kinds of mites (Mangenot and Reisinger, 1976). Species primarily inhabiting deciduous forest soils, primarily inhabiting coniferous forest soils, and characteristic in the A_1 horizons of mixed forests include those listed here.

(a) Deciduous hardwood forests

    Gliocladium roseum
    Paecilomyces carneus
    Penicillium herquei
    P. kapuscinskii
    P. multicolor
    P. palitans
    P. steckii

(b) Conifer forests

    Mucor silvaticus
    Penicillium diversum var. *aureum*
    P. odoratum
    P. radulatum
    P. rolfsii
    Torulopsis aeria

(c) Mixed conifer and deciduous hardwood forests, broad amplitude

    Chalara
    Mortierella macrocystis
    M. nana
    M. vinacea
    Mucor ambiguus
    Oidiodendron (9 species!)

    Penicillium montanense
    P. nalgiovensis
    P. pinetorum
    Thysanophora
    Torulomyces lagena
      (= *Monocillium humicola*)
    Trichoderma (white)

Tolypocladium/Pachybasium is common in conifer forest soils, but is only occasionally listed as a prevalent form (Christensen and Whittingham, 1965; Söderström, 1975), probably because the genera are unfamiliar (Gams, 1971). Several species recorded in Table 6 were about equally prominent in conifer forest soils and heathland soils, namely:

(d) Conifer forests and heathlands, broad amplitude

    Mortierella isabellina
    M. parvispora
    M. ramanniana

    Penicillium spinulosum
    P. terlikowskii
    P. thomii

P. thomii, however, is widely distributed; it is the only species that has been present in all vegetational units sampled in Wisconsin, including the sedge meadows (Tresner et al., 1954; Orpurt and Curtis, 1957; Christensen et al., 1962; Christensen and Whittingham, 1965; Gochenaur and Backus, 1967; Gochenaur and Whittingham, 1967; Christensen, 1969). Singh (1976) recently reported *Mucor ambiguus*, *M. silvaticus*, the *Mortierella isabellina* group, *Oidiodendron* species, *Penicillium montanense*, *P. pinetorum*, *P. rolfsii*, and *P. spinulosum* from conifer forest soils in Newfoundland.

4. Heath

The distinctive assemblage of microfungi in heath soils includes *Penicillium adametzi*, *P. fellutanum*, *P. melinii*, and *P. namyslowskii* along with the conifer-heathland associates mentioned in the preceding paragraph, especially *Mortierella ramanniana*, *M. isabellina*, *M. parvispora*, and *Penicillium spinulosum*, and the broadly distributed and probably ecotypic taxa, *Penicillium nigricans* and *P. brevi-compactum*.

5. *Tundra*

The tundra vegetational unit shared species with the grasslands, especially the New Zealand grasslands, and with certain of the forest communities. It may have been too sparsely sampled to allow recognition of distinctive prevalents. *Chrysosporium*, *Trichocladium*, and *Cylindrocarpon*, however, occurred in both arctic and alpine tundra sites and in just 4-10 of the 23 grasslands and forests sampled.

D. Conclusions

We can conclude based on the foregoing surveys that different but intergrading assemblages of microfungi are characteristic in the soils of deserts, grasslands, deciduous forests, coniferous forests, heathlands, and tundras. Further, all or most of the prevalent species in native soils appear to be pandemic. Thus, similar assemblages occur in regions widely separated geographically. Comparisons of coefficients of similarity using top-ranking species in each survey has revealed striking soil microfungal community similarities for: (1) Polish, Italian, Swiss, and North American *Fagus* forests; (2) Swedish and North American conifer forests; (3) Peruvian, Israeli, and North American deserts; (4) North American, British, and New Zealand grasslands, and (5) British and Australian heathlands. In all of these instances, ecologically similar but taxonomically different vegetations are supporting similar fungi. Could it be that soil mycofloras have remained relatively stable through several million years of speciation in *Fagus*, *Picea*, and *Pinus*? In any event, species composition in soil microfungal communities clearly is a sensitive gauge of similarity in soil environments. In general, species of *Penicillium*, *Fusarium*, *Aspergillus*, *Mucor*, *Mortierella*, and *Oidiodendron* are more habitat specific than are certain primary colonizers, e.g., *Phoma*, *Beauveria*, *Botrytis*, and *Aureobasidium*.

The relation of vegetation to mycoflora reported here is in complete agreement with the conclusion of Badura and Badurowa (1964), "The quantitative and qualitative characters of the mycoflora seem to depend chiefly on the kind of organic matter produced by trees," and that of Apinis (1972), who emphasized the remarkable relationship between phytocenoses, edaphic conditions, and mycocenoses in summing up his more than 20 years of experience with soil fungal communities. Both Apinis (1972) and Gochenaur and Whittingham (1967), however, recognized that occurrence of fungi is related to physical factors of the soil as well as type of vegetation. Christensen (1969), in a survey of the soil microfungi of conifer-hardwood forests in Wisconsin, found that calcium content of the litter superseded vegetation in the first axis of an ordination of microfungal communities; second- and third-axis separations established the correlation between microfungal dominants and tree dominants.

Dilution plating and soil washing appear to provide comparable estimates of principal forms and the relative prominence of each in native soils. Söderström's (1975) species list for a Swedish *Picea* forest (isolation by soil washing) and Morrall and Vanterpool's (1968) and Christensen's (1969) species lists for North American conifer and conifer-hardwood communities (isolation by dilution plating) shared 11, 12, and 12 "top-30" species, respectively. Twelve was the highest number of shared species obtained in 112 comparisons involving forests in different states or countries. Clearly the similarities in soil mycofloras concomitant with soil-vegetation similarity far exceed the dissimilarities attributable to isolation method.

ACKNOWLEDGMENTS

The author thanks Ann C. Worley, Paula Simms, and Ramona Wilson for technical assistance and gratefully acknowledges helpful discussions with Mike and Edie Allen, Ed Beals, Sally Gochenaur, J. H. Lussenhop, and R. H. Thornton.

REFERENCES

Apinis, A. E. (1958). Distribution of microfungi in soil profiles of certain alluvial grasslands. *Angew. Pflanzensoziol. 15*: 83-90.

Apinis, A. E. (1972). Facts and problems. *Mycopathol. Mycol. Appl. 48*: 93-109.

Apinis, A. E., and Chesters, C. G. C. (1964). Ascomycetes of some salt marshes and sand dunes. *Trans. Brit. Mycol. Soc. 47*: 419-435.

Apinis, A. E., Chesters, C. G. C. and Taligoola, H. K. (1972). Colonization of *Phragmites communis* leaves by fungi. *Nova Hedwigia 23*: 113-124.

Badura, L., and Badurowa, M. (1964). Some observations on the mycoflora in the litter and soil of the beech forest in Lubsza region. *Acta Soc. Bot. Poloniae 33*: 507-527.

Badurowa, M., and Badura, L. (1967). Further investigations on the relationship between soil fungi and the macroflora. *Acta Soc. Botan. Pol. 36*: 515-529.

Bazzaz, F. A. (1975). Plant species diversity in old-field successional ecosystems in southern Illinois. *Ecology 56*: 485-488.

Beals, E. W. (1960). Forest bird communities in the Apostle Islands of Wisconsin. *Wilson Bull. 72*: 156-181.

Bisby, G. R. (1943). Geographical distribution of fungi. *Bot. Rev. 9*: 466-482.

Borut, S. (1960). An ecological and physiological study on soil fungi of the northern Negev (Israel). *Bull. Res. Council Israel 8D*: 65-80.

Brown, J. C. (1958). Soil fungi of some British sand dunes in relation to soil type and succession. *J. Ecol. 46*: 641-664.

Christensen, M. (1960). The soil microfungi of conifer-hardwood forests in Wisconsin. Ph.D. thesis, University of Wisconsin, Madison.

Christensen, M. (1969). Soil microfungi of dry to mesic conifer-hardwood forests in northern Wisconsin. *Ecology 50*: 9-27.

Christensen, M., and Whittingham, W. F. (1965). The soil microfungi of open bogs and conifer swamps in Wisconsin. *Mycologia 57*: 882-896.

Christensen, M., Whittingham, W. F., and Novak, R. O. (1962). The soil microfungi of set-mesic forests in southern Wisconsin. *Mycologia 54*: 374-388.

Ciborska, E., and Zadara, M. (1974). Studies on the mycoflora of meadow soil in Kazun. *Acta Mycol. 10*: 159-169.

Clarke, D. C. (1973). The soil microfungi of a western South Dakota grassland. M.S. thesis, University of Wyoming, Laramie.

Curtis, J. T. (1959). *The Vegetation of Wisconsin: An Ordination of Plant Communities*. Univ. of Wisconsin Press, Madison.

Dowding, P., and Widden, P.   (1974).   Some relationships between fungi and their environment in tundra regions.   In *Soil Organisms and Decomposition in Tundra*, A. J. Holding, O. W. Heal, S. F. McLean, Jr., and P. W. Flanagan (Eds.).   Tundra Biome Steering Committee, Stockholm, pp. 123-150.

Dwivedi, R. S.   (1966).   Ecology of the soil fungi of some grasslands of Varanasi. *Trop. Ecol. 7*:   84-99.

England, C. M., and Rice, E. L.   (1957).   A comparison of the soil fungi of a tallgrass prairie and an abandoned field in central Oklahoma.   *Bot. Gaz. 118*:   186-190.

Flanagan, P. W., and Scarborough, A.   (1974).   Physiological groups of decomposer fungi on tundra plant remains.   In *Soil Organisms and Decomposition in Tundra*, A. J. Holding, O. W. Heal, S. F. MacLean, Jr., and P. W. Flanagan (Eds.).   Tundra Biome Steering Committee, Stockholm, pp. 159-181.

Frankland, J. C.   (1966).   Succession of fungi on decaying petioles of *Pteridium aquilinum*.   *J. Ecol. 54*:   41-63.

Franz, H.   (1963).   Biozönotische und synökologische Untersuchungen über die Bodenfauna und ihre Beziehungen zur Mikro- und Makroflora.   In *Soil Organisms*, J. Doeksen and J. van der Drift (Eds.).   North-Holland Publ., Amsterdam, pp. 345-367.

Gams, W.   (1971).   *Tolypocladium*, eine Hyphomycetengattung mit geschwollenen Phialiden.   *Persoonia 6*:   185-191.

Garrett, S. D.   (1955).   Microbial ecology of the soil.   *Trans. Brit. Mycol. Soc. 38*:   1-9.

Gleason, H. A., and Cronquist, A.   (1964).   *The Natural Geography of Plants*.   Columbia Univ. Press, New York.

Gochenaur, S. E.   (1970).   Soil mycoflora of Peru.   *Mycopathol. Mycol. Appl. 42*: 259-272.

Gochenaur, S. E., and Backus, M. P.   (1967).   Mycoecology of willow and cottonwood lowland communities in southern Wisconsin.   II.   Soil microfungi in the sandbar willow stands.   *Mycologia 59*:   893-901.

Gochenaur, S. E., and Whittingham, W. F.   (1967).   Mycoecology of willow and cottonwood lowland communities in southern Wisconsin.   I.   Soil microfungi in the willow-cottonwood forests.   *Mycopathol. Mycol. Appl. 33*:   125-139.

Gochenaur, S. E., and Woodwell, G. M.   (1974).   The soil microfungi of a chronically irradiated oak-pine forest.   *Ecology 55*:   1004-1016.

Griffin, D. M.   (1972).   *Ecology of Soil Fungi*.   Syracuse Univ. Press, Syracuse, N.Y.

Hudson, H. S.   (1968).   The ecology of fungi on plant remains above the soil.   *New Phytol. 67*:   837-874.

Jefferys, E. G., Brian, P. W., Hemming, H. C., and Lowe, D.   (1953).   Antibiotic production by the microfungi of acid heath.   *J. Gen. Microbiol. 9*:   314-341.

Jensen, V.   (1963).   Studies on the microflora of Danish beech forest soils.   V.   The microfungi.   *Zbl. Bakteriol. [Abt. 2] 117*:   167-179.

Joffe, A. Z.   (1967).   The mycoflora of a light soil in a citrus fertilizer trial in Israel.   *Mycopathol. Mycol. Appl. 32*:   209-229.

Kershaw, K. A.   (1973).   *Quantitative and Dynamic Plant Ecology*, 2nd ed.   Elsevier, New York.

Krzemieniewska, H., and Badura, L.   (1954).   Zbadan nad mikoflora lasu bukowego. (Some observations on the microflora of beech woods.)   *Acta Soc. Pol. 23*:   545-587.

Kuehn, H. H.   (1960).   Fungi of New Mexico.   *Mycologia 52*:   535-544.

Lange, Lene.   (1974).   The distribution of macromycetes in Europe.   *Dansk Bot. Ark. 30*:   7-105.

Lindsey, B. I., and Pugh, G. J. F.   (1976).   Succession of microfungi on attached leaves of *Hippophaë rhamnoides*.   *Trans. Brit. Mycol. Soc. 67*:   61-67.

Lisiewska, Maria. (1974). Macromycetes of beech forests within the eastern part of the *Fagus* area in Europe. *Acta Mycol. 10:* 3-72.

Llanos, C., and Kjøller, A. (1976). Changes in the flora of soil fungi following oil waste application. *Oikos 27:* 377-382.

Loucks, O. L. (1970). Evolution of diversity, efficiency, and community stability. *Amer. Zool. 10:* 17-25.

McLennan, E. I., and Ducker, S. C. (1954). The ecology of the soil fungi of an Australian heathland. *Aust. J. Bot. 2:* 220-245.

Mallik, M. A. B., and Rice, E. L. (1966). Relation between soil fungi and seed plants in three successional forest communities in Oklahoma. *Bot. Gaz. 127:* 120-127.

Mangenot, F., and Reisinger, O. (1976). Form and function of conidia as related to their development. In *The Fungal Spore: Form and Function*, D. J. Weber and W. M. Hess (Eds.). Wiley, New York, pp. 789-847.

Martin, J. P. (1950). Effects of fumigation and other soil treatments in the greenhouse on the fungus population of old citrus soil. *Soil Sci. 69:* 107-122.

Meyer, J. A. (1963). Écologie et sociologie des microchampignons du sol de la cuvette centrale congolaise. *Publns. Inst. Nat. Étude Agron. Congo. (I.N.É.A.C.), Sér. Sci. No. 101,* 137 pp.

Morrall, R. A. A. (1974). Soil microfungi associated with aspen in Saskatchewan: Synecology and quantitative analysis. *Can. J. Bot. 52:* 1803-1817.

Morrall, R. A. A., and Vanterpool, T. C. (1968). The soil microfungi of upland boreal forest at Candle Lake, Saskatchewan. *Mycologia 60:* 642-654.

Mosca, A. M. (1960). Sulla micoflora del terreno di un pascolo alpino in val di Lanzo (Alpi Graie). *Allionia 6:* 17-34.

Moubasher, A. H., and Moustafa, A. F. (1970). A survey of Egyptian soil fungi with special reference to *Aspergillus*, *Penicillium* and *Penicillium*-related genera. *Trans. Brit. Mycol. Soc. 54:* 35-44.

Mueller-Dombois, D., and Ellenberg, H. (1974). *Aims and Methods of Vegetation Ecology*. Wiley, New York.

Mueller-Dombois, D., and Perera, M. (1971). Ecological differentiation and soil fungal distribution in the montane grasslands of Ceylon. *Ceylon J. Sci. 9:* 1-41.

Novak, R. O., and Whittingham, W. F. (1968). Soil and litter microfungi of a maple-elm-ash floodplain community. *Mycologia 60:* 776-787.

Orpurt, P. A., and Curtis, J. T. (1957). Soil microfungi in relation to the prairie continuum in Wisconsin. *Ecology 38:* 628-637.

Park, D. (1963). The presence of *Fusarium oxysporum* in soils. *Trans. Brit. Mycol. Soc. 46:* 444-448.

Parker-Rhodes, A. F. (1955). The Basidiomycetes of Skokholm Island. XII. Correlation with the chief plant associations. *New Phytol. 54:* 259-275.

Peyronel, B. (1955). Proposta di un nuovo metodo di rappresentazione grafica della composizione dei Consorzi Vegetali. *Nuova Giorn. Bot. Ital.* [*n.s.*] *62:* 379-383.

Ranzoni, F. V. (1968). Fungi isolated in culture from soils of the Sonoran desert. *Mycologia 60:* 356-371.

Richardson, M. J. (1970). Studies on *Russula emetica* and other agarics in a Scots pine plantation. *Trans. Brit. Mycol. Soc. 55:* 217-229.

Ruscoe, Q. W. (1973). Changes in the mycofloras of pasture soils after long-term irrigation. *N. Z. J. Science 16:* 9-20.

Sappa, F. (1955). The soil fungus population as structural element of the plant communities. *Allionia 2:* 293-345.

Scarborough, A. M. (1970). The soil microfungi of a Colorado grassland. M.S. thesis, University of Wyoming, Laramie.

Sewell, G. W. F. (1959). The ecology of fungi in *Calluna*-heathland soils. *New Phytol. 58*: 5-15.

Singh, P. (1976). Some fungi in the forest soils of Newfoundland. *Mycologia 68*: 881-890.

Söderström, B. E. (1975). Vertical distribution of microfungi in a spruce forest in the south of Sweden. *Trans. Brit. Mycol. Soc. 65*: 419-425.

States, J. S. (1978). Soil fungi of cool-desert plant communities in northern Arizona and southern Utah. *J. Arizona-Nevada Acad. Sci. 13*: 13-17.

Thornton, R. H. (1956). Fungi occurring in mixed oakwood and heath soil profiles. *Trans. Brit. Mycol. Soc. 39*: 485-494.

Thornton, R. H. (1958). Biological studies of some tussock grassland soils. II. Fungi. *N. Z. J. Agr. Res. 1*: 922-938.

Tresner, H. D., Backus, M. P., and Curtis, J. T. (1954). Soil microfungi in relation to the hardwood forest continuum in southern Wisconsin. *Mycologia 46*: 314-333.

Warcup, J. H. (1951a). The ecology of soil fungi. *Trans. Brit. Mycol. Soc. 34*: 376-399.

Warcup, J. H. (1951b). Effect of partial sterilization by steam or formalin on the fungus flora of an old forest nursery soil. *Trans. Brit. Mycol. Soc. 34*: 519-532.

Warcup, J. H. (1960). Methods for isolation and estimation of activity of fungi in soil. In *The Ecology of Soil Fungi*, D. Parkinson and J. S. Waid (Eds.). Liverpool Univ. Press, Liverpool, pp. 3-21.

Whittaker, R. H. (1972). Evolution and measurement of species diversity. *Taxon 21*: 213-251.

Whittaker, R. H. (1975). *Communities and Ecosystems*, 2nd ed. Macmillan, New York.

Wicklow, D. T. (1973). Microfungal populations in surface soils of manipulated prairie stands. *Ecology 54*: 1302-1310.

Wicklow, D. T., and Whittingham, W. F. (1974). Soil microfungal changes among the profiles of disturbed conifer-hardwood forests. *Ecology 55*: 3-16.

Wicklow, M. C., Bollen, W. B., and Denison, W. C. (1974). Comparison of soil microfungi in 40-year-old stands of pure alder, pure conifer, and alder-conifer mixtures. *Soil Biol. Biochem. 6*: 73-78.

Wilkins, W. H., and Patrick, S. H. M. (1940). The ecology of the larger fungi. IV. The seasonal frequency of grassland fungi with special reference to the influence of environmental factors. *Ann. Appl. Biol. 27*: 17-34.

Youssef, Y. A. (1974). On the fungal flora of Libyan soils. *Arch. Microbiol. 99*: 167-171.

Chapter 13

CAUSAL FACTORS OPERATIVE IN DISTRIBUTIONAL PATTERNS
AND ABUNDANCE OF FUNGI: A MULTIVARIATE STUDY

Henry A. H. Wallace

University of Manitoba and
Research Station, Agriculture Canada
Winnipeg, Manitoba, Canada

R. N. Sinha

Research Station, Agriculture Canada and
University of Manitoba
Winnipeg, Manitoba, Canada

I. INTRODUCTION

Ecological studies on the role of fungi and microarthropods in cereal grain agroeco-systems have posed several complex questions pertaining to cause and effect relationships. We have attempted to solve these problems through the use of multivariate statistical methods. Some of the multivariate techniques that can be used to analyze our data are cluster, multiple regression, principal component, factor, canonical correlation, and discriminant function analyses. The nature of the problem, the data structure, and the user's knowledge determine the choice of technique. Each of these analyses has its limitations, some more than others. Sinha (1977) has discussed the assumptions and limitations of the above techniques in brief nonmathematical language. Considerable knowledge of the biological subject matter involved, some help from multivariate statisticians and a programmer, and an access to a digital computer are prerequisites to the application of these methods. The mathematical and statistical basis for these techniques are given by Anderson (1958), Kendall (1965), and Morrison (1967). Various applications and limitations of these analyses in biology are given by Seal (1964), Pearce (1965), Cassie (1969), Bryant and Atchley (1975), Orloci (19-75), Sinha (1977), and Mills et al. (1978b).

Because a niche in which organisms live and interact has many dimensions, our strategy has been to measure as many relevant variables as possible and thereby characterize the role of fungi in man-made field crop or stored grain ecosystems. As our knowledge of the organisms grows, we discard irrelevant or barely relevant variables and incorporate new and more relevant variables. To synthesize the data and to uncover natural interrelations among several variables and develop meaningful hypotheses, we selectively apply one or more multivariate statistical techniques to the same set of data. Most of the mathematical models involved in these techniques, however, are linear; the dynamic aspects of events in the life of an organism or population are often missed. We make systematic comparisons of interrelationship patterns among variables as revealed by the multivariate analysis of the data. Real-life, field, or granary data using the principal component analysis (PCA) technique help us generate hypotheses by deduction rather than by induction alone.

Between 1953 and 1960 the first author spent many months making correlations between cultivars and their seed-borne fungi. Since then, the development of high-speed computers with large memories has enabled us to use sophisticated procedures to begin deciphering the relationships among factors responsible for the organization of fungal communities.

In this study we have used PCA because it seems to be most suited for examining exploratory survey and ecological data. PCA screens the data from either the correlation matrix or from the variance-covariance matrix to determine the main constituents, the underlying structure, if any, and order in the system. An objective examination of PCA should indicate whether most of the information can be expressed by a few variables called principal components. PCA does not assume an underlying structure; it does assume that the variables are measured on random samples from a homogeneous population and that their frequency distribution is normal or linear. Because our variables were measured in different units, we were obliged to use the correlation matrix (variance-covariance requires measurement in the same unit). There is no useful statistical test of significance for the number of principal components when a correlation matrix is used.

Proper interpretation of PCA depends on the user's ability to correctly answer several questions that arise before and during the analysis. For example: How many variables are there, and of what type? What is the size of the sample? What type of transformation should be used to achieve linearity in distribution, and how many components are to be extracted and used for interpretation? PCA is used in this paper for a multivariate analysis of the cereal grain ecosystem.

There are three phases between the initial development of cereal seeds and the final usage of the mature seed by the consumers. The seed development, or *preharvest*, phase includes growth and development of the embryo and maturation of the seed. The *harvest* phase includes cutting, swathing, threshing, and trucking to storage. The *postharvest* phase includes storage and transportation until the grain is processed for consumption or is used for seed. Fungi that characteristically occur in any of these three phases are designated by the name of the phase--preharvest, harvest, and postharvest fungi. The subject of the first PCA is the preharvest fungi associated with freshly harvested barley seed. The characteristics of this field crop ecosystem are the large area involved, the geographic location, the effect of climate, the limited time, aerial dispersal of spores, and seed susceptibility to preharvest fungi. The subject of the second PCA is the postharvest fungi associated with stored wheat seed. The characteristics of this stored crop ecosystem are the restricted area (e.g., within two attached storage bins), the temperature and moisture content when the grain was stored, the effect of time (8 years) and the annual cycle of temperature on migration of moisture, and the consequent effects on insects, mites, and fungi.

Although the first PCA is concerned with barley seed production and the second PCA with storage of wheat, our experience has shown that useful generalizations from such studies can be made for other cereal crops.

II. PRINCIPAL COMPONENT ANALYSIS OF VARIABLES ASSOCIATED WITH BARLEY SEED PRODUCTION: PREHARVEST FUNGI

A. Sources of Data

In Canada the *Western Cooperative Barley Test* is conducted by the Associate Commit-tee of Plant Breeding of the National Research Council and the Canada Department of Agriculture. Within each year, the same cultivars and lines of barley are sown at about 20 locations. Although there are some changes of cultivars and lines from year to year, the overall effects on fungi are small (Table 1). At each location six (oc-casionally four) replicates are sown. During the season many variables are measured on the crop. These include date of heading, days to maturity, and disease ratings for all leaf diseases. Seed obtained from each replicate (where possible) at each location are plated on moist filter paper and germination and the frequency of occur-rence (percentage) of each fungus recorded.

During the years 1953-1960 ten genera of fungi were relatively consistent in frequency of occurrence from year to year (Table 1). Because the data for 1956 were the most complete and compared favorably with the 8-year average, they were used for the principal component analysis.

For PCA of the Western Cooperative Barley Test for 1956, the means of the 4 or 6 replicates of each of the 25 cultivars or lines at 17 locations were used. The number of samples is 425 (25 X 17).

The locations in 1956 were as follows: Agassiz, in British Columbia; Beaver-lodge, Edmonton, Lacombe, and Lethbridge, in Alberta; Swift Current, Scott, Saskatoon,

Table 1 Geographical distribution of fungi associated with barley seed in 1956 and for the years 1953-1960 (in parentheses)[a]

	Mean percentage of occurrence				
Fungi	British Columbia 25 (145)	Alberta 100 (561)	Saskatchewan 175 (841)	Manitoba 100 (671)	Ontario 25 (161)
Alternaria	14.0 (42.0)	45.8 (34.0)	76.7 (66.0)	85.0 (85.0)	82.0 (74.0)
Cephalosporium	0.0 (0.3)	1.7 (1.0)	0.4 (0.7)	10.8 (4.3)	0.1 (0.1)
Cladosporium	15.0 (24.3)	29.5 (36.1)	21.6 (14.9)	8.3 (6.1)	3.0 (11.5)
Cochliobolus	0.3 (0.2)	1.5 (1.1)	3.2 (2.0)	30.6 (21.5)	25.2 (11.1)
Epicoccum	0.5 (2.9)	3.4 (1.8)	1.5 (1.2)	0.4 (0.4)	0.3 (0.3)
Fusarium	11.6 (3.3)	8.9 (3.0)	0.8 (0.7)	2.2 (2.0)	1.9 (1.8)
Penicillium	0.7 (1.2)	0.4 (1.4)	1.0 (3.5)	0.6 (0.4)	0.5 (0.9)
Pyrenophora	0.0 (0.0)	0.2 (0.6)	1.4 (0.6)	0.7 (0.8)	0.0 (0.0)
Streptomyces	1.7 (8.7)	8.3 (15.3)	6.5 (9.6)	21.7 (5.6)	1.7 (1.6)
Trichothecium	20.0 (28.7)	17.4 (17.5)	41.3 (37.7)	9.4 (18.0)	1.1 (6.3)

[a]The number of samples are given below each province.

Melfort, Snowden, Tisdale, and Indian Head, in Saskatchewan; Brandon, Melita, Portage, and Winnipeg, in Manitoba; and Ottawa, in Ontario.

The filter paper method was used to determine germination and microflora of the seed (Wallace and Sinha, 1962).

PCA was carried out with the IBM 360-65 digital computer at the University of Manitoba, Winnipeg. The computer program was written by P. J. Lee, Imperial Oil Company, Calgary, Alberta.

B. Description of Variables

The variables used in this analysis, identified by number, are described below; since the geographical range of the tests was very large, the ranges within variables are large. The range is given in parenthesis.

1. *Temperature*--the mean of daily mean temperature for the 32 days prior to date of harvesting, from the nearest available location to the test (12-20°C). The 32 days are the average period between the date of heading and date of harvesting.

2. *Rainfall*--the total rainfall for the same 32 days (20-130 mm). Data for variables 1 and 2 were obtained from the Canada Department of Transport, Ottawa.

3. *Yield*--mean of each cultivar or line at each location (26.9-108.7 kg/ha).

4. *Bushel weight* (bulk density or test weight)--mean of each cultivar or line at each location (48.6-73.6 kg/hl).

5. *Kernel weight*--mean of each cultivar or line at each location (26-57 g per 1000 kernels).

6. *Growth period*--between date of seeding and harvesting (74-113 days). The wide range is due to presence of early and late varieties and to location.

7. *Germination*--seed plated on moist filter paper and incubated at room temperature (ca. 22°C) for one week (46-99%).

8. *Alternaria*--seed infected with *A. alternata* (Fr.) Keissler (= *A. tenuis* sensu Wiltshire) (8-97%).

9. *Cochliobolus*--seed infected with *C. sativus* (Ito and Kurib.) Drechsl. ex Dastur/conidial state: *Drechslera sorokiniana* (Sacc.) Subram. and Wain = *Bipolaris sorokiniana* (Sacc. in Sorok.) Shoem. = *Helminthosporium sativum* Pamm., King and Bakke (0-85%).

10. *Cladosporium*--seeds infected by this genus, usually *C. cladosporioides* (Fres.) DeVries (1-58%).

11. *Fusarium*--seeds infected by this genus, usually *F. poae* (Pk.) Wr. (0-33%).

12. *Streptomyces*--seeds infected by this genus, usually *S. griseus* (Krainsky) Waksman (0-65%). They are probably epiphytes living on the surface of the seed. Although this genus belongs to the Actinomycetes, for convenience it is included as a fungus.

13. *Trichothecium*--seeds infected by *T. roseum* Link (0-97%).

14. *Cephalosporium*--seeds infected by *Cephalosporium* sp. (0-60%). For a discussion of the problems of identification of species of this fungus see Barron (1968).

15. *Pyrenophora*--seeds infected with *P. teres* (Died.) Drechsl. (conidial state; *Drechslera teres* (Sacc.) Shoem. (= *Helminthosporium teres* Sacc.) (0-10%).

16. *Epicoccum*--seeds infected with *E. nigrum* Link (0-11%).

17. *Penicillium*--seeds infected with *Penicillium* spp. (0-12%).

The frequency distributions for most variables used in both PCA were asymmetrical, usually J-shaped, or positively or negatively skewed. Because few of the variables measured had multivariate normal distribution, a necessary assumption for PCA (Sinha, 1977), apropriate transformations were applied. Only data with asymmetric distributions were transformed; for example, data for most J-shaped distributions were transformed to logarithms, and percentages to arc sines.

C. Interpretation of Data

Six principal components extracted from the 17 X 17 correlation matrix accounted for 72.17% of the total variability (Table 2). An arbitrary cutoff of a 0.25 loading (= component coefficient) was used because this loading enabled us to include all variables, e.g., a 0.30 loading would have eliminated bushel weight (Fig. 1). An experienced experimenter is usually able to determine the level at which natural groupings occur, thus eliminating the arbitrariness of the selection process. Generally, opposite signs before the loadings of two variables in one component was interpreted as an inverse relationship for that component. When two variables had the same signs in one component and opposite signs in another, then the variables behaved in both similar and opposing ways at different times. The above are the literal interpretations of mathematical science, but in practice biologists find that two or more variables with the same signs, whether positive or negative, indicate that some sort of relationship exists between the variables. To reflect correct inverse relationship, two variables must maintain a crucial ratio in quantitative terms. Correct interpretation of these signs in the PCA usually requires a knowledge of the environment and the behavior patterns of the organism which can be established by a priori experiments (Sinha, 1977). A change of sign may not necessarily mean a biological contradiction or a statistical error, because several species of fungi can infect the same seed and their behavior and ecological requirements overlap (Mills et al., 1978a).

The first principal component accounts for 24.9% of the total variability (Table 2). This component is composed of three groups: (1) *Cochliobolus*, *Alternaria*, and *Cephalosporium* with positive loadings; (2) *Cladosporium* and *Epicoccum* with negative loadings; and (3) "growth period" and "bushel weight," also with negative loadings (Fig. 1). All the aforementioned fungi are preharvest fungi.

The salient features of the first group are as follows: *Cochliobolus* has the highest loading and occurs only in this component (Fig. 1); *Cephalosporium*, a mycoparasite of *Drechslera* spp., was frequently associated under our conditions with *Cochliobolus* (*Helminthosporium*); *Alternaria*, *Cochliobolus*, and *Cephalosporium* are most abundant on barley seed in Manitoba (Table 1), where the growth period is gen-

Table 2 Principal component matrix with loadings showing the degree to which the original 17 variables are expressed in the six principal components[a]

Variable	Principal component					
	c_I	c_{II}	c_{III}	c_{IV}	c_V	c_{VI}
1. Temperature	22	44	09	10	0	11
2. Rainfall	24	17	10	-36	26	34
3. Yield	06	-18	21	-45	16	-58
4. Bushel weight	-25	28	-10	02	27	-02
5. Kernel weight	-20	-24	31	-29	06	-03
6. Growth-period	-36	-20	02	02	-04	25
7. Germination	-09	40	27	16	07	-34
8. *Alternaria*	31	0	-44	-16	11	-18
9. *Cochliobolus*	41	02	0	21	-20	08
10. *Cladosporium*	-37	02	-14	22	-26	-07
11. *Fusarium*	-06	-20	51	-02	-01	33
12. *Streptomyces*	22	-30	12	30	44	-10
13. *Trichothecium*	-17	-17	-25	24	54	-04
14. *Cephalosporium*	32	-30	09	29	11	17
15. *Pyrenophora*	-01	-38	-38	-18	-24	03
16. *Epicoccum*	-26	-09	-09	08	33	12
17. *Penicillium*	-02	-21	-21	-39	21	38
Variability accounted for (%)	24.9	14.8	10.9	8.1	7.3	6.2

[a]Component loadings have been multiplied by 100; those below 25 were not interpreted.

erally (e.g., in 1956) relatively short; their occurrence is not affected by temperature or rainfall, and they can all occur on the same seed.

In contrast, the second group of fungi, consisting of *Cladosporium* and *Epicoccum*, are most frequently found in Alberta (Table 1). Their appearance might have been a result of lack of competition from *Alternaria* and *Cochliobolus*, which--unlike *Cladosporium* and *Epicoccum*--usually penetrate the seed pericarp. Furthermore, *Cladosporium* is a slower growing fungus than the other members of the group. Because bushel weight increases during a long growth period, it is possible that the late-maturing cultivars had relatively high bushel weight.

The second principal component accounts for 14.8% of the total variability. The positive loadings include temperature, germination, and bushel weight; the negative loadings include *Pyrenophora*, *Streptomyces*, and *Cephalosporium*. Temperature is

Principal Component   Variable   Principal Component

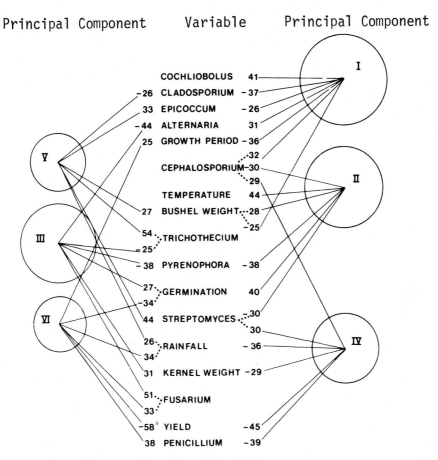

Fig. 1 Diagram illustrating effects of principal components on variables. Numbers at the inner ends of the lines represent the principal component loadings (without the decimal points) of variables. The area of the circles enclosing the principal components is relative to the percentage variability accounted for by each principal component.

the dominant variable in this component. Low temperatures at several locations seem to have caused seed dormancy of some cultivars, and also affected seed germination. Moderate temperatures also facilitate leaf infection by *Pyrenophora*, causing net blotch. Leaf infection and not seed infection by this fungus may be responsible for reduced bushel weight. Apparently, *Cephalosporium* is a mycoparasite of *Pyrenophora* (Kenneth and Isaac, 1964), but we have not observed this phenomenon. Moderate temperatures also appear to favor *Streptomyces*.

The third component accounts for 10.9% of the variability. The positive loadings are *Fusarium*, kernel weight, and germination; the negative loadings are *Alternaria*, *Pyrenophora*, and *Trichothecium*. *Fusarium* was the dominant variable. It was most abundant in British Columbia and northern Alberta. In contrast, *Alternaria*, *Trichothecium*, and *Pyrenophora* were more abundant in Saskatchewan (Table 1). It is

possible that *Alternaria* infection reduced seed size, because the largest kernels
were from British Columbia.

The fourth component accounts for 8.1% of the variability. Again, we have two
natural groups, *Cephalosporium* and *Streptomyces* with positive loadings (compare with
the second component), whereas yield, *Penicillium*, rainfall, and kernel weight have
negative loadings. Rainfall is asociated with kernel weight and yield, and it facil-
itates growth and contamination by *Penicillium*. A possibility of antagonism between
Streptomyces and *Penicillium* exists.

Although the aforementioned four components account for all the variables, components V and VI (Fig. 1) also have an important contribution to make to this ecosystem. Components IV, V, and VI each contain the variable, rainfall, but with contradictory effects.

The fifth component has five variables with positive loadings, viz., *Trichothecium*, *Streptomyces*, *Epicoccum*, bushel weights, and rainfall; and only one negative
loading, *Cladosporium*. This component contains the harvest fungus *Trichothecium*,
Streptomyces and rainfall. We are now faced with the problem of conflicting positive and negative loadings. In component IV the loadings for rainfall and *Streptomyces* were -36 and 30, respectively, but in component V they were 26 and 44, respectively. We suggest that in component IV rainfall encouraged growth of *Penicillium*
to the detriment of *Streptomyces*, but in component V *Penicillium* was absent and
therefore rainfall encouraged growth of *Streptomyces*. In component IV rainfall and
Penicillium have negative loadings, but in component VI they are positive--which
simply means that they respond in the same way.

One of the advantages of PCA is its ability to sort the variables into groups,
e.g., C_I is composed of preharvest fungi; C_{III} and C_{IV} are combinations of several
preharvest, harvest, or postharvest fungi; C_V consists of harvest fungi; and C_{VI},
postharvest fungi (*Fusarium* can occur in stored grain). Likewise, PCA shows the
relative importance of environmental factors, e.g., temperature prevailed in C_{II},
rainfall in C_{IV}, C_V, and C_{VI}. Environmental factors were not involved in C_I and
C_{III}.

V. PRINCIPAL COMPONENT ANALYSIS OF VARIABLES ASSOCIATED
    WITH STORED WHEAT

A. Sources of Data

In November 1959 about 13.6 t* (500 bu) of No. 4 Manitoba Northern wheat of 13.2-
13.8% moisture content were placed in each of two identical attached wooden bins (bins
A and B), the outsides of which were insulated with fiberglass. The bins were 305 X
335 cm and had concrete floors. A parcel of 27.2 kg (1 bu) of wheat with a moisture

---

*Tonnes (or metric tons).

content of 23% wet basis was inserted at a location 74 cm southwest of center at a
depth of 61 cm in bin B.

Samples of grain weighing 200 g were collected monthly with a brass torpedo tube
40 cm long and 3.5 cm in diameter. From the control bin, bin A samples were collected along five vertical lines located at the center and 75 cm diagonally in from each
corner. The samples were taken from the surface, 30, 61, 91, and 152 cm depths (5 X
5 samples). From test bin B, samples were taken along 20 vertical lines spaced equi-
distant apart on a square grid and from the same five depths (20 X 5 samples). From
November 1959 to August 1965, 125 samples were collected each month. The filter paper method was used to determine germination and microflora of the seed. In August
1975 the bins were fumigated with phosphine to determine the effect of this fumi-
gant on mites, insects, and microorganisms. Thereafter, fewer samples were taken,
and less frequently from various locations and depths.

Phosphine had no effect on fungi and therefore the original test was prolonged
to 1968. Altogether 8135 samples were examined and about 250,000 measurements taken.
Detailed description of the materials and methods are given by Sinha and colleagues
(1969a).

B. Description of Variables

The variables used in this analysis, identified by number, are described below.

 1. *Storage period*--indicates time in months elapsed from initial storage of
     grain and between sampling dates (1-97)

 2. *Bin*--involves 2 bins containing 500 bushels of wheat; these bins were
     given the arbitrary numbers of 1 and 2

 3. *Depth*--the depth down from the surface from which the samples were taken;
     these were 0, 30, 61, 91, and 152 cm

 4. *Germination*--the percentage viability of the seed

 5. *Temperature* (°C)--recorded monthly from all depths and at the grain sur-
     face

 6. *Moisture*--percentage moisture in each grain sample

Variables 7-18 and 28 are biological response variables and are expressed in percentages. They refer to species or species groups of fungi or actinomycetes. To avoid
repetition, those variables in the previous PCA are not described in detail here.

 7. *Alternaria*

 8. *Cochliobolus*

 9. *Cladosporium*

10. *Nigrospora sphaerica* (Sacc.) Mason

11. *Penicillium*--includes *P. cyclopium*, *P. funiculosum* and some other unident-
     ified species

12. *Aspergillus versicolor* (Vuill.) Tiraboschi

13. *Aspergillus flavus* Lk.

14. *Streptomyces griseus* (Krainsky) Waksman and Henrici--a common seed-borne
     actinomycete; for convenience, it is referred to as a fungus

15. *Aspergillus* spp.-- includes the *A. glaucus* group, *A. niger* van Tiegh, *A. ochraceus* Wilk., and *A. candidus* Lk.

16. *Absidia*--includes *A. ramosa* (Lindt) Lendner, *A. corymbifera* (Cohn) Sacc. and Trotter, and occasionally *A. orchidis*

17. *Chaetomium*--includes *C. funicola* Cooke and *C. dolichotrichum* Ames.

18. *Rhizopus arrhizus* Fischer

19. *Gonatobotrys simplex* Cda.

Response variables 20-30 are populations of species or group species or arthropods as determined by the Berlese funnel method and expressed as number per 200-g grain sample. Variables 21-27 are species of mites (Acarina).

20. *Lepinotus reticulatus* Endl. (Psocoptera)--a reticulate-winged booklouse common in grain bulks

21. *Lepidoglyphus destructor* (Schr.)--the long-haired mite

22. *Tydeus interruptus* Thor

23. *Cheyletus eruditus* (Schr.)--a predatory mite--the cannibal mite

24. *Tarsonemus* spp. (mainly *T. granarius* Lindquist)

25. *Acarus siro* L.--the grain mite

26. *Aëroglyphus robustus* (Banks)--the warty grain mite

27. *Androlaelaps-Blattisocius*--two genera of mites often associated with rodents; because it is not possible to distinguish between *Androlaelaps casalis* (Berlese) and *Blattisocius keegani* Fox with a stereomicroscope and both genera have a similar predatory consumer role, they are coded as one variable

28. *Spinibdella bifurcata* Atyeo

29. *Cryptolestes ferrugineus* (Steph)--the rust grain beetle

30. *Cryptophagus-Ahasverus*--two genera of fungivorous beetles coded as one variable in this analysis; they include *Cryptophagus varus* Woodroffe and Coombs and *Ahasverus advena* (Waltl)

31. *Location*--sampling locations were divided into two groups, the peripheral and inner locations; the peripheral locations were subject to temperature extremes

32. *Fumigation*--grain bulks were fumigated with phosphine in August 1965 (Sinha et al., 1967); samples were divided into two groups, those taken before fumigation and those taken after fumigation

The effects of 11 principal components on 32 variables, as well as the percentage variability accounted for by each principal component, are shown in Table 3. In the previous PCA (Table 2) loadings of 0.25 were reported, but in this PCA a higher loading (0.30) was used because fewer variables would make the results more interpretable. In Table 3, all loadings below 0.30 have been eliminated and those above are considered "significant" for interpretation of relationships (a workable statistical test of significance for the magnitudes of loadings is not available). In the principal-component matrix (Table 3), *Alternaria* occurs in one component (C_I) and moisture occurs to about the same loading magnitude in four components (C_I, C_{II}, C_{III}, C_{IV}). Component C_I has 16 significant variables; C_X, only 2.

Table 3 Principal-component matrix with loadings showing the effects of 11 principal components on 32 variables, and the percentage of variability explained by each principal component[a]

		C_I	C_{II}	C_{III}	C_{IV}	C_V	C_{VI}	C_{VII}	C_{VIII}	C_{IX}	C_X	C_{XI}
1.	Storage period	89	-32	--	--	--	--	--	--	--	--	--
2.	Bin	--	--	--	--	52	--	--	30	--	--	--
3.	Depth	--	35	36	-41	--	--	--	--	--	--	--
4.	Germination	-77	--	33	--	--	--	--	--	--	--	--
5.	Temperature	--	45	34	43	--	--	--	--	--	--	--
6.	Moisture	34	33	42	-41	--	--	--	--	--	--	--
7.	*Alternaria*	-84	--	--	--	--	--	--	--	--	--	--
8.	*Cochliobolus*	-72	--	--	--	--	--	--	--	--	--	--
9.	*Cladosporium*	--	--	--	--	--	--	44	-31	--	--	--
10.	*Nigrospora*	-59	--	--	--	--	--	--	--	--	--	--
11.	*Penicillium*	51	--	-41	--	--	--	34	--	--	--	--
12.	*Aspergillus versicolor*	48	--	-45	--	--	--	--	--	--	--	--
13.	*A. flavus*	--	--	--	--	34	--	-30	--	--	--	38
14.	*Aspergillus* spp.	61	--	--	--	39	--	--	--	--	--	--
15.	*Streptomyces*	62	--	--	--	--	--	--	--	--	--	-34
16.	*Absidia*	35	--	--	--	53	--	--	--	--	--	--
17.	*Chaetomium*	70	--	--	--	--	--	--	--	--	--	--
18.	*Rhizopus*	43	--	--	--	--	--	--	--	30	-43	47
19.	*Gonatobotrys*	-41	--	--	--	--	31	-31	--	--	--	--
20.	*Lepinotus*	--	55	--	--	--	--	--	--	--	--	--
21.	*Lepidoglyphus*	--	48	35	--	--	-45	--	--	--	--	--
22.	*Tydeus*	--	--	--	--	--	45	--	--	--	--	--
23.	*Cheyletus*	--	66	--	--	--	--	--	--	--	--	--
24.	*Tarsonemus*	36	51	--	--	--	32	--	--	--	--	--
25.	*Acarus*	--	--	58	--	--	--	--	--	--	--	--
26.	*Aëroglyphus*	--	--	--	--	--	--	--	48	43	--	--
27.	*Androlaelaps-Blattisocius*	--	--	--	61	--	--	--	--	--	--	--
28.	*Spinibdella*	--	--	--	--	--	32	--	53	--	--	--
29.	*Cryptolestes*	--	--	--	--	--	--	--	--	--	-67	-37
30.	*Cryptophagus-Ahasverus*	--	--	--	--	--	41	--	31	-54	--	38
31.	Location	--	--	--	-31	-40	--	--	--	--	--	--
32.	Fumigation	60	-58	39	--	--	--	--	--	--	--	--
	Variability explained (%)	19.7	7.9	6.8	5.1	4.8	4.3	3.9	3.7	3.2	3.2	3.0

[a] Component loadings have been multiplied by 100. Only loadings of 30 or larger are reported.

Interpretation of the principal component matrix as given here has been influenced by the results of other multivariate analyses on the same data, e.g., a canonical correlation of seed viability, seed-borne fungi, and environment (Sinha et al., 1969b); effects of phosphine on mites, insects, and microorganisms (Sinha et al., 1967). In addition, the results of a comprehensive descriptive study of the fungi associated with wheat after prolonged storage, based also on the same data (Wallace et al., 1976), have been taken into account in our interpretation.

C. Fungal Components

The first principal component accounts for 19.7% of the variability explained. It has the highest loading on storage period and contains 11 of the 13 fungal variables (Table 3). This component clearly indicates that the duration of storage and the moisture content of the grain are responsible for the decline in frequency of occurrence of the preharvest fungi *Alternaria*, *Cochliobolus*, and *Nigrospora*, as well as for *Gonatobotrys* [which is dependent on *Alternaria* as a source of nutriment (see Whaley and Barnett, 1963)]. Storage period and moisture are also responsible for the increase in the postharvest fungi *Aspergillus* spp., *A. versicolor*, *Penicillium*, *Rhizopus*, and *Streptomyces*. Although *Chaetomium* also occurs in this group, Pelhâte (1968) classified it as an "intermediate" or harvest fungus. However, our experience with *Chaetomium* indicates that it occurs on old stored grain (Wallace et al., 1976) and therefore could be considered a postharvest fungus.

The variable "bin" appears to be responsible for the two fungal variables not occurring in the first component. *Cladosporium* has a negative (C_{III}) and *Aspergillus flavus* a positive (C_V) relation with bin. *Aspergillus* spp. and *Absidia* were also affected by bin. With these exceptions, the bins were not a factor affecting the presence and distribution of the other fungi.

The C_{VII} component contains only four fungi. *Gonatobotrys* is a parasite of *Cladosporium*, a fungus that is not as vigorous as *Alternaria* and hence appears to be retarded by its parasite. This component is the only one in which *Penicillium* and *Aspergillus flavus* occur together. We are not sure which of these fungi is the invader or antagonist, as they often occur on the same seed.

D. Mite and Insect Components

The C_{II} component accounts for only 7.9% of the variability. No fungus occurs in this component, but a psocopteran insect (*Lepinotus*) and also mites (*Lepidoglyphus*, *Cheyletus*, and *Tarsonemus* spp.) were prevalent. Therefore, this component is called "Arthropoda." These arthropods appear to be most affected by temperature and to a lesser extent by moisture. Their numbers were greatly reduced by fumigation--hence, the fumigation and time effect.

C_{VI} component accounts for 4.3% of the variability. This component has been called the "scavenger" component because *Lepidoglyphus*, *Tydeus*, and *Tarsonemus* spp.

are scavenger mites that feed on decomposed organisms and grain debris. *Spinibdella* is a predatory mite, and *Cryptophagus* and *Ahasverus* are beetles that feed on grain debris and fungi.

E. Fungal-Arthropod Components

The C_{III} component with 6.8% variation differs from all the other components in that abiotic variables, postharvest fungi, and mites are all involved. Temperature and particularly moisture favor development of the mites *Acarus* and *Lepidoglyphus*. We note the positive relationship of moisture to *Penicillium* and *Aspergillus versicolor* in the first principal component and the inverse relationship in the third principal component. There are two possible explanations for these superficially contradictory results: (1) in the first component these fungi infected the grain and reduced germination, whereas in the third component they were contaminants from the harvest processes that appeared on plates because the seed was not surface sterilized before plating; (2) *Acarus siro* can feed and breed on *Penicillium* spp., especially *P. cyclopium* (Sinha, 1964; Sinha and Mills, 1968) and *A. gracilis* on *Penicillium* and to some extent on *Aspergillus versicolor* (Sinha, 1966).

Rhizopus (a fungus) and the beetles *Cryptolestes*, *Cryptophagus*, or *Ahasverus*, or all three, occur in the C_{IX}, C_X, and C_{XI} components. The fungus beetles *Cryptophagus* and *Ahasverus* may deter *Rhizopus* (C_{IX}), and the feeding of *Cryptolestes* on the embryo of the seed does facilitate seed infection by *Rhizopus* (C_X); but why the roles of these beetles are reversed when both are present is difficult to explain. The excreta of beetles is a good media for growth of *Streptomyces* (C_{XI}).

F. Abiotic Components

In C_I, storage period and moisture content along with postharvest fungi affect germination; in C_{III}, temperature, moisture, and storage fungi affect germination. Moisture (migration and accumulation) is closely associated with depth in C_{II}, C_{III}, and C_{IV}, as is horizontal location in C_{IV}. In C_I, the effect of fumigation is coincidental since the preharvest fungi had nearly all disappeared and the storage fungi were increasing. As expected, fumigation was effective against mites in C_{II}. The relationship of *Lepidoglyphus* and *Acarus* is not clear, but they may have appeared independently after fumigation.

In both principal component analyses, time (growth period and storage period, or month) and moisture (rainfall) are the predominant variables affecting fungi. Surprisingly, temperature played a secondary role (*Pyrenophora* excepted) in the occurrence of fungi. However, temperature was a requirement for mites. The relationship of mycoparasites and their host fungi was indicated, as was that of predator mites and their prey. Generally, fungi and arthropods appeared in different components; therefore, they tend to be independent of each other. However, *Lepidoglyphus* and *Tarsonemus* may feed on some postharvest fungi. Whether the fungus beetles feed on *Rhizopus* is unknown.

ACKNOWLEDGMENTS

We thank Mr. R. W. Sims for preparing the illustrations; Drs. S. R. Loschiavo, J. T. Mills, W. E. Muir, and F. L. Watters of Winnipeg for critical reviews of the manuscript; and Dr. J. Lacey of Rothamsted, United Kingdom, for helping us develop some of the concepts presented in this chapter.

REFERENCES

Anderson, T. W. (1958). *An Introduction to Multivariate Statistical Analysis.* Wiley, New York.

Barron, G. L. (1968). *The Genera of Hyphomycetes From Soil.* Williams & Wilkins, Baltimore.

Bryant, E. H., and Atchley, W. R. (1975). *Multivariate Statistical Methods: Within-groups Covariation.* Dowden, Hutchinson, and Ross, Stroudsburg, Pa.

Cassie, R. M. (1969). Multivariate analysis in ecology. *Proc. N. Z. Ecol. Soc. 16:* 53-57.

Kendall, M. G. (1965). *A Course in Multivariate Analysis.* Griffin, London.

Kenneth, R., and Isaac, P. K. (1964). *Cephalosporium* species parasitic on *Helminthosporium* (sensu late). *Can. J. Plant Sci. 44:* 182-187.

Mills, J. T., Sinha, R. N., and Wallace, H. A. H. (1978a). Assessment of quality criteria of stored rapeseed: A multivariate study. *J. Stored Prod. Res. 14:* 121-133.

Mills, J. T., Sinha, R. N., and Wallace, H. A. H. (1978b). Multivariate evaluation of isolation techniques for fungi associated with stored rapeseed. *Phytopathology 68:* 1520-1525.

Morrison, D. F. (1967). *Multivariate Statistical Methods.* McGraw-Hill, New York.

Orloci, L. (1975). *Multivariate Analysis in Vegetation Rsearch.* Junk, The Hague.

Pearce, S. C. (1965). *Biological Statistics: An Introduction.* McGraw-Hill, New York.

Pelhàte, J. (1968). Inventaire de la mycoflore des blés de conservation. *Bull. Soc. Mycol. Fr. 84:* 127-143.

Seal, H. (1964). *Multivariate Statistical Analysis for Biologists.* Methuen, London.

Sinha, R. N. (1964). Ecological relationships of stored-product mites and seed-borne fungi. *Acarologia 6:* 372-389.

Sinha, R. N. (1966). Feeding and reproduction of some stored-product mites on seed-borne fungi. *J. Econ. Entomol. 59:* 1227-1232.

Sinha, R. N. (1977). Uses of multivariate methods in the study of stored-grain ecosystems. *Envir. Entomol. 6:* 185-192.

Sinha, R. N., and Mills, J. T. (1968). Feeding and reproduction of the grain mite and the mushroom mite on some species of *Penicillium. J. Econ. Entomol. 61:* 1548-1552.

Sinha, R. N., Berck, B., and Wallace, H. A. H. (1967). Effect of phosphine on mites, insects and microorganisms. *J. Econ. Entomol. 60:* 125-132.

Sinha, R. N., Wallace, H. A. H., and Chebib, F. S. (1969a). Principal component analysis of interrelations among fungi, mites and insects in grain bulk ecosystems. *Ecology 50:* 536-547.

Sinha, R. N., Wallace, H. A. H., and Chebib, F. S. (1969b). Canonical correlations of seed viability, seed-borne fungi, and environment in bulk grain ecosystems. *Can. J. Bot. 47:* 27-34.

Wallace, H. A. H., and Sinha, R. N. (1962). Fungi associated with hot spots in farm stored grain. *Can. J. Plant Sci. 42*: 130-141.

Wallace, H. A. H., Sinha, R. N., and Mills, J. T. (1976). Fungi associated with small wheat bulks during prolonged storage in Manitoba. *Can. J. Bot. 54*: 1332-1343.

Whaley, J. W., and Barnett, H. L. (1963). Parasitism and nutrition of *Gonatobotrys simplex*. *Mycologia 55*: 199-210.

Chapter 14

POPULATION GROWTH AND REGULATION

Ralph Baker and Chandrani Wijetunga
Colorado State University
Fort Collins, Colorado

I. INTRODUCTION

A broad coverage of the subject of population growth and regulation in fungi could encompass almost the entire field of microbial ecology. Obviously, this cannot be done in a single chapter. Instead we shall focus on the mathematical description of the growth, propagule formation, survival, activity and regulation of fungal communities. These have been developed and are applicable to all stages of activity in the life cycle. Mycologists without extensive mathematical training need not be wary, however, as the simple transformations and analyses may be the best. After all, the organisms that mycologists routinely study are the most complex elements of the equation. If we but know their characteristics, they will respond in predictable ways amenable to mathematical description.

II. THE EBB AND FLOW OF POPULATIONS

As found in nature, fungi typically have a period of activity in which they grow on dead and/or living substrates, sporulate (or produce some kind of survival structures), and become dormant awaiting another opportunity to produce thalli. Mathematic models have been suggested for all of these phases of the life cycle.

A. The Vegetative Growth Phase

Literature on this subject usually suggests that an organism behaves as if it had put its thallus (or "capital") in the bank and produced more protoplasm at a fixed rate of growth ("simple interest"). The rate of growth is thus simply proportional to $y_x - y$, where y is the amount of thallus and y_x is the maximum amount of growth. In Van Der Plank's (1963) terms the solution to this becomes:

$$\ln \frac{1}{1-y} = -\ln b + rt \qquad (1)$$

Application of this model to thalli produced in nature has not been accomplished frequently (Mandels, 1965; Robertson, 1968). Perhaps this is mostly due to a certain preoccupation with propagation, that is, generation of propagules, rather than production of the vegetative thallus; for instance, such propagation produces infective units which are of much greater interest to an epidemiologist than growth per se. Moreover, precise measurement of increase in protoplasm in a natural substrate such as the root, stem, or leaf of a host presents certain technical difficulties.

The foregoing simple-interest equation (1) has practical appliation in description of increase in disease resulting from soil-borne pathogens. Van Der Plank (1963) suggested this application and used the data of Ware and Young (1934) as an example for transformation. More recently, Rowe and Powelson (1973) applied this model to cercosporella foot rot of wheat.

B. Generation of Propagules

When fungi are active, they eventualy produce propagules. Under favorable conditions, each propagule produces more of its own kind. In the sense of the terms used by Van Der Plank (1963), this is equivalent to investing thallus capital at compound interest. Thus, Blackman (1919) could formualate the compound-interest law of growth in the logarithmic form:

$$\ln y = \ln y_0 + rt \tag{2}$$

where y_0 is the number of individuals (or capital) at time $t = 0$ and r the relative rate of sporulation (or interest). Multiplication of a fungus, of course, does not proceed exponentially for an indefinite period; the increase is slow at first (y being small), rapid growth follows, rises to a maximum, and finally falls away to zero. The typical S-shaped growth curve generated when increase is plotted over time was delightfully named "lag-log-logy" by Waggoner (1977). The most commonly used mathematical description of the curve is the logistic model first summarized by Yule (1925). The equations take on various forms according to various authors; however, Waggoner's (1977) version should suffice:

$$y = \frac{y_x}{1 + Be^{-y_x rt}} \tag{3}$$

where y_x is the maximum number of individuals produced.

A number of modifications of the logistic model are treated with dexterity by Jowett et al. (1974) and Waggoner (1977). In the Gompertz function, rate is proportional to y and the logarithm of 1/y and the rate of growth is slowed as y approaches unity. Richards (1959) suggested an even more complicated description of a feedback mechanism in which rate of growth eventually is slowed.

After surveying all these models for growth, it is particularly distressing to have Waggoner (1977), at the end of his review, present the logical conclusion that all of these models produce sigmoid curves and that fitting a growth curve derived from nature with any of the models is likely to provide a good correlation. The data of Last et al. (1969) on the buildup of tomato brown rot provide evidence for this. A straight line, simple interest, or logistic curve explain about the same portions of variation in severity of disease.

Examples other than this are now accumulating providing some basis for tests of Waggoner's (1977) conclusion. McCoy (1973) followed the production of conidia of *Cercospora herpotichoides* on straw incubated at 10°C in different soils. Figure 1 graphically displays her data and illustrates the typical lag-log-logy configuration of the curve over time. These data were analyzed to determine best fit using three approaches: (1) ordinate values taken as being simply and directly proportional to conidia increase (nontransformed); (2) ordinate values taken as equal to ln $1/(1 - y)$ (the simple interest transformation); and (3) ordinate values taken as equal to ln $y/(1 - y)$ (the compound interest or logit transformation). To calculate the proportions of population (y), each cumulative value requires division by y_x (the maximum population), an estimate being obtained by averaging the last two maximum values of population. The linear regression equation of the form

$$y = \alpha + \beta t + \gamma t^2 \tag{4}$$

where t is the number of days of incubation, is then fitted. Regression coefficients (α, β, and γ) and the variation explained by regression (R^2) may be calculated for each model (see Table 1). All three models do well in explaining the variations, R^2 values being between 94.7 and 99.9%.

Another example is that contributed by Welch (1975). He followed the increase in inoculum density of *Rhizoctonia solani* in soil under alfalfa (Fig. 2). Again, there is not much to choose from between the models in explaining variation (Table 2). From these examples, we conclude with Waggoner (1977) that the logical course in constructing a model "is to choose a differential equation that is both biologically

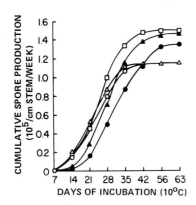

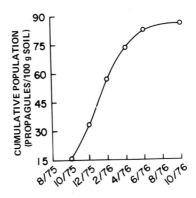

Fig. 1 Production of conidia of *Cercospora herpotrichoides* on straw incubated at 10°C in different soils. Key: ○, Case; △, Summerville; ▲, PES; □, Condon; and ●, TR. (From data of McCoy, 1973.)

Fig. 2 Increase in inoculum density of *Rhizoctonia solani* under alfalfa. (From data of Welch, 1975.)

Table 1 Analyses of the data of McCoy (1973) from Fig. 1 to determine best fit of curves using the linear regression equation, $y = \alpha + \beta t + \gamma t^2$

Soils	Nontransformed				$\ln 1/(1-y)$				$\ln y/(1-y)$			
	α	β	γ	R^2	α	β	γ	R^2	α	β	γ	R^2
PES	-0.327	0.033	0.000	0.947	0.613	-0.102	0.004	0.999	-6.519	0.227	0.000	0.995
Summerville	-0.468	0.058	0.001	0.965	-0.255	-0.009	0.003	0.985	-5.438	0.254	0.000	0.993
TR	-0.205	0.018	0.000	0.964	0.815	-0.115	0.004	0.981	-6.559	0.200	0.000	0.986
Condon	-0.406	0.048	0.000	0.960	0.422	-0.087	0.004	0.992	-4.193	0.127	0.002	0.996
Case	-0.434	0.055	-0.005	0.978	-0.044	-0.0256	0.003	0.993	-4.834	0.191	0.000	0.875

Table 2 Analyses of the data of Welch (1975) from Fig. 2 to determine best fit of curves using various models based on the linear regression equation $y = \alpha + \beta t + \gamma t^2$

Coefficients	Nontransformed	ln 1/(1 − y)	ln y/(1 − y)
α	0.11350	0.51461	−2.12852
β	−0.02726	−0.26088	−0.16564
γ	0.00433	0.02228	0.02558
R^2	0.96623	0.97535	0.99034

appealing and simple and then to use it if it fits the curve of growth reasonably well."

Data provided by McCoy [(1973) see Fig. 1] were generated in a growth chamber. Welch (1975) collected data (Fig. 2) in the field but from a relatively stable subterranean ecological habitat. Nature does not always provide for the generation of such smooth growth curves. The perturbations of a capricious environment influence growth curves so that they become sinous, and models must be modified to take into account the organism's response to fluctuations found in the habitat. Quantitative analyses and elements required for computer simulation can be generated in the physiological laboratory. Thus, such exercises have been named "physiological models" by Waggoner (1977).

C. Survival

Once growth and multiplication of a population have ceased, dormant propagules continue the life line of existence. Intrinsic genetically controlled factors intimately associated with morphology and food storage are basic to determining survival of these propagules over time.

K. F. Baker and Cook (1974) summarize the relation of these factors to persistence of propagules. Fungi such as *Gaeumannomyces graminis* and *Cephalosporium gramineum* are completely dependent on remnant host tissues for survival of mycelia. Sclerotia, as produced by the genera *Sclerotinia*, *Rhizoctonia*, *Sclerotium*, and *Phymatotrichum*, represent an adaptation directed toward improvement in survival; however, their relatively large size and high food reserves present opportunities for attack by antagonists. Some fungi survive in an inactive state for long periods as chlamydospores (species of *Fusarium*, *Cylindrocladium*, and *Thielaviopsis*), teliospores (species of smut fungi), oospores (species of *Aphanomyces*, *Pythium*, and *Phytopthora*), resting spores or sporangia (e.g., *Olpidium brassicae*), and thick-walled hyphae (species of *Rhizoctonia* and *Verticillium*). Factors in the environment, however, may be of great importance in modifying survival characteristics of the organism. However, these qualitative factors have been exhaustively reviewed elsewhere (e.g., by Gar-

rett, 1970; Griffin, 1972; Park, 1965), so we shall press on to the important task
of quantitative description of the observed phenomena.

Yarwood and Sylvester (1959) proposed to measure survival using the half-life
concept. This assumes that propagules die at a logarithmic rate. The half-life
(the time required for half of the propagules to die), may be obtained graphically
by plotting log values of y/y_o against t and recording the time at which half
the propagules in the population have died. This transformation is valid as long
as the death rate is constant, but Dimond and Horsfall (1965) describe exceptions.
Especially if spores are multinucleate or consist of more than one cell, there is a
lag in the survival curve presumably caused by the time rquired to inactivate nuclei
before germination is limited. Thus, departures from the logarithmic law of survival
may be of common occurrence.

Dimond and Horsfall (196) prefer the semilog transformation to describe survival:

$$\log y = \log y_o + rt \log e \qquad\qquad\qquad (5)$$

This is actually the expression of the simple-interest growth curve, but in this case
r is negative since it describes rate of death and y_o is the initial amount of inoculum. In practice transformed values of y/y_o, the proportions of survivors, may be
obtained from tables of $\ln 1/(1 - y)$, as found in Van Der Plank (1963), and plotted
against time. In this graphic solution of survival data, the lag phase is taken into
account and the legitimate assumption that propagules die at a logarithmic rate is
preserved.

The log-probit transformation is also a valid candidate for mathematical descriptions of survival (R. Baker, 1971b). In this transformation, it is assumed that the
tendency of propagules to die follows a normal distribution in time. Time in this
case is the "toxin," and the assumptions of the log-probit transformation (Bliss,
1935) again lead to graphic conversion of the time scale to logarithms. In practice,
then, the proportions of surviving propagules in probit units are plotted against
time on a logarithmic scale.

As concluded in Sec. II.B for models describing growth and multiplication, these
last two survival transformations straighten the reversed S-shape of typical survival
curves despite the different principles used for their construction. Thus Benson and
Baker (1974c) applied the semilog and log-probit transformations to survival data
collected for *Rhizoctonia solani* and found good agreement between TS_{50} values (time
required for 50% of the propagules to die in soil) and the statistical comparison of
slope values among treatments. Thus, these two transformations complement each other
in evaluation of survival data.

The transformations allow computations and objective comparisons of slope values
as they are affected by various intrinsic and extrinsic factors impinging upon ability of the propagules to survive. Interpolations and extrapolations are possible to

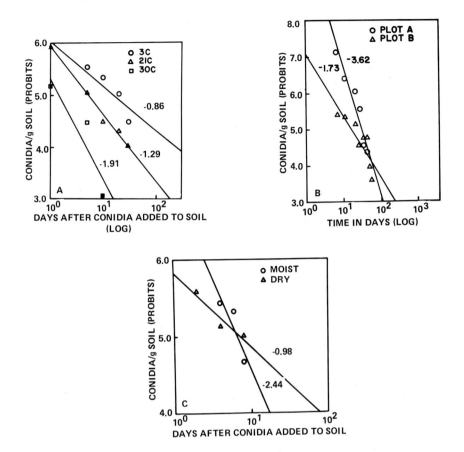

Fig. 3 Survival of conidia of *Botrytis squamosa* as influenced by various environmental factors. Data transformed by log-probit analysis. Numbers adjacent to lines are slope values. (A) Survival of conidia in artifically infested soil stored in plastic bags at 3°, 21°, and 30°C. (B) Survival of conidia in field plots artifically infested with conidia. (C) Survival of conidia in soil maintained at different moisture levels. (From Fig. 1 of Ellenbrock and Lorbeer, 1977.)

determine TS_{50} and final extinction values. As an example, the data of Ellenbrock and Lorbeer (1977) were transformed by log-probit analysis in Fig. 3.

An example in which a biological control agent, chitin (Baker, 1968), added to soil failed to affect survival of *Fusarium oxysporum* f. sp. *pisi* in comparison with nonamended controls was contributed by Guy and Baker (1977). Both the semilog and log-probit transformations of survival over time demonstrated that slope values were not significantly changed when chitin was added to soil. Such observations are valuable in sorting mechanisms of biological control. In this case, the mechanism associated with chitin treatment is not its effect on survival per se.

Transformations, however, may obscure important details. Consider the effect of moisture potential on survival of *R. solani* (cf. Fig. 3 in Benson and Baker, 19-74c). Survival, as measured by transforming the data, was significantly better in

soils with negative metric potential (-540 bars) than in moist conditions. The nontransformed data in Fig. 3A, however, reveal a period after addition of inoculum to
moist soil during which increase in propagules occurred. After 2-6 days propagules
in these systems had a high death rate, but the persistent population left after 10
days apparently was indifferent to the effects of moisture on survival.

Studies on survival of soil-borne fungi suggest what may be the typical machi-
nations of a survival curve (e.g., see Benson and Baker, 1974c; Guy and Baker, 1977;
Ellenbrock and Lorbeer, 1977). When fungi are originally within a host or an aboveground substrate and then added to soil (e.g., during cultivation), they may increase
thalli by vegetative growth and reproduction occasionally (depending on competitive
saprophytic ability). After this brief flurry of activity, a proportion of the now-
dormant propagules may die rapidly resulting in a steep decline in population. Finally, a persistent remnant may still be capable of relatively long-term survival.

D. Arousal

In the rhythmic cycles characteristic of the engineering systems used by fungi, substrates must become available to the dormant stage for the eventual propagation of
new life. For some fungi, there may be a period of inherent dormancy. For all fungi, dormancy may be imposed--by soil fungistasis for underground (and perhaps aquatic) microorganisms or by a variety of environmental factors (e.g., temperature, moisture) unfavorable for resumption of activity for their aerially oriented relatives.
The factor triggering arousal of fungi in soil, then, is almost inevitably a nutrient
stimulus, although examples in which germination occurs with temperature change are
known (Coley-Smith and Cooke, 1971; Trione, 1973). For other fungi that germinate
on substrates or host surfaces aboveground, the triggering mechanisms may be more
complex (e.g., see Waggoner, 1972).

Plant pathologists use the term *inoculum potential* to describe the energy available for the colonization of a substrate at the surface of the substrate to be colonized (Garrett, 1956). This energy, over and above the qualitative genetic factors
(related, for example, to morphological structures required to break host barriers),
is derived from an array of sources. Richly endowed propagules produced on substrates
of high nutrient content may have higher inoculum potential than undernourished ones
do (Phillips, 1965). During germination, prior to the occupation of a substrate,
nutrition profoundly affects the inoculum potential of many fungi (Banttari and Wilcoxson, 1964; Isaac, 1957; Maurer and Baker, 1965; Tousson et al., 1960). Other environmental factors in the ecosystem, including temperature, water potential, hydrogen ion concentration, and antagonism (Henis et al., 1978), may modify inoculum potential. The aforementioned factors are only part of the story, however. The quantity of propagules or the inoculum density is also important in determining inoculum
potential.

Quantitatively, inoculum potential can be mathematically computed in the linear
form for soil-borne organisms (Baker, 1978):

$$\log S = m(\log x + \log v + \log n + \log f) \tag{6}$$

where s is success in invasion of the substrate (or successful infections in a host),
m is the slope of the inoculum density versus successful invasions (or infections)
curve, x is the inoculum density, v is the intrinsic capacity to invade the substrate
(genetically determined), n is the nutritional status of the propagules, and f is the
environmental influences on germination and efficiency of colonization. This equation
could be written to determine the amount of disease (y) if a host was involved by incorporating the multiple infection correction as follows:

$$y = 1 - e^{-(xvnf)^m} \tag{7}$$

Specific values for all elements of the equation have not been determined experimentally. The conventional approach for description of plant-pathogen inoculum potential interrelationships has been to develop the inoculum density-disease curve
(Baker, 1971a; Van Der Plank, 1975). This may be analyzed by semilog (Dimond and
Horsfall, 1965), logarithmic-probability (Horsfall and Dimond, 1963), or log-log
(Baker et al., 1967) transformations. The value of m in Eqs. 6 and 7 may be predicted for soil-borne pathogens according to the geometry of the systems involved. For
fixed-inoculum, fixed-infection courts with a rhizosphere effect the slope should be
one (Baker et al., 1967; Benson and Baker, 1974b; Mitchell, 1975) unless modified by
synergistic effects between propagules (Baker, 1971a). For fixed-inoculum and fixed-
infection courts, the slope should be 0.67 (Byther, 1968; Guy and Baker, 1977; Hanounik et al., 1977; Stienstra and Lacy, 1969). Slope values may change significantly
if the rhizosphere is contracted to a rhizoplane as demonstrated by Rouse and Baker
(1976) when biological control through competition is induced. Usually, environmental influences (Baker, 1971a; Benson and Baker, 1974a,b; Guy and Baker, 1977) or host
substrate resistance (Hanounik et al., 1977) merely shift the position of the inoculum density-infection curve.

III. REGULATION

In soil systems, distances of propagules from each other and from their hosts or organic substrates should have profound ecological significance, especially when inoculum is randomly distributed by cultivation. For example, microorganisms requiring
relatively simple carbon substances, like the sugar fungi (Garrett, 1956), must germinate and grow rapidly and/or produce large quantities of propagules to take advantage of their ephemeral substrates. Such distance relationships, as a function of
inoculum density, were calculated by Baker and McClintock (1965) and corrected by
McCoy and Powelson (1974). The curves generated demonstrate that distance between
propagules decreases dramatically down to 3000, and less rapidly to 10,000 propagules
per gram of soil. It follows that these distances are also proportional to the distances of propagules to their substrates. Thus, it is interesting that values from
3000 to 10,000 propagules per gram of soil are near the maximum for many populations

noticed in the field for pathogens attacking fixed-infection courts in soil (e.g., Nash and Snyder, 1962). Fungi "position themselves" strategically, however, for attack on substrates by means other than adjustments of inoculum density. For example, populations of *Rhizoctonia solani* are quite low in naturally infested soils. Ko and Hora (1971) reported 1-9 propagules per 10 g of dried soil. This fungus compensates by growing through soil at rates up to 1 cm/day (Baker and Martinson, 1965) and responds quickly to infection sites when it is in a rhizosphere (Baker, 1971a). Alternatively, some pathogens with characteristically low inoculum densities are not at a disadvantage because they typically penetrate through root tips invading relatively large volumes of soil. Thus, the host is virtually "disseminated" to the pathogen so that large quantities of inoculum are not necessary. An example of such a system is found in verticillium wilt in which 3.5 microsclerotia per gram of soil, a density typically found in the field, are capable of inducing 100% disease in cotton (Ashworth et al., 1972).

K. F. Baker and Cook (1974) also consider another characteristic of some organisms like *Verticillium albo-atrum*, *Phytophthora cinnamomi*, and *Plasmodiophora brassicae* that have the capacity to produce secondary inoculum from primary survival structures. In these cases a fungus with a relatively low inoculum density may compensate through the production of more propagules, but in soil such examples may be rare.

These developing patterns could mean, in an evolutionary sense, that population levels of fungi in nature resulting from the dual factors of multiplication and survival attain certain levels characteristic of their particular ecological activity. This has been adumbrated rather diffidently as an ecological principle (by R. Baker, 1978): "The extant inoculum density of a given soil-borne pathogen usually is no greater than the exigencies inherent for reasonably efficient positioning in three-dimensional space in relation to the infection."

Developing theory thus indicates that populations do not expand unchecked. The reason(s) for this is not yet completely understood. Certainly exhaustion of nutrient resources (an assumption in growth transformations such as the logistic equation 3) must be considered as a prime mechanism, but there are examples where apparent regulation of population density occurs before this stage is reached. Consider the phenomena developed in our laboratories and graphically displayed in Fig. 4. Successive crops of radish were planted at weekly intervals in soil infested with *R. solani*. The incidence of damping-off reached its peak after 3 weeks and declined thereafter. The inoculum density was almost as low after seven replants as at the beginning of the experiment. At this time the soil was also suppressive and relatively large amounts of inoculum had to be introduced to induce disease (Henis et al., 1978). Thus adequate host substrate was available, but other factors regulated the population. It is significant that no change in suppressiveness of soil to *Rhizoctonia* was observed when crops of radish were planted without the pathogen. Similarly, there was no change in suppressiveness of soil to which only *R. solani* was added. Thus,

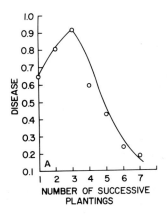
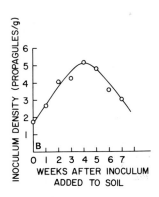

Fig. 4 Disease and population changes when radishes were repeatedly planted at weekly intervals in soil originally infested with *Rhizoctonia solani* at approximately 2 propagules per gram. (A) Incidence of damping-off of radish on a per unit basis. (B) Changes in inoculum density of *Rhizoctonia solani* during the course of the experiment.

both pathogen and host had to be present in the system; the pathogen had to be active to induce the development of postulated antagonists which gained some benefit from the association of host and pathogen. Such population-regulating mechanisms involving antagonism may be of frequent occurrence in soil.

IV. THE MATHEMATICAL MEANS TO AN END

To what advantage is an understanding of these elementary manipulations defining the parameters of population growth and regulation? Certainly epidemiologists vitally concerned with the control of diseases find such a discipline to have great application, as evidenced by recent books on the subject (e.g., Kranz, 1974; Day, 1977). Precise quantitative analyses of manipulations designed to depress pathogens lead to intelligent decisions in the application of integrated controls. To the microbial ecologist there is the satisfaction of understanding how nature behaves in mathematical terms, and this leads to new insights into the mechanisms of community interactions, population dynamics, and the impact of fungal activity in ecosystems. To paraphrase La Fontaine, sensible mycological ecologists will find nothing useless.

REFERENCES

Ashworth, L. J., McCutcheon, O. D., and George, A. G. (1972). *Verticillium alboatrum*: The quantitative relationship between inoculum density and infection of cotton. *Phytopathology 62*: 901-903.

Baker, K. F., and Cook, R. J. (1974). *Biological Control of Plant Pathogens*. Freeman, San Francisco.

Baker, R. (1968). Mechanisms of biological control of soil-borne pathogens. *Ann. Rev. Phytopathol. 6*: 263-294.

Baker, R. (1971a). Analyses involving inoculum density of soil-borne plant pathogens in epidemiology. *Phytopathology 61*: 1280-1292.

Baker, R. (1971b). Simulators in epidemiology--Can a simulator of root disease be built? *Advanced Study Inst.: Epidemiology of Plant Diseases, Wageningen, Netherlands No. 1*: 1-20.

Baker, R. (1978). Inoculum potential. In *Plant Pathology: An Advanced Treatise*, J. G. Horsfall and E. B. Cowling [Eds.], Vol. 2. Academic Press, New York, pp. 137-157.

Baker, R., and McClintock, D. L. (1965). Populations of pathogens in soil. *Phytopathology 55*: 495.

Baker, R., and Martinson, C. A. (1965). Epidemiology of diseases caused by *Rhizoctonia solani*. In *Rhizoctonia solani: Biology and Pathology*, J. R. Parmeter (Ed.). Univ. of California Press, Berkeley, pp. 172-188.

Baker, R., Maurer, C. L., and Maurer, R. A. (1967). Ecology of plant pathogens in soil. VIII. Mathematical models and inoculum density. *Phytopathology 57*: 662-666.

Banttari, E. E., and Wilcoxson, R. D. (1964). Relation of nutrients in inoculum and inoculum concentrations to severity of spring black stem of alfalfa. *Phytopathology 54*: 1048-1052.

Benson, D. M., and Baker, R. (1974a). Epidemiology of *Rhizoctonia solani* preemergence damping-off of radish: Influence of pentacloro-nitrobenzene. *Phytopathology 64*: 38-40.

Benson, D. M., and Baker, R. (1974b). Epidemiology of *Rhizoctonia solani* preemergence damping-off of radish: Inoculum potential and disease potential interactions. *Phytopathology 64*: 957-962.

Benson, D. M., and Baker, R. (1974c). Epidemiology of *Rhizoctonia solani* preemergence damping-off of radish: Survival. *Phytopathology 64*: 1163-1168.

Blackman, V. H. (1919). The compound interest law and plant growth. *Ann. Bot.* (London) *33*: 353-360.

Bliss, C. I. (1935). The calculation of the dosage-mortality curve. *Ann. Appl. Biol. 22*: 134-167.

Byther, R. (1968). Etiological studies on foot rot of wheat caused by *Cercosporilla herpotrichoides*. Ph.D. thesis, Oregon State University, Corvallis.

Coley-Smith, J. R., and Cooke, R. C. (1971). Survival and germination of fungal sclerotia. *Ann. Rev. Phytopathol. 9*: 65-92.

Day, P. R. (Ed.) (1977). *The Genetic Basis of Epidemics in Agriculture*. New York Academy of Science, New York.

Dimond, A. E., and Horsfall, J. . (1965). The theory of inoculum. In *Ecology of Soil-Borne Plant Pathogens*, K. F. Baker and W. C. Snyder (Eds.). Univ. of California Press, Berkeley, pp. 404-415.

Ellenbrock, L. A., and Lorbeer, J. W. (1977). Survival of sclerotici and conidia of *Botrytis squamosa*. *Phytopathology 67*: 219-225.

Garrett, S. D. (1956). *Biology of Root-Infecting Fungi*. Cambridge Univ. Press, New York.

Garrett, S. D. (1970). *Pathogenic Root-Infecting Fungi*. Cambridge Univ. Press, New York.

Griffin, D. M. (1972). *Ecology of Soil Fungi*. Syracuse Univ. Press, Syracuse, N.Y.

Guy, S. O., and Baker R. (1977). Inoculum potential in relation to biological control of fusarium wilt of peas. *Phytopathology 67*: 72-78.

Hanounik, S. B., Pirie, W. R., and Osborne, W. W. (1977). Influence of soil chemical treatment and host genotype on the inoculum-density relationships of cylindrocladium black rot of peanut. *Plant Dis. Rep. 61*: 431-435.

Henis, Y., Ghaffar, A., and Baker, R. (1978). Integrated control of *Rhizoctonia solani* damping-off of radish: Effect of successive plantings, PCNB, and *Trichoderma harzianum* on pathogen and disease. *Phytopathology 68*: 900-907.

Horsfall, J. G., and Dimond, A. E. (1963). A perspective on inoculum potential. In *Maheshwari Commemoration Volume*, T. S. Sadaswan (Ed.). Suppl. to *J. Indian Bot. Soc. 42A*. Bangalore Press, Bangalore, India.

Isaac, I. (1957). The effects of nitrogen supply upon verticillium wilt of *Antirrhinum*. *Ann. Appl. Biol. 45*: 512-515.

Jowett, D., Browning, J. A., and Haning, B. C. (1974). Non-linear disease progress curves. In *Epidemics of Plant Diseases Mathematical Analysis and Modeling*, J. Kranz (Ed.). Springer-Verlag, New York, pp. 115-136.

Ko, W., and Hora, F. K. (1971). A selective medium for the quantitative determination of *Rhizoctonia solani* in soil. *Phytopathology 61*: 707-710.

Kranz, J. (Ed.) (1974). *Epidemics of Plant Diseases: Mathematical Analyses and Modeling*. Springer-Verlag, New York.

Last, F. T., Ebben, M. H., Hoare, R. C., Turner, E. A., and Carter, A. R. (1969). Buildups of tomato brown rot caused by *Pyraenochaeta lycopersici* Schneider & Gerlach. *Ann. Appl. Biol. 64*: 449-459.

McCoy, M. L. (1973). The application of mathematical models in the epidemiology of foot rot of wheat caused by *Cercosporella herpotrichoides* Fron. Ph.D. thesis, Oregon State University, Corvallis, 56 pp.

McCoy, M. L., and Powelson, R. L. (1974). A model for determining spatial distribution of soil-borne propagules. *Phytopathology 64*: 145-147.

Mandels, G. R. (1965). Kinetics of fungal growth. In *The Fungi*, G. C. Ainsworth and A. S. Sussman (Eds.), Vol. 1. Academic Press, New York, p. 599-612.

Maurer, C. L., and Baker, R. (1965). Ecology of plant pathogens in soil. II. Influence of glucose, cellulose, and inorganic nitrogen amendments on development of bean root rot. *Phytopathology 55*: 59-72.

Mitchell, D. J. (1975). Density of *Pythium myriotylum* oospores in soil in relation to rye. *Phytopathology 65*: 570-575.

Nash, S. M., and Snyder, W. C. (1962). Quantitative estimations by plate counts of propagules of the bean root rot *Fusarium* in field soils. *Phytopathology 52*: 567-572.

Park, E. (1965). Survival of microorganisms in soil. In *Ecology of Soil-Borne Plant Pathogens*, K. F. Baker and W. C. Snyder (Eds.). Univ. of California Press, Berkeley, pp. 82-97.

Phillips, D. J. (1965). Ecology of plant pathogens in soil: IV. Pathogenicity of macroconidia of *Fusarium roseum* f. sp. *cerealis* produced on media of high or low nutrient content. *Phytopathology 55*: 328-329.

Richards, F. J. (1959). A flexible growth function for empirical use. *J. Exp. Bot. 10*: 290-300.

Robertson, N. F. (1968). The growth process in fungi. *Ann. Rev. Phytopathol. 6*: 115-136.

Rouse, D., and Baker, R. (1976). Verification of a model for biological control of *Rhizoctonia solani* preemergence damping-off. *Proc. Amer. Phytopathol. Soc. 3*: 220.

Rowe, R. C., and Powelson, R. L. (1973). Epidemiology of cercosporella foot rot of wheat. *Phytopathology 63*: 984-988.

Stienstra, W. C., and Lacy, M. L. (1969). Effect of inoculum density and planting depth on infection of onion by *Urocystis colchici*. *Phytopathology 59*: 1052.

Toussoun, T. A., Nash, S. M., and Snyder, W. C. (1960). The effect of nitrogen sources and glucose on the pathogenesis of *Fusarium solani* f. sp. *phasioli*. *Phytopathology 50*: 137-140.

Trione, E. J. (1973). The physiology of germination of *Tilletia* teliospores. *Phytopathology 63*: 643-648.

Van Der Plank, J. E. (1963). *Plant Diseases: Epidemics and Control*. Academic Press, New York.

Van Der Plank, J. E. (1975). *Principles of Plant Infection*. Academic Press, New York.

Waggoner, P. E. (1972). Weather, space, time and chance of infection. *Phytopathology 52*: 1100-1108.

Waggoner, P. E. (1977). Contributions of mathematical models to epidemiology. In *The Genetic Basis of Epidemics in Agriculture*, P. R. Day (Ed.). New York Academy of Science, New York, pp. 191-206.

Ware, J. D., and Young, V. H. (1934). Control of cotton wilt and "rust." *Arkansas Univ. Agr. Expt. Sta. Bull. No. 308*.

Welch, L. (1975). Interaction of *Rhizoctonia solani* populations and propagule nutrition in agricultural soils. Ph.D. thesis, University of California, Berkeley, 84 p.

Yarwood, C. E., and Sylvester, E. S. (1959). The half-life concept of longevity of plant pathogens. *Plant Dis. Rep. 43*: 125-128.

Yule, G. V. (1925). The growth of populations and the factors which control it. *J. Roy. Stat. Soc. 88*: 1-58.

Chapter 15

RESOURCES AND REGULATORS OF FUNGAL POPULATIONS

Michael W. Dick
University of Reading
Whiteknights, Reading, England

Fungal ecology can provide apposite examples, reveal appropriate systems for experimentation and help to advance theoretical concepts of the ecology of populations and communities. The development of theoretical concepts, and the accumulation of descriptive and experimental evidence supporting these concepts are both facilitated and hindered by the specifically fungal combination of attributes. These attributes include a variable but often rapid generation time; a vegetative phase which is frequently evanescent, of ill-defined extent, and capricious in its survival of isolation procedures; a physiological requirement for heterotrophic nutrition which may be more or less closely circumscribed; and a complexity in the life cycle which may involve more than one vegetative life form and frequently involves several spore life forms.

The Oomycetes have all the physiological attributes of true fungi although they are not fungi in the phylogenetic sense. Within this class there is sufficient variation in morphology, life cycle, physiology, and ecology to provide examples with which to illustrate a discussion of the resources and regulators of fungal populations.

Table 1 provides a framework for this discussion, without attempting to define the terms *resource* and *regulator*: since the limitation of an available resource is in itself a regulator, and regulators of population size may restrict resources, these two terms interrelate and are therefore particularly difficult to circumscribe. Nevertheless, it is helpful to separate those factors which may be designated 'environmental' from those which can be considered as being intrinsic to the species, not least because it enables attention to be drawn to differences between species within a community. The use of a diagram to indicate interactions has been deliberately avoided because these differences are important in comprehending the infrastructure of the population of each species within the community. The use of simple and precise definitions of *community* as "assemblages of populations" and *population* as "progeny within the species" have much to commend them, though the terms *assemblages* and *progeny* often require qualification. Unfortunately, the latter definition especially is inadequate to cover intraspecific clonal differences (groups of progenies), particularly in relation to differences between these clones in their constituent, coexisting life forms. Differences in phenotypic behavior between these life forms

assume a much greater importance in fungal ecology. At its simplest level, the interrelationship between spore forms and vegetative mycelia, which can only be deduced
from indirect or circumstantial evidence, presents to the ecologist a challenge not
experienced in animal or green plant ecology. Yet to understand the role of fungi
in degradation and recycling, this interrelationship must be included in any comprehensive modeling of an ecosystem.

The resources and regulators labeled as environmental in Table 1 may be the more
obvious and probably do not need to be discussed in great detail here. The theoretical concept of the fundamental niche (see McNaughton, Chap. 5) is characterized by
the physical parameters of surface and temperature and the availability of nutrients
and inorganic substances necessary for life. However, for fungal ecologists, it is
often essential to emphasize the fragmentation of this niche in time and space. Nowhere is this more clearly demonstrable than in the freshwater lentic environment,
in which the niche is defined only by the position, at the air-water interface, of
particulate substrates that provide both surface and nutrient source, such as freshly
cast exuviae [*Saprolegnia*, *Aphanomyces* (Dick, 1970)] or freshly incorporated plant
debris [Adenle (1978) has shown that nonsexual aquatic *Phytophthora* species can only
be isolated from freshly submerged leaves within 3 days of submergence]. At least in
the former example, the time factor operates to change the size of the fundamental
niche at two levels: the diurnal fluctuation in insect emergence and the seasonal
change in imago production in temperate lakes. Hallett (1975) and Hallett and Dick
(in press) demonstrated the extensive fluctuations to be found in populations of the
zoospore life form of the major saprobic Oomycetes in the lentic environment. These
populations fluctuate both with the seasonal and with the diurnal changes in the
physical dimensions of the fundamental niche.

In terrestrial situations our concept of the fundamental niche cannot be as precisely delimited. This is because the more closely integrated interaction of the
community, in its all-embracing polyphyletic and multiple magnitudinal definition,
blurs the boundaries of this niche, irrespective of whether these boundaries are conceived as being primarily determined in the spatial sense or in the chemical sense.
Such a community, for a defined ecological niche of spatially indeterminant extent,
may typically be composed of:

1. Parts of an individual (e.g., segments of roots at a given stage of growth
    or senescence)
2. Transients (e.g., the mite that wanders through, browsing, respiring, and
    excreting)
3. An individual (e.g., a fungus mycelium)
4. A collection of individuals (i.e., *parts* of a much larger interbreeding
    population, e.g., sexually capable algae, yeasts, or protozoa)
5. A population (e.g., discrete clonal units of a bacterium, or the asexually
    produced spores of the abovementioned mycelium)

Table 1 Resources and regulators of populations of a species

	RESOURCES	REGULATORS
Environmental (extracellular to the components of the population)	Availability of $\Big\{$ nutrients water supply oxygen supply Existence of a suitable physical substrate	The dimensions of the fundamental niche --particulate with space --intermittent with time Predators, parasites, antagonists, and competitors
Intrinsic (intracellular to the components of the population)	The intraspecific genetic diversity of the population --the degree of adaptability contained within the genetic control of life form and life cycle in a species Phenotypic, morphological, and physiological adaptability of individuals as a result of mutation and/or sexuality and reproduction	The existence of a minimum size to the breeding population Specific constraints related to life cycle --the fundamental requirements of the individual for the production of a potential inoculum of propagules The inoculum potential of the population

In these terrestrial situations, as in aquatic systems, environmental nutritional resources and the regulators of their supply are so interdependent as to be inseparable,
but in the terrestrial situations the significance of surface diminishes in relation
to its reciprocal, *lebensraum*. Thus interstitial space has been shown by Griffin
(1963) to be a potential regulator for *Pythium* populations because of its effect on
one life form--the generation of the sexually formed resting spore.

At the same time, characterization of the fundamental niche will depend on a
nutritional component. However, since natural substrates are complex and provide a
variety of carbon substrates, the same physical unit may simultaneously provide a
nutritional fundamental niche which may be common to more than one species *and* different for other species. Separate communities of species with a genotypic adaptation to limited classes of nutrients coexist. Such distinct heterotrophic communities may be recognized as keratinophilic (H. G. Tribe and A. Abu-El-Souod, personal
communication, 1978), chitinophilic (Sparrow, 1968), or cellulophilic (Park, 1974).
Recognition of these component communities has the practical advantage of channeling
the design of experimental work, though it becomes all too easy to forget the other
coexisting communities and the constraints that they may apply to other aspects of
resource availability.

Likewise, the fundamental niche may be defined by the respiratory enzyme complement of the species, as in the obligately fermentative *Aqualinderella* (Emerson
and Weston, 1967).

Definition of that environmental regulator, the fundamental niche, may therefore
primarily depend either on environmental resources or on intrinsic resources of the
population.

The interaction between the intrinsic resources and regulators of populations
with the environmental resources modifies the fundamental niche into the realizable
niche (niches) of population(s) of a species. The realizable niche has been the subject of vegetational ecology (pattern ecology) of earlier work on Saprolegniaceae
(Dick and Newby, 1961; Dick, 1962, 1966, 1976) which has revealed distribution patterns both at geographic levels (e.g., the absence of *Achlya radiosa* and *Aplanopsis*
spp. from North America) and at local levels, and has demonstrated superimposed mosaic
patterns of differing orders of magnitude compatible with the continuum concept rather
than the community concept. In contrast, the latter concept has been shown to have
more validity for acidophiles (Dick, 1963, 1971).

However, the work just cited takes no account of the life cycles and life forms
of the species which make up the community. The data presume a uniformity between
and within species that may not be justified, even though this community is defined
by a phylogenetic affinity, together with a more or less positive and similar response
to an isolation procedure.

With fungal material, the necessary approaches of "blind" sampling and culture,
which reduce errors resulting from subjective sampling, have the disadvantage of
rarely being able to identify the life form being sampled. The population of a spe-

cies is in reality a group of populations of life forms. Just as isolation procedures are known to be differentially selective for component species of the community,
so they may be differentially selective for different life forms. The difficulty in
both cases is the assessment of the degree to which the selection is differential.
For Oomycetes, these life forms may be mycelium, deciduous sporangia and chlamydospores, zoospores and zoospore cysts, or oogonia and oospores. While it is custom-
ary to attach most importance to the vegetative mycelium because this is presumed to
be metabolically active, respiring, and converting carbon compounds, the ecologist
must also consider the passive role of spore forms in providing either an unoccupied
surface for colonization for other members of the community at large or a nutrient
base for those members of the community which may be functioning either as predators
or as parasites. It is worthwhile to list some interactions which affect one particular life form rather than another and which could thereby affect (if not "regulate")
the balance existing between the life-form populations of a species, either in vivo
or during the culturing procedures necessary for isolation and identification:

1. Hyphosphere effects--bacterial, actinomycete and ciliate communities existing on *Saprolegnia* hyphae
2. Browsing of hyphae by chironomid larvae
3. Endoparasitism of hyphae by *Olpidiopsis*
4. Browsing of zoospores by *Daphnia*
5. Parasitism of oospores and oogonia by *Rhizidiomycopsis*

The biomass of the population is composed of the totals for each of the constituent life forms present, but the relative importance of these components will depend
on a number of factors ultimately controlled by the genetic integrity of the species.
Such factors include generation time, the half-life (-lives) of the spore population(s)
and the interconvertibility of preexisting vegetative biomass to spore biomass. These
are genotypic intrinsic resources of the species, and the degree to which they may be
phenotypically modified provides an additional resource.

Assessment of this biomass depends on the confidence that can be placed on sampling procedures *designed* to isolate only from particular life forms and also on the
numbers and sizes of the individual population units being sampled. The indirect sampling procedures of the mycologist produce a situation, the nearest analogy to which
would be asking the higher plant ecologist to consider the seed populations separately
and together with the populations of growing plants.

How big is a mycelium? The population at any moment in time may be composed of
relatively few large vegetative units, accompanied by--or not accompanied by--small
vegetative units or spore units. The vegetative unit may be extensive, as with a
well-nourished mycelium, or it may be restricted, as with monocentric chytrids or
determinate thalli (e.g., *Rhipidium*). Biomass deductions made from sampling, as
opposed to assay, are dependent on knowledge of the mean size of the population unit
being sampled. Such knowledge is rare. Thus in very few circumstances is it yet

possible to have any confidence in the precision of population assessment in this
sense; discussion of resources and regulators consequently becomes somewhat specula-
tive. Nevertheless, in the case of the community of *Saprolegnia* and *Aphanomyces* at
lentic margins, the works of Dick (1970) and Hallett (1975) have shown that the in
vivo populations of these species during the summer months are predominantly of small
mycelial population units which convert, almost holocarpically, into zoospore population units. The diurnal fluctuation of substrate and the short half-life of the
spore form involved (zoospores or zoospore cysts appear to have a natural half-life
of less than 24 hr) results in a wide fluctuation in the proportions of vegetative
population units to zoospore population units during the day. But these same parameters also enable a biomass calculation to be made on the basis of the maximal values of zoospore counts. The experimental technique which has provided the data for
this argument is the development of continuous flow centrifugation pioneered by
Fuller and Poyton (1964), which has more affinity to an assay than to a sampling procedure. Thus marginal surface temperate-lake waters in summer may provide a daily
production of up to 5000 zoospores/liter (mean value 750 zoospores/liter), from which
a rough calculation of an annual wet biomass production of $0.1-0.3$ cm^3/m^3 can be derived (Hallett, 1975; Hallett and Dick, in press). When it is recalled that the number of zoospores per liter is comparable with equivalent figures for algal unicells,
which may have a longer independent existence, this small and highly specialized
community must be regarded as providing a significant contribution to the economy of
the lake ecosystem.

The identification of two spring and one autumn maxima in assay numbers of zoospores for Saprolegniaceae and Pythiaceae, taken together with the fact that zoospores
of species of both families show the same complex seasonal-diurnal relationship, is
indicative of a general response by the populations of all the species of the community to the environment. These maxima are familiar to limnologists and indicate that
availabilities of the primary environmental resources are most likely to provide the
principal regulators limiting the sizes of these populations. However, a difference
between the phenology of *Pythium* species and Saprolegniaceae, on the one hand, and
Phytophthora species (zoospores of which are available principally only in the autumn),
on the other, provides evidence that specific nutrient factors may affect community
composition at different times of the year.

When the Oomycetes most frequently isolated from exuviae and by direct assay of
zoospores are grown in vitro, they produce a very different, more extensive, and more
persistent eucarpic vegetative life form; zoospores are readily produced, but oospore
production is unpredictable. This emphasizes an adaptability or variability inherent
in the intrinsic resources within these species, an adaptability which may not be as
great in phylogenetically related species. The probable pattern of life-form populations for *Aplanopsis*, though not so thoroughly explored, is likely to be very different from that of the lentic community, since the species depend on oospore production for survival in soil and, at least in culture, the generation time for *Aplanopsis spinosa* is in excess of 10 days. The coexistence of vegetative units of unknown

extent with oospores that appear to be readily capable of germinating does not allow
for analysis of the life-form components of the population, nor therefore for any
assessment of the contribution made by the vegetative population units to the ecosystem.

These comments on life form are in accord with the more general hypothesis of
Stamberg and Koltin (see Chap. 9) that asexuality--though without concomitant inbreeding in *Saprolegnia* and *Aphanomyces* referred to earlier--predominates in a stable environment, that is, a fundamental niche which is narrow in environmental/physiolog-
ical terms but often spatially extensive, such as a lake. It might follow that in
the unstable emergent littoral environment, narrow spatially but embracing a broad
range of environmental/physiological niches, phenotypic life cycles emphasizing sexuality, and with less asexuality, would be found. This appears to be true for community concepts in the Saprolegniaceae insofar as they have been revealed by studies
of pattern. The schematic diagram previously published (Dick, 1968) and redrawn
here (Fig. 1) is as valid for classification of life form as it is for life cycles.
Although life-form categories superficially appear to differ little between Saprolegniaceae, phenotypic life cycles and even genotypic life cycles are markedly different, and the intrinsic regulators associated with a life cycle must be taken into
account when considering the extent of the realizable niche. Specifically, these
regulators relate to the generation time of the vegetative cycle (as discussed earlier) and to the times to onset, and the durations of, asexual and sexual sporula-
tion--factors which certainly vary from species to species and may also vary intra-
specifically. The relative importance of asexual and sexual cycles developed from
the vegetative phase within the generation time is also important, particularly in
relation to nutrient support for spore production in vivo and survival by spores,
thereby fulfilling the fundamental requirements of the individual for the production
of a potential inoculum of propagules formed with adequate endogenous reserves. The
environmental regulators--predation, antagonism, and competition--will also act through
this requirement to further regulate the supply and fitness of the potential inoculum
of the population. Oospore dormancy; oospore germination and its response to environmental stimuli; the degree of diversity in oospore germination behavior; the relationship between oospore production and resultant oospore volume, and the relationship between oospore volume and inoculum potential, all constitute intrinsic resources
of the life cycle, but too little is known about these factors, even in vitro, to
allow an assessment of their importance in the regulation of population size. Such
factors undoubtedly account for the observed interspecific differences in realizable
niches, but experimental procedures designed to provide data to support this statement
have yet to be developed and we can only rely on descriptive work. Intrinsic genotypic regulators other than those controlling life cycle, and possibly linked with morphological characters, must presumably also exist to interact with the environment, further limiting the extent of the realizable niche. For example, *Pythium* species are
known to have differently shaped sporangia (round, torulose, or filamentous), which
apparently function similarly. Hallett (1975) has substantiated the widely held

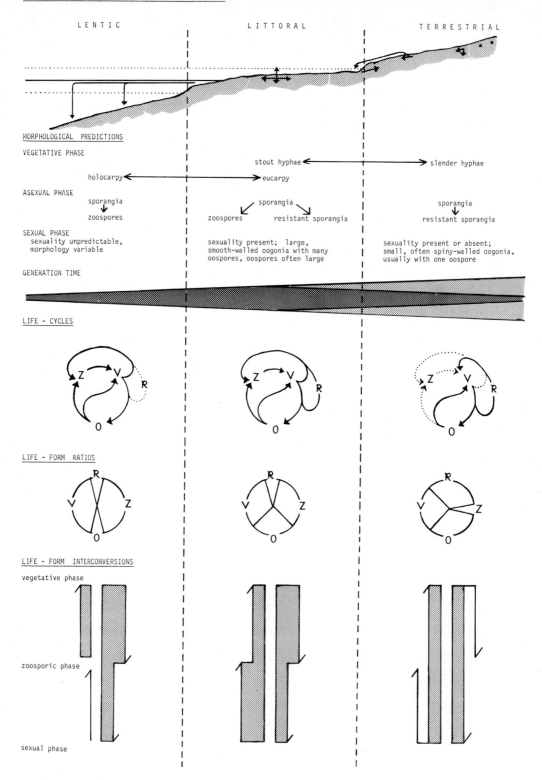

Fig. 1 Diagram to demonstrate the diversity of strategies open to Oomycetes; strategies which will be subject to different regulators and dependent on different resources. Predictions for morphology, life-cycle type, ratios of life forms and life-form interconversions are shown in relation to a continuum of environmental niches. The extent of the realizable niche for a given species may vary either with or without phenotypic or genotypic change in one or more of the three life-cycle/life-form predictions. Key: V, vegetative life form; Z, zoosporangia/zoospore life forms; R, resistant sporangia/chlamydospore life forms; O, oospore life form.

concept that species with round sporangia predominate in terrestrial situations and species with torulose or filamentous sporangia predominate in lentic and littoral zones. Thus a morphological tag may provide a marker for spatial disjunction between niches and their communities.

One of the commonly quoted intrinsic regulators, population size (breeding population), is unlikely to be a major factor for Oomycetes, at least in its obvious application through sexually produced progeny. In the lentic environment recurrent asexual cycles persist, and even in terrestrial and littoral habitats homothallism is predominant. Indirect applications of regulation by population size are possible, and both synergism and clone development should be considered. There appears to be no information on synergistic responses in Oomycetes. Hyphal fusions are almost undocumented, and there are no data on synergistic germination of zoospores or oospores. A single zoospore in a realizable niche with a food base will rapidly generate a potential inoculum of propagules, and population size is unlikely to provide a restriction either way. However, the combination of rapid generation time and a largely asexual, holocarpic life cycle with uninucleate propagules, together with an assumption, at least for several generations, of abundant environmental resources will allow the clonal development of potential inoculum. In the lentic environment all these criteria are met. The realizable niche is wide, and random mix is unlikely to be achieved within the short generation time. Clonal populations, possibly characterized by cryptic or nonlethal mutations in these diploid or polyploid organisms, would develop, albeit with disturbed and indistinct boundaries. Viewed from this aspect, population size may be an important regulator since it is conceivable that the physiological concept of inoculum potential (as opposed to the numerate concept of potential inoculum) may vary between clones. Such a development may underlie the more parasitic propensities of the *Saprolegnia diclina-Saprolegnia parasitica* complex, in which apparently similar individuals of either taxon may be isolated from exuviae or from fly-eating fish. I envisage this as a situation which would be unlikely to stabilize readily, ultimately being regulated by a combination of a failure in environmental resources such as nutrient supply and intrinsic resources such as a phenotypic unsuitability in circumstances changing with season or a physiological restriction of inoculum potential. A similar explanation could be put forward for the observed unpredictability of sexual reproduction in isolates from lentic populations. This unpredictability, or sexual failure, does not constitute a regulator when generation times are short and life cycles are normally recurrently asexual with a tendency to vegetative holocarpy. Very little time difference in achieving the top of a population growth curve would be involved, whether the population at the beginning of the season was derived from a few oospores or many oospores. On the other hand, the number of oospores will directly affect the number and diversity of the clonal populations asexually derived from them.

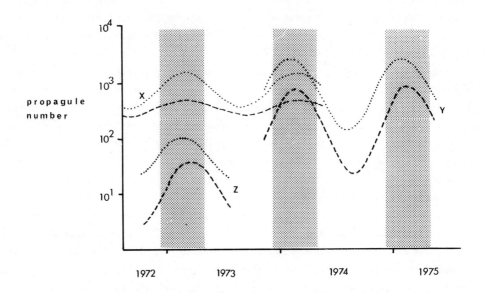

Fig. 2 Computed models of fluctuations in propagule numbers of *Pythium* per gram dry weight in two adjacent soils of different mean water content, using different isolation procedures. Formula used: $\log_e y = a + b \sin[\pi(t + c)/n]$, where y = propagule number recorded; t = date of collection; n = period of fluctuation; e^{a+b} = maximum availability; e^{a-b} = minimum availability; c = phase of fluctuation. Key: X, dilution plate data using a modified Kerr medium; Y, dilution plate data using MPVM medium; Z, direct observation in CMA. Periods from December to April inclusive are stippled. The wetter soil is indicated by the dotted lines; the drier soil, by the dashed lines. (From Ho, 1975.)

In terrestrial environments it is more difficult to ascertain the role of the life cycle as an intrinsic regulator of population levels, partly because the life-form ratios cannot accurately be assessed. This is made apparent by consideration of the phenological data of Ho (1975) (Fig. 2). Irrespective of method, all the data are consistent with a single winter maximum availability of *Pythium* species. It is clear that the two sites, adjacent but differing in water content, yield comparable data. Spot analyses have shown that more than 80% of the MPVM-medium data are derived from sporangial propagules, whereas only 60% of the modified-Kerr-medium data are from this life form. Zoospores and mycelia were much more rarely implicated and therefore appear to indicate only minimal numbers of these population units. There were also some differences in species composition, in that the modified-Kerr-medium revealed a wider range of *Pythium* species with more sexual species; the direct observation data were distinctive in providing a comparison between *Pythium* and *Aplanopsis* isolations, the latter genus being unrepresented on the dilution plates. It is abundantly clear that despite their much higher levels, neither direct plate method is providing a complete picture of the community. The MPVM medium is certainly underscoring in summer (possibly because it fails to activate oospores?), whereas the modified Kerr medium underscores in winter. The interpretation of these phenological data is further complicated when data of Hallett (1975) are considered. From collections made at the same time and in more or less the same area, zoospores were extracted from soil by combining a soil washing technique (Bissett and Widden, 1972) with that of continuous-flow centrifugation. A spring peak and autumn peak in zoospore availability were found for *Pythium* and Saprolegniaceae identical to that established for Saprolegniaceae elsewhere (Dick and Newby, 1961). Taken together these data suggest that a complex phenological interrelationship may exist between the different life forms which constitute the potential inoculum. In particular, there is a need for specifically autecological studies of *Pythium intermedium* and related taxa, which provided the bulk of the data from Hallett (1975) and Ho (1975). One hypothesis would be that the life-form proportions of the potential inoculum change with season from the two life forms (zoospores and sporangia) in autumn, through the single life form (sporangia) in midwinter, and back to the two life forms in the spring. In this case the midwinter isolations may represent persistent sporangia, detached from vegetative population units and related only to a previously active vegetative population, whereas the autumn and spring isolations could have more correspondence with vegetative activity.

It is also open to question whether the zoospore life form is derived in the same manner in autumn and spring. Zoospores could be derived either from freshly formed sporangia, thereby reflecting vegetative activity, or they could be produced only from afterripened sporangia, but as yet there is no way of separating these possibilities.

The inoculum potential, as a function of the potential inoculum, may therefore vary depending on the life-form ratios involved. This, in turn, will act as an intrinsic regulator of the population with season.

There is another inference which can be drawn from these phenological studies in relation to life-form interconversions. It has been assumed that phenological and pattern data for Saprolegniaceae are derived principally from oosporic propagules. It now becomes possible that isolation is via zoospores which may be derived either from vegetative mycelium or from oospores. Oospores are known to vary in the duration of their afterripening dormancy. Therefore conclusions concerning population levels in these studies must be tempered by knowledge of the interconversion patterns of populations of these different life forms in a phenological context. The life cycle may again be seen to be more complex than has been accepted in the past, and it may prove to be a potent intrinsic regulator.

Germination stimuli provided by the environment will affect the numbers of population units of potential inoculum as well as the life form of the potential inoculum, depending on the ability of persistent sporangia, chlamydospores, and oospores to germinate directly or via zoospore production. In the absence of any evidence of synergism, the inoculum potential of the population will depend on the numbers of units constituting the potential inoculum and differences between the inoculum poten-

tial of individual population units of different life forms. It has been convenient
to assume that the population is homogeneous in respect of this intrinsic regulator.
Ho (1975) has shown, for two species of *Pythium*, that this is unlikely to be the
case and that clonal differences may exist in the inoculum potential of population
units. Ho studied population diversity among 60 isolates of two nonsexual species
of *Pythium*. One species with catenulate sporangia was referable to *P. intermedium*,
but the other species could not be identified. Isolates were obtained from the
two sites noted in Fig. 2. Isolations of these representative isolates were made
at different times of the year. The morphological characteristics of these isolates
did not apear to change significantly with time in culture. Morphological data for
the 60 isolates revealed that there were differences in mean sporangial diameters
which enabled the isolates to be placed in a series, or cline. Adjacent isolates in
this cline did not differ significantly from each other, though sporangial diameter
ranges from the extreme ends of the clines for both species either showed no overlap
or were significantly different. There was an approximate but by no means absolute
correlation between isolates with smaller sporangia and the drier soil and those with
larger sporangia and the wetter soil. No correlation was found between sporangial
dimensions and season of isolation. Electrophoretic esterase patterns were also
prepared to see if there were any indications of similarities or differences within
the two taxa. There were slight indications that isolates of the noncatenulate species with smaller sporangia were somewhat different from those with larger sporangia.

Twelve of those isolates (Fig. 3) were also studied for physiological parameters
of growth which could reasonably be related to inoculum potential. The parameters
examined were the hyphal growth unit (HGU) and linear growth rate (LGR). HGU, the
ratio of the length of the juvenile mycelium to the number of hyphal tips, is regarded as a morphological measure of the physiological growth unit. This parameter can
be presumed to reflect some aspects of the inoculum potential of the mycelium source.
The efficiency of deployment of endogenous reserves during the critical period between germination and colonization, which may be termed *establishment*, will also be
related to the dimensions of the physiological growth unit.

LGR is a function of the established vegetative phase. The efficiency of colonization involving the utilization of exogenous reserves, i.e., the competitive ability,
is likely to be related to linear growth rate, though the relationship may well be
complex since rapid lateral spread may enable enhanced primary colonization at the
expense of less efficient utilization of available exogenous reserves. The existence
in the community of competitors and antagonists will regulate the relative importance
of colonization and utilization efficiency. Lateral spread is also related to the
degree of differentiation within the mycelium, and individual physiological unit dimensions may diverge in such differentiated mycelia. This divergence may be developed
more sharply by nutritional resource variation, as shown in the development of peta-
loid growth patterns of *Phytophthora* and *Pythium* in culture.

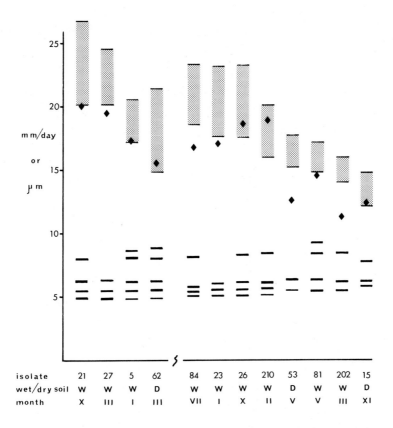

Fig. 3 Clonal differences between 12 of 60 isolates examined of two nonsexual *Pythium* species from the same soils referred to in Fig. 2. Isolate numbers are the same as those used in Fig. 4 (see below). Isolates 21, 27, 5, 62 were of the species with catenulate sporangia. The remaining isolates had noncatenulate sporangia. The stippled bars represent the mean ± 1 standard deviation for sporangial diameter in micrometers; the diamonds refer to mean linear growth rate in millimeters per day, grown on corn meal agar and incubated at 22°C; the y axis figures refer to both these parameters. The bars represent the relative electrophoretic esterase patterns and are drawn as originating from the base line of the x axis. (From Ho, 1975.)

The 12 isolates not only all produced sporangia capable of direct germination but also could be induced to form zoospores. Two propagule life forms of different volume and presumed endogenous reserve capacity could therefore be compared. The scatter diagram, Fig. 4, plots the relationships between these two physiological parameters for the two life forms for each isolate. The same clines can be seen for both species and both propagule types. For sporangia, a linear relationship with positive correlation exists between LGR and HGU: isolates with large sporangia have a high LGU and a high HGU. For zoospores also there exists a linear relationship, but for this life form the relationship is inverse, the zoospores from isolates which form large sporangia having a low HGU. Zoospore dimensions between the isolates do not vary significantly, and the recorded differences between the LGRs of zoospore-

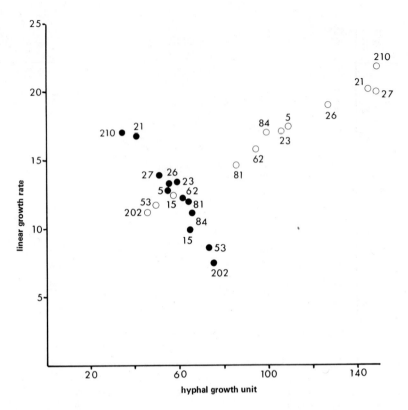

Fig. 4 Scatter diagram of relationships between linear growth rate (LGR) and hyphal growth unit (HGU) from zoosporangial and zoospore inocula for the 12 isolates refer- red to in Fig. 3. Open circles represent zoosporangial inocula; solid circles re- present zoospore inocula. Isolate numbers are listed against each point. Linear growth rates are recorded in millimeters per day; hyphal growth units are calculated in micrometers at 8 hr. Both were from growth on corn meal agar, with incubation at 22°C. (From Ho, 1975.)

derived mycelia and sporangium-derived mycelia are comparable, though it is inter- esting to note that the former is consistently somewhat lower than the latter, even after several days' growth. The fact that isolates show variation in LGR probably indicates that intraspecific variability can be anticipated in competitive ability. When the two linear trends are considered together, it can be seen that for both spe- cies the greatest divergence in HGUs occurs when the difference between mean zoospore volume and mean sporangial volume is greatest. This is interpreted as indicating that the inoculum potential of the two life forms is very different in clones with large sporangia (principally from the wetter soil) whereas the inoculum potential of the two life forms is much the same for clones with smaller sporangia, despite the still considerable difference in volume. In the former situation, and probably only to a much lesser degree in the latter situation, the zoospore life form must be pre- sumed to have a function within the life cycle which is specific and distinct from that of the sporangium. Extrapolated, this may underlie differences in life cycle between clones, which in turn would suggest that the realizable niches for these clones are not identical. The population concept itself is called into question by this conclusion.

Inoculum potential is seen as an intrinsic regulator of population composition separate from competitive ability. Phenotypic variability, coupled with recurrent asexual cycles, can produce physiologically and morphologically distinct clonal sub-populations of the species, each with its own proportions of constituent life forms. This phenomenon must also be considered an intrinsic resource.

At the conclusion of an earlier symposium in fungal ecology in 1960, Burges wrote: "Unless we know their growth patterns, their turn-over rates, and their zones of influence we cannot hope to advance beyond the counting and listing which has occupied us for so long." The counting and listing has enabled community structure to be outlined, and phenological patterns have been established against a background of spatial variation (see Gray and Williams, 1971), so that the zones of influence, or realizable niches, can be described. Problems encountered in determining growth patterns and turnover rates in fungi have been discussed and a few communities have been identified which are particularly suitable for further studies of this kind.

Different regulators are seen to control the size and structure of the population in different ways for different species. Similarly the resources of a population vary depending upon the infrastructure of that population. Population infrastructure, in terms of its life forms, life cycle, phenology, phenotypes, and genotypes, must be recognized in order to interpret the data obtained from the cultural techniques necessary in mycology. When this is done, the rich diversity in clonal development so prevalent in fungi can be incorporated into the general theories concerning the organization of populations.

REFERENCES

Adenle, V. F. O. (1978). Ecological and developmental studies on aquatic *Phytophthora* species. Ph.D. thesis, University of Reading, England.

Bissett, J. D., and Widden, P. (1972). An automatic multichambered soil-washing apparatus for removing fungal spores from soil. *Can. J. Microbiol. 19:* 1399-1404.

Burges, N. A. (1960). Dynamic equilibria in the soil. In *The Ecology of Soil Fungi,* D. Parkinson and J. S. Waid (Eds.). Liverpool Univ. Press, Liverpool, England.

Dick, M. W. (1962). The occurrence and distribution of Saprolegniaceae in certain soils of south-east England. II. Distribution within defined areas. *J. Ecol. 50:* 119-127.

Dick, M. W. (1963). The occurrence and distribution of Saprolegniaceae in certain soils of south-east England. III. Distribution in relation to pH and water content. *J. Ecol. 51:* 75-81.

Dick, M. W. (1966). The Saprolegniaceae of the environs of Blelham Tarn: Sampling techniques and the estimation of propagule numbers. *J. Gen. Microbiol. 42:* 257-282.

Dick, M. W. (1968). Considerations of the role of water on the taxonomy and ecology of the filamentous biflagellate fungi in littoral zones. *Veröffentlichen Inst. Meeresforsch. Bremerhaven, Sonderband 3:* 22-38.

Dick, M. W. (1970). Saprolegniaceae on insect exuviae. *Trans. Brit. Mycol. Soc. 55:* 449-458.

Dick, M. W. (1971). The ecology of Saprolegniaceae in lentic and littoral muds with a general theory of fungi in the lake ecosystem. *J. Gen. Microbiol. 65:* 325-337.

Dick, M. W. (1976). The ecology of aquatic phycomycetes. In *Recent Advances in Aquatic Mycology*, E. B. G. Jones (Ed.). Elek, London.

Dick, M. W., and Newby, H. V. (1961). The occurrence and distribution of Saprolegniaceae in certain soils of south-east England. I. Occurrence. *J. Ecol. 49:* 403-419.

Emerson, R., and Weston, W. H. (1967). *Aqualinderella fermentans* gen. et sp. nov., a phycomycete adapted to stagnant waters. I. Morphology and occurrence in nature. *Amer. J. Bot. 54:* 702-719.

Fuller, M. W., and Poyton, R. O. (1964). A new technique for the isolation of aquatic fungi. *Bioscience 14:* 45-46.

Gray, T. R. G., and Williams, S. T. (1971). *Soil Micro-organisms*. Oliver & Boyd, Edinburgh.

Griffin, D. M. (1963). Soil physical factors and the ecology of fungi. II. Behaviour of *Pythium ultimum* at small water suctions. *Trans. Brit. Mycol. Soc. 46:* 368-372.

Hallett, I. C. (1975). Studies on the zoospores of the Oomycetes. Ph.D. thesis, University of Reading, England.

Hallett, I. C., and Dick, M. W. Seasonal and diurnal fluctuations of oomycete propagule numbers in the free water of a freshwater lake. *J. Ecol.* (in press).

Ho, C.-L. (1975). Population studies on the genus *Pythium*. Ph.D. thesis, University of Reading, England.

Park, D. (1974). Accumulation of fungi by cellulose exposed in a river. *Trans. Brit. Mycol. Soc. 63:* 437-447.

Sparrow, F. K. (1968). Ecology of freshwater fungi. In *The Fungi: An Advanced Treatise*, G. C. Ainsworth and A. S. Sussman (Eds.), Vol. III. Academic Press, New York.

Part IV

INTERACTIONS AMONG FUNGAL POPULATIONS

Chapter 16

INTRODUCTION TO THE THEORY OF SPECIES INTERACTIONS

David C. Culver
Northwestern University
Evanston, Illinois

I. INTRODUCTION

The theory of species interactions forms an integral part of contemporary ecology. Over the past decade this theory of species interactions has become increasingly extensive and sophisticated, but its relevance to experimental and field ecology in general and fungal ecology in particular is often not immediately apparent. Even though contemporary ecological theory often seems excessively abstract and/or developed in ignorance of mycology, it is my contention that it has important insights to offer the mycologist and that it provides a useful basis for considering fungal species interactions. There are, of course, fundamental ecological differences among different kinds of organisms. For example, animal species may typically interact competitively with other animal species while fungi may typically interact with non-fungi. Animals are usually motile; vascular plants and fungi usually are not. It might even be true that competition dominates vascular plant interactions, predation dominates animal interactions, and mutualism dominantes fungal interactions. Nevertheless, whether one's chief interest is mycological, zoological, or botanical, some important aspects of the theory should be common to all groups of organisms. Competition, predation, and mutualism should be fundamental concerns of all ecologists.

Given the relative isolation of mycology and theoretical ecology, that part of ecological theory concerned with general, qualitative results and predictions is most likely to be initially applicable to mycology. Since ecology should be an evolutionary topic, the evolution and co-evolution of interacting species will be given special scrutiny. In this review I have relied heavily on two sources: Hirsch and Smale's (1974) text on differential equations and dynamical systems, and the text on theoretical ecology edited by May (1976a).

The theory of competition, predation, and mutualism is very unevenly developed. The principal question in competition theory is how similár species can be and still coexist, a question intimately tied to the problem of how to measure competition. Much of predation theory concerns the exact form of the equations that describe the dynamics of the interactions. In contrast, mutualism theory is in its beginning stages, as yet without a clear focus. I will attempt, however, to consider these interactions on an equal basis for two reasons. First, the greater emphasis on com-

Table 1 Classification of possible interactions between two species

Effect of species A on species B

	0	−	+
0	−−	Amensalism	Commensalism
−	Amensalism	Competition	Predation
+	Commensalism	Predation	Mutualism

Effect of species B on species A

petition and predation may be accidents of history that do not reflect the actual importance of different kinds of interactions. Second, since the theory has rarely explicitly considered fungi, the emphasis on competition and predation may be totally unjustified for fungi.

The usual classification of species interactions is given in Table 1. The three most important interactions are competition, mutualism, and predation. Mathematically at least, amensalism and commensalism have received far less attention. A priori, it is unlikely that one species is affected by but does not affect another species. What seems more likely is that both species affect each other but that the interaction has very unequal intensities. That is, most cases of amensalism may be competition with asymmetric effects, and most cases of commensalism may be mutualism with very asymmetric effects.

For continuously reproducing populations without time lags we have a fairly complete catalog of the consequences of these two species interactions, and this catalog is a convenient place to begin consideration of species interactions.

II. TWO SPECIES INTERACTIONS

G. F. Gause, in his classic theoretical and experimental study of competition and predation (Gause, 1934), was especially concerned with what he called the struggle for existence. Although his experiments with protozoa are more widely known, it is worth noting that many of his competition experiments involved two species of yeast. Gause felt that two of the major forms that this struggle took were interspecific competition and predation. Competition and predation share the characteristic that at least one species experiences a negative effect as a result of the interaction and that some resource is limited. Gause's implication was that mutualism was not part of this struggle, but rather one possible consequence of this struggle. Thus it is competition and predation that is the template upon which natural selection operates.

What are the short-term dynamic consequences of competition and predation? For continuously reproducing populations without time lags, we have fairly complete answers. In the case of predation, May (1975) has shown that eventually predator and prey settle down to a stable limit cycle if predation is strong enough or show damped oscillations if predation is less intense. However, the limit cycle may have an amplitude that results in extinction. These predictions correspond to the results of laboratory experiments on both predator-prey pairs and host-parasite pairs. These two cases are shown in the top row of Fig. 1, where the "phase portraits" are shown. Fungi may be involved in many predator-prey systems as prey for animals and protists, and in host-parasite systems as parasites of animals and plants.

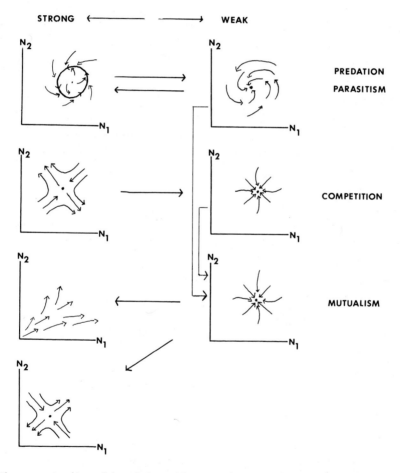

Fig. 1 Phase portraits of two interacting species. Each sketch has as its axes the population sizes of the two species. The small arrows in each sketch indicate the direction of population size change of the two species. For the most part, these sketches correspond to a qualitative classification of critical points of the differential equations. Hirsch and Smale (1974) can be consulted for more details, and Arnold (1973) gives similar sketches for three-dimensional systems. The large arrows connecting the sketches give probable directions of evolution and co-evolution.

Likewise, competition results in either exclusion or in a stable equilibrium. These two cases are shown in the second row of Fig. 1. Laboratory experiments generally confirm the prediction that if competition is strong enough, exclusion results. Experiments with fungi also confirm this (Horn, 1971; Armstrong, 1976). There may be a rough correspondence between interference competition (denial of access to a resource) and strong competition on the one hand, and exploitation competition (co-utilization of a resource without direct interaction) and weak competition. As MacArthur (1972) points out, exploitation competition is the more universal of the two, and interference competition evolves when a species can increase its share of the limiting resource. Most of the ways to measure competition coefficients in natural populations are valid only for exploitation competition (see Abrams, 1976).

What are the evolutionary consequences of competition and predation? One possible consequence of competition, of course, is competitive exclusion, with the allopatric or allotopic separation of the two species as a consequence. More generally, Lawlor and Maynard Smith (1976) show that the co-evolution of two competitors results in reduced competition. Therefore an arrow can be drawn from strong to weak competition to indicate the direction of selection (Fig. 1). The evolution of greater predator efficiency results in oscillations, but the evolution of predator avoidance and perhaps group selection operating aginst oscillating populations (Gilpin, 1975) results in damped oscillations. The relative paucity of documented oscillations (Hutchinson, 1975) indicates that most predator-prey interactions do not result in limit cycles.

Another, and perhaps the major possible consequence of evolution and co-evolution of competitors, or predator and prey, especially for fungal populations, is the evolution of mutualism. The conditions under which the evolution of mutualism is favored have been little studied, but Roughgarden (1975) has established conditions under which at least commensalism is evolutionarily advantageous. In any case, mutualisms involving fungi are certainly widespread (Raper, 1952; Watson, 1965; Tribe, 1966; Dowding, 1973; Hedger and Hudson, 1974; Barras and Perry, 1975). Hirsch and Smale (1974) show that most mutualisms should come to a stable equilibrium point if there is an upper limit to the population sizes of both species. This is shown on the right-hand side of the third row in Fig. 1. However, strong mutualisms per se do not set a bound to population size and result in unbounded growth, as shown on the third row on the left in Fig. 1. Obligate mutualism can result in either extinction or unbounded growth, as shown in the bottom left sketch of Fig. 1. If the co-evolution of pairs of mutualists is towards stronger mutualism and greater dependence (as is likely), then the dynamic consequences are a loss of stability of the pairwise interaction, resulting in either extinction or unbounded growth. The claim from the theory is that the dynamics of mutualism per se do not usually set a bound on population size. But this bound must exist--otherwise extinction or unbounded growth results. The bound might be due to extrinsic factors such as space or to the presence of other species as prey, competitors, or predators. The question of interactions between three or more species is taken up later (in Secs. III and IV), but it is clear that experimental and field work on the dynamics of mutualism are needed both to investigate the validity of the theory and to point to new directions for the theory. Since mutualism is undoubtedly an important kind of fungal interaction (e.g., see DeVay, 1956; Graham, 1967; Culberson, 1970; Cooke, 1977), fungi should play a major role in further research.

The dynamic consequences of species interactions indicated in Fig. 1 hold for a variety of models in addition to the usual Lotka-Volterra equations (see Hirsch and Smale, 1974). At least for competition and predation, experimental data generally agree with the theory. However, two extensions of two-species theory are largely untested, and their importance is unknown. These are (1) discrete generation models of population interactions, and (2) alternate stable equilibria due to "higher-order" interactions.

The extension of two-species models for populations with continuous reproduction to populations with discrete generations introduces additional complexities. The theory itself is not completely worked out even for single-species growth models (see May, 1976b), but the potential consequences of these population growth models are clear: for certain parameter values of predation, competition, or mutualism coefficients, the result is cycles of any length, or population trajectories that are not cycles, but never the extinction of either species. An example of such "chaotic behavior" is shown in Fig. 2, for a predator-prey pair. Whether this is a mathematical curiosity or a pattern of population fluctuation that actually occurs is not known. Many nonlinear difference equation models can generate this kind of population dynamics (May and Oster, 1976), but it does not appear to be common in natural populations (Hassell et al., 1976). Unless experimental and field ecologists recognize that such "chaotic" behavior as shown in Fig. 2 is a real possibility, there may be a tendency to attribute observations of such population fluctuations as due to experimental error or due to random factors affecting the population.

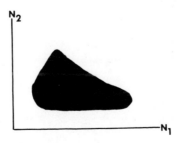

Fig. 2 A possible equilibrium for a predator-prey pair, using a discrete generation model. The example is taken from Beddington et al. (1975) and should be contrasted with the equilibrium points and cycles of Fig. 1.

The final aspect of two-species models to be considered, and potentially the most interesting, is the question of how many equilibrium points or limit cycles exist. The most general treatments of pairs of interacting species (e.g., Hirsch and Smale, 1974) usually give rise to statements that the number of equilibrium points or limit cycles is finite. On the other hand, the Lotka-Volterra equations commonly used in ecology give rise to a single equilibrium point (stable or unstable) or a single limit cycle. The difference between these two results is that the Lotka-Volterra models ignore what are usually called "higher-order" interactions. In Lotka-Volterra models the per capita growth rate, $1/N_i$ (dN_i/dt), where N_i is the population size of species i, is determined by terms of the following type: $a_i N_i$ and $a_{ij} N_i N_j$, where a_i's are coefficients of growth rate and a_{ij}'s are interaction coefficients. Higher-order interactions are ones where the interactions are not only the product of an interaction coefficient and two population sizes but include other terms such as $b_{ijj} N_i N_j^2$ and, when more than two species are present, $b_{ijk} N_i N_j N_k$.

The documentation of multiple stable points in two-species interactions has not been done and remains one of the important unanswered questions in our understanding of such interactions. Vandermeer (1973) gives some examples of multiple equilibria that could arise when either inter- or intraspecific effects change sign at different densities. He also shows that all higher-order interactions do not result in multiple equilibrium points. One hypothetical possibility is sketched out in Fig. 3. Schoener (1976) suggests that alternate stable equilibria are most likely to occur among competitors when there is strong interspecific relative to intraspecific interference and when the two species obtrain a small fraction of their energy from resources for which they do not compete.

Some mention needs to be made as to how either multiple equilibria or chaotic behavior can be detected. The usual procedure of either observing populations near equilibrium in the field or replicating growth experiments with the same starting numbers will not suffice. What is required is that the direction of population change be known for a broad array of population densities, not just those near an equilibrium

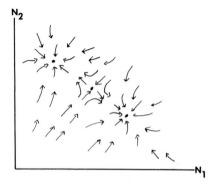

Fig. 3 A hypothetical phase space for two species, with two stable and one unstable equilibrium points.

or with small starting numbers (see Maly, 1969). This could be done either with field perturbations or with short-term growth experiments in the laboratory.

The consequences of two-species interactions as illustrated in Fig. 1 provide a framework for considering some important questions of community structure. For example, if competition is rare, then one might conclude that competition either results in exclusion or evolves to something else. In general, if the relative frequency of competition, predation, and mutualism in natural communities were known, then the relative importance of different evolutionary pathways would be clearer. But, as seen below, the extension from two-species interaction models to three or more species is anything but straightforward.

III. MULTISPECIES INTERACTIONS: LINEAR MODELS

An idea which goes back to Gause is that two-way interactions are the fundamental building blocks of community structure and evolution. In a general sense this is true. The dynamical behavior of three or more populations near an equilibrium point is a straight-forward extension of the dynamical behavior of two populations near equilibrium (Hirsch and Smale, 1974; Arnold, 1973), although the biological interpretation of stable and unstable equilibria becomes more complex.

One hypothesis which is implicit in the two-species models is that interactions set a limit to the number of species in a community. An example for two-species interactions is competitive exclusion (see Fig. 1). In this case we could say that competition limits species diversity. Levins (1968) develops some rules that allow for prediction of the number of species present in a community. Levins shows (see also May, 1975, 1976a) that as the number of species increases, the likelihood of a stable equilibrium point decreases. This hypothesis can explain the very common observation that the number of species that co-occur in any local community is less than the number of species that could potentially occur. For example, three of five cellular slime molds studied may occur at any microsite, but four or five cannot (see Eisenberg, 1976). With three or more species interacting, the community may be unstable because of a few strong interactions or because of many weaker interactions and indirect effects. The second kind of interaction has been called "diffuse competition" for competitive communities (MacArthur, 1972).

In general, the extension from two-species-interaction models to three or more species allows for greater complexity and new biological phenomena. It is important to recognize that there are two qualitatively different kinds of complexity possible when three or more species interact. One is a direct consequence of the increased dimensionality of a community with linear, pairwise interactions. The other is a consequence of multiway interactions and echoes earlier comments about multiple equilibria. The simple increase in dimensionality introduces a surprising amount of complexity, and these will be discussed first.

Implicit in the hypothesis that interactions limit species diversity is the prediction that alternate stable communities can occur. For example, any three cellular slime mold species may occur at any microsite. Which particular three species co-occur at a microsite might have a historical explanation, rather than a dynamical one. Alternate stable communities have been reported for animal communities (Sutherland, 1974; Culver, 1976) and may be important in fungal communities.

A major cause of increased complexity due to higher dimensionality is the occurrence of unstable subcommunities within stable communities. As May (1975) points out, increasing the number of species usually destabilizies a community, but the reverse can occur. For example, E. R. Heithaus and I have found that some pairs of unstable mutualistic species can be stabilized by a predator species. In fact, it may be a general pattern that strong mutualisms occur in rather complex communities, which act to stabilize mutualistic interactions that may be unstable in isolation. Unstable subcommunities appear to be widespread, at least in animal communities. Paine (1966) found that in the marine intertidal zone, starfish predation increases species numbers by stabilizing an unstable subcommunity of competitors. Such keystone predators appear important in many marine intertidal communities. Diamond (1975) found several examples of unstable subcommunities in the avifauna of small islands near New Guinea.

When unstable subcommunities occur, commonsense ideas like perturbation experiments may become very difficult to interpret, even in the linear case. Consider the following hypothetical example. Assume that there are three species of slime molds (A, B, and C) that compete with each other and that there are no higher-order interactions. Assume further that A and B can stably coexist and that A, B, and C can stably coexist. In the two-dimensional case (A and B in the absence of C), species B will decrease in abundance if species A is increased in abundance. But now consider the three-dimensional case where all three species coexist. If species A is increased in abundance, species B may increase, decrease, or stay the same, depending on whether the subcommunity of species B and C is stable or not. These different results can occur because A affects B in two ways. First, there is the direct negative effect of A on the abundance of B due to competition. Second, there is the indirect positive effect of A on the abundance of B because A has a negative effect on C, which in turn has a negative effect on B. That is, A reduces the abundance of competitiors of B, which causes B to increase in abundance. This is diagrammatically shown in Fig. 4. If the direct effect is greater than the indirect effect, then increasing A will decrease B. If the indirect effect is greater, then increasing B will increase A. It is those cases where indirect effects of interactions are stronger than direct effects of interactions that counterintuitive results occur.

Selection within communities of three or more species has consequences that parallel those of the perturbation experiment (Levins, 1975; Roughgarden, 1976). Roughgarden showed that natural selection within a species for increased population size can actually result in reduced population size of the species if the subcommunity ob-

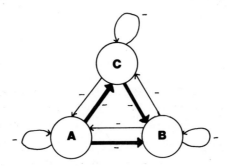

Fig. 4 Illustration of the direct effect and indirect effect of species A on species B. Each arrow indicates a competitive effect. In this example all three species with each other. The arrows from a species back to itself represents intraspecific competition. The thickened arrow from species A to species B is the direct negative effect. The thickened arrows through species C represent the indirect positive effect of species A on species B.

tained by deleting the species being selected is unstable. For example, a keystone predator may select itself to extinction.

It is obviously important to look for unstable subcommunities. One simple way to detect them is to see if some subsets of naturally occurring communities never occur in isolation. Diamond (1975) used this technique on the avifauna of small islands near New Guinea. The necessary data are species lists for different communities. Eisenberg's (1976) data on cellular slime molds can be used in this way. By way of illustration, he found *Dictyostelium purpureum*, *Polysphondylium violaceum*, and *P. pallidum* together, but did not find the first two in the absence of *P. pallidum*. Although sample sizes are too small for statistical testing, *D. purpureum* and *P. violaceum* may be an unstable subcommunity.

Thus far, I have avoided discussing the problem of measurement of interaction coefficients, a problem that is still far from being completely solved. There have been two basic approaches. First, one can impose a perturbation on the community, and observe the return to equilibrium. Since the interaction coefficients can be interpreted as partial regression coefficients in multiple linear regression (Seifert and Seifert, 1976), interaction coefficients of competition, mutualism, and predation can be interpreted in this way. The second method, which is used to calculate competition coefficients, is to find an explicit formula for the competition coefficient, α, usually by expressing α in terms of the characteristics of the competitors' prey populations. For example, MacArthur develops a formula for α involving frequency of prey taken, weight of the prey, etc. Such explicit formulas have two requirements. First, they must be dynamically based, and in fact correspond to partial differentials of the form

$$\alpha_{ij} = \cfrac{\cfrac{\partial(dN_i/dt)}{\partial N_i}}{\cfrac{\partial(dN_i/dt)}{\partial N_i}}$$

where N_i is the population size of species i. Thus, measurements of niche overlap
using information theory do not qualify as α because they are not partial differentials. Second, and most importantly, the assumptions about the nature of competition
must make biological sense. Most explicit measures of α assume that exploitation competition is occurring, and so the greater the overlap between species, the greater the
competition. If, in fact, interference competition is more important, reduced overlap indicates competition. There is simply no substitute for knowing some biology
of the organisms we are studying.

Although it has not been made explicit, I have assumed that any community we observe must be stable. This need not be true. A small spate of theoretical papers
have appeared in the past several years that show that locally unstable communities
may be stable regionally because of continued extinction and migration between communities (see Levin, 1976). This process is especially important for fugitive species, and Armstrong (1976) demonstrates that the process can be important for fungal
communities.

IV. MULTISPECIES INTERACTIONS: NONLINEAR MODELS

If we also allow for the possibility of higher-order interactions, the situation becomes very complex. In fact, even Lotka-Volterra equations, which are not strictly
linear, can result, when three or more species are involved, in patterns not observed
for two interacting species, including cycles of increasing length (May and Leonard,
1975) and equilibrium points that are locally but not globally stable (Goh, 1977).
For very general competition equations (including all possible kinds of higher-order
interactions), Smale (1976) shows that three or more competitors can display almost
any conceivable dynamical behavior (cf. Fig. 2), and it is likely that this will hold
for predators and mutualists as well. Although most ecologists seem to believe that
higher-order interactions are important, the evidence is equivocal. Neill (1974)
found them to be important in zooplankton assemblages in the laboratory, but Vandermeer (1969) found them to be unimportant in protozoan assemblages. In some cases,
the higher-order interactions that do exist "deform" the phase space rather than produce more equilibria or new dynamics (Ayala et al., 1973). Holling (1973) argues
very cogently that multiple equilibria are widespread and that the domain of attraction of a stable equilibrium point is at least as important as the rate of return of
population sizes to equilibrium following a temporary disturbance (i.e., the stability
of the equilibrium). Recent theoretical work (e.g., Goh, 1977) suggests that exploitation competition does not result in multiple equilibria. I know of no mycological

data that bear directly on these points. It is the task of experimental and field ecologists to provide evidence about the importance of higher-order interactions. This can only be done if the consequences of linear interactions are first understood.

If higher-order interactions are important, and if they occur among two species, then the classification of interactions shown in Table 1 is itself in doubt. That is, a pair of species may compete in some circumstances and act as mutualists or predator and prey in others. A classic example of this is the lichen symbiosis, where the fungi can be either parasites or mutualists of the algae (Ahmadjian, 1966) and so the sign of the interaction is ambiguous. There is a great deal of anecdotal information about ambiguous interactions (e.g., different life history stages interacting in different ways), but its overall impact is unknown. It also may be that interactions change with density. The ambiguity of the sign of the interaction may be formally equivalent to higher-order interactions. For example, if the growth rate of algae (dN_1/dt) in a lichen symbiosis contains the terms aN_1N_2 and $bN_1^2N_2$ with $a > 0$, $b < 0$, and N_2 = fungal abundance, the sign of the effect of fungi on algae depends on the density of algae as well as the magnitudes of a and b. In this example, fungi would be mutualists at low algal densities and parasites at high algal densities.

Some theoretical ecologists would claim that I have neglected one crucial ingredient of any realistic ecological model--stochastic processes. And ultimately ecological models must include stochastic elements. Random events do play at least a tangential role in most of the models discussed, i.e., stability is only meaningful if perturbations occur. Stochastic models may play a central role in the problem of competitive indeterminancy, which Mertz et al. (1976) have recently reviewed. At present, deterministic models have not yet been thoroughly developed, and this remains the first priority.

V. SUMMARY

The qualitative dynamics of two-species interactions have been compared: nearly all models of predator-prey interactions result in either limit cycles or damped oscillations; nearly all models of competition result in a stable equilibrium point or the exclusion of one species; nearly all models of mutualism result in stable equilibrium if the mutualism is weak, and unbounded growth or extinction otherwise. Laboratory studies on competition and predation are in general agreement with the models, but the models of mutualism still need experimental confirmation. The evolutionary role of competition and predation differs from that of mutualism. Competition and predation are integral to what G. F. Gause termed the struggle for existence, while mutualism is one of the consequences of the struggle. This does not imply that mutualism is less important, since mutualism may be a major evolutionary consequence of both competition and predation. The other possible consequences of evolution and coevolution of pairs of interacting species are reduced intensity of competition and predation, and increased intensity of mutualism. Increasingly strong mutualism, however, results in instability. It is not clear whether this is a defect of the models or an important ecological insight.

The most general models for two-species interactions predict the kind of equilibria (e.g., limit cycles) but not the number of such equilibria. If higher-order interactions are important, then there may be multiple stable equilibria. This has not yet been demonstrated experimentally for two species. Difference equation models predict the possibility of other dynamical behavior of two interacting species, including cycles of any length and "chaotic behavior." The biological relevance of these findings is unknown.

Linear models can be extended to three or more species, with the possibility of alternate stable equilibria where the same number of species coexist but where the particular species present very. The extension of even linear models of species interactions to more than two species introduces new complexities. Multispecies interactions can result in (1) counterintuitive results of perturbation experiments, (2) stabilization of unstable interactions such as obligate mutualism, and (3) individual selection resulting in reduced population size of the species under selection. Most of this complexity is a consequence of unstable subcommunities within stable communities, which may be a widespread phenomenon in natural communities.

Nonlinear interactions introduce further complexities that have been little studied theoretically, experimentally, or in the field. If higher-order interactions are widespread even in two-way interactions, then interactions will show shifts between competition, predation, and mutualism, depending on density or age structure.

REFERENCES

Abrams, P. (1976). Limiting similarity and the form of the competition coefficient. *Theor. Pop. Biol. 8*: 356-375.

Ahmadjian, V. (1966). Lichens. In *Symbiosis*, S. M. Henry (Ed.), Vol. I. Academic Press, New York, pp. 36-98.

Armstrong, R. A. (1976). Fugitive species: Experiments with fungi and some theoretical considerations. *Ecology 57*: 953-963.

Arnold, V. I. (1973). *Ordinary Differential Equations*. MIT Press, Cambridge, Mass.

Ayala, F. J., Gilpin, M. E., and Ehrenfeld, J. G. (1973). Competition between species: Theoretical models and experimental tests. *Theor. Pop. Biol. 4*: 331-356.

Barras, S. J., and Perry, T. J. (1975). Interrelationships among microorgansisms, bark or ambrosia beetles, and woody host tissue: An annotated bibliography, 1965-1974. Southern For. Exp. Sta., New Orleans, 34 pp. (U.S. Dept. of Agriculture For. Serv. Gen. Tech. Rep. SO-10.)

Beddington, J. R., Free, C. A., and Lawton, J. H. (1975). Dynamic complexity in predator prey models framed in difference equations. *Nature 255*: 58-60.

Cooke, R. (1977). *The Biology of Symbiotic Fungi*. Wiley, New York.

Culberson, W. L. (1970). Chemosystematics and ecology of lichen-forming fungi. *Ann. Rev. Ecol. Syst. 1*: 153-170.

Culver, D. C. (1976). The evolution of aquatic cave communities. *Amer. Natur. 110*: 945-957.

DeVay, J. E. (1956). Mutual relationships in fungi. *Ann. Rev. Microbiol. 10*: 115-140.

Diamond, J. M. (1975). Assembly of species communities. In *Ecology and Evolution of Communities*. Harvard Univ. Press, Cambridge, Mass., pp. 342-444.

Dowding, P. (1973). Effects of felling time and insecticide treatment on the interrelationships of fungi and arthropods in pine logs. *Oikos 24*: 422-429.

Eisenberg, R. M. (1976). Two-dimensional microdistribution of cellular slime molds in forest soil. *Ecology 57*: 380-384.

Gause, G. F. (1934). *The Struggle for Existence*. Williams & Wilkins, Baltimore. (Reprinted, 1971, by Dover, New York.)

Gilpin, M. E. (1975). *Group Selection in Predator-Prey Communities*. Princeton Univ. Press, Princeton, N.J.

Goh, B. S. (1977). Global stability in many species systems. *Amer. Natur. 111*: 135-143.

Graham, K. (1967). Fungal-insect mutualism in trees and timber. *Ann. Rev. Entomol. 12*: 105-126.

Hassell, M. P., Lawton, J. H., and May, R. M. (1976). Patterns of dynamical behavior in single-species populations. *J. Anim. Ecol. 45*: 471-486.

Hedger, J. N., and Hudson, H. J. (1974). Nutritional studies of *Thermomyces lanuginosus* from wheat straw compost. *Trans. Brit. Mycol. Soc. 62*: 129-143.

Hirsch, M. W., and Smale, S. (1974). *Differential Equations, Dynamical Systems, and Linear Algebra*. Academic Press, New York.

Holling, C. S. (1973). Resilience and stability of ecological systems. *Ann. Rev. Ecol. Syst. 4*: 1-24.

Horn, E. G. (1971). Food competition among cellular slime molds (Acrasieae). *Ecology 52*: 575-584.

Hutchinson, G. E. (1975). Variation on a theme by Robert MacArthur. In *Ecology and Evolution of Communities*, M. L. Cody and J. M. Diamond (Eds.). Harvard Univ. Press, Cambridge, Mass., pp. 492-521.

Lawlor, L. R., and Maynard Smith, J. (1976). The coevolution and stability of competing species. *Amer. Natur. 110*: 79-99.

Levin, S. A. (1976). Population dynamic models in heterogeneous environments. *Ann. Rev. Ecol. Syst. 7*: 287-310.

Levins, R. (1968). *Evolution in Changing Environments*. Princeton Univ. Press, Princeton, N.J.

Levins, R. (1975). Evolution in communities near equilibrium. In *Ecology and Evolution of Communities*, M. L. Cody and J. M. Diamond (Eds.). Harvard Univ. Press, Cambridge, Mass., pp. 16-50.

MacArthur, R. H. (1972). *Geographical Ecology*. Harper & Row, New York.

Maly, E. J. (1969). A laboratory study of the interaction between the predatory rotifer *Asplanchna* and *Paramecium*. *Ecology 50*: 59-63.

May, R. M. (1975). *Stability and Complexity in Model Ecosystems*, 2nd ed. Princeton Univ. Press, Princeton, N.J.

May, R. M. (Ed.) (1976a). *Theoretical Ecology: Principles and Applications*. Saunders, Philadelphia.

May, R. M. (1976b). Simple mathematical models with very complicated dynamics. *Nature 261*: 459-467.

May, R. M., and Leonard, W. J. (1975). Nonlinear aspects of competition between three species. *SIAM J. Appl. Math. 29*: 243-253.

May, R. M., and Oster, G.F.   (1976).   Bifurcations and dynamic complexity in simple ecological models. *Amer. Natur. 110*: 573-599.

Mertz, D. B., Cawthon, D. A., and Park, T.   (1976).   An experimental analysis of competitive indeterminancy in *Tribolium*. *Proc. Nat. Acad. Sci. U.S. 73*: 1368-1372.

Neill, W. E.   (1974).   The community matrix and interdependence of the competition coefficients. *Amer. Natur. 108*: 399-408.

Paine, R. T.   (1966).   Food web complexity and species diversity. *Amer. Natur. 100*: 65-75.

Raper, J. R.   (1952).   Chemical regulation of sexual processes in the Thallophytes. *Bot. Rev. 18*: 447-545.

Redmond, D. R., and Cutter, V. M.   (1951).   An example of synergistic growth inhibition between root-inhabiting fungi. *Mycologia 43*: 723-726.

Roughgarden, J.   (1975).   Evolution of marine symbiosis: A simple cost-benefit model. *Ecology 56*: 1201-1208.

Roughgarden, J.   (1976).   Resource partitioning among competing species: A coevolutionary approach. *Theor. Pop. Biol. 9*: 388-424.

Schoener, T. W.   (1976).   Alternatives to Lotka-Volterra competition: Models of intermediate complexity. *Theor. Pop. Biol. 10*: 309-333.

Seifert, R. P., and Seifert, F. H.   (1976).   A community matrix analysis of *Heliconia* insect communities. *Amer. Natur. 110*: 461-483.

Smale, S.   (1976).   On the differential equations of species in competition. *J. Math. Biol. 3*: 5-7.

Sutherland, J. P.   (1974).   Multiple stable points in natural communities. *Amer. Natur. 108*: 859-873.

Tribe, H. T.   (1966).   Interactions of soil fungi on cellulose film. *Trans. Brit. Mycol. Soc. 49*: 457-466.

Vandermeer, J. H.   (1969).   The competitive structure of communities: An experimental appraoch with Protozoa. *Ecology 50*: 362-371.

Vandermeer, J. H.   (1973).   Generalized models of two-species interactions: A graphical analysis. *Ecology 54*: 809-818.

Watson, P.   (1965).   Further observations on *Calcarisporium arbuscula*. *Trans. Brit. Mycol. Soc. 48*: 9-17.

Chapter 17

ECOLOGY OF MYCOPARASITISM

Robert D. Lumsden

U.S. Department of Agriculture
Beltsville, Maryland

I. INTRODUCTION

Microbial interactions in soil involve diverse microorganisms which act on the soil
microflora to influence the behavior and survival ability of individual species or
communities. These interactions constitute the biological activity of a soil--a complex system of biochemical and ecological events that results in checks and balances
among competing populations. This antagonism results in the direct or indirect adverse effect of one individual or population against another. Microbial equilibria
are established in soil as the result of antagonistic processes termed (a) competition; (b) antibiosis, including fungistasis, lysis, and bacterial necrosis; and (c)
mycoparasitism. All of these factors bring about biological buffering in soil and
greatly influence the activity, longevity, and rate of survival of fungal mycelia
and propagules.

Mycoparasitism, also referred to as hyperparasitism and direct or interfungus
parasitism, is probably the least explored and least understood of the phenomena
relating to survival of fungi in soil. Mycoparasitism is the direct interaction of
a fungus (the *mycoparasite*) on another fungus (the *mycohost*, and appears to be of
widespread occurrence. Examples of mycoparasitism can be found among all groups of
fungi from the chytrids to the higher basidiomycetes (DeVay, 1956). Table 1 illustrates selected examples of mycoparasitic combinations that have been studied in vivo
or described from natural materials. This list includes both parasites with very
broad host ranges, such as *Trichoderma* spp., and those with a limited host range or
even a single host species, such as many of the chytrids. Descriptions and studies
of mycoparasitism in culture have been omitted because the emphasis of this review
is on ecological considerations.

The following two broad types of mycoparasitic interactions are recognized
(Barnett and Binder, 1973):

1. *Biotrophic or balanced mycoparasitism* In this type of parasitism, often
obligate, the parasite obtains nutrients from living cells with little or no apparent
harm to the mycohost. Apparently the loss of some nutrient synthetic ability has

Table 1 Selected mycoparasite-mycohost combinations described from natural ecosystems

Mycoparasite	Mycohost	References
Chytridiomycetes		
Various species	Other chytrids	Sparrow (1960)
Phlyctochytrium sp.	*Peronospora tabicina*	Person et al. (1955)
	Sclerospora sorghi	Kenneth et al. (1975)
	Glomus macrocarpus	Ross and Ruttencutter (1977)
Hyphochytriomycetes		
Hyphochytrium catenoides	*Pythium* spp.,	Ayers and Lumsden (1977)
	Aphanomyces euteiches	
	Pythium, Phytophthora, and	Sneh et al. (1977)
	Aphanomyces spp.	
Oomycetes		
Pythium oligandrum,	*Pythium* spp.	Drechsler (1943a)
P. acanthicum, P. periplocum	*Sclerotinia sclerotiorum*	Abawi et al. (1976)
Pythium sp.	*Glomus macrocarpus*	Ross and Ruttencutter (1977)
Pythium sp.		
Zygomycetes		
Mucor sp.	*Sclerotinia sclerotiorum*	Abawi et al. (1976);
		Merriman (1976)
Piptocephalis virginiana	*Mucor hiemalis*	Berry and Barnett (1957)
Rhizopus sp.	*Sclerotinia sclerotiorum*	Abawi et al. (1975)
Syncephalis californica	*Rhizopus oryzae*	Hunter and Butler (1975);
		Hunter et al. (1977)
Ascomycetes		
Mitrula sclerotiorum	*Sclerotinia trifoliorum*	Röed (1954); Ylimäke (1968)
Eudarluca caricis	*Puccinia* spp.	Eriksson (1966)
Didymella exitiales	*Ophiobolus graminis*	Siegle (1961)

Fungi Imperfecti

Acrostalagmus roseus	*Sclerotinia trifoliorum*	Pohjakallio et al. (1956)
Ampelomyces quisqualis	*Erysiphe* spp., *Microsphaera alni,* *Sphaerotheca pannosa,* *Sphaerotheca fuliginea*	Yarwood (1932); Jarvis and Slingsby (1977)
Aspergillus sp.	*Sclerotinia sclerotiorum*	Abawi et al. (1976)
Calcarisporium arbuscula	*Dothiorella quercina*	Barnett (1958)
Cephalosporium sp.	*Helminthosporium teres;* *Sclerotinia sclerotiorum*	Kenneth and Isaac (1964); Abawi et al. (1976)
Cercospora uromycestri	*Uromyces cestri*	Pollack (1971)
Coniothyrium minitans	*Sclerotinia sclerotiorum*	Campbell (1947); Tribe (1957); Erviö (1964); Vörös (1969); Schmidt (1970); Abawi et al. (1976); Turner and Tribe (1976); Huang (1977)
Cylindrocarpon sp.	*Sclerotinia sclerotiorum*	Abawi et al. (1976)
Cylindrocarpon roseum	*Botryosphaeria piceae*	Funk (1973)
Dactylella spermatophaga	*Pythium* spp.	Drechsler (1938)
Darluca filum	*Puccinia recondita*	Swendsrud and Calpouzos (1972)
Engelhardtiella alba	*Cytospora abietis*	Funk (1973)
Fusarium sp.	*Sclerotinia sclerotiorum*	Makkonen and Pohjakallio (1960); Abawi et al. (1976); Merriman (1976)
Fusarium semitectum	*Pythium* spp.	Lumsden and Ayers (1977)
Fusidium parasiticum	*Sclerospora graminicola*	Rao and Pavgi (1976)
Gliocladium sp.	*Xylaria oxyacanthae;* *Sclerotinia sclerotiorum*	Backus and Stowell (1953); Abawi et al. (1976)
Gliocladium catenulatum	*Sclerotinia trifoliorum;* *Sclerotinia sclerotiorum*	Makkonen and Pohjakallio (1960); Huang (1977)
Gliocladium roseum	*Ceratocystis fagacearum;* *Ceratocystis fimbriata;* *Sclerotinia trifoliorum;* *Sclerotinia sclerotiorum;* *Botrytis allii;* *Phomopsis sclerotioides*	Shigo (1958); Barnett and Lilly (1962); Makkonen and Pohjakallio (1960); Karhuvaara (1960); Walker and Maude (1975); Moody and Gindrat (1977)
Gonatobotryum fuscum	*Graphium* sp.	Shigo (1960)
Papulospora stoveri	*Rhizoctonia solani*	Warren (1948)
Penicillium spp.	*Sclerotinia sclerotiorum;* *Sclerotinia borealis*	Rai and Saxena (1975); Abawi et al. (1976); Karhuvaara (1960)

Table 1 (cont.)

Mycoparasite	Mycohost	References
Penicillium vermiculatum	Sclerotium cepivorum	Papavizas (1977)
Sporotrichum carnis	Rhizoctonia solani	Boosalis (1956)
Stephanoma phaeosporum	Sclerotinia sclerotiorum	Erviö (1964)
Trichoderma spp.	Fusarium sp.	Butler and McCain (1968)
	Rhizoctonia solani	Boosalis (1956)
	Typhula incarnata	Harder and Troll (1973)
Trichoderma hamatum	Sclerotium delphinii	Coley-Smith et al. (1974)
Trichoderma harzianum	Sclerotium rolfsii	Wells et al. (1972)
Trichoderma viride	Armillaria mellea	Munnecke et al. (1976)
	Sclerotinia sclerotiorum	Abawi et al. (1976); Huang (1977)
	Sclerotinia trifoliorum	Karhuvaara (1960)
Trichothecium arrhenopus	Pythium gramicolum	Drechsler (1943b)
Trinacrum subtile	Pythium spp.	Drechsler (1938)
Verticillium sp.	Sclerotinia sclerotiorum	Makkonen and Pohjakallio (1960)
Basidiomycetes and Mycelia Sterilia		
Peniophora gigantea	Heterobasidion annosus	Ikediugwu et al. (1970)
Phymatotrichum fungicola	Botryosphaeria piceae	Funk (1973)
Rhizoctonia oryzae	Rhizoctonia solani	
	Sclerotium oryzae	Shahjahan et al. (1976)
Sterile basidiomycete	Macrophomina phaseolina	de la Cruz and Hubbell (1975)

resulted in a dependence on the mycohost for survival. The parasite may merely contact the host cell without penetration, obtaining nutrients through "buffer" or absorptive cells; it may penetrate the host with a peg from an appressorium-like swel-
ling and form haustoria within the host cell; or it may penetrate the host hyphae
and grow internally from cell to cell without an apparent effect in the early stage
of development. Later the association may cause distortion of hyphae and an adverse
effect on host sporulation, as in parasitism of *Rhizopus oryzae* by *Syncephalis californica* (Hunter, et al. 1977).

    2. *Necrotrophic or destructive mycoparasitism* The parasite destroys the host
cell after or slightly before invasion and utilizes nutrients from the dying or dead
host. The infection is initiated by coiling of hyphae around mycohost cells or by
direct penetration and invasion of hyphae or survival structures.

    Most studies relating to ecological aspects of mycoparasitism have dealt with
the destructive mycoparasites and with soilborne plant pathogens. This review,
therefore, centers mainly around the ecology of soilborne plant pathogens in relation
to their mycoparasites and the environmental factors affecting their interaction.
Other reviews have treated the broad aspects of mycoparasitism (Wood and Tveit, 1955;
DeVay, 1956; Boosalis, 1964; Boosalis and Mankau, 1965), the physiology and nutrition
of mycoparasitism (Barnett, 1963a,b, 1964; Barnett and Binder, 1973), and the fine
structure of the host-parasite interaction (Hashioka, 1973).

    Little information is available on the natural occurrence of mycoparasitism in
ecological systems. For example, only a few critical investigations of the interaction in natural soils have been undertaken. Boosalis (1956) and Huber, et. al. (1966)
were among the first to investigate the survival and activity of fungi in natural
soil in relation to mycoparasites. Boosalis (1956) estimated the frequency of parasitism of *Rhizoctonia solani* by *Penicillium vermiculatum* in soil by observing parasitism of *R. solani* hyphae by *P. vermiculatum*. Similarly, Huber and colleagues
(1966), using a soil profile technique in natural root rot soil, observed invasion
of *R. solani* by an unknown fungus, and invasion of *Fusarium solani* f. sp. *phaseoli*
by a soil actinomycete. The actual viability of these hyphae in soil at the time
of the observed parasitism is, however, open to question. Recent investigations by
Ayers and Lumsden (1974, 1977); Hunter et. al. (1977); and Sneh et. al. (1977) indicate that living structures of fungi are parasitized in natural soils. The study by
Hunter and associates (1977) provides the clearest evidence of the occurrence of
mycoparasitism in natural soil. They demonstrated that the biotrophic, nondestruc-
tive parasite *Syncephalis californica* parasitized *Rhizopus oryzae* in naturally and
artificially infested soils. The evidence for mycoparasitism was the occurrence of
swellings within the living hyphae of *R. oryzae*.

    In spite of the evidence for the natural occurrence of mycoparasitism, the ac-
tual ecological importance of the phenomenon is not known. Griffin (1972) stated
that mycoparasitism is relatively rare and is not ecologically important in soil.
In contrast, Drechsler (1938) and Sneh et. al. (1977) suggested that mycoparasitism

may greatly influence soil populations of host fungi by reducing the longevity of
survival structures. Furthermore, Wood and Tveit (1955) believed that investigation
of the ways (including mycoparasitism) in which saprophytes influence the natural
establishment and persistence of pathogens will lead to the development of cultural
methods for disease control.

Documenting the ecological importance of mycoparasitism in nature must await
further detailed systematic studies that define the ecological niches and environmental conditions in which parasitism occurs. Unfortunately, most studies on mycoparasitism have been carried out in vitro or in sterilized soils after empirical or chance
observation of mycoparasitic relationships. Care must be taken in interpreting observations of mycoparasitism in vitro, especially because such relationships may occur
as an artifact of the cultural conditions or the age of the mycohost and the nutrient
status of the medium. The metabolism of an organism is influenced by its nutrition,
and the composition of the mycohost's medium is one of the most important factors
affecting the degree of development of the mycoparasites, especially biotrophic parasites (Barnett, 1963b). Self-invasion of fungal oospores in vitro has been reported
for *Pythium myriotylum* (Drechsler, 1943c) and *Saprolegnia megasperma* (Nolan, 1975).
In these fungi, oospores were penetrated by antheridial branches, and the oospore
contents were destroyed. In addition, formation of appressoria and penetration pegs,
and invasion and destruction of *Pythium myriotylum* and *P. aphanidermatum* oospores
has been observed (Ayers and Lumsden, 1975, and unpublished results). This response
in vitro appears to be related to a nutritional deficiency.

While much of the evidence to date is largely descriptive, it does suggest that
mycoparasitism is widespread and is ecologically important to the survival of fungal
pathogens. Innovative research methods and careful studies on the ecological factors
affecting mycoparasitism can reveal the time, place, and nature of mycoparasitism.

II. TECHNIQUES FOR STUDY OF MYCOPARASITISM IN
       NATURAL SYSTEMS

Procedures previously employed in the study of mycoparasitism place little emphasis
on the study of mycoparasites under natural conditions where they actually influence
the growth and survival of other fungi. Emphasis needs to be placed on methods that
give a more balanced picture of the soil microflora as it occurs naturally.

Direct observation of mycoparasitism in soil has been used in limited cases.
Boosalis (1956) examined parasitism of *R. solani* by *P. vermiculatum* in green-manure
amended and nonamended soils by isolating the *Rhizoctonia* inoculum added previously
to soil in maize meal and sand medium. The inoculum was rinsed in tap water to free
the mycelium from soil and debris, and random samples were examined microscopically
for evidence of mycoparasitism. Three methods were used for isolation of fungi associated with *R. solani*: (1) A trap-fungus culture method consisted of transferring
washed fungus hyphae to petri dish cultures of *R. solani* grown on a mixture of maize
meal and sand. The resulting fungi growing on the *Rhizoctonia* culture were isolated

and further tested for pathogenicity. (2) A hot water method consisted of submerging isolated, invaded *Rhizoctonia* hyphae in water heated to 75°C for 1.5 min and then plating them on water agar. The heat treatment selectively encouraged growth of heat-tolerant mycoparasites. (3) A dilution technique allowed isolation of fungi observed microscopically within hyphae of *R. solani*. Recovered inoculum was comminuted in water, and invaded hyphae were isolated and transferred to water agar, incubated, and the resulting colonies isolated.

A plate profile technique was utilized by Andersen and Huber (1965) and by Huber et. al. (1966) to study the frequency, distribution, and parasitism of *R. solani* and *Fusarium solani* in soil. A plastic plate with 0.5-cm holes spaced at 2.5-cm intervals was used as a trapping device. The holes were filled with sterilized agar medium and covered with plastic electrician's tape in which small holes were punched directly over the agar filled holes. The entire apparatus was buried next to a minimally disturbed soil profile prepared in the field. After 5 days of incubation, the plates were retrieved, the agar plugs removed, and the microflora identified microscopically.

These methods necessitated the introduction of maize meal or agar into the soil, which may have influenced the natural ability of the associated parasites to affect the mycohosts.

More recently, methods have been employed that avoid alteration of the environment. Direct observation of fungal interactions and behavior of fungi in soil by scanning electron microscopy (SEM) offers a valuable tool for observing mycoparasitism in vivo, not only from the point of view of defining the ecological interaction, but also as a means of studying the ultrastructure of the host-parasite interaction in soil. Rovira and Campbell (1975) studied the interaction of *Gaeumannomyces graminis* with microorganisms directly on wheat roots, and Merriman (1976) showed alterations in the structure of the rinds of sclerotia of *Sclerotinia sclerotiorum* buried in soil. He attributed these changes to the activity of the microflora on the sclerotia. Van Eck (1976) used membrane filters as support for fungal spores, buried them in soil, and examined them by SEM. Undoubtedly, this valuable technique will be used more in the future.

Sneh (1977) described the use of polycarbonate membrane filters on which oospores of *Phytophthora megasperma* were applied. The membranes were covered with nylon net and buried in soil. After a period of incubation, the membranes were recovered and placed on top of the cover of a petri dish filled with crushed ice. Glass rings were placed on the membranes, and melted water agar was poured into them. When the agar solidified, the rings were removed, a 1-2 mm thick section of the agar cylinder was cut with a razor blade, and the cylinder was inverted on a glass slide. The membrane was removed, leaving most of the spores adhering to the agar. The oospores were then observed for development of mycoparasites.

Other direct approaches utilized characteristic structures formed by the parasite within cells of the host as a readily observable marker for evidence of mycoparasitism. In soil smears, Ayers and Lumsden (1977) observed the readily stainable sporangia of *Hyphochytrium catenoides* within infected oospores of *Pythium* spp. Ross and Ruttencutter (1977) removed spores of *Glomus macrocarpus* from chopped soybean roots by decanting and wet-sieving. The spores were examined microscopically for evidence of formation of sporangia of *Phlyctochytrium* sp. and a *Pythium*-like fungus. Swollen hyphae of *Rhizopus oryzae*, characteristic of the infections by *Syncephalus californica*, were used as a means of direct assessment of infection by the mycoparasite (Hunter, et. al., 1977). The number of primary infections were estimated by counting swellings when they first became visible to the unaided eye.

Baiting techniques have been employed successfully to trap and isolate mycoparasites in soil. The following were used as baiting methods for isolation of parasites of survival structures of soilborne fungi: (1) Oospore mats of *Pythium* spp., obtained from colonies grown in V-8 juice medium, were washed, dried in petri dishes in order to kill mycelia but not the oospores, and then flooded with nonsterile soil extract. Organisms in the soil extract invaded oospores, which could be observed readily in the clarified extract. The associated organisms were isolated and identified (Ayers and Lumsden, 1974, 1975, 1977). (2) Plant material infected with *Pythium myriotylum* in the greenhouse and buried in field soil was incubated for several months and recovered periodically to test for viability of *P. myriotylum*. After viability was greatly reduced, the plant debris was recovered, comminuted, and dilutions were poured on dried oospore mats. After several days the oospores became infected, the mats were blended, and samples were streaked on water agar plates. The colonies that grew from the oospores were isolated and identified (Lumsden and Ayers, 1977). (3) Oospores were prepared free of mycelium, mixed with 0.3% water agar, and applied to 2 X 2 cm cover slips. The cover slips were placed in soil, retrieved periodically, and examined for evidence of invasion of the oospores still adhering to the glass surface (Ayers and Lumsden, 1975). (4) Sclerotia of *Sclerotium cepivorum* and *Macrophomina phaseolina* were placed on the surface of soil and observed for colonization by soil fungi (Papavizas, 1977). (5) Sclerotia of *Sclerotinia sclerotiorum* were recovered from soil by a combination of ·dry- and wet-sieving and plated on a growth medium or sliced carrot disks to determine their viability (Hoes and Huang, 1975). The microorganisms associated with decaying sclerotia were observed and isolated. Trapping techniques similar to the above have been used by other workers (e.g., Abawi et al., 1976; Jeffries and Kirk, 1976; Sneh, 1977).

Enrichment techniques show promise as methods of enhancing populations of antagonists and thus increasing the probability of detecting mycoparasites. Zogg (1976) grew mycelium of *Gaeumannomyces graminis* in nutrient solution, washed the mycelium thoroughly with water, and repeatedly added it at different times to soil. The intent was to enhance possible accumulation of antagonists or mycoparasites capable of decomposing *G. graminis*. Ayers and Lumsden (1977) added large numbers (500,000 per gram) of *P. myriotylum* oospores to soil to enhance infection and increase the probability of detecting oospores infected by *H. catenoides*.

Indirect approaches to the study of mycoparasitism include estimation of populations of the mycoparasite in soil (Aytoun, 1953; Hunter et al., 1977) and estimation of the population of the host remaining after treatment with the mycoparasite (Huang, 1976; Hunter et al., 1977; Ross and Ruttencutter, 1977; Schmidt, 1970). Numerous isolation media have been developed for assessment of populations of plant pathogens (Tsao, 1970). Estimation of the decrease in plant disease incidence caused by parasitism and destruction of plant pathogens (Ahmed and Tribe, 1977; de la Cruz and Hubbell, 1975; Huang, 1976; Moody and Gindrat, 1977; Swendsrud and Calpouzos, 1972; Turner and Tribe, 1975; Wells et al., 1972; Zogg, 1976) is another indirect method.

Expanded use of these techniques as well as other innovative methods will undoubtedly enhance the probability of isolation, identification, and study of mycoparasites in natural environs. These techniques, coupled with the study of plant disease suppressive soils (Baker and Cook, 1974) and examination of soils where disease organisms occur but where disease is sporadic, will reveal avenues of research to better explain the ecological interactions among associated fungi.

III. ECOLOGICAL FACTORS THAT AFFECT MYCOPARASITISM

Ecological relationships and environmental factors are of equal importance to the inherent factors that determine successful mycoparasitic associations. Inherent factors include the genetic makeup and variability of the parasite and the host; the stage of development, vegetative or sexual phase, and age of the host; the chemical nature of cell walls and thickness of walls; and the presence or absence of growth factors in the mycohost required by the mycoparasite. Many of these factors in mycoparasitic relationships have been studied in vitro, especially in regard to the physiology of the host-parasite interaction (Barnett and Binder, 1973). A discussion of these inherent factors, however, is beyond the scope of this review, which is concerned principally with extrinsic, ecological factors.

A. Population Dynamics and Distribution of Mycoparasites

Populations of mycoparasites in relation to the incidence of mycoparasitism have been determined to a limited extent in natural soils. Aytoun (1953), with the use of a selective medium, studied the quantitative distribution of *Penicillium* spp. and *Trichoderma* spp. in wooded and nonwooded soils as they may be related to parasitism of *Armillaria mellea* and *Polyporus schweinitzii*. The quantity of these fungi in woodland soils did not appear to differ markedly from that in similar soils outside the influence of trees. *Penicillium* spp. were more prevalent than *Trichoderma* spp.

Recently, Huang (1977) demonstrated the seasonal fluctuation in activity of *Coniothyrium minitans* on *Sclerotinia sclerotiorum*. The percentage of sclerotia of *S. sclerotiorum* killed by *C. minitans* in sunflower pith cavities increased and the

percentage of viable sclerotia decreased toward the end of the season. The host-
parasite relationship was clearly demonstrated in which two correlated patterns of
population changes occurred--one for the mycohost and one for the mycoparasite. Also
working with *C. minitans*, Tribe (1957) determined the natural distribution for this
parasite on sclerotia of *S. sclerotiorum* in southern England. The incidence of infected sclerotia varied considerable from 0% infection in some sampling areas to 42%
in others.

The population levels and survival of endomycorrhizal fungi was studied recently. Ross and Ruttencutter (1977) found a high initial buildup of *Glomus macrocarpus*
in soybean roots in fumigated soil. These authors suggested that soil fumigation may
have reduced the populations of the mycoparasites *Phlyctochytrium* sp. and a *Pythium*-
like fungus. In addition, populations of *G. macrocarpus* were much greater in the
fumigated field than in nonfumigated field soils. A decrease in the population of
G. macrocarpus the following year was associated with increases in the incidence of
mycoparasitism. Moreover, an increase during this time in the population of another
mycorrhizal fungus, *Gigaspora gigantea*, suggested that it was not as susceptible as
G. macrocarpus to mycoparasites. Mycoparasites were found in spores of *G. macrocarpus* but not in those of *G. gigantea*. This suggests that mycoparasites may be
detrimental to the survival of beneficial endomycorrhizae.

Hunter and colleagues (1977) quantitatively examined populations of the biotrophic mycoparasite *Syncephalis californica* and its mycohosts, *Rhizopus* spp., in the
soil of a California apricot orchard. The population of *Rhizopus* at a depth of 3 cm
in the soil greatly increased from May to August and remained high through November.
The population of *S. californica* did not, however, show any large changes with time.
Despite the lack of a correlation between populations of hosts and the parasite, the
incidence of parasitism of *Rhizopus* spp. by *S. californica* was fairly frequent in the
field. In fact, six separate fields had significantly high incidence of parasitism
(Hunter and Butler, 1975). The effect of this mycoparasite may be widespread and
result in significant reduction in populations of *Rhizopus* spp. Aside from its obvious effect of causing swellings in the host hyphae, *S. californica* also suppressed
sporulation. Sporulation in nonsterilized soil was greatly reduced by a limited number of infections. Because of the probable importance of sporangia and sporangio-
spores to survival of *Rhizopus*, the authors conjectured that suppression of sporulation of the magnitude caused by *S. californica* very likely influences the population
dynamics of *R. oryzae* in nature. *Syncephalis californica* is a versatile biotrophic
mycoparasite that interferes with the growth and asexual reproduction of its mycohost
over a wide range of natural conditions.

The distribution of *Syncephalis* spp., as well as *Piptocephalis* spp., was studied
by Richardson and Leadbeater (1972). Both mucorine mycoparasites were more frequently isolated from areas of rapid organic matter turnover such as litter and the A
horizon of soil profile samples, especially from woodland and pasture habitats. Distribution of mucoraceous hosts for these mycoparasites was also widespread and of

common occurrence in the soils sampled. In fact, the habitat seemed to have more influence on the parasites than the presence of the host. *Mucor* spp. and *Mortierella*
spp. were present throughout the study area, but *Syncephalis* spp. and *Piptocephalis*
spp. were associated primarily with closed grassland communities--although not commonly associated with nearby sand dunes.

Distribution patterns of mycoparasites have important implications in plant
pathology. Plant disease conducive and suppressive soils have recently received attention (Baker and Cook, 1974). These soils with contrasting incidence of disease
offer excellent opportunities for study of natural disease-resistance mechanisms.
Soils that contain certain types and numbers of mycoparasites and other antagonists
suppress the pathogenic activity of soilborne plant pathogens. In contrast, soils
not containing the antagonists have high incidence of disease. The role that mycoparasitism plays in suppressive soils has not been adequately studied. Sneh and
coworkers (1977), however, compared the parasitism of a *Phytophthora* sp. in soils
collected near healthy and *P. cactorum*-infected apple trees. Infection of oospores
by chytrids in soil near apparently healthy trees was significantly greater than in
soil near trees infected by *P. cactorum*. The greater oospore destruction in the
suppressive soil may have contributed to the health of trees in that soil.

The dynamics of mycoparasitic relationships in natural environments are complex.
Multiple infections of mycohost populations may be quite common. This is suggested
by the array of microorganisms isolated from host survival structures. Abawi and
associates (1976) isolated 12 different fungi from sclerotia of *Sclerotinia sclerotiorum* collected from lettuce field soils. Species of *Trichoderma*, *Fusarium*, *Cylindrocarpon*, and *Penicillium* accounted for more than 95% of the organisms isolated.
Sneh et. al. (1977) isolated 11 fungi, including chytrids, phycomycetes, and hyphomycetes from oospores of *Phytophthora megasperma* var. *sojae* and other oomycetes.
This evidence, along with other reports of multiple isolations of fungi (Warren,
1948; Karhuvaara, 1960; Makkonen and Pohjakallio, 1960; Erviö, 1964; Huber et al.,
1966; Smith, 1972a; Rai and Saxena, 1975; Merriman, 1976) suggests that mycoparasitic
associations may be very complex and difficult to elucidate.

Multiple associations of parasites with soil fungi offers the interesting possibility of ecological succession in populations of soil microorganisms. Sneh, et
al. (1977) suggested that succession of infections of *Phytophthora* oospores occurred as soil moisture conditions changed. Phycomycetes invaded oospores under saturated soil conditions, and hyphomycetes did so under drier soil conditions. In addition, Drechsler (1943a) noted that *Pythium oligandrum* was not usually a primary parasite of plants but was, instead, a secondary invader of diseased tissue and was at
least partly parasitic on other *Pythium* spp. that caused the initial plant infection.
An indirect role in the control of *Gaeumannomyces graminis* infection of wheat by
Phialophora radicicola was suggested for *P. oligandrum* by Deacon (1976). This mycoparasite, as well as others, may increase in population on *P. radicicola* (avirulent
to wheat), then attack the closely related *G. graminis*, and so result in decrease of

wheat infection. This relationship, however, has not been demonstrated in soil. Lumsden and Ayers (1977) described a possible succession of mycoparasites in *Pythium myriotylum*-infected bean tissue buried in soil. The *P. myriotylum* increasingly lost viability in the buried debris, which was increasingly colonized by other *Pythium* spp. in the soil. *Fusarium* spp., also presumably a secondary colonizer of the debris, were subsequently isolated from *P. myriotylum* oospore bait flooded with dilutions of blended bean debris recovered from the soil.

Mycoparasitism may have survival value for higher plant pathogens in soil. Although carried out in vitro, studies of Barron and Fletcher (1970) suggested that mycoparasitism of *Verticillium albo-atrum* and *V. dahliae* on *Rhopalomyces elegans* may enhance the survival of the former in soil in the absence of suitable plant hosts. Mycoparasitism by other plant pathogens on other fungi has also been reported (Butler, 1957; Endo et al., 1973).

B. Effect of the Environment of Mycoparasitism

The environment under which mycoparasitism occurs influences the success of the interaction, the degree to which it occurs, and the populations of mycoparasites. Nutrition, temperature, pH, humidity, water, gas exchange, and soil texture all have been shown to affect mycoparasitism. Only those studies done in natural or living soils or other natural environments will be considered here.

1. Effect of nutrition

Host nutrition is one of the most important factors affecting mycoparasitism (Boosalis, 1964; Barnett and Binder, 1973). Only limited information is available, however, as to the direct effect of nutrition on mycoparasitism in the environment. Boosalis (1956) showed that amendment of soil with dried, ground soybean stems and leaves enhanced the incidence of parasitism of *Rhizoctonia solani* by *Trichoderma* sp. and *Penicillium vermiculatum* when the soil was incubated at 28°C. In contrast, parasitism occurred 50% less in the nonamended soil. Apparently, amending the soil with organic matter provided nutrients that enhanced parasitism of *R. solani* hyphae. Similarly, Wells and associates (1972) attributed success in control of *Sclerotium rolfsii* infection of tomato seedlings to the fact that they provided an adequate food base for vigorous growth of the mycoparasitic *Trichoderma harzianum* used as the biological control agent. A fresh food base for *T. harzianum* seemed to temporarily overwhelm the infection court of *S. rolfsii* and prevent it from utilizing organic debris near the host plants that was required for successful infection.

The nutrition or the prior condition of the mycohost may also indirectly affect mycoparasitism, possibly as it affects the ability of the parasites to invade host structures. Merriman (1976) found that sclerotia of *S. sclerotiorum* buried in soil for 240 days had greater incidence of *Fusarium* spp., *Mucor* spp., and *Trichoderma* spp. when the sclerotia were obtained from infected lettuce than when obtained from cultures on sterilized potato tissue. The physical condition of the sclerotial rind was

considered to be the determining factor in this observation. The structure and resistance of the sclerotia to invasion may have been affected by their nutrition during formation.

Essential growth factors supplied by the living mycohost may affect the ability of some mycoparasites to infect them. Published results have dealt with biotrophic
parasites in vitro (Barnett and Binder, 1973). Recent results of W. A. Ayers and P.
B. Adams of the U.S. Dept. of Agriculture, Beltsville, Md. (personal communication,
and see additional references) however, indicate that a destructive mycoparasite of
S. sclerotiorum required viable sclerotia. No growth of the mycoparasite occurred
on conventional laboratory media or in soil free of sclerotia. However, limited
growth occurred in an agar medium prepared from autoclaved sclerotia.

2. Effect of temperature

In all studies concerned with mycoparasitism in soil, it was found that temperature affected the intensity of mycoparasitism. Boosalis (1956) determined that parasitism of *Rhizoctonia solani* by *Penicillium vermiculatum* and *Trichoderma* sp. was
negligible at 18°C. In contrast, about 18% and 8% of *R. solani* hyphae were parasitized in soybean-tissue-amended soils and -nonamended soils, respectively, at 28°C.
Turner and Tribe (1976) also found increased parasitism of sclerotia of *S. sclerotiorum* by *Coniothyrium minitans* with an increase in temperature. All sclerotia placed
on the surface of soil and incubated at 20°C were infected within 3 weeks; most of
those incubated at 10°C, within 6 weeks. When sclerotia were buried in soil, only
occasional sclerotia were infected at either temperature.

In an indirect manner, temperature also affected survival of *Armillaria mellea*
in relation to parasitism by *Trichoderma* spp. When *A. mellea* was stressed by sub-
lethal treatment, i.e., heating at 40-49°C for 2 days, and then incubated in soil at
23°C, survival was significantly decreased. There was no obvious direct effect on
A. mellea at these temperatures, but the stress may actually have stimulated parasitic activity on the weakened *A. mellea* mycelium. When untreated, *A. mellea* apparently resisted attack by *Trichoderma*.

3. Effect of pH

Aytoun (1953) was unable to detect a difference in the incidence of *Trichoderma*
spp. in soils ranging from pH 4.6-6.5, although increased parasitism of *Armillaria
mellea* was noted at pH values of 3.4 and 5.1--but not at pH 7.0. In contrast, Papavizas (1977) detected more colonization of *Macrophomina phaseolina* by several fungi
and also *Streptomyces* spp. on alkaline soil than on acid soils. Colonization of
Sclerotium cepivorum was as high on acid as on alkaline soils. This meager information on the effect of pH on mycoparasitism is not adequate to permit us to draw
any substantive conclusions.

4. Effect of moisture

Humidity would not be expected to affect mycoparasitism in soil; however, its
effect has been reported on soil and on leaves of wheat infected with *Puccinia*

recondita. Swendsrud and Calpouzos (1972) detected increased percentage of uredinia of *P. recondita* which contained pycnidia of *Darluca filum* by prolonging periods of increased humidity from 3 to 15 days. The longer mist period had a greater effect on the mycoparasite than on the rust fungus, since only slightly more uredinia per wheat plant developed on check plants inoculated with the rust alone. Papavizas (1977) reported that a higher percentage of sclerotia of *M. phaseolina* and *S. cepivorum* were colonized on soil by fungi and *Streptomyces* spp. after exposure to 0% relative humidity than after exposure to 33%, 55%, and 78% relative humidity. The significance of this colonization could not be ascertained but may be related to leakage of nutrients from the dried sclerotia when reexposed to moist soil.

The leakage of nutrients from sclerotia of various fungi previously dried for short periods was studied by Smith (1972a,b) and Coley-Smith et al. (1974). Although somewhat different results were obtained in the two studies, the overall effect was that leakage of nutrients from sclerotia due to drying enhanced invasion and rotting of sclerotia by soil microorganisms. The principle colonizer of sclerotia of *Sclerotium delphinii* was *Trichoderma hamatum* (Coley-Smith et al., 1974). Likewise, *Trichoderma viride* in soil was able to parasitize mycelia of *Armillaria mellea* in infected peach roots after being weakened by drying of the root pieces prior to burial in soil (Munnecke et al., 1976). The fungus was vulnerable to *T. viride* parasitism if root pieces lost more than 60% of their total moisture. If less than 40% of the moisture was lost, no effect by *T. viride* was noted. Soil moisture directly affected the number of infections of *R. oryzae* by *S. californica* (Hunter et al., 1977). Numbers of infected hyphae increased significantly to a maximum as soil water potential decreased from 0 to -0.3 bars. Soil moisture levels between saturation and field capacity, but not including saturation, were favorable for infection. In addition, the extent of sporulation by *R. oryzae* infected by *S. californica* was inversely related to the number of infections.

The type of mycoparasites associated with oospores of *Phythophthora* spp. was affected after 2 weeks in soils at 25%, 50%, and 150% of waterholding capacity (Sneh et al., 1977). Differences were noted in the types of parasites which invaded oospores. In flooded soil, many oospores were parasitized by chytrids, whereas only a few were parasitized by hyphomycetes. Conversely, at soil moistures below saturation there was no infection by chytrids, but many oospores were infected by hyphomycetes.

5. *Effect of CO_2/O_2 levels and soil type*

Although not necessarily related in effect, these parameters can influence mycoparasitic activities as they affect aeration of soils. Gas composition of soil affected sporulation of *Rhizopus oryzae* and its infection by *Syncephalis californica* (Hunter et al., 1977). Concentrations of CO_2 higher than 0.1% significantly reduced infection of *R. oryzae* in soil. Almost no infection occurred in 5% CO_2, even though *R. oryzae* continued to grow and sporulate. Infection decreased with a decrease in

O_2 concentration from 21 to 1%, but differences were not statistically significant. Sporulation of *R. oryzae* was greatly reduced by *S. californica* infections at all CO_2 and O_2 concentrations tested. There seemed to be a more direct effect on the parasite than on the host in this case.

Merriman (1976) obtained greater degradation of sclerotia of *S. sclerotiorum* in sandy clay loam than in a sandy loam. Tribe (1957), however, found no observable relation between soil type and infection of *S. sclerotiorum* sclerotia by *C. minitans*.

Understanding the influence of ecological interactions and the microenvironment on mycoparasitism is extremely important in establishing the true importance of this phenomenon in natural ecosystems, including those in which agriculture has a role. It is clear that the effect of nutrition on mycoparasitism is poorly understood, but it is a factor that can be manipulated in the environment to enhance the incidence of mycoparasitism. Addition of specific organic amendments or growth factors required by mycoparasites may enable modification of the soil environment in favor of antagonists and to the detriment of important soilborne plant pathogens. In turn, these studies have the potential for development of cultural or biological methods of plant disease control.

C. Biological Control of Plant Diseases by Mycoparasites

Antagonism in natural ecosystems is a well-established phenomenon. Existence of natural biological control is also documented (Baker and Cook, 1974) and undoubtedly is instrumental in preventing more widespread and destructive occurrence of plant diseases.

Examples of biological control of plant diseases resulting from mycoparasitism, however, are limited (Boosalis and Mankau, 1965; Baker and Cook, 1974; Papavizas, 1973; Snyder et al., 1976). This minimum documentation of both natural and imposed mycoparasitic biological control can be attributed to inadequate research efforts and methods and the complexity of the ecosystems involved. In addition, success may require greater perseverance by research workers. Long-term studies may be required to determine long-term effects of exposure of survival structures to mycoparasites.

Until adequate knowledge is accumulated to establish the environmental factors that determine the success of fungal interactions in complex natural environments, the possibility of biological control by mycoparasites may well be limited. Progress, nevertheless, is being made in documenting important discoveries of biological control of disease by mycoparasites. The following examples of biological control have been made mostly by chance discovery and empirical methods. Perhaps future examples would not be so slow in coming if more extensive knowledge were available on the ecology of the environments in which diseases occur.

1. Control of Sclerotinia sclerotiorum

S. sclerotiorum and related species (*S. minor* and *S. trifoliorum*) have been studied extensively in relation to the numerous fungi that colonize sclerotia of these fungi (Table 1). Among the colonizers, *Coniothyrium minitans*, *Gliocladium roseum*, and *Trichoderma* spp. have been most frequently isolated and studied to determine their capabilities for control of *Sclerotinia* diseases. In both greenhouse and field studies, (Huang, 1976), *C. minitans* successfully decreased the incidence of sclerotinia wilt of sunflower in infested soil. In the field, the incidence of wilt was 25% in the mycoparasite-treated plots and 43% in the untreated plots. Also, yield of sunflower seed was increased. These results indicate that *C. minitans* has potential in the biological control of sclerotinia wilt of sunflower. Other results (Smith, 1972a; Turner and Tribe, 1975; Huang, 1977) indicate successful reduction in numbers of sclerotia of *S. sclerotiorum* in the field with the use of *C. minitans* and other biological agents. In these cases, however, actual control of plant disease was not assessed. Future use of mycoparasites to control *Sclerotinia* diseases appears to have potential.

2. *Control of Macrophomina phaseolina*

Successful control of black root rot of pine seedlings was demonstrated by de la Cruz and Hubbell (1975). They showed an increase in seedling stand from 57% in soil artificially infested with *M. phaseolina* to 81% in soil simultaneously treated with *M. phaseolina* and an unidentified sterile basidiomycete. This increase compared favorably with the survival rate of 84% in the noninoculated controls. Unfortunately the authors did not utilize naturally infested soil nor soil previously infested with *M. phaseolina* prior to inoculation with the sterile basidiomycete. Inoculum of the pathogen and the mycoparasite were added simultaneously in cultures grown on vermiculite wetted with potato dextrose broth. A more natural system would be the addition of the sterile basidiomycete to nursery beds where *M. phaseolina* was established. This biological control agent is promising because of its potential economical application in intensive nursery practices, especially forest seedling nurseries.

3. *Control of Rhizoctonia solani*

The first effort to compare biological control in sterilized and nonsterilized soil was by Boosalis (1956) with *Penicillium vermiculatum* and *R. solani*. Nearly complete control of damping-off and seedling blight of peas incited by *R. solani* occurred when sterilized soil was infested simultaneously with cultures of *P. vermiculatum* and *R. solani*. Biological control in nonsterilized soil was not successful. About 60% of the peas damped-off in the simultaneously infested soil, and nearly 50% of the surviving plants had symptoms of seedling blight. Control soils infested with *R. solani* also had high levels of disease. This demonstrates the fallacy of studying biological control under grossly unnatural conditions such as sterilized soil and the difficulties encountered in extrapolating results from artificial conditions to that of a more natural ecosystem.

Natural biological control of *R. solani* infection of rice stems and sheaths in the field by *Rhizoctonia oryzae* was reported recently (Shahjahan et al., 1976). This study suggested that *R. oryzae* interferes with normal growth of *R. solani* and other activities of the fungus in causing leaf and stem disease of rice. It also raises the interesting possibility that seemingly closely related fungi in the same environment may compete antagonistically.

4. Control of Sclerotium spp.

Trichoderma harzianum has been shown to effectively control diseases caused by *Sclerotium rolfsii*. The first report of its efficacy was by Wells and coworkers (1972), who demonstrated excellent control of southern blight on lupine, tomato, and peanuts. The investigators attributed this successful biological control to the use of an adequate food base for vigorous growth of the mycoparasite. The food base allowed *T. harzianum* to aggressively colonize plant debris needed by *S. rolfsii* as a source of nutrients for successful infection of its host plant. This suggests that competition for substrate or antibiosis may have been responsible for biological control. The role of mycoparasitism in this instance is not firmly established and should be examined. However, *T. harzianum*, which has both mycoparasitic capabilities and antibiotic activity, seems firmly established as an effective biological control agent. It is considered to be economically feasible for commercial application and also to offer advantages over chemical control (Backman and Rodriguez-Kabana, 1975).

White rot of onions, caused by *S. cepivorum*, has been successfully controlled in greenhouse experiments by the mycoparasite *Coniothyrium minitans* (Ahmed and Tribe, 1977). Pycnidial dust of *C. minitans* was applied to soil and as a seed dressing to protect onion seeds planted in *S. cepivorum*-infested soil. Control of the disease and appearance of the onion seedlings was enhanced by the *C. minitans* treatments compared to a standard mercury fungicide control. This potential biological control agent also has proven effective against other sclerotial fungi.

5. Control of Fomes annosus

The control of *F. annosus*-caused butt rot of pine trees by *Peniphora gigantea* is the first widespread application of a biological method of plant disease control (Greig, 1976). The efficacy of this practice is based on the application of the biological control agent to recently cut pine stumps to prevent invasion of the stumps and infection of adjacent trees by *F. annosus* (Rishbeth, 1963). The mechanism of action is unclear, but it may involve exclusion of *F. annosus* as well as mycoparasitism by *P. gigantea* mycelium. The only study to elucidate the mechansim, by Ikediugwu et al. (1970), describes "hyphal interference" as the antagonistic action. Growth of *F. annosus* was not inhibited at a distance, but when hyphae in opposed cultures contacted each other in vitro the *F. annosus* mycelia became granulated, vacuolated, and were killed.

6. Control of Phomopsis sclerotioides

Moody and Gindrat (1977) describe biological control of *P. sclerotioides*, cause of cucumber black root rot in natural soils in the greenhouse. When a highly organic soil was artificially infested with *P. sclerotioides*, the disease potential was consistently lower than in a mineral control soil that contained equal inoculum. *Glio-cladium roseum* was isolated from mycelial mats of the plant pathogen buried in the organic soil. When *G. roseum* was isolated and added to infested mineral soil, the disease potential was significantly lowered. Although the authors do not describe the mode of action of the antagonist in detail, they indicate that it was destructively mycoparasitic to *P. sclerotioides* in vitro.

7. Control of Puccinia recondita

Swendsrud and Calpouzos (1972) reported negative results in attempts to control wheat leaf rust caused by *P. recondita* with use of pycnidiospores of the rust mycoparasite *Darluca filum*. Even under varied conditions of inoculation sequence and periods of misting of wheat plants, control was not effective. In fact, with prior inoculation of leaves with *D. filum* spores and prolonged misting, uredinia formation by *P. recondita* was enhanced when compared to the control with *P. recondita* alone. On the basis of their study, the authors doubted the success of biological control by spraying wheat leaves with *D. filum* spores before rust infection. In spite of this negative result, the results are briefly mentioned here as an example of the limited research on biological control of a foliage disease by a potentially mycoparasitic antagonist.

IV. SUMMARY AND CONCLUSIONS

The widespread occurrence of mycoparasitism in natural environments is an established fact (Table 1). The importance of this extremely interesting interaction between fungi in natural ecosystems is not well documented, however. Much of this lack of knowledge is due to a reluctance of research workers to study interactions in the very complex soil ecosystem, and because of the lack of reliable, convenient research methods. Imaginative techniques for the study of the interactions in natural soils are beginning to appear in the literature and will no doubt make studies of mycoparasitism more attractive. Direct observation techniques and selective plating methods for both the parasite and the host should stimulate systematic, detailed studies.

Relatively few species of mycoparasites have been studied other than in vitro or in sterilized soil. The study of mycoparasitism in vitro has provided considerable knowledge on the interaction between host and parasite, but it has little relevance to their interaction in the natural environment where other microorganisms and environmental factors exert myriad influences.

Emphasis should be placed in the future on ecological investigations, especially with soilborne mycoparasites. This area of research has been neglected. Consequently, there is little information to show the importance of mycoparasitism relating to the survival and activity of plant pathogens. Limited evidence suggests that necrotrophic mycoparasitism may be the most important type in determining the longevity and survival of mycohosts. Penetration, colonization, and destruction of the sexual and asexual survival units of soilborne pathogens appears to be more important in mycoparasitic relationships than infection of hyphae in soil. Hyphae of many plant pathogens are ephemeral in soil. Biotrophic, or balanced, mycoparasites are dependent on host mycelium for survival and therefore are less prone to affect survival of their hosts.

The success of necrotrophic parasites in destroying host cells or reducing populations of competing organisms, especially in soil, is dependent on the microenvironment. Several factors, studied to a limited extent, appear especially influential in affecting mycoparasitism. Moderately high temperature and adequate nutrition are conducive to successful mycoparasitism. Soil moisture appears to influence the type of microorganisms attacking a host population. These factors, and others even less well understood, need to be more thoroughly studied. Much more needs to be learned about the ecology of mycoparasitism as well as the etiology, physiology, and fine structure of the mycoparasite-host interaction.

Mycoparasitic relationships are interesting and potentially very fruitful areas of research. It is imperative to know what is involved in the host-parasite interaction and what factors determine the outcome of the relationship in order to intelligently exploit this phenomenon for human benefit. Mycoparasitism is probably an important part of the natural biological control system evolved for certain plant diseases. If we can successfully direct the activities of mycoparasites, through an understanding of their ecology and physiology, to the disadvantage of their plant-parasitic mycohosts, we will have accomplished a great deal in our quest for biological control.

REFERENCES[*]

Abawi, G. S., Grogan, R. G., and Duniway, J. M. (1976). Microorganisms associated with sclerotia of *Whetzelinia sclerotiorum* in soil and their possible role in the incidence of lettuce drop. *Proc. Amer. Phytopathol. Soc. 3*: 275 (Abstr. 334).

Ahmed, A. H. M., and Tribe, H. T. (1977). Biological control of white rot of onion (*Sclerotium cepivorum*) by *Coniothyrium minitans*. *Plant Pathol. 26*: 75-78.

Andersen, A. L., and Huber, D. M. (1965). The plate-profile technique for isolating soil fungi and studying their activity in the vicinity of roots. *Phytopathology 55*: 592-594.

Ayers, W. A., and Lumsden, R. D. (1974). Invasion of *Pythium* spp. oospores by *Hypho-chytrium catenoides*. *Proc. Amer. Phytopathol. Soc. 1*: 148 (Abstr.).

Ayers, W. A., and Lumsden, R. D. (1975). Factors affecting production and germination of oospores of three *Pythium* species. *Phytopathology 65*: 1094-1100.

Ayers, W. A., and Lumsden, R. D. (1977). Mycoparasitism of oospores of *Pythium* and *Aphanomyces* species by *Hyphochytrium catenoides*. *Can. J. Microbiol. 23*: 38-44.

Aytoun, R. S. C. (1953). The genus *Trichoderma*: Its relationship with *Armillaria mellea* (Vahl ex Fries) Quel. and *Polyporus schweinitzii* Fr., together with preliminary observations on its ecology in woodland soils. *Trans. Bot. Soc. Edinbergh 36*: 99-114.

Backman, P. A., and Rodriguez-Kabana, R. (1975). A system for the growth and delivery of biological control agents to the soil. *Phytopathology 65*: 819-821.

Backus, M. P., and Stowell, E. A. (1953). A *Fusidium* disease of *Xylaria* in Wisconsin. *Mycologia 45*: 836-847.

---

[*] The literature search for this review was completed in November 1977. A few recent references follow this list.

Baker, K. F., and Cook, R. J. (1974). *Biological Control of Plant Pathogens*. Freeman, San Francisco.

Barnett, H. L. (1958). A new *Calcarisporium* parasitic on other fungi. *Mycologia* *50*: 497-500.

Barnett, H. L. (1963a). The nature of mycoparasitism by fungi. *Ann. Rev. Microbiol.* *17*: 1-14.

Barnett, H. L. (1963b). The physiology of mycoparasitism. In *The physiology of fungi and fungus diseases*. West Virginia Univ. Agr. Exp. Sta. Bull. 488T, pp. 65-90.

Barnett, H. L. (1964). Mycoparasitism. *Mycologia 56*: 1-19.

Barnett, H. L., and Binder, F. L. (1973). The fungal host-parasite relationship. *Ann. Rev. Phytopathol. 11*: 273-292.

Barnett, H. L., and Lilly, V. G. (1962). A destructive mycoparasite, *Gliocladium roseum*. *Mycologia 54*: 72-77.

Barron, G. L., and Fletcher, J. T. (1970). *Verticillium albo-atrum* and *V. dahliae* as mycoparasites. *Can. J. Bot. 48*: 1137-1139.

Berry, C. R., and Barnett, H. L. (1957). Mode of parasitism and host range of *Piptocephalis virginiana*. *Mycologia 49*: 374-386.

Boosalis, M.G. (1956). Effect of soil temperature and green-manure amendment of unsterilized soil on parasitism of *Rhizoctonia solani* by *Penicillium vermiculatum* and *Trichoderma* sp. *Phytopathology 46*: 473-478.

Boosalis, M. G. (1964). Hyperparasitism. *Ann. Rev. Phytopathol. 2*: 363-377.

Boosalis, M. G., and Mankau, R. (1965). Parasitism and predation of soil microorganisms. In *Ecology of Soil-borne Plant Pathogens*, R. R. Baker and W. E. Snyder (Eds.). Univ. of California Press, Berkeley, pp. 374-391.

Butler, E. E. (1957). *Rhizoctonia solani* as a parasite of fungi. *Mycologia 49*: 354-373.

Butler, E. E., and McCain, A. H. (1968). A new species of *Stephanoma*. *Mycologia 60*: 955-959.

Campbell, W. A. (1947). A new species of *Coniothyrium* parasitic on sclerotia. *Mycologia 39*: 190-195.

Coley-Smith, J. R., Ghaffar, A. and Javed, Z. U. R. (1974). The effect of dry conditions on subsequent leakage and rotting of fungal sclerotia. *Soil Biol. Biochem. 6*: 307-312.

Deacon, J. W. (1976). Biological control of take-all by *Phialophora radicicola* Cain. *EPPO Bull. 6*: 297-308.

de la Cruz, R. E., and Hubbell, D. H. (1975). Biological control of the charcoal root rot fungus *Macrophomina phaseolina* on slash pine seedlings by a hyperparasite. *Soil Biol. Biochem. 7*: 25-30.

DeVay, J. E. (1956). Mutual relationships in fungi. *Ann. Rev. Microbiol. 10*: 115-140.

Drechsler, C. (1938). Two hyphomycetes parasitic on oospores of root-rotting oomycetes. *Phytopathology 28*: 81-103.

Drechsler, C. (1943a). Antagonism and parasitism among some oomycetes associated with root rot. *J. Wash. Acad. Sci. 33*: 21-28.

Drechsler, C. (1943b). Another hyphomycetous fungus parasitic on *Pythium* oospores. *Phytopathology 33*: 227-233.

Drechsler, C. (1943c). Two species of *Pythium* occurring in southern states. *Phytopathology 33*: 261-299.

Endo, S., Shinohara, M., Kikuchi, J., and Hirabayashi, Y. (1973). Studies on the soil-born pathogens parasitic to the same host. I. Antagonistic action of *Rhizoc-*

tonia solani Kühn to *Fusarium oxysporum* Schlechtendahl f. *cucumerinum* Owen and its effect on the occurrence of *Fusarium* wilt of cucumber plant. *Bull. Coll. Agr. Vet. Med., Nihon Univ. 30:* 85-95.

Eriksson, O. (1966). On *Eudarluca carici* (Fr.) O. Eriks., comb. nov., a cosmopolitan uredinicolous pyrenomycete. *Bot. Not. 119:* 33-69.

Erviö, L. (1964). Certain parasites of fungal sclerotia. *Maataloustieteellinen Aikak. 36:* 1-6.

Funk, A. (1973). Some mycoparasites of western bark fungi. *Can J. Bot. 51:* 1643-1645.

Greig, J. W. (1976). Biological control of *Fomes annosus* by *Peniophora gigantea*. *Eur. J. For. Pathol. 6:* 65-71.

Griffin, D. M. (1972). *Ecology of Soil Fungi*. Syracuse University Press, Syracuse, N.Y.

Harder, P. R., and Troll, J. (1973). Antagonism of *Trichoderma* spp. to sclerotia of *Typhula incarnata*. *Plant Dis. Rep. 57:* 924-926.

Hashioka, Y. (1973). Mycoparasitism in relation to phytopathogens. *Shokubutsu Byogia Kenkyu (Forsch. Gebiet Pflanzenkrankh.), Kyoto 8:* 179-190.

Hoes, J. A., and Huang, H. C. (1975). *Sclerotinia sclerotiorum:* Viability and separation of sclerotia from soil. *Phytopathology 65:* 1431-1432.

Huang, H. C. (1976). Biological control of sclerotinia wilt of sunflower. *Ann. Conf. Manitoba Agron., 1976*, pp. 69-72.

Huang, H. C. (1977). Importance of *Coniothyrium minitans* in survival of sclerotia of *Sclerotinia sclerotiorum* in wilted sunflower. *Can. J. Bot. 55:* 289-295.

Huber, D. M., Andersen, A. L., and Finley, A. M. (1966). Mechanisms of biological control in a bean root rot soil. *Phytopathology 56:* 953-956.

Hunter, W. E., and Butler, E. E. (1975). *Syncephalis californica*, a mycoparasite inducing giant hyphal swellings in species of Mucorales. *Mycologia 67:* 863-872.

Hunter, W. E., Duniway, J. M., and Butler, E. E. (1977). Influence of nutrition, temperature, moisture, and gas composition on parasitism of *Rhizopus oryzae* by *Syncephalis californica*. *Phytopathology 67:* 664-669.

Ikediugwu, F. E. O., Dennis, C., and Webster, J. (1970). Hyphal interference by *Peniophora gigantea* against *Heterobasidion annosum*. *Trans. Brit. Mycol. Soc. 54:* 307-309.

Jarvis, W. R., and Slingsby, K. (1977). The control of powdery mildew of greenhouse cucumber by water sprays and *Ampelomyces quisqualis*. *Plant Dis. Rep. 61:* 728-730.

Jeffries, P., and Kirk, P. M. (1976). New technique for the isolation of mycoparasitic Mucorales. *Trans. Brit. Mycol. Soc. 66:* 541-543.

Karhuvaara, L. (1960). On the parasites of the sclerotia of some fungi. *Acta Agr. Scand. 10:* 127-134.

Kenneth, R., Cohn, E., and Shahor, G. (1975). A species of *Phlyctochytrium* attacking nematodes and oospores of downy mildew fungi. *Phytoparasitica 3:* 70.

Kenneth, R., and Isaac, P. K. (1964). *Cephalosporium* species parasitic on *Helminthosporium* (sensu late). *Can. J. Plant Sci. 44:* 182-187.

Lumsden, R. D., and Ayers, W. A. (1977). Mycoparasitism of *Pythium* oospores by *Fusarium* spp. *Proc. Amer. Phytopathol. Soc. 4:* 217.

Makkonen, R., and Pohjakallio, O. (1960). On the parasites attacking the sclerotia of some fungi pathogenic to higher plants and on the resistance of those sclerotia to their parasites. *Acta Agr. Scand. 10:* 105-126.

Merriman, P. R. (1976). Survival of sclerotia of *Sclerotinia sclerotiorum* in soil. *Soil Biol. Biochem. 8:* 385-389.

Moody, A. R., and Gindrat, D., (1977). Biological control of cucumber black root rot by *Gliocladium roseum*. *Phytopathology 67*: 1159-1162.

Munnecke, D. E., Wilbur, W., and Darley, E. F. (1976). Effect of heating or drying on *Armillaria mellea* or *Trichoderma viride* and the relation to survival of *A. mellea* in soil. *Phytopathology 66*: 1363-1368.

Nolan, R. A. (1975). Fungal self-parasitism in *Saprolegnia megasperma*. *Can. J. Bot. 53*: 2110-2114.

Papavizas, G. C. (1973). Status of applied biological control of soil-borne plant pathogens. *Soil Biol. Biochem. 5*: 709-720.

Papavizas, G. C. (1977). Survival of sclerotia of *Macrophomina phaseolina* and *Sclerotium cepivorum* after drying and wetting treatments. *Soil Biol. Biochem. 9*: 343-348.

Person, L. H., Lucas, G. B., and Koch, W. J. (1955). A chytrid attacking oospores of *Peronospora tabacina*. *Plant Dis. Rep. 39*: 887-888.

Pohjakallio, O., Salonen, A., Ruokola, A., and Ikaheimo, K. (1956). On a mucous mould fungus, *Acrostalagmus roseus* Bainier, as antagonist to some plant pathogens. *Acta Agr. Scand. 6*: 178-194.

Pollack, F. G. (1971). *Cercospora uromycestri*, hyperparasite of rust on *Cestrum diurnum*. *Mycologia 63*: 689-693.

Rai, J. N., and Saxena, V. C. (1975). Sclerotial mycoflora and its role in natural biological control of 'white-rot' disease. *Plant Soil 43*: 509-513.

Rao, N. N. R., and Pavgi, M. S. (1976). A mycoparasite on *Sclerospora graminicola*. *Can. J. Bot. 54*: 220-223.

Richardson, M. J., and Leadbeater, G. (1972). *Piptocephalis fimbriata* sp. nov., and observations on the occurrence of *Piptocephalis* and *Syncephalis*. *Trans. Brit. Mycol. Soc. 58*: 205-215.

Rishbeth, J. (1963). Stump protection against *Fomes annosus*. III. Inoculation with *Peniophora gigantea*. *Ann. Appl. Biol. 52*: 63-77.

Röed, H. (1954). *Mitrula sclerotiorum* Rostr. and its relations to *Sclerotinia trifoliorum* Erikss. *Acta Agric. Scand. 4*: 78-84.

Ross, J. P., and Ruttencutter, R. (1977). Population dynamics of two vesicular-arbuscular endomycorrhizal fungi and the role of hyperparasitic fungi. *Phytopathology 67*: 490-496.

Rovira, A. D., and Campbell, R. (1975). A scanning electron microscope study of interactions between micro-organisms and *Gaeumannomyces graminis* (Syn. *Ophiobolus graminis*) on wheat roots. *Microbial Ecol. 2*: 177-185.

Schmidt, H. (1970). Untersuchungen über die Lebensdauer der Sklerotien von *Sclerotinia sclerotiorum* (Lib.) de Bary im Boden unter dem Einfluss verschiedener Pflanzenarten und nach Infektion mit *Coniothyrium minitans* Campb. *Arch. Pflanzenschutz 6*: 321-334.

Shahjahan, A. K. M., O'Neill, N. R., and Rush, M. C. (1976). Interaction between *Rhizoctonia oryzae* and other sclerotial fungi causing stem and sheath diseases of rice. *Proc. Amer. Phytopathol. Soc. 3*: 229 (Abstr. 113).

Shigo, A. L. (1958). Fungi isolated from oak-wilt trees and their effects on *Ceratocystis*. *Mycologia 50*: 757-769.

Shigo, A. L. (1960). Parasitism of *Gonatobotryum fuscum* on species of *Ceratocystis*. *Mycologia 52*: 584-598.

Siegle, H. (1961). Über mischinfektionen mit *Ophiobolus graminis* und *Didymella exitialis*. *Phytopathol. Z. 42*: 305-348.

Smith, A. M. (1972a). Biological control of fungal sclerotia in soil. *Soil Biol. Biochem. 4*: 131-134.

Smith, A. M. (1972b). Drying and wetting sclerotia promotes biological control of *Sclerotium rolfsii* Sacc. *Soil Biol. Biochem. 4*: 119-123.

Sneh, B. (1977). A method for observation and study of living fungal propagules incubated in soil. *Soil Biol. Biochem. 9:* 65-66.

Sneh, G., Humble, S. J., and Lockwood, J. L. (1977). Parasitism of oospores of *Phytophthora megasperma* var. *sojae, P. cactorum, Pythium* sp., and *Aphanomyces euteiches* in soil by oomycetes, chytridiomycetes, hyphomycetes, actinomyces, and bacteria. *Phytopathology 67:* 622-628.

Snyder, W. C., Wallis, G. W., and Smith, S. N. (1976). Biological control of plant pathogens. In *Theory and Practice of Biological Control*. C. B. Huffaker and P. S. Messenger (Eds.). Academic Press, New York, pp. 521-539.

Sparrow, F. K. (1960). *Aquatic Phycomycetes*, 2nd rev. ed. Univ. of Michigan Press, Ann Arbor.

Swendsrud, D. P., and Calpouzos, L. (1972). Effect of inoculation sequence and humidity on infection of *Puccinia recondita* by the mycoparasite *Darluca filum*. *Phytopathology 62:* 931-932.

Tribe, H.T. (1957). On the parasitism of *Sclerotinia trifoliorum* by *Coniothyrium minitans*. *Trans. Brit. Mycol. Soc. 40:* 489-499.

Tsao, P. H. (1970). Selective media for isolation of pathogenic fungi. *Ann. Rev. Phytopathol. 8:* 157-186.

Turner, G. J., and Tribe, H. T. (1975). Preliminary field plot trials on biological control of *Sclerotinia trifoliorum* by *Coniothyrium minitans*. *Plant Pathol. 24:* 109-113.

Turner, G. J., and Tribe, H. T. (1976). On *Coniothyrium minitans* and its parasitism of *Sclerotinia* species. *Trans. Brit. Mycol. Soc. 66:* 97-105.

Vörös, J. (1969). *Coniothyrium minitans* Campbell, a new hyperparasite fungus in Hungary. *Acta Phytopathol. Acad. Sci. Hungary 4:* 221-227.

Van Eck, W. H. (1976). Suitability of membrane-filter techniques to study the ultrastructure of *Fusarium solani* in soil. *Can. J. Microbiol. 22:* 1628-1633.

Walker, J. A., and Maude, R. B. (1975). Natural occurrence and growth of *Gliocladium roseum* on the mycelium and sclerotia of *Botrytis allii*. *Trans. Brit. Mycol. Soc. 65:* 335-338.

Warren, J. R. (1948). An undescribed species of *Papulospora* parasitic on *Rhizoctonia solani* Kuhn. *Mycologia 40:* 391-401.

Wells, H. D., Bell, D. K., and Jaworski, C. A. (1972). Efficacy of *Trichoderma harzianum* as a biocontrol for *Sclerotium rolfsii*. *Phytopathology 62:* 442-447.

Wood, R. K. S., and Tveit, M. (1955). Control of plant diseases by use of antagonistic organisms. *Bot. Rev. 21:* 441-492.

Yarwood, C. E. (1932). *Ampelomyces quisqualis* on clover mildew. *Phytopathology 22:* 31 (Abstr.).

Ylimäki, A. (1968). *Mitrula sclerotiorum* Rostr., a parasite on the sclerotia of *Sclerotinia trifoliorum* Erikss. *Ann. Agr. Fenn. 7:* 105-106.

Zogg, H. (1976). Problems of biological soil disinfection. *Soil Biol. Biochem. 8:* 299-303.

ADDITIONAL REFERENCES

Adams, P. B., and Ayers, W. A. (1980). Factors affecting parasitic activity of *Sporidesmium sclerotivorum* on sclerotia of *Sclerotinia minor* in soil. *Phytopathology 70:* 366-368.

Ayers, W. A., and Adams, P. B. (1978). Mycoparasitism of sclerotia of *Sclerotinia* and *Sclerotium* species by *Sporidesmium sclerotivorum*. *Can. J. Microbiol. 25:* 17-23.

Ayers, W. A., and Adams, P. B. (1979). Factors affecting germination, mycoparasitism, and survival of *Sporidesmium sclerotivorum*. *Can. J. Microbiol. 25*: 1021-1026.

Hadar, Y., Chet, I., and Henis, Y. (1979). Biological control of *Rhizoctonia solani* damping off with wheat bran culture of *Trichoderma harzianum*. *Phytopathology 69*: 64-68.

Hoch, H. C., and Abawi, G. S. (1979). Mycoparasitism of oospores of *Pythium ultimum* by *Fusarium merismoides*. *Mycologia 71*: 621-625.

Huang, H. C. (1978). *Gliocladium catenulatum*: hyperparasite of *Sclerotinia sclerotiorum* and *Fusarium* species. *Can. J. Bot. 56*: 2243-2246.

Huang, H. C. (1980). Control of sclerotinia wilt of sunflower by hyperparasites. *Can. J. Plant Pathol. 2*: 26-32.

Papavizas, G. C., and Lumsden, R. D. (1980). Biological control of soilborne fungal propagules. *Ann. Rev. Phytopathol. 18*: 389-413.

Uecker, F. A., Ayers, W. A., and Adams, P. B. (1978). A new hyphomycete on sclerotia of *Sclerotinia sclerotiorum*. *Mycotaxon 2*: 275-282.

Uecker, F. A., Ayers, W. A., and Adams, P. B. (1980). Teratosperma oligocladum, a new hyphomycetous mycoparasite of sclerotia of *Sclerotinia sclerotiorum, S. trifoliorum*, and *S. minor*. *Mycotaxon 10*: 421-427.

Chapter 18

EXPLOITATION COMPETITION[*]

John L. Lockwood
Michigan State University
East Lansing, Michigan

I. INTRODUCTION

It is generally understood that fungi compete for food and perhaps for other components of their environment, yet relatively little definitive work or analytical thought
has been applied either to the mechanisms or consequences of competition among microorganisms, including fungi. In 1961, of 19 papers given at a symposium of the Society of Experimental Biology on *Mechanisms of Biological Competition* (Milthorpe, 1961),
none dealt with microorganisms. This situation does not seem to have changed greatly
since, and it contrasts markedly with that of plants and animals, concerning which a
great deal of attention has been given to competition theory and experimentation, as
any modern ecology text will testify.

Competition among animals has been defined by Milne (1961), after a lengthy consideration of various definitions and concepts, as "the endeavor of two (or more)
animals to gain the same particular thing, or to gain the measure each wants, from
the supply of a thing when that supply is not sufficient for both (or all)." More
applicable to microorganisms, perhaps, is an older definition of Clements and Shelford (1939), later adopted by Clark (1965), that "competition is an active demand in
excess of the immediate supply of a material or condition on the part of two or more
organisms." Competition is generally divided into two types: *exploitation* and *interference*. The former is restricted to the depletion of resources by one organism
or population without reducing the access of another organism or population to the
same resource pool, whereas the latter deals with behavioral or chemical mechanisms
by which access to a resource is influenced by the presence of a competitor (McNaughton and Wolf, 1973). This review will deal with expoitation competition.

A fungus may be involved in competitive interactions at any stage of its life
cycle, e.g., during quiescence of resistant structures, during active exploitation
of a fresh substrate, or during the sequential colonization of a substrate previously occupied. Moreover, competition involves not only the direct confrontation of

---

[*]Journal article No. 8600 from the Michigan Agricultural Experiment Station.

two populations utilizing the same resource but also changes effected in a substrate by one or more populations that limit secondary colonization. If the alterations involve inhibitory chemicals, pH changes, etc., these represent interference competition, but where resource deprivation occurs they are properly appropriate to exploitation competition. Instances of competition among fungal populations isolated from other microoraganisms are probably uncommon, since in most habitats bacteria, actinomycetes, and perhaps soil microfauna also are involved. Therefore, much of the discussion here will of necessity include as coactors in a competitive interaction with a particular fungal population not only other fungi but also other types of microorganisms--indeed, sometimes the total microflora of that population's microhabitat.

II. OBJECTS OF COMPETITION

A. Substrate, Water, Oxygen

It is often stated axiomatically that microorganisms in soil compete for food, water, air, and space. However, in an incisive essay on the subject, Clark (1965) argued convincingly that food is the component most likely to be in short supply and is therefore the prime object of competition. He considered that competition for nitrogen might also occur, though more rarely. Space was considered to be ample, based upon microscopic observations of substrata and the fact that populations greater than the initial maxima often can be achieved if nutritional needs are met. Water was not considered a likely object for competition: even though required for growth, water supply is depleted by evaporation and physical binding rather than by microbial activity, which is more likely to produce than to consume water. Since most microorganisms consume oxygen, this element was suggested to be a possible object of competition, particularly when its supply is limited, such as in flooded conditions or in microsites of intense oxygen demand. I see no reason to question Clark's rationale or conclusions. However, I have reexamined the question of competition for space since more data that can be applied to the question have come to light, and would stress that competition for nitrogen may occur in substrates of high C/N ratio.

The overriding importance of substrate, i.e., carbon source, as the primary object of competition, particularly in soil, has been emphasized by several recent research studies in which the energy requirements for turnover and maintenance of measured microbial populations in terrestrial ecosystems were related to annual input of substrate (Babiuk and Paul, 1970; Clark and Paul, 1970; Gray and Williams, 1971; Shields et al., 1973; Gray et al., 1974). These studies all indicate that the amount of substrate available is sufficient to permit turnover to occur only a few times per year. For example, one such study indicated that one-fifth of the total substrate available annually was required for each generation of new cells produced (Clark and Paul, 1970). Maintenance energy requirements, i.e., energy needed simply to keep the existing population alive without growth and reproduction, may consume one-third or

more of the available substrate, reducing even further that available for growth and
reproduction. In another approach to the problem, CO_2 production from field soils
was compared with that from stationary populations in laboratory cultures (Clark,
1967; Clark and Paul, 1970). The amount of CO_2 evolved from field soils was only
1/35 to 1/70 of that predicted from microbial populations of equivalent biomass in
stationary laboratory cultures. This great discrepancy further indicates that growth
and activity of microorganisms in soil must be severely restricted. It is difficult
to escape the conclusion that competition for substrate, in soil at least, must be
very intense.

B. Space

Competition for physical space in or on dead substrates would not seem to be a likely
possibility based on considerations of respiration losses during trophic transfer,
i.e., the biomass of saprophytic microorganisms would have to be less than that of
the substrate consumed. Possible exceptions might be loci into which nutrients accumulate forming a larger reservoir, such as the rhizosphere or phylloplane. Recent
studies make it possible to assess this possibility. Reference also will be made to
microorganisms other than fungi, since these may contribute to crowding of fungi.

Living plant leaves are often sparsely colonized by microorganisms. Leben and
Daft (1967) found very few or even no microorganisms on cleared and stained leaves of
various plant species in Ohio and Puerto Rico, during dry weather. In wet conditions,
numbers increased but were still considered "few and scattered." Buds of soybean
seedlings exposed to a humid atmosphere had large areas that did not carry any microorganisms, when examined by scanning electron microscopy (SEM) (Leben, 1969). Dickinson (1967) found less than one viable propagule of fungi or yeasts per square centimeter of nail varnish impressions of *Pisum* leaves, until senescence. Dickinson and
Skidmore (1976) have commented that the scarcity of nutrients and frequency of adverse
physical conditions rigorously selects species capable of colonizing the phylloplane
til the onset of senescence. In contrast, larger numbers were recorded on beech
leaves in a forest (Holm and Jensen, 1972). Mean number of bacterial cells for May,
June, and July, as determined by plate counts, was $3940/cm^2$, and that for August and
September was $54,510/cm^2$. The highest value was $120,000/cm^2$. If the two-dimensional
area of a bacterial cell is taken as $1~\mu m^2$, even this highest value would represent
only 0.1% of the available leaf surface. Numbers of bacteria as high as 10^7 to 10^9/g
fresh weight of grass leaves, and of yeasts as high as 10^7/g were recorded by Last
and Deighton (1965). These represent 0.001% to 0.1% of the leaf on a fresh-weight
basis, hardly a crowded condition. However, Ruinen (1961) found a rich and diverse
microflora, including algae, fungi, lichens, and bacteria, on the leaves of *Coffea*,
Pellionia, and *Theobroma* in Indonesia. On some leaves the microbial layer was up to
$22~\mu m$ thick. Bacteria tended to be concentrated along wall junctions or were associated with fungal hyphae or algal filaments. On mature cacao leaves the bacterial
population was estimated by direct microscopy to be 12.5×10^6 cells/cm^2, which is

equivalent to about 10% of the leaf surface area. Whether competition for space exists here is not clear. The growth of algae and lichens on the leaves would present additional surfaces for colonization by other microorganisms.

With senescence, microbial colonization of leaves increases. Dickinson (1967) measured lengths of germ tubes, hyphae, and conidiophores of fungi on the surfaces of *Pisum* leaves in early September when the leaves had yellowed, and again at the end of September when they were dead and the plants had fallen to the ground. Total fungal lengths increased from 42 to 146 mm, respectively, in this short period. If mean hyphal width is taken as 2.5 μm (Rovira et al., 1974), then we can conclude that 0.1% and 0.37% of the leaf surface areas, respectively, were occupied.

Colonization is most extensive on freshly fallen and decomposing leaves, then declines as leaves become mineralized. Mean hyphal lengths in the L, F, H, and A_1 horizons of soil in an oak-beech forest were 1849, 1602, 3685, and 445 m/g dry soil (Nagel-de Boois and Jansen, 1971). For our purposes such data are better expressed on a volume basis, since the bulk densities of the horizons differed widely. For example, the amount of organic matter was 94% in the L layer and only 6.5% in the A_1 layer. Moreover, the proportion of living hyphae as determined by staining declined with soil profile depth; these values were 81, 18, 5 and 4%, respectively. When the appropriate corrections were applied, fungal hyphae in the L, F, H, and A_1 layers occupied 0.8, 0.1, 0.004, and 0.025%, respectively, of the volume available.

Minderman and Daniels (1967) followed the colonization of fallen leaves in an oak forest from leaf fall in October until the following June. Bacteria were the predominant early colonizers, with transient numbers reaching as high as $150-180 \times 10^6$ cells under each square centimeter of three-dimensional space between the upper and lower epidermis. Their numbers declined rapidly, followed by an increase in fungi, then by actinomycetes. These increases were more or less sustained throughout the experimental period; maximum hyphal lengths recorded were 20 m/cm^2 and 30 m/cm^2, respectively. In early December, colonization by all three groups was high: fungal hyphae, 10 m/cm^2; actinomycete hyphae, 20 m/cm^2; and number of bacterial cells, 20×10^2. Using hyphal widths of 2.5 μm for fungi and 1 μm for actinomycetes, and 1 μm^3 as the volume of each bacterial cell, the respective biomasses represented were 0.05, 0.016, and 0.05 mm^3, totaling 0.116 mm^3. Taking the leaf thickness as 60 μm, these microorganisms occupied only 0.2% of the space available.

The root surfaces and the surrounding rhizosphere soil represent other loci wherein microbial activity is enhanced, with the possibility of overcrowding. Rovira and colleagues (1974) grew four grass species and four dicotyledonous herbs in pots of natural soil, and after 12 weeks observed stained root segments microscopically. Bacteria were estimated to cover 4.7-9.3% of the root surfaces. In further experiments, coverage of roots by bacteria and fungi was 7.7% and 3%, respectively, for *Lolium perenne,* and 6.3% and 3%, respectively, for *Plantago lanceolata.* In calculating fungal coverage of roots of 2- and 5-year-old *Pinus nigra* var. *laricio* in a pine forest from data of Parkinson and Crouch (1969), Rovira et al. (1974) found 2.3% to

3.7%, respectively, of the root surfaces covered by fungal mycelium. The 1- to 21-
day-old roots of *Pinus radiata* plants 7-90 days old, growing in a sandy forest soil,
had 0-16% of their surfaces covered by microorganisms (Bowen and Theodorou, 1973).
The greatest coverage, 37%, was on root bases of plants 90 days old. Seedlings of
the same species transplanted into sterile soil had about 5% of their root surfaces
covered by bacteria (Bowen and Rovira, 1973, 1976).

Colonization of rhizosphere soil is likewise incomplete. Lengths of fungal hyphae in rhizospheres of dwarf bean plants in pots of field soil ranged from some 50
to 100 m/g soil (Parkinson and Thomas, 1969). These values represent occupancy of
only 0.018 to 0.036% of the soil volume. The generally low levels of occupancy of
root surfaces and of rhizosphere soil by microorganisms are also shown in other work,
using SEM or transmission electron microscopy (TEM) and isolation procedures (Marchant, 1970; Campbell and Rovira, 1973; Foster and Rovira, 1973; Old and Nicolson,
1975). Some root segments yielded no microorganisms upon isolation, and some portions of roots appeared to be free of microorganisms by EM examination. The rhizosphere appears to represent "a collection of spatially non-interacting communities..
.." (Bowen and Rovira, 1973). The localized occurrence of the root microflora was
attributed to the chance distribution of potential rhizosphere colonizing microorganisms in soil through which the root grows (Bowen and Rovira, 1973, 1976; Bowen
and Theodorou, 1973).

Studies of the soil apart from the influence of plant roots also show minimal
occupancy of the available volume. Bacterial numbers ranged from 10 to 20 X 10^8 and
hyphal lengths from 500 to 1300 m/g in a wheat field soil (Shields et al., 1973).
These values represent, respectively, 2.5-6.5 mm^3 and 1.0-2.0 mm^3 of biomass, which
together would occupy some 0.25-0.65% of the soil volume. Even assuming that bacteria must occupy the water film (0.3 ml) on the available soil surfaces, only 0.4-
1.0% of the space available would be occupied by a bacterial population of 10^9 cells/g
(Clark, 1967). Hyphal lengths in soil tend to decrease with soil depth (Burges and
Nicholas, 1961; Parkinson et al., 1968). For example, in a pinewood soil, mean hyphal
lengths over a 7-month period in the H, A, C_1, and C_2 horizons were 2.7, 25, 6, and
3 m/g dry soil, respectively (Parkinson et al., 1968). The proportion of the soil
occupied ranged from 0.1% downward. Such estimates represent total hyphae, living
and dead. Viable hyphae may be much less (Frankland, 1975; Nagel-de Boois and Jansen, 1971).

The sum of the foregoing review strongly supports the earlier view of Clark
(1965) that microbial competition for space occurs rarely, if at all.

C. Nitrogen and Minerals

Many organic residues contain nitrogen and other minerals in sufficient quantity that
the supply of these materials should not limit decomposition (Clark, 1967). For example, the C/N ratio of herbaceous plant residues may be as low as 10:1 (Burges,

1967; Levi and Cowling, 1969); however, that of woody stem tissue may be on the order of 50:1 or greater (Burges, 1967; Frankland, 1974; Levi and Cowling, 1969), and that of wood, greater than 200:1 (Levi and Cowling, 1969). Competition for nitrogen may be expected to occur during decomposition of tissues of higher C/N ratios, though Clark (1967) does not consider nitrogen to be limiting in plant residues if its content is not less than 1.5% (C/N ratio of ca. 28:1). Supplemental nitrogen is commonly applied to hasten the decomposition of plant residues during composting and will also aid the decomposition of straw and other materials of high C/N ratio in or on soil (Alexander, 1977). An indirect consequence of enhanced decomposition of wheat stem tissue in nitrogen-supplemented soil is more rapid competitive displacement of certain weakly saprophytic root-and foot-rotting pathogens of wheat that had previously colonized the stem pieces (Garrett, 1970). This subject will be discussed further in Sec. IV.

A possible role of nitrogen or mineral competition by mycorrhizal roots resulting in the suppressed litter decomposition in a pine forest is suggested by work of Gadgil and Gadgil (1975). Litter decomposition was enhanced after mycorrhizal roots were severed, suggesting that nutrients had been tied up in such roots. This was supported by laboratory experiments showing that decomposition of litter was more rapid in the presence of nonmycorrhizal roots than in the presence of mycorrhizal roots.

Response to nitrogenous amendments may sometimes be due to factors other than nitrogen content per se. When urea was used to supplement fallen apple leaves, fungi in general were increased as indicated by viable spore counts, and this was attributed to the supplemental nitrogen (Hudson, 1971). However, when pine needles were treated with urea and with ammonium salts to give equivalent concentrations of nitrogen, only urea caused changes in the microfungi (Lehmann and Hudson, 1977). It was considered that the effect of the urea was due to increased pH rather than just to nitrogen itself.

Park (1976b) has pointed out that the generalization that more concentrated nitrogen sources permit more rapid cellulose decomposition may not have universal validity. He evaluated the ability of 45 fungal isolates to decompose disks of cellulose filter paper in liquid media with inorganic nitrogen content varied to give C/N ratios of 0.5:1, 5.0:1, and 22.0:1. Six different patterns emerged, with only eight fungi falling into the expected pattern of increased cellulose decomposition with increased nitrogen. Although the highest C/N ratio used, 22:1, may not have been sufficiently high to limit nitrogen, the results suggest that fungi may have a much wider tolerance of C/N ratio than is often assumed.

The nitrogen content of wood is very low (ca. 0.03-0.1%) and thus presents a very deficient substrate for microorganisms. Fungi adapted for this habitat have a remarkable ability to grow and produce cellulase at low nitrogen concentrations (Levi and Cowling, 1969), to scavenge efficiently for the small amount available, to concentrate

it in sporophores and spores (Merrill and Cowling, 1966), and to recycle the nitrogenous constituents of their own mycelium through autolysis and reutilization (Levi et
al., 1968).

The supply of nitrogen (or other minerals) does not appear to limit microbial
activity in soils, except when the supply of carbon is artificially enhanced, thus
rendering other materials unavailable. For example, in a natural sandy soil increased
CO_2 evolution in the presence of nitrogen, phosphorus, and sulfur could be shown only
following addition of supplemental glucose (Stotzky and Norman, 1961a,b). Similarly,
growth suppression of *Fusarium oxysporum* f. sp. *cubense* by several nonantibiotic-
producing bacteria in sterilized soil was overcome by addition of sucrose but not of
nitrogen and phosphorus (Marshall and Alexander, 1960; Finstein and Alexander, 1962).
These elements gave a slight growth response only after the carbon demand was satisfied.

Competition for nitrogen and phosphorus in the presence of excess carbon also
was suggested by results of Kong and Dommergues (1971) in culture experiments. More
nitrogen and phosphorus were required to obtain maximum decomposition of powdered
cellulose by *Trichoderma* sp. in the presence of other soil microorganisms as competitors than in their absence.

In plant pathology, experimental control of *Fusarium* root rot of bean has been
obtained with organic amendments of high C/N ratio such as barley straw or cellulose,
which presumably result in immobilization of nitrogen required by the pathogen
(Snyder et al., 1959; Maurer and Baker, 1965).

Competition for growth factors could result in suppression of fungi autotrophic
for such factors. Growth of *Phytophthora cryptogea*, which required thiamine, was
suppressed in culture by thiamine-requiring isolates of *Arthrobacter* sp. (Erwin and
Katznelson, 1961). Since soil extracts failed to stimulate growth of *P. cryptogea*
in thiamine-deficient media, it was suggested that competition for thiamine might
limit growth of the fungus in soil, especially in rhizospheres wherein thiamine-
synthesizing bacteria are known to grow.

In water, the absolute amount of nitrogen may be very low. Park (1976a), in a
most significant study, compared the ability of a pythiaceous fungus to decompose
cellulose in a liquid medium containing the same total amount of nutrients but in
different volumes of solution. Decomposition was many times faster in the larger
volume than in the smaller, suggesting that low concentrations of nutrients may be
more favorable than high concentrations for aquatic fungi. When the C/N ratio was
varied from 40:1 to 4000:1, most decomposition occurred at the lower ratios, but
efficiency (amount decomposed/unit of N) was greatest at 400:1. There was some decomposition even at 1600:1.

In summary, it would seem that competition for nitrogen can occur in substrates
where nitrogen concentration is low, such as in woody stem tissues, wood itself, and
water. Some fungi appear to be physiologically well adapted for growth in such habitats, in which they would be expected to have a competitive advantage.

III. COMPETITION AND COEXISTENCE

Competition requires that the niches of competing populations overlap sufficiently
that they share in the utilization of the same resources. Given the great numbers
of individual microorganisms and the diversity of species occurring in nature, competition would be expected to be intense. Since competition theory, based on laboratory models, indicates that one species should eventually displace all others (Pielou, 1974), questions can be raised about the coexistence of so many species of fungi (and other microorganisms), often in close proximity, in the natural world. This
has been a subject of much research and thought in plant and animal ecology (e.g.,
see McNaughton and Wolf, 1973; Ricklefs, 1973; Pielou, 1974) but has not assumed a
dominant theme in microbial ecology. It would seem worthwhile at this juncture to
enumerate and briefly discuss some of the mechanisms of coexistence described for
plant and animal populations that may also apply to microorganisms.

    1. *Spatial separation along an environmental gradient* (Pielou, 1974) *or because
of patchy distribution of environmental factors* The literature of fungal physiology
is replete with examples of different optima and tolerances with respect to environmental factors such as temperature, moisture, pH, or oxygen, many of which are no
doubt expressed in the natural world. [For examples see Chaps. 21-24 in Ainsworth
and Sussman (1965) and Chaps. 1-5 in Ainsworth and Sussman (1968).]

    In an experimental model, spatial separation of two species of slime molds was
shown at opposite ends of a temperature gradient from 15° to 30°C established in a
long culture dish (McQueen, 1971). Whereas both *Dictyostelium discoideum* and *Polysphondylium pallidum* cofruited from 20° to 26°C, agar from 15° to 20°C constituted
a refuge for *D. discoideum* free from competition, and that from 26° to 30°C was a
refuge for *P. pallidum*.

    2. *Temporal separation due to fluctuating environmental conditions* (McNaughton
and Wolf, 1973) Niche parameters may change to favor one competing species, then
another, subsisting on the same substrate. This might occur during the normal wet
and drying cycles on vegetation and in soil. For example *Fusarium roseum* f. sp.
cerealis 'Culmorum' is a more effective saprophyte on wheat residues, as well as a
parasite of living wheat roots, at low water potentials, because many bacterial and
many fungal competitors are not able to grow in dry soil (Cook, 1970; Cook and Papendick, 1970).

    3. *Spatial separation because of patchy resource distribution* (Ricklefs, 1973;
Pielou, 1974) One such example might be the root surface, where exudation is localized and colonization is dependent upon the distribution of microorganisms in the
soil through which the root passes (Bowen and Rovira, 1976). This possibility is
also suggested by the work of Lowe and Gray (1973), who tested competition in sterilized soil among pairs of coccoid bacteria which were phenetically similar or different, based on a large number of morphological and biochemical properties. Two pairs

with high similarity coefficients interacted competitively, but of two pairs with low similarity coefficients, one pair did not interact whereas the other did. Therefore, phenetic similarity was not exclusively associated with competitive interaction. This could mean that bacteria, being thinly distributed in soil, may arrive at specfic microsites by chance, i.e., similar microorganisms can coexist by occupying spatially separated though similar microsites. The results could also be explained by assuming that ecological fitness is determined by smaller variations in characters than those affecting taxonomic relationships. The experimental approach would seem to have great promise for investigating questions of physiological characters in relation to niche and should be adaptable to fungi. However, it might be improved by avoiding the use of sterilized soil which is rich in nutrients readily available to a large number of fungi; this may mask competition for more specific substrates.

4. *Resource specialization, with consequent development of an increased number of niches* (Ricklefs, 1973) This has been elucidated only in the crudest sense in the fungi. Garrett (1970) has recognized four large groups of saprophytic fungi: (1) primary saprophytic sugar fungi, which are pioneer colonizers of fresh plant tissue substrates; (2) secondary saprophytic sugar fungi, associated with cellulose and lignin decomposers; (3) root surface and rhizosphere fungi; and (4) cellulose and lignin decomposers. More specific food specialization undoubtedly occurs. For example, there are quite specific groups of fungi adapted to grow on such substrates as keratin, chitin, and waxes. There also may be quantitative differences: all forms of carbonaceous substrates are not equally suitable for all fungi. Some fungi are obligate or facultative parasites of plants or animals and as a consequence can escape competition while in the parasitic phase. [For detailed discussion of substrate relations in fungi, see Chap. 18 of Ainsworth and Sussman (1965) and Chaps. 7-11 in Ainsworth and Sussman (1968).]

The sequential changes in fungal species and numbers occurring in decomposing plant litter probably reflect, in part, a succession of niches as the substrate changes. Although up to now the main emphasis in this research has been on enumeration and identification, one such interesting niche is the stage when sugars are released late in the succession of fungi decomposing plant litter (Saitô, 1965; Chang, 1967; Hudson, 1968; Frankland, 1969). This stage involves Basidiomycetes, associated with which are Mucorales, *Penicillium*, or *Trichoderma* (Hudson, 1968). Tribe (1966) showed also that several *Pythium* species grow well on cellulose filter paper on sand in the presence of various cellulose-decomposing fungi, apparently from sugars released during cellulose degradation, whereas they grow poorly alone.

Fungi must differ in ability to tolerate reduced resource availability. Those able to utilize nutrients efficiently would have a competitive advantage when resources are scarce.

5. *Divergence in characters related to resource exploitation* (McNaughton and Wolf, 1973; Pielou, 1974) Armstrong (1976) showed that *Aspergillus nidulans* and *Penicillium* sp. could coexist in an experimental agar ecosystem provided that con-

ditions allowed for the complementary expression of the characters each required for
survival. A. *nidulans* was competitively dominant in that it could displace established colonies of *Penicillium* sp., whereas *Penicillium* sp. was the better primary
colonizer in that it produced more colonies and fruited more rapidly. When spores
of donor cultures derived from mixed inoculum of both species of a single generation
were applied to uncolonized agar, only one or the other species was obtained because
the inoculum consisted only of *Penicillium* sp. in younger donor cultures or of *A.*
nidulans in older cultures. However, when the donor cultures were of mixed ages, coexistence was demonstrated. *Penicillium* colonized from the younger patches in which
A. nidulans had not yet become dominant, and thus was considered a fugitive species,
whereas *A. nidulans* dominated the older patches.

A genetic change in reponse to competition was shown experimentally for one of
a pair of slime molds allowed to compete for bacterial food on an agar medium along
a temperature gradient (McQueen, 1971). Initially, in mixed cultures, *Dictyostelium*
discoideum fruited over a temperature range of 15°-26°C and *Polysphondylium pallidum*
from 24° to 30°C. However, with continual competition, *P. pallidum* extended its
fruiting range down to 20°C. The change was not temperature dependent since it also
occurred during incubation at 24°C, did not occur in the absence of competition, and
was permanent. The altered population had a more rapid rate of colony expansion
than did the original culture.

6. *Competitive advantage resulting from sparseness* (Pielou, (1974) In this
situation advantage alternates between two species, such that the temporarily sparser
one is favored at the expense of the temporarily commoner one. One example familiar
to plant pathologists is the tendency for genetically pure stands of a crop to favor
races of pathogens capable of attacking that genotype, hence to provide optimal conditions for their dispersal and increase, whereas mixed genotypes tend to suffer less
damage (Pielou, 1974; Trenbath, 1977). A corollary of this principle is that increased genetic diversity in host populations should tend to reduce pathogen populations and thus result in decreased competition among races.

Noncompetitive limiting factors may also serve to keep populations below the
size at which competition plays an important role (McNaughton and Wolf, 1973).

7. *Exogenous and endogenous dormancy* The question of species coexistence is
perhaps most acute in the soil, in which hundreds of species of bacteria, actinomycetes, fungi, arthropods, and algae all reside within the same relatively small volume of substratum. Though the six mechanisms just enumerated, and possibly others,
play a part in the coexistence of a large number of species, I believe that exogenous
dormancy imposed through chronic deficiency and patchy distribution of substrate is
the main factor restricting competition among species in soil and other occupied
substrates. This will be discussed in detail in Sec. V.

Constitutive dormancy, characteristic of oospores of many of the Pythiaceae and
of ascospores of some carbonicolous Ascomycetes, also provides a mechanism of competition avoidance. Dormancy of the former is commonly terminated in vitro by ex-

posure to light (Ribeiro et al., 1975), but whether this or some alternative factor provides the stimulus for germination in nature is not clear, particularly for root-infecting species. Dormancy of the latter is commonly terminated by heat (from fires) in nature, allowing the development of such fungi in the absence of many potentially competitive microorganisms (Warcup and Baker, 1963).

The problem of species diversity in soil is in some ways similar to that of plankton in which a large number of species coexist in a relatively isotropic and unstructured environment, often under conditions of severe mineral nutrient stress. Hutchinson (1961) believed that mechanisms such as those enumerated in items 1-6 above were involved, but felt that the species diversity of phytoplankton in lakes, at least, was based upon its origin and constant replenishment from species existing in benthic or littoral niches.

IV. COMPETITIVE INTERACTIONS

In this section I will summarize research on competitive interactions involving fungi. Approaches used have included (1) simplified experimental ecosystems, (2) natural environments perturbed for experimental purposes, and (3) mathematical modeling. I will not deal with the third approach out of my own lack of competence in this area. Competitive arenas include (1) leaf surfaces, wounds, and infected tissues of living plants, and (2) plant residues in soil. Some work with simplified experimental eco-systems, such as culture media, were discussed previously in other contexts, and will be treated here selectively if at all.

A. Living Plant Tissues

There are numerous instances of the experimental inhibition of fungal development on plant leaves by other microorganisms, including other fungi (see Leben, 1965; Preece and Dickinson, 1971; Baker and Cook, 1974). In some instances, lesion development by fungal leaf pathogens has been reduced by co- or preinoculation of leaves with saprophytic filamentous fungi or yeasts that were normal residents of the leaf sur-face (van den Huevel, 1971; Fokkema and Lorbeer, 1974; Spurr, 1977).

In most instances the suppressive mechanism, whether exploitation or interfer-ence, is not known. However, Blakeman and colleagues in Aberdeen have provided clear evidence for nutrient competition by associated microorganisms as the mechanism sup-pressing germination of *Botrytis cinerea* on plant leaves (Sztejnberg and Blakeman, 1973; Brodie and Blakeman, 1976; Blakeman and Brodie, 1977). The successful competi-tors were able to compete for amino acids in leaf exudate, and perhaps also for amino acids in exudate from the spores of the leaf pathogen. The mechanism of inhibition will be discussed further in Sec. V.

An interesting interaction involving a parasite, saprophytes, and pollen on plant leaves has been studied in the Netherlands. In the presence of rye pollen, development of lesions caused by *Cochliobolus sativus* was greatly enhanced (Fokkema,

1971a,b). However, this increase was largely canceled by co- or preinoculation with each of several leaf saprophytes, the most effective being *Aureobasidium pullulans* (Fokkema, 1973). The amount of suppression of lesion development was correlated with the amount of reduction in development of superficial mycelium of the pathogen; germination was not affected. The amount of lesion development was unrelated to the presence or absence or degree of antagonism in vitro. Pollen-stimulated enhancement of lesion development by *Phoma betae* on beet leaves also was suppressed by saprophytic fungal flora (Warren, 1972).

Competitive interactions involving fungi also are well known in wounds and cankered lesions on trees. In these instances most interest has attached to the potential for biological control of pathogenic fungi. For example, the failure of the "dieback" fungus, *Eutypa armeniacae*, to become established in pruning wounds of apricot trees was associated with colonization of such tissue by *Fusarium lateritium* (Carter, 1971). Inoculation experiments confirmed that *F. lateritium* was able to partially protect wounds from infection with *E. armeniacae*. The competitive mechanism may be via interference, since the fungus produced a substance inhibitory to the pathogen in culture (Carter and Price, 1974). One such competitive interaction has been employed as a practical means of preventing the entry of the destructive root pathogen *Fomes annosus* into freshly exposed stumps of Scots pine in England (Rishbeth, 1963). The weak pathogen *Peniophora gigantea* is extremely effective in excluding the aforementioned pathogen, whether in natural infestations or artifically applied to pine stumps after cutting. It is also capable of displacing *F. annosus* from infected stump and root tissues. In agar, the cells of *F. annosus* in close contact with *P. gigantea* were killed, but the mechanism of action is not known (Ikediugwu et. al., 1970). Similarly, larch cankers caused by the fungus *Trichoscyphella willkommii* were sequentially invaded by several saprophytic fungi with the concomitant disappearance of the pathogen (Buczacki, 1973). Canker extension also was reduced in artifical inoculations of these fungi with, or even after, establishment of *T. willkommii*. The natural healing of bark cankers was attributed to competitive displacement of the pathogen by these saprophytes. The weak pathogen *Hypoxylon punctulatum* was able to competitively displace the oak wilt fungus, *Ceratocystis fagacearum*, in girdling wounds made in oak trees (Roncadori, 1962). There was no evidence for the formation of an antibiotic substance; instead, *H. punctulatum* apparently exhausted the sapwood of amino acids required for survival of the more pathogenic species.

The competitive success of races of a fungal species, or of related species, in colonizing infected plants has been investigated for *Puccinia graminis tritici*, *Pyricularia oryzae*, *Phytophthora infestans*, and the combination *Tilletia caries-Tilletia foetida*. Plants of one or more varieties of an appropriate host species were inoculated with equal numbers of spores of each race or species, and then spores from the resulting infection were used to reinoculate new plants. At each generation the population densities of the competing members were enumerated by spore counts.

In general, one member of a pair became dominant after several generations, in some cases eliminating the weaker competitor (Watson, 1942; Loegering, 1951; Thurston, 1961; Katsuya and Green, 1967). However, in other cases, an unstable equilibrium was maintained (Rodenhiser and Holton, 1953; Yamanaka, 1974).

In *P. graminis tritici*, temperature affected the outcome (Watson, 1942; Katsuya and Green, 1967). For example, at 20°-25°C, race 56 predominated and race 15-B was nearly eliminated after four or five generations, whereas at 15°C race 15-B predominated (Katsuya and Green, 1967). Race 15-B also prevailed in the field during a series of years with cooler than normal temperatures. Competitive success was often associated with greater vigor (Watson, 1942; Katsuya and Green, 1967; Ogle and Brown, 1971), viz., greater numbers of appressoria penetrating the host, more pustules formed per spore, shorter incubation period for spore production, faster rate of enlargement of uredinia, larger uredinial size, and greater urediospore production. Proportion of spores germinating, forming appressoria, or giving rise to uredinia did not differ (Loegering, 1951; Katsuya and Green, 1967; Ogle and Brown, 1971). In the case of race 56 versus race 15-B, race 56 was superior in many of the aforementioned characters, even at cool temperatures, which favored race 15-B in paired competition experiments. Since race 15-B predominated on one of two varieties with a heavy--but not with a light--inoculum, it was speculated that race 15-B can compete better for host nutrients when stress on the nutrient supply is great (Katsuya and Green, 1967).

In similar experiments with *P. oryzae*, the dominant isolates had slightly higher rates of lesion expansion (Yamanaka, 1974) but no differences in numbers of spore produced on the lesions. In investigations with *P. infestans*, race 0 was dominant over three other isolates and was the more virulent pathogen alone (Thurston, 1961). Competitive advantage could not be related to differences in spore germination or mycelial growth rate (Rodenhiser and Holton, 1953).

In these experiments, it cannot be assumed that competition was restricted to exploitation. Some slight evidence for the operation of interference in *P. infestans* was that water in which sporangia were soaked usually became inhibitory to sporangial germination of the same or other races (Thurston, 1961).

Simultaneous inoculation of plants with virulent and less virulent or nonpathogenic strains of *Fusarium oxysporum* resulted in colonization of host tissues by both, and mitigation of symptoms as compared with those produced by the virulent component alone (Langton, 1969; Meyer and Maraite, 1971). There was no evidence for antagonism in culture (Langton, 1969), and competition for a limited number of penetration sites was considered the most likely explanation for the amelioration (Meyer and Maraite, 1971).

B. Plant Residues in Soil

Dead plant residues are subject to colonization by microorganisms, including fungi. Among the fungi, a well-documented sequence of specific forms occurs as plant litter decomposes (Hudson, 1968; Dickinson and Pugh, 1974). Undoubtedly, competitive interactions are involved in the population changes, but this aspect of the microbiology of plant litter decomposition so far has had little study.

In a more restricted context, plant pathologists have studied the saprophytic colonization of plant residues as a means of survival of root-infecting fungal pathogens. The most notable work in this respect is that of S. D. Garrett and his coworkers, who have studied the saprophytic colonization and survival of several fungal pathogens of the roots and stem bases of wheat in competition with the general soil microflora. This work has been summarized by Garrett (1956, 1970, 1975, 1976).

Competitive ability was assessed by incubating wheat straw pieces in graded mixtures of cornmeal-sand inoculum and soil, prepared so that as inoculum decreased the amount of soil (hence, of competing microorganisms) was increased. *Fusarium roseum* f. sp. *cerealis* "Culmorum" (*F. culmorum*) and *Curvularia ramosa* were the most successful competitors, since they were able to colonize wheat straws at low inoculum concentrations (against high populations of competing microorganisms). *Cochliobolus sativus* was intermediate, and *Gaeumannomyces graminis* and *Cercosporella herpotrichoides* were poor competitors. Garrett earlier (1956) had suggested that competitive success might be related to the following characters: (1) growth rate; (2) antibiotic production; (3) tolerance of antibiotics produced by competitors; or (4) enzyme-producing ability. The first two characters were not found to be related to competitive success (Garrett, 1956, 1970). However, the rate at which the straws were penetrated by *F. culmorum*, *C. ramosa*, and *C. sativus* was much more rapid than that by *G. graminus* and *C. herpotrichoides* (Garrett, 1970, 1975). Straw penetration rates of the fungi in turn were highly correlated with their ability to decompose cellulose filter paper. Despite these correlations, straw penetration rate and cellulolytic ability did not differentiate the two fungi with highest competitive success from *C. sativus*, the intermediate fungus. Such a distinction was made, however, when growth of the fungi in an agar medium over which a layer of soil had been applied 48 hr earlier was compared with that on control agar without soil. The percentage reduction in growth was inversely correlated with competitive ability; reduction of growth was greater for *C. sativus* than for *F. culmorum* and *C. ramosa* (Garrett, 1970, 1972, 1975). Butler (1953) had earlier found that *C. ramosa* was more tolerant of antibiotics and of bacterial antogonism in culture than was *C. sativus*. Though Garrett (1970) attributed the growth reduction on soil-covered agar to inhibitory substances, it seems equally likely that reduced nutrient concentrations resulting from microbial colonization of the agar could have been involved (Hsu and Lockwood, 1969). Thus, the most important factors in competitive saprophytic success among the wheat root and foot pathogens are ability to decompose cellulose and efficiency in absorbing nutrients from a limited supply, or possibly tolerance of microbial metabolites.

The importance of the soil microflora in suppressing saprophytic colonization by such fungi is further stressed in experiments in which the effect of temperature was studied. Saprophytic colonization of sterilized bean root pieces by *Fusarium oxysporum* and *Cylindrocarpon radicicola* (Taylor, 1964) and of sterilized wheat root pieces by *Fusarium culmorum*, *Gibberella zeae*, *Cochliobolus spicifer* and *C. sativus* (Burgess and Griffin, 1967) was most successful at temperatures considerably lower than those optimal for growth of the fungi in pure culture. The suppression at high temperatures was attributed to increased activity of the general soil microflora. Balis (1975) has derived a mathematical expression of the experimental model used by Garrett and others in assessing competitive success in gaining access to substrate; the validity of his equation was established by varying inoculum density and various environmental factors as provided by examples in the literature.

Competition between pairs of fungi for substrates in sterilized soil also has been investigated (Lindsey, 1965; Tribe, 1966; Davet, 1976). Such studies seem to me to be much farther removed from ecological reality than those described earlier in which competitive success against the natural microflora of soil is measured. It is unlikely, particularly in soil, that competition would be restricted to two component populations. Davet (1976) studied five different pathogens involved in the corky root disease complex of tomato, focusing on the relative efficiencies with which they colonized sterilized tomato root pieces at 20° and 28°C. The most effective colonizers were *Fusarium oxysporum*, *F. solani*, and *Rhizoctonia solani*, whereas *Colletotrichum coccoides* and *Pyrenochaeta lycopersici* were relatively ineffective. *R. solani*, though an efficient colonizer, was least able of all to exclude other fungi. Differences in colonization efficiencies at the two temperatures were in accord with optimal growth temperatures. Colonization efficiencies were correlated in part, but not entirely, with linear growth rates.

In a study of competition between two *Fusarium* species for an agar substrate in sterilized soil (Lindsey, 1965), *F. roseum* was dominant over *F. solani*, even when at an inoculum density disadvantage. The competitive superiority of *F. roseum* was attributed, in part, to its faster growth rate in sterilized soil in the absence of competition. However, apparently something more was involved, since it was able to suppress growth of *F. solani* in infested soil. *F. solani* was a successful colonist only when it preceded *F. roseum* in soil by 10 or 20 days. Even in this sequence, *F. solani* was strongly suppressed by glucose amendment. Since this inhibitory effect was partially overcome if KNO_3 also was supplied, suppression of *F. solani* by *F. roseum* was ascribed to competition for nitrogen. Subsequent studies in vitro showed that *F. roseum* was able to germinate better in the presence of inorganic nitrogen sources, and that it utilized inorganic nitrogen more efficiently than *F. solani*, particularly at low carbon (glucose) concentrations (Byther, 1965).

In a similar manner, Tribe (1966) determined the relative competitive abilities of six cellulolytic fungi grown in pairs on sterilized cellophane on sterilized sand. The dominant member bore no relation to relative cellulolytic ability. When the six cellulolytic fungi were paired with *Pythium oligandrum*, a noncellulolytic fungus often associated with decomposing cellophane film on soil, the order of growth stimulation again was not related to cellulolytic ability. Other unidentified factors apparently are involved.

In a novel approach to the study of competition, Kelley and Rodriguez-Kabana (1976) employed assays for the enzymes β-D-glucosidase, phosphatase, and trehalase produced in sterilized soils infested with *Phytophthora cinnamomi* together with either *Trichoderma polysporum* or *T. harzianum*. The success of the method is dependent upon identifying some quantitative or qualitative specificity in enzyme production by each member of a competing pair. Where enzyme production was less in mixed than in pure cultures, competition was assumed. However, interpretation can be complicated by such factors as production of proteases capable of enzyme degradation or by adaptive enzymes degrading the assay substrate. In some experiments enzyme production by *P. cinnamomi* was reduced, but in others it was greater than for either fungus alone. Overall, with the methods used, *P. cinnamomi* apparently was able to compete successfully with the *Trichoderma* species.

Another aspect of competitive saprophytic colonization is the persistence of fungi in tissues already colonized successfully, i.e., their resistance to competitive displacement. This subject also has been of particular interest to plant pathologists concerned with saprophytic survival of root pathogens. These pathogens are often able to escape the need to compete with other microorganisms for access to plant residues by prior parasitic invasion of the root tissues and lower stems of plants. Once occupied in this manner, the tissues often become highly resistant to subsequent invasion by other microorganisms, providing a protected niche in which the primary colonizer is able to resist competitive displacement. Such exclusive proprietorship of previously colonized tissues has been shown experimentally for wheat stem segments inoculated with root- and foot-infecting fungi (Bruehl and Lai, 1966) and for pieces of wood, turnip root, and oak leaves which, after inoculation with specific fungi, or after nonspecific colonization in soil, were resistant to invasion by *Pythium mamillatum* (Barton, 1960, 1961). Questions can be raised about the relative persistence of different fungi in the invaded tissues, factors affecting their longevity, and the nature of the restrictive effects against secondary colonizers.

Garrett and his coworkers also have studied this aspect of the saprophytic colonization of wheat stem bases by cereal root and foot pathogens (see Garrett, 1956, 1970, 1976). He has been particularly interested in the effect of nitrogen on persistence, since wheat residues are highly cellulosic and therefore nitrogen may be limiting. Stem pieces were colonized by incubating them in pure culture inocula for 4 weeks, then burying them in soil, and determining the persistence of the fungi at intervals. *Gaeumannomyces graminis* and *Fusarium culmorum* survived for relatively short periods in unsupplemented straw pieces, but longevity was enhanced by supplemental nitrogen. By contrast, *Phialophora radicicola* and *Cochliobolus sativus* survived for a relatively long period in nonsupplemented pieces, but their longevity was reduced by supplemental nitrogen. *Curvularia ramosa* was indifferent to nitrogen.

Unlike competition for access to fresh stem pieces, longevity was unrelated to absolute cellulose decomposition rate. However, when ability to decompose cellulose was expressed in relation to inherent growth rates of the fungi (Garrett's cellulose adequacy index), *G. graminis* and *F. culmorum* were shown to have relatively inefficient cellulolytic ability in relation to their metabolic needs for growth. These fungi apparently cannot obtain enough energy to satisfy their needs if nitrogen is low and so tend to become starved out; supplemental nitrogen increased cellulolytic efficiency. In contrast, *C. ramosa*, *P. radicicola*, and *C. sativus* were able to obtain sufficient energy through cellulolysis even when nitrogen was in short supply. The decreased longevity of *P. radicicola* and *C. sativus* in the presence of supplemental nitrogen was attributed to competitive displacement by other microorganissms. Direct supporting evidence for this interpretation is scant, but in unrelated experiments growth of *C. sativus* was suppressed to a greater extent than was *C. ramosa* (whose longevity was not decreased at high nitrogen) on agar previously colonized by soil microorganisms (Garrett, 1970).

It has been pointed out (Balis, 1975) that competitive colonization of a fresh substrate and longevity within the substrate appear to be regulated by common factors, e.g., low temperature and supplemental nitrogen enhanced both colonizing efficiency and longevity of *G. graminis*, and nitrogen decreased both for *C. sativus* and *P. radicicola*. This conclusion also is suggested by work with *Rhizoctonia solani* showing better colonization of and longer persistence in nitrogen-supplemented buckwheat and oat stem pieces than in nonsupplemented stem pieces (Papavizas and Davey, 1961).

The tomato root pathogens discussed previously (Davet, 1976) also were evaluated for their ability to colonize sterilized root pieces previously colonized by *Fusarium oxysporum*, *F. solani*, or *Rhizoctonia solani* in sterilized soil. The *Fusaria* excluded the other pathogens with the exception of *R. solani*, but *R. solani* was not effective in excluding any of the others.

The nature of the restrictive effect of already occupied tissues upon secondary colonizers is a matter of importance. One would expect that the utilization of residual soluble energy sources by the primary colonizers should be significant in this respect, inasmuch as secondary colonizers would be deprived of the germination stimulus needed as a first step in colonization. This mechanism of exclusion is well documented for soil and living plant leaves (Lockwood, 1977) but apparently has not been shown experimentally for dead plant tissues. There is circumstantial evidence that antibiotic production by *Cephalosporium gramineum* in the occupied wheat tissues may contribute to excluding secondary colonizers (Bruehl, 1975). All fresh cultures of the fungus from nature produce a broad-spectrum antifungal antibiotic, whereas some older cultures lose this ability. The antibiotic-producing isolates survive longer in inoculated straws buried in soil than do the nonproducers. However, the presence of the antibiotic in the wheat straws themselves has not been demonstrated. Barton (1960, 1961) also felt that inhibitory metabolic products produced by primary colonizers of turnip root and wood pieces were responsible for the restrictive effects against secondary colonization by *Pythium mamillatum*. However, impregnation of wood pieces with staled culture media of various fungi did not limit colonization.

Neither was evidence found for nutrient competition: turnip root peices were first allowed to be colonized naturally in soil, then soaked in glucose; the root pieces then were placed adjacent to glass-fiber tapes containing spores of *P. mamillatum* and returned to the soil. Upon recovery, there was no evidence for stimulation of germination of the spores.

V. RESPONSES TO COMPETITIVE STRESS

Resource competition in a heterotrophic community in which energy is limiting does not result in wholesale elimination of fungal species. In soil, for example, there may be scores of species and thousands of individuals in each gram, which is strong evidence that fungi must be well-adapted to survive competitive stress. The survival strategies employed can conveniently be considered in relation to pre- and post-germination exposure to competitive stress. Emphasis will placed on the role of substrate depletion in inducing quiescence of propagules (mycostasis), lysis of fungal hyphae, and formation of resistant survival structures. Much of the evidence is based upon the use of an experimental model in which fungal spores or mycelia are exposed to a nutrient-deprived environment in axenic conditions. Usually, this involves the percolation of a dilute salt solution through a bed of sand on which propagules borne on membrane filters are incubated. By this means, a steepened diffusion gradient away from the spore is generated that mimics the effect of microbial competition for nutrients in the fungus's vicinity (Lockwood, 1977).

A. Exposure to Stress Before Emergence of the Germ Tube

Germination of fungal spores and other fungal propagules and growth of hyphae are characteristically restricted in soils, apart from nutrient-rich microsites wherein germination and growth occur (Lockwood, 1977). This is the well-known *mycostasis* (*fungistasis*) phenomenon. It is absent in sterilized soil, but is restored more or less nonspecifically by reinoculation with various bacteria, fungi, or actinomycetes, either of antibiotic-producing or nonantibiotic-producing capability. The suppression is alleviated by addition of energy-yielding nutrients. All of this is consistent with an etiology based upon substrate deprivation. The imposed quiescence undoubtedly is of survival value to the affected propagules by avoiding the wastage that would occur were germination to take place spontaneously without adequate substrate. This concept was confirmed experimentally by Chinn and Tinline (1964), who found that isolates of *Cochliobolus sativus* capable of precocious germination in soil disappeared whereas those which were not persisted. The same inhibitory phenomenon also occurs in other previously colonized substrates such as plant litter and sometimes on living plant leaves (Lockwood, 1977). Actually, a more correct conception is that of a general microbiostasis in soil and elsewhere, since there is strong evidence for imposed quiescence of bacteria (Brown, 1973; Ko and Chow, 1977) and actinomycetes (Lloyd, 1968) under the same conditions and with the same general characteristics as those affecting the fungi.

Quantative differences in fungal response to nutrient inputs have been detected by "titrating" soil with nutrient increments (Lockwood, 1977). A small input of energy nutrient induced germination of large spores, which also tended to germinate rapidly, whereas larger inputs were required for germination of small spores, which also tended to germinate slowly (Steiner and Lockwood, 1969). These relationships were explained on the basis of a competitive race between fungal spores and other microorganisms for utilization of nutrients. Fungi with short germination times would be advantaged in this respect. Quiescent fungal hyphae in soil also differed in response to added nutrient (Hsu and Lockwood, 1971). Hyphae with rapid early growth rates, which also tended to be of wide diameter, resumed growth with minimal additions of nutrients; these species also tended to have large, rapid germinating spores. Thus, species with large spores tend to germinate rapidly and to exhibit low sensitivity to mycostasis; they give rise to large hyphae whose early growth is rapid and which also exhibit low sensitivity to mycostasis. The opposite is true for species with small spores.

The relation of germination and growth characters such as these to ecological niche has been considered but little. Dix (1967) found some relation between the site of colonization of bean roots by 12 fungi and their sensitivity to mycostasis. Pioneer colonizers of the surfaces of root tips were the least sensitive (most easily stimulated to germinate); secondary colonizers of the root tissues farther back were more sensitive; and those restricted to rhizosperes were the most sensitive of all. This pattern was related to increasing exudation of amino acids from the root tip backwards.

The low germination of fungi on substrata depleted of soluble energy sources through prior colonization is an obvious corollary of the fact that many fungi require exogenous energy sources for germination (Lockwood, 1977). However, many sclerotia and some larger spores can germinate independently of exogenous nutrients in vitro, i.e., in water alone, yet fail to germinate in soil apart from energy-yielding nutrients. The suppressed germination of such propagules appears to be due to enhanced loss of endogenous food reserves through microbial competition, at a rate sufficient to prevent synthesis of the germ tubes. This inhibition can be mimicked qualitatively and quantitatively by the aforementioned model system (Ko and Lockwood, 1967; Hsu and Lockwood, 1973a; Sztejnberg and Blakeman, 1973; Bristow and Lockwood, 1975a).

Conidia of *Cochliobolus victoriae* and *Curvularia lunata* and sclerotia of *Sclerotium cepivorum* each lost about twice as much exudate during incubation on sand undergoing leaching where germination was suppressed as on water-saturated sand (without percolation) where it was not (Bristow and Lockwood, 1975a). Exudates from spores and sclerotia contained nutrients stimulatory to germination that were also rapidly utilized by microorganisms in the soil (Bristow and Lockwood, 1975b). These results have given rise to the concept of soil as an effective "sink" for energy-yielding nutrients, including those in fungal exudates (Lockwood, 1977). Thus, an interaction has evolved by which the fungal propagule can "sense" the competitive status of its environment: under minimal nutrient competition stress, exudation losses are small and endogenous reserves may be spent for germination. However, in the presence of intense competition, loss of spore nutrients is increased to such an extent that germination is restricted, thereby conserving the energy reserves of the propagule, at least in the short run. This system of inhibition can be overridden by inputs of exogenous nutrients, such as from a plant root. Other evidence for the role of nutrient depletion in mycostasis derives from experiments showing that the restoration of mycostasis to reinoculated sterilized soil was correlated with depletion of added glucose (Steiner and Lockwood, 1970).

The percolation system which has provided much of the data for the foregoing argument appears to be a quantitatively valid model for the imposition of a nutrient-stressed environment. Loss of exudate from ^{14}C-labeled conidia of *C. victoriae* was shown to be greater on each of eight different soils on the model system at flow rates more than sufficient to restrict germination to a low level (Sneh and Lockwood, 1975; Filonow and Lockwood, 1979).

Leaves containing large populations of bacteria and yeasts may also restrict the germination of fungal spores by a similar competitive mechanism (Sztejnberg and Blakeman, 1973; Brodie and Blakeman, 1976; Blakeman and Brodie, 1977). Bacterial numbers increased greatly in droplets containing spores of *Botrytis cinerea* on leaves, and bacteria were concentrated around the spores. Exudate from ^{14}C-labeled spores was taken up by the bacteria. Suppression of condial germination of several fungi in vitro and on leaves was highly correlated with ability of the competing organisms to remove amino acids applied with the spores and competing organisms. Mycostasis was interpreted in terms of uptake of exogenous and endogenous nutrients, particularly amino acids, by bacteria.

If propagules in soil or other occupied substrata are subject to heightened losses of reserve nutrients, then one might expect that originally nutrient independent propagules would eventually become nutrient dependent. This transition has been shown to take place in several days or weeks for several such conidia, both in soil and in the model system (Sztejnberg and Blakeman, 1973; Bristow and Lockwood, 1975a). This change was followed, upon further incubation, by progressive loss in viability (Bristow and Lockwood, 1975a). Thus, the cost of short-term protection against precocious germination may be a more rapid depletion of reserves resulting in reduced longevity. Almost nothing is known concerning the time required for these changes in other fungi, or whether some may have the ability to conserve their reserves through decreased permeability of the spore membrane or by blocking the hydrolysis of insoluble substances to soluble ones.

A quiescent fungal propagule may be exposed to increments of energy which are sufficient to initiate the germination process but which may be exhausted before emergence of the germ tube. The germination status of such propagules was investigated by Yoder and Lockwood (1973). Conidia of nutrient-dependent fungi were first

exposed to sterile soil for several hours to initiate germination, then were transferred to natural soil or the percolation model system, and then were transferred
again to sterilized soil to determine any changes in the time required for germination. If the period of exposure to natural soil or the leaching system was but a
few hours, there was no loss of ther germination progress made. However, incubation
on natural soil for several days caused regression proportional to the time of incubation. Spores that were allowed to take up ^{14}C-labeled glucose for a period short
of germination, then incubated on natural soil, lost radioactivity in proportion to
the amount of regression as determined by change in germination time. In some cases,
75-90% of the radioactivity was lost, divided approximately equally between the products of respiration and exudation. Possibly, fungal spores with long germination
times may be able to progress stepwise toward germination by taking advantage of increments of nutrient separated in time--provided the interval between such increments
is not extended unduly. One might wonder if the tendency towards reversion to the
ground state upon exposure to soil may decrease vulnerability of the propagule to
lysis and death; whether or not this is so remains unknown.

Readers interested in further information concerning mycostasis are referred to
a recent review article (Lockwood, 1977).

B. Exposure to Stress After Emergence of the Germ Tube

Once the germ tube is produced, a fungal propagule is in an extremely vulnerable
position, as this delicate, thin-walled structure is subject to lysis with the onset
of nutrient deprivation stress. Lysis of the germ tube often, but not always, results
in extinction of the propagule. Fungi differ in sensitivity to lysis (Lloyd and
Lockwood, 1966; Bumbieris and Lloyd, 1967a), and those with melanized hyphae are, as a
group, more resistant than those with hyaline hyphae (Bloomfield and Alexander, 19-
67). The mycolytic property of soil is removed by sterilization, and its onset may
be delayed by supplementing soil with energy-yielding nutrients (Lloyd and Lockwood,
1966; Bumbieris and Lloyd, 1967b). Lytic activity has been restored to sterilized
soil by reinoculation with actinomycetes (Lloyd and Lockwood, 1966), and bacteria
and actinomycetes have been associated with mycolysis in vitro (Carter and Lockwood,
1957; Mitchell and Alexander, 1963) and in natural soil (Bumbieris and Lloyd, 1967a).
The aforementioned characteristics of mycolysis are similar to those of mycostasis
and are consistent with an autolytic origin based upon nutrient-deprivation.

Evidence that autolysis of fungal hyphae in soil can occur by autolysis, as
opposed to heterolysis, is of three kinds: (1) its occurrence of hyphae separated
from soil by biodegradation-resistant membrane filters with pore sizes small enough
to exclude enzymes of exogenous origin; (2) its occurrence in hyphae incubated on
a substrate undergoing aqueous leaching similar to that used to impose mycostasis;
and (3) the induction of chitinase and β-D-glucosidase activity in lysing hyphae incubated on natural soil or the model system (Ko and Lockwood, 1970). Nutrient stress
imposed by the competitive activity of microorganisms in the vicinity of the hypha

is believed to activate autolytic enzymes within the hypha resulting in autolysis
and death (Ko and Lockwood, 1970). The products of lysing hyphae, and presumably
hyphal exudates also, are readily utilized by microorganisms in soil (Lloyd and Lockwood, 1966).

Autolysis resulting in death would seem to be a character without adaptive value. However, it may be related to the synthesis of resistant structures of some fungi, as will be discussed. Moreover, in some fungi the original spore or sclerotium
may survive after destruction of the germ tube. Sclerotia of *Verticillium albo-atrum*
were shown to survive as many as nine cycles of germination in moist, nutrient-amend-
ed soil, followed by death of germ tubes upon exposure to dry soil conditions (Farley
et al., 1971). Similarly, sclerotia of *Macrophomina phaseolina* were able to regerminate several times during exposure to water agar followed by drying to kill the germ
tubes (Locke and Green, 1977). Some of a population of conidia of *Cochliobolus sativus* or *C. victoriae* were able to regerminate at least five successive times on an
agar medium following lysis of germ tubes on soil (Hsu and Lockwood, 1971). Even
single-celled sporangia of *Pythium ultimum* germinated at least three times in soil
following pulses of low levels of nutrients (Stanghellini and Hancock, 1971). Upon
exhaustion of nutrients, the protoplasm in the germ tube was observed to retract into
the sporangium.

Another response to the presence of microorganisms is the formation of appressoria by fungal germ tubes on plant leaves (Emmett and Parbery, 1975; Lenne and Parbery, 1976; Blakeman and Brodie, 1977). For example, a *Bacillus* sp. induced lysis
of germ tubes of germinating conidia of *Colletotrichum gloeosporioides* on citrus
leaves, but appressorium numbers were greatly increased and they did not lyse (Lenne
and Parbery, 1976). Though the mechanism was not elucidated, nutrient competition
is at least a likely possiblity.

The formation of some asexually derived resistant structures, such as chlamydospores and sclerotia, often occurs when mycelium has exhausted its substrate, or upon
transfer of mycelium, or in some cases conidia, to soil (Nash et al., 1961; Christias and Lockwood, 1973). This morphogenetic behavior can be mimicked by incubating
spores or fungal hyphae in media deficient or lacking in energy substrates such as
soil extracts (Ford et al., 1970), salt solutions or distilled water (Qureshi and
Page, 1970; Hsu and Lockwood, 1973b), or water-saturated glass beads substrata (Christias and Lockwood, 1973). The degree of nutrient stress required to promote formation of resistant structures appears to be less than that required for mycostasis
and lysis, since exposure to a steepened nutrient depletion gradient was not required
nor did it enhance their formation (Christias and Lockwood, 1973; Hsu and Lockwood,
1973b).

The formation of these resistant structures, both on soil and in substrate-de-
ficient conditions in vitro, is often associated with lysis of hyphae, sometimes of
massive proportions (Christias and Lockwood, 1973), and suggests that lysis may provide a means of recycling mycelial constituents for synthesis of resistant survival

structures when the fungus is confronted by a depleted environment, either through its own growth or through resource competition by other microorganisms. This possibility was verified experimentally by incubating mycelia of four sclerotium-forming fungi on sterile water-saturated glass beads which excludes factors other than substrate depletion. Sclerotial initials appeared within 24 hr, and sclerotia were apparently mature within a week, during which time the mycelia had nearly completely lysed. Dry weight of sclerotia was some 40-60% of that of the original mycelium. Results in soil were in quantitative and qualititative agreement. In view of the high efficiency of conversion, it seems likely that some of the materials utilized in sclerotium formation must have been derived from cell wall materials made available through autolysis.

The formation of chlamydospores from macroconidia of *Fusarium solani* f. sp. *cucurbitae* also involves autolysis, according to work of Schippers and coworkers. Treatments that suppressed chlamydospore formation such as chitin in soil (Schippers and de Weyer, 1972), and other nitrogen sources in soil (Schippers, 1972) and in vitro (Schippers and Old, 1974), also reduced the rate of autolysis of the macroconidia. Ultrastructural studies of chlamydospore formation in vitro (van Eck and Schippers, 1976) and in soil (van Eck, 1976) showed that degradation of the conidial wall accompanied synthesis of the new wall of the chlamydospore. Since lysis in vitro could only have been autolytic, the same changes in soil were presumed to be of autolytic origin. Moreover, the degradation of the conidial wall occurring in soil was not visibly associated with microbial activity (Old and Schippers, 1973).

To the present writer, the evidence is strong for the causal role of substrate depletion, through microbial competition for energy resources, in imposition of exogenous dormancy, induction of lysis, and formation of resistant structures. The experimental evidence discussed is consistent with the growing body of evidence pointing to chronic substrate deficiency, at least in soil. However, this is not to say that other factors of the environment are not responsible for the same behavior in particular circumstances. Though it is beyond the scope of this review to give detailed consideration to these alternatives, it should be mentioned that mycostasis also has been attributed to volatile and nonvolatile inhibitory substances in soil (see Lockwood, 1977; also Watson and Ford, 1972), lysis has been ascribed to cell-wall-decomposing enzymes from other microorganisms (Alexander, 1971), and the formation of resistant structures has been attributed to morphogenetic substances of microbial origin in soil (Ford et al., 1970). It is the view of this reviewer that such factors, where they occur, are superimposed upon the more pervasive substrate deficiency caused by microbial competition for substrate.

REFERENCES

Ainsworth, G. C., and Sussman, A. S. (Eds.) (1965). *The Fungi: An Advanced Treatise*, Vol. I: *The Fungal Cell*. Academic Press, New York.

Ainsworth, G. C., and Sussman, A. S. (Eds.) (1968). *The Fungi: An Advanced Treatise*, Vol. III: *The Fungal Population*. Academic Press, New York.

Alexander, M. (1971). *Microbial Ecology*. Wiley, New York.

Alexander, M. (1977). *Introduction to Soil Microbiology*, 2nd ed. Wiley, New York.

Armstrong, R. A. (1976). Fugitive species: Experiments with fungi and some theoretical considerations. *Ecology 57*: 953-963.

Babiuk, L. A., and Paul, E. A. (1970). The use of fluorescein isothiocyanate in the determination of bacterial biomass of grassland soil. *Can. J. Microbiol. 16*: 57-62.

Baker, K. F., and Cook, R. J. (1974). *Biological Control of Plant Pathogens*. Freeman, San Francisco.

Balis, C. (1975). A theoretical approach to competitive saprophytic colonization of substrates as a function of inoculum density. *Ann. Inst. Phytopathol. Benaki* [*N.S.*] *11*: 73-93.

Barton, R. (1960). Antagonism amongst some sugar fungi. In *The Ecology of Soil Fungi*, D. Parkinson and J. S. Waid (Eds). Liverpool Univ. Press, Liverpool, England, pp. 160-167.

Barton, R. (1961). Saprophytic activity of *Pythium mamillatum* in soils. II. Factors restricting *P. mamillatum* to pioneer colonization of substrates. *Trans. Brit. Mycol. Soc. 44*: 105-118.

Blakeman, J. P., and Brodie, I. D. S. (1977). Competition for nutrients between epiphytic microorganisms and germination of spores of plant pathogens on beetroot leaves. *Physiol. Plant Pathol. 10*: 29-42.

Bloomfield, B. H., and Alexander, M. (1967). Melanins and resistance of fungi to lysis. *J. Bacteriol. 93*: 1276-1280.

Bowen, G. D., and Rovira, A. D. (1973). Are modelling approaches useful in rhizosphere biology? In *Modern Methods in the Study of Microbial Ecology*, T. Rosswall (Ed.), Bull. 17. Swedish Natural Science Research Council (N.F.R.), Stockholm, pp. 443-450.

Bowen, G. D., and Rovira, A. D. (1976). Microbial colonization of plant roots. *Ann. Rev. Phytopathol. 14*: 121-144.

Bowen, G. D., and Theodorou, C. (1973). Growth of ectomycorrhizal fungi around seeds and roots. In *Ecology and Physiology of Ectomycorrhizae*, G. C. Marks and T. T. Kozlowski (Eds.). Academic Press, New York, pp. 107-150.

Bristow, P. B., and Lockwood, J. L. (1975a). Soil fungistasis: Role of spore exudates in the inhibition of nutrient-independent propagules. *J. Gen. Microbiol. 90*: 140-146.

Bristow, P. B., and Lockwood, J. L. (1975b). Soil fungistasis: Role of the microbial nutrient sink and of fungistatic substances in two soils. *J. Gen. Microbiol. 90*: 147-156.

Brodie, I. D. S., and Blakeman, J. P. (1976). Competition for exogenous substrates in vitro by leaf surface micro-organisms and germination of conidia of *Botrytis cinerea*. *Physiol. Plant Pathol. 9*: 227-239.

Brown, M. E. (1973). Soil bacteriostasis limitation in growth of soil and rhizosphere bacteria. *Can J. Microbiol. 19*: 195-199.

Bruehl, G. W. (1975). Systems and mechanisms of residue possession by pioneer fungal colonists. In *Biology and Control of Soil-Borne Plant Pathogens*, G. W. Bruehl (Ed.). American Phytopathology Society, St. Paul, Minn., pp. 77-83.

Bruehl, G. W., and Lai, P. (1966). Prior-colonization as a factor in the saprophytic survival of several fungi in wheat straw. *Phytopathology 56*: 766-768.

Buczacki, S. T. (1973). A microecological approach to larch canker biology. *Trans. Brit. Mycol. Soc. 61*: 315-329.

Bumbieris, M., and Lloyd, A. B. (1967a). Influence of soil fertility and moisture on lysis of fungal hyphae. *Aust. J. Biol. Sci. 20*: 103-112.

Bumbieris, M., and Lloyd, A. B. (1967b). Influence of nutrients on lysis of fungal hyphae in soil. *Aust. J. Biol. Sci. 20*: 1169-1172.

Burges, A. (1967). The decomposition of organic matter in the soil. In *Soil Biology*, A. Burges and F. Raw (Eds.). Academic Press, New York, pp. 479-492.

Burges, A., and Nicholas, D. P. (1961). Use of soil sections in studying amount of fungal hyphae in soil. *Soil Sci. 92*: 25-29.

Burgess, L. W., and Griffin, D. M. (1967). Competitive saprophytic colonization of wheat straw. *Ann. Appl. Biol. 60*: 137-142.

Butler, F. C. (1953). Saprophytic behaviour of some cereal root-rot fungi. II. Factors influencing saprophytic colonization of wheat straw. *Ann. Appl. Biol. 40*: 298-304.

Byther, R. (1965). Ecology of plant pathogens in soil. V. Inorganic nitrogen utilization as a factor of competitive saprophytic ability by *Fusarium roseum* and *F. solani*. *Phytopathology 55*: 852-858.

Campbell, R., and Rovira, A. D. (1973). The study of the rhizosphere by scanning electron microscopy. *Soil Biol. Biochem. 5*: 747-752.

Carter, H. P., and Lockwood, J. L. (1957). Lysis of fungi by soil microorganisms and fungicides including antibiotics. *Phytopathology 47*: 154-158.

Carter, M. V. (1971). Biological control of *Eutypa armeniacae*. *Aust. J. Expt. Agr. Anim. Husb. 11*: 687-692.

Carter, M. V., and Price, T. V. (1974). Biological control of *Eutypa armeniacae*: Studies of the interaction between *E. armeniacae* and *Fusarium lateritium*, and their relative sensitivities to benzimadazole chemicals. *Aust. J. Agr. 25*: 105-119.

Chang, Y. (1967). The fungi of wheat straw compost. II. Biochemical and physiological studies. *Trans. Brit. Mycol. Soc. 50*: 667-677.

Chinn, S. H. F., and Tinline, R. D. (1964). Inherent germinability and survival of spores of *Cochliobolus sativus*. *Phytopathology 54*: 349-352.

Christias, C., and Lockwood, J. L. (1973). Conservation of mycelial constituents in four sclerotium-forming fungi in nutrient-deprived conditions. *Phytopathology 63*: 602-605.

Clark, F. E. (1965). The concept of competition in microbiology. In *Ecology of Soil-Borne Plant Pathogens*, K. F. Baker and W. C. Snyder (Eds.). Univ. of California Press, Berkeley, pp. 339-347.

Clark, F. E. (1967). Bacteria in soil. In *Soil Biology*, A. Burges and F. Raw (Eds.). Academic Press, New York, pp. 15-49.

Clark, F. E., and Paul, E. A. (1970). The microflora of grassland. *Adv. Agron. 22*: 375-435.

Clements, F. E., and Shelford, V. E. (1939). *Bio-ecology*. Wiley, New York.

Cook, R. J. (1970). Factors affecting saprophytic colonization of wheat straw by *Fusarium roseum* f. sp. *cerealis* "Culmorum." *Phytopathology 60*: 1672-1676.

Cook, R. J., and Papendick, R. I. (1970). Soil water potential as a factor in the ecology of *Fusarium roseum* f. sp. *cerealis* "Culmorum." *Plant and Soil 32*: 131-145.

Davet, P. (1976). Étude de quelques interactions entre les champignons associés a la maladie des racines liégeuses de la tomate. I. Phase non parasitaire. *Ann. Phytopathol. 8*: 171-183.

Dickinson, C. H. (1967). Fungal colonization of *Pisum* leaves. *Can. J. Bot. 45*: 915-927.

Dickinson, C. H., and Pugh, G. J. K. (Eds.) (1974). *Biology of Plant Litter Decomposition*, Vol. I. Academic Press, New York.

Dickinson, C. H., and Skidmore, A. M. (1976). Interactions between germinating spores of *Septoria nodorum* and phylloplane fungi. *Trans. Brit. Mycol. Soc. 66*: 45-51.

Dix, N. J. (1967). Mycostasis and root exudation: factors influencing the colonization of bean roots by fungi. *Trans. Brit. Mycol. Soc. 50*: 23-31.

Emmett, R. W., and Parbery, D. G. (1975). Appressoria. *Ann. Rev. Phytopathol. 13*: 147-167.

Erwin, D. C., and Katznelson, H. (1961). Suppression and stimulation of mycelial growth of *Phytophthora cryptogea* by certain thiamine-requiring and thiamine-synthesizing bacteria. *Can. J. Microbiol. 7*: 945-950.

Farley, J. D., Wilhelm, S., and Snyder, W. C. (1971). Repeated germination and sporulation of *Verticillium albo-atrum* in soil. *Phytopathology 61*: 260-264.

Filonow, A. B., and Lockwood, J. L. (1979). Conidial exudation by *Cochliobolus victoriae* on soils in relation to soil mycostasis. In *Soil-borne Plant Pathogens*, B. Schippers and W. Gams (Eds.). Academic Press, New York, pp. 107-119.

Finstein, M. S., and Alexander, M. (1962). Competition for carbon and nitrogen between *Fusarium* and bacteria. *Soil Sci. 94*: 334-339.

Fokkema, N. J. (1971a). Influence of pollen on saprophytic and pathogenic fungi on rye leaves. In *Ecology of Leaf Surface Micro-organisms*, T. F. Preece and C. H. Dickinson (Eds.). Academic Press, New York, pp. 277-282.

Fokkema, N. J. (1971b). The effect of pollen in the phyllosphere of rye on colonization by saprophytic fungi and on infection by *Helminthosporium sativum* and other leaf pathogens. *Neth. J. Plant Pathol. 77*: Suppl. No. 1.

Fokkema, N. J. (1973). The role of saprophytic fungi in antagonism against *Drechslera sorokiniana* (*Helminthosporium sativum*) on agar plates and on rye leaves with pollen. *Physiol. Plant Pathol. 3*: 195-205.

Fokkema, N. J., and Lorbeer, J. W. (1974). Interactions between *Alternaria porri* and the saprophytic mycoflora on onion leaves. *Phytopathology 64*: 1128-1133.

Ford, E. J., Gold, A. H., and Snyder, W. C. (1970). Induction of chlamydospore formation in *Fusarium solani* by soil bacteria. *Phytopathology 60*: 479-484.

Foster, R. C., and Rovira, A. D. (1973). The rhizosphere of wheat roots studied by electron microscopy of ultra-thin sections. In *Modern Methods in the Study of Microbial Ecology*, T. Rosswall (Ed.), Bull. 17. Swedish Natural Science Research Council (N.F.R.), Stockholm, pp. 93-102.

Frankland, J. C. (1969). Fungal decomposition of bracken petioles. *J. Ecol. 57*: 25-36.

Frankland, J. C. (1974). Decomposition of lower plants. In *Biology of Plant Litter Decomposition*, C. H. Dickinson and G. J. F. Pugh (Eds.), Vol. I. Academic Press, New York, pp. 3-36.

Frankland, J. C. (1975). Estimation of live fungal biomass. *Soil Biol. Biochem. 7*: 339-340.

Gadgil, R. L., and Gadgil, P. D. (1975). Suppression of litter decomposition by mycorrhizal roots of *Pinus radiata*. *N.Z. J. For. Sci. 5*: 33-41.

Garrett, S. D. (1956). *Biology of Root-Infecting Fungi*. Cambridge Univ. Press, New York.

Garrett, S. D. (1970). *Pathogenic Root-Infecting Fungi*. Cambridge Univ. Press, New York.

Garrett, S. D. (1972). Factors affecting saprophytic survival of six species of cereal foot-rot fungi. *Trans. Brit. Mycol. Soc. 59*: 445-452.

Garrett, S. D. (1975). Cellulolysis rate and competitive saprophytic colonization of wheat straw by foot-rot fungi. *Soil Biol. Biochem. 7*: 323-327.

Garrett, S. D. (1976). Influence of nitrogen on cellulolysis rate and saprophytic survival in soil of some cereal foot-rot fungi. *Soil Biol. Biochem. 8*: 229-234.

Gray, T. R. G., and Williams, S. T. (1971). Microbial productivity in soil. In *Microbes and Biological Productivity*, D. E. Hughes and A. H. Rose (Eds.). Cambridge Univ. Press, New York, pp. 255-286.

Gray, T. R. G., Hissett, R., and Duxbury, T. (1974). Bacterial populations of litter and soil in a deciduous woodland. II. Numbers, biomass and growth rates. *Rev. Écol. Biol. Sol. 11*: 15-26.

Holm, E., and Jensen, V. (1972). Aerobic chemoorganotrophic bacteria of a Danish beech forest: Microbiology of a Danish beech forest. I. *Oikos 23*: 248-260.

Hsu, S. C., and Lockwood, J. L. (1969). Mechanisms of inhibition of fungi in agar by streptomycetes. *J. Gen. Microbiol. 57*: 149-158.

Hsu, S. C., and Lockwood, J. L. (1971). Responses of fungal hyphae to soil fungistasis. *Phytopathology 61*: 1355-1362.

Hsu, S. C., and Lockwood, J. L. (1973a). Soil fungistasis: Behavior of nutrient-independent spores and sclerotia in a model system. *Phytopathology 63*: 334-337.

Hsu, S. C., and Lockwood, J. L. (1973b). Chlamydospore formation in *Fusarium* in sterile salt solutions. *Phytopathology 63*: 597-602.

Hudson, H. J. (1968). The ecology of fungi on plant remains above the soil. *New Phytol. 67*: 837-874.

Hudson, H. J. (1971). The development of the saprophytic fungal flora as leaves senesce and fall. In *Ecology of Leaf Surface Micro-organisms*, T. F. Preece and C. H. Dickinson (Eds.). Academic Press, New York, pp. 447-455.

Hutchinson, G. E. (1961). The paradox of the plankton. *Amer. Natur. 95*: 137-145.

Ikediugwu, F. E. O., Dennis, C., and Webster, J. (1970). Hyphal interference by *Peniophora gigantea* against *Heterobasidium annosum*. *Trans. Brit. Mycol. Soc. 54*: 307-309.

Katsuya, K., and Green, G. J. (1967). Reproduction potentials of races 15B and 56 of wheat stem rust. *Can. J. Bot. 45*: 1077-1091.

Kelley, W. D., and Rodriguez-Kabana, R. (1976). Competition between *Phytophthora cinnamoni* and *Trichoderma* spp. in autoclaved soil. *Can. J. Microbiol. 22*: 1120-1127.

Ko, W. H., and Chow, F. K. (1977). Characteristics of bacteriostasis in natural soils. *J. Gen. Microbiol. 102*: 295-298.

Ko, W. H., and Lockwood, J. L. (1967). Soil fungistasis: relation to fungal spore nutrition. *Phytopathology 57*: 894-901.

Ko, W. H., and Lockwood, J. L. (1970). Mechanism of lysis of fungal mycelia in soil. *Phytopathology 60*: 148-154.

Kong, K. T., and Dommergues, Y. (1971). Limitation de la cellulolyse dans les sols organiques. III. Compétition entre la microflore celluloytique et la microflore non cellulolytique dans les sols organiques. *Rev. Écol. Biol. Sol. 10*: 45-53.

Langton, F. A. (1969). Interactions of the tomato with two formae speciales of *Fusarium oxysporum*. *Ann. Appl. Biol. 62*: 413-427.

Last, F. T., and Deighton, F. C. (1965). The non-parasitic microflora on the surfaces of living leaves. *Trans. Brit. Mycol. Soc. 48*: 83-99.

Leben, C. (1965). Epiphytic microorganisms in relation to plant disease. *Ann. Rev. Phytopathol. 3*: 209-230.

Leben, C. (1969). Colonization of soybean buds by bacteria: Observations with the scanning electron microscope. *Can J. Microbiol. 15*: 319-320.

Leben, C., and Daft, G. C. (1967). Population variations of epiphytic bacteria. *Can. J. Microbiol. 13*: 1151-1156.

Lehmann, P. F., and Hudson, H. J. (1977). The fungal succession on normal and urea-treated pine needles. *Trans. Brit. Mycol. Soc. 68*: 221-228.

Lenne, J. M., and Parbery, D. G. (1976). Phyllosphere antagonists and appressorium formation in *Colletotrichun gloesporioides*. *Trans Brit. Mycol. Soc. 66*: 334-336.

Levi, M. P., and Cowling, E. B. (1969). Role of nitrogen in wood deterioration. VII. Physiological adaptation of wood-destroying fungi to substrates deficient in nitrogen. *Phytopathology 59*: 460-468.

Levi, M. P., Merrill, W., and Cowling, E. B. (1968). Role of nitrogen in wood deterioration. VI. Mycelial fractions and model nitrogen compounds as substrates for growth of *Polyporus versicolor* and other wood-destroying and wood-inhabiting fungi. *Phytopathology 58*: 626-634.

Lindsey, D. L. (1965). Ecology of plant pathogens in soil. III. Competition between soil fungi. *Phytopathology 55*: 104-110.

Lloyd, A. B. (1968). Behaviour of streptomycetes in soil. *J. Gen. Microbiol. 56*: 165-170.

Lloyd, A. B. and Lockwood, J. L. (1966). Lysis of fungal hyphae in soil and its possible relation to autolysis. *Phytopathology 56*: 595-602.

Locke, J. C., and Green, R. J., Jr. (1977). The effect of various factors on germination of sclerotia of the charcoal rot fungus, *Macrophomina phaseolina*. *Proc. Amer. Phytopathol. Soc. 97* (Abstr. No. 68).

Lockwood, J. L. (1977). Fungistasis in soils. *Biol. Rev. 52*: 1-43.

Loegering, W. Q. (1951). Survival of races of wheat stem rust in mixtures. *Phytopathology 41*: 56-65.

Lowe, W. E., and Gray, T. R. G. (1973). Ecological studies on coccoid bacteria in a pine forest soil. III. Competitive interactions between bacterial strains in soil. *Soil Biol. Biochem. 5*: 463-472.

McNaughton, S. J., and Wolf, L. L. (1973). *General Ecology*. Holt, Rinehart and Winston, New York.

McQueen, D. J. (1971). Effects of continous competition in two species of cellular slime mold: *Dictyostelium discoideum* and *Polysphondylium pallidum*. *Can. J. Zool. 49*: 1305-1315.

Marchant, R. (1970). The root surface of *Amophila arenaria* as a substrate for microorganisms. *Trans. Brit. Mycol. Soc. 54*: 479-506.

Marshall, K. C., and Alexander, M. (1960). Competition between soil bacteria and *Fusarium*. *Plant and Soil 12*: 143-153.

Maurer, C. L., and Baker, R. (1965). Ecology of plant pathogens in soil. II. Influence of glucose, cellulose, and inorganic nitrogen amendments on development of bean root rot. *Phytopathology 55*: 69-72.

Merrill, W., and Cowling, E. B. (1966). Role of nitrogen in wood deterioration: Amount and distribution of nitrogen in fungi. *Phytopathology 56*: 1083-1090.

Meyer, J. A., and Maraite, H. (1971). Multiple infection and symptom mitigation in vascular wilt diseases. *Trans. Brit. Mycol. Soc. 57*: 371-377.

Milne, A. (1961). Definition of competition among animals. In *Mechanisms in Biological Competition*, F. L. Milthorpe (Ed.). Cambridge Univ. Press, New York, pp. 40-61.

Milthorpe, F. L. (Ed.) (1961). *Mechanisms of Biological Competition*. Cambridge Univ. Press, New York.

Minderman, G., and Daniels, L. (1967). Colonization of newly fallen leaves by microorganisms. In *Progress in Soil Biology*, O. Graff and J. E. Satchell (Eds.). North-Holland Publ., Amsterdam, pp. 3-7.

Mitchell, R., and Alexander, M. (1963). Lysis of soil fungi by bacteria. *Can. J. Microbiol. 9*: 169-177.

Nagel-de Boois, H. M., and Jansen, E. (1971). The growth of fungal mycelia in forest soil layers. *Rev. Écol. Biol. Sol. 8*: 509-520.

Nash, S. M., Christou, T. and Snyder, W. C. (1961). Existence of *Fusarium solani* f. *phaseoli* as chlamydospores in soil. *Phytopathology 51:* 308-312.

Ogle, H. J., and Brown, J. F. (1971). Some factors affecting the relative ability of two strains of *Puccinia graminis tritici* to survive when mixed. *Ann. Appl. Biol. 67:* 157-168.

Old, K. M., and Nicolson, T. H. (1975). Electron microscopical studies of the micro-flora of roots of sand dune grasses. *New Phytol. 74:* 51-58.

Old, K. M., and Schippers, B. (1973). Electron microscopical studies of chlamydo-spores of *Fusarium solani* f. *cucurbitae* formed in natural soil. *Soil Biol. Biochem. 5:* 613-620.

Papavizas, G. C., and Davey, D. B. (1961). Saprophytic behavior of *Rhizoctonia* in soil. *Phytopathology 51:* 693-699.

Park, D. (1976a). Cellulose decomposition by a pythiaceous fungus. *Trans. Brit. Mycol. Soc. 66:* 65-70.

Park, D. (1976b). Nitrogen level and cellulose decomposition by fungi. *Intern. Biodet. Bull. 12:* 95-99.

Parkinson, D., and Crouch, R. (1969). Studies on fungi in a pinewood soil. V. Root mycoflora of seedlings of *Pinus nigra* var. *laricio. Rev. Écol. Biol. Sol. 6:* 263-275.

Parkinson, D., and Thomas, A. (1969). Quantitative study of fungi in the rhizo-sphere. *Can. J. Microbiol. 15:* 875-878.

Parkinson, D., Balasooriya, I., and Winterholder, K. (1968). Studies on fungi in a pinewood soil. III. Fungal growth and total microbial activity. *Rev. Écol. Biol. Sol. 5:* 637-645.

Pielou, E. C. (1974). *Population and Community Ecology: Principles and Methods.* Gordon & Breach Sci. Publ., New York.

Preece, T. F., and Dickinson, C. H. (Eds.) (1971). *Ecology of Leaf Surface Micro-organisms.* Academic Press, New York.

Qureshi, A. A., and Page, O. T. (1970). Observations on chlamydospore formation by *Fusarium* in a two-salt solution. *Can J. Microbiol. 16:* 29-32.

Ribeiro, O. K., Zentmyer, G. A., and Erwin, D. C. (1975). Comparative effects of monochromatic radiation on the germination of spores of three *Phytophthora* spp. *Phytopathology 65:* 904-907.

Ricklefs, R. E. (1973). *Ecology.* Chiron Press, Newton, Mass.

Rishbeth, J. (1963). Stump protection against *Fomes annosus.* III. Inoculation with *Peniophora gigantea. Ann. Appl. Biol. 52:* 63-77.

Rodenhiser, H. A., and Holton, C. S. (1953). Differential survival and natural hy-bridization in mixed spore populations of *Tilletia caries* and *T. foetida. Phytopath-ology 43:* 558-560.

Roncadori, R. W. (1962). The nutritional competition between *Hypoxylon punctulatum* and *Ceratocystis fagacearum. Phytopathology 52:* 498-502.

Rovira, A. D., Newman, E. I., Bowen, H. J. and Campbell, R. (1974). Quantitative assessment of the rhizosphere microflora by direct microscopy. *Soil Biol. Biochem. 6:* 211-216.

Ruinen, J. (1961). The phyllosphere. I. An ecologically neglected milieu. *Plant and Soil 15:* 81-109.

Saitô, T. (1965). Coactions between litter-decomposing Hymenomycetes and their assoc-iated microorganisms during decomposition of beech litter. *Sci. Rep. Tôhoku Univ.* [4] *31:* 255-273.

Schippers, B. (1972). Reduced chlamydospore formation and lysis of macroconidia of *Fusarium solani* f. *cucurbitae* in nitrogen-amended soil. *Neth. J. Plant Pathol. 78:* 189-197.

Schippers, B., and deWeyer, W. M. M. M. (1972). Chlamydospore formation and lysis of macroconidia of *Fusarium solani* f. *cucurbitae* in chitin-amended soil. *Neth. J. Plant Pathol. 78*: 45-54.

Schippers, B., and Old, K. M. (1974). Factors affecting chlamydospore formation by *Fusarium solani* f. *cucurbitae* in pure culture. *Soil Biol. Biochem. 6*: 153-160.

Shields, J. A., Paul, E. A., Lowe, W. E., and Parkinson, D. (1973). Turnover of microbial tissue in soil under field conditions. *Soil Biol. Biochem. 5*: 753-764.

Sneh, B., and Lockwood, J. L. (1975). Quantitative evaluation of the microbial nutrient sink in soil in relation to a model system for soil fungistasis. *Soil Biol. Biochem. 8*: 65-69.

Snyder, W. C., Schroth, M. N., and Christou, T. (1959). Effect of plant residues on root rot of bean. *Phytopathology 49*: 755-756.

Spurr, H. W., Jr. (1977). Protective applications of conidia of nonpathogenic *Alternaria* sp. isolates for control of tobacco brown spot disease. *Phytopathology 67*: 128-132.

Stanghellini, M. E., and Hancock, J. G. (1971). The sporangium of *Pythium ultimum* as a survival structure in soil. *Phytopathology 61*: 157-164.

Steiner, G. W., and Lockwood, J. L. (1969). Soil fungistasis: Sensitivity of spores in relation to germination time and size. *Phytopathology 59*: 1084-1092.

Steiner, G. W., and Lockwood, J. L. (1970). Soil fungistasis: Mechanism in sterilized, reinoculated soil. *Phytopathology 60*: 89-91.

Stotzky, G., and Norman, A. G. (1961a). Factors limiting microbial activities in soil. I. The level of substrate, nitrogen, and phosphorus. *Arch. Mikrobiol. 40*: 341-369.

Stotzky, G., and Norman, A. G. (1961b). Factors limiting microbial activities in soil. II. The effect of sulfur. *Arch. Mikrobiol. 40*: 370-382.

Sztejnberg, A., and Blakeman, J. P. (1973). Studies on leaching of *Botrytis cinerea* conidia and dye absorption of bacteria in relation to competition for nutrients on leaves. *J. Gen. Microbiol. 78*: 15-22.

Taylor, G. S. (1964). *Fusarium oxysporum* and *Cylindrocarpon radicicola* in relation to their association with plant roots. *Trans. Brit. Mycol. Soc. 47*: 381-391.

Thurston, H. D. (1961). The relative survival ability of races of *Phytophthora infestans* in mixtures. *Phytopathology 51*: 748-755.

Trenbath, B. R. (1977). Interactions among diverse hosts and diverse parasites. In *The Genetic Basis of Epidemics in Agriculture*, P. R. Day (Ed.). New York Acad. Sci., New York, pp. 124-150.

Tribe, H. T. (1966). Interactions of soil fungi on cellulose film. *Trans. Brit. Mycol. Soc. 49*: 457-466.

van den Huevel, J. (1971). Antagonism between pathogenic and saprophytic *Alternaria* species on bean leaves. In *Ecology of Leaf Surface Micro-organisms*, T. F. Preece and C. H. Dickinson (Eds.). Academic Press, New York, pp. 537-544.

van Eck, W. H. (1976). Ultrastructure of forming and dormant chlamydospores of *Fusarium solani* in soil. *Can. J. Microbiol. 22*: 1634-1642.

van Eck, W. H., and Schippers, B. (1976). Ultrastructure of developing chlamydospores of *Fusarium solani* f. *cucurbitae* in vitro. *Soil Biol. Biochem. 8*: 1-6.

Warcup, J. H., and Baker, K. F. (1963). Occurrence of dormant ascospores in soil. *Nature 197*: 1317-1318.

Warren, R. C. (1972). Interference by common leaf saprophytic fungi with the development of *Phoma betae* lesions on sugarbeet leaves. *Ann. Appl. Biol. 72*: 137-144.

Watson, A. G., and Ford, E. J. (1972). Soil fungistasis: A reappraisal. *Ann. Rev. Phytopathol. 10*: 327-348.

Watson, I. A. (1942). The development of physiological races of *Puccinia graminis tritici* singly and in association with others. *Proc. Linnean Soc. New South Wales* *67*: 294-312.

Yamanaka, S. (1974). Studies on competition among the isolates of rice blast fungus, *Pyricularia oryzae* Cavara. I. Competition on leaf lesions. *Tòhoku J. Agr. Res.* *25*: 125-129.

Yoder, D. L., and Lockwood, J. L. (1973). Fungal spore germination on natural and sterile soil. *J. Gen. Microbiol. 74*: 107-117.

Chapter 19

INTERFERENCE COMPETITION AND THE ORGANIZATION
OF FUNGAL COMMUNITIES

Donald T. Wicklow

Northern Regional Research Center
U.S. Department of Agriculture
Peoria, Illinois

I. INTRODUCTION

There is a good deal of published evidence to support the view that most fungi do
not occur throughout the entire range of environmental conditions that they are genetically capable of occupying (the *fundamental niche*). The actual portion of the
range an organism occupies in nature constitutes the *realized niche* (Hutchinson,
1957). When different fungi share parts of their fundamental niches (Fig. 1), it
may result in *niche overlap*, a condition in which there is simultaneous demand upon
the same resource by two populations. Interspecific competition occurs when the resource is in insufficient supply to meet the demands of both species. The major
forms of competition in fungi are defined by the types of interactions between competing species. In *exploitation competition* (Lockwood, Chap. 18) an individual depletes the resource but does not reduce the probability that another individual can
exploit the remaining resource pool. In other words, competition for the remaining
resource is unaffected. McNaughton and Wolf (1973) suggest that exploitation competitors tend to utilize resources that are potentially unpredictable in time and
space. They argue that this unpredictability makes the value of interference behavior less certain. Instead, reproductive success from searching for more resources

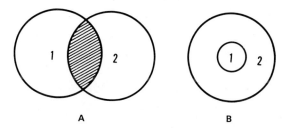

                    A   B

Fig. 1 Diagrammatic representation of niche overlap between competing populations:
(A) competition occurs in overlap portion of niche by reinvasions from the exclusive
subniches of each population; (B) population 1 has a smaller niche but survives by
its ability to outcompete the larger-niched species (population 2) on critical overlap parameters.

or utilization of located resources as quickly as possible may exceed any reproductive gains from the time and energy expended in interference behavior.

Contest for a resource, or *interference competition*, usually involves some form of behavioral or chemical interaction between individuals prior to actual use of the resource. This interaction thus influences the access of a competitor to the resource. An individual that reduces the impact of a competitor on a limited resource increases thereby its own probability of survival and reproduction. For example, in animals, territorial behavior restricts competitor access to a resource in a spatially defined area (Brown, 1964; Miller, 1967). Plants have also evolved mechanisms to ensure that their resources, which are fixed in space, are not exploited by all competitors. By producing chemicals that inhibit the growth of other individuals in the surrounding area, plants are able to maintain distinct zones in excess of their actual area of occupation (Muller, 1966; Whittaker, 1970; Rice, 1974). The chemical substance(s) may be detrimental to individuals of the same species (*autotoxic effects*), as a mechanism for limiting population size, as well as to individuals of other species (*allelopathy*, a form of antibiosis).

Antibiosis is generally recognized as the principal mechanism of interference competition by which fungi may exclude other organisms from resources potentially available to each of the competitors (Brian, 1957; Park, 1967). Resources allocated to territorial defense (antibiotic production) help to ensure that certain initial substrate resources are never so reduced by competitors that the fungal population can no longer survive or sporulate.

Many microbial ecologists believe that antibiotics are active in nature (Brian, 1957; Brock, 1966; Alexander, 1971; Baker and Cook, 1974). Some of the strongest evidence upon which this conclusion is based includes the following, as summarized by Brock (1966): (1) antibiotics can be extracted from nonsterile soil; (2) antibiotic production will occur in nonsterile soil inoculated with antibiotic-producing organisms and supplemented with organic enrichments such as straw; (3) antibiotic-producing microbes can be isolated easily from natural soils; (4) antibiotic-producing organisms are able, under natural conditions, to antagonize other organisms sensitive to antibiotics; (5) conditions unfavorable for the accumulation of antibiotics reduce the antagonistic activity of the organism producing the antibiotic. Brock is careful to emphasize that the proof that a given antagonism is due to the production of an antibiotic can be obtained only if the antibiotic is isolated and purified and the purified material is shown to have the same kind of activity against the test organism. Gottlieb (1976), who has recently reexamined the literature of the late 1940s and early 1950s, concludes that the data still do not allow us to accept the thesis that antibiotics are naturally produced in soil and function there in antagonistic capacities. Gottlieb does not accept the notion that seeds and plant detritus are a natural phenomenon in soil, and therefore does not recognize experimental evidence in which fungal antibiotics are extracted from plant materials (seed, wheat straw) that were buried and later recovered from otherwise unaltered soil (Wright,

1956a,b). How does one explain how soil-inhabiting fungi survive if they don't colonize plant detritus or use as resources the exudates of living plant tissues?

The phenomenon of *hyphal interference* has been identified as a form of interference competition by fungi in which the hyphal tips of one fungus cause the loss
of hydrostatic pressure, vacuolation, and granulation of a competitor's hyphal cells
without the production of any recognizable antibiotic substance (Ikediugwu and Webster, 1970a,b; Ikediugwu et al., 1970).   There is some evidence (see Sec. III below)
to suggest that hyphal interference increases the fitness of slower growing and later
sporulating fungal species (Ikediugwu and Webster, 1970a).

Fungal toxins may also prevent or reduce loss of resource to animals (i.e., insects, mammals, birds).   Janzen (1977) has developed the interesting hypothesis that
fungi which colonize seeds or fruits ("seed- or fruit-eating fungi") render them un-
attractive to seed- or fruit-eating animals by production of toxic metabolites.   According to Janzen this form of interference involving the production of a feeding
deterrent enables fungi that utilize seeds or fruits as a resource to protect this
resource from arthropod or vertebrate predation.   The potential role of these myco-
toxins and antibiotics in predator defense is thoroughly reviewed by Janzen and will
not be considered further in this paper.

While fungal antagonism is a well-documented phenomenon, there are few explanations as to precisely how such biological properties improve the fitness of individual antagonists in specific microhabitats.   Hypotheses are for the most part limited
to analyzing the immediate effects of the confrontation on the populations concerned
and to near-term strategies for survival.   Little attention has been given to the
potential importance of antagonism as a mechanism influencing the organization of
fungal communities.   The evolutionary outcome of fungal interference in a given ecosystem or microcosm is likely to leave its imprint in fungal community structure,
composition, and life history strategies.   The intention of this review is to examine
the role of fungal antagonism as a mechanism contributing to the organization of fungal communities.

II.   FUNGAL ANTAGONISM AS A MECHANISM LIMITING COMPETITOR ACCESS
TO A RESOURCE

Fungal antagonism is usually characterized as the mechanism that enables early colonists of substrates to successfully gain a foothold and then to inhibit the invasion
of other fungi.   The inhibition model of mechansims bringing about successional
change in ecosystems (Connell and Slayter, 1977) rather accurately describes this
recognized pattern of substrate colonization by fungi.   The inhibition model holds
that once earlier colonists secure the space and/or other resources, they inhibit the
invasion of subsequent colonists or suppress the growth of those already present.
Later successional species can invade and grow only when the dominating residents are
damaged or killed, thus releasing resources.   By interfering with further invasion
and colonization, the early colonist prevents further succession.

There is considerable evidence that toxic metabolites (antibiotics) produced by the initial fungal colonist(s) of natural substrates may act to slow or prevent the invasion by other species. The *Penicillium* growth pattern, as described by Sewell (1959), is one in which the fungus densely colonizes small individual substrates with no extension of the mycelium into the surrounding soil and then produces spores heavily over the surface. Under these conditions, the production and accumulation of toxic metabolites in such a well-defined substrate could be significant at the microscale. Wright (1956a,b) established that antibiotic production will take place in pieces of wheat straw or seeds buried in otherwise unaltered soil. The quantities of toxins were large enough for extraction and characterization. Furthermore, the organisms producing such toxins are capable of influencing the pattern of substrate colonization by excluding other microbes. Inoculation of pea seeds with *Trichoderma viride* (Pers.) ex. Fr., *Penicillium frequentans* Westling, and *Penicillium gladioli* Machacek produced gliotoxin, frequentin, and gladiolic acid, respectively, in their seed coats. Evidence has since been presented that Wright's gliotoxin-producing isolate, obtained from P. W. Brian, was not *T. viride* but *Gliocladium virens* Miller, Giddens & Foster (Webster and Lomas, 1964). In other experiments, *T. viride* (= *G. virens*?) growing naturally in soil apparently infected the pea seeds and produced gliotoxin in the seed coat (Wright, 1956b). Wright (1956c) reported that *Pythium ultimum* Thom, a soil-inhabiting pathogenic fungus, was prevented from attacking seeds inoculated with *Trichoderma viride*. Furthermore, the failure to obtain control of *Pythium* damping-off of mustard seedlings by seed inoculation with *Penicillium nigricans* (Bainier) Thom was attributed to the known resistance of *Pythium* spp. to griseofulvin, the antibiotic produced by *P. nigricans*.

Wood and Tveit (1955) questioned whether inoculation of the seed with antagonists was really necessary to protect the seed from fungal attack, since the seed surface is likely to carry a varied microflora. During routine isolations of fungi from oat seed produced in Brazil, it was noted that *Chaetomium* colonies strongly inhibited other fungi such as *Helminthosporium* and *Fusarium* growing from nearby seeds on the same plate (Tveit and Moore, 1954). The *Chaetomium* isolates protected oat seedlings from the pathogen *Helminthosporium victoriae* (= *Cochliobolus victoriae* Nelson) when both were inoculated into sterilized soil. Furthermore, when seeds naturally infected with *H. victoriae* were inoculated with *Chaetomium* and planted in nonsterile soil in greenhouse pots, germinating oat seedlings were also protected from Victoria blight while most of the controls were heavily blighted or killed. The authors suggest that lack of any important Victoria blight in Brazil might be attributed, in part, to the natural presence on oat seed or in soil of antagonistic strains of *Chaetomium*. Isolates of *C. cochlioides* Pall. and *C. globosum* Kunze provided a larger measure of disease control against *Fusarium* species causing seedling blight of oat seedlings. Control of disease was obtained when *Chaetomium inoculum* was applied as a seed dressing (dusted) or when oat seeds were soaked in cell-free culture filtrate of *Chaetomium* isolates that reduced the disease in soil naturally infested

or artificially inoculated with pathogenic fusaria (Tveit and Wood, 1955). Using basically the same approach, Chang and Kommedahl (1968) reported that inoculation of
corn seed with *Bacillus subtilis* or *Chaetomium globosum* before planting in a field
shown to have relatively large amounts of inoculum of *Fusarium roseum* f. sp. *cerealis*
"Graminearum" (= *Fusarium graminearum* Schwabe), a pathogen of corn, controlled pathogen attack. This was believed to be the result of antibiotic production by the seed
coat microflora. Remarkable ecological insight was shown by Baker and Cook (1974),
who advocated that antagonists used in seed coatings or soil amendments should be
complementary, not competitive; that is, they should grow best at different temperatures, water potentials, or pH, or on different nutrients so that some will be operative under most environmental situations. Microbial ecologists should determine
whether in native ecosystems the microfloral colonists of native seeds have in fact
co-evolved to provide this form of natural defense.

Prolonged saprophytic survival of certain root-infecting fungi in dead host
tissues that were invaded during the parasitic phase of growth involves more than
just dormant survival in the form of spores or sclerotia. Ludwig (1957) suggested
that *Helminthosporium sativum* may persist in plant debris as a primary colonist by
production of a fungistatic toxin. *Cephalosporium gramineum* (= *Hymenula cerealis*
Ell. and Ev.), which causes a vascular wilt of winter wheat, occupies the vascular
bundles of wheat stems and reduces competition among potential colonists by producing a broad-spectrum antifungal antibiotic (Bruehl et al., 1969). Wild-type isolates
of the fungus (antibiotic producers) were better able to exclude invasion by other
fungi than were laboratory mutants that had lost the ability to produce the antibiotic. The authors concluded that antibiotic-producing isolates possessed survival
value and that the character is preserved by natural selection in the wild-type population. Cook and Bruehl (1968) made the interesting observation that wheat straw
which had been heavily colonized by fungi when the wheat was left standing in the
field for extended periods in wet cool weather was much less available as a substrate
for colonization by *Fusarium roseum* (= *F. graminearum*) during soil burial than straw
that had not been similarly weathered. It was not determined, however, whether toxin
production or nutrient depletion rendered the substrate unsuitable to subsequent
colonists. Macer (1961a,b) observed that *Cercosporella herpotrichoides* Fron, a pioneer colonist on wheat straw but found to be unable to attack cellulose, could substantially reduce the rate of decomposition of that resource when it was buried in
soil, as contrasted with wheat straw not colonized by *Cercosporella*. In addition,
following soil burial, a more varied fungal flora developed on the unoccupied straw
than on straw first colonized by *Cercosporella*. Once again, the mechanism by which
Cercosporella was able to exclude potential colonists was not determined.

It is important to remember that pioneer colonists of plant detritus include
the weak parasites and/or saprophytes that invade senescing plant tissues and metabolize the available sugars and carbon compounds simpler than cellulose (Garrett,
1970). One must consider the extent to which later arriving colonists are control-

led or limited by the pioneer colonist that has already exploited most of the more easily available substrate nutrients (Lockwood, Chap. 18).

The case for antibiosis in natural and man-managed ecosystems is based, in part, on assumptions about soil saprophyte/root pathogen interactions. Because the root pathogen is able to escape competition with soil saprophytes for a part of its life below ground, Garrett (1956) theorized that this group of organisms might be less competitive in the presence of saprophytic microbes from the soil or rhizosphere. In turn, saprophytic fungi in the rhizosphere must be able to contend with the antibiotics produced by other microorganisms. Jefferys et al. (1953) wanted to determine whether antibiotic-sensitive or nonsensitive fungi are likely to be found in the same microenvironment (ecological niche) as antibiotic-producing molds. They examined the interactions between root-infecting fungi during the saprophytic phase of their life history and common soil saprophytes. Nine antibiotic-producing fungi were highly tolerant of antibiotics they themselves produced, whereas the root pathogens examined were generally more sensitive to the antibiotics. Plant pathologists have provided numerous examples and much data correlating the diminished growth or lack of pathogenicity of a soilborne root-infecting fungus to the antagonistic properties of certain saprophytic fungi from the rhizosphere of a potential host plant (Meyer, 1972; Baker and Cook, 1974; Gottlieb, 1976). An excellent illustration of this phenomenon in native soils was published by Rishbeth (1950, 1951), who studied the spread of *Fomes annosus* (Fr.) Cooke, a root pathogen of pine, in alkaline and acid heath and woodland soils. *Trichoderma viride* was shown to be significant in preventing colonization of tree stumps by *F. annosus*, the increased severity of this disease on alkaline soils being attributed to the absence of the antagonist under these conditions (Rishbeth, 1951). Superficial growth of *F. annosus* was abundant on roots in the alkaline soils but sparse or absent in the more acid soils. Roots in the acid soils were found to be colonized by *T. viride*, which was demonstrated to have a marked in vitro antibiotic effect on *F. annosus*. Saprophytic fungi living near the root surface compete for the sugars and amino acids that are excreted by the plant. Some of these fungi are known to excrete antibiotic substances that may enable them to keep competitors at a distance. The rapid development of saprophytic microorganisms on plant roots may provide a barrier against root pathogens. However, the production of antibiotics by these microbes in the rhizosphere in sufficient quantities to inhibit a fungal pathogen has not been firmly established. Baker and Cook (1974) have theorized that if a number of niches in the rhizosphere are unfilled or if those organisms filling the niches are ineffective as antagonists to root pathogens, then root infection may take place.

A good deal of research has been conducted which shows that ectomycorrhizal fungi reduce a pathogen's access to the feeder roots. Zak (1964) recognized several possible mechanisms by which mycorrhizal fungi may protect plant roots from fungal disease. These included (1) utilization of excess carbohydrate nutrients produced by the root, thus reducing the amount of nutrients available to stimulate propagule germination among potential pathogenic fungi; (2) presentation of a physical barrier to penetration by the pathogen; (3) secretion of antibiotics inhibitory to pathogens; and (4) contribution to the support of a protective microbial rhizosphere community. Numerous basidiomycetes that produce antifungal antibiotics in pure culture have been experimentally proven to be mycorrhizal symbionts or have been associated with mycorrhizae (Marx, 1973). The antifungal compounds produced by mycorrhizal fungi were also shown to be inhibitory to different fungi pathogenic on feeder roots by demonstrating that these antibiotics are produced in mycorrhizal association with the host root. Marx and Davey (1969) extracted diatretyne nitrile and diatretyne III from mycorrhizae formed by *Leucopaxillus cerealis* var. *piceina* nomen nudum and from the rhizosphere of septically grown mycorrhizal roots of *Pinus echinata* Mill. seedlings. In experiments designed to determine the susceptibility or resistance of mycorrhizas to infection by a pathogen, Marx and Davey also demonstrated that the diatretynes present in mycorrhizae formed by *L. cerealis* var. *piceina* protected seedlings from *Phytophthora cinnamomi* Rands.

Antibiotics produced in the rhizosphere are ideally located to inhibit pathogen access to mycorrhizal and possible adjacent nonmycorrhizal roots by translocation or diffusion. Marx (1973) argues that in this situation the presence of a few mycorrhizae formed by an antibiotic-producing fungal symbiont may be as effective in controlling pathogenic root-infecting fungi as the presence of considerably more mycorrhizae produced by a non-antibiotic-producing mycorrhizal fungus. It should also be noted that microbial populations of mycorrhizal rhizospheres are different from those of other mycorrhizae, nonmycorrhizal rhizospheres, and nonrhizosphere soil, both quantitatively and qualitatively (Marx, 1973). Marx points out that while the value of these different microbial populations in the control of feeder root diseases has yet to be firmly established, one can logically conclude that they do have an effect on microbial competition and probably influence root pathogens.

There are numerous but fragmentary observations recorded in the literature that suggest antibiosis is important in the inhibition of a competitor's growth in woody substrates. For example, Ricard and Bollen (1968) surveyed the fungi in heartwood of Douglas-fir utility poles attacked by the wood-rotting basidiomycete *Poria carbonica* Overh. and reported that an unidentified species of *Scytalidium* was isolated from sound wood as a unifungal culture. They suggested that poles colonized by *Scytalidium* escape decay by *P. carbonica*. Strong antagonism of *Scytalidium* to *P. carbonica* has been attributed, in part, to the production of an antibiotic substance, "scytalidin" (Strunz et al., 1972; Stillwell et al., 1973).

Hulme and Shields (1972) argue that the growth inhibition of wood-rotting basidiomycetes in wood blocks by certain wood-inhabiting saprophytes is due largely to the depletion of more accessible nutrients and not to the excretion of toxic metabolites or mycoparasitism. Such inhibition would be expected if the initial colonist removes traces of more accessible nutrients that are necessary for rapid colonization by a competitor. While acknowledging that antibiosis may be involved in the inhibition of

a competitor's growth in wood blocks, Hulme and Shields reason that this would mean
that production of antifungal antibiotics is a common phenomenon, since the wood-inhabiting fungi (competitors) used in their experiments were for the most part selected
at random, and a good number of them were successful inhibitors of wood-rotting basidiomycetes on culture media.

Fungal succession in wood is a complex process involving microbial populations
that have co-evolved over a long time, and it must be remembered that utility poles,
wood stakes, fresh and autoclaved wood blocks, wood chips, and even cut stump surfaces
do not necessarily present the indigenous fungal community with the number and sequence of niche options provided by a living, dying, or dead tree. In most instances
where a wood-inhabiting fungus is reported to inhibit the colonization and decay by
wood-rotting fungi, the former was first allowed to become established. It is also
difficult to relate these observed antagonisms to a fungal colonization strategy,
since the combination of events determining patterns of fungal colonization in nature
no longer applies. It is important to stress that the significance of competition
can only be recognized if the natural context within which it occurs can clearly be
envisaged. Failure to realize this has led to much fruitless speculation and argument
(e.g., doubt about whether antibiotic production on substrates of high energy and
nutrient composition is realistic in agricultural soils). "We must remember that
competition only occurs when two organisms attempt to occupy the same resource in the
same place at the same time. Many supposed examples of competition have not been
unequivocally shown to fulfill these criteria, and arguments about competitive mechanisms may thus not be relevant" (M. J. Swift, personal communication, 1978).

III. FUNGAL ANTAGONISM AS A MECHANISM OF SPECIES REPLACEMENT

Successional ecosystems and individual resources within ecosystems can be visualized
as islands, the species composition of the biota at any time representing the outcome
of two separate processes, immigration and extinction (MacArthur and Wilson, 1967).
Immigration is the means by which individual species of fungi reach a substrate, either passively through environmental carriers such as wind, water, invertebrates and
vertebrates or actively by hyphal extension. Extinction is the process by which individual fungal genotypes are permanently eliminated from a resource when the habitat
becomes modified (i.e., the niche parameters change) or when subsequent colonists are
better competitors. Early successional species have evolved adaptations for continual
colonization of newly available areas, whereas later successional species are adapted
for invasion of existing communities (McNaughton and Wolf, 1973). The term *competitive saprophytic ability* was coined by Garrett (1956) to describe the ability of a fungus to successfully colonize a substrate in the presence of competitors (see Frankland, Chap. 22). As outlined by Garrett, the share of the substrate obtained by a
fungus will be determined: (1) directly by its competitive saprophytic ability; (2)
directly by its *inoculum potential* (propagule density and total energy available for
colonization) at the surface of the substrate; (3) inversely by the inoculum potential

of its competitors. Four of the physiological traits or characteristics that favor a
high competitive saprophytic ability include: (1) high growth rate and rapid germinability of spores; (2) good enzyme production; (3) production of antibiotics; (4)
tolerance of antibiotics produced by other organisms. It should be noted that the
first of these characteristics fits the description of an r-selected trait, while the
latter three might be considered traits of species responding to K-selection (Mac-
Arthur and Wilson, 1967). For this reason, Prof. J. Lussenhop (personal communication, 1978) points to the potential danger in characterizing fungi, as either r- or
K-selected species (Swift, 1976), since fungi often combine traits associated with
r- or K-strategists among large organisms. Garrett (1970) introduced a scheme for
examining the various attributes of fungal substrate groups from soil that contribute
to their competitive saprophytic ability. Primary saprophytic sugar fungi are described (p. 116) as "pioneer colonists" which "occupy their substrate ahead of competitors so that neither production of nor tolerance to antibiotic substances is
likely to give them any significant advantage." Garrett's other three fungal substrate groups (i.e., cellulose and lignin decomposers; root surface and rhizosphere
fungi; secondary saprophytic sugar fungi) are said to live in an environment in which
they are exposed to a high degree of microbial competition; therefore, both the production of and tolerance to antibiotic substances will confer an additional degree of
fitness. It has been theorized that, in microbial successions, pioneer colonists lose
their selective advantage as they make the habitat increasingly unfavorable to themselves by removal of nutrients and by formation of acidic or autoinhibitory substances
(Alexander, 1971; Park, 1967). Such colonists are said to exhibit a low tolerance to
biologically formed inhibitors since few are present early in the succession. Furthermore, it is sensitivity to antagonism that may restrict some fungi to the role of
pioneer colonist.

Mechanisms determining patterns of saprophytic colonization of pine logs or
stumps by fungi were studied by Gibbs (1967). The succession, involving *Leptographium
lundbergii* Lagerb. & Melin. as an initial colonist, followed by *Fomes annosus*, *Peniophora gigantea* (Fr.) Massee, and eventually *Trichoderma viride*, is believed to be
based on (1) a decreasing ability to tolerate toxic substances in the freshly cut
pine log which decrease with time, and (2) increasing tolerance of antibiotics produced by earlier members of the succession. Gibbs (1967) did not indicate whether
later appearing fungi interfere in some way with earlier colonists. The distinction
is important; tolerance to the metabolites of earlier colonists is one thing, but the
antagonistic behavior of later colonists so as to replace an earlier colonist would
be an entirely different proposition. The latter implies that a "competive hierarchy"
(Horn, 1977) exists based on increasingly aggressive antagonists which, not insignifi-
cantly, are also less sensitive to the metabolites of earlier appearing species.

Horn (1977) has identified *competitive hierarchy*, a type of succession in which
later successional plants are increasingly capable of dominating earlier successional
species, as a regulatory process in determining the sequence of colonists in a successional sere. There are numerous fragmentary observations recorded in the mycologi-

cal literature which suggest that (1) fungal antagonism may be an important mechanism of species replacement during substrate colonization and (2) such a competitive hierarchy may be operative in delimiting patterns of fungal colonization and sporocarp appearance. Let us now consider some of this evidence.

In a forest soil profile, substrate and microclimatic heterogeneity in litter horizons, particularly of mixed forest tree composition, give way progressively to fewer and fewer substrate choices and a less variable microclimate in humus and mineral soil zones. Moreover, the richness of microfungal communities also declines progressively with depth and biodegradation of the primary resource (Hering, 1965; Swift, 1976; Wicklow and Whittingham, 1974). Fungal dominants in the fermentation, humus, and mineral soil horizons would have to be particularly competitive since, as these primary resources are converted to terminal resources (humus), total niche space is progressively reduced (Swift, 1976) and niche overlap among members of the component communities colonizing specific litter types becomes more important. Furthermore, since litter falling onto the soil surface is already colonized by fungi, subsequent change in species composition usually involves some degree of species replacement, presumably by more aggressive colonists, and the extinction of an earlier colonist(s). Fungi in most strata are in the continual fix of having to invade previously colonized substrates lying above them. It is not suprising then that *Trichoderma* spp., *Penicillium nigricans*, and other fungi long recognized for their ability to produce antifungal antibiotics or for their potential as antagonists, become increasingly important components of the humus/A_1, horizons of forest soils having well-defined organic profiles (Kendrick and Burges, 1962; Christensen, 1969; Widden and Parkinson, 1973; Wicklow and Whittingham, 1974). Equivalent examples can be found in studies on the microfungal colonization and decomposition of specific forms of plant litter (Hering, 1965; Saitô, 1966, Hudson, 1968).

The ability then of many fungi in the organic soil horizons to coexist in an ecosystem depends on the regular input of new primary substrate. As substrate diversity and the total pool of energy become reduced during decomposition, competitive interactions increase because of niche overlap. This ultimately results in the exclusion of certain fungi. Rapid mycelial growth or rapid production and dispersal of fungal spores is important for survival in or escape from potentially unfavorable environments (Hawker, 1957). Sporulation patterns of many fungi are undoubtedly an outcome of competitive interactions among members of the fungal community. More recent studies by Brasier (1975) have correlated the ability of certain *Trichoderma* isolates to induce a sexual response in *Phytophthora* (oospore formation) with their ability to produce volatile antibiotics. Root-inhabiting species of *Phytophthora* appear to be more ecologically specialized in their response to *Trichoderma* than others. *Trichoderma* is often noted to replace primary pathogens such as *Phytophthora* in woody material, and Brasier suggests that the "*Trichoderma* effect" may be a specifically evolved defense mechanism that provides both a means of survival (oospores are more resistant and longer lived than the hyphae) and a potential source of genetic variation. Brasier's study provides the best evidence to date that, in nature, organisms sensitive to antibiotics have acquired mechanisms allowing for their survival and competitive coexistence.

Recognizing that wounded or dying plant tissue sooner or later becomes colonized by a saprophytic flora that does not normally spread to undamaged tissue and that under natural conditions the microflora on damaged tissue may exclude a particular pathogen, Wood and Tveit (1955) suggested that it might be possible to isolate the organisms responsible and introduce them to the surface of fresh wounds as a means of biological control. This was accomplished by Rishbeth (1963), who demonstrated that artificial inoculation of freshly cut pine stumps by oidia of *Peniophora gigantea* brought about considerable reduction of stump colonization by *Fomes annosus*, the causal fungus of butt-rot disease of pines; *F. annosus* may be replaced under natural conditions by *P. gigantea* (Rishbeth, 1950; Meredith, 1960). It has been shown that *P. gigantea* promotes symptoms of hyphal interference in *F. annosus* when the two are grown next to one another on culture media (Ikediugwu et al., 1970). Moreover, *P. gigantea* was found to have a slower growth rate in fresh pine logs than *F. annosus* (Gibbs, 1967), fitting precisely the model of an organism that operates via mechanisms of hyphal interference (Ikediugwu and Webster, 1970b). From an ecological point of view, what Rishbeth (1963) did was reverse the order of appearance of fungi in a natural succession on wood, placing a fungus capable of aggressive secondary invasion (*P. gigantea*) in the role of a pioneer colonist. There have been numerous field or laboratory studies of fungal antagonism on woody substrates that might better be interpreted by first examining the ecological status of the interacting pair.

Fungal antagonism may be involved in species replacement within fungal cankers of trees. In a comprehensive study of the microflora of European larch (*Larix decidua* Mill.) cankers, Buczacki (1973) has shown that *Trichoscyphella willkommii* (Hart.) Nannf. was the only fungus consistently associated with canker extension tissue. Sequential colonization of 208 cankers by a total of 83 other fungal species was demonstrated. The species were mostly well known as bark- and wood-inhabiting species and included some ubiquitous saprophytes. The spread within the canker of other species coincided with the disappearance of *T. willkommii*. The four most frequently isolated canker-invading fungi, *Cryptosporiopsis abietina* Petrak, *Zalerion arboricola* Buczacki, *Tympanis laricina* (Fuckel) Sacc., and *Phialophora* sp., were shown to be able to replace *T. willkommii* in bark blocks. Furthermore, *Zalerion arboricola* and *Phialophora* sp. brought about the death of its hyphae on agar plates. The manner in which the process might operate was not determined. Reports by Stillwell (1966), Stillwell et al. (1969), and Strunz et al. (1969) show that an unidentified species of *Crytosporiopsis* isolated from *Betula alleghaniensis* Britt. was antagoistic to 31 species of wood-destroying basidiomycetes. A broad-spectrum antibiotic was isolated and termed "cryptosporiopsin." The mechanism by which *Cryptosporiopsis abietina* naturally replaces the larch canker fungus *T. willkommii* might be related to the production of this antibiotic. Some canker-producing fungi may themselves produce antifungal antibiotics. Funk and McMullan (1974) report that *Potebniamyces balsamicola* Smerlis var. *boycei* Funk produces a fungal growth inhibitor, "phacidin," which restricts the growth of a wide variety of wood-inhabiting and wood-rotting fungi.

An outstanding example of competitive exclusion among fungal pathogens of jack pine is provided by Basham (1975). Intense antagonism exhibited by *Ascocoryne sarcoides* (Jacq. ex Gray) Groves toward *Peniophora pseudo-pini* Weres. and Gibson in laboratory tests is related to the ecology of each of these species. In young trees, *P. pseudo-pini* exists in the basal regions of the stem because of the scarcity of *A. sarcoides* in trees of this age. As the jack pine ages, however, *A. sarcoides* becomes increasingly dominant in the lower third of the stems. In older trees *P. pseudo-pini* gradually disappears from the lower third of the tree, and in "overmature" trees it is limited to the upper stem.

Data presented by Rayner (1975, 1977a,b) can also be applied to the general model of a competitive hierarchy among wood-inhabiting fungi. *Chondrostereum purpureum* (Pers. ex Fr.) Pouz. is well known as a parasite of fruit trees and as a predominant early colonizer of other freshly exposed living or recently dead wood tissues. However, it is a poor agent of decay, disappears rapidly as stumps age (Rayner, 1977b), and was shown to be highly susceptible to competition from other organisms (Rayner, 1975). According to Rayner (1977b), *C. purpureum* is more likely to be at an advantage in tissues possessing at least some host resistance, an amount sufficient to prevent the establishment of other less-specialized fungi. Moreover, *C. purpureum* may be replaced in birch (*Betula pendula* Roth) stumps by *Phlebia merismoides* Fr., which is not noted for a capacity to invade living tissues but is highly competitive toward other fungi in the nonliving wood (Rayner, 1977b). Rayner (1977a) also observed that a species of *Scytalidium* (close to *S. album* Klingström and Beyer) was widely distributed through white-rotted wood in which the basidiomycetes responsible for the decay were completely replaced by *Scytalidium*. The potential importance of *Scytalidium* species as producers of antifungal antibiotics has been treated earlier.

The role of fungal antagonism in species replacement on dung is illustrated by the work of Webster and associates. Harper and Webster (1964) experimentally demonstrated that the appearance of fruiting bodies on rabbit dung is not related to changes in the nutritional characteristics of the substratum but to the minimum time required for each fungus to develop such structures. On sterile rabbit dung, these fungi required the same number of days to produce mature spores as did fungi colonizing freshly collected dung from the field. Significantly, it was noted that in the absence of certain fungal colonists other species of coprophilous fungi are capable of producing sporocarps over longer intervals (Harper and Webster, 1964; Ikediugwu and Webster, 1970a,b). Hyphal interference and antibiotic activity were identified as mechanisms contributing to the early cessation of fruiting of these fungi (Ikediugwu and Webster, 1970a; Singh and Webster, 1973). Harper and Webster (1964) proposed that, while competition had little effect on the initial appearance of fungal fruiting bodies on rabbit feces, competition is important in limiting the duration

and intensity of fruiting of particular species. In the presence of the basidiomycete *Coprinus heptemerus* M. Lange & A. H. Smith premature cessation of fruiting of
the zygomycetes *Pilobolus crystallinus* (Wiggers) Tode and *Pilaira anomala* (Cesati)
Schröter occurred, along with a drastic reduction in numbers of sporangia produced
by *Pilobolus* and *Ascobolus viridulus* Phill. & Plowr. (= *A. crenulatus* P. Karst.)
This phenomenon was further explored by Ikediugwu and Webster (1970a). When the hyphae of *C. heptemerus* and most of the other coprophilous basidiomycetes tested came
in contact with hyphae of *Ascobolus crenulatus*, the cells of the latter were observed
to undergo a loss of hydrostatic pressure, vacuolation, and granulation of the cytoplasm. Because antagonism was observed only after contact between hyphae had been
made, Ikediugwu and Webster termed the phenomenon "hyphal interference." Nearly all
of the fungi Ikedeguwu and Webster (1970b) tested were sensitive to *C. heptemerus*.
Of particular interest was their observation that those basidiomycetes with the
greatest capacity for hyphal interference had the lower rates of hyphal extension on
dung agar. Ikediugwu and Webster explain this relationship by noting that the intensity of the interference action was found to be greatest in the growing hyphal tips of
the antagonist. A slow-growing species would have this particular zone of hyphae in
contact with the hyphae of a sensitive competitor for a longer time than would a fast-
growing antagonist, thereby amplifying this effect.

Webster (1970) observed that when *Stilbella erythrocephala* (Ditm.) Lindau was
found on rabbit feces collected from the field, relatively few other fungi developed
sporocarps on the fecal surface. It was later determined (Singh and Webster, 1973)
that *S. erythrocephala* was strongly inhibitory to *C. heptemerus* and all other coprophilous fungi through the production of a diffusible antibiotic. *Stilbella erythrocephala* is not an obligate coprophilous fungus in that it does not require gut passage
for spore germination. It probably colonizes rabbit feces secondarily after a period
of field exposure. Therefore, in order to invade a substrate already colonized by
coprophilous fungi, the late arrival would have to be a particularly aggressive invader.

In a study designed to contrast the structure, composition, and development of
coprophilous fungal communities in semiarid to mesic grasslands in the western United
States, Angel (1977) determined that species richness and diversity were highest for
fecal collections from the driest site. This phenomenon was characterized as an evolutionary response to the long-term survival of fecal substrates in these semiarid
grasslands (approximately 10 years) and the increased spatial heterogeneity resulting
from the greater variety of age classes of dung. Three functional groups can be
identified in the long-term successional sere: (1) early sporulating colonists; (2)
later sporulating colonists; and (3) early successional species that are able to persist (sporulate) through later states. Wicklow and Hirschfield (1979) examined
cultural antagonism among representative fungi from each of these groups and determined that interference competition becomes increasingly more important as the substrate ages. *Poronia punctata* L. ex Fr. which appeared on feces exposed 18 months

or longer in field sites was antagonistic to all earlier appearing and co-occurring species tested. Coprophilous fungi might be described by and large as food and habitat generalists with a large amount of overlap in resource requirements among coexisting species. In older feces, the resources in short supply include available carbohydrate and available nitrogen. Since the chemical composition of these older feces remains approximately the same as the substrate aged beyond 18 months of field exposure (Angel, 1977), competitive interactions are believed to account primarily for resource partitioning.

Usually, once a fungus is already in a substrate, it will retain possession of that substrate even in the presence of other potential saprophytic colonists. Slower-growing or later-arriving fungal colonists are faced with the problem of invading a substrate that is already occupied by fungi and partially depleted of more readily available nutrients in that the latter are tied up in fungal biomass. Once a mycelial network becomes established, it may continue to produce spores until either all resources are exhausted or the physiochemical environment is altered sufficiently so that conditions required for continued hyphal activity or sporocarp production are no longer adequate. While later successional species may be able to utilize a wider array of carbon sources, competition with earlier-appearing forms may nonetheless be for a limiting factor(s) required by both organisms. A competitive hierarchy would enable later species to transfer nutrient resources, by lysis, from the mycelium of earlier-appearing and faster-growing colonists. It might even be disadvantageous for the slower-growing antagonist to prevent entirely the development and sporulation of an earlier colonist. The latter are capable of competing successfully in fresh substrate and retaining substrate nutrients against losses to the environment. In time, these nutrients may be released to other organisms by lytic mechanisms. From the evidence summarized herein, we might conclude that interference competition can be of considerable significance in the organization of fungal communities. The suppression of niche breadth of a few colonists through different forms of interferenmce should also result in a greater degree of niche partitioning and, correspondingly, increased richness and diversity. The situation may be analogous to that of a plant community in which strong dominance and intense allelopathic effects contribute to low species diversity, whereas a variety of chemical accommodations are responsible, in part, for the high species diversity of others (Whittaker, 1970). Indeed, some antagonistic or antibiotic-producing fungi may be analogous to predators that kill their prey up close and, like predators, help to maintain diversity in a community.

Consideration thus far has mainly been given to the role of fungal interference in the colonization of natural substrates. Let us now examine present views on the relative importance of microbial antagonism in disturbed (species-poor) versus mature (species-rich) ecosystems. The mature stages of ecosystems, as compared to developing ones, show the following trends regarding the partitioning of available energy by the species composing those systems (Odum, 1969). Mature stages are said to (1)

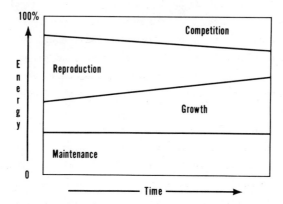

Fig. 2 Trends in the partitioning of energy flow during ecosystem maturation.

have a large and diverse organic structure within the limits set by the available energy input and the prevailing abiotic or physical conditions; (2) partition greater amounts of energy to competition and maintenance among the organisms present rather than reproduction (Fig. 2); (3) have high species diversity (for exceptions see Connell, 1978); (4) have narrow niches; (5) show greater importance of species interactions (i.e., mutualism, parasitism, predation, commensalism); and (6) have high biochemical diversity. As succession progresses, organic extra-metabolites are believed to serve increasingly important functions as regulators that stabilize the growth and composition of the ecosystem. Moreover, selection pressure is said to favor species with lower growth potential but better capabilities for competitive survival under the equilibrium density of late stages in succession. In contrast, species with high rates of reproduction and growth are more likely to survive in the early uncrowded stages of colonization.

     How well does this theoretical argument based on "large organism" ecology apply to fungal communities in ecosystems? One could find general support for the argument that in mature ecosystems there are more fungal species than in immature or recently disturbed ones. If so, it might be anticipated that there would be a greater chance of finding antagonistic microbes in mature systems, simply by virtue of the fact that there are more species to screen for microbial interference. Whether the proportion of antagonistic fungal species varies according to the stage of maturity of an ecosystem has not been studied, although indirect evidence that microbial interference in virginal soils may be greater than in recently disturbed soils has surfaced from time to time in the phytopathology literature. Baker and Cook (1974) have theorized that the more varied and numerous the soil microbes, the greater the chance of biological control of a plant pathogen. This notion is based upon the ecological idea that the more complex the biological community, the greater will be its stability. According to Baker and Cook, initial irrigation and cultivation of pathogen-susceptible crops in desert soils is believed to allow for the development of some patho-

gens, because these virgin desert soils supposedly have fewer species of potential
microbial competitors. They argue that the microbial community of desert soils is
"relatively simple, qualitatively if not quantitatively, until such time as natural
introductions of saprophytes fill the biological vacuum and the organic matter content can be increased to support a more complex microbiota." They also suggest that
the same phenomenon occurs when a native grassland is first plowed and planted to
wheat. Disease is said to be more prevalent because there are supposedly fewer species of microbes, more unfilled niches, and fewer antagonists in the recently plowed
field soil. However, as the wheat field becomes "established," one supposedly finds
more species, more interactions, more antagonism, and therefore less substrate available for soilborne pathogens in their saprophytic phase. This conclusion seems to
be based entirely on the results of a single study by Clark et al. (1960), in which
it was shown that the nitrifying capacity of a virgin desert soil remains low until
under cultivation for a few years. Ranzoni (1968), Scarborough (1970), and States
(1978) have examined the microfungi of numerous soil types from semiarid to arid regions of the western United States, finding a considerable number of species. Prof.
M. Christensen (personal communication, 1978) observes that although fungal population densities may sometimes be low, fungal communities are often remarkably rich
and diversified. It could be argued that a shift from semiarid to more mesic soil
conditions, a result of irrigation, increases the importance of a few antagonists,
some of which may already be residing, albeit in low numbers, in the virginal soil.
These antagonists might progressively increase their propagule density and relative
importance in the soil community under conditions of cultural enrichment associated
with particular cultivation practices. This could occur without adding species to
the microbial community and most probably contributes to a reduction in species. I
agree with Baker and Cook that there must be a period of readjustment before the soil
microflora can become "balanced for the new conditions imposed by cultivation." However, since there have not been any comprehensive studies on microbial community response to cultivation of virginal soils, it is premature to suggest that cultivation
increases microbial species richness. Certainly in the case of larger organisms
(i.e., insects) cultivation of virginal ecosystems dramatically lowers species richness and diversity (Bey-Biyenko, 1961).

The notion that cultivation of soils provides greater environmental heterogeneity, a greater number of different niches, and thus greater numbers of fungal species
(Park, 1965) seems to be based entirely upon a limited study by England and Rice (19-
57). Curiously, in the same review, Park also recognizes the importance of Parkinson
and Kendrick's (1960) demonstration that different fungi occur in different zones of
the soil organic horizons in a pine forest. Presumably, the stratification of soil
organic horizons contributed to the environmental heterogeneity, number of niches,
and fungal species richness of that soil profile. Cultivation of a native grassland
of forest soil might initially be expected to increase species richness among soil
microfloral components, since the uppermost soil horizons and their microfungi would

become "homogenized." However, since the diversity of higher plant species and their remains would be eliminated, both above and below ground, we should eventually expect that there would be fewer substrate choices, less variation in microhabitat, and therefore fewer niches for microfloral colonists. A comprehensive study on the impact of cultivation of virginal soils on the soil microbiota would be of considerable interest and importance.

IV. REGULATION OF ANTIBIOTIC PRODUCTION IN NATURAL SUBSTRATES

A critical question relating to the role of fungal antibiotics in improving the fitness of initial substrate colonists, or in enabling a fungus to replace an earlier arrival, concerns the mechanism(s) by which toxin production is regulated in nature. Microbial toxins are classed as secondary metabolites in that their production is generally associated with the stationary phase of growth. According to Bu'Lock et al. (1974), the production of secondary metabolites by fungi can be viewed as a chemical mode of differentiation, which is minimal when growth is fastest and greatest when growth is minimal or has ceased. Growth and production of secondary metabolites therefore are not mutually exclusive processes. It is generally recognized that antibiotics are produced most rapidly on a rich substrate providing high levels of available energy and when the growth of the fungus is slow or has ceased because some other essential growth requirement (e.g., nitrogen) becomes limiting in the enviroment (Bu'Lock, 1967; Demain, 1972). Under such conditions, the fungus continues to absorb simple carbohydrates from the substrate and, instead of making metabolically active cell constituents such as nucleic acids or producing structural components of the cytoplasm and cell walls, it converts these carbohydrates to a shunt metabolite that may have toxic properties. Antibiotics are excreted by living cells or released upon autolysis of senescent cells within the mycelial network (Fig. 3). Antibiosis works rapidly, often resulting in the lysis of the competitor's hyphae (Brian, 1949) and by preventing spore germination (Stanley and English, 1965). After they are released in soil, antibiotics are either enzymatically degraded (Jefferys, 1952; Brian, 1957) or rendered inactive through adsorption to clays (Pinck et al., 1961a,b; Soulides et al., 1962) and, therefore, the rate of antibiotic production and inactivation will determine the concentration of an antibiotic in any microhabitat.

Microbiologists have argued that antibiosis can take place only on nutrient-rich substrates. At low levels of available organic carbon, antibiotic synthesis is unlikely to be significant enough to influence microbial competitors. Specific soil microhabitats in which nutrients are abundant, such as those associated with freshly incorporated organic detritus and seed or root exudates, have long been recognized as sites where antibiosis may occur in nature (Brian, 1957). It has also been demonstrated that antibiotic production can be detected when a fungus first invades fresh plant tissue, such as is the case for patulin production by *Penicillium expansum* Link during its colonization of mature apple fruits (Brian et al., 1956). One can

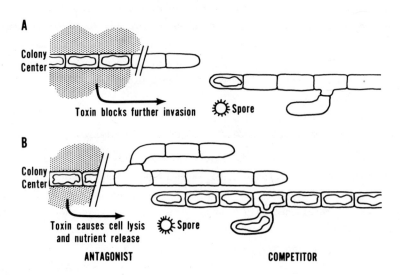

Fig. 3 Fungal antibiosis as a mechanism of interference competition: (A) by preventing competitor access to a substrate; (B) by enabling the antagonist to replace a fungal colonist already in a substrate.

easily envision a model of fungal interference in which a fungal colonist exploits a nutrient-rich substrate, partially depleting substrate nitrogen or some other growth-limiting nutrient, and then produces toxins (secondary metabolites) from the remaining sugars in sufficient quantities to restrict the invasion of potential competitors.

Wood and Tveit (1955) proposed that an antagonist might be effective in an environment of limited nutrients because fungi are generally more susceptible to toxins under poor nutrient conditions. However, the circumstances require that there be close contact between the antagonist and target organism, since the antibiotics produced would probably be of insufficient quantity to act at a distance. It is well known that toxins causing membrane permeability that are host specific make nutrients available to the organisms in closest proximity (Gardner et al., 1972). Therefore, it should not be surprising that antibiotics are not found in large quantities in nature. In fact, it is probably to an organism's advantage not to produce large quantities. Were an antibiotic to cause cell lysis beyond the zone in which the antagonist could readily retrieve the available nutrients, advantage from producing the toxin would be lost. Moreover, these nutrients could be used by a potential competitor. Therefore, it would make more sense for antibiotic-producing organisms to limit their antagonism to close-range encounters.

Let us consider a situation in which two species of fungi (one of which is a toxin producer) possessing basically the same nutrient requirements invade opposite sides of a nutrient-rich substrate and grow toward one another, meeting at some hypothetical midpoint (Fig. 3). Nutrient reserves would be equally depleted by each organism, and both the antibiotic-producing antagonist and the competitor would be equally weakened by low nutrition. However, since the antagonist would likely be less sensitive to its antibiotics than would the competitor, competitor invasion would be halted and the antagonist might even enter the zone between the opposing colonies and eventually replace the competitor entirely throughout the substrate.

The process by which a secondary colonist might, with the aid of toxins, invade a substrate partially depleted of nutrients and already occupied by a primary colonist(s) suggests a steady-state model of interference (Fig. 4). It can be expected that a primary colonist of plant material will utilize the simple carbohydrates that are initially available and, beyond that, might enzymatically attack plant cell walls or starch granules to obtain additional carbon. Nitrogen is one essential nutrient that would become more limited under these circumstances, since fungi have been shown

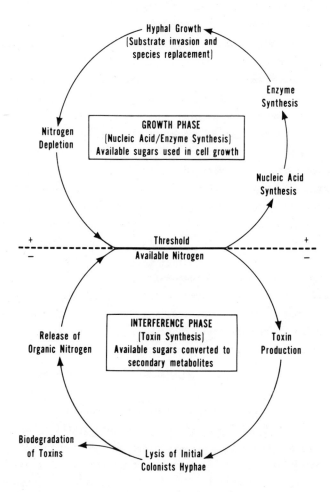

Fig. 4 Steady-state model of alternating growth and interference phases during fungal invasion and species replacement under conditions of limited nitrogen availability.

to retain or conserve nitrogen against losses to the environment by reallocating it
from the older portions of the mycelium to the younger and actively metabolizing hyphae (Cowling and Merrill, 1966; Levi et al., 1968). Toxin production by a secondary
colonist would promote the lysis of the initial colonists hyphae, thereby providing
a readily available source of organic nitrogen. Additional nitrogen can be obtained
through enzymatic biodegradation of chitinous cell walls. In this instance, the production of antibiotics might be viewed as the evolutionary outcome of competition
with other secondary invaders, in that the more successful secondary invader is able
to procure the least costly form of nitrogen. Using this nitrogen, the fungal antagonist can shift to the biosynthesis of the so-called essential (primary) metabolites,
such as amino acids and nucleotides, which are critical for growth and further substrate invasion. The process repeats itself as nitrogen reserves become depleted
and there is a return to metabolic pathways leading to the synthesis of secondary
metabolites (Fig. 4). This model, in which changes in the level of available nitrogen regulates, in part, the nature of fungal biosynthetic pathways, provides a steady-
state form of antagonism that results in the progressive replacement of an initial
colonist by a later successional species (Fig. 3). Since the microbial succession
progressively depletes the capacity of the habitat (substrate) rather than building
it up (Garrett, 1955), one might envision a hierarchy of fungal competitors (colonists), each of which in turn became progressively weakened by low nutrition and subjected to the toxins of later members of the sere. Garrett (1970) views it as unlikely that the sugars released by cellulolysis of residual celluose during the later stages of saprophytic mycelial survival in mature plant tissue will reach a high
enough concentration to permit a significant production of antibiotic. He suggests
that the rate of cellulolysis would be self-limiting, because accumulation of sugar
above a certain concentration temporarily inhibits further production of cellulases
as adaptive enzymes. Moreover, the external digestion of cellulose through the excretion of cellulase offers other microorganisms, i.e., "secondary saprophytic sugar
fungi," an ecological niche (Tribe, 1966). The enzymatic machinery of secondary invaders, however, is often more diversified, and such organisms might be utilizing as
energy the carbon from lignocellulosic complexes, the chitinous walls of dead mycelium, and the organic matter released from lysed fungal cells.

V. CONCLUSIONS

I have made an effort in this chapter to examine some possible relationships between
the antagonistic properties of fungi and their ecological status in various fungal
communities. If fungal communities are organized by the same processes that regulate
plant or animal communities (physical factors, competition, and predation), then we
might expect eventually to be able to make predictions as to when and where fungal
antagonists are likely to affect the reproductive potential of the microbes around
them. Such an approach would certainly be important in developing effective strate-

gies for biological control. Moreover, while the antibiotic-producing capabilities of various microbes have been studied intensively during the past 40 years, use of ecological principles to identify potential antagonists (antibiotic producers) has been virtually neglected. Our current repertoire of antibacterial and antifungal antibiotics is based primarily on (1) chance observations of cultural antagonism; (2) efforts to identify potential antagonists from microhabitats in which a particular target organism is either absent or unimportant; (3) general screening programs; (4) specialized screening programs based upon discoveries made in categories 1-3. For those who have read Janzen (1977), it should come as no suprise that the famous penicillin-producing strain of *Penicillium chrysogenum* Thom (NRRL 1951) was isolated from a moldy cantaloupe (Raper and Thom, 1949). In the present chapter, I have attempted to show that there are numerous examples of antibiosis/antagonism which might be examined or even predicted within the framework of certain ecological models of community development.

ACKNOWLEDGMENTS

I am grateful to Drs. Karl Smiley, Michael Swift, Robert Detroy, and John Lussenhop for their helpful discussions. In addition, Prof. Lussenhop both developed and kindly contributed Fig. 2.

REFERENCES

Alexander, M. (1971). *Microbial Ecology*, Wiley, New York.

Angel, K. (1977). *Structure, Composition, and Development of Coprophilous Fungal Communities in Three Grassland Ecosystems*. Ph.D. thesis, University of Pittsburgh, Pittsburgh, Pa.

Baker, K. F., and Cook, R. J. (1974). *Biological Control of Plant Pathogens*, Freeman, San Francisco.

Basham, J. T. (1975). Heart rot of jack pine in Ontario. IV. Heartwood-inhabiting fungi, their entry and interactions within living trees. *Can. J. For. Res.* 5: 706-721.

Bey-Biyenko, G. Ya. (1961). Some pecularities of the formation of the wheat agrobiocenosis fauna under cultivation of virgin steppe. *Proc. 11th Int. Congr. Entomol.*, *Vienna*, Vol. 2, Secs. XII-XIV, pp. 5-7.

Brasier, C. M. (1975). Stimulation of sex organ formation in *Phytophthora* by antagonistic species of *Trichoderma*. II. Ecological inplications. *New Phytol.* 74: 195-198.

Brian, P. W. (1949). Studies on the biological activity of griesofulvin. *Ann. Bot.* (London) 13: 59-77.

Brian, P. W. (1957). The ecological significance of antibiotic production. In *Microbial Ecology*, R. E. O. Williams and C. C. Spicer (Eds.). Cambridge Univ. Press, New York, pp. 168-188.

Brian, P. W., Elson, G. W., and Lowe, D. (1956). Production of patulin in apple fruits by *Penicillium expansum*. *Nature 178*: 263-264.

Brock. T. D. (1966). *Principles of Microbial Ecology*. Prentice-Hall, Englewood Cliffs, N.J.

Brown, J. L. (1964). The evolution of diversity in avian territorial systems. *Wilson Bull. 76*: 160-169.

Bruehl, G. W., Millar, R. L., and Cunfer, B. (1969). Significance of antibiotic production by *Cephalosporium gramineum* to its saprophytic survival. *Can. J. Plant Sci. 49*: 235-246.

Buczacki, S. T. (1973). A microecological approach to larch canker biology. *Trans. Brit. Mycol. Soc. 61*: 315-329.

Bu'Lock, J. D. (1967). *Essays in Biosynthesis and Microbial Development*. Wiley, New York.

Bu'Lock, J. D., Detroy, R. W., Hostalek, Z., and Munim-Al-Sharkarchi, A. (1974). Regulation of secondary biosynthesis in *Gibberella fujikuroi*. *Trans. Brit. Mycol. Soc. 62*: 377-389.

Chang, I-P., and Kommedahl, T. (1968). Biological control of seedling blight of corn by coating kernels with antagonistic microorganisms. *Phytopathology 58*: 1395-1401.

Christensen, M. (1969). Soil microfungi of dry to mesic conifer-hardwood forests in northern Wisconsin. *Ecology 50*: 9-27.

Clark, F. E., Beard, W. E., and Smith, D. H. (1960). Dissimilar nitrifying capacities of soils in relation to losses of applied nitrogen. *Proc. Soil Sci. Soc. Amer. 24*: 50-54.

Connell, J. H. (1978). Diversity in tropical rain forests and coral reefs. *Science 199*: 1302-1310.

Connell, J. H., and Slayter, R. O. (1977). Mechanisms of succession in natural communities and their role in community stability and organization. *Amer. Natur. 111*: 1119-1144.

Cook, R. J., and Bruehl, G. W. (1968). Relative significance of parasitism versus saprophytism in colonization of wheat straw by *Fusarium roseum* "Culmorum" in the field. *Phytopathology 58*: 306-308.

Cowling, E. B., and Merrill, W. (1966). Nitrogen in wood and its role in wood deterioration. *Can J. Bot. 44*: 1539-1554.

Demain, A. (1972). Cellular and environmental factors affecting the synthesis and excretion of metabolites. *J. Appl. Chem. Biotechnol. 22*: 345-362.

England, C. M., and Rice, E. L. (1957). A comparison of the soil fungi of a tall grass prairie and of an abandoned field in central Oklahoma. *Bot. Gaz.* (Chicago) *118*: 186-190.

Funk, A., and McMullan, E. E. (1974). Phacidin, a fungal growth inhibitor from *Potebniamyces balsamicola* var. *boycei*. *Can J. Microbiol. 20*: 422-425.

Gardner, J. M., Mansour, I. S., and Scheffer, R. P. (1972). Effects of the host specific toxin of *Periconia circinata* on some properties of sorghum plasma membranes. *Physiol. Plant Pathol. 2*: 197-206.

Garrett, S. D. (1955). Microbial ecology of the soil. *Trans. Brit. Mycol. Soc. 38*: 1-9.

Garrett, S. D. (1956). *Biology of Root-Infecting Fungi*. Cambridge Univ. Press, New York.

Garrett, S. D. (1970). *Pathogenic Root-Infecting Fungi*. Cambridge Univ. Press, New York.

Gibbs, J. N. (1967). A study of the epiphytic growth habit of *Fomes annosus*. *Ann. Bot.* (London) *31*: 755-774.

Gottlieb, D. (1976). The production and role of antibiotics in soil. *J. Antibiot. 29*: 987-1000.

Harper, J. E., and Webster, J. (1964). An experimental analysis of the coprophilous fungus succession. *Trans. Brit. Mycol. Soc. 47*: 511-530.

Hawker, L. E. (1957). Ecological factors and the survival of fungi. In *Microbial Ecology*, R. E. O. Williams and C. C. Spicer (Eds.). Cambridge Univ. Press, New York, pp. 238-258.

Hering, T. F. (1965). Succession of fungi in the litter of a Lake District oakwood. *Trans. Brit. Mycol. Soc. 48*: 391-408.

Horn, H. S. (1977). Succession. In *Theoretical Ecology*, R. M. May (Ed.). Saunders, Philadelphia, pp. 187-204.

Hudson, H. J. (1968). The ecology of fungi on plant remains above the soil. *New Phytol. 67*: 837-874.

Hulme, M. A., and J. K. Shields (1972). Interaction between fungi in wood blocks. *Can. J. Bot. 50*: 1421-1427.

Hutchinson, G. E. (1957). Concluding remarks. *Cold Spring Harbor Symp. Quant. Biol. 22*: 415-427.

Ikediugwu, F. E. O., and Webster, J. (1970a). Antagonism between *Coprinus heptemerus* and other coprophilous fungi. *Trans. Brit. Mycol. Soc. 54*: 181-204.

Ikediugwu, F. E. O., and Webster, J. (1970b). Hyphal interference in a range of coprophilous fungi. *Trans. Brit. Mycol. Soc. 54*: 205-210.

Ikediugwu, F. E. O., Dennis, C., Webster, J. (1970). Hyphal interference by *Peniophora gigantea* against *Heterobasidion annosum*. *Trans. Brit. Mycol. Soc. 54*: 307-309.

Janzen, D. (1977). Why fruits rot, seeds mold, and meat spoils. *Amer. Natur. 111*: 691-713.

Jefferys, E. G. (1952). The stability of antibiotics in soils. *J. Gen. Microbiol. 7*: 295-312.

Jefferys, E. G., Brian, P. W., Hemming, H. G., and Lowe, D. (1953). Antibiotic production by the microfungi in acid heath soils. *J. Gen. Microbiol. 9*: 314-341.

Kendrick, W. B., and Burges, A. (1962). Biological aspects of the decay of *Pinus silvestris* leaf litter. *Nova Hedwigia 4*: 313-342.

Levi, M. P., Merrill, W., and Cowling, E. B. (1968). Role of nitrogen in wood deterioration. VI. Mycelial fractions and model nitrogen compounds as substrates for growth of *Polyporus versicolor* and other wood-destroying and wood-inhabiting fungi. *Phyopathology 58*: 626-634.

Ludwig, R. A. (1957). Toxin production of *Helminthosporium sativum* P. K. & B. and its significance in disease development. *Can. J. Bot. 35*: 291-303.

MacArthur, R. H., and Wilson, E. O. (1967). *The Theory of Island Biogeography*. Princeton Univ. Press, Princeton, N.J.

Macer, R. C. F. (1961a). Saprophytic colonization of wheat straw by *Cercosporella herpotrichoides* Fron and other fungi. *Ann. Appl. Biol. 49*: 152-164.

Macer, R. C. F. (1961b). The survival of *Cercosporella herpotrichoides* Fron in wheat straw. *Ann. Appl. Biol. 49*: 165-172.

McNaughton, S. J., and Wolf, L. L. (1973). *General Ecology*, Holt, Rinehart and Winston, New York.

Marx, D. H. (1973). Mycorrhizae and feeder root diseases. In *Ectomycorrhizae: Their Ecology and Physiology*, G. C. Marks and T. T. Kozlowski (Eds.). Academic Press, New York, pp. 351-382.

Marx, D. H., and Davey, C. B. (1969). The influence of ectotrophic mycorrhizal fungi on the resistance of pine roots to pathogenic infections. III. Resistance of aseptically formed mycorrhizae to infection by *Phytophthora cinnamomi*. *Phytopathology 59*: 549-558.

Meredith, D. S. (1960). Further observations on fungi inhabiting pine stumps. *Ann. Bot.* (London) [N.S.] *24*: 63-78.

Meyer, J. A. (1972). The ecological significance of toxin production by microorganisms. In *Phytotoxins in Plant Diseases*, R. K. S. Wood, A. Ballio, and A. Graniti (Eds.). Academic Press, New York, pp. 331-343.

Miller, R. S. (1967). Pattern and process in competition. *Advan. Ecol. Res 4*: 1-74.

Muller, C. H. (1966). The role of chemical inhibition (allelopathy) in vegetational composition. *Bull. Torrey Bot. Club 93*: 332-351.

Odum, E. P. (1969). The strategy of ecosystem development. *Science 164*: 262-270.

Park, D. (1965). Survival of microorganisms in soil. In *Ecology of Soil-borne Plant Pathogens*, K. F. Baker and W. C. Snyder (Eds.). Univ. of California Press, Berkeley, pp. 82-97.

Park, D. (1967). The importance of antibiotics and inhibitory substances. In *Soil Biology*, A. Burges and F. Raw (Eds.). Academic Press, New York, pp. 435-447.

Parkinson, D., and Kendrick, W. B. (1960). Investigations of soil microhabitats. In *The Ecology of Soil Fungi*, D. Parkinson and J. S. Waid (Eds.). Liverpool Univ. Press, Liverpool, England, pp. 22-28.

Pinck, L. A., Holton, W. F., and Allison, F. E. (1961a). Antibiotics in soil. I. Physico-chemical studies of antibiotic-clay complexes. *Soil Sci. 91*: 22-28.

Pinck, L. A., Soulides, D. A., and Allison, F. E. (1961b). Antibiotics in soil. II. Extent and mechanisms of release. *Soil Sci. 91*: 94-99.

Ranzoni, F. V. (1968). Fungi isolated in culture from soils of the Sonoran Desert. *Mycologia 60*: 356-371.

Raper, K. B., and Thom, C. (1949). *A Manual of the Penicillia*. Williams & Wilkins, Baltimore.

Rayner, A. D. M. (1975). *Fungal Colonization of Hardwood Tree Stumps*. Ph.D. thesis, University of Cambridge, England.

Rayner, A. D. M. (1977a). Fungal colonization of hardwood stumps from natural sources. I. Non-basidiomycetes. *Trans. Brit. Mycol. Soc. 69*: 291-302.

Rayner, A. D. M. (1977b). Fungal colonization of hardwood stumps from natural sources. II. Basidiomycetes. *Trans. Brit. Mycol. Soc. 69*: 302-312.

Ricard, J. L., and Bollen, W. B. (1968). Inhibition of *Poria carbonica* by *Scytalidium* sp., and imperfect fungus isolated from Douglas-fir poles. *Can. J. Bot. 46*: 643-647.

Rice, E. L. (1974). *Allelopathy*. Academic Press, New York.

Rishbeth, J. (1950). Observations on the biology of *Fomes annosus*, with particular reference to East Anglian pine plantations. I. The outbreaks of disease and ecological status of the fungus. *Ann. Bot.* (London) *14*: 365-383.

Rishbeth, J. (1951). Observations on the biology of *Fomes annosus*, with particular reference to East Anglian pine plantations. II. Spore production, stump infection, and saprophytic activity in stumps. *Ann. Bot.* (London) [N.S.] *15*: 1-21.

Rishbeth, J. (1963). Stump protection against *Fomes annosus*. III. Inoculation with *Peniophora gigantea*. *Ann. Appl. Biol. 52*: 63-77.

Saitô, T. (1966). Sequential pattern of decomposition of beech litter with special reference to microbial succession. *Ecol. Rev. Sendai 16*(4): 245-254.

Scarborough, A. M. (1970). The soil microfungi of a Colorado grassland. M.S. thesis, University of Wyoming, Laramie.

Sewell, G. W. F. (1959). Studies of fungi in a *Calluna*-heathland soil. II. By the complementary use of several isolation methods. *Trans. Brit. Mycol. Soc. 42*: 354-369.

Singh, N., and Webster, J. (1973). Antagonism between *Stilbella erythrocephala* and other coprophilous fungi. *Trans. Brit. Mycol. Soc. 61*: 487-495.

Soulides, D. A., Pinck, L. A., and Allison, F. E. (1962). Antibiotics in soils. V. Stability and release of soil-adsorbed antibiotics. *Soil Sci.* 94: 239-244.

Stanley, V. C., and English, M. P. (1965). Some effects of nystatin on the growth of four *Aspergillus* species. *J. Gen Microbiol.* 40: 107-118.

States, J. S. (1978). The soil fungi of cool-desert plant communities in northern Arizona and southern Utah. *J. Ariz.-Nev. Acad. Sci.* 13: 13-17.

Stillwell, M. A. (1966). A growth inhibitor produced by *Cryptosporiopsis* sp., an imperfect fungus isolated from yellow birch, *Betula alleghaniensis* Britt. *Can. J. Bot.* 44: 259-267.

Stillwell, M. A., Wood, F. A., and Strunz, G. M. (1969). A broad-spectrum antibiotic produced by a species of *Cryptosporiopsis*. *Can. J. Microbiol.* 15: 501-507.

Stillwell, M. A., Wall, R. E., and Strunz, G. M. (1973). Production, isolation, and antifungal activity of scytalidin, a metabolite of *Scytalidium* species. *Can. J. Microbiol.* 19: 597-602.

Strunz, G. M., Court, A. S., Komlossy, J., and Stillwell, M. A. (1969). The structure of cryptosporiopsin, a new antibiotic substance produced by a species of *Cryptosporiopsis*. *Can. J. Chem.* 47: 2087-2094.

Strunz, G. M., Kakushima, M., and Stillwell, M. A. (1972). Scytalidin: A new fungi toxic metabolite produced by a *Scytalidium* species. *J. Chem. Soc. Perkin Trans.* 1: 2280-2283.

Swift, M. J. (1976). Species diversity and the structure of microbial communities in terrestrial habitats. In *The Role of Terrestrial and Aquatic Organisms in Decomposition Processes*, J. M. Anderson and A. Macfadyen (Eds.). Blackwell Sci. Publns., Oxford, England, pp. 185-222.

Tribe, H. T. (1966). Interactions of soil fungi on cellulose film. *Trans. Brit. Mycol. Soc.* 49: 457-466.

Tveit, M., and Moore, M. B. (1954). Isolates of *Chaetomium* that protect oats from *Helminthosporium victoriae*. *Phytopathology* 44: 686-689.

Tveit, M., and Wood, R. K. S. (1955). The control of *Fusarium* blight in oat seedlings with antagonistic species of *Chaetomium*. *Ann. Appl. Biol.* 43: 538-552.

Webster, J. (1970). Coprophilous fungi. *Trans. Brit. Mycol. Soc.* 54: 161-180.

Webster, J., and Lomas, N. (1964). Does *Trichoderma viride* produce gliotoxin and viridin? *Trans. Brit. Mycol. Soc.* 47: 535-540.

Whittaker, R. H. (1970). The biochemical ecology of higher plants. In *Chemical Ecology*, E. Sondheimer and J. B. Simeone (Eds.). Academic Press, New York, pp. 43-70.

Wicklow, D. T., and Hirschfield, B. J. (1979). Evidence of competitive hierarchy among coprophilous fungal populations. *Can. J. Microbiol.* 25: 855-858.

Wicklow, D. T., and Whittingham, W. F. (1974). Soil microfungal changes among the profiles of disturbed conifer-hardwood forests. *Ecology* 55: 3-16.

Widden, P., and Parkinson, D. (1973). Fungi from Canadian coniferous forest soils. *Can. J. Bot.* 51: 2275-2290.

Wood, R. K. S., and Tveit, M. (1955). Control of plant diseases by use of antagonistic organisms. *Bot. Rev.* 21: 441-492.

Wright, J. M. (1956a). The production of antibiotics in soil. III. Production of gliotoxin in wheatstraw buried in soil. *Ann. Appl. Biol.* 44: 461-466.

Wright, J. M. (1956b). The production of antibiotics in soil. IV. Production of antibiotics in coats of seeds sown in soil. *Ann. Appl. Biol.* 44: 561-566.

Wright, J. M. (1956c). Biological control of a soil-borne *Pythium* infection by seed inoculation. *Plant and Soil* 8: 1-9.

Zak, B. (1964). Role of mycorrhizae in root disease. *Ann. Rev. Phytopathol.* 2: 377-392.

Part   V

FUNGAL COMMUNITY DEVELOPMENT

Chapter 20

COMMUNITY PATTERNS AND PROCESSES:
A MULTIVARIATE PROBLEM

K. A. Kershaw
McMaster University
Hamilton, Ontario, Canada

I. INTRODUCTION

It came as something of a surprise to plant ecologists to discover that plants were not distributed at random (Gleason, 1920; Svedberg, 1922), and very few ready explanations were available for this phenomenon. There is now, however, a considerable body of evidence documenting the ubiquitous nature of nonrandomness, or pattern, in vegetation. Greig-Smith (1952) provided the statistical technique which utilized density data, to estimate the scale of pattern present and focused attention on the actual physical dimensions of a pattern. The extension of pattern analysis to percentage cover data (Kershaw, 1957) effectively allowed pattern analysis of vegetation types, normally intractable to density analysis, and the causality mechanisms of pattern were subsequently extensively and intensively studied for the next decade. It is now relevant to examine pattern and its dimensionality in the multivariate ecological framework of the late 1970s and early 1980s, especially in relation to the current topic of this text. Mycologists have not traditionally accepted lichens simply as ascomycetes (despite the protestations of lichenologists), so it is convenient here to use lichen-dominated heath in northern Ontario as illustrative material of pattern dimensionality.

II. THE PHYSICAL DIMENSIONS OF PATTERN

A typical primary scale of pattern which is classed as a morphological pattern is shown in Fig. 1, where the fruiting bodies of a species of white rust infecting *Ledum groenlandicum* are contagiously distributed. Morphological pattern in vegetation usually represents the primary pattern scale exhibited by a species (Kershaw, 1973) and is a function of the form of the rhizome branching and especially the physical limits of vegetative spread. In the example shown here (Fig. 1) it simply represents the extent of the subepidermal hyphal development of the *Albugo* sp., although within such a pattern scale there may be smaller patterns reflecting, for example, the actual mode of hyphal penetration into the leaf and its subsequent pathway of extension. Primary morphological scales of pattern in vegetation can also reflect the vegetative vigor of the individuals of a species. Thus the vegetative

Fig. 1 A clumped distribution of the fruiting bodies of an *Albugo* sp. infecting *Ledum groenlandicum*. This primary pattern scale is typical of many morphological patterns found in plants.

Fig. 2 Secondary and tertiary pattern scales of *Albugo*, reflecting the physiographic range of pattern scales of *L. groenlandicum*.

spread of an individual plant is dependent, for instance, on the nutrient status of the soil as well as the habitat microclimate, and of course on the intrinsic response of the species to such parameters. Similarly the extent of infection of the *Ledum* host (Fig. 1) reflects at least the chemical climate, subepidermal microclimate, and general leaf microclimate of the *Ledum* host as it interacts with the vigor of the *Albugo* sp. This implies that beyond these primary morphological pattern scales there are a further secondary series of patterns. These are termed *environmental patterns* (Kershaw, 1973) and reflect the control by a complex of environmental parameters. In our example here, nonrandom environmental complexes at the habitat level lead to the correlated and equally nonrandom distribution of *L. groenlandicum* (Fig. 2). At this much larger scale, there will be an associated and secondary pattern scale for *Albugo*. This very obvious patterning of *L. groenlandicum*, usually below a shallow raised-beach ridge, is a widespread feature of the coastal tundra in northern Ontario and reflects the overall control of the growth and performance of *Ledum* by a number of patterned environmental parameters.

III. THE MULTIVARIATE ASPECT OF PATTERN
     AND THE TIME DIMENSION

Similar environmental patterns are equally evident at an equivalent scale in the other associated species in this heath vegetation, many of them lichens characteristic of the tundra zone in northern Ontario. Thus there is usually a zone of *Cladonia stellaris* (Opiz.) Pouz & Vezda on the lower ridge slopes, with *Alectoria nitidula* (Th. Fr.) Vain forming a distinct zone on the ridge crest (Fig. 3). The distinctness of these two zones is particularly evident as a color contrast between the pale green thallus of *C. stellaris* and the black thallus of *A. nitidula*. The spatial pattern is thus amplified by the marked color contrast. An analysis of the environmental parameters interacting with these two lichen species over the beach profile reveals a

Fig. 3 Marked and contrasting physiographic pattern scales in lichen species growing on raised beaches in the taiga of northern Ontario.

typical multivariate situation: The ridge crest is considerably more xeric in its characteristics with higher rates of evaporation from the lichen thalli, largely as a function of higher average wind speeds during low-energy conditions, which contrasts markedly with the relatively mesic *C. stellaris* zone. The ridge crest is also cooler, with both thallus and air temperature on average being several degrees below the more sheltered and warmer lower slopes. The physiological response of *A. nitidula* and *C. stellaris* in terms of net photosynthetic rate, and hence carbohydrate supplies to the mycobiont, clearly reflects these differences in microclimate. Maximum rates of net photosysthesis are only developed in *A. nitidula* at low levels of thallus moisture and at thallus temperatures below 20°C. Conversely, *C. stellaris* has an appreciable net photosynthetic rate of thallus saturation which is maintained up to 25°C (Fig. 4).

In addition to these physical variables, however, there is an additional and important time component. In winter and early spring the ridge microclimate is quite different in the two zones, with some 2 m of snow over the lower slopes of the ridge but only a few centimeters over the ridge crest. The *Alectoria* zone is exposed quickly in early spring (Fig. 5), and probably a considerable proportion of its metabolic activity occurs during this spring melt period. Under the still air conditions ocurring in the snow melt pockets which have developed during thaw periods, the dark thallus of *A. nitidula* is several degrees warmer than an equivalent but pale-colored thallus would be (e.g., *Alectoria ochroleuca*; see Fig. 6). Thus, even at air temperatures of -4°C the dark *Alectoria* thallus remains above freezing point and hence is metabolically active.

The time component in this multivariate situation is also important in yet another sense; the temperature optimum of net photosynthesis acclimates markedly on a

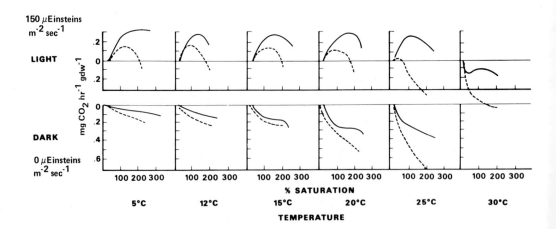

Fig. 4 The contrasting net photosynthetic and respiratory response of *C. stellaris* (———) and *A. nitidula* (- - - -) to moisture and temperature, under 150 and 0 μEinsteins illumination.

Fig. 5 Late winter physiographic conditions, with secondary but temporal environmental patterns clearly visible.

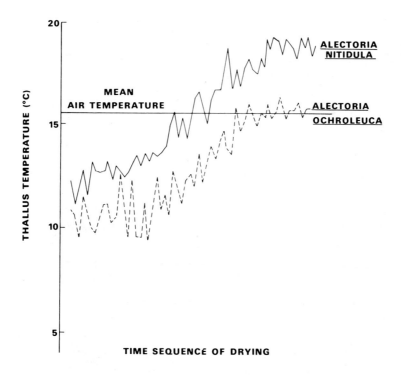

Fig. 6 Contrasting thallus temperatures under identical drying conditions for the dark brown thalli of *A. nitidula* and the pale green thalli of *A. ochroleuca*.

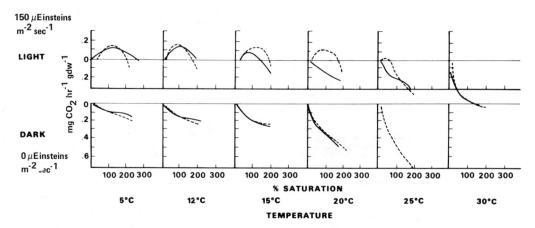

Fig. 7 The summer (----) and winter (——) pattern of net photosynthetic and respiratory response of *A. nitidula* to temperature and moisture at 150 and 0 µEinsteins illumination.

seasonal basis (Fig. 7). The summer response shows almost optimal levels of net photosynthesis at 20°C, but in winter the temperature optimum is apparently 5-10°C and at 20°C the net photosynthetic response is totally respiration dominated.

It is apparent, then, that every scale of pattern in vegetation is usually--and at least potentially--a multivariate situation. Some of the variables will be static, but many of the variables will be dynamic. Likewise, the metabolic response of a species may be quite plastic and time may become a dominant variable.

IV.   ABSTRACT DIMENSIONS AND PATTERNS

The extent of the defined correlation between the environmental parameters operating over the beach ridge and the pattern of *A. nitidula* or *C. stellaris* complements the idea of an ultimate level of species adaptedness. Implicit is the interchange of an ultimate level of species vector with an environmental vector, both in terms of the vector direction (trend of correlation) and the vector length (species abundance/ operative level of the environmental factor). Although each macropattern situation is controlled by multiple environmental parameters, it is equally reflected by a number of plant species, not just one. Thus, for example, a group of species vectors can be replaced by one generalized component vector that effectively summarizes those species which are largely correlated with the xeric, cool ridge top. The inverse of the foregoing statement is also implicit: A single generalized component vector can replace a group of correlated environmental variables. Furthermore, noncorrelated environmental or species variables (vectors) are orthogonal, and thus a plant association which represents the ultimate pattern scale can in fact be defined dimensionally, where the dimensions are either the species abundances or conversely environmental measures. These macropatterns are in a sense abstractions and can be more conveniently handled by ordination or classification methods than by traditional pattern analysis. Conceptually the numerical variability of the specific composition of each replicate stand of an association leads to the nodum (Poore, 1955). Thus a plant association has an average composition, and replicates from this population of stands with a composition close to this average occur more frequently than do replicates near the two tails of the distribution. This realization parallels a similar concept in pattern analysis; it is the nodal patterns which reflect the intrinsic patterns of the community.

V. SUMMARY

Community patterns exist at multiple scales; they are controlled by mirror images of time-space environmental patterns and are also reflected in species physiological response patterns. These physiological patterns can also contain a time component. The concept of the substitution of a species vector with an environmental vector underlines both the dimensionality of plant associations and the concept of complete species adaptedness.

REFERENCES

Gleason, H. A. (1920). Some applications of the quadrat method. *Bull. Torrey Bot. Club 47*: 21-33.

Greig-Smith, P. (1952). The use of random and contiguous quadrats in the study of the structure of plant communities. *Ann. Bot.* (London) [N.S.] *16*: 293-316.

Kershaw, K. A. (1957). The use of cover and frequency in the detection of pattern in plant communities. *Ecology 38*: 291-299.

Kershaw, K. A. (1973). *Quantitative and Dynamic Plant Ecology*, 2nd ed. Arnold, London.

Poore, M. E. D. (1955). The use of phytosociological methods in ecological investigations. IV. General discussions of phytosociological problems. *J. Ecol. 44*: 28-50.

Svedberg, T. (1922). Ettbidrag till de statiska metodernas användning inom Växtbiologien. *Svensk Bot. Tidskr. 16*: 1-8.

Chapter 21

PATTERNS OF PHENOLOGY AMONG FUNGAL POPULATIONS

Paul Widden

Loyola Campus, Concordia University
Montreal, Quebec, Canada

I. INTRODUCTION

A. Definition of Phenology

In the study of plant ecology, phenology forms an important discipline as it is an essential part of the adaptive strategy employed by plants to survive in their particular habitats. As defined by a plant ecologist, "The study of biologic periodicity in relation to the seasonal sequence of climatic factors is called phenology" (Daubenmire, 1968). Daubenmire further states that "events such as the development of new foliage, pollination, root growth and seed dissemination ... are all aspects of the phenology of plants." This definition has been expanded by Mueller-Dombois and Ellenberg (1974) who state: "Phenological changes do not involve only the flowering, fruiting, leaf emergence and leaf fall of trees or other perennials but also the periodic appearance and disappearance of therophytes (annuals) and geophytes (plants perennating from rhizomes, bulbs or tubers)."

B. Phenology of Plants

In the temperate deciduous forests of the northeastern region of North America phenological changes in flowering plants are quite obvious, even to the casual observer. In one of my Quebec study sites the forest floor in April and early May produces a spectacular display of spring flowers, such as the large-flowered trillium (*Trillium grandiflorum* Michx, Salisb.), bloodroot (*Sanguinaria canadensis* L.), and the dog's-tooth violet (*Erythronium americanum* Ker-Gawl). These flowers appear before the trees forming the upper canopy [mainly the sugar maple, *Acer saccharum* Marsh., bitternut hickory, *Carya cordiformis* (Wang.) K. Koch and the hop-hornbeam, *Ostrya virginiana* (Mill.) K. Koch] have leafed out and therefore occluded a major portion of the available light energy. By the time the trees are in full leaf, the majority of these flowers have finished blooming and set seed, and by July even the leaves of some of these species (notably *E. americanum* have disappeared. During the summer and fall, very few herbaceous plants are seen flowering, though some such as the indian pipe (*Monotropa uniflora* L.), wood aster (*Aster acuminatus* Michx.), and broad-leaved goldenrod (*Solidago flexicaulis* L.) are found. The only other plants found on the forest floor at this time of the year are mosses, tree seedlings, and a variety of ferns, all of which are shade tolerant.

The foregoing examples illustrate at least three strategies that are applied by plants, based on phenology, whereby they are surviving in the deciduous forest habitat. The deciduous trees, by means of their tall stature and large mass, are competing with each other to form part of the canopy and to intercept the light directly from the sun. The spring flowers, by storing food over winter, either in fleshy rhizomes (*T. grandiflorum*, *S. canadense*) or in corms (*E. americanum*), are able to develop in spring when the ground warms up but before the trees have leafed out. They flower, set seed, and perform much of their photosynthetic activity in the brief period when light intensity on the forest floor is high, before the tree leaves reduce the light intensity. The other plants do not store large quantities of food over winter but instead appear to be adapted to low light intensities and therefore develop slowly during spring and flower or form spores (the ferns) later in the year. *M. uniflora* is well adapted to low light intensities, as it relies not on photosynthesis but on saprophytism for energy.

Some useful generalizations that emerge from phenological studies of higher plants which may be applicable to fungal populations are as follows:

1. Phenological information gives valuable clues to the strategies which organisms employ in order to survive in their particular habitat.
2. The form of the organism dictates to some extent the possible phenological responses; for example, the production of a rhizome by *T. grandiflorum* enables it to store food during winter and therefore to leaf out and flower early in the spring.
3. Larger organisms (such as trees) have to put much of their energy into maintaining structure and are therefore less likely to be able to respond rapidly to environmental change than are smaller organisms (such as herbaceous plants and microorganisms).
4. When phenological responses are an important part of the organism's adaptive strategy, responses to the environment will be genetically programmed in a manner that ensures an appropriate response to environmental cues.

It is the purpose of this chapter to examine the extent to which phenology of fungi has been studied and to determine the applicability of the above generalizations to fungal populations.

C. Events in Fungal Life Cycles

There are many significant events in fungal life cycles, the timing of which in relation to climatic change may be important. These include the production of asexual and sexual spores, the production of resting structures or resistant bodies, germination of spores, periods of growth and assimilation, fruit body initiation and development, sexual events, and--for parasites--the infection of hosts or the trans-

fer from one host to another. The study of phenology in fungi should therefore include the timing of these events within naturally occurring field populations as well
as laboratory studies of both the environmental triggers that cue these events and
their genetic control. Thus, the study of the effects of temperature, light, and
nutrient status on sporulation, sclerotium formation, etc. as well as fungistasis
and the breaking of dormancy in spores would be important to an understanding of fungal phenology. In keeping with Mueller-Dombois and Ellenberg's (1974) definition of
phenology quoted earlier, the cyclic appearance and disappearance of microfungi in
the soil is also an important aspect of the phenology of fungal populations.

II. PHENOLOGY OF PARASITIC FUNGI

Though this chapter deals mainly with saprophytes, some mention of parasitic fungi
should be made. Parasitic fungi often have complex life cycles involving many morphological phases. It is clear that the life cycles of these fungi are synchronized
with those of their hosts. As the host (particularly plant hosts) exhibits periodic
behavior in response to the environment, so must the parasitic fungus. I am sure
that when Austwick (1968) wrote "The slow, inexorable seasonal changes are closely
bound with fungal life cycles and hence are almost part of the taxa," he was thinking mainly of the parasitic fungi. The relationship between fungal life cycles and
the host plant is reviewed by Wheeler (1968). In parasitic fungi it is difficult to
separate the direct effects of the environment on development from those of the host
physiology. It is clear, however, that the environment plays an important direct
role, both through dormancy mechanisms (Sussman, 1966) and in determining the type
of spore produced. Thus, in cool summers it appears that *Puccinia antirrhini* only
produces urediospores but that in hot summers it may produce teliospores (Hawker,
1966).

III. PHENOLOGY OF SAPROPHYTIC FUNGI

A. Comparison between Macro- and Microfungi

In comparison with the parasitic fungi, it appears that very little work has been
done on the phenology of saprophytic fungi. Indeed, from a computer search of the
literature which I performed before writing this chapter, it would appear that the
word *phenology* is not generally part of the vocabulary of mycologists studying saprophytes. There is, however, a body of literature that is relevant to an understanding
of the phenology of saprophytic fungi. For convenience of discussion I wish to divide
the fungi into two basic groups--the *macrofungi*, which are usually detected by their
formation of macroscopic sporocarps, and the *microfungi*, usually dectected and enumerated by some form of plating and isolation method. It could further be said that
many of the macromycetes have a comparatively long-lived mycelium, which is often
perennial, and utilize "hard" substrates such as lignin and humic compounds, whereas
most micromycetes are comparatively short-lived and utilize "soft" substrates such as

simple sugars. In a sense, those macrofungi which have a perennial mycelium are analogous to the geophytes of Raunkiaer (1934), whereas those microfungi with a short-
lived mycelium and heavy spore production are analagous to the annuals or therophytes.

B. Phenology of Macromycetes

Macrofungi with a perennial mycelium must undergo a number of phases during the annual cycle. In periods of adverse climate such as the hard winters of continental
America or Europe, some form of dormancy occurs, accompanied by the production of
specialized structures such as sclerotia and/or the onset of metabolic arrest in the
vegetative hyphae. When conditions are favorable, growth must be initiated. At some
time the initiation of fruit-body production must occur, followed by differentiation,
maturation, sporulation, and eventually senescence. It is evident that these different phases must occur under different conditions. For many fungi, growth and sporulation usually have different temperature optima (see the review by Hawker, 1966).
Frankland (1966) showed that the period of maximum mycelial biomass of basidiomycetes
in an English deciduous woodland occurred during the summer but that maximum fruiting
occurred in the fall.

    The timing of fruiting varies considerably for different macrofungi. Many dis-
comycetes such as *Morchella* and *Sarcoscypha* fruit regularly in the spring, whereas
many agarics (such as *Amanita* and *Agaricus* species) fruit in the fall. In my own
study area in southern Quebec, the giant puffball *Calvatia gigantea* (Batsch ex Pers.)
Lloyd fruits regularly in the last two weeks of August. While the regular fruiting
of these fungi appears to be part of the common knowledge of mycologists and is mentioned in the numerous popular field guides to the identification of mushrooms, there
are very few studies that examine in depth the seasonal nature of macromycete sporocarp production.

    The paper by Matveev (1972) is the only recent report that I am aware of which
attempts to explain in any detail the regular annual appearance of fruit bodies in
the field. Matveev examined the records of fruiting of basidiomycetes in the Leningrad area over a number of years and divided the mushrooms into early, summer, and
late species. This investigator concluded that the timing of fruit-body production
depended on both the initiation of mycelial growth in the spring and the time taken
for mycelial growth before fruit-body development could occur. The initiation of
mycelial growth was found to depend on the heat store in the soil and the availability of moisture. Once the heat store in the soil (calculated as the temperature sum
of degrees above 0°C) reached a critical level, a rainfall in excess of 10 mm would
initiate mycelial growth. The critical temperature sum for early mushrooms was 500°C,
for summer mushrooms 800°C, and for late mushrooms 1000°C. Once growth was initiated,
there was a characteristic time period for development of fruit bodies for each species. For *Russula claroflava*, an early mushroom, the growth period was 28 days; for
the summer mushroom *Boletus edulis*, it was 36 days; and for the fall mushroom

B. luteus, it was 46 days. From these data Matveev concludes that if one allows for a longer growth time when temperature or rainfall drop below the mean and a shorter growth time when temperature or precipitation exceed the mean, it is possible to predict fairly accurately the fruiting of many mushrooms.

It would be of interest to duplicate the kinds of study just described in other geographical regions to see how fixed the behavior of individual species might be. There is some evidence that mushroom species can show phenological adaptations for growth in extreme climates. Thus Miller (1977) points out that in the Arctic mushrooms have smaller fruit bodies and fruit over a shorter period of time than they do in more favorable climates. Savile (1972) has noted that on Somerset Island, in the Canadian Arctic, the puffball *Calvatia tatrensis* may take more than 1 year for maturation of the fruit body. Savile also noted that many agarics fruit irregularly in the Arctic, though some at least (*Agaricus arvensis*, on Ellesmere Island) produce normal-sized sporocarps. I have observed on Devon Island that *Russula* sp. produced normal-sized fruit bodies, though they do not fruit every year. Saville suggested that this behavior may imply that the mycelium of some fungi in the Artic may grow in the soil for several seasons, accumulating nutrients until such time as a normal-sized fruit body can be produced.

Frankland's (1975) calculation at 0.05 µg/ha annual sporocarp production by 1.19 kg/ha vegetative mycelium suggests that the allocation of resources to reproduction can be small and that even for macromycetes a preoccupation with the phenology of fruiting alone may be unjustified. The importance of the growth of mycelium before fruiting is illustrated by Matveev's work (1972); already cited. The formation of sclerotia underground is also of great importance to macrofungi. According to Austwick (1968), "Sclerotia are perhaps the triumph over adversity of the soil-inhabiting larger fungi, for they act as collecting foci for nutrients obtained by far-reaching hyphae and allow a protracted period of grace before the fungus gathers the momentum to form a fruiting body." In spite of this, there is little information concerning either the timing of sclerotium formation in the field by macromycetes or which species are capable of forming sclerotia. It is significant, however, that the discomycetes, many of which fruit in the spring, are a group in which sclerotia are commonly formed. Presumably, in these cases the ability to carry an extensive food supply in the form of the sclerotium would permit early fruiting.

In the case of spring-fruiting agarics the fungi may carry a food reserve over the winter; alternatively, they may remain metabolically active under a cover of snow and derive substantial energy from decomposition of substrates during the winter months. For example, Stark (1972) has shown that 80% of the first-year weight loss of surface litter in a stand of Jeffrey pine (*Pinus jeffreyi*) at 2013 m in Nevada occurs during the winter months when the litter is colonized by basidiomycetous fungi with dark hyphae.

C. Laboratory Studies of Fruit Body and Sclerotium Production

In comparison with the paucity of data on the production of fungal structures in the
field, there is a relatively extensive literature on the conditions required to initiate the fruiting of macrofungi in the laboratory. More recently, there has also
been extensive research into the factors involved in sclerotium formation, especially by plant pathogens.

    Reviews by Taber (1966), Volz and Beneke (1969), and Smith and Berry (1974)
stress the importance of light, temperature, humidity, pH, and CO^2 tension for the
production of basidiomycete fruit bodies. Generally, fruiting is favored by low CO^2
tensions, high humidity, and temperatures in the mesophilic range, whereas response
to light varies greatly according to the species being studied. Studies by McLaughlin (1970) showed that *Boletus rubinellus* initiates primordia in the absence of light
but requires light for pileus and stipe development. In *Coprinus domesticus* (Chapman
and Fergus, 1973) and in *Favolus arcularius* (Kitamoto et al., 1974), light is required
for primordium initiation, stipe elongation, and differentiation of the pileus. In
Agaricus bisporus, however, Koch (1958) showed that light inhibits the initiation and
development of fruit bodies. The relevance of these observations to the behavior of
fungi in the field has not generally been discussed, though the statement by Buller
(1924) that soil mushrooms are not phototropic whereas lignicolous and coprophilous
ones are may be of ecological importance and therefore worthy of further investigation.

    A recent review by Chet and Henis (1975) indicates that sclerotium formation is
influenced by factors similar to those affecting fruit-body formation. The conclusions appear to be that light varies in its effects from species to species, that temperature and pH are optimal for sclerotium production when they are optimal for mycelial growth, and that staling products and increase in carbohydrates stimulate
sclerotial production. Low O^2 concentration and high CO^2 concentration apparently
inhibit sclerotium production. These data, however, come mainly from studies of
Sclerotium spp. and *Verticillium* spp., both plant pathogenic micromycetes. It would
therefore by foolhardy to draw any conclusions concerning the behavior of soil macromycetes.

D. Phenology of Micromycetes in Soil

The microfungi are generally short lived; most of them are Fungi Imperfecti and therefore produce no fruit bodies. Some do, however, produce resistant structures such as
chlamydospores and microsclerotia, and those microfungi which have a sexual phase may
have sexual spores that are more persistent and resistant than the conidia. Due to
various methodological problems, however, seasonal studies of soil micromycetes have
been concerned more with changes in propagule density than with changes in the form
of the organism in the soil.

1. Quantitative changes in the fungal community

Studies designed to evaluate quantitative changes in the total community of soil fungi, using either the dilution plate method or direct hyphal measurements, usually indicate clear, seasonal trends (Wright and Bollen, 1961; Frankland, 1966; Nagel-de-Boois and Jansen, 1966, 1971; Prakash and Khan, 1971; Nair, 1973; Widden and Parkinson, 1973; Miller and Laursen, 1974; Toth et al., 1975). These trends may show maximum propagule or hyphal densities in the summer, or separate spring and fall maxima, or--in hot climates--the maxima may be in the cooler months with a decline in propagules during the hot, dry summers (Moubasher and El-Dohlob, 1970).

In a Hungarian oak forest, Toth et al. (1975) used the dilution plate method simultaneously with the nylon mesh method of Waid and Woodman (1957) to study changes in the fungal community. Toth and colleagues found that the maximum fungal plate counts occurred during November, whereas the maximum hyphal activities occurred during the summer. This suggests that the fungal community as a whole in these forests is highly active during the summer, existing in the mycelial phase, whereas during fall sporulation is occurring; presumably these fungi overwinter as spores which will germinate in spring. The work of Nagel-de-Boois and Jansen (1971), using the nylon mesh method in deciduous woods in Holland, suggests a different pattern. They found that hyphal activity, particularly in the organic layers, was highest in the spring and the fall. The authors suggest that the summer decline in hyphal activity was probably due to a lack of nutrients rather than growth inhibition by climatic factors. One could speculate that in Holland, with its mild winters, decomposition is occurring, though at a slower rate, during winter, resulting in an exhaustion of nutrients by midsummer. This would result in increased activity in the fall when litter drops to the ground, a reduction in winter due to low temperature, and then an increase in spring with the increase in temperature. Under these conditions, it is reasonable to postulate that sporulation would occur in late spring, with the spores remaining dormant during summer while nutrients are low. The observation of Dobbs et al. (1960) that in English woodlands, which also have mild winters, fungistasis is at a maximum in midsummer would then take on a increased ecological significance. Work of a similar nature to that of Toth et al. (1975), i.e., using the two aforementioned methods simultaneously, would give strong indications as to whether these speculations have any validity.

2. Seasonal change in fungal populations

Given that seasonal trends in the microfungal community of the soil have been established, it would seem reasonable to assume that such trends could also be established for individual species populations.

There have been a number of studies of microfungi occurring on litter, and regular, repeating patterns of occurrence have been observed that can be related to the change in litter quality and to competetive pressure during decay (see reviews by Hudson, 1968; Bell, 1974; Frankland, 1974; Jensen, 1974; Katz and Lieth, 1974; Millar,

1974). Kendrick and Burges (1962), in their study of pine-forest litter in England, looked for seasonal changes in litter fungi and could not detect any regular seasonal patterns which could be related directly to climate, though on the needles it could be seen that *Aureobasidium pullulans* was at a maximum just after leaf fall and then declined during the late fall-early winter period. A recent study by McKenzie and Hudson (1976) of the change in fungus flora over a 7-month period on detached leaves showed that some fungi (*Cladosporium* spp., *Aureobasidium pullulans*, and *Eipcocum purpurascens*) on wheat and poplar leaves had high initial occurrences in the fall, which rapidly declined, whereas others (*Phomopsis perniciosa*, *Pleospora phaeocomoides*, and *Phialophora* spp.) had higher occurrences in late winter-early spring. It is not clear, however, whether these changes relate more to changes in nutritional status and competetive pressures than to climatic factors.

A number of workers have attempted to demonstrate seasonal fluctuations in populations of soil fungi in temperate regions, but few clear trends have been observed (Gams and Domsch, 1969; Parkinson and Balasooriya, 1969; Dickinson and Kent, 1972). Mabee and Garner (1974) have shown seasonal trends of soil fungi in Tennessee, using a strip-baiting method. In Egyptian soils, Moubasher and El-Dohlob (1970), using the dilution plate method, showed that *Penicillium* spp. occurred with the highest frequency in the fall and were supressed during the summer whereas *Aspergillus* spp. occurred with a low frequency in winter and with highest frequency in summer, except during conditions of extreme drought. In the continental climate of Canada, where seasonal trends may be accentuated, Widden and Parkinson (1973) were unable to establish any clear cyclical trends in forest soils in spite of the fact that there was a marked midsummer peak in the fungal biomass, as measured by the agar film method of Jones and Mollison (1948).

One reason for the seeming inability of workers to demonstrate seasonal variations in species populations of microfungi may lie in the ability of these organisms to grow and complete their life cycles very rapidly. This may enable them to respond to immediate changes in the environment which would mask any cyclical, seasonal changes. The great amount of spatial heterogeneity in population distributions may also mask the effects of seasonal change (Parkinson and Balasooriya, 1969). In order to overcome these problems it is therefore essential, if one wishes to relate population density to environmental change, to take replicated samples and to measure all environmental parameters that one is interested in at the time of sampling and in the same soil samples used for the population study. One can then relate changes in the population directly to changes in the environment rather than to a time sequence.

It would be useful to have some information regarding the timing of life cycle events of microfungi in the soil. The simultaneous use of different methods such as dilution plates and soil washing may indicate the timing of active mycelial growth versus survival in the soil as spores. The use of direct observations in conjunction with fluorescent antibodies, as suggested by Frankland (1975), would also help in identifying the form of the organism in the soil.

There is as yet no information that I am aware of concerning the possible modifications of the phenology of soil microfungi in order to survive in extreme environments, although Savile (1972) has given some attention to this problem with reference to Arctic fungi. He mentions the simplification of the life cycle and the breeding system as a possible response to cold, short summers. It is possible that the extraordinarily large percentage of sterile fungi reported from tundra soils by many workers (Cooke and Lawrence, 1959; Cooke and Fournelle, 1960; Bunt, 1965; Heal et al., 1967; Hayes, 1973; Dowding and Widden, 1974; Widden, 1977) represents an extreme example of a modification of fungal phenology in relation to an extreme environment.

E. Phenology of Aquatic Microfungi

The seasonal occurrence of aquatic fungi has been examined by a number of workers, and this work has been reviewed by Sparrow (1968) and more recently by Clausz (1974). Many of the water molds appear to be highly seasonal in their occurrence. Roberts (1963) reported that in England many were commonest in the winter months, others in summer, and some showed no seasonal pattern. Sparrow (1968) in his review states that "Seasonal occurrence would certainly seem to be intimately associated with temperature and its effects." Willoughby (1962), however, studied the seasonal abundance of zoospores of the Saprolegniales in lakes in England and concluded that the major correlation was between high zoospore counts and periods of high rainfall. The work of Clausz (1974) on the water molds of Muskrat Lake in North Carolina shows that the periods of major activity of the Saprolegniales occur in the spring and again in the fall, with *Achlya* mainly active in late summer, *Saprolegnia* spp. active from spring through fall, and *Leptolegnia* active from fall through spring. Clausz concluded that these fungi may either be responding directly to climatic changes or indirectly through changes in food abundance.

The relationship between aquatic fungi and food supplies has been studied more directly by workers interested in fungal populations on dead leaves immersed in streams. In these running-water systems, the dominant fungi appear to be aquatic hyphomycetes rather than oomycetes (Barlocher and Kendrick, 1974; Ingold, 1975; Suberkropp and Klug, 1976). The work of Barlocher and Kendrick (1974) shows clearly that there is a seasonal pattern of microfungi on submerged leaves in streams. The aquatic Hyphomycetes such as *Alatospora*, *Anguillospora*, *Heliscus*, *Tetracladium*, and *Tricladium* species predominate during the colder months in southern Ontario, whereas in the summer the geofungi such as *Alternaria*, *Cladosporium*, *Penicillium*, and *Trichoderma* become predominant. It is clear that temperature is as important in determining this behavior as substrate availability. The work of Suberkropp and Klug (1976) in a stream in Michigan also demonstrated the fact that the aquatic hyphomycetes occur at a maximum frequency in winter; they also showed that aquatic hyphomycetes had a low temperature optimum (25°C) for growth.

F.   Phenology of Leaf Surface Fungi

There has been a rising interest recently in the fungi occurring on the surfaces of
living leaves. This is reflected in the publication of two volumes of proceedings
of symposia on leaf surface microorganisms held in the last decade (see Preece and
Dickinson, 1971; Dickinson and Preece, 1976). A number of these fungi appear to be
nonpathogenic epiphytes which live as saprophytes on the leaf surface such as *Sporobolomyces*, *Aureobasidium*, *Cladosporium*, *Alternaria*, *Epicoccum,* and *Stemphylium* (Dickinson, 1976). A number of people have shown that the occurrence of these organisms
is seasonal in nature. Using a leaf plating method, Pugh and Mulder (1971) showed
that, at the leaf tips of *Typha latifolia* L., *Cladosporium herbarum* and *Aureobasidium
pullulans* increased during the period of May through September whereas *Sporobolomyces*
spp. had a summer peak and declined in the fall. Using similar methods, Lindsay and
Pugh (1976) showed that, on attached leaves of *Hippophae rhamnoides* L., populations
of *Aureobasidium pullulans*, *Cladosporium herbarum*, and *Epicoccum purpurascens* increased from May through November whereas *Botrytis cinerea*, *Fusarium culmorum*, and
Cephalosporium acremonium have a peak occurrence in August. Using leaf impressions,
it was also shown that the population of *Sporobolomyces roseus* increased throughout
the summer and fall. Using a herbage washing technique, Latch and McKenzie (1977)
showed that on *Lolium perenne* L. and *L. multiflorum* Lam., in Welsh pastures, *Acremonium* spp., *Cladosporium* spp., and *Tricellula aquatica* had peak occurrences in midsummer (June, July) whereas *Phoma* spp. occurred with a maximum peak in September.

     The main explanation for the seasonal patterns in fungi on leaves may be nutri-
tional, related to changes in leaf exudates and senescence of the leaf. It is clear,
however, that phylophane fungi are exposed to high levels of radiation, dessication,
and--in winter--to very cold conditions (Dickinson, 1976). It has been suggested
by Dickinson (1976) that *Aureobasidium* and *Cladosporium* may form microsclerotia
which enable these organisms to overcome adverse conditions, but I am not aware of
any investigations that have examined the conditions under which microsclerotia may
be formed by these fungi.

G.   Effects of Environment on Sporulation in Microfungi

There have been some studies of the effects of environmental parameters such as light
and temperature on the production of spores by microfungi, some of which are reviewed
by Hawker (1966) and more recently by Turian (1974). It is known that light influences sporulation in many fungi and that some responses are affected by the wavelength
of light. Thus, in *Botrytis cinerea*, near-ultraviolet light promotes sporulation and
red and blue light inhibit sporulation (Tan, 1974, 1975a,b). In *Phoma medicaginis*
light promotes both the formation of pycnidia and conidia (Chung and Wilcoxson, 1971),
and light is essential for sporulation in *Aspergillus ornatus* (Hill, 1976). In *Phy-
comyces blakesleeanus* dwarf sporangiophore primordia were abundantly formed in dark
cultures and suppressed in the light, whereas the reverse was true for the giant

sporangiophore primordia. In the field, *Sporobolomyces* produces maximum numbers of ballistospores during dark conditions (Pady, 1974). Temperature often has an influence on sporulation; thus, Chung and Wilcoxson (1971) showed that in *Phoma medicaginis* pycnidial formation is maximal at 30°C whereas maximal conidium formation occurs at 20°C. Commonly, optimal temperatures for growth and sporulation are not the same (Hawker, 1966).

The aforementioned studies may not be directly relevant to this discussion, as no attempt was made to relate laboratory studies to field conditions. A paper by Koske and Duncan (1973), however, relates the effects of temperature on growth and sporulation to seasonal changes in populations of "aquatic" hyphomycetes isolated from terrestrial habitats. They concluded that species with a narrow temperature range for sporulation in culture had a more strictly seasonal occurrence than those with wide ranges. This paper is one of the few examples of a study relating changes in seasonal population densities of fungi with their responses in the labratory to environmental parameters.

IV. CONCLUSIONS

It is clear that the attention of students of the saprophytic fungi has not generally been focused on phenology. Nevertheless, there is a quantity of information relevant to the topic. The information available indicates that both microfungi and macrofungi do employ a number of adaptive strategies in their phenological responses. The macrofungi, because they tend to utilize "hard" substrates such as lignin and cellulose, generally have substrate available at all times of the year, as decomposition of such substrates takes place over periods of years (Swift, 1977). Therefore, some of these organisms can have a perennial mycelium and show periodicity in their fruiting. The timing of events in these organisms is fairly fixed and therefore predictable, though there is still much to be learned about events such as hyphal growth and the formation of resistant structures. It is known that different macrofungi fruit at different times of the year, but the basis for these differences in phenology is little understood. The ability, however, to produce large resistant storage structures for overwintering must have some significance to spring-fruiting fungi. Microfungi, because of their morphology, more ephemeral food sources, and lack of large structures, appear and disappear in their habitat in a much less predictable manner. Nevertheless, there is some evidence to show that at least some of these organisms show a periodicity related to seasonal change. In these organisms, however, an ability to respond rapidly to physical and chemical changes in the environment, particularly temperature, moisture, and substrate availability, may mask any long-term annual cycles.

The form of a fungus therefore does affect the phenological strategies available to it, and in general the larger fungi do have more predictable behavior than the microfungi. The extent to which fungal responses to environmental cues are genetically fixed and the way in which these cues ensure appropriate responses are not fully understood. This is mainly because, with few exceptions, laboratory studies of fungal behavior have not been correlated with observations in the field. Park's (1968) statement that "few detailed studies of the conditions necessary for different phases of the life histories (of fungi) have been made" remains as true now as it was then. It is my opinion that a greater attention to the form of the fungus in its natural environment, coupled with a study of the relation between the environment and morphogenesis, would lead to a much greater understanding of the ecology of saprophytic fungi.

I would also like to emphasize the fact that the division of the fungi into macromycetes and micromycetes is artificial; moreover, the assumptions I have made here regarding substrates and longevity of the mycelium of these organisms are most probably too generalized. There are many saprophytic microfungi that do not fit neatly into either of these categories and whose phenology it would be of interest to study.

ACKNOWLEDGMENTS

I wish to thank Don Wicklow for suggesting this topic to me, as it has opened up a number of avenues in my own mind for further investigation.

REFERENCES

Austwick, P. K. C. (1968). Effects of adjustment to the environment on fungal form. In *The Fungi*, G. C. Ainsworth and A. S. Sussman (Eds.), Vol. III. Academic Press, New York, pp. 419-445.

Barlocher, F., and Kendrick, W. B. (1974). Dynamics of fungal populations on leaves in a stream. *J. Ecol. 62*: 761-791.

Bell, M. K. (1974). Decomposition of herbaceous litter. In *Biology of Plant Litter Decomposition*, C. H. Dickinson and G. J. F. Pugh (Eds.), Vol. I. Academic Press, New York, pp. 37-68.

Buller, A. H. R. (1924). *Researches on Fungi*, Vol. III. Longmans, Green, New York, 146 pp.

Bunt, J. S. (1965). Observations on the fungi of Macquarie Island. *ANARE, Sci. Repts. Ser. B*, No. 78.

Chapman, E. S., and Fergus, C. L. (1973). An investigation of the effects of light on basidiocarp formation of *Coprinus domesticus*. *Mycopathol. Mycol. Appl. 51*(4): 315-326.

Chet, I., and Henis, Y. (1975). Sclerotial morphogenesis in fungi. *Ann. Rev. Phytopathol. 13*: 170-192.

Chung, H. S., and Wilcoxson, R. D. (1971). Effects of temperature, light, carbon and nitrogen nutrition on reproduction in *Phoma medicaginis*. *Mycopathol. Mycol. Appl. 44*(4): 297-308.

Clausz, J. D. (1974). Periods of activity of water molds in a North Carolina lake. In *Phenology and Seasonality Modeling*, H. Lieth (Ed.). Springer-Verlag, New York, pp. 191-203.

Cooke, W. B., and Fournelle, H. T. (1960). Some fungi isolated from an Alaskan tundra area. *Arctic 13*: 266-270.

Cooke, W. B., and Lawrence, D. B. (1959). Soil mould fungi isolated from recently glaciated soils in South-East Alaska. *J. Ecol. 47*: 529-549.

Daubenmire, R. (1968). *Plant Communities: A Textbook of Plant Synecology*. Harper & Row, New York.

Dickinson, C. H. (1976). Fungi on the aerial surfaces of higher plants. In *Microbiology of Aerial Plant Surfaces*, C. H. Dickinson and T. F. Preece (Eds.). Academic Press, New York, pp. 293-324.

Dickinson, C. H., and Kent, J. W. (1972). Critical analysis of fungi in two sand-dune soils. *Trans. Brit. mycol. Soc. 58*: 269-280.

Dickinson, D. H., and Preece, T. F. (1976). *Microbiology of Aerial Plant Surfaces*. Academic Press, New York.

Dobbs, C. G., Hinson, W. H., and Bywater, J. (1960). Inhibition of fungal growth in soils. In *The Ecology of Soil Fungi*, D. Parkinson and J. S. Waid (Eds.). Univ. of Liverpool Press, Liverpool, England, pp. 130-147.

Dowding, P., and Widden, P. (1974). Some relationships between fungi and their environment in tundra regions. In *Soil Organisms and Decomposition in Tundra*, A. J. Holding, O. W. Heal, S. F. MacLean, and P. W. Flanagan (Eds.). Tundra Biome Steering Committee, Stockholm, pp. 123-150.

Frankland, J. C. (1966). Succession of fungi on decaying petioles of *Pteridium aquilinum*. *J. Ecol. 54*: 41-63.

Frankland, J. C. (1974). Decomposition of lower plants. In *Biology of Plant Litter Decomposition*, C. H. Dickinson and G. J. F. Pugh (Eds.), Vol. I. Academic Press, New York, pp. 3-36.

Frankland, J. C. (1975). Fungal decomposition of leaf litter in a deciduous woodland. In *Biodégradation et humification*, G. Kilbertus, O. Reisinger, A. Mourey, and J. A. Cancela da Fonseca (Eds.). Rapport du 1er Colloque International, 1974, Nancy University, Pierron, France, pp. 33-40.

Gams, W., and Domsch, K. H. (1969). The spatial and seasonal distribution of microscopic fungi in arable soils. *Trans. Brit. mycol. Soc. 52*: 301-308.

Hawker, L. E. (1966). Environmental influences on reproduction. In *The Fungi*, G. C. Ainsworth and A. S. Sussman (Eds.), Vol. II. Academic Press, New York, pp. 435-469.

Hayes, A. J. (1973). Studies on the microfungi occurring at Stordalen and Njulla, 1972. Swedish IBP Tundra Biome Project Technical Report, p. 15.

Heal, O. W., Bailey, A. D., and Latter, P. M. (1967). Bacteria, fungi and protozoa in Signy Island soils compared with those from a temperate moorland. *Phil. Trans. Roy. Soc. London, Ser. B 252*: 191-197.

Hill, E. P. (1976). Effect of light on growth and sporulation of *Aspergillus ornatus*. *J. Gen. Microbiol. 95*: 39-44.

Hudson, H. J. (1968). The ecology of fungi on plant remains above the soil. *New Phytol. 67*: 837-874.

Ingold, C. T. (1975). Convergent evolution in aquatic fungi: The tetraradiate spore. *Biol. J. Linnean Soc. 71*: 1-25.

Jensen, V. (1974). Decomposition of angiosperm leaf litter. In *Biology of Plant Litter Decomposition*, C. H. Dickinson and G. J. F. Pugh (Eds.), Vol. I. Academic Press, New York, pp. 69-104.

Jones, P. C. T., and Mollison, J. E. (1948). A technique for the quantitative estimation of soil microorganisms. *J. Gen. Microbiol. 2*: 54-69.

Katz, B. A., and Lieth, H. (1974). Seasonality of decomposers. In *Phenology and Seasonality Modeling*, H. Lieth (Ed.). Springer-Verlag, New York, pp. 163-184.

Kendrick, W. B., and Burges, A. (1962). Biological aspects of the decay of *Pinus sylvestris* leaf litter. *Nova Hedwigia 4*: 313-342.

Kitamoto, Y., Horikoshi, T., and Suzuki, A. (1974). An action spectrum for photo-induction of pileus formation in a basidiomycete, *Favolus arcularius*. *Planta* (Berlin) *119*: 81-84.

Koch, W. (1958). Untersuchungen uber Mycelwachstum und Fructkorperbildung bei einigen Basidiomyceten (*Polystictus versicolor*, *Polyporus annosus*, *Pleurotus ostreatus*, und *Psaliota bispora*). *Arch. Mikrobiol. 30*: 409-432.

Koske, R. E., and Duncan, I. W. (1973). Temperature effects on growth, sporulation and germination of some "aquatic" hyphomycetes. *Can. J. Bot. 52*: 1387-1391.

Latch, G. C. M., and McKenzie, E. H. C. (1977). Fungal flora of ryegrass swards in Wales. *Trans. Brit. mycol. Soc. 68*: 181-184.

Lindsay, B. I., and Pugh, G. J. F. (1976). Succession of microfungi on attached leaves of *Hippophae rhamnoides*. *Trans. Brit. mycol. Soc. 67*: 61-67.

Mabee, H. F., and Garner, J. H. B. (1974). Seasonal variations of soil fungi isolated from the rhizosphere of *Liriodendron tulipifera* L. In *Phenology and Seasonality Modeling*, H. Lieth (Ed.). Springer-Verlag, New York, pp. 185-190.

McKenzie, E. H. C., and Hudson, H. J. (1976). Mycoflora of rust-infected and non-infected plant material during decay. *Trans. Brit. mycol. Soc. 66*: 223-238.

McLaughlin, D. J. (1970). Environmental control of fruit-body development in *Boletus rubinellus* in axenic culture. *Mycologia 62*: 307-331.

Matveev, V. A. (1972). [Prognosis of fruiting of edible mushrooms.] *Lesnoe Khozyaistivo 9*: 27-28.

Millar, C. S. (1974). Decomposition of coniferous leaf litter. In *Biology of Plant Litter Decomposition*, C. H. Dickinson and G. J. F. Pugh, Vol. I. Academic Press, New York, pp. 105-128.

Miller, O. K. (1977). Fungi in cold environments. (Abst.) *Proc. 2nd Int. Mycol. Congr.*, *Tampa, Fla.*, p. 439.

Miller, O. K., and Laursen, G. A. (1974). Belowground fungal biomass on U.S. Tundra Biome sites at Barrow, Alaska. In *Soil Organisms and Decomposition in Tundra*, A. J. Holding, O. W. Heal, S. F. McLean, and P. W. Flanagan (Eds.). Tundra Biome Steering Committee, Stockholm, pp. 151-158.

Moubasher, A. H., and El-Dohlob, S. M. (1970). Seasonal fluctuations in Egyptian soil fungi. *Trans. Brit. mycol. Soc. 54*: 45-51.

Mueller-Dombois, D., and Ellenberg, H. (1974). *Aims and Methods of Vegetation Ecology*. Wiley, New York.

Nagel-de-Boois, H. M., and Jansen, E. (1966). Hyphal activity in mull and mor of an oak forest. In *Progress in Soil Biology*, O. Graff and J. E. Satchell (Eds.). Vieweg, Braunschweig, Germany, pp. 27-36.

Nagel-de-Boois, H. M., and Jansen, E. (1971). The growth of fungal mycelium in forest soil layers. *Rev. Ecol. Biol. Soc. 4*: 509-520.

Nair, P. K. R. (1973). Quantitative changes in soil microorganisms under rice-based multiple cropping in northern India. *Soil Biol. Biochem. 5*: 387-389.

Pady, S. M. (1974). Sporobolomycetaceae in Kansas. *Mycologia 66*: 333-338.

Park, D. (1968). The ecology of terrestrial fungi. In *The Fungi*, G. C. Ainsworth and A. S. Sussman (Eds.), Vol. III. Academic Press, New York, pp. 5-39.

Parkinson, D., and Balasooriya, I. (1969). Studies on fungi in a pine wood soil. IV. Seasonal and spatial variations in the fungal populations. *Rev. Ecol. Biol. Sol. 6*: 147-153.

Prakash, D., and Khan, A. M. (1971). Fungal population in sugarcane soils. *J. Ind. Bot. Soc. 50*: 153-157.

Preece, T. F., and Dickinson, C. H. (Eds.) (1971). *Ecology of Leaf Surface Microorganisms*. Academic Press, New York.

Pugh, G. J. F., and Mulder, J. L. (1971). Mycoflora associated with *Typha latifolia*. *Trans. Brit. mycol. Soc.* *57*: 273-282.

Raunkiaer, C. (1934). The life forms of plants and statistical plant geography: Being the collected papers of C. Raunkiaer, translated into English by H. G. Carter, A. G. Tansley and Miss Fausboll. Oxford Univ. Press (Clarendon), New York.

Roberts, R. E. (1963). A study of the distribution of certain members of the Saprolegniales. *Trans. Brit. mycol. Soc.* *46*: 213-224.

Savile, D. B. O. (1972). Arctic adaptations in plants. *Can. Dept. Agr. Monogr. No. 6.* Ottawa, Canada.

Smith, J. E., and Berry, D. R. (1974). *An Introduction to Biochemistry of Fungal Development.* Academic Press, New York.

Sparrow, F. K. (1968). Ecology of freshwater fungi. In *The Fungi*, G. C. Ainsworth and A. S. Sussman (Eds.), Vol. III. Academic Press, New York, pp. 41-93.

Stark, N. (1972). Nutrient cycling pathways and litter fungi. *Bioscience 22*: 355-360.

Suberkropp, K., and Klug, M. J. (1976). Fungi and bacteria associated with leaves during processing in a woodland stream. *Ecology. 57*: 707-719.

Sussman, A. S. (1966). Dormancy and spore germination. In *The Fungi*, G. C. Ainsworth and A. S. Sussman (Eds.), Vol. II. Academic Press, New York, pp. 733-764.

Swift, M. J. (1977). The ecology of wood decomposition. *Sci. Progr.* (Oxford) *64*: 175-199.

Taber, W. A. (1966). Morphogenesis in Basidiomycetes. In *The Fungi*, G. C. Ainsworth and A. S. Sussman (Eds.), Vol. II. Academic Press, New York, pp. 387-412.

Tan, K. K. (1974). Blue-light inhibition of sporulation in *Botrytis cinerea*. *J. Gen. Microbiol. 82*: 191-200.

Tan, K. K. (1975a). Interaction of near-ultraviolet, blue, red and far-red light on sporulation of *Botrytis cinerea*. *Trans. Brit. mycol. Soc. 64*: 215-222.

Tan, K. K. (1975b). Recovery from the blue-light inhibition of sporulation in *Botrytis cinerea*. *Trans. Brit. mycol. Soc. 64*: 223-228.

Toth, J. A., Papp, L. B., and Lenkey, B. (1975). Litter decomposition in an oak-forest ecosystem (*Quercetum petraeas cerris*) of northern Hungary studied in the framework of "Sikfokut Project." In *Biodégradation et humification*, G. Kilbertus, O. Reisinger, A. Mourey, and J. A. Cancela da Fonsca (Eds.). Rapport du 1er Colloque Internatinal, 1974, Nancy University, Pierron, France, pp. 41-58.

Turian, G. (1974). Sporogenesis in fungi. *Ann. Rev. Phytopathol. 12*: 129-137.

Volz, P. A., and Beneke, E. S. (1969). Nutritional regulation of basidiocarp formation and mycelial growth of Agaricales. *Mycopathol. Mycol. Appl. 37*: 225-253.

Waid, J. S., and Woodman, M. J. (1957). A method of estimating hyphal activity in soil. *Pedologie 7*: 155-158.

Wheeler, B. E. J. (1968). Fungal parasites of plants. In *The Fungi*, G. C. Ainsworth and A. S. Sussman (Eds.), Vol. III. Academic Press, New York, pp. 179-211.

Widden, P. (1977). Microbiology and decomposition on the Truelove Lowland. In *Truelove Lowland, Devon Island, Canada: A High Artic Ecosystem*, L. C. Bliss (Ed.). Univ. of Alberta Press, Edmonton, Alberta, Canada, pp. 505-530.

Widden, P., and Parkinson, D. (1973). Fungi from Canadian coniferous forest soils. *Can. J. Bot. 51*: 2275-2290.

Willoughby, L. G. (1962). The ecology of some lower fungi in the English Lake District. *Trans. Brit. mycol. Soc. 45*: 121-136.

Wright, E., and Bollen, W. B. (1961). Mycoflora of Douglas-fir forest soil. *Ecology 42*: 825-828.

Chapter 22

MECHANISMS IN FUNGAL SUCCESSIONS

Juliet C. Frankland

Merlewood Research Station
Institute of Terrestrial Ecology
Grange-over-Sands, Cumbria, England

I. INTRODUCTION

Community life for a fungus is dynamic. The terms *niche* and *organization of popula-tions*, discussed in previous chapters, have a somewhat static tone. They suggest comfortable permanency, whereas a battleground image is much nearer the truth. Niches are lost or gained and the population levels alter as living organisms interact with each other and their environment. Species succession follows, and the community pattern changes.

Fungal successions are familiar facts of mycology, and our ignorance is centered not so much on the species composition of successive communities as on the changing structure of these communities and the actual mechanisms whereby one species replaces another. Without this knowledge, we can neither understand nor manipulate a succes-sion. This chapter does not provide the answers; it can only set the scene for fur-ther investigations of fungal successions within the context of general plant ecology. It refers, in particular, to the fungal saprophytes of plant litter and soil.

II. SUBSTRATE AND SERAL SUCCESSIONS

Fungal successions have been classified into two types, *substratum* and *seral* (Park, 1968), the latter being *primary* if it starts on a site which has not been previously occupied by organisms, and *secondary* if it follows disruption of a primary succession (see Chap. 24). Any plant, animal, or man-made substrate that a fungus can colonize is a potential site for a succession of species, unless the substrate is too small or simple to support a true succession. On a human hair, for example, secondary colo-nists may grow on mycelium of the primary colonists and not on the substrate itself (Griffin, 1972). In this chapter the term *substrate*, the less ponderous anglicized version of *substratum*, is used loosely for any material substance acting as a fungal food base. It is convenient to use this familiar term here, but more precise expres-sions, i.e., *primary* and *secondary resource*, have been proposed by Swift (1976) to avoid the biochemical connotation of a chemical entity that reacts in a specific manner with an enzyme.

Many substrate successions of fungi have been described--some substrates being as simple as a cellulose film, others as complex as a leaf. These successions include

those on plant remains above the soil, compared in detail by Hudson (1968). The
higher-plant ecologist must make a considerable mental adjustment to the scales involved when, as Griffin (1960) said (quoting R. Y. Stanier): "A single cellulose
fibre provides a specialized environment with its own characteristic microflora, yet
may occupy a volume of not more than a cubic millimetre." Figure 1 illustrates a
cellulose film which has been buried in soil and colonized by a fungus in a substrate
succession described by Tribe in some classic studies (1957, 1961). The hyphal
branches of *Botryotrichum*, a cellulose decomposer, can be seen "rooting" in the film.
The exact succession of species depended on the type of soil in which the film was
buried. Sometimes a noncellulolytic species, *Pythium oligandrum* Drechsler, parasitized fungi already established. As a succession proceeds the growth of mycelium
tends to increase the spatial complexity, providing more opportunities for parasites
and predators; as Southwood said (1977), "elephants and oak trees provide more niches
for other organisms than mites and mosses."

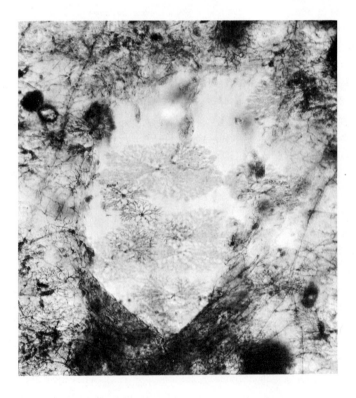

Fig. 1 Mycelium of the *Botryotrichum* state of *Chaetomium piluliferum* J. Daniels on
a cellulose film, showing "rooting branches" where the surface mycelium has been
pulled back. The film was buried 27 days in a loamy sand soil. Magnification:
X150. (Photograph by H. T. Tribe.)

A fungal succession on a more complex substrate, leaf litter of *Betula pendula* Roth, decomposing on a woodland mull (J. C. Frankland, unpublished data), is illustrated in outline in Table 1. Host-restricted fungi parasitized the leaves before leaf fall. They were followed by three overlapping waves of saprophytic fungi, each wave characterized by certain abundant species, and after 1 year including fungi more typical of soil. This succession with its wave pattern conforms with Hudson's generalized scheme of fungal succession on nonwoody plant debris. The duration of these successions, however, varies considerably with the type of substrate and its situation. For example, on *Quercus* leaves and *Pteridium* petioles decaying on a similar mull, soil fungi did not predominate until the second and sixth year, respectively (Frankland, 1966, 1975, 1976).

The succession on *Betula* litter, as described, is on a macroscale; it is an "average" succession for the total heterogeneous substrate, since each type of tissue in the leaf could carry a specific fungal flora. Pugh and Buckley (1971), for example,

Table 1 A substrate succession of fungi isolated commonly[a] from leaves of *Betula pendula* decomposing on a mull humus, Meathop Wood, United Kingdom

	Leaf state		
Living	Senescent	Dead	
		6	12 months after leaf fall
Leaf parasites $\longrightarrow$	Primary saprophytes $\longrightarrow$	Secondary saprophytes $\longrightarrow$	Soil fungi
Venturia ditricha *Gnomonia intermedia* $\Big\}$ $\longrightarrow$			
	Aureobasidium sp. *Epicoccum nigrum* *Cladosporium herbarum* Sphaeropsidales $\Big\}$ $\longrightarrow$		
		Basidiomycetes $\rightarrow$	
		$\longrightarrow$	*Penicillium* spp. *Trichoderma* spp. *Mortierella ramanniana*

[a] Species are listed at the first stage in decomposition at which they became "dominant," i.e., occurrence in ≥25% of whole leaves incubated in damp chambers, or production of ≥50 colonies from 500 1- to 2-mm^2 washed leaf fragments in modified Czapek-Dox agar. Indicator lines denote presence of a species or fungal group before or after this event.

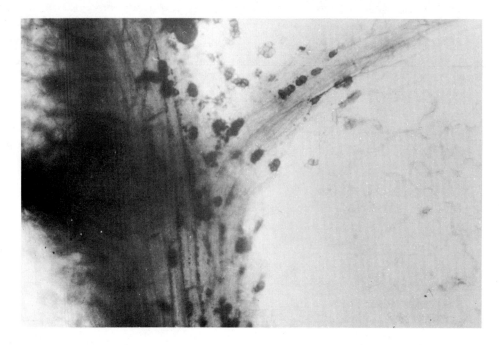

Fig. 2 Chlamydospores and microsclerotia, mainly of *Aureobasidium*, arranged linear-
ly along the veins of cleared leaves of *Acer pseudoplatanus*. Magnification: Approx.
X24. (From Pugh and Buckley, 1971.)

found that *Aureobasidium* was significantly associated with the leaf veins of *Acer*
pseudoplatanus L. (Fig. 2). Similarly, Hering (1965) found that *Dactylaria purpur-*
ella (Sacc.) was associated exclusively with the vascular tissue of the leaf litter
of *Quercus petraea* (Matt.) Liebl. The sampling techniques used by mycologists in the
past have often been too ham-handed to preserve these microhabitats. All too often
the substrate was broken up and dispersed in a culture medium, so that the patterns
of species within communities, familiar to higher-plant ecologists, were seldom recorded. Kendrick and Burges (1962), however, gave an excellent close-up view of a
substrate succession in their description of fungi colonizing pine needles. One quotation alone indicates the intricate information to be found on a few square millimeters of a pine needle: "*Desmazierella* conidiophores, arising from the substomatal
cavities, were unaffected by the distribution of *Helicoma* or *Sympodiella*, but were
strongly influenced by the presence of *Lophodermium* diaphragms."

Turning to seral successions of fungi, few have been described, but they can be
illustrated by the distribution of soil fungi across some developing dune systems.
A gradient series of fungal populations, from the foreshore and mobile dunes to the
semifixed and fixed dunes of grassland, heath, or woodland, was found by the author
in eight coastal systems of the British Isles (Brown, 1958a). Classic seral successions of dune vegetation were mirrored by changes in the soil mycoflora, so that the

center of a seral stage could be recognized as readily by the fungal population of a culture plate, inoculated with only 5 mg of sand, as by a community of higher plants. Table 2 lists the fungi most commonly isolated from one of these systems, where *Ammophila* grass was succeeded by *Calluna*, *Pteridium*, and *Betula*. Apparent anomalies in the final stages were caused by wind blowouts of sand around decomposing *Pteridium* plants, followed by secondary succession.

Higher-plant succession does not fit a universal pattern, so that attempts to generalize have been conflicting and argued at length (Odum, 1969; Drury and Nisbet, 1973; Horn, 1974; Whittaker, 1975), but some analogies between seral successions of vegetation and these dune fungi can be drawn. Vegetation becomes increasingly stratified with progressive development of a soil profile and increasing organic matter. Similarly, "fungal horizons" developed in these dunes with different species compositions and, as expected, were most differentiated in the fixed dunes with the most developed soil profile. Here this analogy stops, because the fungal succession was not accompanied by obvious changes in the dominant growth form to longer-lived species. Agarics with macroscopic fruit bodies occurred at all stages of the sere. The greatest fungal diversity or richness (number of species) occurred at an intermediate stage in the succession, in 100-year-old semifixed dunes where pioneer and late successional plants overlapped. This pattern is unlike that of another decomposer succession described by Usher and Parr (1977), in which arthropod diversity appeared to increase as plants became less diverse, a situation which, as they point out, could have important implications in the management of an ecosystem. In higher-plant succession, however, a diversity pattern like that of the dune fungi has often been described (Horn, 1974), although many authors claim that plants tend to increase in diversity from simple to mature communities. The giant redwood forests of America are a spectacular example of mature communities poor in plant species. A further generalization that productivity and biomass increases through succession is also not consistent among higher plants. There are no productivity data for the dune fungi, but--as may seem obvious where a community is progressively closing in--fungal biomass (abundance of mycelium) reaches a maximum in fixed stable dunes (Brown, 1958b). Another common feature of successions of higher plants and other organisms appears to be the slowing of species replacement (proportion of species lost per unit time) through succession after the initial stages of colonization (Shugart and Hett, 1973). There is some fungal evidence for this in the dune example, where there was a mean net gain of 0.35 species per year in the first century of dune development and a mean net loss of 0.03 species per year in the second century.

Finally, a higher-plant succession is said to reach a so-called climax or steady state, where changes in the species composition are undetectably slow, provided the physical environment remains constant. The dune-heath mycoflora, in which casual fungi had decreased from 91% to 37% of the population, appeared to be approaching this stage. If community stability is measured as resistance to disturbance, and not in

Table 2 A seral succession of soil fungi isolated commonly[a] from an acid dune system, Studland, United Kingdom

| | | | Seral stage | | | | |
| | | | Semifixed | | | Fixed | |
	Beach $\rightarrow$	Fore dunes $\rightarrow$	"Dune grass" $\rightarrow$	Dune heath $\rightarrow$	Dune heath $\rightarrow$	Dune scrub
Age (approx. years)			100	150	200	250
Soil pH (mean; 0–2.5 cm)	6.7	6.8	4.7	3.9	3.7	3.7
Soil organic matter (mean % loss on ignition; 0–2.5 cm)	0.2	0.2	0.7	2.0	11.9	2.2

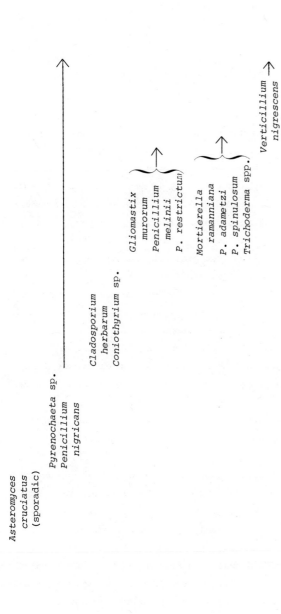

[a]Present on $\geq$50% of culture plates prepared from 150 sand samples in any one seral stage.

Source: After Brown (1958a).

terms of species composition, stability can be said to decrease during succession
(Horn, 1974). Again, this was illustrated by the effect of blowouts on the communities of dune fungi; late stages were set back further than early stages. There is no
mycological evidence to support Clement's (1916) extreme view that succession leads to
a recognizable organismal entity in the climax community, which can grow, mature, and
die. The opposite extreme of continuously variable associations of organisms with individual properties appears to be more realistic (Gleason, 1926; Kershaw, 1973), but
more fungal data are needed before mycologists can contribute substantially to the
argumer.ts of these and intermediate theories. It would be interesting, for instance,
to examine the types of transition between the communities of dune fungi, as done for
dune vegetation, using tests of heterogeneity, by Morrison and Yarranton (1974), who
found some support for the organismal concept.

The evidence for succession, even of higher plants, is often largely circumstantial (Greig-Smith, 1964). The fungal succession in old dune systems may seem self-evi-
dent, but it is not proven when the sampling of each stage was contemporary. The
sampling methods were also, as usual, destructive, so that the succession in dunes
(Table 2) is again an "average," this time derived from many substrate successions.
The overall picture is of a seral succession equivalent to that of higher plants,
but the details at species level are missing. Exactly which species has ousted another cannot be said. A decomposing leaf of *Ammophila* may be colonized by fungus
A → B → C in one dune, and by Al → Bl → Cl in the next dune with a different environment; Al replaces A not C. These substrate successions are discontinuous, and usually not much is known about the origin of fungal colonists, the extent to which
they live in the soil and grow through it from one substrate to another, or the degree of competition between, say, A and Al when colonizing a new substrate.

The root surface of *Ammophila arenaria* (L.) Link, a dominant plant in most of
the dunes examined, provided a good opportunity of comparing the mycoflora of a known,
even if variable, substrate at different stages in the succession (Brown, 1957). The
commonest isolates of fungi attached or closely adhering to roots of mixed ages and
sizes after washing are listed in Table 3. The association between the roots and
fungi was not a highly selective one. Few species were found exclusively on the
roots, and the population varied across and between dune systems. The divergence
between the populations of the root surface and free soil was greatest in the fore
dunes and then decreased until in the fixed dunes the root mycoflora could be classed
as "acid" or "alkaline" according to soil type. A seral pattern of communities therefore emerged whether a single substrate or sand was examined, but differences in timing occurred. *Trichoderma,* for example, was common on roots in the fore dunes but
infrequent in the sand until the dunes were 100 years older. It is difficult to interpret such differences when most methods of sampling soil fungi do not distinguish
between spores and mycelium. Both the microhabitat and the state of the fungus must
be known before the mechanisms of a succession can be understood.

Table 3 Fungi isolated commonly[a] from a specific substrate, the root surface of
Ammophila arenaria growing in the seral stages of an acid dune system,
Studland, United Kingdom

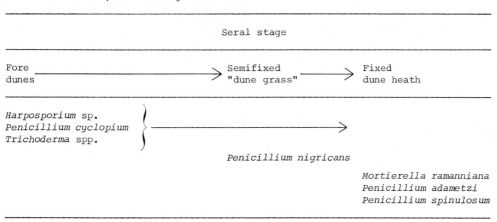

Seral stage

| Fore dunes | → | Semifixed "dune grass" | → | Fixed dune heath |

Harposporium sp.
Penicillium cyclopium   ⎫ ————————————→
Trichoderma spp.   ⎭

Penicillium nigricans

Mortierella ramanniana
Penicillium adametzi
Penicillium spinulosum

[a]Present on $\geq 10\%$ of 300 2-mm root segments at any one soil depth.

Substrate successions differ from seral successions not only in scale but also
in direction. Instead of moving toward a state of relative equilibrium, they will
in theory decline to zero when the substrate is totally exhausted (Fig. 3). In practice, descriptions of succession on decaying organic matter usually fall well short
of this endpoint, and little is known in particular about succession on the refractory remains of plants once they are incorporated in soil. A substrate succession
may be said to be cyclical, but only in the sense that it can start again on a fresh
substrate. Somewhat similar cycles are exhibited by higher plants. A *Calluna* community, for example, may degenerate and be replaced by lichens and other plants, finally
returning to the bare soil (Watt, 1947). Initially, decomposition of a substrate often opens up new microhabitats and fungal diversity increases, but diversity has been

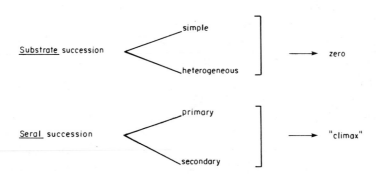

Fig. 3 Types of fungal succession.

shown to decrease as decay proceeds (Swift, 1976), so the diversity pattern of fungi
in the two types of succession can be similar.

All fungal successions, seral or substrate, will tend to be more flexible than
those of higher plants, because fungi are so adaptable. Fungi do not appear to possess unique biochemical and physiological mechanisms of adjustment (Waid, 1968), but
they are very plastic forms of life. Their lack of a truly cellular organization,
their ability as hyphae to grow, fuse, and exploit substrates rapidly, their versatile methods of nutrition and propagation, and their tolerance of adverse conditions
are all likely to contribute to their ecological success. The genetic adjustments
of fungi again, in principle, are not unlike those of other organisms, but there are
various alternative mechanisms to sexual reproduction that may influence the survival
of a saprophyte in the wild (Person, 1968). Nuclear ratios within a heterokaryon,
for instance, have been shown in the laboratory to adjust to changes in the environment.

III. ARRIVAL, SURVIVAL, AND ESTABLISHMENT

Before one species can replace another, it must arrive, survive, and establish itself.
Fungal spores are notoriously ubiquitous, and it is sometimes suggested that the potential mycofloras of various soils are essentially the same. Arrival of spores or
hyphae on a substrate is likely to depend largely on random selection, operating on
spore dispersal and hyphal growth as on seed dispersal and root growth in higher-plant
succession. The supply of propagules will in turn depend on life cycles, dispersal
mechanisms, season, and distance of the fungus from the site. The secondary succession of horseweed, aster, and broomsedge on abandoned fields in North Carolina (Keever, 1950) illustrates particularly well the influence of life cycle and season on
the entry of plants to a succession. Mycological successions are not as well documented, although the influence of these factors on the arrival of specific fungi are
sometimes known. The distribution of "tar spot," *Rhytisma acerinum* (Pers.) Fr., on
Acer crowns, declining with height up the tree, is a showy example of the effect of
the distance between a leaf and the source of inoculum on litter at ground level.
Similarly, Carruthers and Rayner (1979) found that the succession of fungi on decaying hardwood in the field varied according to its exposure to fungi in the air,
litter, and other well-rotted wood. The relative time of arrival of different fungi
on a natural virgin substrate is not often known, but Meredith (1959) obtained some
evidence of when fungi landed on pine stumps by recording fungal growth on the wood
when it was sampled and incubated at various time intervals after felling (Table 4).
Small relatively isolated colonies were taken as evidence of colonization by single
spores. He also deduced from his observations that the relative proportions of
spores in the air above the stumps influenced the incidence of infection by *Heterobasidion annosum* (Fr.) Bref. and *Peniophora gigantea* (Fr.) Massee.

Table 4 Arrival of fungi on pine stumps: Average number of colonies on wood samples removed and incubated at intervals after felling

	Interval between felling and sampling			
	0	1 hr	3 hr	5 days
Peniophora gigantea	0.3	2.3	3.5	>10.0
Blue stain fungi	0.3	1.0	1.8	1.0
Heterobasidion annosum	0	0	0.8	1.8

Source: After Meredith (1959).

The many factors, besides competition, which can affect the germination and survival of a fungus after its arrival on a new substrate are discussed by Sussman (1966,
1968). Both stress, such as shortage of nutrients, and disturbance, such as grazing,
may limit further growth before true competition occurs. The pioneers of a succession
may not in fact meet competition immediately, and it will be seen later that they may
be less well adapted to it than are later colonists.

As a fungal community closes in on a substrate, hardships can arise from the
nearness of neighbors and the overlap of niches, i.e., there may be competition between fungal species for limited resources. Nutrients, water, air, heat, light, and
space are often listed as the requirements of organisms in general. Grime (1973,
1977) succinctly defined competition among higher plants as "the tendency of neighboring plants to utilize the same quanta of light, ions of mineral nutrients, molecules of water or volume of space (1973, p. 311). The list of requirements, if not
the order of priority, is similar for fungi. Some would delete light, although it
has effects on sporulation and growth, and, in general, soil fungi do not appear to
compete for water or space. Microbial activity is more likely to produce water than
consume it, so that in Clark's view (1965) fungi are "struggling" with the environment, rather than with one another, for available water.

Several workers have commented on the microbial emptiness of soil, so space in
itself is probably not important in this habitat, but the logistics of mycelia in
the soil need further investigation. It is often assumed that mycelia of the same
species will fuse on meeting to form a single ecological unit, with potentially un-
limited vegetative spread, so that the population concepts pertaining to truly sessile organisms would not always apply in fungal successions. Recent field evidence
from studies of rotting stumps suggests, however, that individual dikaryotic mycelia
of some basidiomycetes are often delimited by intraspecific antagonism (Rayner and
Todd, 1977). The extent of this phenomenon and the effect of density on the mortal-

ity and plasticity of the mycelia of single species in natural substrates, all of
consequence in succession, are not known.

The success of a fungus in establishing itself on a new substrate against competition results from a whole web of interactions. Garrett (1950, 1956, 1970) summed
up the intrinsic factors of a saprophytic fungus which contribute to its success in
two terms, originally proposed for root-infecting fungi: *competitive saprophytic
ability* and *inoculum potential*. The first is "the summation of physiological characteristics that make for success in competitive colonization of dead organic substrates"
(1956; p. 130). It includes such properties as high growth rate, rapid germination,
and the production of specific enzymes and antibiotics. It is not a fixed character,
but substrate specific, its expression depending on the immediate conditions. In
plant ecology it is matched by the term *competitive ability* (or *competitive capacity*).
Mueller-Dombois and Ellenberg (1974, p. 352), writing of higher-plant communities,
stated that in general "the competitive ability of a species depends on its genetic
potential which is manifested in its morphological structure and physiological requirements." Properties that they list as important in competitive ability, such as
growth rate, overlap those given by Garrett, although a few such as duration of photosynthesis obviously do not apply to fungi. Quoting from them again, competitive ability "varies with habitat factors, it is not a constant species property" (p. 353).

Inoculum potential of a saprophytic fungus is defined by Garrett (1956, p. 41)
as "the energy of growth of a fungus available for colonization of a substrate at the
surface of the substrate to be colonized." It is governed by such factors as propagule density, hyphal age, and (sensu Garrett) conditions in the enviroment, including
grazing pressure, diffusion of nutrients from the substrate, etc. The concept is more
familiar in fungal pathology; a single conidium of *Botrytis cinerea* Fr. will not usually infect a strawberry petal, a normal inoculum threshold being about 160 conidia
in a droplet of water (Chou and Preece, 1968). The pathogen must overcome the resistance of an active host, whereas the resistance to a colonizing saprophyte comes
mainly from competitors. Success will depend partly on the balance between the competitive saprophytic ability and the inoculum potential. A poor competitive saprophyte may gain a foothold by having a relatively high inoculum potential. Higher-
plant ecologists do not appear to have a universal term for this invasive force of a
colonist, but examples of the combined effect of the competitive ability of a plant
and its age, performance, or population level on community patterns are frequently
quoted. Kershaw (1973) stated that the causal factors of sociological pattern are not
only dependent on the competitive ability of an individual plant but also on the possible presence of toxins exuded by an individual, as well as its age. Similarly, "the
phasic development of *Calluna* affects other associated species through its changing
levels of competitive ability" and "the variation of density of *Salix herbacea* as it
is invaded by an 'advancing front' of *Eriophorum angustifolium* ... confirm the relationship between age and competitive ability" (Kershaw, 1973, pp. 84, 162).

IV. REPLACEMENT OF SPECIES

In every ecosystem the competitive balance and species composition are constantly changing, even if imperceptibly. Various mechanisms have been suggested for this species replacement, but they have rarely been verified, even by mycologists with their advantage of dealing with short-lived organisms. Possible interactions will be discussed, but all mechanisms of replacement in a saprophytic succession can be summed up in one general statement: a decomposer can replace another species when changes in the substrate (or site) have interacted with changes in its *relative* competitive saprophytic ability and inoculum potential giving it a decisive advantage. The macroenvironment is an overall constraint, as shown in Fig. 4.

Cause and effect are difficult to disentangle, but modification of the habitat, violent or gradual, is probably the principal driving force of a succession. Thus, in seral successions, changes in the green plant community regulate the decomposer community. Reactions between the two are reciprocal, but the accumulation of litter and organic matter tends to "drag" decomposer succession in the wake of higher-plant succession (McNaughton and Wolf, 1973). Mere occupation of a site by a microorganism alters the microenvironment and may even modify the species niche out of existence.

On a substrate such as a decomposing petiole of *Pteridium* (Frankland, 1966, 1969) a species might improve the site for a successor or impair it for itself by, for example, accumulation of nitrogen, exhaustion of pectin, removal of tannins, or changes in physical structure. *Trichoderma* did not become prominent until the second year of the succession, when it parasitized live mycelium of *Mortierella*, grew over moribund mycelium of its predecessors, and also penetrated deeply into xylem vessels and fibers. The decomposition rate of bracken by this species in vitro was doubled by

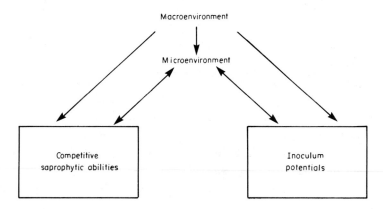

Fig. 4 The network of changes which can be involved in the replacement of a saprophytic fungus.

addition of nitrogen, so its field performance may have depended on a supply of fungal nitrogen. Levi et al. (1968) showed that certain wood-destroying basidiomycetes can utilize their own mycelium or that of other species as a sole source of nitrogen. The ability of fungi to tap nutrient sources beyond the immediate substrate inevitably complicates the interpretation of replacement processes and also the definition of the boundaries of a substrate or site.

By-products of cellulose decomposition may also prime a substrate for species replacement. The "sugar fungus" *Mucor hiemalis* Wehm. followed the cellulose decomposer *Mycena galopus* (Pers. ex Fr.) Kummer in the *Pteridium* succession when the initial sugar content of the substrate was likely to have been exhausted. The soluble carbohydrate content of the petioles increased significantly when they had been incubated for 1 year with *M. galopus*, suggesting that the activity of this basidiomycete could be a source of sugars for fungi late in the succession (Frankland, 1974). McNaughton and Wolf (1973) cite parallel examples in higher-plant succession of site priming, such as mats of *Dryas* acting as loci for invasion by cottonwood seedlings on deglaciated sites.

Accumulation of nutrients is probably more important in succession than exhaustion of the food supply. In fact there is little evidence that a species dies out in fungal succession either from starvation or senility. *Aureobasidium pullulans* (de Bary) Arn. was a pioneer decomposer of *Pteridium* petioles, but growth ceased abruptly in the first year, although there is evidence that this fungus can utilize cellulose as well as pectin (Siu, 1951; Wieringa, 1955; Seifert, 1964); it attacks the former weakly and probably insufficiently to maintain itself under competition, but in axenic culture it continued to grow on *Pteridium* fibers for several months. Performance as a decomposer, however, cannot be correlated with survival time in a substrate succession. *Hypoxylon multiforme* (Fr.) Fr., for example, capable of only moderate decay of hardwood, can persist for long periods in dead stumps among competitors, whereas *Coriolus versicolor* (L. ex Fr.) Quel., capable of the complete breakdown of woody tissues, is often replaced in the sapwood by more competitive fungi at a relatively early stage (A. D. M. Rayner, personal communication, 1978).

Inhibitors may also have a major influence on fungal successions. When fungi from *Quercus* leaf litter were tested for inhibition by tannins, some early colonists were found to be tolerant, whereas some later species including *M. galopus*, which grew on the litter when the concentration of tannins would be lower, were inhibited (Harrison, 1971). It is tempting to "explain" the succession in this way, but the effect of tannins on several other members did not fit the hypothesis, emphasizing the need for a multivariate approach to succession.

The effects of changes in the physical character of a substrate, such as particle size, are probably overlooked more often than the chemical aspects of a succession. The initial breakdown of a plant organ by pioneer fungi is likely to increase aeration of the internal tissues and access for the less diffusible fungal enzymes.

The secondary colonist, *M. galopus*, entered the *Pteridium* petioles when there were breaks in the epidermis and longitudinal tunnels in the interior; it then formed, around its hyphae, characteristic "bore-hole" cavities in the fiber walls. It appears to be equipped to decompose chunks of hard plant tissues, rather than finely divided litter where the increased surface area could favor species with more diffusible enzymes. Nilsson (1974, 1976) has suggested that the cellulase production of some wood-attacking microfungi is stimulated by the physical structure of a substrate and that, in some cases, enzymes are produced only when a hypha in a plant cell wall is growing parallel to the cellulose microfibrils.

The mechanism suggested, in which earlier species "pave the way" for later species (mechanism 1), corresponds with Horn's (1976) *obligatory succession* and Connell and Slatyer's (1977) *facilitation model* for higher-plant succession. The mechanism could apply to many heterotrophic successions where penetration or growth is initially hindered, and there is evidence for it in some autotrophic successions in extreme conditions--for example, the plant colonization of soils newly exposed by a retreating glacier (Crocker and Major, 1955; Reiners et al., 1971).

Alternative mechanisms of replacement in higher-plant succession are described by Connell and Slatyer's *tolerance* and *inhibition models*. In the former, late plant colonists replace early species because they can grow better at lower levels of resources; they are successful whether or not earlier species have preceded them (mechanism 2). In the latter, early occupants of the site modify the environment so that it becomes less suitable for subsequent colonization of both early and late species, and the early species are killed not by competition as in the first two models but by some local disturbance such as grazing or frost (mechanism 3).

Any of these three mechanisms could operate in a fungal decomposer succession, but there is a dearth of direct evidence. *M. galopus* appears in the field to depend on changes brought about by earlier colonists (mechanism 1). In monoculture, however, it grows equally well on recently fallen and old weathered litter (J. C. Frankland, unpublished data), so mechanism 2 could be operating, but the priming effect of phylloplane fungi and the season in which this basidiomycete actually enters the succession is not known. The basidiomycete life form, relatively long-lived with more elaborate hyphal systems than many microfungi, equivalent perhaps to that of higher plants of late succession, may fit it to survive beyond the demise of other colonists (mechanism 3). There is evidence from both plants and animals of the advantage of larger size in many interactions between species (Grime, 1977). Mechanism 3 could be tested by showing that *M. galopus* grows better in the absence of earlier members of the succession.

The actual process of a fungal replacement, like the cause, will vary with the species combination, and a population may be extinguished or merely reduced in size. The details again are usually destroyed by the fungal sampling techniques used in successional studies. Initially, one fungus may colonize literally on top of another,

or two mycelia growing on adjacent sites may meet and interact at their margins.
Replacement can follow when the growth of one fungus is inhibited and the other grows
over or through it. The process can vary greatly in speed and can be described as
passive, or active if it involves parasitism or some type of antagonism, such as
"hyphal interference" (Ikediugwu and Webster, 1970). These reactions have often been
observed in dual cultures on agar media (e.g., see Rypáček, 1966; Rayner and Todd, 19-
79), but there are few observations of the process of replacement on natural substrates.

Rayner recorded the replacement process when fungal decomposers of hardwood
stumps and logs were grown in pairs on 3% malt agar, sawdust, and wood blocks. Usually the growth of the "aggressor" slowed before it overgrew the partner, sometimes
lysing it, e.g., the replacement of *Bjerkandera adusta* (Willd. ex Fr.) Karst. by
Phlebia merismoides Fr., but some species were replaced much more rapidly and appeared to be parasitized, e.g., *B. adusta* by *Pseudotrametes gibbosa* (Pers.) Bond.
In the field he detected replacements by such evidence as: a relict zone line within another interaction area; a succession of zone lines in front of an advancing mycelium; or a fungus growing only at the periphery of a decay column occupied by a
second species. Interactions in the field and laboratory corresponded closely, indicating the importance of the innate competitive abilities of the species. The
mechanisms underlying these interactive characters are obscure. Natural selection
favors self-replacement which slows succession, and it can be argued that there are
no adaptations to succession itself (Drury and Nisbet, 1973; Horn, 1976).

Patterns of change in the biology of successive fungal decomposers may help to
explain some of the mechanisms of species replacement and have been discussed by
Swift (1976). Garrett's generalized hypothesis (1963) of a nutritional sequence on
plant debris of "sugar fungi," followed by cellulose and lignin decomposers, channeled mycologists' thoughts for some time, but further evidence (Webster, 1970; Swift,
1976) shows that the relationship between substrate composition and fungal function
is not as clear cut. However, there appears to be some pattern in competitive features. On nonwoody plants many early colonists, e.g., phylloplane fungi, have a
relatively short life span with one flush of sporing, which "avoids" competition.
They are specialist pioneers, or "opportunists," similar to the so-called r-selected
species of other groups of organisms (Pianka, 1970; Swift, 1976). Like annual weeds,
they develop rapidly, have a high reproductive output, a relatively small thallus,
and are noncompetitive. Later colonists include more powerful antibiotic producers
such as penicillia, or basidiomycetes with life spans often longer than 1 year. The
latter can be compared with perennial plants or K-selected species, often developing
slowly with delayed reproduction, buffered against climatic changes by their relatively large thallus, and often strongly interactive. Annual types which appear late
in a succession are often exploiting new pioneer situations such as fungal mycelium
and animal feces.

On woody substrates colonized by a succession of basidiomycetes, similar competitive trends can be seen within the one taxonomic group of fungi. Rayner and Todd

Table 5 Competitive reactions between paired wood-rotting fungi on malt agar at 24°C[a]

	Species	"Wins"	"Losses"	"Draws"
Parasites and heart rots	*Fomes fomentarius*	3	16	8
	Ganoderma applanatum	6	16	5
	Heterobasidion annosum	1	9	17
	Phellinus igniarius	3	14	10
Total		13	55	40
Sapwood saprophytes	*Bjerkandera adusta*	19	6	2
	Coniophora puteana	24	1	2
	Coriolus versicolor	16	3	8
	Stereum hirsutum	16	4	7
Total		75	14	19

[a]The results of pairing 27 species with each of the eight species listed are shown. See text for method of scoring.

Source: After Rypáček (1966).

(1979), for example, found that the species present at the later stages of wood decay are more competitive in terms of ability to replace, e.g., *Hypholoma fasciculare* (Huds. ex Fr.) Kummer and *Phlebia merismoides* on hardwood stumps, whereas an early colonist such as *Chondrostereum purpureum* (Pers. ex Fr.) Pouz. grows rapidly, fruits rapidly and is quickly replaced. Rypáček (1966) recorded the reaction of 28 wood-rotting fungi when paired on malt agar. Table 5 gives the results of a selection of these pairings: suppression or overgrowth of one species by another is recorded as a "win" or "loss"; and deadlock, where neither fungus can grow past the other, or an indecisive reaction, is recorded as a "draw." It shows that the attackers of living wood had more losses than wins and that the sapwood saprophytes or secondary colonizers had the reverse, suggesting again that the latter are better adapted to compete with other fungi.

V. THE FUTURE

Can the mycologist advance beyond the compiling of species lists in the study of succession? Few studies have done so. Basically there seem to be two possible approaches—either to break the process down or to build it up. The succession can be "kicked" by adding chemicals, irradiation, and various other interventions. Spectacular changes can occur in this way. A succession of specific fruit bodies on pine needles was dramatically altered by adding urea (Lehmann, 1975); see Table 6. The system, however, remains complicated, and the changes are difficult to interpret.

Table 6 Succession of fungal fruit bodies on needles of *Pinus sylvestris* treated
with urea in woodland plots[a]

	Months after application	Treatment	
		Water control	Urea (20 gN/m^2)
Marasmius androsaceus	0-5	+	-
Desmazierella acicola	4-5	+	-
Lophodermium pinastri	4-5	+	-
Ascobolus denudatus	1-2	-	+
Pseudombrophila deerata	1-3	-	+
Mycena sp.	2-4	-	+
Tephrocybe tesquorum	2-5	-	+
Coprinus echinosporus	3-4	-	+

[a]Key: +, present; -, absent.

Source: After Lehmann (1975).

A clear warning against false interpretations was given by Harper and Webster's (1964)
evidence that the appearance of fruit bodies in a coprophilous succession corresponded
to a characteristic minimum time of fruiting and was not, as might be expected, the
result of competition.

Close observation of successional communities in the field, or extremely critical
sampling on a microscale, as opposed to disruptive macrosampling, can still be rewarding as in Rayner's recent studies of wood decomposition (see Rayner and Todd, 1979).
Not much is known, for instance, of the spatial relationships, or "pattern," of fungal communities, either inter- or intraspecific. Does the scale of pattern alter
through succession, and is it related to the stability of the mycoflora? There is
evidence from statistical measurements that the association between both species
and individuals is at a minimum in stable vegetation (Greig-Smith, 1961; Kershaw,
1973). Maximum pattern of *Ammophila arenaria*, i.e., maximum level of association
between individuals, was found in stabilizing, not stable, dunes.

The possibilities of building a predictive Markovian model (Horn, 1976) for a
substrate succession of fungi would also be worth exploring if sufficient quantitative data, perhaps on an area basis, could be obtained and if various theoretical constraints could be satisfied. Several higher-plant ecologists have represented succession as a Markovian replacement chain, in which the current composition of a community
can be projected into the future by constructing a matrix showing the probability that
a given species will be replaced in a specified time. Usher and Parr (1977) obtained
some success in applying this type of model to a succession of decomposer termites on
wood. The succession must be a continuum in which all species of the final stage are
present at the start of the calculations, without discontinuities occurring during

the succession. Discontinuities in fungal successions are frequently reported (Swift, 1976) but are probably more often apparent than real, due to shortcomings of the sampling technique. The relatively short life of many substrates and rapid fungal replacement rates have advantages in that predictions can be verified, not by inference, as in many seral successions of higher plants, but by actual observation of the changes taking place on one substrate.

Mathematical modeling is currently fashionable because of the progress it has made in many fields, but in general the soil mycologist, perhaps unlike the plant pathologist, is barely ready for it in successional studies. Even higher-plant ecologists do not appear to have applied it to a succession as a whole (Fogg, 1977). Modeling of fungal succession and also the use of modern methods of gradient analysis are hampered by the still sketchy information on the autecology of many common soil and litter species and by taxonomic problems. Species and strains occurring at different stages in a succession are often incorrectly lumped together in one taxonomic unit. *Trichoderma viride* agg. was recorded in every stage of the dune succession described, but it may have represented a series of different genotypes. Soil fungi are not known by mycologists as higher plants are known by plant ecologists. Fungi are also notoriously difficult to quantify in situ, especially in a mixed community, when the concept of an "individual" has no precise meaning, although such parameters as annual extension of mycelia or fruit body size may be useful measures of performance in some successions.

It is virtually impossible to disentangle the causal factors of fungal succession without resorting to laboratory experiments and simple model systems. Experimenters in the past have concentrated on single organisms and the effects of isolated variables. More studies are needed in which the various species combinations of a succession are built up and subjected to interacting factors. Higher-plant ecologists have shown that the "physiological optimum" of a species in monoculture is rarely the same as its "ecological optimum" and that mixed cultures often exhibit a response approaching the response in the field (Mueller-Dombois and Ellenberg, 1974). This is likely to be true of all organisms, but the mycological evidence is sparse. Garrett's classic experiments with root-disease fungi in which he recorded the competitive colonization of "units" of wheat straw, buried in soil containing fungi at various inoculum levels under various conditions, show how the gap between the agar plate and the natural situation can be bridged. His stepwise development (with coworkers) of simple model systems, readably described in his 1970 review, is an approach which could well be extended more widely in successional studies.

Fungal successions, like most of the ecological processes described in this volume, are intimately tied in with the activities of other organisms and other trophic levels, but they are not often investigated together. Winston's study in 1956 of the complex interrelated succession of fungi and animals on decomposing acorns (Fig. 5) still has a relevant message. These are well-worn sentiments, but there

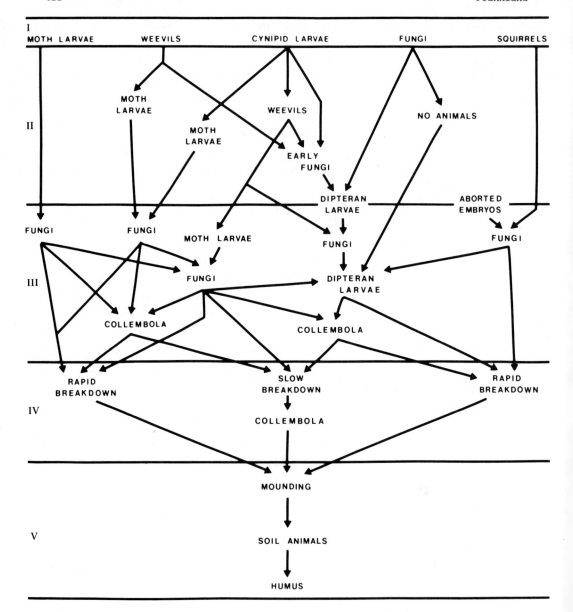

Fig. 5 Interrelated succession of fungi and animals on decaying acorns of *Quercus rubra* L. Stages of decay: I, damage to living nut; II, destruction of embryo; III, breakdown of inner shell layers; IV, final decay of outer shell and cup; V, collapse of softened shell into a mound and incorporation into humus. (After Winston, 1956.)

are recent signs that more zoologists and mycologists are combining their efforts in the model-systems approach to succession, as in the interesting experiments by Parkinson et al. (1979) on the effects of collembolan grazing on fungal colonization of leaf litter. However, "succession is the natural integration of all ecosystem processes" so it will remain "among the most complex and difficult to unravel of ecological phenomena" (McNaughton and Wolf, 1973, pp. 392, 390).

ACKNOWLEDGMENTS

I am grateful to Dr. H. T. Tribe for Fig. 1 and to Professor G. J. F. Pugh for Fig. 2. I thank also Dr. T. F. Hering (the arranger of Table 5), Miss P. M. Latter, Miss K. Newell, and Dr. A. D. M. Rayner for valuable discussions; Mr. A. D. Bailey, Mrs. S. Benson, Miss P. L. Costeloe, and Mrs. G. Howson for their technical assistance, and Mrs. A. Killalea for her help in obtaining literature.

REFERENCES

Brown, J. C. (1957). An ecological study of the soil fungi of some British sand dunes. Ph.D. thesis, University of London.

Brown, J. C. (1958a). Soil fungi of some British sand dunes in relation to soil type and succession. *J. Ecol. 46*: 641-664.

Brown, J. C. (1958b). Fungal mycelium in dune soils estimated by a modified impression slide technique. *Trans. Brit. mycol. Soc. 41*: 81-88.

Carruthers, S. M., and Rayner, A. D. M. (1979). Fungal communities in decaying hardwood branches. *Trans. Brit. mycol. Soc. 72*: 283-289.

Chou, M. C., and Preece, T. F. (1968). The effect of pollen grains on infections caused by *Botrytis cinerea* Fr. *Ann. Appl. Biol. 62*: 11-22.

Clark, F. E. (1965). The concept of competition in microbial ecology. In *Ecology of Soil-borne Plant Pathogens*, K. F. Baker and W. C. Snyder (Eds.). Murray, London, pp. 339-347.

Clements, F. E. (1916). Plant succession: An analysis of the development of vegetation. *Carnegie Inst. Wash. Publ. No. 242*.

Connell, J. H., and Slatyer, R. O. (1977). Mechanisms of succession in natural communities and their role in community stability and organization. *Amer. Natur. 111*: 1119-1144.

Crocker, R. L., and Major, J. (1955). Soil development in relation to vegetation and surface age at Glacier Bay, Alaska. *J. Ecol. 43*: 427-448.

Drury, W. H., and Nisbet, I. C. T. (1973). Succession. *J. Arnold Arbor. 54*: 331-368.

Fogg, G. E. (1977). Physiology under the worst possible conditions. *J. Inst. Biol. 24*: 73-79.

Frankland, J. C. (1966). Succession of fungi on decaying petioles of *Pteridium aquilinum*. *J. Ecol. 54*: 41-63.

Frankland, J. C. (1969). Fungal decomposition of bracken petioles. *J. Ecol. 57*: 25-36.

Frankland, J. C. (1974). Decomposition of lower plants. In *Biology of Plant Litter Decomposition*, C. H. Dickinson and G. J. F. Pugh (Eds.). Academic Press, New York, pp. 3-36.

Frankland, J. C. (1975). Fungal decomposition of leaf litter in a deciduous woodland. In *Biodégradation et humification*, G. Kilbertus, O. Reisinger, A. Mourey, and J. A. Cancela da Fonseca (Eds.). Rapport de 1er Colloque International, 1974, Nancy University, Pierron, France, pp. 33-40.

Frankland, J. C. (1976). Decomposition of bracken litter. *Bot. J. Linnean Soc. 73*: 133-143.

Garrett, S. D. (1950). Ecology of the root-infecting fungi. *Biol. Rev. 25*: 220-254.

Garrett, S. D. (1956). *Biology of Root-infecting Fungi*. Cambridge Univ. Press, New York.

Garrett, S. D. (1963). *Soil Fungi and Soil Fertility*. Pergamon Press, Elmsford, N.Y.

Garrett, S. D. (1970). *Pathogenic Root-infecting Fungi*. Cambridge Univ. Press, New York.

Gleason, H. A. (1926). The individualistic concept of the plant association. *Bull. Torrey Bot. Club. 53*: 7-26.

Greig-Smith, P. (1961). Data on pattern within plant communities. II. *Ammophila arenaria* (L.) Link. *J. Ecol. 49*: 703-708.

Greig-Smith, P. (1964). *Quantitative Plant Ecology*. Butterworths, Washington, D.C.

Griffin, D. M. (1960). Fungal colonization of sterile hair in contact with soil. *Trans. Brit. Mycol. Soc. 43*: 583-596.

Griffin, D. M. (1972). *Ecology of Soil Fungi*. Chapman & Hall, London.

Grime, J. P. (1973). Competition and diversity in herbaceous vegetation: A reply. *Nature 244*: 310-311.

Grime, J. P. (1977). Evidence for the existence of three primary strategies in plants and its relevance to ecological and evolutionary theory. *Amer. Natur. 111*: 1169-1194.

Harper, J. E., and Webster, J. (1964). An experimental analysis of the coprophilous fungus succession. *Trans. Brit. Mycol. Soc. 47*: 511-530.

Harrison, A. F. (1971). The inhibitory effect of oak leaf litter tannins on the growth of fungi in relation to litter decomposition. *Soil Biol. Biochem. 3*: 167-172.

Hering, T. F. (1965). Succession of fungi in the litter of a Lake District oakwood. *Trans. Brit. Mycol. Soc. 48*: 391-408.

Horn, H. S. (1974). The ecology of secondary succession. *Ann. Rev. Ecol. Syst. 5*: 25-37.

Horn, H. S. (1976). Succession. In *Theoretical Ecology: Principles and Applications*, R. M. May (Ed.). Blackwell Sci. Publns., Oxford, England, pp. 187-204.

Hudson, H. J. (1968). The ecology of fungi on plant remains above the soil. *New Phytol. 67*: 837-874.

Ikediugwu, F. E. O., and Webster, J. (1970). Hyphal interference in a range of coprophilous fungi. *Trans. Brit. Mycol. Soc. 54*: 205-210.

Keever, C. (1950). Causes of succession on old fields of the Piedmont, South Carolina. *Ecol. Monogr. 20*: 299-250.

Kendrick, W. B., and Burges, A. (1962). Biological aspects of the decay of *Pinus sylvestris* leaf litter. *Nova Hedwigia 4*: 313-342.

Kershaw, K. A. (1973). *Quantitative and Dynamic Plant Ecology*. Arnold, London.

Lehmann, P. F. (1975). Changes in the fungal succession after application of urea to litter of Scots Pine (*Pinus sylvestris* L.). In *Biodégradation et humification*, G. Kilbertus, O. Reisinger, A. Mourey, and J. A. Cancela da Fonseca (Eds.). Rapport du 1er Colloque International, 1974, Nancy University, Pierron, France, pp. 470-476.

Levi, M. P., Merrill, W., and Cowling, E. B. (1968). Role of nitrogen in wood deterioration. VI. Mycelial fractions and model nitrogen compounds as substrates for growth of *Polyporus versicolor* and other wood-destroying and wood-inhabiting fungi. *Phytopathology 58*: 626-634.

McNaughton, S. J., and Wolf, L. L. (1973). *General Ecology*. Holt, Rinehart and Winston, New York.

Meredith, D. S. (1959). The infection of pine stumps by *Fomes annosus* and other fungi. *Ann. Bot.* (London) [*N.S.*] *23*: 455-476.

Morrison, R. G., and Yarranton, G. A. (1974). Vegetational heterogeneity during a primary sand dune succession. *Can. J. Bot. 52*: 397-410.

Mueller-Dombois, D., and Ellenberg, H. (1974). *Aims and Methods of Vegetation Ecology*. Wiley, New York.

Nilsson, T. (1974). Microscopic studies on the degradation of cellophane and various cellulosic fibres by wood-attacking microfungi. *Studia forestalia Suecica No. 117.*

Nilsson, T. (1976). Soft-rot fungi-decay patterns and enzyme production. In *Organismen und Holz*, G. Becker and W. Liese (Eds.). Internationales Symposium, Berlin-Dahlem, 1975. Duncker & Humblot, Berlin, pp. 103-112.

Odum, E. P. (1969). The strategy of ecosystem development. *Science 164*: 262-270.

Park, D. (1968). The ecology of terrestrial fungi. In *The Fungi*, G. C. Ainsworth and A. S. Sussmann (Eds.), Vol. 3. Academic Press, New York, pp. 5-39.

Parkinson, D., Visser, S., and Whittaker, J. B. (1979). Effects of collembolan grazing on fungal colonization of leaf litter. *Soil Biol. Biochem. 11*: 529-535.

Person, C. (1968). Genetical adjustment of fungi to their environment. In *The Fungi*, G. C. Ainsworth and A. S. Sussman (Eds.), Vol. 3. Academic Press, New York, pp. 395-415.

Pianka, E. R. (1970). On r- and K-selection. *Amer. Natur. 104*: 592-597.

Pugh, G. J. F., and Buckley, N. G. (1971). The leaf surface as a substrate for colonization by fungi. In *Ecology of Leaf Surface Micro-organisms*, T. F. Preece and C. H. Dickinson (Eds.). Academic Press, New York, pp. 431-445.

Rayner, A. D. M., and Todd, N. K. (1977). Intraspecific antagonism in natural populations of wood-decaying basidiomycetes. *J. Gen. Microbiol. 103*: 85-90.

Rayner, A. D. M., and Todd, N. K. (1979). Population and community structure and dynamics of fungi in decaying wood. *Adv. Bot. Res. 7*: 334-420.

Reiners, W. A., Worley, I. A., and Lawrence, D. B. (1971). Plant diversity in a chronosequence at Glacier Bay, Alaska. *Ecology 52*: 55-69.

Rypáček, V. (1966). *Biologie holzzerstörender Pilze*. Fischer, Jena.

Seifert, K. (1964). Die Veränderung der chemischen Holzzusammensetzung durch den Bläuepilz *Pullularia pullulans* (de Bary) Berkhout [= *Aureobasidium pullulans* (de Bary) Arnaud]. *Holz Roh-u, Werkstoff 22*: 405-409.

Shugart, H. H., and Hett, J. M. (1973). Succession: Similarities of species turnover rates. *Science 180*: 1379-1381.

Siu, R. G. H. (1951). *Microbial Decomposition of Cellulose*. Reinhold, New York.

Southwood, T. R. E. (1977). Habitat, the templet for ecological strategies. *J. Anim. Ecol. 46*: 337-365.

Sussman, A. S. (1966). Dormancy and spore germination. In *The Fungi*, G. C. Ainsworth and A. S. Sussman (Eds.), Vol. 2. Academic Press, New York, pp. 733-764.

Sussman, A. S. (1968). Longevity and survivability of fungi. In *The Fungi*, G. C. Ainsworth and A. S. Sussman (Eds.), Vol. 3. Academic Press, New York, pp. 447-486.

Swift, M. J. (1976). Species diversity and the structure of microbial communities in terrestrial habitats. In *The Role of Terrestrial and Aquatic Organisms in Decomposition Processes*, J. M. Anderson and A. Macfadyen (Eds.). Blackwell Sci. Publns., Oxford, England, pp. 185-222.

Tribe, H. T. (1957). Ecology of micro-organisms in soils as observed during their development upon buried cellulose film. In *Microbial Ecology*, R. E. O. Williams and C. C. Spicer (Eds.). Cambridge Univ. Press, New York, pp. 287-298.

Tribe, H. T. (1961). Microbiology of cellulose decomposition in soil. *Soil Sci. 92*: 61-77.

Usher, M. B., and Parr, T. W. (1977). Are there successional changes in arthropod decomposer communities? *J. Envir. Management 5*: 151-160.

Waid, J. S. (1968). Physiological and biochemical adjustment of fungi to their environment. In *The Fungi*, G. C. Ainsworth and A. S. Sussman (Eds.), Vol. 3. Academic Press, New York, pp. 289-323.

Watt, A. S. (1947). Pattern and process in the plant community. *J. Ecol. 35*: 1-22.

Webster, J. (1970). Coprophilous fungi. *Trans. Brit. Mycol. Soc. 54*: 161-180.

Whittaker, R. H. (1975). *Communities and Ecosystems*. Macmillan, New York.

Wieringa, K. T. (1955). Der Abbau der Pektine: der erste Angriffe der organischen Pflanzensubstanz. *Z. PflErnähr. Düng. Bodenk. 69*: 150-155.

Winston, P. W. (1956). The acorn microsere, with special reference to arthropods. *Ecology 37*: 120-132.

Chapter 23

RELATIONSHIPS BETWEEN MACROMYCETES AND THE DEVELOPMENT
OF HIGHER PLANT COMMUNITIES

Roy Watling

Royal Botanic Garden
Edinburgh, Scotland

I. INTRODUCTION

Undoubtedly the most desirable way to conduct a study of the successional pattern of
higher fungi in any plant association is to relate those mycorrhizal species present
to the development of the community as a whole. Mycorrhizal species in climax and
most proclimax vegetation usually make up a high percentage of the constituent fungi
and are more related to the phanerogamic dominant than the humicolous species found
in the same areas.

When the vascular plants form a complex community, difficulties often arise as
to determining which fungi grow with which host. Therefore the British Isles probably rank among the best places in which to study successional patterns in macrofun-
gi--for, although there is a huge variety of trees introduced from abroad, the native
flora itself is depauperate when compared with other areas of the world at the same
latitude. The fungus flora of Britain is consequently rather limited.

Several problems are inherent in the study of the succession of larger fungi,
the major one of which is the dependence on the successful recording of basidiomes.
Anyone who has collected fungi in a single locality for many years will appreciate
that mycorrhizal species may appear suddenly in places where they may have not been
seen previously but where they have surely been all the time in the vegetative state.
Even in the absence of basidiomes one might suggest that a certain taxon is present
as a mycelial network within the substrate, but in ecological studies if the fungus
does not fructify it cannot be regarded as present.

Fruiting of agarics and boletes is erratic--of that there is no doubt even in
common species--and often brief, the individual basidiomes being so ephemeral that
much depends on the presence of an interested party at the right place at the right
time.

We know about the factors controlling initial mycorrhizal infection and the
metabolic differences betwen various host species, indeed provenances within a single
host (Lobanow, 1960). Is it that certain fungi are tied to a particular host because
of physiological characteristics of the host or of the fungus, or both? Could it be

that there is a hierarchy of mycorrhizal competence within the Agaricales and that
competition is experienced for colonization of a single root system? No doubt climate
influences the stability of any relationship just as fruiting is a response to particular climatic factors, and when in stress one relationship could well be replaced
by another more favorable partner under the new circumstances. According to Trappe
(1977), the ecology of mycorrhizal fungi hinges on those biochemical pathways which
fungi have evolved in order to survive environmental variation at the tree root.

Kauffman (1906) has shown that when mycorrhizal fungi are present some can associate with more than one host. Therefore is not the boreal woodland, with its characteristic mycorrhiza-dependent trees, operating as a complex reticulum of mycelium,
in some cases connecting individuals of different vascular plant taxa? Thus when one
host species is removed, could some of the mycorrhizal fungi present continue their
existence with a second host? It has been shown (Hayes, 1967) from pollen data that
numerous areas within the ancient pine woodland at Rannoch in Scotland experienced
oscillations of first birch then pine, then birch and pine again, and so on. Thus
a changing pattern of higher fungi can be envisaged paralleling these fluctuations
not only in such pine-birch ecosystems but in other communities.

These questions should not be asked in isolation. Consideration must be paid
to the substrate characteristics in which the dual organism is growing. There is
some evidence to suggest that some mycorrhiza only form in a given range of soil conditions (Trappe, 1977). Thus the characteristics offered to explain the fructification of a given species may operate over quite a different range of parameters to
the actual mycorrhizal relationship and, indeed, to the initial infection. Grainger
(1946) discussed the fruiting of larger fungi in relation to physical factors and
suggested that it is dependent on a favorable soil carbon/nitrogen ratio, but this
is perhaps a rather coarse way of monitoring the situation even when this ratio is
modified by reference to soil moisture, temperature, and pH. He claimed that accumulation of free nitrogen compounds in spring and autumn were partly responsible for
the timing of the fructification of larger fungi. Hora (1972) applied a nitrogen-
containing fertilizer to woodland plots and, although the yield of basidiomes increased, most of this could be attributed to the single mycorrhizal agaric *Paxillus
involutus*; in some species the yield decreased, but it can be hypothesized that a
spectrum of species reaction exists in the field even to the evolution of ecotypic
variation, (Laiho, 1970; Trappe, 1977). Fructification is the end product of a process which may take in a particular mycorrhizal species not just one but several seasons for completion. This end product is the result of a series of interconnecting
factors: magnitude of the mycelial reserves, initiation of primordia, and supply of
nutrients and water to the developing basidiomes.

When the tree has unrestricted growth and its roots are in fairly uniform condition, a complete or disjunct fairy ring of basidiomes at the edge of the tree canopy may be produced. But most naturally occurring trees do not grow in this fashion.
In grazing areas of the British Isles trees are often restricted to rocky ledges grow-

ing with their aerial parts out of the reach of sheep, some roots exposed or certainly in thin soils, while others, often those supporting the tree, push deep into the surrounding soil above the ledges. Even a casual examination shows the morphology of the mycorrhizal shoots in these two ecological situations to be very different: on the soils prone to drying out, the shoots are frequently dark colored and velvety in appearance; in the deeper soils they are pale colored and fluffy. This indicates that under field conditions mature trees can probably support more than one mycorrhizal fungus, a fact partially confirmed by the appearance of basidiomes of different species at different points around the tree.

In the future, studies should provide a much more complete picture of the entire fungal community, not only by a careful examination of those fructifications appearing in August and September, but also of those appearing well into the winter months. This implies a failure in our observational data, emphasized in my mind by the results we obtained in Edinburgh after an examination of Ascomycetes collected in a single season by Jack Warcup during a mycorrhizal study in Scotland. The possible role played by such Discomycetes as members of the *Peziza badia* complex, *Inermisia* spp., and *Sepultaria* spp. should not be ignored.

In addition we should look further afield within our taxonomic framework to assess the vast array of resupinate fungi as potential symbionts. This should not be surprising when many species within the *Thanatephorus-Ceratobasidium* complex are proven mycorrhizals. Reexamination of field records of Aphyllophorales and reassessment of the part played by such fungi in the woodland economy leads one to believe that *Amphinema byssoides*, *Tylosperma fibrillosum*, and several *Tomentella* spp. are worthy of more attention.

Richardson (1970) has demonstrated after careful examination of set quadrats in a pine woodland over a 5-year period that with a week between visits about a quarter of the basidiomes produced would not be recorded, so that after 5 years species new to the quadrats were still being noted. He also drew attention to the fact that little is known about the longevity of individual basidiomes and the precise nature of the response to temperature, rainfall, and other climatic factors. Even less is known of the effect of removal of basidiomes on the subsequent production of the mycelium, or the reason for so-called good and bad years of particular species, years which do not always appear to coincide with similar fluctuations in other taxa.

Ronald Rayner (personal communication) has analyzed the species collected on various fungus forays in Britain, and his results show that the picture obtained is rather different to that with vascular plants, there being an overemphasis of the rare species and an underemphasis of the common ones. Thus with the study of fruiting Basidiomycetes in contrast to other organisms, rather different interpretations of the field data must be applied before a final picture can be assembled.

Mycorrhizal species undoubtedly play a dominant role both up to the timber line and in the mountains, and many species found apparently possess a rather narrow host spectrum. It has been suggested (Lange, 1974) that by strong mycorrhizal specialization a fungus can satisfy its requirements through the symbiotic relationship and survive as a dual organism in localities with extreme climatic conditions. However, as we rely on presence or absence of basidiomes it is possible that we have rather a biased picture of the potential range of mycorrhizal fungi associated with a particular tree species.

In the *Picea sitchensis* plantings in Britain, *Lactarius rufus* is frequently collected and obviously is now an important component of the fungal flora. This species is not infrequent in the natural communities of Sitka spruce in the United States, and in Michigan it grows abundantly under spruce and sometimes in the north under pine. In Britain it characterizes our native pine woodland. In pure culture Ian Alexander has achieved a successful inoculation of *Picea sitchesis* with *Lactarius rufus*, and he communicated to me that in cultures primoridal basidiomes had formed. *Lactarius rufus*, although typical of native pine woodland in Britain, may infrequently be seen in hill birch scrub on base-poor soils derived from acid millstone grit rocks in northern England, indicating that even in Britain it must have a much wider range in the vegetative state than is generally considered.

Some kind of succession of species can be observed in the field, indeed, can probably be derived from analysis of foray records if enough field data are available. These observations might be indicative of trends and one such trend would appear to indicate that *Laccaria laccata* and *Hebeloma* spp. are pioneer fungi colonizing newly planted woodland and open land invaded by sapling trees. These two genera have also been found in either or both pot and nursery experiments [*Laccaria* growing with *Quercus petraea*; *Hebeloma fragilipes* with *Pinus* spp., *H. populinum* with *Picea sitchensis* (Seaby, personal communication) and *H. versipelle* with *Tilia cordata*] as has *Suillus grevillei* with *Larix europea*.

II. FIELD OBSERVATIONS

A. Creeping Willow Communities

Despite the drawbacks just described, a preliminary attempt is offered here to analyze the possible succession in communities containing the mycorrhizal shrub *Salix repens* ("creeping willow"): communities have been studied in mainland hill pasture, coastal dunes, and cliff turf in northern areas of the British Isles. These areas have been compared with similar communities containing *S. repens* on St. Kilda, a group of small Atlantic islands, (due west of Harris in the Outer Hebrides) where tree species have long since vanished. *S. repens* is an ideal subject, for it is thought to form mycorrhiza with a whole range of larger fungi generally considered associated with other vascular plant species.

Cliff turf *S. repens* communities have been examined at Farr Bay, Skerrary, and Terrisdale, all near Bettyhill in Sutherland, and on St. Kilda. Coastal dune communities have also been examined at Strathnaver in Sutherland; Forvie Sands, Aberdeenshire; Culbin Sands, Morayshire; Rhum, Inner Hebrides; Tentsmuir, Fife; St. Cyrus,

Table 1 Location of study areas in the British Isles[a]

I. Communities with creeping willow (*Salix repens*)

    A. Grassland/heath communities

        1. Mainland grassland (hill pasture)
            Straloch, Perthshire 1

        2. Maritime
            St. Kilda, Outer Hebrides 2
            Mingulay, Outer Hebrides 9

    B. Sea cliff-turf communities
        St. Kilda 2
        Farr Bay, Bettyhill, Sutherland 3
        Skerrary, near Bettyhill 6

    C. Coastal dune communities
        Farr Bay, Bettyhill 4
        Strathnaver, Bettyhill, Sutherland 5
        Forvie Sands, Aberdeenshire 7
        Rhum, Inner Hebrides 8
        St. Cyrus, Angus 10
        Tentsmuir, Fife 11
        Loch Gairloch, Argyllshire 12
        Culbin Sands, Morayshire 15
        Ross Links, Northumberland 13

II. Main study area with arborescent willows
        Whitlaw Moss, Selkirkshire 14

III. Montane study areas
        Cairnwell, Perthshire 16
        Ben Vrackie, Perthshire 17
        Beinn Eighe, Wester Ross 18

---

[a]Numerals refer to areas labeled on the map in Fig. 1.

Angus; Gairloch, Argyllshire; and Ross Links in Northumberland. (See Table 1 and Fig. 1.) These communities have been compared with an area of *S. repens* colonizing hill pasture and acidic heathland on a morainic complex at Straloch near Kirk Michael in Perthshire and a community developed on a raised beach on St. Kilda.

Straloch and the Bettyhill localities have been surveyed at least once annually for 15 and 8 years, respectively. St. Kilda and Rhum were each intensely surveyed for 9- and 14-day periods in 1964 and 1967, respectively. Forvie Sands, Culbin Sands, and Tentsmuir have been visited on several occasions at irregular intervals and the other localities on single occasions only.

From field observations it bacame apparent that the fungal flora which had developed at Straloch and the other sites studied depended not only on climatic and edaphic conditions but also on the plant community which the *S. repens* replaced. In Britain *S. repens* has twice been recorded from deposits (Godwin, 1956) of the Lower Pleistocene; other material has been derived from Upper Pleistocene deposits of southeastern England in which leaves and wood have been found. Although more recent records are sparse and uncertain, they indicate that *S. repens* has survived in the Brit-

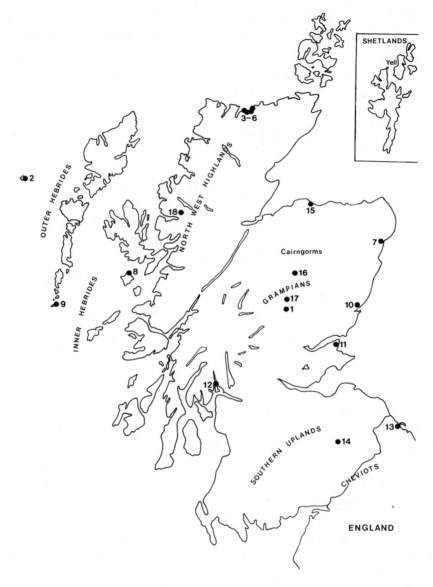

Fig. 1 Map showing Scottish mainland and island localities cited in the text.
(For key to the numerals, see Table 1.)

ish Isles since the Upper Pleistocene, if not longer, to possess at the present day
a wide, though somewhat scattered, range. Apart from the British sites it is restricted to the southern part of Scandinavia except for scattered localities between 60°N
and 66°N; a belt extends around the Baltic Coast.

In contrast to many other species of willow, *S. repens* will colonize soils of
all degrees of productivity in the absence of other tree competitors. It can be an
understory plant but usually forms marginal communities to birchwoods and the like,

rapidly colonizing the areas after removal of the trees. It probably invaded the bare ground along with pioneer birch after the last glaciation (8500 B.C.), theoretically similar to some ecological niches which can be still seen on some of the Scottish mountains. It is difficult, however, to confirm this suggestion without the finding of macrofossils, as willow pollen is much the same for a whole range of species of *Salix* (Birks, 1973).

Kreisel (1965) has attempted an analysis of the larger fungi associated with *S. repens* in East Germany.

1. Mainland grassland/heath: Straloch

The locality is hill pasture grassland on a complex of terminal moraines of a small glaciation after the Great Ice Age and is situated in the Ardle Valley on mica and mica/quartzite schists with chlorite bands formed from igneous activity. A continuous pollen diagram is available from man's present activities back to the Palaeolithic, indicating that like much of the rest of Britain the site passed from tundra through park tundra, then from *Pinus/Betula* to mesophytic forest of the climatic optimum, and later was burnt by humans to clear the land for agriculture (see Table 2).

The natural woodland which was cleared probably consisted of mixed scrub birch and in the damper areas willow and alder with perhaps some oak, which may have resembled the trees still to be seen at Kirkmichael nearby. This destruction of the woodland allowed *S. repens* to spread, and whereas the regenerating oak and birch were grazed out the prostrate *S. repens* remained.

A "standing stone" (monolith) only 6 km from Straloch indicates early man's presence in the area. Indeed, iron age activity can also be traced, and it is known from historic records that the valley was later attacked by the Danes. Grazing animals were an important factor in the Ardle Valley. Kirkmichael, only 3.2 km from Straloch, became the area's principal market, especially for cattle; cattle were ultimately replaced by sheep, which are still present today.

The Straloch locality involves a very extensive development of *S. repens* on marshy land, acidic heath, and fairly base-rich soils all in close proximity; the population demonstrates the great variability found in this species. Some plants which are considered typical have very long prostrate stems, small narrowly elliptic subglabrous leaves, and glabrous capsules. However, others are relatively stout and ascending with broadly oblong, shortly acute sericeous leaves and pubescent capsules; some of these plants even approach the morphology of *S. repens* subsp. *argentea*.

S. repens at Straloch is found in *Agrostis/Festuca* grassland intermixed with *Nardus*, the actual constituents depending on the base status of the glacial till which covers the base rock. In some areas it is quite calcareous and supports *Helianthemum chamaecistus*, whereas on leached podzolic areas *Salix* is codominant with *Calluna vulgaris*. In the marshy areas at the base of the moraines the *Salix* is found associated with *Menyanthes trifoliata*, *Narthecium ossifragum*, and *Juncus effusus*.

Table 2 Sequence of events leading to mainland communities dominated by *Salix repens*

Palaeolithic man	Mesolithic man	Neolithic man	Bronze age	Iron age/Roman Empire	Norman Conquest onwards
Tundra					
→ Park tundra			Dry		
→ *Pinus/Betula*	*Ulmus* increase			*Betula* cropping/grazing → grassland	
	Corylus				
	→ Mesophytic forest				
	Alnus/Betula/Corylus *Quercus/Pinus/Ulmus* with *Salix* in wet areas			*Quercus* felling/burning/grazing → grassland	
		Appearance of charcoal and weeds	Wet	*Salix/Alnus* → *Salix/Alnus*	
Late glacial period	Postglacial period			1000 A.D.	
	5000 B.C.				
	Climatic optimum	Climatic deterioration			

Salix repens ---

Table 3 Site 1--Mainland hill pasture: Straloch, near Kirkmichael, Perthshire

Present community	Possible mycorrhizal origin
Agrostis/Festuca grassland with *Salix repens:*	Mixed scrub birch/birch woodland:
Amanita muscaria	*Amanita muscaria*
A. pantherina	*A. rubescens*
A. rubescens	*Boletus edulis*
Boletus aestivalis	*B. piperatus*
B. edulis	*B. subtomentosus* agg.
B. piperatus	*Cantharellus cibarius*
B. subtomentosus	*Clitocybe infundibuliformis*
Cantharellus cibarius	*Cortinarius anomalus*
Clitocybe clavipes	*C. biformis*
C. infundibuliformis	*C. callisteus*
Clitopilus prunulus	*C. cinnamomeus*
Collybia butyracea	*Laccaria laccata*
C. confluens	*Russula persicina*
C. dryophila	*R. vesca*
Cortinarius anomalus	*Thelephora terrestris*
C. biformis	
C. callisteus	Mixed oak/alder woodland:
C. cinnamomeus grp.	*Amanita pantherina*
C. obtusus	*A. rubescens*
C. uliginosus	*Boletus calopus*
Cystoderma amianthinum	*B. edulis*
C. carcharias	*B. erythropus*
Entoloma porphyrophaeum	*Cantharellus cibarius*
E. prunuloides	*Collybia butyracea*
E. turbidum	*C. clavipes*
Hygrocybe cantharella	*Cortinarius* spp.
H. glutinipes	*Laccaria laccata*
H. splendidissima	*Russula albonigra*
Inocybe lacera (subsp. *subsquarrosa*)	*R. atropurpurea*
Laccaria laccata	*R. fragilis*
Lactarius lacunarum	*R. illota*
Leptonia atromarginata	*R. xerampelina*
L. caerulea	
L. griseocyanea	Marshy woodland:
L. sericella	*Cortinarius uliginosus*
Mycena aetites	*Laccaria laccata*
M. filopes	*Lactarius lacunarum*
M. galopus	
M. leptocephala	Montane:
M. pura	*Laccaria laccata*
M. sanguinolenta	*Russula alpina*
M. swartzii	
Nolanea staurospora	Major litter-decomposing
N. verna	woodland fungi:
Russula albonigra	*Collybia confluens*
R. alpina	*C. dryophila*
R. atropurpurea	*Mycena galopus*
R. fragilis	*M. leptocephala*
R. illota	*M. sanguinolenta*
R. persicina	
R. vesca	*Lycoperdon perlatum*
R. xerampelina (incl. var. *barlae*)	
Tricholoma carneum	

Table 3 (cont.)

Present community	Possible mycorrhizal origin

Thelephora terrestris

Lycoperdon foetidum
L. perlatum

Calluna vulgaris with
S. repens:
  Boletus calopus
  B. spadiceus
  B. subtomentosus

Marshland:
  Hypholoma elongatum
  H. ericaeum
  Mycena bulbosa
  Omphalina oniscus
  O. sphagnorum
  Pholiota myosotis

These last-named areas at Straloch are characterized by typical peat bog fungi. However, the notable feature of the Straloch site is the several populations of *Amanita muscaria* and *Boletus edulis*; the former one might suspect to be mycorrhizal with *Betula*, which inhabits the drier tops to the moraines to the west, but a search for *Betula* within the community has proved unsuccessful. The species found at this locality are listed in Table 3; *Russula albonigra* and *R. alpina* were both collected. The first was quite unexpected, although it may be associated with a wide range of frondose and sometimes coniferous trees. *R. alpina* is usually found at higher elevations, although Dennis (1964) records it at low levels on the island of Canna in the Inner Hebrides.

Cortinarius is a frequent genus in the *S. repens* community in many areas of Britain and is present to some degree at the Perthshire site; *Boletus piperatus*, usually associated with *Betula* in *Betula/Pinus* woodland in the Highlands (Watling, 1963) has also been found.

Table 3 indicates the species collected, many of which are suspected mycorrhizals: some, such as *Collybia confluens*, are saprophytic members of the flora; others, e.g., *Nolanea staurospora*, are equally at home in woodland and hill pasture and so could have colonized the area from a whole spectrum of sites within the Ardle Valley.

Salix aurita is scattered in the adjacent heath, but the hybird with *S. repens*, *S. x ambigua*, was only located at the edge of the woodland tracks on the southern side of the valley where the two species are in juxtaposition. This may indicate a possible origin of the variability found in the Straloch *S. repens*, i.e., hybridization with other species, and also the much larger range of suspected mycorrhizals found at this one site. By hybridization and backcrossing to the parental *S. repens* over many

generations some degree of morphological variability would be expected. Hybridization might also be expected, as when woodland was cleared by human activity different
species of *Salix* were brought closer together.

At Straloch typical birch fungi, e.g., *Amanita crocea*, *Cortinarius crocolitus*,
Lactarius turpis, and *Russula velenovskyi*, are found in woodlands quite near, although
never within, the *S. repens* community. Also at Straloch typical grassland agarics
have colonized the margins of the community or invaded the turf containing the *Salix
repens*; these include *Hygrocybe* spp., some preferring the more acidic soils, others
the company of *Calluna vulgaris*. The two puffballs *Lycoperdon perlatum* and *L. foetidum* were recorded; the former, although found in grasslands, is more typically a
woodland fungus, whereas *L. foetidum*, as the *Hygrocybe* spp., is very characteristic
of hill pastures.

2. *Maritime grassland/heath: St. Kilda*

St. Kilda is a group of islands situated 241 km from the mainland of Scotland
and 80 km due west of Harris in the Outer Hebrides. The pollen diagram indicates
that, as elsewhere in Scotland, a climatic optimum was reached during Boreal and Early
Atlantic times with *Alnus*, *Betula*, *Corylus*, *Quercus*, *Pinus*, and *Ulmus* present. *S.
repens* survived as a member of this community because twigs, dating to the early Atlantic period (3500 B.C.), have been collected on Hirta, the main island of the St.
Kilda group. The vegetation of St. Kilda would appear to be a detached unit of Highland vegetation, although since the optimum period there has been a greater deterioration in the climate than in mainland Scotland, with excessive sea spray eliminating
woodland on these Atlantic islands and replacing some types of vegatation by *Plantago
maritima* communities.

Sheep were introduced soon after the permanent settlement of St. Kilda; in fact,
this island group is still the home of the primitive breed known as Soay sheep. St.
Kilda once supported 1300 Scottish Blackface sheep, but the grazing factor brought
about by the presence of thse domestic animals was removed in 1930 when sheep and
humans were evacuated, leaving only the Soay sheep.

McVean (1961) has compared the dense prostrate mat of *Calluna vulgaris* at a
height of about 65 m on Hirta with the dwarf *Calluna* communities of intermediate
elevation on mainland Scotland. In this plant association *Calluna* is dominant, and
there are small amounts of *Trichophorum caespitosum* and *Molinia caerulea*; on Hirta,
S. repens is scattered throughout the mat. The associated agarics, twelve in four
genera, were not found anywhere else on the island, and all are thought to be mycorrhizal with the willow. Collecting over a period of time would no doubt increase the
list of associates of *S. repens* on St. Kilda. Attention was drawn to certain higher
plant communities when distinctive mycorrhizal fungi were found, e.g., *Cortinarius
uliginosus*, a member of the *Cortinarius cinnamomeus* complex, but only after careful
searching were several small but woody and obviously old plants of *S. repens* located.
Three distinct fungal associations were observed, probably limited by edaphic condi-

tions because as at Straloch the *Salix repens* was found in marshy areas on the raised
beach and among sea-cliff turf.

In one limited area two species of *Cortinarius* were collected; at another, five
species of the same genus accompanied by three members of the Russulaceae, including
Lactarius lacunarum. The last species has been found with *S. aurita* at Kilmory on
Rhum (Watling, 1965) and on Eigg, both in the Inner Hebrides (Dennis, 1964).

A third locality is characterized by the presence of two major Russulaceae: (1)
Lactarius controversus, of which some basidiomes reach 155 mm in diameter; and (2)
Russula persicina, reported earlier as a form of *R. pulchella* (Watling and Richardson, 1971). This species is probably also the same as that which Kreisel (1965) in
his study called *R. pulchella*; it has been found in the Netherlands (C. Bas; personal
communication, 1973). It is a widespread species, collections being taken from as
far apart as Kilmory on Rhum to Ross Links in Northumberland. *Lactarius controversus* has also been found at Ross Links with *Salix repens* and at Mingulay in the Outer
Hebrides, although its normal mycorrhizal associates in England, central Europe, and
North America are *Populus* spp., particularly *P. tremula* and *P. alba* (Trappe, 1962b).
Kreisel (1965) records this particular milk-cap from Germany, and Komirnaya and Fursayev (1953) record it along with *Paxillus involutus* with *Salix caspica*. Terkelsen
(1956) has described a new species in Greenland differing only slightly from *L. controversus*.

Hebeloma leucosarx, which is characteristic of willow carrs on mainland Britain,
was not found on St. Kilda, but a fungus resembling it was collected from both ends
of the *Salix* zone. Several species of *Hebeloma* have been found with *Salix repens*,
but difficulty in separating taxa has been experienced.

In addition to members of the Cortinariaceae and Russulaceae, *Leptonia caerulea*,
often associated with *S. repens* in hollows among maritime sand dunes (slacks) on
the mainland, was found on St. Kilda. The relationship between this agaric and the
Salix species is at present unknown, but it might be suggested that whenever the former is recorded from "grassland" careful search should be made for the creeping willow. The Entolomataceae are not thought to contain mycorrhizal species, members
probably only being humicolous within the community. *Mycena aetites* and *M. leptoce-
phala*, also found in the *Salix* zone on St. Kilda, probably colonized from outside
the community, as both were recorded from the surrounding grasslands. Table 4 lists
all those species recorded for the St. Kilda communities.

Luzula sylvatica, normally found as a member of British woodland communities,
is found on St. Kilda, although it is doubtful whether trees have been present there
in historic time. Nevertheless woodland fungi, e.g., *Nolanea cetrata*, have also been
found. Does this reflect earlier vegetational times in the same way that the presence
of *Scilla non-scripta* and *Anemone nemorosa* on heath in northern Britain reflect former
woodland cover? Or is the sum total of the ecological factors operating at these
damp, cool sites on islands and mountains so buffered that they resemble the woodland

Table 4 Site 2--St. Kilda

Present community	Possible mycorrhizal origin
Cliff top turf:	Birch woodland:
Cortinarius cinnamomeus	*Cortinarius* spp.
C. lepidopus	*Russula fragilis*
C. obtusus complex (three microspecies)	*R. persicina*
C. pseudosalor	
Cortinarius sp. A.	Poplar woodland:
Hebeloma hiemale	*Lactarius controversus*
H. cf. *leucosarx*	
Lactarius controversus	Marshy woodland:
L. lacunarum	*Cortinarius uliginosus*
Leptonia caerulea	*Hebeloma* cf. *leucosarx*
L. serrulata	*Lactarius lacunarum*
Mycena aetites	
M. leptocephala	Montane:
Russula alpina	*R. alpina*
R. fragilis	
R. persicina	
Marshy areas:	
Cortinarius uliginosus	

ecosystem? In parallel, several phanerogams, e.g., *Dryas octopetala*, normally considered high-mountain plants in Britain, grow at lower altitudes as one moves across the country in a northwesterly direction (McVean and Ratcliffe, 1962), reflecting the increase in oceanic impact from east to west and the latitudinal decrease in temperature from south to north; larger fungi appear to show a similar pattern.

   3. *Sea-cliff turf/coastal dune*

   Four quite distinct sites have been studied in Sutherland on the northern coast of Scotland: two at Farr Bay near Bettyhill, one at Skerrary, and a unique community of *S. repens* and *Dryas octopetala* on the west of the estuary of the River Naver. The first three are sea-cliff turf exposed to sea spray, often for long periods.

   Evidence supplied by Kenworthy (1976) indicates that humans were active in the Bettyhill area of Sutherland as much as 8000 to 10,000 years ago. Bones of Arctic mammals have been found, split for extraction of marrow, etc., but lack of domestic animals' bones indicates that man was then a hunter. The presence of scorched stones and charcoal suggests that even at this early period at least some burning of woodland had commenced.

   Pollen analysis of deep peats and examination of plant remains in Chambered Cairns* indicate that changes were taking place in the vegetation--on the one hand, the replacement of the widespread coniferous woodland by peat and birch scrub because

---

   *
   Burial places.

of climatic changes and, on the other, burning of woodland and elimination of regenerating trees by grazing. While grazing associated with early villages was undoubtedly of some considerable importance, it was not solely restricted to sheep, as a
wide variety of grazing animals were involved, including deer, cattle, Kerry sheep,
goats, and horses. Cattle are recorded as having annually cropped birch shoots in
autumn and winter, and such shoots as survived to a second year were used by the vil-
lagers. Thus, prior to clearance of the land, people for thousands of years have had
adverse effects upon birch regeneration, an effect which was accentuated by the introduction of the Cheviot sheep in the late eighteenth century during the extensive
Highland clearances.

Where the grasslands at Bettyhill extend to the sea cliffs, *S. repens* invades
the turf to even become dominant, and although parallel communities are found in
many other areas of Sutherland those overlooking Farr Bay are particularly rich.

Table 5 Sites 3-6 near Bettyhill and Site 9 at Mingulay

(a) Sites 3-6 (Bettyhill)

Present communities	Possible mycorrhizal origin
Coastal dune grassland with *Salix repens* often accompanied by *Dryas octopetala*:	Oak/hazel woodland:
Agrocybe subpediades	*Boletus luridus*
Boletus luridus[a]	*Entoloma jubatum*
Clitocybe infundibuliformis	
C. sericella	Birch/pine woodland:
Collybia dryophila	*Cortinarius callisteus*
Conocybe dunensis	*Inocybe pyriodora*
Cortinarius biformis	*Laccaria laccata*
C. callisteus	*Lactarius hysginus*
C. trivialis	
Entoloma ameides	Wet woodland:
E. jubatum	*Cortinarius trivialis*
Hygrocybe conicoides	*Inocybe dulcamara*
Hebeloma mesophaeum	*Lactarius lacunarum*
H. sinuosum	
Inocybe dulcamara	
I. dunensis	
I. pyriodora	
I. serotina	
Laccaria laccata	
Lactarius hysginus	
L. lacunarum	
Lepiota alba	
Leptonia caerulea	
Marasmius oreades	
Melanoleuca grammopodia	
Nolanea staurospora	
Pluteus lepiotoides	

Table 5 (cont.)

Coastal dune grassland
on dung (d) and highly
nitrogenous material:
  Panaeolus rickenii
  P. sphinctrinus (d)
  Psilocybe semilanceata
  Stropharia pseudocyanea
  S. semiglobata (d)

(b) Sites 3-6 (Bettyhill) and 9 (Mingulay)

Present community	Possible mycorrhizal origin

Sea-cliff turf with
Salix repens:
  Amanita rubescens
  Amanita nov. sp.
  Boletus edulis
  Cantharellus cibarius[c]
  Clitocybe infundibuliformis
  Cortinarius anomalus
  C. fasciatus[b]
  C. largus
  C. obtusus complex[c]
  C. pseudosalor[b]
  Entoloma ameides
  E. jubatum
  E. madidum[b,c]
  E. porphyrophaeum
  E. prunuloides
  E. turbidum
  Inocybe cookei
  I. dulcamara
  I. fastigiata
  Lactarius controversus[d]
  L. hysginus
  L. lacunarum
  L. torminosus
  Leccinum salicola[b]
  Leptonia atromarginata
  L. chalybea[d]
  L. fulva[d]
  L. serrulata
  Mycena alcalina
  Nolanea sericea[d]
  N. staurospora[b,c]
  Russula fallax
  R. fragilis
  R. persicina
	Birch scrub/birch woodland:
	  Amanita rubescens
	  Boletus edulis
	  Lactarius torminosus
	  Russula fallax
	  R. fragilis
	  R. persicina
	
	Birch/pine woodland:
	  Cantharellus cibarus
	  Lactarius hysginus
	
	Marshy woodland:
	  Inocybe dulcamara
	  Lactarius lacunarum

[a]Found in parallel community among *Dryas* and *Salix repens* Co. Clare (I. Alexander, personal communication).

[b]Also recorded for Skerrary (site 6).

[c]Also recorded for Mingulay (site 9).

[d]Mingulay only.

The grasslands in which the willow grows are dominated by *Festuca rubra* and usually
on slightly acidic soils other prostrate shrubs may be associated with the willow,
e.g., *Empetrum nigrum*, especially on the thinner, less base-rich soils. Many cliff
tops are much richer, and here the *Salix* becomes replaced by *Festuca-Agrostis-Antho-
xanthum* grassland with *Primula scotica* and *Scilla verna*. In contrast, on the immature
coastal dune grasslands (Gaelic: machair) above Farr Bay *S. repens* is associated with
a very rich herb flora containing *Antennaria dioica, Arabis hirsuta, Campanula rotun-
difolia, Oxytropis halleri*, and *Draba incana*. At Strathnaver, in coastal dunes, the
same creeping willow is associated with *Dryas octopetala*. The codominants with *Dryas*
range from *E. nigrum* to *Pteridium aquilinum* as well as to *S. repens*, or the *Salix*
may be scattered throughout the community. In some areas it grows in juxtaposition
with *Juniperus communis*. According to Dominik et al. (1954a) *Dryas* has been shown
to form four types of mycorrhiza whereas *J. communis* forms endotrophic mycorrhiza
(Nespiak, 1953). One dominant agaric in the community shown in Table 5(a) is *Inocybe
serotina* which occurs with *Salix glauca* (Lange, 1957) and with *S. acutifolia* (Zerova,
1955). The possible community which dominated parts of the area before grazing is
indicated by the presence of birch woodland supporting typical birch fungi on inaccessible rocky debris close by.

In Farr Bay communities the normal *S. repens* associates are intermixed with
various *Hygrocybe* spp. and *Leptonia* spp., members of an assoication recognized in
various modifications in grasslands found throughout the British Isles (Watling,
1967). It is uncertain whether the presence of these species is a result of weeding
out and selection of normal woodland constituents or reflects a community capable of
colonizing proclimax vegetation whether woodland or grassland. In support of the
latter argument, it should be mentioned that a similar and parallel agaric flora is
found in second growth woodland in the Great Lakes region of North America; but, to
uphold a hypothesis that it is possibly a derived flora, the same species can all be
found in small outliers of woodland on thin limestone soils in northern Britain.

Typical grassland species have been found at several localities both within and
outside of the *Salix* communities, and where the *S. repens* grows in sand interspersed
with *Ammophila arenaria*, several typically psammophilous (i.e., sand-loving) fungi
are found including *Lepiota alba*; the related *L. subgracilis* is mycorrhizal with *S.
repens* in the Ukraine (Zerova, 1955). Those species recorded are not considered
typically mycorrhizal. Similar communities have been found on Rhum, Forvie Sands,
etc., and are listed in Table 6.

Dung is found in most creeping willow communities for, as indicated earlier,
grazing is a means by which the community is maintained. On such highly nitrogenous
substrates *Stropharia semiglobata* is found where the dung is of an acidic nature,
and *Panaeolus* spp., especially *P. sphinctrinus*, where the dung is of a highly basic
nature. Where nitrogenous material leaches out into the surrounding soils, other
species are recorded.

Table 6 Larger fungi of coastal dune grassland other than Bettyhill sites[a,b]

Species	1	2	3	4	5	6
Agaricus campestris	x	x	x	x	St. Cyrus[c]	Tent.
A. devoniensis	x	x	x	x	x	Tent.
A. semotus	x	x	x	x	St. Cyrus	x
Agrocybe praecox	x	x	x	x	x	Tent.
A. semiorbicularis incl. *A. subpediades*	Cul.	x	x	Rhum	St. Cyrus	Tent.
*Amanita citrina	x	x	Ross	x	x	x
*A. rubescens	x	x	Ross	x	x	x
Bolbitius vitellinus	Cul.	x	x	x	St. Cyrus	Tent.
*Cantharellus cibarius	x	Fov.	x	x	x	x
Clitocybe rivulosa	x	x	x	x	x	Tent.
Conocybe dunensis	x	Fov.	x	x	x	Tent.
C. lactea	x	x	x	x	x	Tent.
*Cortinarius biformis	Cul.	x	x	Rhum	x	x
*C. pseudosalor	x	Fov.	x	x	x	x
*C. trivialis	x	Fov.	Ross	x	x	x
Cystoderma amianthinum	x	Fov.	x	x	x	x
C. carcharias	x	x	Ross	x	x	x
Entoloma jubatum	x	x	x	Rhum	x	x
E. turbidum	x	x	x	Rhum	x	x
Galerina vittaeformis	Cul.	Fov.	x	x	St. Cyrus	x
*Hebeloma hiemale	x	x	x	Rhum	x	x
*H. cf. leucosarx	x	Fov.	x	x	x	x
*H. pumilum	x	x	x	Rhum	x	x
*H. vaccinum	x	x	Ross	x	x	x
Hygrophoropsis aurantiaca	Cul.	Fov.	x	x	x	Tent.
Hygrocybe conicoides	Cul.	Fov.	x	x	x	Tent.
H. psittacina	x	x	x	x	x	Tent.
*Inocybe agardhii	x	x	Ross	x	x	x
*I. devoniensis	x	x	Ross	x	x	Tent.
*I. dulcamara	Cul.	x	Ross	x	x	x
*I. halophila	Cul.	x	x	Rhum	x	x
*I. lacera	x	Fov.	x	x	x	x
Laccaria laccata	x	x	Ross	x	x	Tent.
L. maritima	Cul.	x	x	x	x	x
*Lactarius controversus	x	x	Ross	x	x	x
*L. lacunarum	x	x	x	Rhum	x	x
Lepiota alba	x	x	x	x	St. Cyrus	x
L. (Leucoagaricus) leucothites	x	x	x	x	x	Tent.
L. procera	x	x	x	x	St. Cyrus	x
Leptonia chalybea	x	Fov.	x	x	x	x
L. fulva	x	Fov.	x	x	x	x
L. sarcitula	x	Fov.	x	x	St. Cyrus	x
L. serrulata	x	Fov.	x	x	x	x
Lycoperdon foetidum	x	x	x	x	x	Tent.
Marasmius oreades	Cul.	Fov.	x	x	St. Cyrus	Tent.
Mycena leptocephala	x	Fov.	x	x	x	x
M. pura	x	Fov.	Ross	x	St. Cyrus	x
M. swartzii	x	x	x	x	x	Tent.
Nolanea cucullata	x	x	x	x	St. Cyrus	x
N. sericea	x	x	Ross	x	St. Cyrus	Tent.
N. staurospora	Cul.	x	Ross	x	St. Cyrus	x
Omphalina ericetorum	x	x	x	x	x	Tent.
O. pyxidata	x	x	x	Rhum	x	Tent.
Panaeolina foenisecii	Cul.	x	x	x	St. Cyrus	Tent.
Panaeolus ater	x	x	x	x	x	Tent.
P. rickenii	x	x	x	x	St. Cyrus	x
P. sphinctrinus	x	x	x	Rhum	x	Tent.

Table 6 (Cont.)

Phaeomarasmius erinaceus	Cul.	x	x	x	x	x
Phallus hadriani	x	x	x	x	x	Tent.
Psathyrella ammophila	Cul.	Fov.	x	x	x	Tent.
*Russula laccata	x	x	Ross	x	x	x
*R. nigricans	x	x	Ross	x	x	x
*R. persicina	x	x	Ross	Rhum	x	x
Stropharia albocyanea	x	x	x	x	x	Tent.
S. coronilla	x	x	x	x	St. Cyrus	Tent.
S. semiglobata	Cul.	Fov.	Ross	Rhum	x	Tent.

[a]Abbreviations: x = not present; Cul. = Culbin sands, Morayshire; Fov. = Forvie Sands, Aberdeenshire; Ross, Ross Links, Northumberland; Rhum = Isle of Rhum, Inner Hebrides; Tent. = Tentsmuir, Fife.

[b]Suspected mycorrhizal species are asterisked.

[c]St. Cyrus, Forfar, is taken as a parallel locality lacking Salix repens.

Plants of S. repens with 60 growth rings have been located in the Bettyhill area, and a 12 year-old tree has been shown to possess a root-system covering several square meters. Thus, although the aerial part of the creeping willow in grazed turf may superficially appear restricted, the plants colonize considerable areas of sur-rounding soil. The root tips are typically covered in mycorrhizal sheaths, but some plants, particularly those in sandy soils, have a Cenococcum-like felt: Dominik and Pachlewski (1955) have recorded Cenococcum graniforme with S. repens, and Dominik (1951) has described endotrophic mycorrhiza in S. repens subsp. argentea in Pinetum empetrosum communities. Those taxa found at the Bettyhill sites are listed in Table 5(b) and include several interesting fungi, e.g., Boletus luridus, and Leccinum sal-icola.

4. Sand dune communities

Creeping willows in coastal dunes generally have relatively stout and ascending shoots, although these may be appressed by wind, with broadly oblong, shortly acute, sericeous leaves and pubescent capsules. This taxon is considered a subspecies by some, i.e., S. repens subsp. argentea, or as a distinct species (S. arenaria); both these epithets refer to the grayish leaves.

The flowering and dehiscence of the capsule of subsp. argentea is earlier than for the more typical subspecies. The creeping willow found in sand dunes usually grows where the water table has just reached the surface, particularly in blowouts; in the more extensive marshy areas (dune slacks) other species of willow invade the edges of any permanent water.

S. repens subsp. argentea occurs in Sutherland, in coastal dune grassland con-ditions on the Isle of Rhum, and in dunes at Tentsmuir near St. Andrews, at St. Cy-rus, etc., in Scotland, and at Ross Links in Northumberland. At Culbin Sands, Moray-shire, it grows intimately in some places with Laccaria maritima, and on Rhum with Cortinarius biformis and Inocybe halophila. Laccaria trullisata, often considered conspecific with *L. maritima*, grows with *Salix glauca* in Greenland (Lange, 1957) and with *Salix daphnoides* in Scandinavia (Andersson, 1950). South of Loch Gairloch *S. repens* forms sheer-sided platforms 1-1.5 m high above the recently colonized sandy base of wind eroded areas (Gimmingham, 1970). Fungi have as yet not been found in this highly dissected community, probably because of its exposed nature. *Psilocybe muscorum* is found most frequently among lichen on podzolized heath on sand along with *Omphalina hepatica*; neither is thought to be specifically associated with the *Salix*.

The fungi recorded from the coastal dune systems are listed in Table 6. The entries should be compared with those Bon and Gehu (1973) tabulated from continental *S. repens* subsp. *argentea* communities (*Salicion arenariae*). *Inocybe tarda* var. *subulosa* was described by M. Beller and Bon (in Bon, 1975) specifically from this community.

B. Arborescent Willows

Salix caprea is soemtimes found in close proximity to *S. repens*, and where the two species grow close together the hybrid *S.* x *laschiana* may be found. Whereas *S. repens* has a wide range of associates, the other parent, *S. caprea*, which is much more widely distributed, rarely appears to be accompanied by larger fungi. Indeed the whole genus *Salix* can be separated into the prostrate, creeping, and dwarf willows, particularly the montane forms, which undoubtedly are mycorrhizal with larger fungi, and the tree and shrub willows, which although often accompanied by characteristic fungi are thought not to be obligately ectotrophic mycorrhizal.

The agarics typically recorded with arborescent willows in Britain are listed in Table 7. *Tricholoma cingulatum* frequently occurs with several arboresant salices including *S. repens* subsp. *argentea*; this same species is indicated by J. Lange (1938-1940) as being characteristic of willow carrs (i.e., marshy woods) in Denmark along with *Russula atrorubens*; this author also describes *R. graveolens* var. *subrubens* as being characteristic associates of willow. Other species are *R. fragilis*, *R. gracillima* [=*R. gracilis* var. *altaica* fide M. Lange (1957)], and *R. olivaceoviolascens* (Delzenne-van Haluwyn, 1971c). *R. graveolens* has been described by Favre (1955) from the Alps. *Lactarius aspideus*, *L. cyathula*, *L. lacunarum*, *L. sphagneti*, and *L. tabidus* are all recorded by Delzenne-van Haluwyn (1971d) with *S. aurita* and *S. cinerea*, and *L. uvidus* and *L. violascens* for unspecified willow communities. Two boletes, *Leccinum duriusculum* and *Boletus chrysenteron*, have been recorded as mycorrhizal with *S. caprea* and *S. couteriana*, respectively (Becker 1956; Trappe, 1960; Bruchet, 1970).

Cenococcum graniforme has been recorded for *S. repens* and *S. caprea* (Dominik, 1961; Dominik et al., 1954a,b) and several alpine willows (see below); in addition it has been recorded in association with at least twelve other species of *Salix* (Dominik and Pachlewski, 1955; Trappe, 1962a,b). In addition Fontana (1962) has analyzed 14 species and varieties of *Salix*; she found that in some species of *Salix*, even in a single plant up to 6 kinds of ectotrophic mycorrhiza might be present. *S. purpurea* was found to be endotrophic in two localities, and it was shown that some individuals

Table 7   Species found with arborescent *Salix* spp. and *Alnus glutinosa* in the
British Isles

Fungus	Associate
Boletus subtomentosus	*Salix aurita* (Dennis, 1955)
Cortinarius cinnamomeus	*S. atrocinerea* (Wallace, 1954)
C. decoloratus	*S. atrocinerea* (and *Myrica gale*) (Dennis, 1955)
C. triformis	*S. atrocinerea* (Wallace, 1954)
C. uliginosus	*S. atrocinerea* (Wallace, 1954
Galerina salicicola	*Salix* spp.
Hebeloma crustuliniforme	*S. atrocinerea* (Wallace, 1954, 1956)
H. mesophaeum	*S. atrocinerea* (Wallace, 1957)
Inocybe acuta	*S. atrocinerea* (Wallace, 1954)
I. dulcamara	Widespread; *Salix* spp.: *S. atrocinerea* (Wallace, 1956, 1957)
I. halophila	*S. atrocinerea* (Wallace, 1954, 1956)
Naucoria celluloderma	*Alnus glutinosa*
N. escharoides	*A. glutinosa*
N. permixta	*Salix* spp.
N. salicis (?= *N. macrospora* J. Lange)	*Salix atrocinerea* and *S. cinerea*
N. scolecina	*Alnus glutinosa*
N. striatula	*A. glutinosa*
N. subconspersa	*A. glutinosa*
Russula atrorubens[a]	*Salix atrocinerea* (Wallace, 1956)
R. atropurpurea	*S. atrocinerea* (Wallace, 1954)
Tricholoma cingulatum	Widespread: *S. cinerea*

[a]Also recorded by Kreisel (1965) with *Salix repens*.

of arborescent *Salix* examined were not mycorrhizal at all. She identified 16 ectotrophic mycorrhizal forms of which 7 were of the 12 morpho-types documented by Dominik (1959).

Examination of *Salix cinerea* communities in the Border Counties of Scotland,
e.g., Whitlaw Moss, Selkirkshire, indicate that the fungus flora is totally different
to the *Salix repens* communities. These eutrophic marshes ("mosses"), despite their
small aggregate size, are of outstanding importance, representing a northern example
of rich calcareous fen. Their flora includes a rich variety of phanerogamic species,
but probably because of the high water table they command a very limited fungus flora;
where *Alnus glutinosa* occurs the species mentioned in Table 7 have been recorded. In
parallel to other British plant communities, indeed northern boreal woodland communities in general, the arborescent constituents in these marshes are extremely limited
in number and diversity and consequently in fungi. These fungi are generally components of the hydroseral succession.

Delzenne-van Haluwyn (1971a) has listed the species of *Hebeloma* she considers
hygrophilous and growing with *Salix* spp. under these conditions. She also (Delzenne-
van Haluwyn, 1971b) gives *Inocybe cucculata* and *I. salicis* as components of the same
flora.

Bruchet (1970) has recorded *Lactarius lacunarum* from mountain areas with *Hebeloma hiemale* under *S. aurita*. The first species has been found in *S. repens* communities, and the second with *S. aurita* on St. Kilda and Rhum, respectively (Watling, 1970; Watling and Richardson, 1971). Bruchet also records *Hebeloma pusillum* var. *longisporum* and *H. sacchariolens* with *S. aurita*; Singer (1950) records the last named *Hebeloma* species (as *H. austroamericanum*) with *Salix humboldtiana* in South America.

Darimont (1973) in his mycosociological groups mentions both *Salix caprea* and *S. aurita* in the "hautes forêts" of Belgium, but unfortunately his listings of the fungi cannot be related to our understanding of the flora of British willow communities.

Bon and Gehu (1973) indicate that *Inocybe lanuginosa* and *I. trechispora* are not tied to--but typify with *Lactarius lacunarum*--wet places containing *S. aurita* and *S. cinerea*. However, *Cortinarius cinnamomeoluteus*, *C. pulchripes*, *C. uliginosus*, *Inocybe salicis*, and *Galerina salicola* may all be obligately dependent on *Salix*. The same authors indicate diverse species of *Inocybe* with a range of *Salix* spp.

The relationships just described contrast with those of *Salix* spp. in alpine areas of the Scottish mountains and also with the rather extensive *Salix/Betula nana* scrub of Finmark. It is surprising to find in the latter community *Boletus piperatus*, a species normally associated with mesophytic communities, and *Rozites caperata* charachertistic both of the remmant Caledonian woodland of Scotland, composed of *Pinus sylvestris* subsp. *scotica*, and *Quercus* woodlands of Central Europe (Watling, 1974).

The bolete *Leccinum rotundifoliae* is found mycorrhizal with *Betula nana* in these damp areas; however, *L. scabrum* subsp. *tundrae*, recently described by Kallio (1975), might well be assoicated with the *S. glauca* and not with the birch.

C. Arcto-Alpine Willows

The mountain salices of Scotland cover two subgenera: *Salix reticulata*, *S. herbacea*, and *S. myrsinites* in subgenus *Chamaeota*; and *S. lanata* and *S. arbuscula* in subgenus *Caprisalix*. *S. repens* also belongs in the latter subgenus, along with *S. glauca*, *S. cinerea*, and *S. lapponum*. The arctic mountain willows have been shown by Hesselman (1960) to form ectotrophic mycorrhiza.

Salix reticulata, an arctic to subarctic plant although widespread in high mountains in Europe, in the British Isles is confined to Scotland except for a single locality in Wales. *S. herbacea* is confined to Scandinavia, Iceland, and Britain; however, it is widespread in the British Isles, being found in Wales, Ireland, and the Lake District of England as well as in several Scottish localities. *S. myrsinites* is more restricted in distribution in Scotland, having been recorded only to the north of the Central Valley.

Salix lanata, found in arctic to subarctic Europe, occurs in northern Scotland; *S. arbuscula*, found in Fennoscandia and northern Russia, has more westerly distribution than other species in Scotland; and *S. lapponum*, with a northerly distribution in Europe, is confined to Scotland except for one locality in the Lake District of England.

Salix lapponum and *S. reticulata* are considered historically northern arcto-alpine species, whereas *S. arbuscula*, *S. herbacea*, and *S. myrsinites* are historically Tertiary arcto-alpine plants (Mathews, 1937). That is, the first two species migrated from the north during the Ice Age, retreating as the climate improved, and therefore are now relic in the British Isles, whereas the other three have inhabited mountains in Central Europe since Tertiary times and have reached their present northerly position since the retreat of the ice (Mathews, 1937). Perhaps therefore the British populations of the last species are of alpine origin and their present distribution is controlled by climatic factors.

Salix lanata belongs to an arcto-subarctic element in the British flora and was originally of boreal origin, but not all ancestral types in this group may themselves have originated in the Arctic; perhaps the British populations are derivative types which migrated north from continental Europe subsequent to the maximum invasion of northwestern Europe.

Considerable numbers of the sites in Scotland with montane willows are either small in size or rather inaccessible; many sites are on mountain slopes where the plants are confined to ledges and therefore not favorable to the fructification of agarics.

Many of the British sites of *S. myrsinites* and *S. reticulata* would encourage the development of *Cenococcum*-like mycorrhiza, the habitats having thin soils prone to drying out. Dominik et al. (1954a) record *Cenococcum graniforme* with both *S. myrsinites* and *S. reticulata* in the Tatra Mountains of Poland, but its presence has not been confirmed for specimens from Scottish localities. In *S. myrsinites* Fontana (1962) described six distinct mycorrhizas.

In *S. arbuscula*, a montane species in continental Europe, Fontana (1962) found a distinct type of mycorrhiza with starlike hyphal projections not appearing in Dominik's scheme (Dominik, 1959). This mycorrhiza was also found in *S. caesia*; in *S. retusa* var. *serpyllifolia* three mycorrhizas were found. *S. caesia* and *S. retusa* are not recorded for Britain.

Salix glauca occurs in Iceland, the Faeroes, northern and western Fennoscandia, and arctic Russia but is not found in the British Isles; this species is also found in Greenland. There is a closely related taxon in the North American Rockies. M. Lange (1957) records 31 species of agaric mycorrhizal with *S. glauca* and 9 with unidentified salices in Greenland. The rich fungus flora of *Betula nana/Salix glauca* has been examined and documented by M. Lange (1957). *S. glauca* also grows in sand dunes in parallel to *Salix repens* subsp. *argentea* in Scotland, commanding there a similar series of mycorrhizals.

Salix herbacea is a basiphilous species favoring fresh and open soils, and forms communities characterized by their exposed locations; it is often abundant in regions of solifluction or scree formation. It is also strongly characteristic of late snow

patches and in arctic areas it has a wide range of habitats. *S. herbacea* apparently
formed similar communities during two periods in the Upper Palaeolithic some 12,000
to 14,000 years ago and 10,500 years ago (Godwin, 1956). During Full Glacial and
Late Glacial times it was a highly successful colonizer, and judging from the distribution of macrofossils it had a wide range in Britain. However, it suffered an early
and severe restriction to higher altitudes in the Postglacial period. It is very
susceptible to competition from taller plants, and there is little doubt that many of
the alpine *Salix* communities in Scotland are maintained at their present sites by the
grazing activity of both sheep and deer.

During our own study concentration has been paid to the *Salix herbacea* community on the Cairnwell, where it forms a fairly extensive turf, not being confined to
ledge communities. The Cairnwell straddles the borders of Perthshire and Aberdeenshire, at an altitude of about 900 m. The area largely consists of quartzitic and
quartz-mica schistose rocks intruded with both lamprophyre and quartz porphyry dikes.
Limestone appears in patches within the system and apparently is contemporaneous
with similar outcrops in the central massif of Scotland.

Soils derived from the limestone contain considerable quantities of iron. Much
of the outcrops of limestone are covered in peat deposits, although some boulder clay
is also present; heavily leached calcareous drift is present in the immediate vicinity of the three limestone outcrops (Coker, 1970).

The sites studied are open habitats continually maintained in an immature state
by physical factors, i.e., wind, rain, frost, and biotic factors; grazing and tramp-
ling by deer and sheep are also major factors. *Dryas octopetala* is found within the
comunity, although *S. herbacea* may in some places become dominant, being associated
in such areas with *Empetrum nigrum*. Plants of the latter are often stunted and
chlorotic and are apparently growing in pockets of soil in which the calcium status
has been reduced by leaching and slight humas enrichment. The small shrub community
may also contain *Calluna vulgaris*, *Vaccinium uliginosum*, *V. vitis-idaea*, and *Lois-
eleuria procumbens*. Control of grasses and sedges is provided by the grazing activities of mountain hares and voles, which are present in considerable numbers (Coker,
1970). The surrounding areas are flanked by agarics typical of heath with peaty
soils (see Table 8).

Amanita nivalis, first described from the Scottish mountains by Greville (1823),
is characteristic of *S. herbacea* beds on Cairnwell. Those fungi recorded at Cairnwell appear in Table 8 along with other important finds; several are suspected mycorrhizals, but others are undoubted humicoles. The appearance of the latter at high
altitudes resembles the finds of *Nolanea cetrata* on Ben Loyal at 701 m (Dennis, 1955)
and on St. Kilda, and of *Mycena filopes* on the Isle of Eigg, Inverness-shire (Dennis,
1964). Several of the species tabulated have been recorded from other scattered
sites in Scotland, Iceland (Christiansen, 1941), and Greenland (Lange, 1957; Watling,
1978).

Table 8 Communities on Cairnwell and important records from other Scottish
mountains

Rhacomitrium heath and peat banks:
   Galerina praticola grp.
   Hygrocybe lilacina
   Leptonia atromarginata
   Omphalina ericetorum
   O. luteovitellina
   O. hudsoniana

Salix herbacea:
  Suspected mycorrhizal elements
   Amanita nivalis
   Boletus spadiceus
   Cortinarius pseudosalor
   C. obtusus agg.
   Inocybe dulcamara
   I. praetervisa
   Russula alpina
   R. pascua

  Humicolous elements
   Clitocybe sericella
   Collybia confluens
   C. dryophila
   Cystoderma amianthinum
   C. carcharias
   C. granulosum
   Entoloma cordae
   Galerina mniophila
   G. paludosa[a]
   Hygrocybe cantharella[a]
   H. coccineocrenata
   H. nigrescens
   Hypholoma elongatum[a]
   H. udum[a]
   Laccaria laccata
   Leptonia catalunica

Salix herbacea (cont.):
  Humicolous elements
   L. lampropus
   Mycena amicta
   M. filopes
   Nolanea cetrata
   N. hebes
   N. staurospora[a]
   Omphalina oniscus[a]
   Pholiota myosotis[a]

   Calvatia utriformis

Records from other Scottish mountains:
   Amanita nivalis (Ben Macdui,
    Lochnager, Meall nan Ptarmachan,
    Ben Lawers, all Perthshire;
    Applecross, Wester Ross)
   Boletus spadiceus (Ben Vrackie)
   C. actus var. *striatulus* (Dennis,
    1964)
   C. favrei (Beinn Eigh, Wester Ross)
   C. pertrisis (Henderson, 1958)
   C. tabularis (Dennis, 1964)
   Russula alpina (Ben Vrackie, Perth-
    shire)

---

[a]Marshland species.

   Russula alpina, which grows both at Stralock and on the Cairnwell, has been
found accompanying *Boletus spadiceus* on a small deposit of peat near the summit of
Ben Vrackie, Perthshire, among *S. herbacea*, *Lycopodium selago*, and *Rubus chamaemorus*;
this area is a relatively low (840 m) outlying hill towards the southern fringes of
the Highlands. In Britain *Russula alpina* has not often been critically studied, and
it may be necessary to reexamine collections of this species to ensure that *Russula
norvegica*, described by Reid (1972), is not also present; similar problems are experienced with the *Cortinarius cinnamomeus* complex (Peyronel, 1931).

   Kühner alone and with colleagues has documented many of the montane and arcto-
subarctic agarics, including the mycorrhizal genera *Lactarius* (Kühner, 1976a) and
Russula (Kühner, 1976b), and *Clitocybe* spp. (Lamoure, 1972); many of these records
are from *Salix* communities. Bruchet (1970) has documented the *Hebeloma* spp. from

the Alps, and Delzenne-van Haluwyn (1971a) has documented both the alpine-subalpine, and arcto-subarctic species. In a companion publication Delzenne-van Haluwyn (1971b) listed the *Inocybe* spp. found in the same communities. Bon and Gehu (1973) have listed all those agarics which occur in the Braun-Blanquet units *Salicetalia herbacea* (acidic associations) and *Arabidetalia caeruleae* (neutral associations. Fifteen and twelve species were recorded for *Lactarius* and *Russula* respectively for the arcto-subarctic localities.

Nespiak (1953) has discussed the *S. herbacea* communities in the Tatra Mountains. The list of species he recorded is rather limited but includes *Laccaria laccata*, *Russula puellaris*, and *Amanita vaginata* (cf. Bas, 1977); the last probably being the alpine variety now considered to be a different species. The record of *Russula puellaris* is perplexing; in the British Isles this species grows in both pine and frondose-woodland usually on siliceous soils but never in high mountains. As Nespiak calls this a "dwarf form," it may in fact be an autonomous species. The *Russula* was described from *Trifidi-Distichetum* (i.e., *Juncus-Sesleria*) and the *Amanita* species from *Luzula-nodum* (*Luzula spadicea*); surprisingly only the first community was indicated as containing *S. herbacea*.

The agaric flora of the montane willow communities in Scotland had to be critically compared with that in the Alps (Favre, 1955; Horak, 1960; Kühner, 1976a,b), the Faeroes (Moeller, 1945), Iceland (Christiansen, 1941), and Greenland (Lange, 1957), although several of these localities possess a wider range of willow species. It is impossible at the moment to correlate British and continental European floras in minutia, but this problem is experienced even when studying the phanerogamic flora.

III. CONCLUSIONS

It is quite clear from the field data that *Salix repens* acts in every way as a forest tree, with the full spectrum of Russulaceae (*Russula* and *Lactarius*), Cortinariaceae (*Inocybe, Cortinarius*, and *Hebeloma*), Boletaceae (*Leccinum* and *Boletus*), and Amanitaceae expected of such communities. Similar woodlands are to be found on European mountains but composed of montane willows *S. herbacea*, *S. lapponum*, *S. myrsinites*, *S. reticulata*, etc.; only the first- and the last-named species of dwarf willow have been studied in Britain in detail, although parallel communities have been examined in the northern Rocky Mountains in Canada and the United States, as well as in continental Europe.

The arborescent willows, on the other hand, are "wetland" plants apparently commanding a rather restricted range of Agaricales. It would appear that these willows, which usually grow in moist and often base-rich soils, are not as dependent on mycorrhizal fungi for their nutrition as are the dwarf montane species. Meyer (1973) would consider them facultatively mycorrhizal.

Mycorrhizal fungi fructify toward the end of the year when the number of active mycorrhizal root tips are apparently declining. The length of the growing season for fungi with *S. repens* is similar to that for typical forest trees, but the alpine willows have a telescoped season and therefore the mycorrhizal partners are on a very fine balance awaiting favorable weather conditions to induce fructification; it is very likely that for long periods fructification is never experienced.

The *S. repens* site on St. Kilda parallels locations found on the Faeroes where the vegetation is equally dictated by the proximity to the sea. Moeller (1945) considers that the richness of the fungus flora of the Faeroes results from the fact that the soil and climate there produces an environment corresponding to that found by the same or related species in montane or woodland communities in continental Europe.

An alternative suggestion would be that these mycorrhizal fungi are actually associated with dwarf willows, and a careful search should always be made in future studies. The significance of this has been indicated in the work on St. Kilda (Watling and Richardson, 1971), and the record of *C. cinnamomeus* in pastures in Yell, Shetlands (Dennis and Gray, 1954) suggests the possible presence of *S. repens*.

Those fungi growing on twig and leaf debris, as might be expected, are found in all dwarf *Salix* communities, and as their substrate is constantly present they can fruit (if in the mountains) at any period between the snow melting and the first snowfall the following winter. This is in contrast to the phenomenon found with *S. repens*, where the pattern of fruiting of humicoles is more normal and follows the climatic pattern experienced by lowland vegetation. In the *S. repens* beds at Straloch, Perthshire, even spring activity is indicated by the fruiting of *Nolanea verna*.

Thus the net result of basidiome cataloging in all willow communities is as follows. The can be split into two major groupings: (1) mycorrhizal species fruiting in a response to a rhythm not fully explained, and (2) those species decomposing leafy and twiggy remains of *Salix*.

From the field observations it would appear that *S. repens* communities in the British Isles can be split into those which are rather specialized, i.e., coastal communities characterized by *Lactarius controversus* and probably having had woodland areas containing *Populus tremula*, and those communities which have replaced other woodland dominants such as *Quercus* or *Betula*. In the latter communities a whole range of fungal species might be expected and are indeed found. After a fairly extensive survey it appears that it is possible to correlate the succession of flowering plants and its modification by man, with changes in the mycorrhizal flora--the individual fungal flora of an area being somewhat related to the woodland which the *S. repens* replaced; variable and fairly extensive floras replacing fairly rich mesophytic woodland, e.g., Straloch, Perthshire; and more specialized flora being found in exposed areas where the original vegetation was restricted, e.g., Farr Bay, Bettyhill.

Although much of the foregoing speculation is based at most only on fairly intensive field observations, it would appear that the mycologist just like the flowering plant ecologist must think not only in terms of edaphic and climatic factors in his study of succession of larger fungi but also the part played by biotic and historic factors.

Agarics, although equipped with easily dispersible diaspores, appear not to be as widely distributed as might at first be expected, e.g., comparison of European and North American floras; however, if Kemp's arguments, expressed at the Lausanne symposium (Clémençon, 1977), are correct, then the evolution of agarics is composed of two components: (1) geographical isolation and later genetic drift away from the parent stock over a period of time, and (2) speciation at the hyphal level, so well exhibited and documented in members of the Coprinaceae, etc. Although the families so far examined by Kemp are composed on nonmycorrhizal fungi, he considers the same pattern of speciation might be expected in mycorrhizal agarics.

Historical studies have shown that the vegetation of Britain became dissected after the retreat of the last glaciation and consequently plant species became irregularly distributed over the country; but these perhaps were derived stocks, and it may be now possible by analysis not only of the plants but also of the cryptogams including the agarics to solve some of the phytogeographical problems observed. We should not be surprised if the two groups of montane willows had different mycorrhizal fungi accompanying them, but the possibility is only now being explored in detail. Perhaps in the future new information may be offered to support Godwin's aforementioned hypothesis (1956).

Similarities are found between Favre's and Kühner's studies and those discussed here, but as indicated in Sec. I the British phanerogamic flora is rather depauperate with some species found only in small restricted pockets at the margin of their European distribution. A somewhat restricted agaric flora might therefore be expected. Perhaps similarities can be drawn between these mycorrhizal fungi and the distribution of certain rust fungi whose host plants are also at the limit of their distribution (Savile, 1963).

Any study, however, depends entirely on recording the appearance of fruit bodies and their subsequent accurate determination. The former requires years of patient field observations, and the latter considerable exposure to agaric systematics. Ecological studies of the higher fungi usually come up wanting in each of these requirements.

APPENDIX

All entries for the Agaricales, unless otherwise stated, follow the check list prepared by Dennis et al. (1960). Authorities for species not appearing in this publication along with nonagaric names are given below. Names of vascular plants follow *Flora Europea*.

Agrocybe subpediades (Murr.) Watling, Bolbitiaceae
Amphinema byssoides (Pers. ex Fr.) J. Erikss., Corticiaceae; Aphyllophorales
Boletus spadiceus Quélet, Boletaceae

Botrydina vulgaris Bréb. ex Meneghini, Lichenes: incertae sedis

Calvatia utriformis (Bull. per Pers.) Jaap, Lycoperdaceae; Lycoperdales

Cenococcum graniforme (Sow.) Ferd. & Winge, Agonomycetes

Clitocybe sericella Kühn & Romagn. (= *C. subalutacea* Fr. s. Lge.)

Cortinarius acutus var. *striatulus* Henry, Cortinariaceae

C. pertrisis Favre

C. pulchripes Favre

Hebeloma austroamericana Singer, Cortinariaceae

H. fragilipes Romagn., Cortinariaceae

H. populinum Romagn.

H. pusillum var. *longisporum* Bruchet

H. vaccinum Romagn.

Hygrocybe conicoides (Orton) Orton & Watling, Hygrophoraceae (= *Hygrophorus* auct.
     brit.)

Inocybe cucculata Martin, Cortinariaceae

I. tarda var. *subulosa* Bon

Laccaria maritima (Theod.) Singer, Tricholomataceae (= *L. trullisata* auct. brit.)

Lactarius groenlandicus Terkel., Russulaceae

L. sphagneti (Fr.) Neuh.

L. violascens (Otto ex Fr.) Fr.

Leccinum rotundifoliae (Singer) Smith, Thiers & Watling, Boletaceae

L. salicola Watling

Lycoperdon foetidum Bon, Lycoperdaceae; Lycoperdales

L. perlatum Pers.

Naucoria macrospora J. Lge., Cortinariaceae

Nolanea verna Lundell, Entolomataceae

Omphalina hudsoniana Jenning, Tricholomataceae (= *O. luteolilacina* (Favre) Henderson)

Peziza badia Pers. ex Mérat, Pezizaceae; Pezizales

Psilocybe muscorum (P.D. Orton) P.D. Orton, Strophariaceae (= *Deconica* auct. brit.)

Russula atrorubens Quél., Russulaceae

R. graveolens Romell

R. graveolens var. *subrubens* J. Lge.

R. illota Romagn.

R. laccata Huijsman

R. norvegica Reid

R. olivaceoviolascens Gillet

R. pascua (Moell. & Schaeff.) Kühner

R. persicina Krombh. (= *R. pulchella* Borszczov auct. pl.)

Suillus grevillei (Klotsch) Singer, Boletaceae (= Boletus)

Thelephora terrestris (Ehrh.) Fr., Thelephoraceae; Aphyllophorales

Tylosperma fibrillosum (Burt) Donk, Corticiaceae; Aphyllophorales

REFERENCES

Andersson, O. (1950). Larger fungi on sandy grass heaths and sand-dunes in Scandinavia. *Bot. Not. 2* Suppl. 2: 1-89.

Bas, C. (1977). Species concept in *Amanita*, Sect. *Vaginatae*. In *Species Concept in Hymenomycetes*, H. Clémençon (Ed.). Cramer, Vaduz, Liechtenstein, pp. 79-103.

Becker, G. (1956). Observations sur l'écologie des champignons supérieurs. *Ann. Sci. Univ. Besançon* [Ser. 2, Bot.] 7: 15-128.

Birks, H. J. B. (1973). Modern pollen rain studies in some arctic and alpine environments. In *Quaternary Plant Ecology*, H. J. B. Birks and R. G. West (Eds.). Blackwell Sci. Publns., Oxford, England, pp. 143-172.

Bon, M. (1975). Agaricales de la côte Atlantique Française. *Docum. Mycol. 4*: 1-40.

Bon, M., and Gehu, J. M. (1973). Unités supérieurs de végétation et récoltes mycologiques. *Docum. Mycol. 2*: 1-40.

Bruchet, G. (1970). Contribution à l'étude du genre *Hebeloma* (Fr.) Kummer: Partie speciale. *Bull. Mens. Soc. Linn. Lyons 39*: 1-132.

Christiansen, M. P. (1941). *The Botany of Iceland*, Vol. III, Pt. II: *Studies in the larger Fungi of Iceland*. Munksgaard, Copenhagen, pp. 191-225.

Clémençon, H. (Ed.) (1977). *Symposium Herbette: Species Concept in Hymenomycetes*. Cramer, Vaduz, Liechtenstein, pp. 1-444.

Coker, P. (1970). Some observations on the flora of the Cairnwell, Perthshire. I. Limestone areas. *Trans. Proc. Bot. Soc. Edinburgh 40*: 592-603.

Darimont, F. (1973). *Recherches mycosociologiques dans les forêts de haut Belgique*. *Inst. Roy. Sci. Nat. Belg. Mem. 170*: 1-220.

Delzenne-van Haluwyn, Ch. (1971a). Notes écologiques sur les champignons supérieurs. I. Le genre *Hebeloma*. *Docum. Mycol. 1*(1): 7-18.

Delzenne-van Haluwyn, Ch. (1971b). Notes écologiques sur les champignons supérieurs. II. Le genre *Inocybe*. *Docum. Mycol. 1*(1): 19-31.

Delzenne-van Haluwyn, Ch. (1971c). Notes écologiques sur les champignons supérieurs. IV. Le genre *Russula*. *Docum. Mycol. 1*(2): 15-30.

Delzenne-van Haluwyn, Ch. (1971d). Notes écologiques sur les champignons supérieurs. V. Le genre *Lactarius*. *Docum. Mycol. 1*(2): .33-44.

Dennis, R. W. G. (1955). Larger fungi in the North West Highlands of Scotland. *Kew Bull. 10*: 111-126.

Dennis, R. W. G. (1964). The fungi of the Isle of Rhum. *Kew Bull. 19*: 77-127.

Dennis, R. W. G., and Gray, E. (1954). A first list of the fungi of Zetland (Shetland). *Trans. Proc. Bot. Soc. Edinburgh 36*: 215-223.

Dennis, R. W. G., Orton, P. D., and Hora, F. B. (1960). The new check list of British agarics and boleti. *Trans. Brit. Mycol. Soc. 43* (Suppl.).

Dominik, T. (1951). Badanie mykotrolizmu róslinnósa wydm radmorskich i srodladowych. *Acta Soc. Bot. Pol. 21*: 125-164.

Dominik, T. (1959). Synopsis of a new classification of the ectrotrophic mycorrhizae established on morphological and anatomical characteristics. *Mycopathol. Mycol. Appl. 11*: 359-367.

Dominik, T. (1961). Badanie mikotrofizmu zespolów róslinnych W Parku Narod wym, W Pieninach i na skalce nad Lysa Polona W Tatrach ze szczególnym u wzglednieniem mikotrofizina sosny reliktowey. *Inst. Badawczy Lésn. Prace 208*: 31-58.

Dominik, T. and Pachlewski, R (1955). Investigations on mycotrophism of plant associations in the lower timber zone of the Tatra. *Acta Soc. Bot. Pol. 25*: 3-26.

Dominik, T., Nespiak, A., and Pachlewski, R. (1954a). Badanie mykotrofizmy rostin-
ńości zespołów na skálkach wapiennych W Tatrach. *Acta Soc. Bot. Pol. 23*: 471-485.

Dominik, T., Nespiak, A., and Pachlewski, R. (1954b). Investigations of mycorrhizae
in plant communities of the Higher Tatrach Mountains. *Acta Soc. Bot. Pol. 23*: 487-
504.

Favre, J. (1955). Les champignons supérieurs de la zone alpine du Parc National
Suisse. *Ergebn. Wiss. Unters. Schweiz. Nat.* [5, N.S.] *33*: 1-212.

Fontana, A. (1962). Richerche sulle micorrize del genre *Salix*. *Allionia 8*: 67-85.

Gimmingham, C. H. (1970). Maritime Zone 4. Maritime and submaritime communities.
In *Vegetation of Scotland*, J. H. Burnett (Ed.). Oliver & Boyd, Edinburgh, pp. 67-
125.

Greville, R. K. (1823). *Scottish Cryptogamic Flora*, Vol. 1. Maclachan & Stewart,
Edinburgh.

Godwin, H. (1956). *The History of the British Flora*. Cambridge Univ. Press, New
York.

Grainger, J. (1946). Ecology of larger fungi. *Trans. Brit. Mycol. Soc. 29*: 52-63.

Hayes, A. J. (1967). Palaeo-ecology of Blackwood of Rannoch. *Scott. For. 21*: 153-
162.

Henderson, D. M. (1958). New and interesting Scottish fungi. I. *Notes Roy. Bot.
Garden Edinburgh 22*: 593-597.

Hesselman, H. (1900). Om Mykorrhizabilndingar hos Arktiska växter Bihang Till.
K. Svenska Vetensk.-Acad. Handl. (Stockholm) *No. 26* (Ajd. III) 1-46.

Hora, F. B. (1972). Quantitative experiments on toadstool production in woods.
Trans. Brit. Mycol. Soc. 42: 1-14.

Horak, E. (1960). Die Pilzvegetation im Gletschervorfeld (2290-2350 m.) des Rotmoos-
ferners in den Otztaler Alpen. *Nova Hedwigia 2*: 487-507.

Kallio, P. (1975). *Leccinum scabrum* (Fries) S. F. Gray subsp. *tundrae* Kallio, a new
subspecies from Lapland. *Rep. Kevo Subarctic Res. Sta. 12*: 25-27.

Kauffman, D. H. (1906). *Cortinarius* as a mycorrhiza-producing fungus. *Bot. Gaz.
42*: 208-214.

Kenworthy, J. B. (1976). *John Anthony's Flora of Sutherland*. Botanical Society of
Edinburgh, Edinburgh, pp. 1-201.

Komirnaya, O. N., and Fursayev, A. D. (1953). Forest fungi of a tree plantation of
semidesert Zavolzh and questions of mycorrhiza formation. *Bot. Zh. Kyyiv 38*: 426-
428.

Kreisel, H. (1965). Ectotrophbildende Pilze als Begleiter der Kriechweide *Salix
repens* L. *Westf. Pilze 5*: 135-139.

Kühner, R. (1976a). Agaricales de la zone Alpine. Genre *Lactarius* D. C. ex S. F.
Gray. *Bull. Soc. Mycol. Fr. 91*: 5-69.

Kühner, R. (1976b). Agaricales de la zone Alpine. Genre *Russula* Pers. ex S. F.
Gray. *Bull. Soc. Mycol. Fr. 91*: 313-390.

Laiho, O. (1970). *Paxillus involutus* as a mycorrhizal symbiont of forest trees.
Acta For. Fenn. 106: 1-72.

Lamoure, D. (1972). Agaricales de la zone Alpine: Genre *Clitocybe*. *Trav. Sci.
Parc Nat. Vanoise 2*: 107-152.

Lange, J. (1938-1940). *Flore Agaricina Danica*. Recato, Copenhagen.

Lange, L. (1974). The distribution of macromycetes in Europe. *Dansk Bot. Ark. 30*:
1-105.

Lange, M. (1957). Macromycetes. Part III. 1. Greenland Agaricales (Pars), *Macromycetes caeteri*. 2. Ecological and plant geographical studies. *Meddel Grønland
148*: 1-126.

Lobanow, N. W. (1960). *Myxotrophie der Holzpflanzen*. Berlin (Orig. Ed.: State Publ. Sov. Sci., Moscow, 1953), pp. 1-232.

McVean, D. N. (1961). Flora and vegetation of the Islands of St. Kilda and North Rona in 1958. *J. Ecol. 49:* 39-54.

McVean, D. N., and Ratcliffe, D. (1962). *Plant Communities of the Scottish Highlands*. H.M.S.O., London.

Mathews, J. R. (1937). Geographical relationships of the British flora. *J. Ecol. 25:* 1-90.

Meyer, F. H. (1973). Distribution of Ectomycorrhizae in nature and man-made forests. In *Ectomycorrhizae: Their Ecology and Physiology*, G. C. Marks and T. T. Kozlowski (Eds.). Academic Press, New York, pp. 79-105.

Moeller, F. H. (1945). *Fungi of the Faeroes*, Pt. 1: *Basidiomycetes*. Munksgaard, Copenhagen.

Nespiak, A. (1953). Researches on mycotropism of the alpine vegetation in the granite parts of the Tatra mountains over the limit of *Pinus mughus. Acta Soc. Bot. Pol. 22:* 97-125.

Peyronel, B. (1931). Simbiasi micorizici tra plante alpine e basidiomiceti. *Nuova G. Bot. Ital.* [N.S.] *37:* 655-663.

Reid, D. A. (1972). *Fungorum rariorum icones coloratae*. Cramer, Vaduz, Liechtenstein.

Richardson, M. J. (1970). Studies on *Russula emetica* and other agarics in a Scots Pine plantation. *Trans Brit. Mycol. Soc. 55:* 217-229.

Savile, D. B. O. (1963). Mycology in the Canadian Arctic. *Arctic 16:* 17-25.

Singer, R. (1950). Die höheren Pilze Argentiniens. *Schweiz. Z. Pilzk. 28:* 181-196.

Terkelsen, F. (1956). *Lactarius groenlandicus* sp. nov. *Friesia 5:* 417-418.

Trappe, J. M. (1960). Some probable mycorrhizal associations in the Pacific Northwest. II. *Northwest Sci. 34:* 113-117.

Trappe, J. M. (1962a). *Cenococcum graniforme*: Its distribution, ecology, mycorrhiza formation and inherent variation. Ph.D. dissertation, University of Washington, Seattle.

Trappe, J. M. (1962b). Fungus associates of ectotrophic Mycorrhizae. *Bot. Rev. 28:* 539-606.

Trappe, J. M. (1977). Selection of fungi for mycorrhizal inoculation in nurseries. *Ann. Rev. Phytopath. 15:* 203-222.

Wallace, T. J. (1954). The plant ecology of Dawlish Warren. Part II. The larger fungi. *Trans. Dev. Assoc. Advancement Sci. Lit. and Arts 86:* 201-210.

Wallace, T. J. (1956). The larger fungi of the South Haven Peninsula, Studland Heath, Dorset. *Proc. Dorset Nat. Hist. Archaeol. Soc. 77:* 113-122.

Wallace, T. J. (1957). Some larger fungi of the Berrow Dunes. *Proc. Somerset Archaeol. Nat. Hist. Soc. 98:* 127-129.

Watling, R. (1963). Fungi of the Garth area. *Rep. Scott. Field Stud. Assoc. 1962:* pp. 15-26.

Watling, R. (1965). Notes on British boleti. IV. *Trans. Proc. Bot. Soc. Edinburgh 40:* 100-120.

Watling, R. (1967). Fungi of the Kindrogan area. *Rep. Scott. Field Stud. Assoc. 1966:* pp. 28-46.

Watling, R. (1970). Check list of plants of Rhum. III. *Trans. Proc. Bot. Soc. Edinburgh 40:* 497-535.

Watling, R. (1974). Macrofungi in the oak woods of Britain. In *The British Oak*, M. G. Morris and F. H. Perring (Eds.). Classey, Farringdon, pp. 222-235.

Watling, R. (1978). Larger fungi from Greenland. *Astarte 10*: 61-71.

Watling, R., and Richardson, M. J. (1971). The agarics of St. Kilda. *Trans. Proc. Bot. Soc. Edinburgh 41*: 165-188.

Zerova, M. Ya. (1955). Mycorrhiza formation by forest species in Ukrainian S.S.R. *Trudy Konf. po Mikotrofic Rast. Akad. Nauk. S.S.S.R., Moscow, 1953*, pp. 43-62.

Chapter 24

RESPONSE OF SOIL FUNGAL COMMUNITIES TO DISTURBANCE

S. E. Gochenaur

Adelpni University
Garden City, New York

I. INTRODUCTION

The concept of community is usually restricted to populations of one or more individuals with similar life habits and similar resource demands co-occurring in time and
space (McNaughton and Wolf, 1973). By this definition, soil and litter contain several microfloral and microfaunal communities. Among the microfungi whose presence
in soil and litter is revealed through plating, washing, and dilution techniques,
temporal and spatial relationships exist as well as similar nutritional, reproductive,
and survival strategies. These fungi can be grown in vitro on ordinary mycological
media where they have no or only minimal requirements for organic compounds other
than a carbon-energy source and exhibit relatively broad environmental tolerances.
Although sexual stages may occur, in the majority reproduction and dissemination are
accomplished by asexually produced, small aplanospores. Their principal mechanism
for survival is the production of large numbers of propagules including memnospores.
An assemblage of these microfungi isolated by indirect methods (Parkinson et al.,
1971) from the same habitat at the same time constitute a community that will herein
be designated the opportunistic decomposers (OD). Such a connotation is useful because it helps one focus on these microfungi as a distinct community type separate
from the other microfloral and microfaunal communities with which they coexist. In
this respect it is somewhat analogous to the term unit-community proposed by Swift
(1976) for microbial communities (fungi plus bacteria) occurring on branches, roots,
fruits, leaves, and other resource types. This chapter will concentrate on OD communities in the upper horizons of soil.

    With an assimilative stage that is usually very transitory (Old, 1967; El-Abyad
and Webster, 1968a; Meyer, 1970) the opportunistic decomposers occur primarily as
spores (Warcup, 1955; Christensen, 1969; Griffin, 1972). In studying their populations, the dilution method has proved to be especialy valuable since it provides the
investigator with a means of determining the numerically superior fungi within an
assemblage. Because most if not all members of the OD community are capable of
sporulating abundantly, those that achieve numerical dominance over the others are
regarded as the most ecologically fit irrespective of whether this is due to the pro-

duction of more recalcitrant propagules or because the organisms during their assimilative stage are more competitively successful, or both. Such "dominants" form u-
nique ecosystem-specific assemblages since their selection will depend upon the response of their propagules and assimilative stages to the particular combination of
biotic and abiotic factors that occur.

The following account deals first with the community parameters that characterize OD assemblages isolated from unperturbed "steady-state" ecosystems. Particular
reference will be made to data for a Long Island oak-birch forest where these attributes have been studied in detail (Gochenaur, 1978). The second part examines modification of these community characteristics by acute disturbance. The latter is
herein defined as a severe disordering of a soil ecosystem usually caused by destruction of part or all of the vegetation whether natural or man induced but excluding
chemical pollutants. Such perturbations include cultivation of fallow land, periodic
flooding, clear-cutting, chronic ionizing radiation, fire, and wind erosion. Finally,
I shall attempt to extrapolate from the experimental data generalizations that within broad limits may be used to predict possible patterns of change in any OD community following ecosystem disturbance.

II. THE UNDISTURBED COMMUNITY

Before examining the effects of disturbance upon OD communities, it is necessary to
consider what attributes characterize these assemblages in their unperturbed state.
In doing this, I shall concentrate on OD communities from late successional forest
ecosystems because they have been the most intensively studied. Examination of reports that deal with populations obtained from several sites per forest and that present data in terms of frequency (sites of occurrence as a percentage of total sites
sampled) and relative density (isolates of a species as a percentage of total isolates) show that communities from distinctly different plant associations of boreal
(Morrall and Vanterpool, 1968; Söderström, 1975), temperate (Tresner et al., 1954;
Christensen et al., 1962; Gochenaur and Whittingham, 1967; Novak and Whittingham,
1968; Christensen, 1969; Gochenaur and Woodwell, 1974; Gochenaur, 1978), and tropical (Gochenaur, 1975) origins hold many features in common. Among these features,
those related to biomass, organization and structure, species and biochemical diversity, and stability are most relevant to the discussion.

A. Biomass

Biomass is a measure of the ability of an ecosystem to provide conditions that are
suitable for the development of the opportunistic decomposers' propagative and assimilative stages. Ideally, it estimates that portion of the total energy (calories)
or mass of the system stored at any time in their propagules and mycelium.

Hyphal biomass is used to assess assimilative productivity (Nagel-deBoois and
Jansen, 1967; Hanssen and Goksoyr, 1975). Because the technique includes all mycelial fungi irrespective of their community membership, it is not strictly applicable

to the opportunistic decomposers. This is not the case with propagative biomass, from here on to be called simply biomass. Calculated directly from the isolation plates that yield representatives of the OD community, it is the number of their propagules per volume or weight of soil. Biomass, in the sense it is used in this chapter, reflects the cumulative and integrated effects of several ecosystem functions on the members. These effects include the quality of the resources to be utilized in producing assimilative stages and the suitability of environmental conditions that allow these stages to produce or differentiate into propagules of reproductive, survival, and/or disseminative functions; and also include the inherent resistance of these propagules to biotic and abiotic factors that cause their destruction or transport from an area (Gochenaur, 1978).

OD biomass is high in late successional temperate forests (Tresner et al., 1954; Christensen et al., 1962; Christensen, 1969; Gochenaur, 1978) where the quasi-steady-state or "climax" conditions encountered by the members produce fluctuations around an average rather than directional change (Whittaker, 1953). The results for an oak-birch forest are representative (Gochenaur, 1978). In the A horizon, propagule numbers show annual cyclic fluctuations, being lowest in the summer, remaining essentially constant at 340,000 per gram of dry soil in the fall and winter, and increasing to well over 1 million per gram in the early spring.

B. Organization and Structure

OD communities from late successional forests have certain organizational and structural features in common. First, their species show a Raunkiaer distribution (McNaughton and Wolf, 1973). That is, a few taxa couple wide distribution (high frequency) with abundance (high density), whereas most combine limited occurrence with rarity (Tresner et al., 1954; Christensen et al., 1962; Gochenaur and Whittingham, 1967; Novak and Whittingham, 1968; Gochenaur and Woodwell, 1974; Gochenaur, 1975; Gochenaur, 1978). Second, populations from the same ecosystem are compositionally alike. For instance, when Sørensen's index of similarity (Mueller-Dombois, 1974), weighted by frequency and/or relative density, is used to compare populations isolated from different soil samples of the same forest, the similarity can exceed 70% (Christensen et al., 1962; Novak and Whittingham, 1968; Christensen, 1969; Gochenaur and Woodwell, 1974; Gochenaur, 1978). Third, the assimilative stages of the members are hyaline, thin-walled, and ephemeral. The majority of the isolates, often more than 70%, belong to the Mucorales and/or Moniliaceae. The former are most common in boreal and other conifer-dominated woods (Morrall and Vanterpool, 1968; Christensen, 1969; Söderström, 1975), whereas the latter are characteristic of broad-leaf forests of temperate regions (Tresner et al., 1954; Christensen et al., 1962; Gochenaur and Woodwell, 1974; Gochenaur, 1978). The major genera typically include *Mortierella*, *Oidiodendron*, *Penicillium*, *Trichoderma*, *Acremonium*, and *Torulomyces*. Fourth, most have reproductive structures that are erect and often intricate and highly branched,

although single cells that produce blastospores and prostrate structures of intermediate complexity may be present (Gochenaur and Woodwell, 1974).

C. Species Diversity

As usually employed, diversity measures compositional heterogeniety among members of
a collection. It consists of a varietial component, or species richness, and an equitability component that gauges how evenly isolates are distributed among the taxa
(Peet, 1974).

Intuitively, one expects large numbers of species to occur in late successional
forests due to the moderate conditions, the array of resources, and the variety of
microhabitats encountered. Indeed, many reports support this idea. For example,
115 species were obtained from a green ash forest in Oklahoma (Mallik and Rice, 1966),
90 species were recorded from a Swedish spruce forest (Söderström, 1975), 113 species
were isolated from a boreal forest (Morrall and Vanterpool, 1968), and 89 taxa occurred in a single mineral horizon of an oak-birch forest (Gochenaur, 1978).

Although informative, these figures are of little value when we are comparing
the diversity of microfungi in different ecosystems, due to the dependence of such
data upon sample size. In other words, one can expect the number to increase when
larger populations are examined and conversely to decrease when populations of smaller size are employed. This problem is overcome by basing diversity (i.e., species
richness) on the mean number of species occurring in populations of fixed size derived
from samples of known weight. Such figures have the advantage of being comparable
but have their own limitations because they give no indication of the actual number of
microfungi in an ecosystem. However, this is not a serious limitation since it is
doubtful that such a number could ever be experimentally determined for any natural
ecosystem.

Again using the oak-birch forest study as an example (Gochenaur, 1978), diversity
is found to fluctuate throughout the year. It is highest in the spring, significantly
lower during the summer, and equal to the yearly average of 16 (per 100 isolates) in
the fall and winter. An identical average has been recorded for two other late successional forests (Gochenaur and Woodwell, 1974; Gochenaur, 1975). Such comparable
data on richness are rare, and consequently in this chapter diversity will be employed
in a more qualitative sense indicating a state of variety rather than a quantitative
measure of either richness or equitability.

D. Biochemical Diversity

In forested ecosystems, a majority of the organic input comes from root production
and decomposition phenomena that supply the microfungi with an incredible array of
resources in the form of suberized or unsuberized fine and course roots, sloughed cells
and root hairs, and exudates and mucigels (Coleman, 1976). Although some of these
materials are soluble, the majority are biopolymers that must be enzymatically degraded

to their constituent monomers before permeation can occur. This is accomplished by extracellular hydrolases that cleave peptide, ester, or glycosidic bonds. Since individual organisms are known to produce both exo- and endo-isoenzymes, cleavage of many of these substrates is probably accomplished by a combination of extracellular hydrolases (Matile, 1975).

Information on the distribution of these enzymes among the members of a community provides a measure of the community's biochemical diversity as well as a better understanding of how its resources may be partitioned. For the past several years, I have been routinely screening microfungi for their ability to hydrolyze 17 substrates (see Table 1, footnote a), using methods summarized elsewhere (Holding and Collee, 1971; Hankin and Anagnostakis, 1975). While this in vitro study provides conditions and substrates quite different from what the organisms encounter in nature, it nevertheless is useful in estimating potential hydrolyzing abilities. Since pure culture studies do not allow for synergistic interactions between decomposers of differing biochemical potential (Swift, 1976), the results probably underestimate rather than overestimate an organism's capacity to utilize compounds in the soil. A portion of my data (unpublished) for OD communities from a Long Island oak-birch ecosystem and the Irradiated Forest, Brookhaven Laboratory, Upton, New York, will be summarized here. When the test fungi are divided into groups based upon the number of substrates hydrolyzed and the percentage of the total population represented by each group is determined, the data (Table 1) reveal two aspects of the communities' biochemical diversity. First, a significant proportion of each community displays a great capacity for depolymerization. Over 45% of each OD community could hydrolyze between 60% and 88% of the test substrates. Similar versatility, but on fewer substrates, has been recorded for other microfungal populations (Domsch and Gams, 1969; Flanagan and Scarborough, 1974). Second, OD communities from the undisturbed ecosystems exhibit high biochemical diversity with taxa ranging in ability from "generalists" that can degrade all or nearly all of the substrates to "specialists" with an extremely limited capacity, and including all variations in between. This implies, as Swift (1976) has so cogently pointed out, a considerable potential difference in the "biochemical strategy" for obtaining carbon and energy on the part of the decomposer as well as a great potential for resource partitioning among its members.

E. Stability

Ecosystems show increased stability as succession proceeds (McNaughton and Wolf, 1973), and this pattern appears to extend to the microfungi as well. Seasonal cycles observed during the first year of study of an oak-birch forest's OD community were faithfully repeated during the next 28 months (Gochenaur, 1978). Not only did the same species reoccur, but their numerical importance in the community did not change. Similar stability but over a shorter time span has been reported for populations from other late successional forests (Novak and Whittingham, 1968; Christensen, 1969). These results suggest that the opportunistic decomposers are best viewed

Table 1 Biochemical diversity of OD communities from undisturbed and disturbed
forest ecosystems as measured by the ability of representative isolates
to hydrolyze in vitro one or more of seventeen substrates[a]

	Number of substrates utilized (percentage of total community)					
	1-3	4-6	7-9	10-12	13-15	16-17
Undisturbed						
Oak-birch forest OD community[b]	0.3	18	34	25	22	0.3
Oak-pine forest OD community[c]	1	14	36	38	11	0
Disturbed						
Oak-pine forest OD community[d]	0	1	39	25	30	5

[a]Test substrates were aesculin, amylose, amylopectin, arbutin, casein, cellulose,
chitin, deoxyribonucleic acid, gelatin, glycogen, keratin, lipase reagent (Bacto),
ovalbumen, ribonucleic acid, pectin, phenolphthalein diphosphate, and xylan.

[b]See Gochenaur (1978) for a description of the OD community; 280 isolates representing 97% of the population and 78% of the species were tested.

[c]See Gochenaur and Woodwell (1974) for a description of the OD community of the
Control Zone at 100 m from the radiation source; 51 isolates representing 98% of the
population and 81% of the species were tested.

[d]See Gochenaur and Woodwell (1974) for a description of the OD community of the
Devastated Zone at 10 and 15 m from the radiation source; 37 isolates representing
97% of the two populations and 83% of their species were tested.

not as an assemblage of individuals but as an integrated, structured community consisting of self-perpetuating species populations that are in dynamic equilibrium with
their environment and that show a remarkable resilience to annual changes in temperature, in soil moisture content, and in the addition and withdrawal of nutrients.

As the following discussion will show, productivity, structure and organization,
diversity, and stability of the community are not immutable parameters but are altered through ecosystem perturbation, the extent of the change being directly related
to the level of destruction of the macrophytes and the subsequent modification of the
soil environment.

III. THE DISTURBED COMMUNITY

Acute disturbance modifies an ecosystem through damage to or elimination of the primary producers. Although not all the consequences of perturbation are necessarily
negative, the effects on the plants directly influence the quality and quantity of

the resources available to the decomposers and indirectly often lead to a more rigorous soil environment. The maximum response of the OD community to these changes
is usually not immediately evident but takes place over a period of time. It is typically the end product of a sequence of events rather than the stages leading up to
it that has been investigated and will form the basis for much of what follows. The
various disturbances described in this section are presented in order of increasing
overall damage to the soil ecosystem. For each, initial remarks concentrate on the
effects of disturbance to the environment. These are followed by the opportunistic
decomposers' response.

A. Cultivation

When virgin or fallow land with its mixed flora is converted to land supporting a
single crop plant, a reduction in the diversity of resources and microhabitats as
well as total carbon and nitrogen content occurs (Lemaire and Jovan, 1966; Chu and
Stephen, 1967; Ayanaba et al., 1976). These detrimental effects are greatly out-
weighed by the benefits that accrue to the opportunistic decomposers. Through till-
age, insolation, irrigation, and fertilization, the soil environment becomes more
homogeneous, warmer, with improved texture, humidity, and gas exchange, as well as
increased levels of phosphorus, potassium, magnesium, and calcium (Lemaire and Jovan,
1966; Chu and Stephen, 1967). Similar positive effects occur when recently submerged land or desert soil is reclaimed (Taha et al., 1970; Tichelaar and Vruggink, 1975)
and when areas normally devoid of higher vegetation such as cut-away peat beds are
colonized by plants (Dooley and Dickinson, 1970). Of the perturbations to be discussed here, only cultivation improves rather than lowers the overall quality of the
environment.

Productivity of the opportunistic decomposers is always greater under cultivation (Lemaire and Jovan, 1966; Chu and Stephen, 1967; Dooley and Dickinson, 1970;
Taha et al., 1970; Tichelaar and Vruggink, 1975), but the effect on species diversity
varies. In comparing microfungal populations from a fallow and a cropped subtropical
soil, Chu and Stephen (1967) report that the former yielded nearly twice the number
of genera and species as did the latter. They suggest that in the fallow soil the
absence of competition from markedly dominant species allows many taxa to coexist
whereas the more favorable and uniform environment in the cultivated area encourages
a few strongly competitive fungi to develop at the expense of high diversity. In contrast to these results, Lemaire and Jovan (1966) find the reverse to be true for a
cropped and fallow soil in Brittany.

Characteristic patterns of change in the composition of the OD community emerge
following cultivation. There occurs a dramatic increase in the abundance of *Trichoderma* and conconmitantly many native penicillia and mortierellas decrese in number
or disappear entirely (Lemaire and Jovan, 1966; Chu and Stephen, 1967; Tichelaar and
Vruggink, 1975). Certain species rare in or absent from fallow soil appear. These
include various sphaeropsidaceous taxa (Lemaire and Jovan, 1966; Gams and Domsch,

1969; Tichelaar and Vruggink, 1975), cellulolytic decomposers (Lemaire and Jovan, 1966; Taha et al., 1970), *Gliocladium roseum* Bainier (Lemaire and Jovan, 1966; Gams and Domsch, 1969), *Verticillium nigrescens* Pethybr. and *Aureobasidium bolleyi* (Sprague) von Arx (Gams and Domsch, 1969). Finally, with time a variety of phyto-pathogens including *Helminthosporium atrovirens* (*H. solani* Dur. and Mont.) (Santerre, 1966), *Ophiobolus graminis* Sacc. and *Streptomyces scabies* (Thaxt.) Waksman and Henrici (Lemaire and Jovan, 1966), and species of *Pythium*, *Fusarium*, *Rhizoctonia* (Chu and Stephen, 1967) and *Cylindrocarpon* (Gams and Domsch, 1969) may become established. From the above it appears that the most noticeable effect of cultivation, aside from increasing productivity, is alteration in composition of the OD community.

B. Flooding

Periodic disturbance due to inundation of stream bank communities by flood waters (Mitchell and Alexander, 1960; Gochenaur, 1964; Novak and Whittingham, 1968), alluvium deposition on river deltas (Gill, 1973a; Gill, 1974), impoundment of water from winter rains in green-tree reserves (Filer, 1975) and irrigation (Ruscoe, 1973; Ioannou et al., 1977) results in no or at best only minor damage to the macrophytes but for a time can significantly alter the soil ecosystem. When flooding is coupled with rap-idly moving water, an initial response may be lowered nutrient levels as litter is swept away or increased levels when litter is buried by alluvium (Gill, 1973a). Sometimes enrichment of poor soils through deposition of nutrient-rich sediments oc-curs, as was reported for the sand of a Wisconsin river bar that showed a fourfold increase in organic matter after flooding (Gochenaur, 1964). These positive effects are overshadowed by negative ones induced through saturation. The importance of this lies not in the absolute water content of the soil but in indirect effects on nitrate level and on the gaseous environment. Denitrification by microbial action and leach-ing are enhanced (Alexander, 1977), oxygen is rapidly depleted (Scott and Evans, 1955; Ioannou et al., 1977; Miller and Burke, 1977), and concomitantly the levels of hydrogen sulfide (Alexander, 1977), carbon dioxide, and ethylene (Griffin, 1963; Ioannou et al., 1977) may be increased.

An obvious response of the microfungi to the alteration of their environment is a decrease in biomass but not a complete elimination of their propagules (Mitchell and Alexander, 1960; Gochenaur, 1964; Filer, 1975). This decrease may be due in part to negative effects exerted on certain members of the soil populations by the accumulation of ethylene at concentrations above 1 ppm (Smith, 1976; Ioannou et al., 1977) and by oxygen depletion (Griffin, 1963). The latter is not necessarily dele-terious to all species, as quite a few molds and yeasts possess facultative anaerobic abilities (Tabak and Cooke, 1968), and forms such as these could be expected not only to survive but also to grow and possibly germinate in waterlogged soils.

Inundation alters the composition of the OD community. At the generic level, the response varies. Some species of *Botrytis* (Ruscoe, 1973), *Penicillium* (Menon and Williams, 1957; Gochenaur, 1964; Filer, 1975), and *Cladosporium* (Gochenaur, 1964) decrease in or are eliminated from saturated soils, while the density of *Trichoderma* remains the same (Filer, 1975) or increases (Gochenaur, 1964; Ruscoe, 1973), as it also does for *Verticillium* (Ruscoe, 1973) and *Gliocladium* (Gochenaur, 1964). It is not uncommon for organisms apparently absent from an ecosystem initially to appear in high numbers following flooding. For example, in a willow-cottonwood soil in Wisconsin, *Aspergillus niger* van Tiegh. and species of *Fusarium*, *Phoma*, and *Cylindrosporium* appear only after the area has been inundated for approximately 50 days and then quickly disappear from the ecosystem as it returns to normal in the ensuing months (Gochenaur, 1964).

Certain macrofungi show an interesting adaptation to spring flooding. *Inocybe* sp., *Omphalina* sp., and *Cortinarius* sp. grow in cracks formed upon drying of alluvium in *Salix alaxensis-Equisetum arvense* communities in the Mackenzie River Delta (Gill, 1973a). Their occurrence in this unusual location is thought to be due to the exposure of buried lignicolous material that they use as a substrate.

The probable sequence of changes that occur in the OD community following flooding is a decrease in numbers of propagules, a transient elimination of some species, a temporary influx of a few alien forms, and the eventual reconstruction of the community both quantitatively and qualitatively within as short a period as 3 or 4 months after recession of the water (Gochenaur, 1964; Novak and Whittingham, 1968).

C. Clear-Cutting

Clear-cutting, i.e., stripping the land of all aboveground vegetation during logging operations or to render it suitable for agriculture, produces with time a radical change in the soil environment (Borman et al., 1967; Marks and Borman, 1972; Gill, 1973; Ayanaba et al., 1976). Nutrient inventory is decreased as materials are flushed from the soil in response to reduced transpiration that no longer retards passage of water through the system and increased mineralization that lowers total root area, reducing the amounts of nutrients that can be sequestered. Losses of nitrate, calcium, sodium, potassium, and magnesium ions (Borman et al., 1967; Marks and Borman, 1972) and biomass carbon (Ayanaba et al., 1976) are severe. Total destruction of the vegetation can also be expected to produce a soil environment that is both warmer and supports greater evaporation rates than the undisturbed system, as was shown for the soil of a forest destroyed by sulfur dioxide (Witkamp et al., 1966). Other changes that have been recorded after clear-cutting include higher soil moisture in a tropical ecosystem (Pacheco de Lira, 1971) and increased soil acidity in what was originally a boreal forest (Gill, 1973b).

Although experimental evidence documenting the effects of recent clear-cutting on the OD community is missing, one can predict that simplification of the plant community through reduction in total standing crop and nutrient inventory will produce a similar response in the opportunistic decomposers. Reduction in fungal biomass and in species diversity should occur. There is the likelihood that for a time this will be accompanied by an increase in certain segments of the community, as was reported for the genus *Absidia*, whose numbers increased in a tropical ecosystem when competing organisms--thought to be ill-adapted to the highly humid soils that occurred after deforestation--were reduced (Pacheco de Lira, 1971). In addition, we can expect clear-cutting to affect members of other fungal communities, especially those dependent on specific host plants. For example, a forest that was logged yielded far fewer sporocarps of the endomycorrhizal fungus *Endogone* than did the adjacent undisturbed area (Kessler and Blank, 1972).

D. Chronic Ionizing Radiation

The response of a late-successional oak-pine forest to chronic ionizing radiation is representative of the types of change induced in an ecosystem by this disturbance (Woodwell and Rebuck, 1967). Although diversity, primary production, total respiration, and nutrient inventory are altered (Woodwell, 1967, 1970; Horrill and Woodwell, 1973), the most dramatic transformation occurs in the forest's structure (Woodwell and Rebuck, 1967). Under irradiation an upright growth form is a disadvantage and consequently systematic dissection of the forest takes place. Strata are removed layer by layer, producing five well-defined zones of modification of the vegetation: (1) a central devasted zone where exposures exceed 200 R/day and no higher plants survive; (2) a sedge zone where *Carex pennsylvanica* endures and ultimately forms a continuous cover under radiation doses in excess of 150 R/day; (3) a shrub zone at levels above 40 R/Day; (4) an oak zone, the pines having been eliminated, at exposures above 16 R/day; and (5) the oak-pine forest where exposures are less than 2 R/day and damage is minimal.

Although the direct consequences of ionizing radiation on the physiochemical nature of the soil are slight, changes in the vegetation are followed by slow but progressive changes in the soil, especially noticeable in the devastated zone (Horrill and Woodwell, 1973). A reduction in litter and nutrient content occurs, and organic matter--measured as percent loss on ignition--declines in most horizons. Other indirect effects produced by removal of the canopy include warmer soil temperatures, lower moisture levels due to increased evaporation, and greater diurnal fluctuations of both parameters (Horrill and Woodwell, 1973; Gochenaur and Woodwell, 1974).

Chronic ionizing radiation both directly and indirectly affects the fungal, algal, and lichen communities of the forest. Initially, irradiation causes defoliation (Woodwell, 1967) and possibly increased exudation of material from the roots in response to damage to the aboveground portions. This flush in nutrients can be expected to produce a similar but transitory flush in microbial numbers, as has been recorded for a tropical ecosystem (Witkamp, 1970). As damage to the forest progresses, resources become poorer in quality and lower in quantity, and this coupled with the deleterious effects of chronic ionizing radiation on the resident populations results in a decrease in biomass as measured by number of organisms and their propagules (Woodwell and Gannutz, 1967; Franz and Woodwell, 1973; Gochenaur and Woodwell, 1974).

Using the microfungi as an example, numbers in the devastated zone are 35 times lower than those recorded for the vegetated areas of the forest (Gochenaur and Woodwell, 1974). Similar decreases in biomass are documented for other irradiated ecosystems (Davis et al., 1956; Skou, 1962; Kashkina and Abaturov, 1969).

Chronic ionizing radiation alters the biochemical diversity of the OD community and appears to diminish the capacity of the ecosystem to support enzymatically restricted organisms. With this disturbance, populations of generalists able to hydrolyze a variety of materials seem to increase at the expense of the specialists with more limited abilities (Table 1). This idea is further supported by the results of the two OD communities on a single substrate, cellulose (unpublished data from my laboratory). Approximately 40% of the community is cellulolytic in that section of the forest undamaged by radiation, while in the devastated zone these forms increase to 61%. It is tempting to conclude that this increase is a direct response of the community to the higher ratio of woody to soluble substrates that develops after disturbance.

Diversity of the lichens decreases in direct response to increasing radiation. This suggests that niches opened by death of sensitive species are not filled by radioresistant forms (Woodwell and Gannutz, 1967). In contrast, algal diversity remains steady or increases slightly at the highest exposures (Franz and Woodwell, 1973). For the microfungi (Gochenaur and Woodwell, 1974), diversity based upon the number of taxa per 100 isolates is reduced but constant throughout the devastated zone as shifting groups of more and more resistant forms occur. Radioresistance appears to be linked to melanization. Under increasing exposures, populations become progressively dominated by dematiaceous species and yeasts, culminating at the maximum level in a population consisting primarily of intensely melanized, extremely slow-growing isolates. In vitro studies suggest that the latter are modified colonies of the black "yeast" *Aureobasidium pullulans* (de Bary.) Arnaud, since irradiation of aqueous solutions of its blastospores at doses in excess of 150 krad produced upon plating colonies that mimicked those isolated from the forest (unpublished data from my laboratory). Skou (1964) is one of the first workers to study the survival ratio of *A. pullulans* to gamma radiation. His results reveal that it is among the most radioresistant organisms known, with decimation of the culture requiring up to 500 krad. Also, its different propagules vary in their sensitivity. Young unpigmented blastospores (LD_{50}, 100 krad) are more sensitive than older unpigmented blastospores (LD_{50}, 180 krad), which are in turn more sensitive than melanized chlamydospores (LD_{50}, 250 krad). Other workers as well have demonstrated in vitro that melanin appears to protect cellular organelles from harmful ionizing radiation so that melanized fungi are typically more radioresistant than their unpigmented counterparts (Mirchink et al., 1972; Zhdanova and Pokhodenko, 1973; Zhdanova et al., 1974; Ellis and Griffiths, 1975).

A well-established characteristic of radiation damage to higher plants, sorting by size, has parallels among the cryptogams. Under increasing exposure, stature and morphological complexity decline. Among the fungi, basidiomycetes such as *Armillaria mellea* (Vahl ex Fr.) Kummer (Holt, 1972) are the most sensitive, whereas unicells and forms with a rudimentary mycelium are the least sensitive (Gochenaur and Woodwell, 1974). The lichens show a similar progression, with crustose species replacing foliose or fruitcose kinds at the higher radiation levels (Woodwell and Gannutz, 1967). Among the soil algae, increasing exposure results in the substitution of procaryotes for eucaryotes in the community (Franz and Woodwell, 1973).

Chronic radiation disturbance produces a simplification of the OD community: indirectly, it reduces the quantity and quality of the resources, thus limiting the size of the heterotrophic population that can be supported; directly, it destroys the more sensitive species, while at the same time selecting for forms with protective pigmentation and simple morphology. As a consequence of this selection, diversity shows essentially no decrease as radiation levels increase.

E. Fire

Probably every factor in the environmental complex is changed in some way by fire (Christensen and Muller, 1975), with the degree of alteration regulated by the season of the year in which the fire occurs (Kelting, 1957) and the depth and intensity of the burn (Ahlgren, 1974). During a severe fire, temperatures may for a time exceed 980°C at ground level and reach 315°C some 2-3 cm below the surface (Ahlgren and Ahlgren, 1960). Such intense heat will destroy much of the vegetation, partially sterilize the upper layers of the soil, and mineralize the accumulated organic matter. With the latter, fire accomplishes rapidly what the decomposers normally take many seasons to do (Walter, 1975).

The white ash produced when combustion is complete enriches the surface with a variety of oxides, hydroxides, and salts (Ahlgren and Ahlgren, 1960; Petersen, 1970; Christensen and Muller, 1975). Water-soluble anions, CO_3^{2-}, HCO_3^-, SO_4^{2-}, Cl^-, and cations, K^+, Na^+, Mg^{2+}, are increased; ammonium concentrations are significantly higher (Christensen and Muller, 1975); and the level of acid-soluble P can exceed 15,000 ppm immediately after the fire (Petersen, 1970). The soil often becomes very alkaline due to the addition of carbonate, hydrogen carbonates, and hydrogen phosphates, and a pH of 9 or higher has been recorded by some workers (Jalaluddin, 1969; Petersen, 1970). These direct fire-induced changes persist, sometimes for more than 1 year (Petersen, 1970) or until rain leaches the water-soluble materials from the soil (Ahlgren and Ahlgren, 1960; Christensen and Muller, 1975).

Fire, by destroying the vegetation, indirectly influences several edaphic parameters. Moisture content commonly is reduced as erosion, surface-runoff and evaporation increase (Ahlgren and Ahlgren, 1960; Viro, 1974; Christensen and Muller, 1975). Temperatures of the upper layers are higher, since opening of the canopy allows greater light penetration and insolation. When combustion is incomplete, charred and blackened organic matter covers the surface, further increasing the amount of heat absorbed (Ahlgren and Ahlgren, 1960; Christensen and Muller, 1975). Higher nutrient levels occur immediately after the fire through mineralization and the addition of organic matter in the form of dead roots. With time, nutrients decrease both quantitatively and qualitatively in response to microbial activities, leaching, and the absence of root exudates, litter leachates, and plant runoff (Christensen and Muller, 1975; DeBoois and Jansen, 1976).

The effects of these changes on the fungal community are striking. Analysis of samples collected soon after the surface of the soil has cooled reveals a great reduction in species diversity (Jalaluddin, 1969; Wicklow, 1973) and numbers of propagules (Wright and Bollen, 1961; Jalaluddin, 1969; Rambelli et al., 1973) due to partial sterilization of the upper few centimeters of soil. This decrease is followed, usually after the first rain, by a dramatic increase in density so that total counts often exceed those recorded prior to burning by tenfold (Ahlgren and Ahlgren, 1965; Wicklow, 1973; Ahlgren, 1974; Christensen and Muller, 1975). Such a "postfire bloom" may be due to the selective multiplication of surviving fungi or of early colonizers from unburned areas. Most of these organisms are "pyrophilous" in the sense that they become abundant in an area only after it has burned, being rare in or absent from the prefire population. These fungi can include *Coniochaeta discospora* (Auersw.) Cain, *C. tetraspora* Cain, and *Rhodotorula* sp. (Wicklow, 1973), *Trichophaea abundans* (Sow. ex Fr.) Sacc. (El-Abyad and Webster; 1968a), *Gelasinospora* sp. (Widden and Parkinson, 1975), *G. calospora* (Mout.) C & M Moreau (Wicklow, 1975), and a variety of macromycetes (Jalaluddin, 1968; Jalaluddin, 1969; Petersen, 1970; Sagara, 1973; Petersen, 1975; Widden and Parkinson, 1975). The macromycetes often follow a distinct succession, beginning with species of *Anthracobia*, followed by *Ascobolus carbonarius* Karst. and *Coprinus angulatus* Lloyd among others, and culminating with species of *Octospora* (Petersen, 1970).

Evidence derived from autecological studies suggests that such pyrophilous fungi may thrive because of their response to one or more of four fire-induced modifications. First, heat decreases or completely eliminates the normal mycoflora. As a consequence, pyrophiles that are otherwise often poor competitors can flourish (El-Abyad and Webster, 1968a; Widden and Parkinson, 1975). Second, heat pretreatment greatly stimulates spore germination in some of these fungi. Often this is accompanied by a rapid growth rate permitting them to fill open niches ahead of the remaining slower-growing residents (Jalaluddin, 1967; El-Abyad and Webster, 1968a,b; Petersen, 1970; Widden and Parkinson, 1975). Third, increased alkalinity of the soil may reduce the intensity of the competition by acidophilous species, may select for forms that tolerate a high pH, or may enhance germination, subsequent outgrowth, and/or fruiting (El-Abyad and Webster, 1968a; Petersen, 1970). Fourth, such pyrophilous fungi may be tolerant of or stimulated by products produced through incomplete combustion of organic matter that are toxic to the remaining normal mycoflora. Widden and Parkinson (1975) demonstrated that aqueous extracts of burned material reduce or inhibit spore germination and growth in liquid culture of many test fungi including species of *Penicillium* and *Trichoderma* whereas others such as *Cylindrocarpon destruc-*

tans (Zins.) Scholten grow faster in their presence. These investigators further suggest that recolonization of the upper soil layers by resident fungi will be slowed or prevented until these toxic materials are reduced through leaching.

The magnitude of the response of OD communities to burning is directly related to the severity of the fire. When the heat is held in check by environmental factors, only minor and very transitory changes can be expected to occur. In contrast, an intense fire alters the soil habitat and reduces the number of resident fungi, thus providing conditions that encourage the successional development of communities of uniquely adapted pyrophilous forms. These will eventually be replaced as regeneration of the vegetation from subterranean buds, colonization of the area by herbaceous plants, and leaching ameliorates the edaphic environment and permits restoration of the original mycoflora--usually only after several growing seasons have elapsed (Wright and Bollen, 1961; Jalaluddin, 1969; Jorgensen and Hodges, 1970; Petersen, 1970).

F. Wind Erosion

On suitable sites, movement of sandy soil by wind results in the development of new vegetational types in either blowouts from which the sand is removed or on dunes or deserts upon which it is deposited. In Wisconsin, sand barrens may result from the plowing of dry-mesic prairies and sand blowouts (holes in the plant cover) may develop under excessive wind erosion (Curtis, 1959). These blowout hollows continue to increase in area and depth as erosion proceeds. The parent material often contains, in addition to sand, small stones and gravel that are too heavy to be picked up by the wind. After the hollow has reached a certain depth, a sufficient quantity of these stones is concentrated on the surface, protecting the underlying sand from further erosion. These blowout areas differ significantly from the undisturbed xeric prairie. Vegetationally, the heavy cover of native grasses and herbs has been replaced by a sparsely distributed group of plants dominated by the xerophytic evergreen shrub *Hudsonia tomentosa* and to a lesser extent by the three-awn grass (*Aristida basiramea*) and several sedges in the genus *Cyperus* (Curtis, 1959). Edaphically, the environmental conditions are similar to those encountered in the deserts of the southwestern United States and northern Africa. The blowout areas have an almost complete lack of available nutrients, a very low surface water supply, and extremely high evaporation rates. Soil temperatures at the surface in the summer may reach values of 60° to 70°C (Curtis, 1959; Phelps, 1973).

Biomass production by the OD community from undisturbed xeric prairies is moderate and species diversity somewhat lower than in the mesic prairies (Orpurt and Curtis, 1957). The community is dominated by members of the Moniliaceae. Dematiaceous fungi account for less than 10% of the isolates, and ascomycetes are very uncommon. Several taxa, including *Aspergillus fumigatus* Fres., *A. terreus* Thom, *Penicillium funiculosum* Thom, *P. lilacinum* Thom, and *Spicaria violacea* (*Paecilomyces marquandii* (Massee) Hughes), reach their maximum frequencies in this segment of the prairie continuum. Other species, such as *Fusarium oxysporum* Schlecht. emend Snyder & Hansen, also are widespread (Orpurt and Curtis, 1957).

The prairie OD community is greatly modified in the wind-eroded sand blowout (Phelps, 1973). Biomass is reduced tenfold and species diversity is low. Dematiaceous and sphaeropsidaceous taxa (55% of the isolates) are abundant. Ascomycetes, especially species of *Chaetomium*, increase to about 10%, and mucoraceous fungi are conspicuous by their absence. Several taxa, including *Aspergillus fumigatus*, *Alternaria humicola* Oud., *Penicillium restrictum* Gilman & Abbott, *Cladosporium* sp., and *Arthrinum phaeospermum* (Corda) M. B. Ellis, are widely distributed, and the populations when compared exhibit a high degree of similarity (60-73%). In composition, the community resembles those isolated from edaphically similar areas including bare soil (Wicklow, 1973), sand dunes (Brown, 1958), and sandbars (Gochenaur and Backus, 1967).

The severity of the environmental conditions appears to favor microfungi with certain morphological and/or physiological features. First, taking the population as a whole, less energy and nutrients are expended on reproductive structures since these are on the average simpler and more reduced than those observed in the undisturbed OD community. Second, propagule mass is high even though the number is reduced. As a consequence, cell walls can be thicker and storage materials greater in quantity. Third, the majority produce assimilative and/or reproductive structures with melanized walls that serve to protect cellular organelles from ultraviolet-radiation damage (Sussman, 1968; Zhdanova et al., 1974) and possibly dehydration (Zhdanova and Pokhodenko, 1973). Finally, certain taxa, notably *A. fumigatus*, can tolerate the high temperatures that develop in insolated soils (Gochenaur, 1975).

IV. SUMMARY

When "climax" ecosystems are severely disturbed, as occurs with clear-cutting, ionizing radiation, fire, and wind erosion, a sequence of events transpires that is predictable in broad outline. Typically, damage or elimination of the higher plants creates a more rigorous soil environment. Humidity and moisture decrease as erosion, surface runoff, and evaporation increase. Temperatures on the average become warmer as opening or removal of the canopy allows greater light penetration and insolation. As dead roots and other materials become available, nutrient levels increase. Such positive effects are short lived, however. Eventually, nutrients are both qualitatively and quantitatively reduced as enrichment of the soil with root and foliage exudates and litter leachates ceases and readily hydrolyzable substrates are utilized, leaving behind more recalcitrant materials.

Although the response of individual microfungi to these environmental modifications varies, at a community level certain generalizations can be drawn and predictions made since the different disturbances appear to produce parallel modifications in the structure, diversity, and composition of the community. Some or all of the following alterations may occur after acute disturbance of an ecosystem.

1. *Community biomass will change.* Immediately after perturbation it is common for progagule numbers to be lower. This will be followed by an increase of variable magnitude as reduction in competition permits survivors and/or immigrant colonizers to exploit the higher nutrient levels occurring in the area. Normally such an increase is transitory and biomass can be expected to decline as the variety of resources dwindle and quantity is reduced. It will remain low until plant recolonization begins and then can be expected to return to its original level with restoration of the area.

2. *Community organization will not be altered.* Probably all ecosystems contain a few microfungi that will respond favorably to perturbation and as a result come to dominate the new community. As a consequence, it seems likely that distribution of the isolates among the taxa will still follow a Raunkiaer pattern and replicate populations when compared will still show a high degree of similarity.

3. *Community composition and structure will be modified.* With deterioration of the environment, different species assume dominance and taxa new to the area appear. A decline in the importance of mucoraceous and moniliaceous isolates is expected, with their place being taken by ascomycetes and melanized taxa. Some of these may possess physiological and/or nutritional adaptations allowing them to compete more successfully or possibly merely survive for longer periods of time in the modified environment. A trend towards reduction in the stature and complexity of the individual may also appear. Such a progression is logical, since it represents a decreased investment in energy and nutrients on the part of the fungus while the reduction in surface area that accompanies reduced stature will minimize water loss by evaporation.

4. *Other survival strategies will assume importance.* The production of large numbers of small unicellular spores seems to be the principal survival strategy employed by members of the undisturbed community. Under perturbation the reverse is often true. That is, the mass of a propagule is increased with a corresponding decrease in the number produced. Greater mass may be accompanied by thicker, less permeable cell walls and larger deposits of reserved materials. As a consequence of this trend, individuals producing multicellular conidia, bulbils, sclerotia, and resting spores may become more common in the community.

5. *Species diversity will decrease.* As environmental quality, microhabitats, and nutrient diversity are reduced, it is likely that fewer taxa will be supported and species diversity will therefore be less.

6. *Biochemical diversity of the community will decrease.* Under disturbance, nutritionally flexible decomposers able to degrade a variety of biopolymers are likely to increase at the expense of the more fastidious members of the community. These generalists will possess a distinct advantage over the specialists because they can consume a diversity of nutrients, including more recalcitrant ones.

7. *Community stability will be reduced.* The existence of the new community that develops under perturbation will be transitory since it represents but a first

stage in a succession back to the original community. It will persist only as long as the conditions that brought it into being remain.

As Woodwell (1970) points out, the potential of a site for supporting life depends heavily upon the pool of nutrients supplied by the vegetation. When the latter is destroyed reduction of the system as a whole occurs. The evidence reported here strongly suggests that that part of the system represented by the opportunistic decomposers suffers a similar reduction and loss of structure. These changes, while variable at the level of the individual, produce broadly predictable ecological patterns when the community as a whole is considered.

REFERENCES

Ahlgren, I. F. (1974). The effect of fire on soil organisms. In *Fire and Ecosystems*, T. T. Kozlowski and C. E. Ahlgren (Eds.). Academic Press, New York, pp. 47-72.

Ahlgren, I. F., and Ahlgren, C. E. (1960). Ecological effects of forest fires. *Bot. Rev. 26*: 483-533.

Ahlgren, I. F., and Ahlgren, C. E. (1965). Effects of prescribed burning on soil microorganisms in a Minnesota jack pine forest. *Ecology 46*: 304-310.

Alexander, M. (1977). *Introduction to Soil Microbiology*, 2nd ed. Wiley, New York.

Ayanaba, A., Tuckwell, S. B. and Jenkinson, D. S. (1976). The effects of clearing and cropping on the organic reserves and biomass of tropical forest soils. *Soil Biol. Biochem. 8*: 519-525.

Borman, F. H., Likens, G. E., Fisher, D. W., and Pierce, R. S. (1967). Nutrient loss accelerated by clear-cutting of a forest ecosystem. *Science 159*: 882-884.

Brown, J. C. (1958). Soil fungi of some British sand dunes in relation to soil type and successions. *J. Ecol. 46*: 641-664.

Christensen, M. (1969). Soil microfungi of dry to mesic conifer-hardwood forests in northern Wisconsin. *Ecology 50*: 9-27.

Christensen, M., Whittingham, W. F., Novak, R. O. (1962). The soil microfungi of wet-mesic forests in southern Wisconsin. *Mycologia 54*: 374-388.

Christensen, N. L., and Muller, C. H., (1975). Effects of fire on factors controlling plant growth in *Adenostoma* chaparral. *Ecol. Monogr. 45*: 29-55.

Chu, M., and Stephen, R. (1967). A study of the free-living and root-surface fungi in cultivated and fallow soils in Hong Kong. *Nova Hedwigia 14*: 301-311.

Coleman, D. C. (1976). A review of root production processes and their influence on soil biota in terrestrial ecosystems. In *The Role of Terrestrial and Aquatic Organisms in Decomposition Processes*, J. M. Anderson and A. Macfayden (Eds.). Blackwell Sci. Publns., Oxford, England, pp. 417-434.

Curtis, J. T. (1959). *The Vegetation of Wisconsin*. Univ. of Wisconsin Press, Madison.

Davis, R. J., Sheldon, V. L., and Auerbach, S. I. (1956). Lethal effects of gamma radiation upon segments of a natural microbial population. *J. Bacteriol. 72*: 505-510.

DeBoois, H. and Jansen, E. (1976). Effects of nutrients in throughfall rainwater and of leaf fall upon fungal growth in a forest soil layer. *Pedobiologia 16*: 161-166.

Domsch, K. H., and Gams, W. (1969). Variability and potential of a soil fungus population to decompose pectin, xylan and carboxymethyl-cellulose. *Soil Biol. Biochem. 1*: 29-36.

Dooley, M., and Dickinson, C. H. (1970). The microbiology of cut-away peat. II. The ecology of fungi in certain habitats. *Plant Soil 32*: 454-467.

El-Abyad, M. S. H., and Webster, J. (1968a). Studies on pyrophilous discomycetes. I. Comparative physiological studies. *Trans. Brit. Mycol. Soc. 51*: 353-367.

El-Abyad, M. S. H., and Webster J. (1968b). Studies on pyrophilous discomycetes. II. Competition. *Trans. Brit. Mycol. Soc. 51*: 369-375.

Ellis, D. H., and Griffiths, D. A. (1975). Melanin deposition in the hyphae of a species of *Phomopsis*. *Can. J. Microbiol. 21*: 442-452.

Filer, T. H., Jr. (1975). Mycorrhizae and soil microflora in a green-tree reservoir. *Forest Sci. 21*: 36-39.

Flanagan, P., and Scarborough, A. (1974). Physiological groups of decomposer fungi on tundra plant remains. In *Soil Organisms and Decomposition in Tundra*, A. Holding, O. Heal, S. Maclean, Jr., and P. Flanagan (Eds.). Tundra Biome Steering Committee, Stockholm, pp. 159-181.

Franz, E., and Woodwell, G. (1973). Effects of chronic gamma irradiation on the soil algal community of an oak-pine forest. *Rad. Bot. 13*: 323-329.

Gams, W., and Domsch, K. H. (1969). The spatial and seasonal distribution of microscopic fungi in arable soils. *Trans. Brit. Mycol. Soc. 52*: 301-308.

Gill, D. (1973a). A restricted habitat for mushrooms in the MacKenzie River Delta, Northwest Territories. *Can. Field Nat. 87*: 53.

Gill, D. (1973b). Ecological modifications caused by the removal of tree and shrub canopies in the Mackenzie Delta. *Arctic 26*: 95-111.

Gill, D. (1974). Influence of waterfowl on the distribution of *Beckmannia syzigachne* in the Mackenzie River Delta, Northwest Territories. *J. Biogeog. 1*: 63-69.

Gochenaur, S. E. (1964). The soil microfungi of willow-cottonwood forests in southern Wisconsin. Ph.D. thesis, University of Wisconsin.

Gochenaur, S. E. (1975). Distributional patterns of mesophilous and thermophilous microfungi in two Bahamian soils. *Mycopathologia 57*: 155-164.

Gochenaur, S. E. (1978). Fungi of a Long Island oak-birch forest. I. Community organization and seasonal occurrence of the opportunistic decomposers of the A horizon. *Mycologia 70*: 975-994.

Gochenaur, S. E., and Backus, M. P. (1967). Mycoecology of willow and cottonwood lowland communities in southern Wisconsin. II. Soil microfungi in the sandbar willow stands. *Mycologia 59*: 893-901.

Gochenaur, S. E., and Whittingham, W. F. (1967). Mycoecology of willow and cottonwood forests. *Mycopathol. Mycol. Appl. 33*: 125-139.

Gochenaur, S. E., and Woodwell, G. M. (1974). The soil fungi of a chronically irradiated oak-pine forest. *Ecology 55*: 1004-1016.

Griffin, D. M. (1963). Soil moisture and the ecology of soil fungi. *Biol. Rev. 38*: 141-166.

Griffin, D, M. (1972). *Ecology of Soil Fungi*. Syracuse Univ. Press, Syracuse, New York.

Hankin, L., and Anagnostakis, S. (1975). The use of solid media for detection of enzyme production by fungi. *Mycologia 67*: 597-607.

Hanssen, J. F., and Goksoyr, J. (1975). Biomass and production of soil and litter fungi at Scandinavian tundra sites. In *Fennoscandian Tundra Ecosystems*, Pt. 1: *Plants and Microorganisms*, F. W. Wielgolaski (Ed.). Springer-Verlag, New York, pp. 239-243.

Holding, A. J., and Collee, J. G. (1971). Routine biochemical tests. In *Methods in Microbiology*, J. R. Norris and D. W. Ribbons (Eds.), Vol. 6, Pt. A. Academic Press, New York, pp. 1-32.

Holt, B. R. (1972). Radiosensitivity and distribution of common macrofungi in a chronically-irradiated oak-pine forest. *Rad. Bot. 12*: 339-342.

Horrill A. D., and Woodwell, G. M. (1973). Structure and cation content of a podzolic soil of Long Island, N.Y., seven years after destruction of the vegetation by chronic gamma irradiation. *Ecology 54*: 439-444.

Ioannou, N., Schneider, R. W., and Grogan, R. G. (1977). Effect of flooding on the soil gas composition and the production of microsclerotia by *Verticillium dahliae* in the field. *Phytopathology 67*: 651-656.

Jalaluddin, M. (1967). Studies on *Rhizina undulata*. I. Mycelial growth and ascospore germination. *Trans. Brit. Mycol. Soc. 50*: 449-459.

Jalaluddin, M. (1968). Fire as an agent in the establishment of a plant disease. *Pakinstan J. Sci. 20*: 42-44.

Jalaluddin, M. (1969). Microorganic colonization of forest soil after burning. *Plant Soil 30*: 150-152.

Jorgensen,, J. R., and Hodges, C. S. (1970). Microbial characteristics of a forest soil after twenty years of prescribed burning. *Mycologia 62*: 721-726.

Kashkina, G., and Abaturov, Y. (1969). Influence of chronic irradiation of the microflora of the soil of a gamma field. *Radiobiology 9*: 185-188.

Kelting, R. W. (1957). Winter burning in central Oklahoma grassland. *Ecology 38*: 520-522.

Kessler, K. J., and Blank, R. W. (1972). Endogone sporocarps associated with sugar maple. *Mycologia 64*: 634-638.

Lemaire, J. M., and Jovan, B. (1966). Modifications microbiologiques entraînées par la mise en culture de sols nouvellement défriches: Incidences sur l'installation de l'*Ophiobolus graminis* Sacc: (= *Linocarpon cariceti* B. et Br.) et du*Streptomyces scabies* (Thaxt.) Waksman et Henrici. *Ann. Epiphytol. 17*: 313-333.

McNaughton, S. J., and Wolf, L. L. (1973). *General Ecology*. Holt, Rinehart and Winston, New York.

Mallik, M. A. B., and Rice, E. L. (1966). Relation between soil fungi and seed plants in three successional forest communities in Oklahoma. *Bot. Gaz. 127*: 120-127.

Marks, P. L., and Borman, F. H. (1972). Revegetation following forest cutting: Mechanisms for return to steady-state nutrient cycling. *Science 176*: 914-915.

Matile, P. (1975). *The Lytic Compartment of Plant Cells*. Springer-Verlag, New York.

Menon, S. K., and Williams, L. E. (1957). Effect of crop, crop residues, temperature and moisture on soil fungi. *Phytopathology 47*: 559-564.

Meyer, F. H. (1970). Abbau von Pilzmycel im Boden. *Z. Pflanzen. Bodenkunde 127*: 193-199.

Miller, D. E., and Burke, D. W. (1977). Effect of temporary excessive wetting on soil aeration and *Fusarium* root rot of beans. *Plant Disease Rep. 61*: 175-179.

Mirchink, T. G., Kashkina, G. B., and Abaturov, Y. D. (1972). The resistance of fungi with various pigments to gamma radiation. *Microbiology 41*: 83-86.

Mitchell, R., and Alexander, M. (1960). Microbiological changes in flooded soils. *Bacteriol. Proc.* p. 28.

Morrall, R. A. A., and Vanterpool, T. C. (1968). The soil microfungi of upland boreal forest at Candle Lake, Saskatchewan. *Mycologia 60*: 642-654.

Mueller-Dombois, D. (1974). *Aims and Methods of Vegetation Ecology*. Wiley, New York.

Nagel-deBoois, H. M., and Jansen, E. (1967). Hyphal activity in mull and mor of an oak forest. In *Progress in Soil Biology*, O. Graff and J. Satchell (Eds.). Vieweg Braunschweig, Germany, pp. 27-35.

Novak, R. O., and Whittingham, W. F. (1968). Soil and litter microfungi of a maple-elm-ash floodplain community. *Mycologia 60*: 776-787.

Old, K. (1967). Effects of natural soil on survival of *Cochliobolus sativus*. *Trans. Brit. Mycol. Soc. 50*: 615-624.

Orpurt, P. A., and Curtis, J. T. (1957). Soil microfungi in relation to the prairie continuum in Wisconsin. *Ecology 38*: 628-637.

Pacheco de Lira, Nelly. (1971). Especies de *Absida* do solo do Maranhao: Aspectos ecologicos. *Univ. Fed. Pernambuco Inst. Micol. Publ. 657*: 3-6.

Parkinson, D., Gray, T. R. G., and Williams, S. T. (1971). *Methods for Studying the Ecology of Soil Micro-organisms*. Blackwell Sci. Publns., Oxford, England.

Peet, R. K. (1974). The measurement of species diversity. *Ann. Rev. Ecol. Syst. 5*: 285-307.

Petersen, P. M. (1970). Danish fireplace fungi. *Dansk Bot. Ark. 27*: 1-97.

Petersen, P. M. (1975). Fireplace fungi in an arctic area: Middle West Greenland. *Friesia 10*: 270-280.

Phelps, J. W. (1973). Microfungi in two Wisconsin sand blows. *Trans. Brit. Mycol. Soc. 61*: 386-390.

Rambelli, A., Puppi, G., Bartoli, A., and Albonetti, S. G. (1973). Deuxième contribution à la connaissance de la microflore fongique dans les sols de Lamto en Côte d'Ivoire. *Rev. Ecol. Biol. Sol 10*: 13-18.

Ruscoe, Q. W. (1973). Changes in the mycofloras of pasture soils after long-term irrigation. *N. Z. J. Sci. 16*: 9-20.

Sagara, N. (1973). Proteophilous fungi and fireplace fungi: A prelimary report. *Trans. Mycol. Soc. Japan 14*: 41-46.

Santerre, J. (1966). Absence apparente de l'organisme de la tache argentée des pommes de terre, *Helminthosporium atrovirens*, dan les sols nouvellement défriches. *Can. J. Plant Sci. 46*: 647-652.

Scott, A. D., and Evans, D. D. (1955). Dissolved oxygen in saturated soil. *Proc. Soil Sci. Amer. 19*: 7-12.

Skou, J. P. (1962). Studies on the microflora of soil under chronic irradiation. *Gen. Microbiol. 28*: 521-530.

Skou, J. P. (1964). *Aureobasidium pullulans*: A common and very radio resistant fungus in fresh fruit and vegetables. *IVA Medd. (Ingenjoersvetenskapsakademien) 138*: 63-70.

Smith, A. M. (1976). Ethylene production by bacteria in reduced microsites in soil and some implications to agriculture. *Soil Biol. Biochem. 8*: 293-298.

Söderström, B. E. (1975). Vertical distribution of microfungi in a spruce forest soil in the south of Sweden. *Trans. Brit. Mycol. Soc. 65*: 419-425.

Sussman, A. S. (1968). Longevity and survivability of fungi. In *The Fungi*, G. C. Ainsworth and A. S. Sussman (Eds.), Vol. III. Academic Press, New York, pp. 447-486.

Swift, M. J. (1976). Species diversity and the structure of microbial communities in terrestrial habitats. In *The Role of Terrestrial and Aquatic Organisms in Decomposition Processes*, J. M. Anderson and A. Macfayden (Eds.). Blackwell Sci. Publns., Oxford, England, pp. 185-222.

Tabak, H. H., and Cooke, W. B. (1968). The effects of gaseous environments on the growth and metabolism of fungi. *Bot. Rev. 34*: 126-252.

Taha, S., Mahmoud, S., and Moubarek, M. (1970). Effect of reclamation of sandy soil on some chemical and microbiological properties. *Plant Soil 32*: 282-292.

Tichelaar, G., and Vruggink, H. (1975). A survey of fungi and actinomycetes colonizing the newly reclaimed soil of the polder Zuidelijk Flevoland. *Plant Soil 42*: 241-254.

Tresner, H. D., Backus, M. P., and Curtis, J. T. (1954). Soil microfungi in relation to the hardwood forest continuum in southern Wisconsin. *Mycologia 46*: 314-333.

Viro, P. J. (1974). Effects of forest fire on soil. In *Fire and Ecosystems*, T. T. Kozlowski and C. E. Ahlgren (Eds.). Academic Press, New York, pp. 7-45.

Walter, H. (1975). Besonderheiten des Stoffkreislaufes einiger terrestrischer Oeko-systeme. *Flora 164*: 169-184.

Warcup, J. H. (1955). On the origin of colonies of fungi developing on soil dilution plates. *Trans. Brit. Mycol. Soc. 38*: 298-301.

Whittaker, R. H. (1953). A consideration of climax theory: The climax as population and pattern. *Ecol. Monogr. 23*: 41-78.

Wicklow, D. T. (1973). Microfungal populations in surface soils of manipulated prairie stands. *Ecology 54*: 1302-1310.

Wicklow, D. T. (1975). Fire as an environmental cue initiating ascomycete develop-ment in a tallgrass prairie. *Mycologia 67*: 852-862.

Widden, P., and Parkinson, D. (1975). The effects of a forest fire on soil micro-fungi. *Soil Biol. Biochem. 7*: 125-138.

Witkamp, M. (1970). Aspects of soil microflora in a gamma-irradiated rain forest. In *A Tropical Rain Forest: A Study of Irradiation and Ecology at El Verde, Puerto Rico*, H. T. Odum and R. F. Pigeon (Eds.). U.S. Atomic Energy Commission, Div. Tech. Information Exten., Oak Ridge, Tenn., pp. F29-F33.

Witkamp, M., Frank, M., and Shoopman, J. (1966). Accumulation and biota in a pioneer ecosystem of kudzu vine at Copperhill, Tennessee. *J. Appl. Ecol. 3*: 383-391.

Woodwell, G. M. (1967). Radiation and the patterns of nature. *Science 156*: 461-470.

Woodwell, G. M. (1970). Effects of pollution on the structure and physiology of eco-systems. *Science 168*: 429-433.

Woodwell, G. M., and Gannutz, T. (1967). Effects of chronic gamma irradiation on lichen communities of a forest. *Amer. J. Botany 54*: 1210-1215.

Woodwell, G. M., and Rebuck, A. (1967). Effects of chronic gamma radiation on the structure and diversity of an oak pine forest. *Ecol. Monogr. 37*: 53-69.

Wright, E. W., and Bollen, W. B. (1961). Microflora of Douglas-fir forest soil. *Ecology 42*: 825-828.

Zhdanova, N. N., and Pokhodenko, V. D. (1973). The possible participation of a mel-anin pigment in the protection of the fungus cell from desiccation. *Microbiology 42*: 753-757.

Zhdanova, N. N., Vasilevskaya, A. I., and Gavryushina, A. I. (1974). Radioresistance of dark-colored hyphomycetes (Dematiaceae). *Mikol. Fitopatol. 8*: 8-12. [In Rus-sian.]

Chapter 25

ECOLOGICAL APPROACHES TO THE USE OF LICHENIZED FUNGI
AS INDICATORS OF AIR POLLUTION

T. H. Nash III and L. Sigal [*]

Arizona State University
Tempe, Arizona

I. INTRODUCTION

One of the major challenges of modern ecology is to develop quantitative measures of community response to perturbations. Some perturbations allow for the maintenace of greater species richness in a community than would occur without perturbations, whereas other perturbations may limit community richness. For example, hurricanes that cause windfalls in a forest create a patchy habitat in which more species exist because successional species occupy the patches in close proximity to members of the more mature and unperturbed community. In such cases the intensity of the perturbation is reflected by the frequency and size of such patches, except in the extreme case, such as those created by a crown fire, where the perturbation leads to the retrogression of the whole community. In the case of adverse perturbations, such as the application of posticides, community response may include simplification of community structure by the elimination of one or more species, and also more subtle alterations such as shifts in reproductive potential or growth patterns of individuals composing the community. Thus cognizance of community response to perturbations is essential to an ecologist in delimiting and quantifying the integrated effects of environmental factors on organisms and communities.

With the advent of modern technology there has been a proliferation of man-caused perturbations which have had significant impacts on ecosystems. Air pollution is a prominent example of such a technology-related perturbation. Collectively the lichenized fungi are probably the group of organisms most sensitive to air pollution, although differential sensitivity among different species is generally recognized (Nash, 1976a). Because of this differential sensitivity to air pollution, it is possible to recognize areas receiving different levels of air pollution on the basis of lichen community characteristics. Thus the use of lichens as air pollution indicators should involve a community ecology approach as opposed to single-species studies. However, such ecological studies need to be carefully stratified in a statistical sense to differentiate among a number of potential causes of community variation.

Present affiliation: Environmental Sciences Division, Oak Ridge National Laboratory, Oak Ridge, Tennessee

The evidence which supports the hypothesis that lichens are sensitive indicators of air pollution has accumulated over the last 130 years. In the 1800s several lichenologists noted that lichens were disappearing from urban areas, and they suggested that the cause might be due to the smoke (air pollution) associated with industrialization (Nylander, 1866; Leighton, 1868; Arnold, 1892). Subsequently the phenomenon was demonstrated with varying degrees of precision in cities in many parts of the world. Similar patterns were also noted around isolated factories (Skye, 1958; Rao and LeBlanc, 1967; Nash, 1972), where factors other than air pollution are less likely to be present and thus account for part of the observed pattern. Over the last two decades a series of laboratory studies has been performed to document the response of various lichens to specific air pollutants. Although at first unrealistically high concentrations were employed, more recent studies have actually measured lichen sensitivity to air pollutant concentrations in urban and industrial environments. It is also of practical importance that in areas where air pollution emissions have ceased, renewed lichen growth is demonstrable (Schönbeck, 1969; Skye and Hallberg, 1969). Also renewed growth of at least some species has been demonstrated in urban environments after a reduction of emissions (Jürging, 1975; Seaward, 1976).

II. THE NATURE OF LICHENS AND THEIR IMPORTANCE
     TO AIR POLLUTION

Lichenized fungi are symbiotic organisms, usually described as a mutualistic relationship between algae and fungi. Because of the prevalence of fungal sexual reproduction among lichens, classification is based on the fungus. Almost all lichens are Ascomycetes although a few belong to the Basidiomycetes and Fungi Imperfecti (Hale, 1967). More than half of all the described Ascomycetes are lichens, which indicates the success of symbiosis in that group of organisms. Lichens are predominantly terrestrial organisms, occurring both as epiphytes on trees and shrubs and as colonizers of mineralized soil, detritus, and bare rocks. In contrast to higher plants, lichens possess no root absorptive system and consequently are assumed to be largely dependent on atmospheric sources of moisture and nutrients for their nutrition and growth. Metabolic activity is confined to periods when the lichens are moist (Kershaw, 1972; Lange et al., 1975), and alternation of periods of wetness (activity) and dryness (inactivity) is apparently prerequisite for the maintenance of the symbiotic state (Ahmadjian and Heikkilä, 1970). Because of their dependence on the low levels of atmospheric nutrients, lichens have developed an efficient mechanism for retaining nutrients (Brown, 1976), and it is partially for this reason that lichens are sensitive to atmospheric pollutants. In contrast, higher plants which gain their nutrients from soils are at least partially sheltered from air pollution fallout due to the buffering capacity of the soil and the relative immobility of many pollutants in soils. Another important factor is that lichens accumulate pollutants over a number of years because they are relatively long-lived organisms, with life spans of tens to hundreds of years being common and some even living for thousands of years. Their perennial habit and slow growth means that they have no mechanism, such as leaf shedding in deciduous higher plants, for discarding pollution-laden parts.

III. ECOLOGICAL APPROACHES TO THE USE OF LICHENS AS AIR POLLUTION MONITORS

From an historical perspective, the sensitivity of lichens to air pollution was initially documented around large pollution sources where air pollution had been present for many years. In such situations a zonation effect was readily demonstrable by the absence of most lichens near the pollution source. As one proceeded centrifugally from the pollution source, lichen communities became increasingly rich until a normal flora was encountered. Serander (1912, 1926) observed such a phenomenon around Stockholm, Sweden, and he proposed the terminology *lichen desert* for the area around the pollution source, *struggle zone* for the area where differential lichen response was most clearly recognizable, and *normal zone* where the expected or normal complement of species was found. This terminology was rapidly adopted and used in scores of subsequent studies around city centers and other pollution sources. The delimitation of such zones is largely based on the presence or absence of species and thus reflects a measure of species richness in the lichen community. The use of species richness is a simple, direct approach which lends itself to graphic presentation. In recent years such studies have been expanded to cover large geographic regions. For example, Hawksworth and Rose (1970) established a semiquantitative scale relating the presence and absence of common British lichens to seasonal sulfur dioxide values. Although the scale was initially established on an intuitive approach based on extensive field experience, it was subsequently largely confirmed by a multivariate regression analysis (Creed et al., 1973). Today detailed distribution maps of critical species are available from the British Lichen Society. Similarly, a large regional study for the Netherlands by de Wit (1976) is based on species richness studies.

Reliance on species richness alone does not provide a particularly sensitive measure of community response. For example, a species' abundance or vitality may be affected prior to its outright elimination. By quantifying such parameters, one may examine in greater detail the relation of community variation to the perturbation under study. To achieve this goal, species diversity calculations, phytosociological analyses, or standard plant community analyses may be used.

The modern concept of species diversity includes the concept of relative abundance of individuals within species categories (species evenness) combined with the number of species present (species richness). For example, Nash (1972) used the Shanon-Wiener index (see MacArthur, 1965), a commonly used measure of species diversity, and found values of 0.00 to 0.84 for the most polluted area and values of 2.01 to 3.73 in the control area. Part of the same data set is presented graphically in Fig. 1, using the dominance diversity curves developed by Whittaker (1965). In constructing these curves, species are ordered with respect to their abundance. By

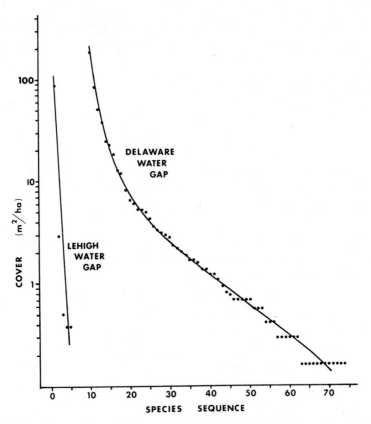

Fig. 1 Dominance diversity curves (Whittaker, 1965) for lichens sampled in ground quadrats in Delaware Water Gap (control) and Lehigh Water Gap (near a zinc plant). [Reproduced from Nash (1975) with permission of the Ecological Society of America.]

glancing at the graphs, one can see not only that there are more species in the control area (Delaware Water Gap) than in the polluted site (Lehigh Water Gap) but also that there are more common, intermediately abundant, and rare species in the control area than in the polluted area.

Two phytosociological approaches have been developed which provide a more rigorous definition of zonal boundaries around air pollution sources. The first was developed by De Sloover and LeBlanc (1968) and is called the index of atmospheric purity (IAP). Before calculating the IAP, all the epiphytes present at a site are listed and their frequency coverage is estimated. Then the IAP is calculated according to the following formula:

$$IAP = \sum_{n}^{1} \frac{Qf}{10}$$

In this case n is the total number of epiphyte species, f is the frequency-coverage score of each species expressed by a number on a numerical scale determined by the investigator, and Q is the "resistance" factor or ecological index of each species.

The Q value for a species is found by adding the numbers of its companion species present at all the investigated sites and then dividing this total by 10 to reduce it to a small manageable number. Zones which reflect different degrees of epiphytic richness may then be mapped by constructing isometric lines among the plotted sites which were sampled. In contrast to the assumption originally made by the authors, such IAP zones do not necessarily reflect the long-range effect of air pollution on the species studied. Rather, these zones reflect underlying variability in the epiphytic populations. In terms of air pollution studies, the technique has its greatest credibility when the zones can be directly related to ambient levels of air pollutants, as has been done by LeBlanc et al. (1972a) in Canada around the nickel smelters at Sudbury, Ontario, and by Lötschert and Köhn (1977) in a German industrial area. Furthermore, the technique should not be applied casually, as the underlying assumption is that the sites had been carefully stratified before sampling to reflect ecological homogeneity. For example, the tree species should be the same throughout the study area, and the sampled trees should occur in similar habitats and have similar structures. Even when the study has been stratified as carefully as possible, the investigators should be congnizant of other possible causes which might explain part of the observed variation pattern.

A second phytosociological approach involves use of the index of poleotolerance (IP) developed by Trass (1968, 1973). This index has been used extensively in eastern Europe (Liiv, 1975; Martin, 1976). In Estonia, Trass has been able to relate his zones to approximate annual levels of ambient sulfur dioxide. IP is calculated according to the following formula:

$$IP = \sum^{n} \frac{a_i - c_i}{c_i} \qquad i = 1$$

In this case a is the poleotolerance value of each species on a scale of 1 to 10 based on field experience, and c stands for the cover degree of the given species. The C value is the sum of the cover degree of the species forming the community. Because of the subjectivity inherent in the estimation of the a values, other more objective approaches to the study of lichen communities are preferable.

Skorepa and Vitt (1976) discovered that, although the IAP approach does incorporate an estimate of frequency and cover in the calculations, ultimately the delimitation of zones is largely dependent on species differences among the zones. In other words the IAP zones are primarily a reflection of differences in species richness. Skorepa and Vitt studied the distribution of lichens in 21 stands around two gas plants in western Alberta, Canada. Because the gas plants had only operated for 3 years, distinct patterns or zones of lichen richness had not developed, although the lichens near the power plant were clearly impoverished. There were no differences among stands in lichen species richness, and the applicaiton of the IAP technique led to the assumption that there were few differences among stands. Consequently these investigators developed a procedure by which they estimated both the luxur-

iance and density of the species in each stand. An LD value for each species was
calculated as follows:

$$LD_j = \Sigma \ \frac{L_j D_j}{N} \ 10$$

Here LD_j is the luxuriance-density value for species j; L is the relative luxuriance
value on a scale of 1 to 5; D is the relative density on a scale of 1 to 5; and N is
the number of quadrats sampled. The summed value is multiplied by 10 to obtain a
manageable value, and the LD values of all the species in a given stand are summed.
With the LD values in hand Skorepa and Vitt ran a standard Bray and Curtis (1957)
ordination of the stands, using Beals's (1960) equations for axis distances and Sø-
renson's (1948) coefficient of community similarity. The ordination yielded a plot
of stands which primarily reflected a horizontal gradient in the luxuriance-diversity
index. Furthermore, use of the minimum variance algorhithm for constructing a den-
drogram according to Pritchard and Anderson's (1971) cluster analysis program yielded
a plot reflecting distinct differences among the stands. Thus by two independent
techniques differences were detectable among the stands where no differences were
detectable using the IAP procedure. Furthermore, by plotting the stands in relation
to the gas plants and by plotting a scaled value reflecting the stand's LD value
(Fig. 2), a pattern of low luxuriance-density near the gas plants was readily detectable.

The absence of a species from an area may, of course, be due to a variety of
reasons. In the case of rare species, chance distribution of reproductive propagules
may create apparent patterns in communities. Transplant experiments provide a means
for assaying whether the species is capable of surviving in areas where it was observed to be absent. If the species does survive in such an area, then the observed
pattern is probably spurious or at least unimportant in relation to pollution. On
the other hand, death of the transplants may result from a variety of causes. It is
absolutely essential to make transplants in control areas to validate the transplan-
tation procedure. Death or marked decline of such controls would, of course, invali-
date any further interpretation. However, if the only injured transplants occurred
in the heretofore void area, then it is reasonable to seek a causal mechanism. If
the pattern of injury corresponds to specific symptoms induced by control fumigations
with the air pollutant, then the air pollutant is reasonably implicated. However,
a number of injury symptoms are nonspecfic. The argument in relation to pollution
as a cause may be strengthened by demonstrating the presence of toxic ambient levels

Thus in future studies it is recommended that a more precise definition of patterns in lichen communities may be obtained by using the standard multivariate procedures employed by plant ecologists over the last two decades. Once the pattern has
been defined rigorously, the problem of assigning probable cause becomes of para-
mount importance. In this regard transplant studies as well as fumigation studies
with the species present will provide important data.

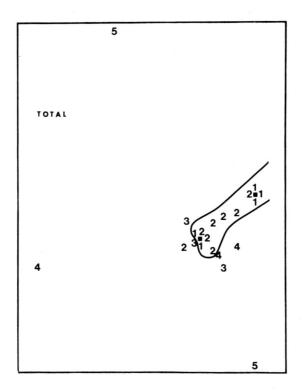

Fig. 2 Geographic relation of sampled stands to two gas plants (squares) in south-central Alberta, Canada. Each number corresponds to a luxuriance-density (LD) scale of 1 through 5, as derived by Skorepa and Vitt (1976), with 5 representing the greatest abundance and luxuriance. The solid line gives an estimate of the area of severe lichen damage. (Reproduced with permission of the aforementioned authors.)

of the air pollutant and, if possible, demonstrating accumulation of toxic materials by the declining transplants. Procedures for transplanting corticolous lichens were developed by Brodo (1961) and have been successfully applied in a number of areas (Brodo, 1966; Schönbeck, 1969; Nash, 1971). In the case of saxicolous lichens, Seaward (1976) has conducted the most careful studies.

IV. RESPONSE OF LICHENS TO SPECIFIC AIR POLLUTANTS

In the final sections of this paper, some of the evidence in relation to specific pollutants is briefly summarized to facilitate access to the literature and to provide examples of the application of the techniques.

A. Sulfur Dioxide

Sulfur is a ubiquitous element present in coal and many petroleum products. Combustion of these products results in the release of sulfur dioxide, a gas long known

for its phytotoxicity. Because of the need for winter heating, sulfur dioxide release is particularly high in northern latitudes, and in these areas the gas frequently has been implicated as the major cause of lichen decline (Hawksworth and Rose,
1976). One of the classic field studies was conducted by Rao and LeBlanc (1967) in
Wawa, Ontario, a small isolated town on the north shore of Lake Superior. An iron
sintering plant was a major source of sulfur dioxide. Because of the prevailing
north-northeasterly winds, the distribution of the gas, as reflected by soil sulfate
concentrations, was confined to a narrow corridor to the north-northeast where a
strong zonation pattern in lichen distribution was demonstrable (Fig. 3). Other studies have shown that lichens accumulate sulfur when present in atmospheres containing
sulfur dioxide (Olkkonen and Takala, 1975; Laaksovirta and Olkkonen, 1977). Sulfur
dioxide may affect lichens indirectly through acidification of the lichens' substrate
or directly by disrupting a variety of physiological processes. Skye (1968) and
others have shown that the buffering capacity of urban tree bark is markedly reduced,
as is reflected by a pH decrease of over 2.0 units. Under such acidic conditions,
lichens are demonstrably more sensitive to sulfur dioxide (Türk and Wirth, 1975). A
variety of physiological processes including photosynthesis, respiration, and maintenance of cell membrane integrity are adversely affected by sulfur dioxide (Nieboer
et al., 1976). Importantly, many of the recent lichen experimental studies with sulfur dioxide have been conducted at realistic sulfur dioxide concentrations and consequently the evidence implicating sulfur dioxide as a major cause of lichen decline
is very strong.

B. Hydrogen Fluoride

Although hydrogen fluoride is phytotoxic at much lower concentrations than sulfur
dioxide, it is not as universally important because its release is largely confined
to specific industrial point sources such as fertilizer plants, glassworks, ceramic
manufacturing, and aluminum and rare earth metal refining. The effects on the lichen
flora around aluminum factories (Martin and Jacquard, 1968; Gilbert, 1971; LeBlanc
et al., 1972b; Horntvedt, 1975), around a glass works (Steubing et al., 1976) and
around a titanium plant (Nash, 1971) are well demonstrated. One example of such a
pattern is given in Fig. 4, where the pattern of randomly chosen trees without *Physcia millegrana* corresponds closely to the presence of atmospheric fluoride, as indicated by an 8 month study with limed filter papers (Nash, 1969). Subsequent transplant studies repeatedly demonstrated death of *Physcia* and *Parmelia* species associated with fluoride uptake. From a combination of fumigation and transplant studies,
a critical fluoride accumulation level of 20-80 ppm was defined as toxic to the species utilized, and this threshold level was independently confirmed by Gilbert (1971)
for a very different area.

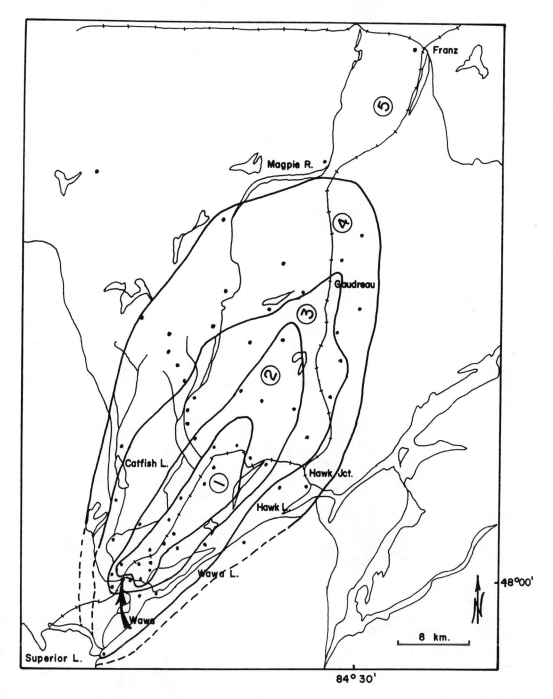

Fig. 3 Approximate boundaries of pollution zones in the Wawa, Ontario area as established by the number of epiphytic species and the concentration of soil sulfate (Rao and LeBlanc, 1967). Zone 1 has no epiphytes and sulfate exceeding 1.4 meq/100 g; zone 2, 1-5 epiphytes per site and 0.9-1.4 meq sulfate/100 g; zone 3, 5-15 epiphytes per site and 0.7-0.9 meq sulfate/100 g; zone 4, 15-30 epiphytes per site and 0.4-0.7 meq sulfate/100 g; zone 5, more than 30 epiphytes per site and soil sulfate less than 0.4 meq/100 g. (Reproduced with permission of the American Bryological and Lichenological Society and the aforementioned authors.)

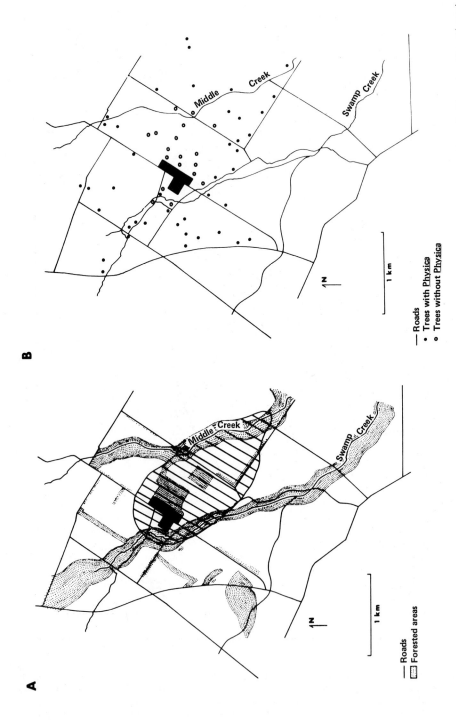

Fig. 4 Elevated fluoride levels (A, diagonal hatching), as determined by limed filter papers, in relation to the distribution of *Physcia millegrana* (B) adjacent to a titanium plant near Boyertown, Pennsylvania. [Adapted from Nash (1969) and subsequent unpublished work.]

C. Heavy Metals

Trace metals are released from a variety of industrial, manufacturing, and municipal sources. In recent years, the potential effect of long-term accumulation of trace metals is of increasing concern. Because of their well-established ability to accumulate elements (Lounamaa, 1956), lichens are useful organisms to study in relation to metal "fallout" patterns as well as for toxicity studies to the lichens themselves. The mechanism of the accumulation phenomenon is documented by Nieboer et al. (1977). Preliminary studies on the degree of toxicity of metal ions have been conducted by Nash (1975) and Puckett(1976).

A number of studies have shown accumulation of lichen metal contents in urban and industrial areas: Nieboer et al. (1972) for nickel near the Sudbury nickel smelters in Ontario, Canada; Laaksovirta et al. (1976) for lead in southern Finland; Gilbert (1976) for calcium around a cement plant; Schutte (1977) for chromium in the United States Midwest; Steinnes and Krog (1977) for mercury around a Norwegian industrial complex. However, none of these authors have clearly demonstrated toxicity of the elements accumulated. In contrast, Nash (1975) has demonstrated toxic concentrations of zinc and secondarily cadmium around a zinc factory in Pennsylvania. In this area a lichen impoverished zone extended 15 km west and 10 km east of the factory along an overlooking mountain ridge (Fig. 5). Although detectable, potentially toxic

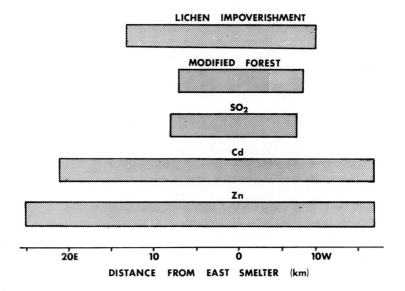

Fig. 5 The distribution of the lichen-impoverished zone (Nash, 1972) at Palmerton, Pennsylvania, in relation to the zones of modified forest (Jordan, 1975), to detectable levels of ambient sulfur dioxide (Nash, 1975), and to elevated levels of cadmium and zinc in surface soils (Buchauer, 1973). (Reproduced with permission of Springer-Verlag, New York.)

levels of sulfur dioxide were found adjacent to the factory, the distribution of ambient sulfur dioxide did not extend as far as the lichen-impoverished zone. The lack
of importance of sulfur dioxide was further substantiated by the lack of acidification of tree bark. In contrast heavy metal concentrations in the soil duff as high
as 135,000 ppm zinc and 1750 ppm cadmium (Buchauer, 1973) were found in areas where
no lichens grew. The concentration of zinc and cadmium decreased exponentially with
distance from the factory until background concentrations were found at 20-25 km east
and 16 km west of the factory. Physiological studies demonstrated that zinc and cadmium were toxic at 300-500 ppm. Since zinc was present in concentrations 100 times
that of cadmium and since zinc is approximately as toxic as cadmium, zinc which decreased to nontoxic levels at the boundaries of the lichen-impoverished zone was assumed to be the most important pollutant.

D. Oxidants

With the widespread use of motor vehicles, oxidant air pollution has become a major
problem. The term *oxidant* refers to a group of air pollutants which, in contrast to
the reducing agent sulfur dioxide, are strongly oxidizing agents. Ozone, which is
formed by a phytochemically mediated reaction between nitrogen oxides and various
hydrocarbons, is the most important phytotoxic air pollutant in the United States.
In addition, peroxyacetylnitrate (PAN) and its various derivatives have caused considerable agricultural damage. In contrast, nitrogen oxides are relatively unimportant at current concentrations.

Initial fumigation studies with ozone (Brown and Smirnoff, 1978; V. Ahmadjian,
personal communication, 1977) and nitrogen dioxide (Nash, 1976b) have not shown marked injury to lichens. However, additional studies which better correspond to field
conditions are needed.

At present we know of only one field study, our own (Sigal and Nash, unpublished), which pertains to the probable effects of ozone. In southern California, stands
of *Quercus kelloggii* and conifers were sampled for lichens among five mountain ranges
(Fig. 6). The Cuyamaca area, which represents the best control site, was approximately twice as rich in species as the "polluted" sites. Many of the species which
are absent from the San Bernardino Mountains today are known to have grown there
abundantly in the early 1900s (Hasse, 1913). Of 41 foliose and fruticose lichens
reported by Hasse, only 20 have been found in the present study. Thus it is clear
that the lichen flora has declined. It is reasonable to hypothesize that ozone is
the probable cause of the decline. Sulfur dioxide emissions are extremely low in
Los Angeles, and plant injury from sulfur dioxide is rare in the region (P. R. Miller,
personal communication, 1977). In contrast, the high volume of motor vehicular traf-
fic has led to the generation of high levels of oxidants, particularly ozone. In the
San Bernardino Mountains in areas where we have studied lichens there has been severe
injury to ponderosa pine by ozone (Miller and McBride, 1975; Parmeter and Miller,

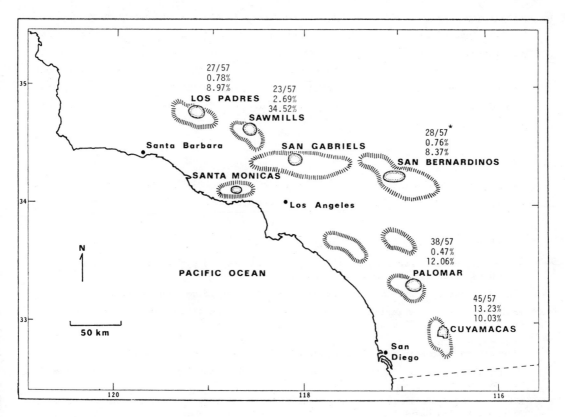

Fig. 6 Distribution of the major sampling areas in relation to the Los Angeles Basin. Number of species sampled is given for each mountain system, as well as the percentage of cover of macrolichens on conifers and the percentage of cover of selected species on *Quercus kelloggii*. The Cuyamaca area has the lowest oxidant levels today. [*An additional five species are found in vestigial amounts in a "sheltered area" in the San Bernardinos.] (Based on unpublished research by L. Sigal and T. H. Nash III.)

1968). Experimental studies are currently being performed to ascertain if the missing lichens are actually sensitive to ozone.

V. SUMMARY

Lichenized fungi are probably the group of terrestrial organisms most sensitive to air pollution. In areas where pollution sources have existed for a number of years, distinct patterns reflecting differential sensitivity to the pollutants are apparent on the basis of species richness considerations. Thus sensitive species are excluded from areas with high air pollutant concentrations; these niches are then often filled by tolerant species. Similar patterns are clearly detectable when species diversity measures or phytosociological techniques are employed. Finer resolution of the patterns may be obtained by quantitatively documenting the abundance and/or relative vitality of species. By employing this approach together with modern multivariate analysis procedures it is possible to detect patterns developing in lichenized fungal populations in areas where air pollution has been present for a comparatively short time. Experimental studies together with field observations at locations throughout the world have documented that lichenized fungi are particularly sensitive to sulfur dioxide and to a lesser extent to hydrogen fluoride, "trace" metals, and oxidant air pollutants.

ACKNOWLEDGMENT

Funds provided by National Science Foundation (Grant DEB-7610244) in developing part of the data employed are gratefully acknowledged.

REFERENCES

Ahmadjian, V. and Heikkilä, H. (1970). The culture and synthesis of *Endocarpon pusillum* and *Staurothele clopima*. *Lichenologist 4*: 259-267.

Arnold F. (1892). Zur Lichenflora von München. *Ber. Bayer Bot. Ges. 2*: 1-76.

Beals, E. (1960). Forest bird communities in the Apostle Islands of Wisconsin. *Wilson Bull. 72*: 156-181.

Bray, J. R., and Curtis, J. T. (1957). An ordination of upland forest communities of southern Wisconsin. *Ecol. Monogr. 27*: 325-349.

Brodo, I. M. (1961). Transplant experiments with corticolous lichens using a new technique. *Ecology 42*: 838-841.

Brodo, I. M. (1966). Lichen growth and cities: A study on Long Island, New York. *Bryologist 69*: 427-449.

Brown, D. H. (1976). Mineral uptake by lichens. In *Lichenology: Progress and Problems*, D. H. Brown, D. L. Hawksworth, and R. H. Bailey (Eds.). Academic Press, New York, pp. 419-439.

Brown, D. H., and Smirnoff, N. (1978). Observations on the effect of ozone on *Cladonia rangiformis*. *Lichenologist 10*: 91-94.

Buchauer, M. (1973). Contamination of soil and vegetation near a zinc smelter by zinc, cadmium, copper and lead. *Environ. Sci. Technol. 7*: 131-135.

Creed, E. R., Lees, D. R., and Duckett, J. G. (1973). Biological method of estimating smoke and sulphur dioxide pollution. *Nature 244*: 278-280.

De Sloover, J. and LeBlanc, F. (1968). Mapping of atmospheric pollution on the basis of lichen sensitivity. In *Proceedings of the Symposium on Recent Advances in Tropical Ecology, Varanasi, India*, R. Misra and B. Gopal (Eds.), pp. 42-56.

de Wit, T. (1976). *Epiphytic Lichens and Air Pollution in the Netherlands*. Cramer, Vaduz, Liechtenstein.

Gilbert, O. L. (1971). The effect of airborne fluorides on lichens. *Lichenologist 5*: 26-32.

Gilbert, O. L. (1976). An alkaline dust effect on epiphytic lichens. *Lichenologist 8*: 173-178.

Hale, M. E., Jr. (1967). *The Biology of Lichens*. Arnold, London.

Hasse, H. E. (1913). The lichen flora of southern California. *Contrib. U.S. Nat. Herb. 17*: 1-132.

Hawksworth, D. L., and Rose, F. (1970). Qualitative scale for estimating sulfur dioxide air pollution in England and Wales using epiphytic lichens. *Nature 227*: 145-148.

Hawksworth, D. L., and Rose, F. (1976). *Lichens as Pollution Monitors*. Arnold, London.

Horntvedt, R. (1975). Epiphytic macrolichens on Scots pine related to air pollution from industry in Odda, Western Norway. *Medd. Norsk Inst. Skogforsk. 31*: 581-604.

Jordan, M. J. (1975). Effects of zinc smelter emissions and fire on a chestnut-oak woodland. *Ecology 56*: 78-91.

Jürging, P. (1975). *Epiphytische Flechten als Bioindikatoren der Luftverunreinigung*. Cramer, Vaduz, Liechtenstein.

Kershaw, K. A. (1972). The relationship between moisture content and net assimilation rate of lichen thalli and its ecological significance. *Can. J. Bot. 50*: 543-555.

Laaksovirta, K., and Olkkonen, H. (1977). Epiphytic lichen vegetation and element contents of *Hypogymnia physodes* and pine needles examined as indicators of air pollution at Kokkola, W. Finland. *Ann. Bot. Fenn. 14*: 112-130.

Laaksovirta, K., Olkkonen, H., and Alakuijala, P. (1976). Observations on the lead content of lichen and bark adjacent to a highway in southern Finland. *Envir. Pollut. 11*: 247-255.

Lange, O. L., Schulze, E.-D., Kappen, L., Buschbom, U., and Evenari, M. (1975). Adaptations of desert lichens to drought and extreme temperatures. In *Environmental Physiology of Desert Organisms*, N. F. Hadley (Ed.). Dowden, Hutchinson and Ross, Stroudsburg, Pa., pp. 2-37.

LeBlanc, F., Rao, D. N., and Comeau, G. (1972a). The epiphytic vegetation of *Populus balsamifera* and its significance as an air pollution indicator in Sudbury, Ontario. *Can. J. Bot. 50*: 519-528.

LeBlanc, F., Rao, D. N., and Comeau, G. (1972b). Indices of atmospheric purity and fluoride pollution pattern in Arvida, Quebec. *Can. J. Bot. 50*: 991-998.

Leighton, W. A. (1868). Notulae lichenologicae No. XXIII. *Ann. Mag. Nat. Hist. 4*: 245-249.

Liiv, S. (1975). Linnade õhu saastatuse astme bioindikatsiooni võimalusi. *Eesti NsV Tead. Akad. Juvres asuva Loodusuurijate selsti Aastaraamat 63*: 82-89.

Lötschert, W., and Köhn, H.-J. (1977). Characteristics of tree bark as an indicator in high immission areas. *Oecologia 27*: 47-64.

Lounamaa, K. J. (1956). Trace elements in plants growing wild on different rocks in Finland: A semi-quantitative spectrographic survey. *Ann. Bot. Soc. "Vanamo" 29*: 1-196.

MacArthur, R. H. (1965). Patterns of species diversity. *Biol. Rev. 40*: 510-533.

Martin, J. F., and Jacquard, F. (1968). Influences des fumées d'usines sur la distribution des lichens dans la vallée de la Romanche (Isère). *Pollut. Atmos. 19*: 95-99.

Martin, L. N. (1976). Detalbnoe likhenoindikatsionnoe kartirovanie zagryazneniya vozdukha v parke kadriorg (Tallin). In *Indikatsiya Pripodnykh Prostessov i Sredy, Matericaly Respublikanskoy Konferentsii* pp. 61-62.

Miller, P. R., and McBride, J. R. (1975). Effects of air pollutants on forests. In *Response of Plants to Air Pollution*, F. B. Mudd and T. T. Kozlowski (Eds.). Academic Press, New York, pp. 195-235.

Nash, T. H., III (1969). The influence of hydrogen fluoride on lichens. M.S. Thesis, Rutgers--The State University, New Brunswick, N.J.

Nash, T. H., III (1971). Lichen sensitivity to hydrogen fluoride. *Bull. Torrey Bot. Club 98*: 103-106.

Nash, T. H., III (1972). Simplification of the Blue Mountain lichen communities near a zinc factory. *The Bryologist 75*: 315-324.

Nash, T. H., III (1975). Influence of effluents from a zinc factory on lichens. *Ecol. Monogr. 45*: 183-198.

Nash, T. H., III (1976a). Lichens as indicators of air pollution. *Die Naturwissenschaften 63*: 364-367.

Nash, T. H., III (1976b). Sensitivity of lichens to nitrogen dioxide fumigations. *The Bryologist 79*: 103-106.

Nieboer, E., Ahmed, H., Puckett, K. J., and Richardson, D. H. S. (1972). Heavy metal content of lichens in relation to distance from a nickel smelter in Sudbury, Ontario. *Lichenologist 5*: 292-304.

Nieboer, E., Richardson, D. H. S., Puckett, K. J., and Tomassini, F. D. (1976). The phytotoxicity of sulphur dioxide in relation to measurable responses in lichens. In *Effects of Air Pollutants on Plants*, T. A. Mansfield (Ed.). Cambridge Univ. Press, New York, pp. 61-85.

Nieboer, E., Puckett, K. J., Richardson, D. H. S., Tomassini, F. D., and Grace, B. (1977). Ecological and physiochemical aspects of the accumulation of heavy metals and sulfur in lichens. *Symposium Proceedings of the International Conference on Heavy Metals in the Environment* Vol. 2, pp. 331-352.

Nylander, W. (1866). Les lichens du Jardin du Luxembourg. *Bull. Soc. Bot. Fr. 13*: 364-372.

Olkkonen, H., and Takala, K. (1975). Total sulphur content of an epiphytic lichen as an index of air pollution and the usefulness of the x-ray fluorescence method in sulphur determinations. *Ann. Bot. Fenn. 12*: 131-134.

Parmeter J. R., Jr., and Miller, P. R. (1968). Studies relating to the cause of decline and death of ponderosa pine in southern California. *Plant Disease Reporter 52*: 707-711.

Pritchard, N. M., and Anderson, A. J. B. (1971). Observations on the use of cluster analysis in botany with an ecological example. *J. Ecol. 59*: 727-747.

Puckett, K. J. (1976). The effect of heavy metals in some aspects of lichen physiology. *Can. J. Bot. 54*: 2695-2703.

Rao, D. N., and LeBlanc, F. (1967). Influence of an iron sintering plant on corticolous epiphytes in Wawa, Ontario. *Bryologist 70*: 141-157.

Schönbeck, H. (1969). Eine Methode zur Erfassung der biologischen Workung von Luftverunreinigungen durch transplantierte Flechten. *Staub 29*: 14-18.

Schutte, J. A. (1977). Chromium in two corticolous lichens from Ohio and West Virginia. *Bryologist 80*: 279-283.

Seaward, M. R. D. (1976). Performance of *Lecanora muralis* in an urban environment. In *Lichenology: Progress and Problems*, D. H. Brown, D. L. Hawksworth, and R. H. Bailey (Eds.). Academic Press, New York, pp. 323-357.

Serander, T. (1912). Studier öfver lafvarnas biologi. I. Nitrofila lafver. *Svensk Bot. Tidskr. 6.*

Serander, T. (1926). *Parmelia tiliacea*, en kustlav och marin islandsvelikt i Skandinavien. *Svensk. Bot. Tidskr. 20*: 352-365.

Skorepa, A. C., and Vitt, D. H. (1976). A quantitative study of epiphytic lichen vegetation in relation to SO_2 pollution in western Alberta. *Envir. Can., Can. For. Serv., Northern For. Res. Cent. Inf. Rep.* NOR-X-161.

Skye, E. (1958). Luftföroeneningars inverkan på busk-och bladlarfloran kring skifferoljeverket i Närkes kvarntorp. *Svensk. Bot. Tidskr. 52*: 33-190.

Skye, E. (1968). Lichens and air pollution. *Acta Phytogeog. Suec. 52*: 1-114.

Skye, E., and Hallberg, I. (1969). Changes in the lichen flora following air pollution. *Oikos 20*: 547-552.

Sørensen, T. (1948). A method of establishing groups of equal amplitude in plant sociology based on similarity of species content, and its application to analysis of the vegetation of Danish commons. *K. Danska Vidensk. Selsk. Biol. Skr. 5*: 1-34.

Steinnes, E., and Krog, H. (1977). Mercury, arsenic and aluminum fall-out from an industrial complex studied by means of lichen transplants. *Oikos 28*: 160-164.

Steubing, L., Kirschbaum, U., and Gwinner, M. (1976). Nachweis von Fluorimmissionen durch Bioindikatoren. *Angew Bot. 50*: 169-185.

Trass, H. (1968). In deks samblikurü hmituste kasutamiseks õhu saastatuse määramisel. *Eesti Loodus 11*: 628.

Trass, H. (1973). Lichen sensitivity to the air pollution and index of poleotolerance (I.P.). *Fol. Crypt. Est. 3*: 17-24.

Türk, R., and Wirth, V. (1975). The pH dependence of SO_2 damage to lichens. *Oecologia 19*: 285-291.

Whittaker, R. H. (1965). Dominance and diversity in land plant communities. *Science 147*: 250-260.

Part VI

FUNGAL BIOMASS AND PRODUCTIVITY IN ECOSYSTEMS

Chapter 26

FUNGAL PRODUCTION ON LEAF SURFACES

Barry J. Macauley and John S. Waid
La Trobe University
Bundoora, Victoria, Australia

I. INTRODUCTION

The surfaces of living or dead leaves appear at first sight to provide a readily defined habitat in which the growth of hyphae or yeast cells may be observed and measured. In theory, in order to identify the fungi which live in this habitat, the application of direct observation coupled with a variety of isolation techniques should
provide fruitful results. In practice, however, the study of fungal production on
leaf surfaces is fraught with many of those technical problems which have hindered
progress in the study of fungal production in natural habitats.

Some of these problems include determining whether the hyphae or cells are alive,
dormant, or dead; isolating and identifying the organisms; relating the nutrition
provided by the leaf, or obtained from elsewhere, to the activity of the mycoflora;
and understanding the interactions between the microfungi themselves and between the
microfungi and other inhabitants of the leaf surface.

It is important to determine the significance of microfungi of the leaf surface,
or phylloplane (Last and Deighton, 1965), in the ecosystem. One measure would be to
determine their productivity, and by this we mean an estimation of the rate of energy
flow through the phylloplane microorganisms to obtain a picture of the energy exchanges occurring between the host plant and the inhabitants on its leaf surface.

Most workers in this field, while being aware of and applying themselves to
solving the problems of measuring productivity, have--perhaps reluctantly--expressed
their data in terms of numbers of species or numbers of cells, length of mycelium per
area of leaf, and percentage cover of fungal structures, all of which can be roughly
extrapolated to give an estimate of standing crop at the time of sampling. But we
have little information on the dynamics, interactions, and energy exchanges occurring
between the host plant and the inhabitants on its leaf surfaces. To appreciate these
important ecological factors it is necessary to obtain better estimates of biomass
and productivity, and with our present technology this appears to be a formidable
task.

Conceivably such estimates would reveal that the proportion of energy flow within
the leaf ecosystem channeled through the microfungi of the phylloplane is minuscule
and ecologically insignificant. What then, we could ask, is the ecological importance

of the phylloplane mycoflora? There is evidence to show that they may influence their host by the production of plant growth substances or by preventing or modifying the development of leaf pathogens. Thus the productivity of the phylloplane micro-fungi measured in terms of energy flow may prove to be insufficiently sensitive or comprehensive to permit objective assessment of their significance in the ecosystem.

In this chapter we have examined the methodology which may be valuable in deter-mining fungal biomass and productivity on leaf surfaces. Most of these methods have been employed in the study of decomposition of leaf litter and thus suffer from the deficiency of being unable to distinguish surface from internal colonizers, though attempts have been made to rectify this problem.

We consider the factors which affect, regulate, or control fungal production, such as nutrition, environmental factors, and fungicides. The availability of nut-rients appears to be an important factor and may influence the change in the myco-flora which occurs as the leaf ages. The interactions on leaf surfaces of nonpatho-genic fungi with pathogenic fungi in relation to biological control is another topic which has attracted considerable interest.

We have restricted ourselves to considering studies on the leaf-surface micro-fungi in terrestrial ecosystems and have omitted discussion on the considerable lit-erature concerning such microfungi in aquatic habitats, e.g., aquatic hyphomycetes.

The subject of fungi on plant surfaces has been covered in great detail in two recent books, *Ecology of Leaf Surface Micro-organisms* (edited by Preece and Dickin-son, 1971) and *Microbiology of Aerial Plant Surfaces* (edited by Dickinson and Preece, 1976), which are essential reading for those interested in this field. We have at-tempted to give a broad coverage of the topic and consequently the references cited are, in many cases, ones which refer to many other papers that we have been unable to incorporate into this chapter but are nonetheless pertinent and important contri-butions to the study of fungal production on leaf surfaces.

II. METHODS

There are two broad groups of techniques, direct and indirect methods, commonly em-ployed to study the microfungi on leaf surfaces. With direct methods, the fungi are observed in situ with the minimum of disturbance or modification to the substrate. When indirect techniques are used, the fungi are either allowed to develop on the substrate or attempts are made to separate them from the substrate by growing them out on artificial media, collecting dispersed spores or dislodging progagules by washing the substrate vigorously, or measuring specific products of their metabolism.

Lindsey (1976) has pointed out that the techniques used in early studies of microfungal succession on leaf surfaces were mainly indirect, such as plating (Pugh, 1958), spore fall (Last, 1955), and humidity-chamber incubation (Hudson and Webster, 1958; Pugh, 1958). Since 1960, there has been an increasing use of observational techniques employing impression films (Dickinson et al., 1974; Dickinson and Wallace,

1976; Dickinson and O'Donnell, 1977), cutical and epidermal stripping (Preece, 1962, 1963), clearing (Shipton and Brown, 1962; Hering and Nicholson, 1964; Daft and Leben, 1966; Minderman and Daniels, 1967; Lamb and Brown, 1970; Eyal and Brown, 1976; Lindsey and Pugh, 1976a,b; McBride and Hayes, 1977) or scanning electron microscopy (Barnes and Neve, 1968; Leben, 1969; Pugh and Buckley, 1971; Lindsey and Pugh, 1976-a,b). Increasing emphasis is also being placed on direct quantitative determinations of numbers of fungal cells (Dickinson, 1967; Apinis et al., 1972; Bainbridge and Dickinson, 1972; Dickinson and Wallace, 1976; McBride and Hayes, 1977), hyphal length (Dickinson, 1967; Minderman and Daniels, 1967; Bainbridge and Dickinson, 1972; Waid et al., 1973; Witkamp, 1974; Visser and Parkinson, 1975; Dickinson and Wallace, 1976; Dickinson and O'Donnell, 1977; McBride and Hayes, 1977), areas covered by fungal structures (Macauley, 1972; Diem, 1974; Eyal and Brown, 1976; Bernstein and Caroll, 1977), or biomass (Ride and Drysdale, 1972; Swift, 1973a; Nagel-de Boois, 1976; Harrower, 1977; Frankland et al., 1978; Carroll, 1979). This brings the study of fungi on leaf surfaces into line with the techniques recommended in International Biological Program (IBP) studies for estimating populations of soil fungi (Parkinson et al., 1971a,b; Parkinson, 1973).

A. Direct Techniques

1. *Impression films and surface stripping*

Beech and Davenport (1971) have reviewed the development of the impression film technique. Initially acetate adhesive tape was applied, stripped off, stained, and examined with the light microscope. To ensure complete sampling of the phylloplane microorganisms for counting, the tape must have a uniform adhesive layer and the same pressure must be applied for each sampling. Beech and Davenport (1971) describe a simple lever-operated device for taking Sellotape impressions at known pressures. In addition to examining the impressions directly, the tape can be applied to an agar medium, incubated, then examined to distinguish between dead and living propagules or cells.

In later studies a variety of substances have been employed to remove the phylloplane microorganisms, such as clear nail polish (Bainbridge and Dickinson, 1972), celloidin strip (Norse, 1972), cellulose acetate in amyl acetate (Dickinson et al., 1974), and Bedacryl 122xGurr (Lindsey and Pugh, 1976b).

The impression films have the advantage that there is minimal preparation of the leaf material and hence little disturbance of phylloplane microorganisms. The arrangement of the epidermal cells and stomata can also be clearly observed, and this aids interpretation of the spatial distribution of the mycoflora (Dickinson, 1967).

Some workers have obtained satisfatory material for direct observation by first stripping off the cuticle or the epidermis. The enzyme pectinase is used to bring about softening and erosion of the cuticle and the walls of the cells of the epidermal layer. The cuticle floats free, leaving behind the phylloplane microflora more or less intact on what remains of the outer surfaces of the epidermal layer (Preece, 1962, 1963).

2. Leaf clearing

The direct examination of phylloplane microorganisms in situ gives valuable information about the distribution of the population. With the light microscope, the structure and contents of the leaf reduce optical resolution to an unsatisfactory level. Thus developing methods to clear the leaf and improve the resolution have been the aim of a number of workers.

These techniques include soaking the leaf in such agents as hydrogen peroxide (Minderman and Daniels, 1967), potassium hydroxide (Eyal and Brown, 1976) or chloral hydrate (Preece, 1959; Pugh and Buckley, 1971; Godfrey, 1974). In some situations an opaque background is desirable and methanol (Preece, 1971a; Warren, 1972c) achieves this effect. Boiling leaves in alcoholic lactophenol cotton blue and then clearing in chloral hydrate demonstrated rust on wheat leaves (Shipton and Brown, 1962); although this would appear likely to disrupt the distribution of the nonparasitic mycoflora, the technique proved successful in indicating the presence of these fungi (Lamb and Brown, 1970), as did a modification of the technique (Lindsey and Pugh, 1976a,b).

A technique which is less likely to dislodge fungal cells or hyphae from the leaf surface is to bleach the leaf in chlorine, usually liberated from bleaching powder (Daft and Leben, 1966; Warren, 1972a; Skidmore and Dickinson, 1973; McBride and Hayes, 1977). Hering and Nicholson (1964) found that soaking leaves in a bleaching solution gave satisfactory results, but Daft and Leben (1966) using such techniques obtained erratic results.

The main criticism of the use of leaf-clearing techniques is the possibility of disturbance to the phylloplane so as to cast doubt on the quantitative results obtained by these techniques. Quantitative studies comparing clearing with other techniques are required to determine their relative merits.

3. Stains

An important component of techniques for direct observation with the light microscope is the effectiveness of a stain to differentiate fungal structures from other material, including the leaf itself.

Fungi in plant tissue may be differentiated by lactophenol containing cotton blue (Shipton and Brown, 1962; Warren, 1972c), aniline blue (Warren, 1972a), chlorazol black (Minderman and Daniels, 1967), or trypan blue (Boedijn, 1956; Dickinson et al., 1974). Staining with Ziehl's carbol fuchsin (Daft and Leben, 1966), phenol acetic aniline blue, or the periodic Schiff method (Preece, 1971a) has also been recommended. Staining methods which are much discussed but little used are those based on fluorescent techniques. Preece (1971b) discusses autofluorescence and the various dyes, stains, and fluoresence brighteners that give a brilliant picture when excited with ultraviolet radiation. The fluorescent-antibody technique (Preece, 1971c; Schmidt, 1973) would be well worthwhile developing particularly for phylloplane studies because it allows direct observation and identification of specific vegetative hyphae.

Two techniques that do not rely on stains but can be considered in this area of improving the visualization of fungal hyphae in plant tissue are infrared photography and autoradiography. Sporulating *Phytophthora infestans* emerging from stomata has been successfully photomicrographed using infrared film (Purnell and Farrell, 1969). Autoradiographic techniques (Waid et al., 1971, 1973) have been used to detect meta- bolically active microorganisms in leaf litter habitats. Labeled [^{14}C] glucose is applied to the experimental samples and controls, and the surface hyphae are removed by painting with dissolved polystyrene, then stripping off. Liquid photographic emul- sion was used for the autoradiographs which showed those hyphae that had actively tak- en up the labeled glucose. A drawback in the technique is that if dead or dormant hyphae were colonized by mycolytic bacteria, they might be recorded as active. Thus the usefulness of the technique may be limited to the time until fungal growth has reached a plateau.

4. *Scanning electron microscope (SEM)*

Direct observations of plant surfaces with the SEM has advantages over the light microscope in that it provides greater magnifications and a much greater depth of field, together with less disturbance of the natural distribution of the microorgan- isms as there is less preparation and handling of material (Barnes and Neve, 1968; Leben, 1969). The SEM also has advantages in less preparation of the specimen, being shadowed with gold (Pugh and Buckley, 1971) or aluminum (Lindsey and Pugh, 1976a), and also in that it can accommodate specimens up to 2.5 cm (Barnes and Neve, 1968). As Diem (1974) has suggested, however, the SEM has given no information additional to that obtained with the light microscope.

B. Indirect or Cultural Techniques

1. *Agar film*

Jones and Mollison (1948) developed the agar film technique, wherein measured amounts of soil are suspended in molten agar and small drops are removed and allowed to solidify in thin films on a hemocytometer slide of known depth. For IBP studies (Parkinson et al., 1971a) when only one technique to estimate microbial biomass in soil is to be used, the agar film method is the one recommended. It was also sug- gested (Parkinson et al., 1971a,b) that the method could be adapted to estimate bio- mass in leaf litter. The method does not, however, distinguish between living and dead fungal hyphae, nor does it indicate the distribution of the hyphae in relation to the leaf tissue. Also, for determining surface-inhabiting species only, this tech- nique is of limited value. In senescent leaves the internal tissue is extensively colonized, while even in healthy leaves species of *Aureobasidium* have been observed as internal colonizers (Lindsey and Pugh, 1976a). If investigators can recognize the type of cell to which the hyphae are firmly attached, say, an epidermal, mesophyllic, or vascular cell, then they may be able to distinguish between internal and surface colonizers.

2. *Dilution plate and leaf washing*

The dilution plate and leaf washing techniques employed to determine the phylloplane population have been adapted from techniques used to study rhizosphere microorganisms. In particular, modifications of the washing technique of Harley and Waid (1955) have been extensively employed (Voznyakovskaya and Khudyakov, 1960; Dickinson, 1965, 1967; Hogg and Hudson, 1966; Pugh and Buckley, 1971; Stott, 1971; Apinis et al., 1972; Bainbridge and Dickinson, 1972; Lindsey and Pugh, 1976a; Warren, 1976).

When dilution plates are prepared from the washings, numbers of easily detachable cells or propagules can be estimated. It is generally accepted (Dickinson, 1971) that bacteria and yeast cells together with fungal spores will be isolated most frequently by this method. While those bacteria, yeasts, and actively sporulating fungi which are growing actively on the leaf surface will be represented in the leaf washings, so too will many casual residents that have been deposited on the leaf.

Caution has been advised in the interpretation of results obtained when plating leaf washings. When using detergents or surfactants to improve the efficiency of washing (Dickinson, 1967; Kendrick and Burges, 1962; Macauley and Thrower, 1966), their fungitoxic properties must be determined. In a study of the effect of nine surfactants on the growth of fungi (Steiner and Watson, 1965), it was found that a number of cationic surfactants inhibited colony growth and were toxic at lower concentrations than were other types of surfactants. M. A. Stott (cited in Dickinson, 1971) recorded lower counts of fungi on dilution plates using Tween 80 as a surfactant. Thus the difficulty arises in deciding whether improved release of propagules from the leaf surface is justified, considering the decrease in colony development on the growth medium. This problem may be overcome by washing in 0.1% agar solution (Pugh and Mulder, 1971), which reduces surface tension and removes most surface spores and debris.

A similar problem has been reported by Hislop and Cox (1969), who studied the microflora on leaves sprayed with the fungicide captan. Occasionally larger numbers of colonies developed on plates prepared from the more dilute suspensions than on those from suspensions of higher concentrations. Also, different fungi were affected to different degrees by these dilution effects. These effects may be due to dilution of inhibitory materials from the leaves or to the fungistatic effects of the captan in the more concentrated suspensions.

Following the removal of detachable cells and spores, the colonies which develop from the leaf material plated on agar are assumed to have originated from hyphae (Dickinson, 1971). These hyphae may have been surface or internal colonizers of the leaf. Thus surface sterilization techniques have been used in an attempt to differentiate between these two groups in studies of succession on decomposing leaves (Kendrick and Burges, 1962; Hering, 1965; Macauley and Thrower, 1966; Ruscoe, 1971), although the technique was not suitable for *Typha* leaves (Pugh and Mulder, 1971).

To ensure adequate sampling, large numbers of small even-sized pieces or disks of leaf should be plated (Kendrick and Burges, 1962; Hogg and Hudson, 1966; Dickinson, 1967; Nagel-de Boois, 1976). Complete leaves can be cut into small sections and plated in sequence to obtain a distribution map, as has been done to determine the distribution and interaction of two fungal parasites which produce a leaf spot disease of *Eucalyptus* seedlings (Ashton and Macauley, 1972).

Another technique involves maceration of leaf tissue (Hislop and Cox, 1969; Warren, 1972a; Visser and Parkinson, 1975; McBride and Hayes, 1977). This technique suffers from many of the disadvantages of both the dilution plate and washed leaf methods in relation to determining the surface mycoflora, but for the total leaf flora it ensures the release of more propagules than the washed leaf method (Davenport, 1976). Other disadvantages of the method include possible release of toxic or inhibitory components from the leaf and the effect of maceration on the viability of certain microorganisms, especially filamentous fungi (Dickinson, 1971), although this does not always occur (Visser and Parkinson, 1975).

3. *Impression plate*

Another method used to determine those species in the phylloplane which are readily detachable is the leaf-impression plate technique first developed by Potter (1910). The leaf is lightly pressed on a solidified nutrient medium, then removed, and the plate incubated (Lamb and Brown, 1970; Apinis et al., 1972). The main disadvantage of this method is that one cannot know how many of the total detachable organisms on the leaf are represented on the impression plate. This is offset by the highly advantageous fact that this method enables samples to be taken in situ and thus avoids disturbance and changes in the physiology of the leaf which occur after it has been detached (Dickinson, 1971). Another advantage is the pattern of distribution of the phylloplane species evident in the leaf imprint on the agar plate; this permits results to be expressed, for example, as the number of germinating spores per square centimeter of leaf surface (Apinis et al., 1972).

Diem (1974) modified Potter's method by pouring cooled agar on the leaf, then removing the thin pellicle which formed. The embedded, epiphytic microorganisms were counted, and the areas of the fungal colonies were determined by outlining their perimeters.

If the epiphytic microflora is particularly dense, the balloon print method can be useful and allows the detection of slow-growing species among the fast growers (Rusch and Leben, 1968). A sterilized, deflated rubber balloon is pressed on the leaf, then inflated and applied to agar in large flat vessels. The technique is most appropriate for single-celled microorganisms rather than filamentous types.

Impression plate methods are reviewed by Beech and Davenport (1971), who indicate the availability of a variety of plates, containers, or agar sausages suitable for these techniques.

4. *Spore fall method*

The spore fall method, used extensively by Last (1955, 1970; Last and Deighton, 1965) and others (Hogg and Hudson, 1966; Lamb and Brown, 1970; Pugh and Buckley,

1971; Apinis et al., 1972; Lindsey and Pugh, 1976a), is a selective method which favors the isolation of members of the Sporobolomycetaceae, which are common phylloplane inhabitants of the green leaf. On agar plates inoculated with dilutions or leaf washings, species such as *Sporobolomyces* are usually overgrown by other fungal species, but separation can be effected by allowing these leaf-inhabiting organisms to discharge their spores so that they fall on the culture medium. Another aspect of the technique is that it is thought that the actively growing and sporulating fungi are more likely to liberate spores than those species present by chance which are not actively growing (Lamb and Brown, 1970). Norse (1972), however, believes that the technique does not differentiate between the actual residents and those species which are very common members of the air spora.

The sample leaf is suspended above the lower half of a petri dish containing the culture medium and held in place by replacing the upper lid, or the leaf may be attached to the lid with a sticking agent (Last, 1955; Hogg and Hudson, 1966; Lamb and Brown, 1970). The whole leaf can be cut in half and each half attached to the petri dish lid with either the adaxial or abaxial surface toward the agar (Pugh and Buckley, 1971; Bainbridge and Dickinson, 1972; Lindsey and Pugh, 1976a).

The investigator can obtain additional data from this method by noting the area of the sample leaf and the venation configuration so numbers of colonies developing per area of leaf and the microdistribution of the sporogenic cells on the leaf surface can be determined.

In studies of microfungal succession on leaves (Hogg and Hudson, 1966) the spore fall method has proved useful in isolating those species which produce few, minute ascocarps, such as *Mycosphaerella*, *Guignardia*, and *Leptosphaeria*, and are difficult to observe directly. From senescent and moribund tissue, *Cladosporium* (Hogg and Hudson, 1966; Pugh and Buckley, 1971) and *Alternaria* species can be isolated, with *Aureobasidium pullulans* (Pugh and Buckley, 1971) occurring infrequently.

Last and Deighton (1965) emphasized the need for selective techniques by pointing to the example of contrasting mycofloras on old oak leaves: in one study a washing technique showed the dominant species to be a black *Pullularia*, whereas in another the spore fall method demonstrated the dominance of *Sporobolomyces*.

5. *Selective media*

Selective media can be used to reinforce the use of manipulative techniques to preferentially isolate phylloplane fungi which may be outstripped in competition on a nutrient medium by fast-growing or antagonistic species. Dickinson (1971) points out that the use of such media has been limited, but he suggests that development and improvement of such media may well provide more data on the relationships of leaf organisms to their natural substrates.

Selective media can provide data on the relationships of leaf surface colonizers to their substrates. While the technique has been used only for phylloplane bacteria, the detection of lipolytic fungi, which indicates the presence of species able to degrade the cuticle, should be feasible using the medium of Sierra (1957). Very few selective media have been used in phylloplane studies, but the wide choice of media developed for soil research such as those to isolate and count pectolytic (Wieringa, 1947) or cellulolytic (Eggins and Pugh, 1962) microorganisms could be adopted to investigate leaf surface microfungi.

While selective media have definite advantages over nonselective media when the goal is to isolate specific fungi, Beech and Davenport (1971) have stressed the need to carefully define each isolation medium. They have shown an agar medium based on apple juice and yeast extract is a valuable means to isolate yeasts from apple leaves. When, for example, different yeast extracts were incorporated in the same basic medium, different yields of the same selection of species of yeast were obtained. Similarly, when liver extract instead of yeast extract was used in a grape-juice basic medium to isolate yeasts from leaves, counts dropped from 2.7 X 10^6 to 3.8 X 10^4 g^{-1} tissue (Davenport, 1976). So unless care is taken to ensure that the same composition of medium is maintained from experiment to experiment, variation in the composition of the yeast microflora isolated from natural substrates may be due solely to variation in the composition of the medium.

6. Humidity chamber

The humidity--or moist-chamber--technique (Keyworth, 1951), wherein leaf samples are incubated, for example, on moist filter paper in a petri dish (Bainbridge and Dickinson, 1972), allows the development of certain species to the stage where fungal colonies can be picked off for culturing or even identified in situ (Dickinson, 1967; Pugh and Mulder, 1971; Apinis et al., 1972; Diem, 1974; Lindsey and Pugh, 1976a). This technique is valuable in that it permits the development of fungal species actively growing on the leaf surface, but it may also allow the germination of spores of casual residents which may otherwise have remained dormant. Dickinson (1971) has questioned the use of filter paper in humidity chambers and has suggested using an inert water reservoir. Kendrick and Burges (1962) set up chambers containing saturated asbestos cord.

While fungal colonies develop in humidity chambers, the physiological condition of the green leaf changes to one of induced senescence (Baddeley, 1971). Attempts have been made to maintain the physiological condition of the detached leaf by floating leaf disks on kinetin (Blakeman and Dickinson, 1967) or benzimidazole (Wolfe, 1965) solution, or inducing the petiole to root in naphthalene acetic acid solution and then growing the rooted, detached leaf in a mineral salt solution (Collins, 1976). Attached leaves have been enclosed by humidity chambers in order to maintain an environment favorable for the development of phylloplane microorganisms (Fokkema, 1971b; Blakeman, 1972).

7. Measurement of fungal products

Measurements of fungal metabolism or products of metabolism have potential in estimating fungal biomass or production on leaf surfaces. The important criterion is that the component being measured is characteristic of fungal populations only, and not of the substrate or other organisms in the phylloplane.

A technique developed recently to estimate biomass of filamentous fungi in plant tissue is the hexosamine (chitin) assay (Ride and Drysdale, 1972; Swift, 1973a,b; Wu and Stahmann, 1975). It has been used to estimate the mycelial content of *Septoria tritici* (Harrower, 1977) and stem rust (Mayama et al., 1975) in wheat and the mycelial biomass in leaf litter (Swift, 1973b; Frankland et al., 1978). The growth of a fungal pathogen during the infection process in melon seedlings (Toppan et al., 1976) as estimated by the hexosamine assay gave good correlation with proteolytic activity within the host induced by the infection.

Because the chitin content of a fungus varies depending on the species, age of mycelium, growth medium, temperature, and oxygen tension (Swift, 1973a,b; Sharma et al., 1977), the assay is applicable only to studies concerning single species (Swift, 1973a,b; 1978; Frankland et al., 1978). It may, however, prove to be a valuable means for determining, in vitro, the ability of monocultures to colonize selected organic substrates, such as in studies on decomposition and pathogenicity. Disadvantages of the technique are that it estimates standing biomass, including dead hyphal walls, rather than productivity; and as insects also contain chitin, care has to be taken to exclude their contaminating influence. It cannot be used to estimate hyphal production by Oomycetes, as they do not form chitin (Frankland et al., 1978).

Ergosterol concentrations have been used to estimate fungal biomass in stored grain (Seitz and Paukstelis, 1977; Seitz et al., 1977a,b). Since ergosterol is a predominant and unique sterol in nearly all fungi, one might expect it to prove a more useful indicator of total fungal biomass than chitin. Unlike the chitin determination, ergosterol assays are not subject to erroneous high readings because of the presence of insects in the substrate. Further, the sterol can be determined reliably and rapidly (1 hr) with high-pressure liquid chromatography, while chitin assays require 4-6 hr. The requirement for sophisticated and expensive chromatography equipment and for specially trained personnel to operate the equipment and run the assay may be viewed as the only drawback to the method. Definitive studies on variation in ergosterol content among various fungal species and with mycelial age have yet to be performed.

Measurements of respiration (Howard, 1967) are useful in litter decomposition studies in that the substrate per se is not contributing to CO_2 evolution, but they do not separate fungal activity from that of other organisms in the decomposer system. Similarly ATP assays (Ausmus, 1973; Holm-Hansen, 1973) give an estimate of the activity of the total decomposer organisms in litter. ATP meets the important criteria of (1) being present in all living cells and absent from dead--including non-living detrital--material; (2) being present in uniform concentrations in all cells regardless of environmental stresses; and (3) being quickly and easily measured in submicrogram quantities (Holm-Hansen, 1973).

Dehydrogenase activity has been measured as an indicator of biological activity in soils and has been found to correlate well with proteolytic activity, nitrification, and respiration but not with microbial numbers (Skujins, 1973).

The problems associated with the use of any of the aforementioned assay techniques suggests that it is most unlikely that any one specific metabolite could be utilized for determination of fungal biomass.

C. Comparison of Techniques in Relation to Fungal
    Biomass and Productivity

The results obtained with the methodology we have outlined range from a list of species present, i.e., the floristic approach, to an estimate of fungal productivity. Such results have a sound ecological basis and are of particular interest to some fungal ecologists, depending upon their area of expertise. The observation that *Alternaria*, a common inhabitant of temperate regions, appears to be less common in leaf litter in the tropics while species of *Curvularia* and *Nigrospora* increase in frequency (Hudson, 1971) is an important observation to the ecologist interested in the distribution of fungi on a global basis. The fact that there is a large increase in the standing crop of mycelium during summer and early autumn in aspen litter (Visser and Parkinson, 1975) is of vital interest to the ecologist studying the energetics of litter decomposition.

For those interested in the floristics of the phyllophane fungi, cultural techniques will be of most benefit. Washed leaf plating, dilution plating of the washings, maceration, spore fall, humidity chamber, and agar film techniques will provide information on the frequency of isolation of each species in relation to the amount of leaf material samples; such techniques can yield data on dry or fresh weight, area of surface, and the like (Fokkema, 1971a; Apinis et al., 1972; Dickinson and Wallace, 1976). Ideally, more than one cultural technique should be employed with a variety of selective media to ensure maximum isolation of the various species depending on their cultural requirements (Dickinson, 1971; Last and Warren, 1972).

Some of the cultural techniques provide data other than floristic information. For example, the leaf impression plate and the balloon print methods show distribution patterns in relation to the leaf surface, e.g., distribution in relation to venation. There is the added advantage of being able to identify or isolate the species concerned, which is not always possible by direct observation.

It is generally accepted that cultural techniques allow the identification of fungi present on or in the leaf, but whether they are true inhabitants or casual residents is difficult to know with certainty. Vigorous washing techniques to remove the casual residents may also remove the less tightly adhering true inhabitants such as yeasts. The time and degree of shaking, together with the type of leaf surface (Fokkema, 1971a) can influence the amount and type of propagules washed off. Plating leaf pieces and macerations after washing gives an estimation of the total mycoflora, both surface and internal colonizers, but surface sterilization techniques can help to distinguish between these two groups.

But the important point is that colony counts obtained by cultural methods represent the number of viable propagules which are able to grow on particular media and under the specific incubation conditions employed. Thus, they are not necessarily estimates of the colonies capable of active growth on the leaf at the time of sampling (Last and Warren, 1972).

While it has been stated that microscopic techniques are generally too laborious for quantitative assessment of fungal biomass or productivity, it would appear that these techniques are the best available at present. During the last 15 years, studies involving a combination of direct observational and indirect cultural techniques have increased (Lindsey, 1976).

Data available from direct observation include measurements of hyphal lengths or areas of colonies (leaf spots) as well as cells and spore counts expressed in relation to the area or weight of the leaf material. Thus results such as 3.5 km mycelium g^{-1} leaf material (Minderman and Daniels, 1967) and 1 km cm^{-2} leaf surface (Witkamp, 1974) appear in the literature, but both these estimates include both surface and internal colonizers. As the average diameter of fungal hyphae can also be obtained by direct microscopy, the volume of fungal mycelium can be calculated. From this, standing biomass has been calculated using the average specific gravity of a variety of fungi obtained from Saitô (1955) to obtain figures such as 42.8 mg mycelium g^{-1} oven-dry weight leaf (Frankland et al., 1978) and 0.91 g wet-weight mycelium m^{-2} live leaves (Visser and Parkinson, 1975).

But how realistic are these estimations? In order to determine the validity of their estimates of biomass and to enable comparisons of biomass between different studies, Frankland and colleagues (1978) compared the effectiveness of a variety of methods. In a study on the estimation of the combined mycelial biomass of both surface and internal colonizers in leaf litter, they compared the agar film technique (Jones and Mollison, 1948) and the hexosamine (chitin) assay (Swift, 1973a). The data were analyzed statistically to determine the degree of precision, sensitivity, and accuracy of the two methods as well as which method was most efficient in terms of man-hours. For this particular experiment sterilized leaf material was inoculated with a pure culture of *Mycena galopus* and incubated under defined environmental conditions. The authors found the agar film method was faster and more sensitive but, even with a correction factor (X3) applied to compensate for hyphae invisible by bright-field illumination, gave biomass volumes only one-half to one-third of those obtained by the hexosamine assay. The conclusions were that it is necessary to obtain more accurate conversion factors such as moisture content and relative density of the hyphae in order to convert the experimental data to biomass values; moreover, it is important to choose the intensity of maceration more carefully. While the agar film technique appears to underestimate biomass, it has an advantage over the hexosamine assay in that it is not restricted to studies of monocultures.

One objective of a study made by Witkamp (1974) was to compare direct microscopic observations of microbial populations on white oak (*Quercus alba*) and yellow poplar (*Liriodendron tulipifera*) leaf litter with estimates from colony counts on dilution plates. Fungal populations were determined on bleached leaf disks by counting the times mycelium was observed to cross a number of parallel lines in a microscope field (20 samples). These data were converted into meters of mycelium per square centimeter of leaf surface. By the direct observation method, estimations of fungi increased significantly on both leaf species with time and there was a net gain in the standing crop of fungal mycelia in the first year of colonization. Witkamp points out that this might indicate only that the average production showed a gain, not necessarily that fungal activity had increased, as some of the standing crop may have been dead, resting, or senescent. The dilution plate count, on the other hand, indicated no significant increase of fungi on yellow poplar leaves, while there was a significant increase on oak leaves. This lack of correlation emphasizes the difficulty of interpreting two sets of data which have been determined by direct and indirect methods. However, Witkamp (1974) stressed that the differences in the data are probably related to what was being measured. For example, the dilution plate technique estimates the population capable of competitive growth under the incubation conditions employed and does not necessarily represent those species actively involved in the decomposition process. Assuming that when a fungal species produces mycelium on and within a substrate it has been able to utilize that substrate or by-products from its decomposition, then direct observation permits estimates of both past and present activity of fungal populations in relation to the decomposition process on or within substrates.

Rather than comparing indirect with direct methods to estimate fungal production, other workers have compared their assessment of biomass with the degree to which the fungus affects its substrate as validation of a meaningful estimate, e.g., of weight loss. Toppan and associates (1976) found a good correlation between fungal growth curves obtained by the hexosamine assay and proteolytic activity which developed in the infected stems and leaves of melon seedlings infected with *Colletotrichum lagenarium*.

In another comparative study, Warren (1976) found numbers of yeast on *Tilia* were underestimated by microscopy compared with dilution plating and concluded that this was due to superimposed cells and difficulty in observing the midrib area.

Lindsey and Pugh (1976a) undertook an extensive comparison of techniques, both observational and cultural, and their results exemplify the problem the fungal ecologist has to face when deciding which methods are best to employ for each particular study. They showed the necessity of using a range of direct and indirect methods in making the observation that the most common species recorded by most techniques, *Aureobasidium pullulans*, was rarely found in humidity chambers while other important species, e.g., *Sporobolomyces roseus*, *Penicillium* spp., and *Phoma* spp., were record-

ed by only a few of the techniques. Thus using a narrow range of techniques, important species may not be detected.

III. FACTORS AFFECTING FUNGAL PRODUCTION ON LEAF SURFACES

When considering fungal production on leaf surfaces, an analysis of the factors likely to affect the distribution and quantity of fungi is required. For heterotrophic
organisms an important factor is an adequate supply of nutrients. The ability of
microorganisms to survive in an exposed environment depends on their resistance to
the extremes to which they are exposed, such as starvation, drought, high and low
temperatures, or ultraviolet radiation. Additional factors likely to affect fungal
production are the presence of fungicides, the population of grazing and mycolytic
organisms living in association with the phylloplane fungi, and the reaction of the
host plant to the presence of the fungi, e.g., production of fungal inhibitors.

A. Nutrients on the Leaf Surface

The available nutrients on leaf surfaces come from three sources: (1) exudates from
the leaf; (2) particles being deposited from the atmosphere or dissolved in precipitation; and (3) inhabitants of the phylloplane (Ruinen, 1965; Last and Warren, 1972;
Godfrey, 1976).

The fact that substances leak from the surface of leaves is demonstrated by
analysis of exudates and leachates. The collection of throughfall under the plant
canopy following rain or inducing leaching by artificial rain or soaking indicates
the presence of all essential minerals, organic compounds, including free sugars,
pectin substances, sugar alcohols, amino acids, gibberellins, vitamins, and phenolic
substances (Long et al., 1956; Tukey, 1971; Godfrey, 1976). Carlisle (1965) found
melezitose, probably from insect honeydew, in precipitation under an oak canopy.
Cooper and Carroll (1978) have shown ribitol to be the principal constituent of leachates from the lichen *Lobaria oregana* (Tuck) Müll Arg.; ribitol has also been found in
throughfall from coniferous stands in which *L. oregana* is the predominant canopy lich-
en. So the leachates can be either promoters of fungal growth, e.g., carbohydrates
and amino acids, or inhibitors thereof, e.g., gallic acid (Godfrey, 1976).

Witkamp (1970) provides data suggesting that phylloplane organisms can retain
substances which are deposited on leaves in leachates, dry fallout, or rainfall.
Leaves from which phylloplane organisms had been removed retained from 5 to 50% of
the quantity of radioactive substances retained by leaves with a normal phylloplane
microflora. This suggests that the phylloplane organisms may indeed act as a sink
for nutrients present on the leaf surface.

The quantity of organic nutrients, such as carbohydrates, leached from the leaf
surface is substantial. Tukey (1971) gives examples of losses of carbohydrate up to
800 kg ha^{-1} year^{-1} from apple trees and up to 6% of the dry weight equivalent could
be leached from young bean leaves in 24 hr. Carlisle (1965) demonstrated losses of

large quantitites of soluble organic matter of up to 86 kg ha^{-1} month^{-1} from oak canopies. Carlisle also mentions the very low nitrogen content of the leachate (0.1 mg per gram on a dry-weight basis), and this may be a factor limiting the development of the phylloplane organisms.

Tukey (1971) has reviewed the factors affecting the quality and quantity of leachates and the mechanism of leaching and should be referred to for discussion of these aspects. Godfrey (1976) has surveyed the literature published after 1971.

The observation that nutrients are more easily leached from old than from young leaves may have important implications in relation to successional changes in the phylloplane population on living leaves. While we can find no documented evidence, we assume that in the healthy leaf there is a continual steady supply of readily available nutrients leaking onto the leaf surface or being deposited as leachate from other leaves and presumably available to the mycoflora. As the leaf ages, analysis of the leachates suggests that greater quantities of these substances become available as nutrients. This may explain the switch from a population of yeast and yeast-like fungi to a population dominated by fast-growing filamentous fungi as the leaf senesces. After the death of the leaf, as the membrane integrity of the leaf cells breaks down there may be a rapid increase in the release of nutrients available to surface colonizers. Then, with the gradual decrease in available nutrients, the initial surface colonizers are replaced by those fungi capable of decomposing waxes, pectins, and cellulose. These aspects will be discussed in Sec. IV.A.

Of the potential nutrients deposited from the atmosphere, pollen has received the most attention (Fokkema, 1971a,b; Norse, 1972). Frequently, the development of saprophytic fungi shows a sudden increase after flowering, and this has been attributed to the deposition of pollen on the leaf surface (Fokkema, 1971b; Warren, 1972a). The effect of pollen on the incidence of disease caused by fungal pathogens and on the nonpathogenic fungi which may inhibit the development of the pathogen (Warren, 1972a,b,c) is a subject of interest.

Some normally mild pathogens become destructive in the presence of pollen (Warren, 1972c; Mansfield and Deverall, 1971). With *Botrytis cinerea* on leaves, Mansfield and Deverall (1971) concluded that pollen reduces the sensitivity of spores to wyerone acid, an antifungal agent found in infected beans. Warren (1972c) found that the stimulatory effect of pollen on germ-tube production of *B. cinerea* and *Phoma betae* could be reproduced with a mixture of hexose sugars plus boric acid.

Little information is available on nutrient production from the phylloplane inhabitants themselves. Certainly a great variety of substances leak from hyphae such as amino acids and carbohydrates (Last and Warren, 1972) and would be available to other fungi. On the death of fungal hyphae, lysis releases additional nutrients and so a cycle of decomposition must then commence on the leaf surface. However, evaluation of the flow of energy or nutrients in such a process would be extemely difficult. A potential source of nutrition is from the propagules deposited on the leaf surface.

If they are able to germinate but cannot develop into phylloplane inhabitants, the dead hyphae would provide a valuable source of nutrients.

B. Environmental Factors

Apart from meeting the nutritional requirements of phylloplane fungi, the microenvironment of the leaf surface itself is an important factor in determining the survival and distribution of these fungi. The physical nature of the leaf surface has a major influence on whether fungi can successfully colonize. The cuticle consists of a number of layers, generally with an external coat of wax. The waxy layer is important in restricting diffusion of water and solutes and thus has an influence on the humidity and nutrient concentration at the leaf surface, both important factors affecting the phylloplane inhabitants. In addition the wax surface has a variety of configurations from flat sheets to projecting tubes that influence the movement of air currents and thus influence the thickness of the still-air boundary layer. Again, this has implications in maintaining high humidity. On the epidermis of many plant species there are a variety of structures, such as hairs and trichomes, which project above the leaf surface. These also influence air movements and help to maintain high humidity.

Thus the leaf surface will provide a suitable niche for phylloplane fungi if they are able to adhere to the surface and the humidity levels are adequate for them to survive. Some species have been shown to possess lipase activity (Ruinen, 1963), which would enable them to become embedded into the wax layers. Some species appear to be preferentially distributed on the leaf surface in the vicinity of trichomes (Lindsey and Pugh, 1976b), at the base of leaf hairs (Leben, 1965), or associated with veins (Last, 1970; Pugh and Buckley, 1971; Lindsey and Pugh, 1976b); other species have no particular distribution pattern (Last, 1970). In some studies (Bainbridge and Dickinson, 1972; Warren, 1976; Bernstein and Carroll, 1977; Carroll, 1979) most phylloplane development occurred on the abaxial surface, whereas in others (Apinis et al., 1972) there were no differences in colonization on abaxial and adaxial surfaces. Generally the cuticle is thinner on the abaxial surface (Last and Warren, 1972) but protection from ultraviolet radiation, more efficient spore trapping, or protection from rain washing may be possible explanations (Bainbridge and Dickinson, 1972). Pugh and Buckley (1971) discuss why some fungi may be restricted to veins, and they suggest that fungal propagules may be washed into channels over the veins where their growth is supported by the presence of exudates.

A major environmental factor in maintaining a viable phylloplane population appears to be high relative humidity (Dickinson, 1967, 1976; Dickinson and O'Donnell, 1977) and may be one explanation for the extensive phylloplane development on leaves in the tropics (Ruinen, 1961). The properties of the cuticle and possession of hairs and trichomes may also be important factors in maintaining a high humidity at the leaf surface. The presence of a moisture film on the leaf surface increases exudation of nutrients from the leaf and possibly from pollen and hyphae, thus favoring

the growth of the phylloplane microorganisms--although there may also be increased
exudation of growth inhibitors (Norse, 1972).

  The leaf surface is generally an exposed microenvironment with wide fluctuations
of humidity, light, and temperature. Exposure to ultraviolet radiation is an obvious
hazard to phylloplane inhabitants, and the possession of pigmented cells in many species may be an evolutionary adaptation to counter this hazard (Last and Deighton,
1965). As well as the direct effect of light on fungi (Leach, 1971), the response
of the host to sunlight can also affect the phylloplane fungi. For example, high
light intensity has been shown to increase loss of carbohydrate from leaves (Tukey
et al., 1957), thus increasing the nutrient concentration on the leaf surface while
ultraviolet radiation may be increasing the susceptibility of leaf tissue to pathogens (Blakeman and Dickinson, 1967).

  The effect of temperature on phylloplane microorganisms has been little studied.
There are the obvious effects of the mean temperature on rate of growth, and the extremes in temperature on survival of hyphae and propagules. Stott (1971) tested on
artificial media the temperature responses of three common phylloplane inhabitants,
Alternaria, *Aureobasidium*, and *Cladosporium*, for vegetative growth and spore germination but found no correlation between the optimum temperature for growth and maximum
population on the leaf surface. The temperature for optimum log phase growth for the
three species was 25°C, but peak populations on the leaf surface occurred in December
and January when the minimum grass temperature was usually below 0°C.

C. Fungicides

Agricultural practice has led to a use of fungicides which could have serious implications in relation to nonpathogenic phylloplane inhabitants. Nonspecific fungicides
are applied usually as a preventative measure for two reasons: tolerant strains of
pathogens are more likely to develop when highly specific sprays are used, and the
broad-spectrum spray can control a number of diseases (Crowdy, 1971). These fungicides, however, can affect the nontarget microorganisms on the leaf surface. On potato, Captafol caused a large reduction in nonpathogenic fungal populations while
Maneb plus fentin acetate was less harmful (Bainbridge and Dickinson, 1972). Another
wide-spectrum spray, captan, reduced numbers of yeasts and filamentous fungi (Hislop
and Cox, 1969; Stott, 1971) as did Fungex (Stott, 1971). Two days after spraying
with captan or Fungex, numbers of viable propagules decreased by about 60% (Stott,
1971). In most studies, numbers soon returned to normal levels (Hislop and Cox, 1969;
Stott, 1971; Dickinson, 1973). Tridemorph, a fungicide relatively specific against
Erysiphe, did not have as wide a spectrum of activity as did Zineb and Benomyl
(Dickinson and Wallace, 1976).

  The possible effects of wide-spectrum fungicides in causing an imbalance between
nonpathogenic and pathogenic fungi have been considered by a number of authors (Hislop and Cox, 1969; Dickinson, 1973; Dickinson and Wallace, 1976; Fokkema, 1976).
Reports of increasing pathogenicity of species tolerant to a fungicide have been ex-

plained as being triggered by a reduction in population of the antagonistic nonpatho-
gen mycoflora.

An interesting side effect of fungicides has been documented by Firman (1970).
Mycosphaerella musicola, causing Sigatoka disease of bananas, has been replaced by
Mycosphaerella sp., causing a much more serious disease, black leaf streak, in Fiji
and other Pacific islands. *M. musicola* can be controlled by oil spraying, which apparently reduces the number of sporodochia and conidiophores. As *Mycosphaerella* sp.
produces few condiophores but abundant ascospores, the oil treatment may be favoring
the development of the new, more virulent pathogen in these regions.

While fungicide spraying may be considered unfavorable in a number of situations,
the observation made by Dickinson (1973) in relation to leaf senescence must be considered: Active growth of several phylloplane species can increase the rate of senescence (Skidmore and Dickinson, 1973), and some reports (see Dickinson, 1973) have
indicated the life of the leaf can be prolonged for several days by fungicide spraying. In situations where it is advantageous to prolong the photosynthetic capacity
of a leaf, as is true for flag leaf of wheat, any disadvantage of disrupting the balance between pathogenic and nonpathogenic fungi may be overridden.

D. Grazing and Lysis of Hyphae

As well as the inherent problems of determining fungal production on leaf surfaces,
an added complication is the reduction in the standing biomass by grazing microarthropods and lysis of hyphae by other microorganisms.

Neither of these aspects has been referred to in the literature on the phylloplane of the living leaf, but studies concerning leaf litter are presumably relevant.
Many species of microarthropods can be cultured on fungi or decaying organic matter
containing fungal hyphae, and examination of gut contents indicates that fungi are
the most important components of the food of mites and collembolans (Harding and
Stuttard, 1974). It has been suggested that the microflora is the important food
source of animals consuming litter (Minderman and Daniels, 1967), and the feeding
pattern of microarthropods in phylloplane populations may be similar.

In soil and on leaf litter, mycelium has been observed to be lysed by the activity of bacteria and actinomycetes and possibly by chemical or enzymatic action (Nagel-
de Boois and Jansen, 1967), and we assume that these processes occur on living leaf
surfaces also.

E. Plant Growth Regulators

Production of the auxins indole-3-acetic acid (IAA) and indole-3-acetonitrile (IAN)
has been demonstrated for *Cladosporium herbarum* (Valadon and Lodge, 1970) and probably for *Aureobasidium pullulans* and *Epicoccum nigrum* (Buckley and Pugh, 1971). This
could account for the presence of plant growth regulators in throughfall under tree
canopies (Tukey, 1971), although it is thought that growth regulators are also leached
from trees without their usual microflora (Good, 1974). If such substances are pro-

duced on leaf surfaces, then the fungi may be playing an important role in plant growth regulation in the ecosystem (Buckley and Pugh, 1971).

While many substances secreted by the leaf promote the growth of fungi, there is evidence that some leaves produce antifungal inhibitory substances. Some of these are produced in response to attack by a pathogen, e.g., phytoalexins (Wood, 1967); others are by-products of the normal metabolism of the plant, e.g., gallic acid (Dix, 1974). Mace and Veech (1973) have reported a response to a pathogen apparently explained by a secretion of organic acids which reduce the pH to levels inhibitory to spore germination. Godfrey (1976) cites a number of examples of the presence of antifungal substances in leaf leachates.

IV. ROLE OF THE FUNGI INHABITING LEAF SURFACES

The importance of the green leaf in primary production in terrestrial ecosystems is unquestioned. Until recently, however, there was very little known about the interface, the leaf surface, where the incident energy from sunlight penetrates to be converted into the organic energy on which the living world depends. A great deal is known about leaf pathogens and how their presence can affect productivity; but do the nonparasitic members of the fungal population inhabiting the leaf surface influence productivity in any way? In an attempt to answer this, we will look at two major areas of study on phylloplane fungi--the succession of fungi as the leaf ages and the interactions between surface colonizers.

A. Fungi Colonizing Leaf Surfaces

One aspect of determining how fungi inhabiting leaf surfaces interact with their host is to know something of the types of fungi present on leaves during their development from green leaf tissue through senescent to moribund tissue.

On the young green leaf, the major fungal inhabitants are yeasts, members of the Sporobolomycetaceae, yeastlike imperfect fungi such as *Aureobasidium pullulans*, and *Cladosporium* species (Ruinen, 1963; Leben, 1965; Last and Warren, 1972; Dickinson, 1973, 1976; Godfrey, 1974; Dickinson and Wallace, 1976). The presence of filamentous fungi appears to be less frequent on young green leaves than on older ones (Dickinson, 1976). Dickinson (1976) lists 100 genera of fungi which occur in the phylloplane of 35 different higher plants, but most species occur infrequently compared with the aforementioned species. As the leaf ages, hyphal development increases rapidly (Pugh and Mulder, 1971; Ruscoe, 1971; Dickinson, 1976; McBride and Hayes, 1977) to the point where, at abscission, the leaf is extensively colonized. These species are the primary colonizers of the dead tissue of the leaf as it becomes incorporated into the decomposing leaf litter (Chesters, 1949; Hudson, 1968; Dickinson, 1976).

The phylloplane species share a number of properties which may explain their presence on the leaf surface. They grow on a wide range of host plants so they probably do not require specific micronutrients (Last and Deighton, 1965). Thus popula-

tions and species, particularly in temperate regions in the Northern and Southern
Hemispheres, are very similar (Last and Deighton, 1965; Stott, 1971). Most phylloplane fungi are found more frequently on leaves than in the soil (Last and Deighton,
1965) suggesting they are well adapted to the microenvironment of the leaf. In the
case of *Aureobasidium* and *Cladosporium*, one such adaptation is the formation of microsclerotia (Pugh and Buckley, 1971; Ruscoe, 1971), which enable these species to survive the adverse periods of moisture supply on the leaf surface. *Sporobolomyces* is
well adapted to the life in the phylloplane due to its rapid rate of growth under
favorable conditions and its ability to reproduce effectively by budding and sporulating and then disperse to new hosts (Dickinson, 1976). A number of phylloplane inhabitants, both fungi and bacteria, possess pigments which could be an adaptation enabling them to survive the high light intensity at the leaf surface (Last and Deighton,
1965).

One important aspect of the presence of a considerable mycoflora on the leaf
surface is the effect of these organisms on the leaf itself. It has been observed
that the mycoflora changes as senescence occurs--but does the mycoflora in any way
affect the rate of senescence? Fungicide application appeared to delay senescence
for a few days, presumably due to reduced development of the nonpathogenic phylloplane (Dickinson, 1973; Dickinson and Wallace, 1976). Other studies wherein phylloplane and pathogenic fungi were inoculated on barley leaves have supported this observation (Skidmore and Dickinson, 1973). Using chlorophyll content to estimate senescence (see Baddeley, 1971), Skidmore and Dickinson (1973) observed that *Alternaria*
and *Stemphylium* (nonpathogens) accelerated senescence but *Cladosporium*, *Aureobasidium*,
and *Botrytis* had little effect.

Dickinson and Wallace (1976) have proposed that saprophytic fungi may penetrate
older leaves, thereby hastening leaf death, or they may accelerate the loss of materials from cells as they lose their membrane integrity on aging. Thus at the stage
of abscission an extensive mycoflora is present on the leaf. Examples of these primary saprophytes are *Ascochytula*, *Leptosphaeria*, *Pleospora*, and *Phoma* (Dickinson,
1976). In addition to the primary saprophytes, parasitic fungi which may or may not
be host specific (Hudson, 1968) inhabit the moribund leaf.

After abscission, the role of the litter colonizers cha¬ges to what Witkamp (1973,
p. 179) has described as "the two most important aspects of microbial life in the
environment--(1) the breakdown of organic matter to prevent its accumulation to levels
harmful to primary production and (2) the mineralization of essential elements out of
the organic debris in order to maintain fertility and the sustained productivity of
the ecosystem."

The role of fungi in the decomposition process is well established (Hudson and
Webster, 1958; Kendrick and Burges, 1962; Hayes, 1965; Hering, 1965; Hogg and Hudson,
1966; Macauley and Thrower, 1966; Minderman and Daniels, 1967; Nagel-de Boois, 1976).
A floristic approach has been the object of most of such studies, though in some
there are attempts to relate the occurrence and the succession of fungi to changes

in the nutritional status of the litter (Hering, 1967; Hudson, 1971). The ecology
of fungi colonizing senescent and fallen leaves is the subject of extensive reviews
(Hudson, 1968, 1971).

It is possible that the succession of fungi which occurs as the leaf ages is related to the changing nutrient status of the leaf. Analysis of leachates indicates
there is a continual exudation of nutrients from the living leaf onto the leaf surface. The initial colonizers, the yeasts and yeastlike fungi, must be capable of
utilizing these simple carbon compounds. As the leaf ages, the amount of exudate increases, and this may explain the increase in frequency of filamentous fungi such as
Cladosporium, *Alternaria*, and *Botrytis* observed by Hogg and Hudson (1966). These
fungi presumably utilize the leaf exudates until they are exhausted or are inaccessible. At this stage those fungi capable of utilizing other available substrates
(e.g., cellulose) persist, and other species also start colonizing the now dead tissue (Hudson, 1971).

Once invasion of the dead leaf tissue begins, many of the problems in estimating
fungal biomass are magnified. Direct observation is difficult, except to examine the
surface colonizers, so again a variety of techniques is required. Cultural techniques associated with maceration or some degree of comminution of the leaf material
are most frequently employed, with the result that many of the slow-growing species
such as basidiomycetes are rarely isolated.

The occurrence of the initial phylloplane fungi such as yeasts, *Sporobolomyces*,
Aureobasidium, and *Cladosporium* appears to have little effect on the leaf; but as the
leaf ages and colonization by filamentous fungi increases, the effect may be to induce
a more rapid rate of senescence. As far as productivity of the leaf is concerned,
there are insufficient critical data to evaluate the effect of the phylloplane fungi.
After abscission, the role of the fungi in the release of nutrients via decomposition
is an essential one in maintaining productivity in the ecosystem.

B. Interactions Among Components of the Phylloplane Mycoflora

Wood and Tveit (1955) and Leben (1965) have reviewed literature on epiphytic microorganisms in relation to plant disease from the time the subject was raised by Potter
(1910) up to 1965.   Leben points out that whether the epiphytes are casual or resident, their presence could affect the pathogenic potential of other organisms present
in the phylloplane. For example, epiphytes may have a direct antagonistic effect on
the pathogen, they may compete successfully against the pathogen for nutrients, or
they may restrict the growth and development of the pathogen by simultaneously occupying the invaded tissue. In addition, effects such as nitrogen fixation on the leaf
surface (Ruinen, 1965; Jones, 1970) or the production of plant growth substances
(Buckley and Pugh, 1971) may change the host's physiology in such a way as to alter
its susceptibility to attack (Leben, 1965). Crosse (1971) has reviewed the literature
on interactions between saprophytic and pathogenic bacteria in plant disease.

One of the first extensive studies on interactions between saprophytic and pathogenic fungi was the control of *Botrytis* attack on lettuce by the direct use of antagonistic organisms (Wood, 1951). A large number of bacteria, actinomycetes, and fungi were tested in culture in lettuce extract agar for antagonism to *Botrytis*. In detached leaves the pathogen was not able to penetrate areas of dead tissue which had been colonized by certain antagonists, and control of the rot was also achieved on leaves of growing plants. Bier (1963) has reported control of a leaf rust by the application of an inoculum of nonpathogenic epiphytes.

While many leaf pathogens can be controlled by aerial spraying with fungicides, Fokkema (1976) notes that the increasing awareness of possible deleterious effects of fungicides on the ecosystem and the consumer interest in pesticide-free foods may mean that biological control has a role to play. Thus there is a need for continued research into the control of leaf pathogens by nonpathogenic microorganisms, and Fokkema reviews much of the work done up to 1975.

Warren (1972b) studied the effects of common saprophytic fungi on the growth and formation of leaf lesions by *Phoma betae* on sugar beet. Leaves of glasshouse-grown sugar beet were inoculated with drops containing *P. betae* with and without pollen, plus propagules from each of four phylloplane fungi, *Sporobolomyces pararoseus*, *Torulopsis candida*, *Aureobasidium pullulans*, and *Cladosporium cladosporioides*, applied at different concentrations. These concentrations were related to the development of the phylloplane population: the lowest concentration was that found on leaves of field-grown sugar beet when little pollen was present; the middle concentration represented the numbers of propagules found when pollen enhances development of the phylloplane fungi; and the high concentration was approximately the sum of the middle concentrations of all four fungi.

Without pollen, low concentrations of the saprophytes strongly decreased the number of expanding lesions; with pollen medium concentrations were required to substantially decrease numbers while high concentrations allowed very few lesions to develop. Thus under experimental conditions closely approximating observed field situations, nonpathogenic saprophytes can restrict this pathogen.

Some studies have been designed to determine the mechanisms of antagonism between saprophyte and pathogen. Warren (1972b) found that a heat-killed mixed mycoflora had no effect on the development of lesions of *P. betae*, suggesting that the mechanism was either by nutrient competition or antibiosis. Assuming that some leakage of nutrients would occur from the dead fungal cells, these results suggest the pathogen was not inhibited by being competitively deprived of nutrients, so some form of antibiosis seems likely.

Norse (1972) found that cell-free filtrates of fungi isolated from tobacco inhibited germ-tube growth of *Alternaria longipes* although germination was unaffected. Such results suggest that fungistatic compounds had been produced by the nonpathogenic fungi.

Fokkema (1976) and Skidmore (1976) have reviewed the literature on possible mechanisms of antagonism of saprophytic phylloplane inhabitants towards leaf pathogens. They discuss the production of antibiotics, toxic metabolites or staling compounds, inhibitory changes in the pH induced by the antagonist, competition for nutrients, physical exclusion of the pathogen by extensive saprophytic growth, hyperparasitism ranging from hyphal interference to direct parasitism and lysis of hyphae, and immunization effects induced in plants by nonpathogenic fungi, e.g., by production of phytoalexins (Wood, 1967).

If the phylloplane fungi do limit the development of pathogens on the leaf, then they must be playing a role in increasing primary productivity. A great deal more research is required in both the laboratory and the field, but it would appear to be a worthwhile goal to attempt to quantify the effect these natural inhabitants on the leaf surface may have on their hosts.

V. CONCLUSIONS

In terms of total energy flow in ecosystems, the phylloplane fungi probably play a minor role, except in the disease situation when virulent pathogens can reduce primary productivity drastically (Klinkowski, 1970). On the healthy leaf, the fungi may reduce the amount of light reaching the photosynthetic tissue, but this is only of consequence in the case of extensive coverage of mycelium such as sooty molds living on honeydew produced by insect feeding. Typical estimates of percentage cover by hyphae range from 0.8% on barley leaves (Diem, 1974) to ca. 20% of 8-year-old Douglas fir needles (Bernstein and Carroll, 1977; Carroll, 1979); assuming hyphal diameter of 5 μm, data from Waid et al. (1973) suggest a hyphal coveral of ca. 1% during early states of leaf decay. The role of the phylloplane fungi as antagonists to pathogens, as well as in increasing the rate of senescence of leaves, may have important implications in relation to primary productivity.

The quantification of the energy flow via the heterotrophic fungi is a difficult task. From studies of leachates it is clear that the supply of energy compounds such as carbohydrates is plentiful, hence presumably fungal production is not limited due to lack of an energy source. We have no data, however, on the amount of available energy on the leaf surface, so an appreciation of the efficiency of conversion of energy by the phylloplane fungi is lacking.

By making assumptions on the average diameter and specific gravity of fungi, data as to the number of cells, hyphal lengths, or volume of hyphae or cells can be converted to approximate fungal biomass values, (Witkamp, 1974; Visser and Parkinson, 1975; Frankland et al., 1978); however, most of the data available are those from studies on decomposition of leaf litter, so the contribution of the surface colonizers is not known. Moreover, such estimates are of limited value in terms of energy flow and the dynamics of the phylloplane community. Studies on the change of biomass with time can give some estimate of productivity, but the problem of whether the meas-

ured hyphae are living, dormant, or dead still remains. The development of autoradiographic (Waid et al., 1973) and fluorescent (Preece, 1971c) techniques should be of
great value.

Until new technology is developed, the best approach appears to be a combination
of techniques. Direct observation with the light microscope of suitably cleared and
stained material gives data on cell and hyphal dimensions which can be converted to
biomass. This may have to be coupled with an indirect technique, such as the agar
film technique (Jones and Mollison, 1948), where by incubating the material it may
be possible to obtain a measure of viability (Jones and Mollison, 1948; Beech and
Davenport, 1971) as well as an estimate of biomass. The problem of testing viability
of all fungi in a limited number of culture media under specific incubation conditions
still remains, however, as does the problem of removing all the fungal propagules
which are not strictly phylloplane species (Last and Warren, 1972).

Cultural techniques have their place in determining the species of fungi inhabiting leaf surfaces. Their geographic distribution (Hudson, 1968) and, to some degree, their distribution on the leaf surface (Last, 1970) can be determined. The
cosmopolitan inhabitants found by most workers are *Aureobasidium pullulans*, members
of the Sporobolomycetaceae, yeasts, and *Cladosporium* spp. (Dickinson, 1973), although
many other species have been recorded (Dickinson, 1976). As the leaf ages, the yeast
and yeastlike fungi are replaced by filamentous fungi--perhaps in response to the
increased availability of nutrients on the leaf surface.

Compared to the soil ecosystem, species diversity in the phylloplane is lower,
which may make the task of direct observation less formidable than it first appears.
If satisfactory techniques can be worked out to identify species (e.g., immunofluor-
escence techniques), to measure cell dimensions, and to determine hyphal viability,
then the study of fungal production on leaf surfaces will undoubtedly develop apace.

REFERENCES

Apinis, A. E., Chesters, C. G. C., and Taligoola, H. K. (1972). Colonization of
Phragmites communis leaves by fungi. *Nova Hedwigia 23*: 113-124.

Ashton, D. H., and Macauley, B. J. (1972). Winter leaf spot disease of seedlings of
Eucalyptus regnans and its relationship to forest litter. *Trans. Brit. Mycol. Soc.
58*: 377-386.

Ausmus, B. S. (1973). The use of the ATP assay in terrestrial decomposition studies.
Bull. Ecol. Res. Comm. (Stockholm) *17*: 223-234.

Baddeley, M. S. (1971). Biochemical aspects of senescence. In *Ecology of Leaf Surface Micro-organisms*, T. F. Preece and C. H. Dickinson (Eds.). Academic Press, New
York, pp. 415-429.

Bainbridge, A., and Dickinson, C. H. (1972). Effect of fungicides on the microflora
of potato leaves. *Trans Brit. Mycol. Soc. 59*: 31-41.

Barnes, G., and Neve, N. F. B. (1968). Examination of plant surface microflora by
the scanning electron microscope. *Trans. Brit. Mycol. Soc. 51*: 811-812.

Beech, F. W., and Davenport, R. R. (1971). A survey of methods for the quantitative
examination of the yeast flora of apple and grape leaves. In *Ecology of Leaf Surface
Micro-organisms*, T. F. Preece and C. H. Dickinson, (Eds.). Academic Press, New York,
pp. 139-157.

Bernstein, M. E., and Carroll, G. (1977). Microbial populations on Douglas fir needle surfaces. *Microbial Ecology 4*: 41-52.

Bier, J. E. (1963). Tissue saprophytes and the possibility of biological control of some tree diseases. *Forestry Chronicle 39*: 82-84.

Blakeman, J. P., (1972). Effect of plant age on inhibition of *Botrytis cinerea* spores by bacteria on beetroot leaves. *Physiol. Plant Pathol. 2*: 143-152.

Blakeman, J. P., and Dickinson, C. H. (1967). The effect of ultraviolet and visible light on infection of host leaf tissue by four species of *Ascochyta*. *Trans. Brit. Mycol. Soc. 50*: 385-396.

Boedijn, K. B. (1956). Trypan blue as a stain for fungi. *Stain Technol. 31*: 115-116.

Buckley, N. G., and Pugh, G. J. F. (1971). Auxin production by phylloplane fungi. *Nature 231*: 332.

Carlisle, A. (1965). Carbohydrates in the precipitation beneath a sessile oak *Quercus petrae* (Mattushka) Liebl. canopy. *Plant Soil 22*: 399-400.

Carroll, G. C. (1978). Needle microepiphytes in a Douglas fir canopy: biomass and distribution patterns. *Can. J. Bot. 57*: 1000-1004.

Chesters, C. G. C. (1949). Concerning fungi inhabiting soil. *Trans. Brit. Mycol. Soc. 32*: 197-216.

Collins, M. A. (1976). Colonization of leaves by phylloplane saprophytes and their interactions in this environment. In *Microbiology of Aerial Plant Surfaces*, C. H. Dickinson and T. F. Preece (Eds.). Academic Press, New York, pp. 401-418.

Cooper, G., and Carroll, G. C. (1978). Ribitol as a major component of water-soluble leachates from *Lobaria oregana*. *Bryologist 81*: 568-572.

Crosse, J. E. (1971). Interactions between saprophytic and pathogenic bacteria in plant disease. In *Ecology of Leaf Surface Micro-organisms*, T. F. Preece and C. H. Dickinson (Eds.). Academic Press, New York, pp. 283-290.

Crowdy, S. H. (1971). The control of leaf pathogens using conventional and systemic fungicides. In *Ecology of Leaf Surface Micro-organisms*, T. F. Preece and C. H. Dickinson (Eds.). Academic Press, New York, pp. 395-407.

Daft, G. C., and Leben C. (1966). A method for bleaching leaves for microscope investigation of microflora on the leaf surface. *Plant Dis. Rep. 50*: 493.

Davenport, R. R. (1976). Ecological concepts in studies of micro-organisms on aerial plant surfaces. In *Microbiology of Aerial Plant Surfaces*, C. H. Dickinson and T. F. Preece (Eds.). Academic Press, New York, pp. 199-215.

Dickinson, C. H. (1965). The mycoflora associated with *Halimione portulacoides*. III. Fungi on green and moribund leaves. *Trans. Brit. Mycol. Soc. 48*: 603-610.

Dickinson, C. H. (1967). Fungal colonization of *Pisum* leaves. *Can. J. Bot. 45*: 915-927.

Dickinson, C. H. (1971). Cultural studies of leaf saprophytes. In *Ecology of Leaf Surface Micro-organisms*, T. F. Preece and C. H. Dickinson (Eds.). Academic Press, New York, pp. 129-137.

Dickinson, C. H. (1973). Interactions of fungicides and leaf saprophytes. *Pestic. Sci. 4*: 563-574.

Dickinson, C. H. (1976). Fungi on the aerial surfaces of higher plants. In *Microbiology of Aerial Plant Surfaces*, C. H. Dickinson and T. F. Preece (Eds.). Academic Press, New York, pp. 293-324.

Dickinson, C. H., and O'Donnell, J. (1977). Behaviour of phylloplane fungi on *Phaseolus* leaves. *Trans. Brit. Mycol. Soc. 68*: 193-199.

Dickinson, C. H., and Preece T. F. (Eds.) (1976). *Microbiology of Aerial Plant Surfaces*. Academic Press, New York.

Dickinson, C. H., and Wallace B. (1976). Effects of late applications of foliar fungicides on activity of micro-organisms on winter wheat flag leaves. *Trans. Brit. Mycol. Soc. 67*: 103-112.

Dickinson, C. H., Watson J., and Wallace B. (1974). An impression method for examining epiphytic micro-organisms and its application to phylloplane studies. *Trans. Brit. Mycol. Soc. 63*: 616-619.

Diem, H. G. (1974). Micro-organisms of the leaf surface: Estimation of the mycoflora of the barley phyllosphere. *J. Gen. Microbiol. 80*: 77-83.

Dix, N. J. (1974). Identification of a water-soluble fungal inhibitor in the leaves of *Acer platanoides* L. *Ann. Bot.* (London) *38*: 505-514.

Eggins, H. O. W., and Pugh, G. J. F. (1962). Isolation of cellulose-decomposing fungi from the soil. *Nature 193*: 94-95.

Eyal, Z., and Brown, M. B. (1976). A quantitative method for estimating density of *Septoria tritici* pycnidia on wheat leaves. *Phytopathology 66*: 11-14.

Firman, I. D. (1970). Possible side effects of fungicides on banana and coffee diseases. *Nature 225*: 1161.

Fokkema, N. J. (1971a). The effect of pollen in the phyllosphere of rye. *Neth. J. Plant Pathol. 77* (Suppl. 1): 1-60.

Fokkema, N. J. (1971b). Influence of pollen on saprophytic and pathogenic fungi on rye leaves. In *Ecology of Leaf Surface Micro-organisms*, T. F. Preece and C. H. Dickinson (Eds.). Academic Press, New York, pp. 277-282.

Fokkema, N. J. (1976). Antagonism between fungal saprophytes and pathogens on aerial plant surfaces. In *Microbiology of Aerial Plant Surfaces*, C. H. Dickinson and T. F. Preece (Eds.). Academic Press, New York, pp. 487-506.

Frankland, J. C., Lindley, D. K., and Swift, M. J. (1978). A comparison of two methods for the estimation of mycelial biomass in leaf litter. *Soil Biol. Biochem. 10*: 323-333.

Godfrey, B. E. S. (1974). Phylloplane mycoflora of bracken *Pteridium aquilinum*. *Trans. Brit. Mycol. Soc. 62*: 305-311.

Godfrey, B. E. S. (1976). Leachates from aerial parts of plants and their relation to plant surface microbial populations. In *Microbiology of Aerial Plant Surfaces*, C. H. Dickinson and T. F. Preece (Eds.). Academic Press, New York, pp. 433-439.

Good, J. E. G. (1974). Naturally-occurring growth regulators in leaf washings of *Picea sitchensis* (Bong.) Carr and *Betula pendula* Roth. *Planta 116*: 45-54.

Harding, D. J. L., and Stuttard, R. A. (1974). Microarthropods. In *Biology of Plant Litter Decomposition*, C. H. Dickinson and G. J. F. Pugh (Eds.), Vol 2. Academic Press, New York, pp. 489-532.

Harley, J. L., and Waid, J. S. (1955). A method of studying active mycelia on living roots and other surfaces in the soil. *Trans. Brit. Mycol. Soc. 38*: 104-118.

Harrower, K. M. (1977). Estimation of resistance of three wheat cultivars to *Septoria tritici* using a chemical method for determination of fungal mycelium. *Trans. Brit. Mycol. Soc. 69*: 15-19.

Hayes, A. J. (1965). Studies on the decomposition of coniferous leaf litter. II. Changes in external features and succession of microfungi. *J. Soil Sci. 16*: 242-257.

Hering, T. F. (1965). Succession of fungi in the litter of a Lake District oakwood. *Trans. Brit. Mycol. Soc. 48*: 391-408.

Hering, T. F. (1967). Fungal decomposition of oak leaf litter. *Trans Brit. Mycol. Soc. 50*: 267-273.

Hering, T. F., and Nicholson, P. B. (1964). A clearing technique for the examination of fungi in plant tissue. *Nature 201*: 942-943.

Hislop, E. C., and Cox, T. W. (1969). Effects of captan on the non-parasitic microflora of apple leaves. *Trans Brit. Mycol. Soc. 52*: 223-235.

Hogg, B. M., and Hudson, H. J. (1966). Micro-fungi on leaves of *Fagus sylvatica*. I. The micro-fungal succession. *Trans. Brit. Mycol. Soc. 49*: 185-192.

Holm-Hansen, O. (1973). The use of ATP determinations in ecological studies. *Bull. Ecol. Res. Comm.* (Stockholm) *17*: 215-222.

Howard, P. J. A. (1967). A method for studying the respiration and decomposition of litter. In *Progress in Soil Biology: Proceedings of the Colloquium on Dynamics of Soil Communities*, O. Graff and J. E. Satchell (Eds.). North-Holland Publs., Amsterdam, pp. 464-472.

Hudson, H. J. (1968). The ecology of fungi on plant remains above the soil. *New Phytol.* *67*: 837-874.

Hudson, H. J. (1971). The development of the saprophytic fungal flora as leaves senesce and fall. In *Ecology of Leaf Surface Micro-organisms*, T. F. Preece and C. H. Dickinson (Eds.). Academic Press, New York, pp. 447-455.

Hudson, H. J., and Webster, J. (1958). Succession of fungi on decaying stems of *Agropyron repens*. *Trans. Brit. Mycol. Soc.* *41*: 165-177.

Jones, K. (1970). Nitrogen fixation in the phyllosphere of the Douglas fir, *Pseudotsuga douglasii*. *Ann. Bot.* (London) *34*: 239-244.

Jones, P. C. T., and Mollison, J. E. (1948). A technique for the quantitative estimation of soil micro-organisms. *J. Gen. Microbiol.* *2*: 54-69.

Kendrick, W. B., and Burges, A. (1962). Biological aspects of the decay of *Pinus silvestris* leaf litter. *Nova Hedwigia 4*: 313-342.

Keyworth, W. G. (1951). A petri-dish moist chamber. *Trans. Brit. Mycol. Soc.* *34*: 291-292.

Klinkowski, M. (1970). Catastrophic plant diseases. *Ann. Rev. Phytopathol.* *8*: 37-60.

Lamb, R. J., and Brown, J. F. (1970). Non-parasitic microflora on leaf surfaces of *Paspalum dilatatum*, *Salix babylonica*, and *Eucalyptus stellulata*. *Trans. Brit. Mycol. Soc.* *55*: 383-390.

Last, F. T. (1955). Seasonal incidence of *Sporobolomyces* on cereal leaves. *Trans. Brit. Mycol. Soc.* *38*: 221-239.

Last, F. T. (1970). Factors associated with the distribution of some phylloplane microbes. *Neth. J. Plant Pathol.* *76*: 140-143.

Last, F. T., and Deighton, F. C. (1965). The non-parasitic microflora on the surfaces of living leaves. *Trans. Brit. Mycol. Soc.* *48*: 83-99.

Last, F. T., and Warren, R. C. (1972). Non-parasitic microbes colonizing green leaves: Their form and functions. *Endeavour 31*: 143-150.

Leach, C. M. (1971). A practical guide to the effects of visible light and ultraviolet light on fungi. In *Methods in Microbiology*, C. Booth (Ed.), Vol. 4. Academic Press, New York, pp. 609-664.

Leben, C. (1965). Epiphytic micro-organisms in relation to plant disease. *Annu. Rev. Phytopathol.* *3*: 209-230.

Leben, C. (1969). Colonization of soybean buds by bacteria: Observations with the scanning electron microscope. *Can. J. Microbiol.* *15*: 319-320.

Lindsey, B. I. (1976). A survey of methods used in the study of microfungal succession on leaf surfaces. In *Microbiology of Aerial Plant Surfaces*, C. H. Dickinson and T. F. Preece (Eds.). Academic Press, New York, pp. 217-222.

Lindsey, B. I., and Pugh, G. J. F. (1976a). Succession of microfungi on attached leaves of *Hippophae rhamnoides*. *Trans. Brit. Mycol. Soc.* *67*: 61-67.

Lindsey, B. I., and Pugh, G. J. F. (1976b). Distribution of microfungi over surfaces of attached leaves of *Hippophae rhamnoides*. *Trans Brit. Mycol. Soc.* *67*: 427-433.

Long, W. G., Sweet, D. V., and Tukey, H. B. (1956). The loss of nutrients by leaching of the foliage. *Quart. Bull. Mich. State Univ. Agr. Exp. Sta.* *38*: 528-532.

Macauley, B. J. (1972). A quantitative technique for assessing colonization of leaf litter of *Eucalyptus regnans* by *Penicillium lapidosum*. *Trans. Brit. Mycol. Soc. 59*: 173-175.

Macauley, B. J., and Thrower, L. B. (1966). Succession of fungi in leaf litter of *Eucalyptus regnans*. *Trans. Brit. Mycol. Soc. 49*: 509-520.

McBride, R. P., and Hayes, A. J. (1977). Phylloplane of European larch. *Trans. Brit. Mycol. Soc. 69*: 39-46.

Mace, M. E., and Veech, J. A. (1973). Inhibition of *Helminthosporium turcicum* spore germination by leaf diffusates from Northern leaf blight-susceptible or -resistant corn. *Phytopathology 63*: 1393-1394.

Mansfield, J. W., and Deverall, B. J. (1971). Mode of action of pollen in breaking resistance of *Vicia faba* to *Botrytis cinerea*. *Nature 232*: 339.

Mayama, S., Rehfeld, D. W., and Daly, J. M. (1975). A comparison of the development of *Puccinia graminis tritici* in resistant and susceptible wheat based on glucosamine content. *Physiol. Plant Pathol. 7*: 243-257.

Minderman, G., and Daniels, L. (1967). Colonization of newly fallen leaves by micro-organisms. In *Progress in Soil Biology: Proceedings of the Colloquium on Dynamics of Soil Communities*, O. Graff and J. E. Satchell (Eds.). North-Holland Publs., Amsterdam, pp. 3-9.

Nagel-de Boois, H. M. (1976). Fungal development on oak leaf litter and decomposition potentialities of some fungal species. *Rev. Ecol. Biol. Sol. 13*: 437-448.

Nagel-de Boois, H. M., and Jansen, E. (1967). Hyphal activity in mull and mor of an oak forest. In *Progress in Soil Biology: Proceedings of the Colloquium on Dynamics in Soil Communities*, O. Graff and J. E. Satchell (Eds.). North-Holland Publs., Amsterdam, pp. 27-36.

Norse, D. (1972). Fungal populations of tobacco and their effect on the growth of *Alternaria longipes*. *Trans. Brit. Mycol. Soc. 59*: 261-271.

Parkinson, D. (1973). Techniques for the study of soil fungi. *Bull. Ecol. Res. Comm.* (Stockholm) *17*: 29-36.

Parkinson, D., Gray, T. R. G., Holding, A. J., and Nagel-de Boois, H. M. (1971a). Heterotrophic microflora. In *Methods of Study in Quantitative Soil Ecology*, IBP Handbook No. 18, J. Phillipson (Ed.). Blackwell Sci. Publns., Oxford, England, pp. 34-50.

Parkinson, D., Gray, T. R. G., and Williams, S. T. (Eds.) (1971b). *Methods for Studying the Ecology of Soil Micro-organisms*, IBP Handbook No. 19. Blackwell Sci. Publns., Oxford, England.

Potter, M. C. (1910). Bacteria in their relation to plant pathology. *Trans. Brit. Mycol. Soc. 3*: 150-168.

Preece, T. F. (1959). A staining method for the study of apple scab infection. *Plant Pathol. 8*: 127-129.

Preece, T. F. (1962). Removal of apple leaf cuticle by pectinase to reveal the mycelium of *Venturia inaequalis* (Cooke) Wint. *Nature 193*: 902-903.

Preece, T. F. (1963). Micro-exploration and mapping of apple scab infections. *Trans. Brit. Mycol. Soc. 46*: 523-529.

Preece, T. F. (1971a). Some environmental and microscopic procedures useful in leaf surface studies. In *Ecology of Leaf Surface Micro-organisms*, T. F. Preece and C. H. Dickinson (Eds.). Academic Press, New York, pp. 245-253.

Preece, T. F. (1971b). Fluorescent techniques in mycology. In *Methods in Microbiology*, C. Booth (Ed.), Vol 4. Academic Press, New York, pp. 509-516.

Preece, T. F. (1971c). Immunological techniques in mycology. In *Methods in Microbiology*, C. Booth, (Ed.), Vol 4. Acedemic Press, New York, pp. 599-607.

Preece, T. F., and Dickinson, C. H. (Eds.) (1971). *Ecology of Leaf Surface Micro-organisms*, Academic Press, New York.

Pugh, G. J. F. (1958). Leaf litter fungi found on *Carex paniculata*. *Trans. Brit. Mycol. Soc.* 41: 185-195.

Pugh, G. J. F., and Buckley, N. G. (1971). The leaf surface as a substrate for colonization by fungi. In *Ecology of Leaf Surface Micro-organisms*, T. F. Preece and C. H. Dickinson (Eds.). Academic Press, New York, pp. 431-445.

Pugh, G. J. F., and Mulder, J. L. (1971). Mycoflora associated with *Typha latifolia*. *Trans. Brit. Mycol. Soc.* 57: 273-282.

Purnell, T. J., and Farrell, G. M. (1969). Photomicrography of unstained fungi using infra-red film. *Lab. Pract.* 18: 1185.

Ride, J. P., and Drysdale, R. B. (1972). A rapid method for the chemical estimation of filamentous fungi in plant tissue. *Physiol. Plant Pathol.* 2: 7-15.

Ruinen, J. (1961). The phyllosphere. I. An ecologically neglected milieu. *Plant Soil* 15: 81-109.

Ruinen, J. (1963). The phyllosphere. II. Yeasts from the phyllosphere of tropical foliage. *Antonie van Leeuwenhoek J. Microbiol. Serol.* 29: 425-438.

Ruinen, J. (1965). The phyllosphere. III. Nitrogen fixation in the phyllosphere. *Plant Soil* 22: 375-394.

Rusch, V., and Leben, C. (1968). Epiphytic microflora: The balloon print isolation technique. *Can. J. Microbiol.* 14: 486-487.

Ruscoe, Q. W. (1971). Mycoflora of living and dead leaves of *Nothofagus truncata*. *Trans. Brit. Mycol. Soc.* 56: 463-474.

Saitô, T. (1955). The significance of plate counts of soil fungi and the detection of their mycelia. *Ecol. Rev.* 14: 69-74.

Schmidt, E. L. (1973). Fluorescent antibody techniques for the study of microbial ecology. *Bull. Ecol. Res. Comm.* (Stockholm) 17: 67-76.

Seitz, L. M., Mohr, H. E., Burroughs, R., and Sauer, D. B. (1977a). Ergosterol as an indicator of fungal invasion in grains. *Cereal Chem.* 54: 1207-1217.

Seitz, L. M., Mohr, H. E., Burroughs, R., Sauer, D. B., and Hubbard, J. D. (1977b). Growth of *Alternaria alternata* on solid substrate monitored by ergosterol, chitin, and secondary metabolites. *Abstracts, Second International Mycological Congress, Tampa, Fla.* p. 604.

Seitz, L. M., and Paukstelis, J. V. (1977). Metabolites of *Alternaria alternata*: Ergosterol and ergostera-4,6,8(14),22-tetraen-3-one. *Agr. Food Chem.* 25: 838-840.

Sharma, P. D., Fisher, P. J., and Webster, J. (1977). Critique of the chitin assay technique for estimation of fungal biomass. *Trans. Brit. Mycol. Soc.* 69: 479-483.

Shipton, W. A., and Brown, J. F. (1962). A whole-leaf clearing and staining technique to demonstrate host-pathogen relationships of wheat stem rust. *Phytopathology* 52: 1313.

Sierra, G. (1957). A simple method for the detection of lipolytic activity of microorganisms and some observations on the influence of the contact between cells and fatty substances. *Antonie van Leeuwenhoek J. Microbiol. Serol.* 23: 15-22.

Skidmore, A. M. (1976). Interactions in relation to biological control of plant pathogens. In *Microbiology of Aerial Plant Surfaces*, C. H. Dickinson and T. F. Preece (Eds.). Academic Press, New York, pp. 507-528.

Skidmore, A. M., and Dickinson, C. H. (1973). Effect of phylloplane fungi on the senescence of excised barley leaves. *Trans. Brit. Mycol. Soc.* 60: 107-116.

Skujins, J. (1973). Dehydrogenase: An indicator of biological activities in arid soils. *Bull. Ecol. Res. Comm.* (Stockholm) 17: 235-241.

Steiner, G. W., and Watson, R. D. (1965). The effect of surfactants on growth of fungi. *Phytopathology 55*: 1009-1012.

Stott, M. A. (1971). Studies on the physiology of some leaf saprophytes. In *Ecology of Leaf Surface Micro-organisms*, T. F. Preece and C. H. Dickinson (Eds.). Academic Press, New York, pp. 203-210.

Swift, M. J. (1973a). The estimation of mycelial biomass by determination of the hexosamine content of wood tissue decayed by fungi. *Soil Biol. Biochem. 5*: 321-332.

Swift, M. J. (1973b). Estimation of mycelial growth during decomposition of plant litter. *Bull. Ecol. Res. Comm.* (Stockholm) *17*: 323-328.

Swift, M. J., (1978). Growth of *Stereum hirsutum* during long-term decomposition of oak branchwood. *Soil Biol. Biochem. 10*: 335-337.

Toppan, A., Esquerré-Tugayé, M. T., and Touzé, A. (1976). An improved approach for the accurate determination of fungal pathogens in diseased plants. *Physiol. Plant Pathol. 9*: 241-251.

Tukey, H. B., Jr. (1971). Leaching of substances from plants. In *Ecology of Leaf Surface Micro-organisms*, T. F. Preece and C. H. Dickinson (Eds.). Academic Press, New York, pp. 67-80.

Tukey, H. B., Jr., Wittwer, S. H., and Tukey, H. B. (1957). Leaching of carbohydrates from plant foliage as related to light intensity. *Science 126*: 120-121.

Valadon, L. R. G., and Lodge, E. (1970). Auxins and other compounds of *Cladosporium herbarum*. *Trans. Brit. Mycol. Soc. 55*: 9-15.

Visser, S., and Parkinson, D. (1975). Fungal succession on aspen poplar leaf litter. *Can. J. Bot. 53*: 1640-1651.

Voznyakovskaya, Y. M., and Khudyakov, Y. P. (1960). Species composition of the epiphytic microflora of living plants. *Mikrobiologiya 29*: 97-103.

Waid, J. S., Preston, K. J., and Harris, P. J. (1971). A method to detect metabolically-active micro-organisms in leaf litter habitats. *Soil Biol. Biochem. 3*: 235-241.

Waid, J. S., Preston, K. J., and Harris, P. J. (1973). Autoradiographic techniques to detect active microbial cells in natural habitats. *Bull. Ecol. Res. Comm.* (Stockholm) *17*: 317-322.

Warren, R. C. (1972a). The effect of pollen on the fungal leaf microflora of *Beta vulgaris* and on infection of leaves by *Phoma betae*. *Neth. J. Plant Pathol. 78*: 89-98.

Warren, R. C. (1972b). Interference by common leaf saprophytic fungi with the development of *Phoma betae* lesions on sugarbeet leaves. *Ann. Appl. Biol. 72*: 137-144.

Warren, R. C. (1972c). Attempts to define and mimic the effects of pollen on the development of lesions caused by *Phoma betae* inoculated onto sugarbeet leaves. *Ann. Appl. Biol. 71*: 193-200.

Warren, R. C. (1976). Microbes associated with buds and leaves: Some recent investigations on deciduous trees. In *Microbiology of Aerial Plant Surfaces*, C. H. Dickinson and T. F. Preece (Eds.). Academic Press, New York, pp. 361-374.

Wieringa, K. T. (1947). A method for isolating and counting pectolytic microbes. *4th Int. Congr. Microbiol., Copenhagen, Report of Proc.*, pp. 482-483.

Witkamp, M. (1970). Mineral retention by epiphytic microorganisms. In *A Tropical Rain Forest*, H. T. Odum and R. F. Pigeon (Eds.). U.S. Atomic Energy Commission, Washington, D.C., pp. H177-H179.

Witkamp, M. (1973). Compatibility of microbial measurements. *Bull. Ecol. Res. Comm.* (Stockholm) *17*: 179-188.

Witkamp, M. (1974). Direct and indirect counts of fungi and bacteria as indexes of microbial mass and productivity. *Soil. Sci. 118*: 150-155.

Wolfe, M. S. (1965). Physiologic specialization of *Erysiphe graminis* f. sp. *tritici*, in the United Kingdom. *Trans Brit. Mycol. Soc. 48*: 315-326.

Wood, R. K. S (1951). The control of diseases of lettuce by the use of antagonistic organisms. I. The control of *Botrytis cinerea* Pers. *Ann. Appl. Biol. 38*: 203-216.

Wood, R. K. S. (1967). *Physiological Plant Pathology*, Blackwell Sci. Publns., Oxford, England.

Wood, R. K. S., and Tveit, M. (1955). Control of plant diseases by use of antagonistic organisms. *Bot. Rev. 21*: 441-492.

Wu, L. C., and Stahmann, M. A., (1975). Chromatographic estimation of fungal mass in plant material. *Phytopathology 65*: 1032-1034.

Chapter 27

ESTIMATION OF LICHEN BIOMASS AND PRODUCTION
WITH SPECIAL REFERENCE TO THE USE OF RATIOS

Lawrence H. Pike
University of Oregon
Eugene, Oregon

I. INTRODUCTION

The thallus of a lichen bears more similarity, both functionally and structurally,
to the body of a green plant than it does to the mycelium of most fungi. Consequently, methods used for estimating biomass and production of lichenized fungi bear more
similarity to those used for the study of photosynthetic plants than those used for
the study of heterotrophic fungi. Many lichens are easily distinguished from their
substrate; they can be photographed and measured and/or removed and weighed. It is
thus considerably easier to estimate biomass and production for lichenized fungi than
it is for free-living fungi immersed in their substrata.

Lichens occupy a range of surfaces which vary greatly in geometrical complexity.
Terrestrial lichens commonly occupy more or less horizontal two-dimensional surfaces,
whereas epiphytic lichens occupy complex three-dimensional surfaces. Certain terrestrial habitats, for example, rock outcrops, present surfaces imtermediate in structural complexity. The ease with which lichen populations can be sampled is largely
determined by the nature of the surface on which they grow.

II. ESTIMATION OF BIOMASS

In the past numerous studies have been carried out producing order-of-magnitude estimates of the standing crop of lichen biomass; thus these estimates have usually
been presented without accompanying error statistics. It is anticipated that future
workers will desire both more precise and more accurate estimates. Because of the
strong relationships between lichen weight and easily measured variables such as cover, height, length and width (e.g., see Bergerud, 1971; Verseghy and K.-Láng, 1971;
Rhoades, 1977), it seem natural to increase the precision of biomass estimates by
gathering information which decribes lichen abundance in many units of the population, processing in detail the lichens of only a few of the units, and estimating
population totals through ratio or regression techniques. In fact some recent investigations have utilized ratios (Nash et al., 1974; Becker et al., 1977) or both ratios and regression (Forman, 1975). Although ratios are relatively simple to compute
and are appropriate for small samples, estimates made from them may be strongly

biased. The investigator can choose among a variety of sampling schemes and ratio estimators depending on desired accuracy, precision, time spent in field sampling, and ease of computations. Over the past several years I have been involved in estimating biomass of lichens in western Oregon for a variety of field situations; data from these studies will be drawn upon to discuss the use of ratio statistics. It should be emphasized that the statistical methods discussed here, though appropriate for use in studies of lichen biomass, have potential for use in other kinds of biological studies.

A. Terrestrial Lichens

Biomass of terrestrial lichens may be estimated through use of quadrats with a sampling design such as would be used for the study of herbaceous vegetation. Lichens may be cleared from a series of quadrats and total biomass estimated from the mean weight of lichens found in the quadrats. This approach has been used by various investigators (e.g., Verseghy and K.-Láng, 1971; Prince, 1974; Crittenden, 1975; Wein and Speer, 1975). Except for small areas of homogeneous lichen cover, standard deviations for these biomass estimates are likely to be very high. Sampling a large number of quadrats will lead to appropriately narrow confidence intervals, but without some change in sampling strategy the cost in terms of time spent picking, cleaning, and weighing lichens becomes very high.

Since there is a high correlation betwen lichen biomass and perceived lichen abundance, it is possible to develop a sampling plan allowing a large number of quadrats to be analyzed while minimizing the time which must be spent processing lichen samples. If some measure of lichen abundance x_i (say, total area or percent cover) is gathered for each of a large number of quadrats and if, in addition, from a randomly selected subset of quadrats a lichen weight y_i is determined, total lichen biomass (Y) may be estimated as

$$\hat{Y} = \overline{X}\hat{R} \tag{1}$$

where $\overline{X}$ is the mean of the x_i's for all N quadrats surveyed for lichen cover and $\hat{R}$ is a ratio estimator based on the relationship between weight y and area x for lichens removed from the n quadrats selected for subsampling.

Several different ratio estimators have been formulated (Kendall and Stuart, 1966); the simplest are R_1, the mean of the ratios for individual quadrats,

$$R_1 = n^{-1} \Sigma' \frac{y_i}{x_i} \tag{2}$$

and R_2, the ratio of the means for all quadrats,

$$R_2 = \frac{\overline{y}}{\overline{x}} \tag{3}$$

where Σ' indicates summation over the subsampled quadrats, $\overline{x} = n^{-1} \Sigma' x_i$, $\overline{y} = n^{-1}$

Σ' y_i; and where $x_i > 0$ for subsampled quadrats. Both R_1 and R_2 are likely to lead to biased estimates of Y (Raj, 1968). Of these estimators, R_1 may appear more favorable since computation of a variance is straight-forward. However, R_2 should be favored since its bias is approximately 1/n that of R_1 and therefore decreases with increased sampling while the bias of R_1 does not (Kendall and Stuart, 1966).

R_1 may be corrected for its bias (Hartley and Ross, 1954), producing R_3--an unbiased ratio-type estimator which is useful in assessing the magnitude of the bias of other ratios:

$$R_3 = R_1 + \frac{(\alpha N - 1)n}{N(n - 1)} \left[\frac{\overline{y} - R_1 \overline{x}}{\overline{x}} \right] \tag{4}$$

Although computation of an exact variance for R_3 is possible (Goodman and Hartley, 1958), the procedure is complex and cumbersome, making general use of R_3 unlikely. Replacement of the population mean ($\overline{X}$) in the denominator of the right-hand term of Eq. 4 by the sample mean ($\overline{x}$) results in an approximately unbiased ratio-type estimator (R_4) which has a smaller variance than R_3:

$$R_4 = R_1 + \frac{(N - 1)n}{N(n - 1)} (R_2 - R_1) \tag{5}$$

When N is large, R_4 is identicial with r*, an estimator proposed by Nieto de Pascual (1961). The precision of R_4 is discussed later in this chapter.

While conducting studies of the distribution of terrestrial lichens in the Cascade Mountains of Oregon, I have investigated the use of the above ratios. Mean lichen area $\overline{X}$ (square meters of lichen per hectare) was estimated from N = 50, 1/2 m^2 quadrats following a systematic random sampling scheme. From the N' quadrats in which lichens were present, a subset of n quadrats was chosen at random with p = 0.5. The ratio R of lichen weight to lichen area (kilograms per square meter) was estimated from these n quadrats. Because of a slight curvilinearity in the relationship between weight and area, R_1 is consistently lower than the other ratios in the more than 50 cases examined. R_2, R_3, and R_4, on the other hand, are very similar in all cases and on the average are nearly identical. The close similarity among R_2, R_3, and R_4 is illustrated by results for litter of arboreal, nitrogen-fixing lichens, as shown in Table 1. (Once they have fallen from the canopy, arboreal lichens present sampling problems identical to those of true terrestrial lichens.) Since R_2 is essentially unbiased (in this situation), and since both R_2 and its error statistics are relatively simple to compute, it is the ratio of choice. It should be noted that even when the correlation betwen x and y is close to one, estimates based on R_1 may be substantially biased.

Since the lichen biomass, $\hat{Y}$, is computed as the product of two quantities ($\overline{X}$ and $\hat{R}$) each estimated with uncertainty, the standard error (SE) of $\hat{Y}$ is computed as follows (Goodman, 1960):

Table 1 Ratio and error statistics for estimates of standing crop of litter of epiphytic lichens at six sites in the Lookout Creek drainage, H. J. Andrews Experimental Forest, Blue River, Oregon[a],[b]

Site	Elevation (m)	Aspect	n	r^2	Ratio (kg/m² lichen)				Standing crop (kg ha^{-1})		
					R_1	R_2	R_3	R_4	$Y_2 = \bar{X}R_2$	SE (Y_2) (Eq. 6)	SE (Y_2) (Eq. 7)
1	730	S	14	0.957	0.1448	0.1617	0.1558	0.1625	9.54	2.3	2.2
2	640	S	29	0.847	0.1331	0.1354	0.1355	0.1354	24.34	3.8	3.7
3	580	S	22	0.906	0.1575	0.1774	0.1763	0.1778	36.33	7.2	6.8
4	610	N	30	0.941	0.1878	0.2042	0.2078	0.2044	64.44	8.9	8.6
5	730	N	23	0.924	0.1612	0.1885	0.1878	0.1891	14.63	2.8	2.7
6	850	N	15	0.997	0.1671	0.2001	0.2016	0.2015	9.45	2.3	2.3
Mean					0.1586	0.1779	0.1774	0.1785			

[a]Location: 44°N, 122°W.

[b]The coefficient of determination (r^2) is for the relationship between observed lichen weight (y) and measured lichen area (x). For each site N = 50 quadrats (½ m²), from which n quadrats were selected for lichen removal. Estimates are corrected to correspond to a horizontal surface.

$$SE(\hat{Y}) = \{\bar{X}^2 SE^2(\hat{R}) + \hat{R}^2 SE^2(\bar{X}) - SE(\hat{R}) SE(\bar{X})\}^{1/2} \qquad (6)$$

where $SE(\hat{R})$ is computed for the n quadrats in the subsample and $SE(\bar{X})$ is computed for the N quadrats in the main sample. If the coefficient of variation of R is much smaller than that of X, the first and third terms under the radical in Eq. 6 may be negligible when compared to the second term, and a much simpler approximation to the standard error (Eq. 7) will yield results nearly identical to those of Eq. 6:

$$SE(\hat{Y}) \simeq \hat{R} SE(\bar{X}) \qquad (7)$$

If most of the uncertainty in $\hat{Y}$ is due to uncertainty in $\bar{X}$ rather than $\hat{R}$, the standard error may be decreased more efficiently by increasing N (the number of quadrats in which lichen area estimates are made) than by increasing n (the number of quadrats from which lichens are removed for weighing).

If R_2 is used in the computation of $\hat{Y}$, the standard error of Y_2 is computed by substituting the mean square error of R_2 for $SE^2(R_2)$ in Eq. 6. The mean square error (MSE) incorporates both variance and bias and is identical with the square of the standard error when the estimator is unbiased. The mean square error of R_2 is estimated as follows (Raj, 1968, p. 93):

$$MSE(R_2) \triangleq \frac{(1 - n/N)(\Sigma' y_i^2 + R_2^2 \Sigma' x_i^2 - 2R_2 \Sigma' x_i y_i)}{n(n - 1)\bar{x}^2} \qquad (8)$$

The above approximation of $MSE(R_2)$ may be an underestimate; computation of bounds on the MSE is discussed by Raj (1968). For the examples in Table 1 the simplified formula for $SE(Y_2)$ (Eq. 7) yields results very similar (on the average understating by about 3%) to those obtained from Eq. 6, indicating that nearly all of the uncertainty in Y_2 may be attributed to uncertainty in $\bar{X}$.

Estimator R_2 varies from 0.135 to 0.204 kg/m^2 of lichen for the examples presented in Table 1. Part of this variability results from combining data for several different species of lichens. However, even when species are treated separately there are relatively large differences in R_2 from site to site which are probably in large part due to observer bias, differing size-frequency distributions for the lichens, and variation in health of lichen thalli. Thus the weight-to-area ratio should not be treated as a constant which can be determined once and applied to all sets of area data; its determination should be part of the sampling plan for each site.

Results of biomass sampling may be expressed as weight per unit area of actual land surface or per unit area on a horizontal plane. The latter is desirable to permit comparison with other ecological data. As long as the slope of the ground surface is less than 15° there is little difference between biomass per unit of actual surface and biomass per unit of horizontal surface. The difference between biomass values expressed in these two manners increases rapidly for slopes greater than 20° and at 40° is a factor of 1.3. Since slopes are commonly in the 20-40° range for our study sites in the Oregon Cascades, the angle θ by which each quadrat deviated from the horizontal was measured and both $\hat{Y}$ and $SE(\hat{Y})$ were multiplied by $N (\Sigma \cos \theta)^{-1}$ to correct estimates to a horizontal surface.

B. Arboreal Lichens

Estimates of lichen biomass obtained for a subset of trees can be expanded to stand estimates utilizing information such as number and species composition of trees in the stand and trunk diameters. Access to the crown of the sampled trees is gained through felling, climbing, or use of ladders. Although it is possible to strip lichens completely from trees, the number of trees sampled is likely to be very small if this is done. It is more satisfactory to remove only a small proportion of the lichens from as large a number of trees as possible. Sampling techniques discussed here assume a monopodial or "conifer" growth form, and modifications must be made for many broadleaf trees which branch throughout the canopy.

Within individual trees biomass is estimated separately for epiphytes of trunk and branches. Trunk lichens are easily sampled using a series of *cylindrats*, i.e., belt transects running completely around the stem (Pike et al., 1972). Total lichen biomass for the trunk is then estimated as the total weight of lichens removed from cylindrats multiplied by the sampling interval (distance from the center of a cylindrat to the center of adjacent cylindrats). Branch lichens are removed and weighed from a sample of branches from throughout the crown. If branches are selected with equal probabilities, total lichen biomass for all branches on the tree may then be estimated by multiplying mean biomass per branch by the number of branches on the tree.

By analogy with methods described for sampling terrestrial lichens, this basic sampling scheme can be modified into one which improves the precision of the estimates obtained for a given amount of lichen processing if, in addition to enumerating the branches, information is gathered which serves as an index of lichen abundance on each branch. This auxiliary information could be diameter at the base of the branch, length of the branch, or an estimate of the total area occupied by lichens on the branch. A ratio is formed expressing the relationship between weight of lichens y and the index of lichen abundance x; total lichen biomass on all branches of the tree is estimated in a manner paralleling that discussed earlier for terrestrial lichens.

To keep at a minimum both sample processing time and destruction of branches, it may be desirable to remove a relatively small number of branches from the tree. However, since the bias of R_2 increases with decreasing sample size, expansion using R_2 may not be appropriate. The precision and accuracy of several estimators when sample size is small have been investigated for a population simulated to resemble the observed distribution of epiphytes on branches. These populations can be characterized as follows: (1) Logarithms of epiphyte biomass are approximately normally distributed. (2) Epiphyte weight (y) and the index of lichen abundance (x) are highly correlated. (3) Variance of y increases as a function of x. (4) The relationship

Table 2 Estimates of Y, the population total, for a set of x's and y's which simulate predicted and actual weights of epiphytes on branches of a tree[a,b]

Estimator	Simple random sampling			Stratified random sampling after ordering by size		
	Mean Ŷ	var (Ŷ)	MSE (Ŷ)	Mean Ŷ	var (Ŷ)	MSE (Ŷ)
Ȳ	92.6	5361	5361	92.6	1873	1873
R₁	68.1***	320	928	68.1***	114	722
R₂	83.7**	421	502	91.4	281	282
R₃	91.6	1005	1006	99.0*	503	545
R₄	88.0	508	530	97.8*	393	421

x	y	x	y	x	y	x	y	x	y	x	y
0.25	0.01	0.30	0.14	0.50	0.75	0.70	1.15	1.50	1.63	5.00	11.64
0.25	0.16	0.35	0.37	0.50	0.54	0.85	2.05	2.00	3.74	7.50	14.81
0.25	0.00	0.40	0.31	0.65	0.77	1.00	0.66	2.50	2.72	10.00	20.18
0.25	0.33	0.50	0.66	0.65	0.64	1.25	1.15	3.50	5.21	15.00	22.98

[a]Mean, variance, and mean square error (MSE) are given for 60 estimates based on five estimators under two sampling regimes with $n = 4$. The actual value of Y is 92.6. Asterisks indicate that the probability that Ŷ equals 92.6 is <0.05 (*), <0.01 (**), or <0.001 (***) by a 2-tailed t-test.

[b]Numbers used to simulate the population of epiphytes on branches of a small tree are given in the lower table (x = index of lichen abundance; y = corresponding lichen weights). Units are arbitrary. Linear regression equation $y = -0.230 + 1.763x$, $r = 0.9814$.

between x and y may be slightly curvilinear. The degree of curvilinearity varies
from tree to tree and cannot be estimated accurately from a small sample.

Repeated sampling from the simulated population (Table 2) has shown that for any
given estimator both the bias and the MSE vary depending on the way the sample is
chosen. Sampling schemes investigated included the two following methods: (1) *Simple random sampling*--Each of the 24 units in the population was randomly assigned to
one of six samples of n = 4. This procedure was repeated 10 times, resulting in 60
samples. Since each unit occurred in precisely 10 samples, the mean population totals ($\hat{Y}$) computed by expanding on the basis of the means ($\overline{y}$'s) and the R_1's are exact. Means of $\hat{Y}$ derived from R_2, R_3, and R_4 are estimates. (2) *Stratified random sampling after ordering by size*--Units in the population were divided from four strata after ordering them by values of x. The first stratum contained the six units
with the lowest values of x_i. The second stratum contained the six units with the
next higher values of x_i, and so on. The six units in each stratum were randomly
assigned to one of six samples of n = 4. Again, this procedure was repeated 10 times,
resulting in 60 samples. Also, again, means of the $\hat{Y}$'s based on $\overline{y}$ and R_1 are exact;
those based on R_2, R_3, and R_4 are estimates.

Under conditions of random sampling, R_1 and R_2 are both significantly biased,
underestimating the known value of Y; R_3 and R_4 give more accurate estimates of Y.
The MSE of estimates based on R_4 is an order of magnitude lower than that of estimates based on the sample mean and about one-half that of estimates based on R_3.
The increase in efficiency of R_4 as an estimator over R_3 has been obtained by using
$\overline{x}$ rather than $\overline{X}$ in the second term of Eq. 3. This result is consistent with that of
Nieto de Pascual (1961), who has shown that r* (nearly identical with R_4) is a more
efficient estimator than R_3. Sampling with larger n would increase the precision of
all estimates and would improve the accuracy of estimates based on R_2 but not those
based on R_1. Under conditions of stratified random sampling after ordering by size,
estimates based on all of the ratios except R_2 are significantly biased. Stratified
random sampling gave smaller MSEs for all estimators than simple random sampling, and
smaller MSEs for estimates based on R_2 than for those based on other estimators. For
a low MSE and simplicity of computation, stratified or systematic sampling after ordering by size and use of R_2 should be favored over simple random sampling and use
of R_4.

Additional sampling from the simulated population has shown that bias and MSE
resulting from systematic sampling without ordering by size are similar to those obtained for simple random sampling, and bias and MSE resulting from systematic sampling after an ordering by size are similar to those obtained for stratified random
sampling after ordering.

This strategy of collecting supplementary information on all branches and removing a subsample of a few branches has been used for estimating biomass of epiphytes on branches of a series of relatively small understory trees [10-30 cm diameter at breast height (dbh), 8-20 m height] in watershed 10 of the H. J. Andrews Experimental Forest (L. H. Pike and W. Denison, unpublished data). Basal diameter in centimeters and epiphytic cover in square decimeters were recorded for each branch of trees chosen at random from strata defined by vascular plant community type. Experience has shown that biomass of epiphytes is positively correlated with both estimated cover of epiphytes (C) and the square of branch diameter (d^2). When epiphytic cover is low, biomass of epiphytes is correlated more strongly with branch diameter than it is with epiphytic cover; the converse is true when epiphytic cover is high. Consequently an index of lichen abundance formed from a product of both types of information works well on the wide variety of trees investigated. Since the relationship between $d_i^2 C_i$ and epiphytic biomass is strongly curvilinear but can be made more or less rectilinear by taking the square root, we have used $d_i C_i^{\frac{1}{2}}$ as an index of lichen abundance.

A systematic random sample of four or five branches was removed from each of these trees, and all epiphytes (including bryophytes) were picked from them, cleaned of debris, dried, and weighed. Since branches were included in the sample with equal probabilities and were not ordered by size before the sample was drawn, Y can be computed from R_4 and also from the average biomass of epiphytes removed from the sampled branches. The standard error of R_4 was computed using the approximation for SE(r*) from Nieto de Pascual (1961) and multiplying by the finite population correction, $1 - n/N$ (Cochran, 1963).

$$SE(R_4) \simeq \left\{ (1 - \frac{n}{N}) \left[\frac{Z}{n-1} + \frac{1}{(n-1)^2} (var(R_1) - 3Z) \right] \right\}^{1/2} \tag{9}$$

where $var(R_1) = (n-1)^{-1} \Sigma' (y_i/x_i - R_1)^2$
$$Z = R_4^2 CV^2(y) + CV^2(x) - 2\rho CV(y) CV(x)$$
Σ' = summation over the sample of n branches
ρ = correlation coefficient
CV = coefficient of variation

Since X (the total of the indices of epiphytic abundance for all branches) is known with certainty, the standard error of $Y_4 = R_4 X$ is computed as $SE(Y_4) = X\ SE(R_4)$. The estimated total epiphytic biomass based on the sample mean is $Y_m = N\bar{y}$. The standard error of Y_m with finite population correction is

$$SE(Y_m) = \{ N(N-1)n^{-1} var(y) \}^{1/2} \tag{10}$$

For these small trees R_4 consistently estimates Y more efficiently than the mean (Fig. 1). In general, some eight to ten branches would have had to be removed to make the coefficient of variation based on the mean equal to that found for a sample size of four or five branches when estimates are based on ratios.

Although the sampling schemes outlined above may be satisfactory for the study of epiphytes on small trees, their use on large trees where quantities of epiphytes on branches may vary over many orders of magnitude would require a considerable in-

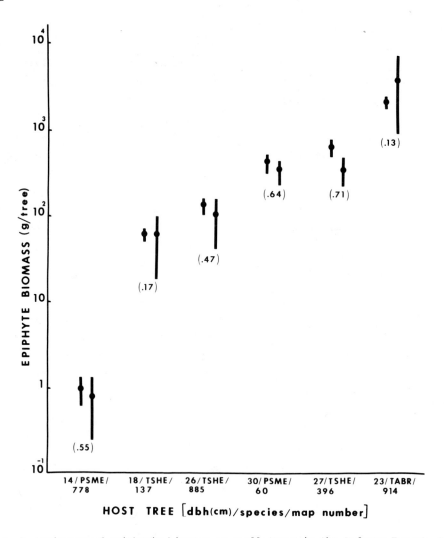

HOST TREE [dbh(cm)/species/map number]

Fig. 1 Estimates of epiphytic biomass on small trees in the Andrews Forest, Oregon. For each tree Y_4, based on the ratio R_4, is on the left and Y_m, the estimate based on the mean weight of epiphytes on four or five branches, is on the right. Vertical lines represent 1 SE on each side on the mean, and the number in parentheses is the ratio of the coefficients of variation ($CV(Y_4)/CV(Y_m)$). Species of host tree are *Pseudotsuga menziesii* (PSME), *Tsuga heterophylla* (TSHE), and *Taxus brevifolia* (TABR).

crease in sample size for a desired level of precision. Analysis of a large sample requires not only a great deal of sample processing but also considerable destruction of the tree crown. Clearly, destruction must be kept to a minimum if long-term studies of these crowns are anticipated. We (my colleagues at Oregon State University and the University of Oregon and myself) have been faced with this problem while studying epiphytes on old-growth Douglas fir trees (*Pseudotsuga menziesii*) in the Andrews Experimental Forest. These trees are commonly 70-80 m tall and 1.5 m dbh.

Methods developed for sampling these trees (Pike et al., 1977) involve establishing relations for predicting weights of tree and epiphytic components and then using these predicted weights as the "sizes" on which sample units are chosen with selection probability proportional to size (pps). The probability of a branch being included in the sample is

$$P_i = \frac{nx_i}{X} \tag{11}$$

where x_i is the "size" of branch i, n is the desired sample size, and $X = \Sigma\ x_i$. The estimated total weight of a component for the entire tree is then

$$\hat{Y} = \Sigma' \frac{y_i}{P_i} \tag{12}$$

The efficiency of pps sampling can be illustrated by sampling the simulated population in Table 2 in this manner. Units in the population were placed in a random order, and population estimates were computed for all possible systematic samples of n = 4. (Actually the largest unit was included in all samples since its x was greater than X/n; three units were selected with pps from the remaining population.) See Raj (1968), Hartley (1966), or Pike et al. (1977) for the procedure for drawing a systematic sample with pps. The mean $\hat{Y}$ for all possible samples from a random starting order of the simulated population of Table 2 weighted for their expected frequency under infinite sampling is 92.6, exactly the same as the known Y. Since expansions involving R_1 in Eq. 1 or P_i in Eq. 12 are identical under pps sampling, Raj's statement that R_1 is exactly unbiased under pps sampling is supported for populations with the aforementioned characteristics. A dramatic result is that the MSE for the population estimates based on R_1 after pps sampling is 118, only 42% of the lowest MSE for any estimator under equal probability sampling (Table 2). Sampling with pps must be carried out in two stages: in the first stage, units of the population are described so that P_i's may be computed; in the second, selected units of the population are sampled. For large complex populations use of a computer is a virtual necessity. Thus the tremendous gain in precision from pps sampling must be weighed against various practical considerations.

Branches on old-growth Douglas fir trees naturally form fan-shaped arrays which we call branch systems. Within crowns of individual trees we have chosen branch systems for sampling with pps where the measure of size was a composite index including information from both epiphyte and tree components (Pike et al., 1977). Since individual branches may be comparable in size to some of the small trees described earlier, branch systems were subsampled in a manner parallel to that described for small trees: *Axes* (stems more than 4 cm in diameter) were sampled using cylindrats. *Branchlets* (systems of twigs less than 4 cm in diameter) are analogous to branches of small trees and were ordered by size and sampled systematically. Weights of epiphytes on branchlets were expanded to the whole branch system using R_2, and estimates of epi-

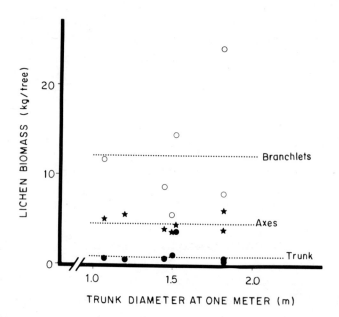

Fig. 2 Estimates of lichen biomass on trunk, axes, and branchlets of old-growth *Pseudotsuga menziesii* in the Andrews Forest, Oregon. Dotted horizontal lines represent several-tree means.

phytic weights on all branchlets of sampled branch systems were expanded to tree totals using a relationship similar to that of Eq. 14. Weights of epiphytes on sampled axes were expanded to total tree estimates in a manner which paralleled that described for branchlet epiphytes.

Estimates for epiphytes on old-growth trees indicate that biomass is fairly consistent on trunks and axes but more variable on branchlets. Epiphyte biomass shows no consistent variation as a function of trunk diameter (Fig. 2), nor does it appear to be related to other tree parameters such as surface area. Thus expanding from sample trees to all trees in the stand by multiplying the mean weight of epiphytes per tree by tree density may be appropriate.

III. ESTIMATION OF PRODUCTION

With the exception of the economically important reindeer lichens (*Cladina* spp.), production estimates have been made for relatively few lichen populations. Although estimates of lichen production can be based on determination of photosynthetic rates through studies of gas exchange, most estimates have been based on changes in biomass employing methods similar to those used for vascular plants (Milner and Hughes, 1967; Newbould, 1967; Šesták et al., 1971). Changes in biomass can be estimated directly by harvesting lichens at appropriate time intervals from randomly selected plots or indirectly by measuring lichens over a period of time and estimating weights on the basis of parallel measurements on other lichens. Furthermore, if lichen ages can be determined on the basis of substrate age or morphology (e.g., the branching pattern), production can be estimated from samples taken at one time. Several examples are discussed in this section to illustrate the diversity of approaches available for estimating lichen production.

If a lichen population is reasonably stable rrom year to year but growth and attrition (due to grazing and decomposition) are out of phase with one another, standing crop will oscillate around some average value and the difference between maximal and minimal standing crops gives a direct estimate of production. The data of Verseghy and K.-Láng (1971) for three grassland communities in Hungary dominated by *Cladonia* spp. can be analyzed in this manner. In all three of the communities studied the fall season appears to account for virtually all of the year's biomass increase. For example, in one community, a perennial open grassland on sandy soil, lichen biomass decreased from an air-dry weight of 1900 kg ha^{-1} in spring to approximately 800 kg ha^{-1} in summer and increased again to 1900 kg ha^{-1} during the fall. The increase of 1100 kg ha^{-1} during the fall represents a relative production of 140% when compared with the summertime standing crop. Estimates made in this manner are expected to be conservative since losses due to decomposition or consumption during the periods of growth are not considered. Since standing crop of one species may be increasing while that of another species is declining, higher estimates of total production are likely if estimates are computed separately for each species and then summed (Wiegert and Evans, 1964).

Increases in thallus area or volume as determined from sequential photography or other measurements may provide the basis for making indirect estimates of lichen production. Additional thalli must be harvested and weighed so that regression equations can be established for predicting weights of measured thalli. Although growth rates of individual thalli for various lichen species have been determined by numerous authors (e.g., see Hale, 1973; Armstrong, 1976), such data have been little explored as a basis for making production estimates for lichen populations. Through the analysis of photographs, Lange and Evenari (1971) have determined the change in area of *Caloplaca aurantia* thalli; they have also related the change in area to the initial thallus area. If it is assumed that the thalli were circular and that each grew at a constant radial rate over the 5-year period, annual relative production for each thallus can be computed (Fig. 3B). If data on size-frequency distribution of thalli in the population were available, it would be possible to compute annual relative production of new area for the entire population. However, inspection of the relative production values for individual lichens suggests that production for the population is unlikely to be more than some 4-6% of the initial area per year. If thallus density increases somewhat as a function of age, as seems likely, estimates of relative production based on areas will be slight underestimates of the relative production of biomass. According to Kappen et al. (1973), estimates of production obtained from measuring gas fluxes for *C. aurantia* agree with those determined from photographs.

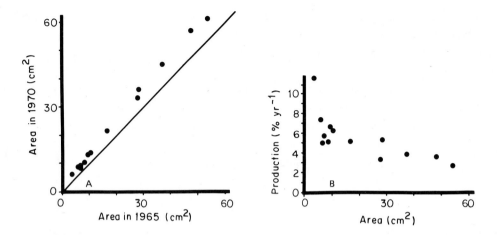

Fig. 3 Growth and production of *Caloplaca aurantia* in the Negev Desert [based on data in Lange and Evenari (1971)]. (A) Area of thalli in 1970 compared with area in 1965. The diagonal represents positions points would occupy if there were no change in thallus area during the interval. (B) Relative production calculated for the year 1965-1966.

Similar methods can be used to make indirect estimates of production for epiphytic lichens; however, the problem of estimating the size-frequency distribution for lichen thalli on tree surfaces is likely to be much more complex than it is for terrestrial lichens. Production estimates have recently been made for *Lobaria oregana* populations growing epiphytically on old-growth Douglas fir in western Oregon by Rhoades (1977, 1978). The relationship between annual increase in area and initial size of a thallus was determined through sequential photography; the relationship between thallus weight and thallus area was determined by harvesting and weighing a large number of measured thalli; and following pps sampling, the size-frequency distribution of the *Lobaria* populations for several host trees was estimated using methods outlined earlier in this chapter and described by Pike and colleagues (1977). Assuming a steady-state for the *Lobaria* population and making corrections for mortality, Rhoades (1978) estimated relative production of four *Lobaria* populations to range from 0.30 to 0.34 g g^{-1} year^{-1}; absolute production was estimated to be 158 kg ha^{-1} year^{-1}.

In situations where thallus age is strongly related to substrate age, accumulation of biomass as a function of substrate age provides a basis for estimating lichen production. I have used this approach for estimating annual production by lichens epiphytic on twigs of *Quercus garryana* in the Willamette Valley of Oregon (Pike, 1971), where the lichen flora is dominated by fruticose species of the genera *Usnea*, *Ramalina*, and *Evernia* that are nearly restricted to twigs less than 20 years old (Pike, 1973). Biomass accumulation was described by removing and weighing lichens from a total of 241 twig sections, each section corresponding to a year's growth.

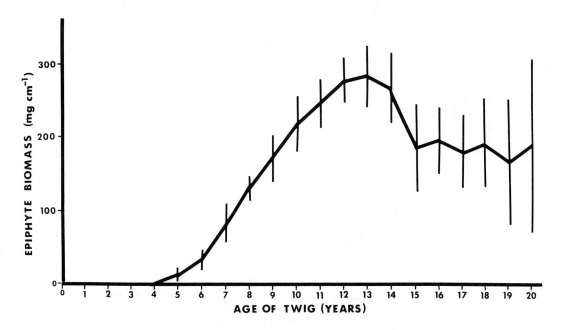

Fig. 4 Epiphyte biomass on *Quercus garryana* twigs in the Willamette Valley, Oregon, as a function of twig age. Vertical lines represent 2 SE on each side of the mean.

The sample consisted of 15-17 sections from each year for age 0 (= current year) to 10, and 5-15 sections from each year for ages 11 to 20. Lichen biomass increased as a function of twig age and reached a peak at age 13 (Fig. 4). The rapid decline after age 13 represents loss of much of the *Usnea* population (fungal decomposition?) and gradual replacement by other species.

An estimate of total net production by all lichens on a twig of age i during the interval i to i + 1 can be made by summing biomass increases during the interval for all lichens on the twig. A steady state was assumed for the lichen population, and biomass increases were estimated from separate age-specific distributions of biomass for each lichen species. In estimating production by lichens on all 0- to 20-year-old twigs, the fact that not all of the twigs of age i remain on the tree at age i + 1 and that the total length of all twigs of age i is greater than that for i + 1 (Fig. 5) must be taken into account. Again these net production estimates are probably conservative since decomposition of lichens on young twigs may lead to underestimates. Although only age-specific distributions of both lichen biomass and relative twig length are necessary for computing relative production (g g^{-1} $year^{-1}$), to compute absolute production (kg ha^{-1} $year^{-1}$) total lichen biomass in the stand must also be estimated. For the *Quercus garryana* forest this was done by multiplying lichen biomass per centimeter of twig by estimates of total twig length per hectare of forest for each twig age. The estimated total length of current-year twigs (11.5 X

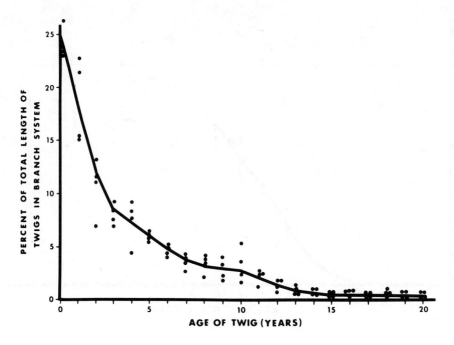

Fig. 5 Relative distribution of total twig length by age in four 20-year-old *Quercus garryana* branch systems. A smoothed curve has been drawn through the points.

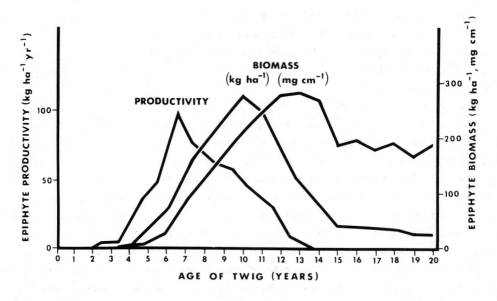

Fig. 6 Biomass and production of epiphytes on 0- to 20-year-old *Quercus garryana* twigs as a function of twig age. Epiphyte biomass is expressed both per unit length of twig and per unit area of forest.

10^6 cm ha^{-1}) was computed as CAB^{-1}, where

A = weight of leaves in annual litterfall (3.4 X 10^3 g ha^{-1}, SE = 0.18)

B = mean weight of litterfall leaf (0.207 g per leaf, SE = 0.0222)

C = number of leaves per centimeter of current-year twigs (1.43 leaves cm^{-1}, SE = 0.09)

Total length of twigs for remaining ages was estimated from relationships illustrated in Fig. 5.

Although lichen biomass per centimeter of twig peaks at age 13, lichen biomass per hectare of forest peaks at age 10 (Fig. 6). Production reaches a peak during the intervals between years 6 and 7 and declines steadily thereafter. Total lichen production during the interval i to i + 1 was computed as

$$P_i = \frac{L_i + L_{i+1}}{2} \sum_{j=1}^{n} (B_{j,i+1} - B_{j,i}) \tag{13}$$

where L is total twig length (cm ha^{-1}) and B_j is biomass per centimeter of twig for species j, and where $B_{j,i+1} - B_{j,i} > 0$. Absolute production by twig lichens was estimated to be 480 kg ha^{-1} year^{-1}, and relative production was estimated to be 0.27 g g^{-1} year^{-1}.

In forests where the bulk of lichen biomass is present on young twigs, it may in general be more satisfactory to sample lichens by relating them to the number of leaves in the canopy since the relationship between leaves and stems is relatively constant and since the number of leaves in the canopy can be estimated quite precisely by sampling the litterfall. It must be pointed out that though such biomass estimates may be relatively precise they will always be underestimates. This bias results from the fact that dead twigs do not have leaves attached. In branch systems destructively analyzed for this study, ages of dead twigs were estimated and were included in the determination of the age-specific distribution of twig length. Although minimized, bias was not removed entirely since completely dead branch systems were not included in the sampling.

A few lichen species exhibit growth patterns analagous to annual rings of a tree, thereby enabling one year's growth to be separated from the growth of other years (Hale, 1973). Thus it is possible to make measurements at one time which enable the reconstruction of the past growth history of a lichen thallus, and this information can provide the basis for making estimates of production. In studying growth rates of species of *Cladina*, various researchers have taken advantage of the fact that certain species produce an "internode" and a whorl of "branches" each year. Thus it is possible to determine the age of a podetium by counting internodes. In a recent study, Prince (1974) estimated biomass and production for a *Cladina* population in Scotland. This population was apparently at steady state with growth at podetial apices keeping up with decay at their bases. Prince estimated biomass at 5400 kg ha^{-1} and an average podetium age of 11 years for an annual production of 470 kg ha^{-1}

or a relative production of 0.087 g g^{-1} year^{-1}. This estimate of production must be viewed as conservative since decay need not be restricted to bases of podetia; data from a more detailed analysis of the distribution of podetium weight as a function of age could help reduce this bias.

Although available data are too incomplete to make precise statements, it is nonetheless interesting to ask how estimates of standing crop and relative production for lichens compare with those for green plants and for free-living fungi. Maximal biomass appears to be on the order of 2-3 t ha^{-1} for epiphytic lichens (Edwards et al., 1960; Scotter, 1962) and 5-8 t ha^{-1} for terrestrial lichens (Prince, 1974; Wein and Speer, 1975). Biomass estimates for free-living soil fungi are few, but two recent estimates give a range of roughly 0.5-1 t ha^{-1} (Gray and Williams, 1971; Parkinson, 1973), which is well within the range of lichens. Biomass of vascular plants is, on the other hand, well studied: for major vegetational units its upper limit (on the order of 400-500 t ha^{-1}) is found in certain forest ecosystems (Olson, 1975; Rodin et al., 1975) and is well above the limits for lichen biomass. Relative production values for vascular vegetation in general range between 3% and 50% per year (Rodin et al., 1975), well within the range of values cited here for lichens; relative production for vascular plant communities would, of course, often be much higher were it not for large amounts of wood and other nonliving tissues. Relative production of soil fungi is presumably much higher than that of lichens and can probably be at least as high as 1000% per year (Nagel-de Boois and Jansen, 1971).

ACKNOWLEDGMENTS

This research was supported in part by NSF grant DEB76-21401 to the author, and in part by NSF grant GB-20963 to the Coniferous Forest Biome, US/IBP. This is contribution No. 324 from the Coniferous Forest Biome.

REFERENCES

Armstrong, R. (1976). Studies on the growth rates of lichens. In *Lichenology: Progress and Problems*, D. Brown, D. Hawksworth, and R. Bailey (Eds.). Academic Press, New York, pp. 309-322.

Becker, V., Reeder, J., and Stetler, R. (1977). Biomass and habitat of nitrogen fixing lichens in an oak forest in the North Carolina Piedmont. *Bryologist 80*: 93-99.

Bergerud, A. (1971). Abundance of forage on the winter range of Newfoundland Caribou. *Can. Field Natur. 80*: 39-52.

Cochran, W. (1963). *Sampling Techniques*, 2nd ed. Wiley, New York.

Crittenden, P. (1975). Nitrogen fixation on the glacial drift of Iceland. *New Phytol. 74*: 41-49.

Edwards, R., Soos, J., and Ritcey, R. (1960). Quantitative observations on epidendric lichens used as food by caribou. *Ecology 41*: 425-431.

Forman, R. (1975). Canopy lichens with blue-green algae: A nitrogen source in a Colombian rain forest. *Ecology 56*: 1176-1184.

Goodman, L. (1960). On the exact variance of products. *J. Amer. Stat. Assoc. 55*: 708-713.

Goodman, L., and Hartley, H. (1958). The precision of unbiased ratio-type estimators. *J. Amer. Stat. Assoc. 53*: 491-508.

Gray, T., and Williams, S. (1971). Microbial productivity in soil. *Symposia Soc. Gen. Microbiol. 21*: 255-286.

Hale, M. (1973). Growth. In *The Lichens*, V. Ahmadjian and M. Hale (Eds.). Academic Press, New York, pp. 473-492.

Hartley, H. (1966). Systematic sampling with unequal probability and without replacement. *J. Amer. Stat. Assoc. 61*: 739-748.

Hartley, H., and Ross, A. (1954). Unbiased ratio estimators. *Nature 174*: 270-271.

Kappen, L., Lange, O., Schulze, E., Evenari, M., and Buschbom, U. (1973). Primary production of lower plants in the desert and its physiological bases. In *Photosynthesis and Productivity in Different Environments*, J. P. Cooper (Ed.). Cambridge Univ. Press, New York.

Kendall, M., and Stuart, A. (1966). *The Advanced Theory of Statistics*, Vol. 3. Griffin, London.

Lange, O., and Evenari, M. (1971). Experimentell-ökologische Untersuchungen an Flechten der Negev-Wuste. IV. Wachstumsmessungen an *Caloplaca aurantia* (Pers.) Helb. *Flora* (Jena) *160*: 100-104.

Milner, C., and Hughes, R. (1968). *Methods for the Measurement of the Primary Production of Grassland*, IBP Handbook No. 6. Blackwell Sci. Publns., Oxford, England.

Nagel-de Boois, H., and Jansen, E. (1971). The growth of fungal mycelium in forest soil layers. *Rev. Ecol. Biol. Sol 8*: 509-520.

Nash, T., White, S., and Nash, J. (1974). Composition and biomass contribution of lichen and moss communities in the hot desert ecoystems. US/IBP, Desert Biome Res. Memo. No. 74-19, pp. 160-165.

Newbould, P. (1967). *Methods for Estimating the Primary Production of Forests*, IBP Handbook No. 2. Blackwell Sci. Publns., Oxford, England.

Nieto de Pascual, J. (1961). Unbiased ratio estimators in stratified sampling. *J. Amer. Stat. Assoc. 56*: 70-87.

Olson, J. (1975). Productivity of forest ecosystems. In *Productivity of World Ecosystems*, D. Reichle, J. Franklin, and D. Goodall (Eds.). National Academy of Sciences, Washington, D.C., pp. 33-43.

Parkinson, D. (1973). Techniques for the study of soil fungi. In *Modern Methods in the Study of Microbial Ecology*, T. Rosswall (Ed.), Ecol. Res. Comm. Bull. No. 17, Swedish Natural Science Research Council, Stockholm.

Pike, L. (1971). The role of epiphytic lichens and mosses in production and mineral cycling of an oak forest. Ph. D. dissertation, University of Oregon, Eugene.

Pike, L. (1973). Lichens and bryophytes of a Willamette Valley oak forest. *Northwest Science 47*: 149-158.

Pike, L., Tracy, D., Sherwood, M., and Nielsen, D. (1972). Estimates of biomass and fixed nitrogen of epiphytes from old-growth Douglas fir. In *Proceedings: Research on Coniferous Forest Ecosystems--a Symposium*, J. Franklin, L. Dempster, and R. Waring (Eds.). Pacific Northwest For. & Range Exp. Sta., Portland, Ore., pp. 177-187.

Pike, L., Rydell, R., and Denison, W. (1977). A 400-year-old Douglas fir tree and its epiphytes: Biomass, surface area, and their distributions. *Can. J. For. Res. 7*: 680-699.

Prince, C. (1974). Growth rates and productivity of *Cladonia arbuscula* and *Cladonia impexa* on the Sands of Forvie, Scotland. *Can. J. Bot. 52*: 431-433.

Raj, D. (1968). *Sampling Theory*. McGraw-Hill, New York.

Rhoades, F. (1977). Growth rates of the lichen *Lobaria oregana* as determined from sequential photographs. *Can J. Bot. 55*: 2226-2233.

Rhoades, F. (1978). Growth, production, litterfall and structure in populations of the lichen *Lobaria oregana* (Tuck.) Müll. Arg. in canopies of old-growth Douglas fir. Ph.D. dissertation, University of Oregon, Eugene.

Rodin, L., Bazilevich, N., and Rozov, N. (1975). Productivity of the world's main ecosystems. In *Productivity of World Ecosystems*, D. Reichle, J. Franklin, and D. Goodall (Eds.). National Academy of Sciences, Washington, D.C., pp. 13-26.

Scotter, G. (1962). Productivity of arboreal lichens and their possible importance to barren-ground caribou (*Rangifer arcticus*). *Arch. Soc. Bot. Fenn. "Vanamo" 16*: 155-161.

Šesták, Z., Čatský, J., and Jarvis, P. (1971). *Plant Photosynthetic Production: Manual of Methods*. Junk, The Hague.

Verseghy, K., and K.-Láng, E. (1971). Investigations of production of grassland communities of sandy soil in the IBP area near Csévharaszt (Hungary). I. Production of lichens. *Acta Biol. Acad. Sci. Hung. 22*: 393-411.

Wein, R., and Speer, J. (1975). Lichen biomass in Acadian and boreal forests of Cape Breton Island, Nova Scotia. *Bryologist 78*: 328-333.

Wiegert, R., and Evans, F. (1964). Primary production and the disappearance of dead vegetation in an old-field in southeastern Michigan. *Ecology 45*: 49-63.

Chapter 28

QUANTIFICATION OF SPOROCARPS PRODUCED BY HYPOGEOUS FUNGI

Robert Fogel[*]
Oregon State University
Corvallis, Oregon

I. INTRODUCTION

Fungi are difficult to study in natural habitats because their minute hyphae ramify throughout their substrate. Most fungi cannot be identified by vegetative character-istics. In the field, even sporocarp-producing fungi whose taxonomy is based on mor-phology often are impossible to identify because of the important role of microscopic characteristics.

Despite the problems in defining individuals and identifying species as well as the ephemeral and seemingly irregular appearance of sporocarps, much can be learned by studying sporocarp production. For example, mycorrhizal fungi--difficult to iso-late by dilution plating or other methods--can be identified, and those species from sites where reforestation is difficult can be used to artificially inoculate planting stock for similar sites. Fungi eaten by mycophagists can be determined. Sporocarps can be used to compare different plant associations. Possible specific or obligate associations, such as that of *Fuscoboletinus ochraceoroseus* (Snell) Pomerlau and Smith with its mycorrhizal host *Larix*, can be inferred. Finally, more can be learn-ed about the reproductive biology of a species, range of variation in morphological characters, and environmental factors controlling sporocarp production.

The hypogeous fungi, those producing sporocarps underground, include members of the Ascomycetes, Basidiomycetes, and a few Endogonaceae. This phylogenetically di-verse group not only produces macroscopic subterranean sporocarps but also generally lacks active spore discharge. Sterile tissue completely encloses sporogenous tissue, and many species are approximately spherical. Most hypogeous fungi are presumed to be ectomycorrhizal (Trappe, 1962, 1971). As mycorrhizal fungi, they play an import-ant role in forest tree nutrition as extensions of the root system, directly draining photosynthates and contributing to soil respiration and nutrient mobilization (Marks and Kozlowski, 1973; Trappe and Fogel, 1977).

Epigeous fungi, those producing sporocarps aboveground, are easier to study than hypogeous fungi. Consequently, sporocarp production by epigeous fungi has been re-

---

[*]*Present address*: Herbarium, University of Michigan, Ann Arbor, Michigan

searched more extensively. Cooke (1948, 1953), Hueck (1953), and Hering (1966) reviewed much of the literature published prior to 1960, but much more has appeared
since then, especially in Europe and Japan. Most of the research can be divided into
three categories (Hering, 1966). First, the epigeous floras of different vascular
plant communities have been summarized by species and by the numbers of sporocarps
each produces (e.g., see Hofler, 1937; Arnolds, 1976). The associated biomass of
epigeous sporocarps has rarely been reported (Richardson, 1970). Second, the structure of the fungus community has been analyzed, usually in relation to the vascular
flora (e.g., see Haas, 1932; Cooke, 1955). Third, sporocarp phenology has been explained by graphs relating it to soil nitrate concentration, temperature, precipitation, and occasionally other environmental factors (e.g., see Wilkins and Harris,
1946; Lange, 1948; Guminska, 1962; Endo, 1972; Thoen, 1976). No statistical or regression analyses of sporocarp production have apparently been reported.

     The ecology of hypogeous fungi has not been comparably researched. The fungal
floras of different communities have been summarized in species lists (Ceruti et al.,
1967; Gross, 1969), and associated biomass has been reported in one study (Fogel, 1976).
Sporocarp production has been related to climatic or edaphic factors (Setchell and
Watson, 1926; Ceruti et al., 1967; Montacchini and Caramiello, 1968; Fogel, 1976).
Mycophagy also has been studied in very little detail (Fogel and Peck, 1975; Fogel
and Trappe, 1978), although the subterranean habit, lack of active spore discharge,
and sterile tissue enclosing spores make mycophagy crucial in spore dispersal. In
turn, mycophagists have become dependent on hypogeous fungi. For example, fungi comprised from 1 to 72% of the food volume consumed yearly by nine species of rodents
(Fogel and Trappe, 1978), and hypogeous fungi comprised 88% of the fungi in the stom-
achs of some small mammals (Maser et al., 1978). The importance of mycophagy to
small mammals also is illustrated by squirrels who dry and cache sporocarps, including hypogeous species (Hardy, 1949).

II. SAMPLING CONSIDERATIONS

The subterranean habit of hypogeous fungi creates several problems in any systematic
sampling scheme. Because the sporocarps are not visible, simply locating the fungi
can be difficult. Ectomycorrhizal hosts on a study site indicate the possible hypogeous flora. Other clues include small pits excavated by rodents and small mounds
raised by sporocarps fruiting at the soil-litter interface. Sporocarps also fruit
in rotten logs that have been invaded by roots of ectomycorrhizal hosts. Patches of
dead grass and the presence of small flies (Mycetophilidae) hovering over ripe sporocarps are clues for Italian truffle collectors (Singer, 1961). The much publicized
use of pigs and dogs to locate truffles depends on their ability to smell ripe sporocarps of a few truffle species, and the animals probably could not efficiently locate
immature sporocarps in quantitative studies.

Once the sporocarps are located, sampling requires removing the forest floor and raking the top 10 to 20 cm of mineral soil. This seriously disturbs quadrats, and new quadrats are needed for each sampling date.

Other sampling considerations include species differences in spatial distribution and sporocarp phenology. Sporocarp distribution is aggregated, not random, and sporocarp fruiting occurs in three distinct patterns (Fogel, 1976): (1) single sporocarps or a few widely scattered over a large area (e.g., 25 m^2); (2) large (2–4 m^2), loose clusters; and (3) arcs and partial arcs ("fairy rings") formed by tight clusters of sporocarps (Fig. 1).

The fruiting of individual species and of the population as a whole varies seasonally, and it apparently depends on the climate of the study area. In western Oregon, for instance, production is bimodal (Fig. 2) with winter and summer valleys due to summer drought and cold winter temperatures (Fogel, 1976). In England (Fig. 3) and Scotland, where sporocarp production is not limited by summer drought, production of epigeous and presumably hypogeous species peaks in the fall (Grainger, 1946; Richardson, 1970).

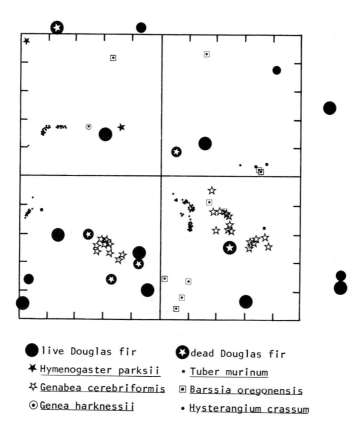

● live Douglas fir   ✪ dead Douglas fir

✱ <u>Hymenogaster parksii</u>   • <u>Tuber murinum</u>

✩ <u>Genabea cerebriformis</u>   ▣ <u>Barssia oregonensis</u>

⊙ <u>Genea harknessii</u>   • <u>Hysterangium crassum</u>

Fig. 1 Distribution of hypogeous sporocarps on a 10-m X 10-m quadrat subdivided into 5-m X 5-m subplots in relation to Douglas-fir stems larger than 4 cm dbh (diameter at breast height) at Woods Creek, Oregon. (From Fogel, 1976.)

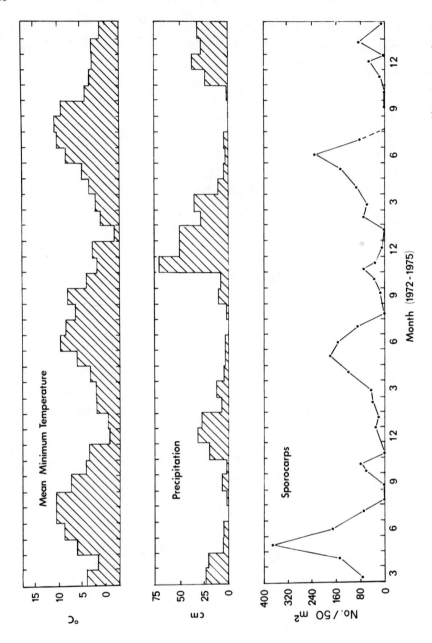

Fig. 2 A 3-year comparison of sporocarp production by hypogeous fungi with corresponding temperature and precipitation in western Oregon. (From Fogel, 1976.)

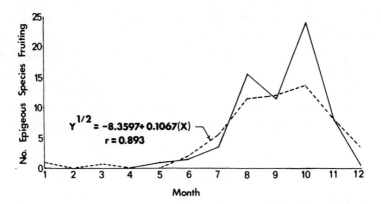

Fig. 3 Simulation of the number of epigeous species fruiting in England. Solid line is the observed number fruiting, and the broken line is the predicted number. (After Grainger, 1946.)

Seasonal and spatial differences in sporocarp production make the minimal area for sampling difficult to determine. Mueller-Dombois and Ellenberg (1974) defined minimal area for population sampling either as the quadrat size that contains 90-95% of one species or as the species number asymptote derived from a species-area curve. The former definition is inoperable for hypogeous fungi because the community represented by sporocarps constantly changes. The asymptote derived for hypogeous populations depends on the seasonal abundance of sporocarps. For example, species-area curves constructed from data collected on 10-m X 10-m quadrats in western Oregon indicate that the minimal area ranges from 0 to 100 m^2 betwen minimum and maximum production. Unfortunately, the time factor (4-6 man-hours needed to sample 100 m^2) severely restrains the number of quadrats that can be thoroughly searched during maximum sporocarp production.

III. COMMUNITY AND POPULATION ATTRIBUTES

Data obtained by irregular collecting of sporocarps can be used to compile species lists for different areas, but the usefulness of such information is limited. Much more can be learned by simply sampling quadrats at periodic intervals. This section begins by comparing hypogeous floras, then explores several attributes of the hypogeous community that can be examined if data are collected periodically.

A. Flora

A given area probably has more epigeous than hypogeous species, although this has not been substantiated. Floristic lists contain far more epigeous than hypogeous species. Epigeous floras in Europe and Japan include 28 to 205 species (Parker-Rhodes, 1951; Hering, 1966; Richardson, 1970; Endo, 1972; Smarda, 1973; Thoen, 1976). Maas and

Stuntz (1970) collected 134 epigeous species from a nonserpentine mixed conifer (*Pseudotsuga-Abies-Pinus*) stand in the Cascade Mountains of Washington. Over a 3-year period, I collected 24 hypogeous species (11 ascomycetes, 13 basidiomycetes) in a young stand of Douglas-fir in western Oregon (Fogel, 1976); however, during the past year, I found only 11 species in a nearby stand, possibly the result of drought last year in the Pacific Northwest. Seventeen hypogeous species have been reported for an Italian oak stand, 12 for a German red beech stand, and 17 for a German spruce stand (Ceruti et al., 1967; Gross, 1969). Because hypogeous fungi appear mycorrhizal, hypogeous and epigeous floras can be compared better if only epigeous mycorrhizal species are considered. Of the 28 species listed by Richardson (1970), 12 (42.9%) were listed by Trappe (1962) as possible mycorrhizal associates. Of the 134 species listed by Maas and Stuntz (1970), 22 (16%) presumably were mycorrhizal.

Sampling for more than 1 year may be necessary to reliably estimate total species number. For example, at Woods Creek, Oregon, I collected 68% of the theoretical hypogeous flora during the first year, 88% the second year, and 96% by the end of the third year (Fig. 4) (Fogel, 1976). Parker-Rhodes (1951) presented a statistical method useful for reducing the number of collections needed to estimate the total number of species per site.

B. Abundance: Number of Sporocarps and Biomass

Yearly production of hypogeous sporocarps ranged from 11,052 to 16,753 ha^{-1} in a young Douglas-fir forest in western Oregon (Fogel, 1976). Other hypogeous studies (Ceruti et al., 1967; Gross, 1969) did not specify plot size, so the three studies cannot be compared. Estimates of epigeous sporocarp numbers range from 7000 to 489,000 sporocarps ha^{-1} year^{-1} (Hering, 1966; Richardson, 1970). Epigeous sporocarp numbers closely approximate hypogeous numbers if mycorrhizal species are compared.

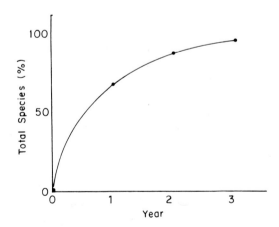

Fig. 4 Percentage of total number of hypogeous species fruiting in each year of a 3-year study at Woods Creek, Oregon. (Author's unpublished data.)

For example, 8750 to 20,250 (1.8-8.5%) of the 239,000 to 489,000 epigeous sporocarps produced annually per hectare in Richardson's (1970) study are mycorrhizal compared to the 11,052 to 16,753 hypogeous sporocarps per hectare reported by Fogel (1976).

The dry weight produced by hypogeous sporocarps ranges from 2.3 to 5.4 kg ha^{-1} year^{-1} (Fogel, 1976). Hering (1966) estimated that epigeous species would produce 0.19 to 19.2 kg dry weight ha^{-1} year^{-1} [using Richardson's (1970) conversion factor of 6.36% to convert fresh weight to dry weight]. Dry weight of epigeous mycorrhizal species (9.8-19.4 kg ha^{-1} year^{-1}) accounted for slightly more than half of the total sporocarp biomass (16-30 kg ha^{-1} year^{-1}) reported by Richardson (1970).

C. Frequency

Frequency of individual species is difficult to measure due to the large minimal area required for sampling and the constant change in fruiting species. Total frequencies of all species ranged from 0 to 68% and species number ranged from 0 to 10 when 50 new 1-m X 1-m quadrats were sampled monthly; when four new 5-m X 5-m quadrats were sampled monthly, the total frequency ranged from 0 to 100%, representing from 0 to 5 species per month (R. Fogel, unpublished data). Therefore, large quadrats with a higher frequency of sporocarps should be used to quantify biomass or numbers, and a large number of small quadrats should be used in floristic studies where the goal is to maximize species number.

D. Diversity

A floristic list also reflects a site's species richness. Diversity can be estimated by the number of species in a sample of standard size, by the steepness of the importance-value sequence (i.e., the Simpson index), or by an index of the relative evenness of the importance values through a sequence (i.e., the Shannon-Wiener index) (Whittaker, 1972). Shannon-Wiener index values (base 2) for hypogeous fungi from the Fogel (1976) site (Woods Creek) and for a new site located 3 km from the original (Dinner Creek) are 3.14 and 2.17. The value for Dinner Creek epigeous fungi was 3.67. For the first year of each study, Woods Creek had 13 hypogeous species, and Dinner Creek had 12 hypogeous and 55 epigeous species. Apparently, a larger proportion of the epigeous species are infrequently encountered. The low values might reflect the low diversity of the vascular plant flora, i.e., four tree species per site. Beta diversity, defined by Whittaker (1972) as the differentiation of communities along gradients, has not been estimated.

E. Coefficient of Community

The similarity in species composition between two sites can be calculated using the coefficient of community, c_c:

$$c_c = \frac{200(s_{xy})}{s_x + s_y} \qquad (1)$$

Table 1 Middates of fruiting by year for common species of hypogeous fungi in western Oregon[a]

Species	Year 1	Year 2	Year 3	Mean	Standard deviation	Fruiting period
Barssia oregonensis	24 May	21 May	12 Apr.	9 May	23	March–July
Genabea cerebriformis	--	12 Feb.	22 May	2 Apr.	70	March–July
Genea harknessii	9 May	18 Mar.	14 Dec.	15 Mar.	56	Nov.–June
Hymenogaster parksii	--	18 Jan.	9 Nov.	14 Dec.	50	Oct.–June
Hysterangium crassum	29 Apr.	28 Mar.	15 Apr.	14 Apr.	16	Sept.–July
Hysterangium separabile	27 Apr.	9 Apr.	30 Apr.	22 Apr.	11	Sept.–Aug.
Truncocolumella citrina var. *citrina*	10 Oct.	1 Oct.	17 Nov.	20 Oct.	25	Sept.–Dec.
Tuber murinum	28 May	6 May	10 June	22 May	15	Feb.–July
Spring peak	--	23 Apr.	11 May	2 May	13	--
Fall peak	10 Oct.	30 Oct.	8 Dec.	5 Nov.	30	--

[a]From Fogel (1976).

Table 2 Species accounting for 5% or more of productivity of hypogeous fungi in western Oregon[a]

	Percentage of total weight		
Species	Year 1	Year 2	Year 3
Gautieria sp.	--	14.5	--
Hymenogaster parksii	8.0	12.2	--
Hysterangium crassum	14.6	17.4	53.2
Hysterangium separabile	9.3	19.2	13.0
Truncocolumella citrina			
var. *citrina*	43.2	--	--
Tuber gibbosum	7.0	--	--
Tuber murinum	6.8	8.3	7.1
Total	88.9	71.6	73.3

	Percentage of total sporocarp no.		
Species	Year 1	Year 2	Year 3
Barssia oregonensis	5.6	5.1	--
Genabea cerebriformis	--	6.3	--
Genea harknessii	13.6	--	--
Hymenogaster parksii	5.8	10.6	--
Hysterangium crassum	26.1	25.9	65.8
Hysterangium separabile	18.7	11.0	14.0
Truncocolumella citrina			
var. *citrina*	9.6	--	--
Tuber murinum	10.0	18.7	6.1
Total	89.4	77.6	85.9

[a]From Fogel (1976).

where s_x and s_y are the number of species in quadrats x and y (Pielou, 1975). In the first year of the Fogel (1976) study and the Dinner Creek study, 64.5% of species collected at each site occurred at both. Normally I would expect more similarity because the sites, only 3 km apart, have comparable aspect and stand structure, but the 1976-1977 drought in the Pacific Northwest may have reduced the number of species fruiting at Dinner Creek.

F. Middate of Fruiting

The middate of fruiting, which permits comparison of different sites and prediction of species abundance on a given date, is calculated by the formula:

$$m = \frac{d(n)}{N} \tag{2}$$

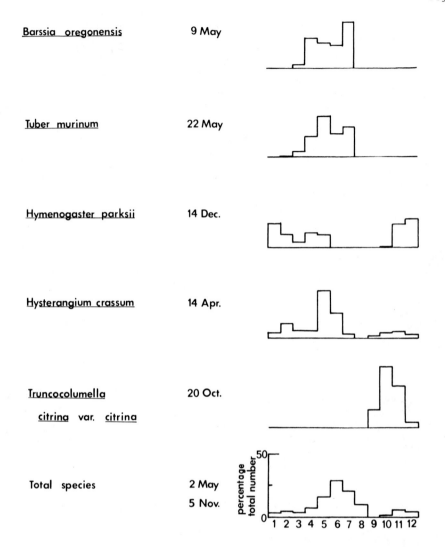

Fig. 5 Sporocarp phenology and midpoint of fruiting for major species and the total species population of hypogeous fungi at Woods Creek, Oregon. (From Fogel, 1976.)

where m is the midpoint in days after the starting point, n is the number of sporo-carps collected d days after the starting point, and N is the total number of sporo-carps collected (Richardson, 1970). Richardson's (1970) epigeous population midpoints for a 5-year study in Scotland range from September 11 to 26, with a standard devia-tion of 6 days; production peaked in the fall. The mean midpoints for a 3-year study of hypogeous production in Oregon (Table 1) were May 2 and November 5, with standard deviations of 13 and 30 days. The two dates reflect the bimodal production of sporo-carps due to summer drought and winter cold, and the standard deviations might reflect the greater variability of western Oregon's climate. The fall middate in Oregon, for instance, is determined by the timing of the first heavy fall rain and the elapsed time to the first freezing temperatures (Fig. 5).

G. Sporocarp Longevity

Richardson (1970) also reported a linear relationship between mean epigeous sporocarp longevity and the slope of the population decay curve for sporocarps of *Russula emetica* (Schaef. ex. Fr.) S. F. Gray. The population decay curve was a function of the percentage of total sporocarps observed for a species and the mean time interval between observations. The calculated sporocarp longevity closely corresponded to observed maximum life of marked sporocarps. Hypogeous species have not been similarly analyzed.

H. Major Species

Hering (1966) classified as major species those contributing 5% to the total number of species or biomass production (Table 2). The major species by weight or number generally differed among the 3 years of the Oregon hypogeous study (Fogel, 1976). Major species by weight comprised some 16-24% of the total species, but many commonly occurring species--especially those with small, hollow ascocarps--did not contribute greatly to total sporocarp biomass.

IV. PHENOLOGY

Most mushroom collectors intuitively sense that sporocarp production is related in some way to temperature, precipitation, or both. Fruiting and environmental factors have been graphically compared many times, but apparently only one study has attempted to predict production. Matveev (1972) developed a method to predict the date for mass fruiting of epigeous species in a birch forest near Leningrad. The method, basically a heat-sum approach, also uses the average period for development of sporocarps for each species and the dates of warm ($\geq 12°C$) spring rains. Unfortunately, Matveev did not quantitatively define mass fruiting (numbers or biomass), and he did not compare observed and predicted fruiting dates.

Visual comparison of hypogeous sporocarp production versus mean monthly temperature and total monthly precipitation (Fig. 2) shows that summer drought and cold winter temperature appear to control sporocarp production in western Oregon (Fogel, 1976). Time-series analysis of my Oregon data failed to detect any pattern (R. H. Strand, Oak Ridge National Laboratory, Oak Ridge, Tennessee, personal communication, 1975), and simple linear regressions of sporocarp biomass, mean monthly temperature, and total monthly precipitation were not significant (Table 3). I then stratified the data by assuming that precipitation was limiting during the summer (mean monthly air temperature > 14°C) and that precipitation and temperature were limiting during the rest of the year. The correlation between biomass and temperature was highly significant ($r = -0.818$, $p > 0.01$), as was the multiple correlation between biomass

Table 3 Correlation between biomass of hypogeous sporocarps, mean monthly air temperature, and total monthly precipitation[a]

Variable	$\overline{X}$ monthly air temperature (°C)	r	n	Site[b]
Temperature	≤14	0.646**	20	WC
	≤14	0.711*	8	DC
	>14	−0.818**	12	WC
	>14	−0.754*	6	DC
	1.7 to 19.9	−0.034	36	WC
	3.0 to 19.9	0.079	14	DC
Precipitation	≤14	−0.417*	24	WC
	≤14	−0.333	8	DC
	>14	0.676*	12	WC
	>14	−0.520	6	DC
	1.7 to 19.9	−0.157	36	WC
	3.0 to 19.9	−0.221	14	DC
Temperature and precipitation	≤14	0.646**	20	WC
	>14	−0.828**	11	WC

[a]The data have been analyzed as collected and stratified into summer (temperature > 14°C) and winter (temperature ≤ 14°C) periods.

[b]Site abbreviations are WC for Woods Creek, Oregon (Fogel, 1976) and DC for Dinner Creek, Oregon (Fogel, unpublished data).

and temperature plus precipitation ($r = -0.828$, $p > 0.01$). The correlation between biomass and total precipitation of the previous month was slightly less significant ($r = 0.676$, $p > 0.05$). During the rest of the year, when the temperature was equal to or less than 14°C, the correlation between biomass and temperature was highly significant ($r = 0.646$, $p > 0.01$), as was the multiple correlation with temperature and precipitation ($r = 0.665$, $p > 0.01$) (Table 3). The correlation between biomass and precipitation was slightly less significant ($r = -0.417$, $p > 0.05$).

I then used a simple model incorporating two of the linear regression equations to predict hypogeous sporocarp biomass in western Oregon (Fig. 6). The correlation between the data used in the regressions and the predicted production was highly significant ($r = 0.648$, $p > 0.01$, $n = 32$), and the visual fit of the two curves was good, although the amplitude of the predicted production was consistently lower than that observed. The model was then used to predict the sporocarp production of Dinner Creek. The correlation was highly significant between observed and predicted production for Dinner Creek, but the fit was poor (Fig. 7). This might be due to the severe drought during the sampling period, because the only significant correlation

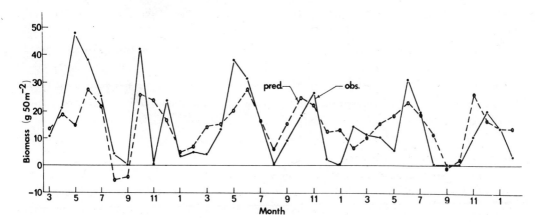

Fig. 6 Simulation of Woods Creek hypogeous sporocarp production from 1972 to 1975. (Author's unpublished data.)

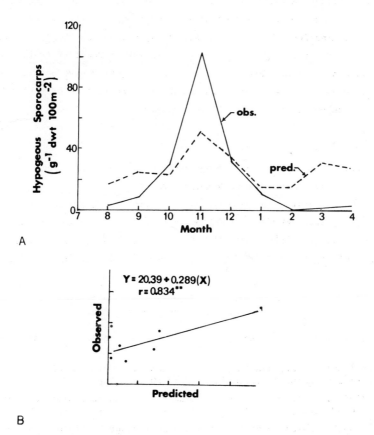

Fig. 7 (A) Simulation of Dinner Creek hypogeous sporocarp production by the Woods Creek model. (B) Correlation between predicted and observed sporocarp biomass. (Author's unpublished data.)

was between biomass and temperature (r = -0.754, p > 0.05) (Table 3). Site factors such as cover and aspect might also have had an effect.

Grainger's (1946) data of sporocarp production and environmental data for a different climatic area had a highly significant correlation (r = 0.893, p > 0.01) between the square root of the number of epigeous species fruiting and mean monthly air temperature (Fig. 3). The correlation between production and precipitation was not significant. Precipitation was not limiting in Grainger's study because of the characteristic high level of summer precipitation in England, contrasting markedly with western Oregon's dry summers.

In summary, sporocarp production by hypogeous and epigeous fungi apparently correlates with environmental factors, and those that are limiting vary with season and climatic region. Other factors, perhaps site or floristic, will have to be incorporated into any general sporocarp production model.

V. SUMMARY

Systematic and periodic sampling of hypogeous sporocarps can yield more information than can collection methods conventionally used by mycologists. Systematic collection permits determination of spatial distribution, phenology, flora, abundance, frequency, species diversity, coefficient of community, middate of fruiting, sporocarp longevity, and major species. Most of these attributes have been investigated only in one western Oregon stand, and different climatic regions or variation along gradients cannot be compared without further research.

ACKNOWLEDGMENTS

Figures 1, 2, and 5 and Tables 1 and 2 were used by permission of the *Canadian Journal of Botany* (National Research Council of Canada).

The preparation of this paper was supported by National Science Foundation Grant DEB 76-10188 and done in cooperation with the Pacific Northwest Forest and Range Experiment Station, Forestry Sciences Laboratory, Corvallis, Oregon. This is contribution No. 1223 from the Forest Research Laboratory, Oregon State University, Corvallis.

REFERENCES

Arnolds, E. J. M. (1976). Sociology and ecology of mushrooms in grasslands. *Coolia 19*: 72-85.

Ceruti, A., Montacchini, F., and Duployez, C. (1967). Ecologia dei funghi ipogei dell' "Arboretum Taurinense." *Allionia 13*: 89-105.

Cooke, W. B. (1948). A survey of the literature on fungus sociology and ecology. I. *Ecology 29*: 376-382.

Cooke, W. B. (1953). A survey of the literature of fungus sociology and ecology. II. *Ecology 34*: 211-222.

Cooke, W. B. (1955). Fungi, lichens and mosses in relation to vascular plant communities in Eastern Washington and adjacent Idaho. *Ecol. Monogr. 25*: 119-180.

Endo, M. (1972). Plant sociological observations on the terrestrial larger fungi in an evergreen broad-leaved forest. (In Japanese.) *Japan. J. Ecol. 22*: 51-61.

Fogel, R. (1976). Ecological studies of hypogeous fungi. II. Sporocarp phenology in a western Oregon Douglas-fir stand. *Can. J. Bot. 54*: 1152-1162.

Fogel, R., and Peck, S. B. (1975). Ecological studies of hypogeous fungi. I. Coleoptera associated with sporocarps. *Mycologia 67*: 741-747.

Fogel, R., and Trappe, J. M. (1978). Fungus consumption (mycophagy) by small animals. *Northwest Sci. 52*: 1-31.

Grainger, J. (1946). Ecology of the larger fungi. *Trans. Brit. Mycol. Soc. 29*: 52-63.

Gross, G. (1969). Einiges über die Hypogäenuche. *Z. Pilzkunde 35*: 13-20.

Guminska, B. (1962). The fungi of the beach forests of Rabsztyn and Maciejawa. *Monogr. Bot. 13*: 3-38.

Haas, H. (1932). Die bodenbewohnenden Grosspilze in den Waldformationen einiger Gebiete von Wurttemberg. *Beih. Bot. Zbl. 50*: 35-134.

Hardy, G. A. (1949). Squirrel cache of fungi. *Can. Field Natur. 63*: 86-87.

Hering, T. F. (1966). The terricolous higher fungi of four Lake District woodlands. *Trans. Brit. Mycol. Soc. 49*: 369-383.

Hofler, K. (1937). Pilzsoziologie. *Ber. Deut. Bot. Ges. 55*: 606-622.

Hueck, H. J. (1953). Myco-sociological methods of investigation. *Vegetatio 4*: 84-101.

Lange, M. (1948). The agarics of Maglemose. *Danske Bot. Ark. 13*: 1-141.

Maas, J. L., and Stuntz, D. E. (1970). Mycoecology on serpentine soil. *Mycologia 61*: 1106-1116.

Marks, G. C., and Kozlowski, T. T. (1973). *Ectomycorrhizae*. Academic Press, New York, 444 pp.

Maser, C., Trappe, J. M., and Nussbaum, R. A. (1978). Fungal-small mammal interrelationships with emphasis on Oregon coniferous forests. *Ecology 59*: 799-809.

Matveev, V. A. (1972). Forecasting fruiting of edible mushrooms. [In Russian.] *Lesnoe Khozyaistivo 9*: 27-28.

Montacchini, F., and Caramiello, R. (1968). Ecologia del *Tuber magnatum* Pico in Piemonte. *Allionia 14*: 1-29.

Mueller-Dombois, D., and Ellenberg, H. (1974). *Aims and Methods of Vegetation Ecology*. Wiley, New York, 547 pp.

Parker-Rhodes, A. F. (1951). The basidiomycetes or Skokholm Island: Some floristic and ecological calculations. *New Phytol. 50*: 227-243.

Pielou, E. C. (1975). *Ecological Diversity*. Wiley, New York, 165 pp.

Richardson, M. J. (1970). Studies on *Russula emetica* and other agarics in a Scots pine plantation. *Trans. Brit. Mycol. Soc. 55*: 217-229.

Setchell, W. A., and Watson, M. G. (1926). Some ecological relationships of the hypogeous fungi. *Science 63*: 313-315.

Singer, R. (1961). *Mushrooms and Truffles*. Wiley (Interscience), New York, 265 pp.

Smarda, F. (1973). The fungal communities of several spruce forests of Moravia. [In German.] *Prirodoved Pr. Ustavu. Cesk. Akad. Ved. Brne. 7*: 1-44.

Thoen, D. (1976). An inquiry into the myocoenoses of the fir tree plantings in the southern Ardennes. *Coolia 19*: 67-71.

Trappe, J. M. (1962). Fungus associates of ectotropic mycorrhizae. *Bot. Rev. 28*: 538-606.

Trappe, J. M. (1971). Mycorrhiza-forming ascomycetes. In *Mycorrhizae*, E. Hacskaylo (Ed.). U.S. Dept. of Agriculture, Misc. Publ. No. 1189, pp. 19-37.

Trappe, J. M., and Fogel, R. D. (1977). Ecosystematic functions of mycorrhizae. In *The Belowground Ecosystem: A Synthesis of Plant-Associated Processes*, J. Marshall (Ed.). Sci. Ser. No. 26, Range Sci. Dept., Colorado State University, Fort Collins, Colo., pp. 205-214.

Whittaker, R. H. (1972). Evolution and measurement of species diversity. *Taxon 21*: 213-251.

Wilkins, W. H., and Harris, G. C. M. (1946). The ecology of the larger fungi. V. An investigation into the influence of rainfall and temperature on the seasonal production of fungi in a beechwood and a pinewood. *Ann. Appl. Biol. 33*: 179-188.

Chapter 29

FUNGAL TAXA, PHYSIOLOGICAL GROUPS, AND BIOMASS:
A COMPARISON BETWEEN ECOSYSTEMS

Patrick W. Flanagan

Institute of Arctic Biology
University of Alaska
Fairbanks, Alaska

I. INTRODUCTION

Fungi in the form of yeasts, mushrooms, toadstools, and molds have occupied man's general interest for a long time. However, scientific investigation of fungi, especially invisible soil fungi, is a relative newcomer to biological sciences, the work of Adametz (1886) being the starting point for serious soil mycological studies. The communications of Jensen (1912) and Waksman (1916a,b, 1917) ask the question as to whether there is any fungus flora of the soil and further attest to the recency of soil mycological studies. Adametz and several others including Reinitzer (1900) and Nikitinsky (1902) studied their isolates for clues to their biochemical or ecological role in soil.

As soon as soil fungi were accepted as natural entities in soils, their role therein became of interest, especially to those investigating soil fertility. Thaysen and Bunker (1927) presented a comprehensive review of the activities of fungi in hydrolysis of cellulose, hemicellulose, pectin, and gums.

During more recent years numerous publications have proven the ubiquitous occurrence of fungi in soils. Lists of taxonomic groups within and between different soils are common to the literature of fungal ecology, i.e., mycoecology. But taxonomic listings tell us little of the role played by soil fungi in ecosystem development and maintenance and nothing of the composite fungal niche in ecosystems.

From the beginning of the twentieth century fungal activities in the soil have been discussed and investigated in relation to their role in ammonification, nitrification, nitrogen transformation, cellulose decomposition, and humification. These investigations concerned themselves with organisms isolated from the soil. The methods and approaches used in different sites by different workers varied, thus forestalling comparative studies among different soils. Until Jones and Mollison (1948), soil fungi were quantified in terms of the number of their occurrences on soil dilution agar plates. The inadequacy of such quantification need not be stressed here, but it should be pointed out that the use of the soil dilution isolation technique for fungi probably led to the generally held concept that soil bacteria were the

major components of soil microfloras and were the main agents in organic decay and
nutrient cycling processes.

The range and magnitude of metabolic capabilities for invisible soil microorganisms have been constructively reviewed recently by Clark (1975). As the number of
reports multiply it is becoming established that fungi, not bacteria, are the major
constituents of soil microfloras (Anderson and Domsch, 1973; Satchell, 1974; Flanagan
and Van Cleve, 1977). Because of the high relative activity of microorganisms [i.e.,
the ratio of energy flow to biomass, which has been estimated to be as high as 1000
(Odum, 1968)], it is obvious that study of major soil microorganisms is essential to
an understanding of ecosystem function and soil fertility.

The biomass of fungi in soil is not easily measured. It is possible to determine the length of fungal filaments per gram of mineral soil or litter, but the proportion of live and dead hyphae in substrates is not so easily resolved. These problems have been discussed most recently by Frankland (1974, 1975) and Söderström
(1977). The problems of "ghost hyphae" (Frankland, 1974) and the resolution of live
hyphae may be overcome by the addition of 2% aceto-orcein, a nuclear stain, to the
material being examined prior to using either the method of Hanssen et al. (1974) or
that of Jones and Mollison (1948) (see Sec. II).

Fungal biomass expressed as meters of hyphae per gram of soil or as grams per
square meter shows wide spatiotemporal variation within ecosystems (Flanagan and Van
Cleve, 1977; Söderström, 1977; Bunnell et al., 1978). Temporal variation is greatest
in the spring and fall in high latitude sites, presumably through stimulation from
incoming nutrients and nonstructural carbon compounds. Studies which did not use
some method to distinguish live from dead fungal hyphae do not show the seasonal
fluctuation in fungal biomass (Nagel-de Boois and Jansen, 1971). Expressing fungal
biomass in meters per gram of dry weight soil (DWS) without regard to the quantity of
organic matter involved must be misleading.

In the following discussion I will attempt to deal with the detectable fungal
components from seven widely separated and different soils, both in terms of their
potential to decay substrates and in terms of their biomass in relation to stated
variables such as the quantity of organic matter. I hypothesize that different microbial taxa from various soils do not show significant differences in their capabilities for attack of indigenous plant residues. Thus, a universal decomposer system
exists whose member microorganisms differ from place to place but whose specific
catabolic capabilities enable it to decay all natural organic residues, limited only
by climatic extremes and by substrate availability. The latter constraint will be
reflected in the dimensions of the decomposer community, i.e., the microbial biomass
and its turnover rate. Turnover rate, or generation times, will not be discussed
here, but work on this important aspect of microbial ecology has begun and is reported in Babiuk and Paul (1970), Gray and Williams (1971), Flanagan and Bunnell (1976),
and Flanagan and Van Cleve (1977).

II. STUDY SITES AND METHODOLOGY

A. Study Areas

The sites were selected more by coincidence than by design. Investigations were carried out in Alaska concurrent with U.S. International Biological Program (US/IBP) tundra investigations, National Science Foundation (NSF) sponsored research on taiga ecosystems, and oil industry sponsored research on taiga forest areas. Two temperate sites, a grassland and a sphagnum-peat bog site in western Ireland were examined during 1974-1975 while the present author was at Galway University.

The site at Pt. Barrow, Alaska (71°18'N, 156°40'W), was in the US/IBP Tundra Biome intensive research area. This site is described in detail by Bunnell et al. (1975). The study plots were covered by an almost pure stand of *Eriophorum angustifolium*.

The Alaskan Eagle Summit site, also a US/IBP Tundra Biome site, situated at 65°28'N and 145°23'W, has been described by Anderson (1974). The microbiological investigations were performed in a pure stand of *Dryas octapetala* at 1200 m in an alpine tundra environment.

Study sites in interior Alaska included a black spruce forest (*Picea mariana* [Mill] B.S.P.), an aspen forest (*Populus tremuloides*), and a paper birch forest (*Betula papyrifera*).

The black spruce forest floor is covered with *Hylocomium spendens* and frequent patches of *Pleurozium schreberi* and *Sphagnum* sp. Emergent from moss layers, *Ledum groenlandicum* and an occasional *Rosa acicularis* compose the shrub layer. *Vaccinium vitis-idaea* and *Oxycoccus macrocarpus* are the dominant herbaceous forms.

The aspen stand was typical of those on south-facing slopes in interior Alaska. Bear berry (*Arctostaphylos uva-ursi*) was dominant in the understory and occasional willows in lower reaches of the stand. *Viburnum* sp. occurred occasionally.

The paper birch stand contained *Alnus* sp. intermingled with the major trees. Open areas between trees were dominated by *Calamagrostis*, while the entire forest floor was covered by *Equisetum arvense*. Patches of *Hylocomium splendens* occurred among the *Equisetum*.

The temperate sites studied were both in western Ireland. A limestone subtended grassland site was selected on the western bank of Lough Corrib, about a half mile (ca. 0.8 km) north of Galway University. *Festuca* spp. accounted for approximately 70% of plant cover, while *Trifolium repens*, *Ranunculus repens*, and *Agrostis tenuis* made up the remainder. In this site flushes of *Pteridium aquilinium* occurred occasionally where the meadow abutted walls or alder thickets.

A temperate peat bog (blanket bog) approximately 30 miles (ca. 48.6 km) west of Galway City in the Maam Valley was studied between April and November. The peat, which never freezes to a significant depth or for much time, is permanently waterlogged, averages 2-5 m depth, and is frequently interrupted by small (1-10 m diameter) bog pools containing acidic water. Ericaceous shrubs including *Calluna vulgaris*,

Erica tetralix, and *E. cinerea* predominate. *Eriophorum angustifolium*, *E. vaginatum*
and *Scirpus caespitosus* are the primary graminoids on the less wet areas, while wetter spots and pool margins contain *Juncus effusus*, *Schoenus nigricans*, and *Molinia
caerulea*. The bryophyte community is composed predominately of *Sphagnum* spp.

The study sites, all in the Northern Hemisphere, occur between the latitudes
55° and 72°. There is a wide range in annual patterns of air temperature between
sites (Table 3), and annual precipitation across sites varies between 1300 mm (temperate peat bog) and 124 mm (Pt. Barrow tundra), the mean between extremes, 675 mm,
occurring at the Eagle Summit site.

B. Isolation of Fungi and Enzyme Hydrolytic Capability

Fungi were isolated from standing dead gramineous leaves and from L, F, H, and A_1
soil horizons, where such were discernible, using the particle-washing technique of
Parkinson and Williams (1961). Fungi were not isolated from tree leaves until they
had fallen as litter. In order to detect seasonal colonization patterns, fungal isolations were made at each site from all plant residues and soil horizons at four separate times throughout at least one field season.

Identification of fungal isolates was carried out in our own laboratories, and
further identification and confirmation was provided through the help of the staff
of the Centraalbureau voor Schimmelcultures, Baarn, Netherlands. Percent frequency
of an isolate was calculated as follows:

$$\frac{\text{number of plated particles producing a species}}{\text{total number of particles plated}} \times 100$$

Fungal enzyme content and capability to degrade cellulose, pectin, starch, humic
acid, and gallic acid were measured as described by Flanagan and Scarborough (1974).

Lignocellulose dowels were powdered by shaving them in pencil sharpeners and
then passing them through a Wiley mill. The powdered lignocellulose was the sole
carbon source in a mineral salt agar layer poured as a thin overlay on plain agar
medium. Organisms capable of degrading lignin and cellulose in combination will
clear a zone among the closely crowded wood particles. Ability to utilize xylan was
measured in the manner described by Domsch and Gams (1969). Organisms giving positive responses at room temperature were examined for the same activity at 2-5°C.

Species which together accounted for >80% of all species were considered the
major species components (exclusive of basidiomycetes). In-site potential in a decomposition characteristic, e.g., cellulose decomposition ability was calculated:

$$\frac{\text{frequency sum of species with a given activity}}{\text{frequency sum of major species components}} \times 100$$

C. Biomass Estimates

Site fungal biomass was measured using several techniques. From 1970 to 1974 the
methods of Jones and Mollison (1948) were used at the Barrow, Eagle Summit, and Fairbanks sites. At the temperate grassland and bog sites, fungal biomass was measured
using the technique of Hanssen and colleagues (1974). To distinguish live from dead

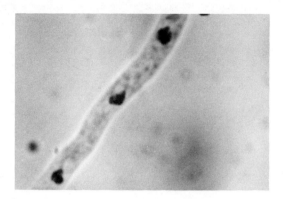

Fig. 1 Fungal hypha in slide stained with aceto-orcein. Width of hypha = 1.8 μm.

hyphae an aceto-orcein stain was included in the application of the two techniques used. Aceto-orcein (2.0 g in 100 ml of 10% acetic acid) was added to the fresh soil sample prior to conventional application of the Jones and Mollison (1948) and Hanssen et al. (1974) techniques. Aceto-orcein stains fungal mycelium a light reddish color, and nuclei are seen as densely stained spheroids (Fig. 1). The assumption was that the presence of cytoplasm and especially cytoplasm with nuclei in hyphae indicated live mycelium. Results with the aceto-orcein technique as applied to tundra soils were compared with estimation of live fungal biomass in the same soils using Frankland's (1974) technique. Results of both techniques were significantly correlated (P < 0.05).

D. Cellulose Decomposition Studies

At each of the study sites cellulose strips, 10 cm long, composed of acid-washed Watman No. 1 filter paper, were placed vertically in the upper 10 cm of earth. Cellulose decay rates within and differences between latitudinally separated sites were measured by determining weight losses in 20-30 strips collected at each site after a 90-day interval.

III. RESULTS

A. Mycofloral Taxonomy, Enumeration, and Distribution

It is not my purpose here to describe or compare the fungal taxa recovered in each study site. However, examples of taxonomic data are shown in Tables 1 and 2 for the reader's information and to clarify the means by which the data presented in the figures were calculated. Lists of species and their enzyme activities for the Pt. Barrow and Eagle Summit sites may be found in Flanagan and Scarborough (1974) and Bunnell et al. (1980). The isolates obtained in the Fairbanks aspen forest and those from the temperate bog site were not as carefully identified as those reported presently and by Flanagan and Scarborough (1974). Dowding and Widden (1974) report in detail on the fungi in a similar peat bog in western Ireland, and the taxa of fun-

Table 1 Taxonomy, utilization potential and frequency (%) of fungi in a temperate-grassland site[a,b]

Organisms	Standing dead in phyllosphere	Litter	Soil 0–10 cm	Soil 10–20 cm	Activity on:						
					Pectin	Starch	Xylan	Cellulose	Humic acid	Gallic acid	Wooden dowels (lignocellulose)
Phycomycetes											
Circinella sp.	1.0	1.7			+	+	+	–	–	–	–
Mortierella elongata	2.3				+	+	+	–	–	–	–
M. pusilla	2.6		4.6		–	+	+	–	–	–	–
M. (cf.) ramanniana	2.8	3.0			–	+	+	–	–	–	–
Mucor spinescens	3.4	2.7		2.3	+	+	+	–	–	–	–
M. corticolus	1.7				+	+	+	–	–	+	–
Phycomyces nitens	0.5	3.2	1.6	0.5	+	+	+	–	–	–	–
Rhizopus nigricans	2.7			1.5	+	+	+	–	–	–	–
R. stolonifer	3.4				+	+	+	–	–	+	–
Zygorhynchus moelleri	0.5	1.7			+	+	+	[+]	–	–	–
Ascomycetes											
Chaetomium sp.	1.8				+	+	+	+	–	–	–
Nectria sp.	OC	OC	OC		+	+	+	+	–	+	–
Penicillium luteum (series)	1.5	OC	1.6	1.6	+	+	+	–	–	–	–
Sordaria sp. (GL 32)	1.0	0.5			+	+	+	+	–	–	–
Sporormia sp. (GL 37)	OC	OC	0.5	OC	+	+	–	+	–	–	–
Fungi Imperfecti											
Alternaria tenuis (var.)	1.0	3.2	2.8		+	+	+	–	–	–	–
Aspergillus clavatus	2.3		1.6		+	+	+	+	–	–	–
A. flavus	OC	1.6	1.6		+	+	+	+	–	–	–
A. niger	2.3	2.3	1.6		+	+	+	–	–	–	–
A. ochraceous	1.6		0.5	1.5	+	+	+	–	–	–	–
A. versicolor	4.6		0.5	1.6	–	+	+	–	–	–	–
Aureobasidium pullulans				0.5	+	+	+	–	–	–	–
Botrytis cinerea	3.3	0.5	1.6	3.2	–	–	+	+	–	+	–
Cercosporidium graminis		1.5	OC		+	+	+	–	–	–	–
Cephalosporium acremonium		1.6	1.6	OC	+	+	+	+	–	–	+
Chalaropsis thielavioides		1.3	0.5	0.5	+	+	+	–	–	+	–

Taxon											
Chrysosporium pannorum	1.2		1.7	2.3	+	−	+	+	−	−	−
C. pruinosum	9.5	OC			+	+	+	(+)	−	−	−
Cladosporium herbarum	2.8	1.6	5.6	1.7	+	+	+	−	[+]	[+]	−
C. cladosporoides		2.3	OC		+	+	+	−	−	+	−
C. macrocarpum (var.)		1.0	OC	1.0	−	+	−	−	−	+	−
Cylindrocarpon sp. (GL 23)		1.3	0.5		−	−	−	+	+	+	+
Cylindrocephalum sp. (GL 7)	2.7	OC			−	+	−	+	−	−	+
Epicoccum purpurascens	OC	1.0			+	+	+	+	−	−	−
Fusarium anguioides	OC	1.6	2.3	1.6	+	+	−	−	−	−	−
F. avenaceum	OC	OC	1.6	OC	+	+	+	+	−	−	−
F. culmorum	OC	1.7	1.6	OC	+	+	−	+	−	−	−
F. gramineum	OC	OC	OC	1.0	+	+	+	+	−	−	−
F. nivale	OC	1.7	1.0	4.6	+	+	+	+	−	−	−
F. oxysporum		5.6	1.7	1.7	+	+	+	−	−	−	−
F. peae			OC	2.3	−	+	+	+	−	−	−
F. roseum			OC	5.6	+	+	+	+	+	+	−
F. sambucinum	OC	1.0	2.8	OC	+	+	+	+	−	−	+
F. solani (?)		OC	OC	OC	−	−	−	−	−	−	−
Fusarium sp.		OC	1.3	OC	+	+	+	−	−	−	−
Fusidium sp.			4.3	OC	−	+	−	+	+	+	−
Gliomastix sp.			OC		+	+	−	−	−	−	−
Monilia sp.		1.0	0.5	1.7	+	−	−	−	−	−	−
Nigrospora sp.		OC	1.0	1.0	+	−	+	−	−	−	−
Penicillium canescens	OC	1.3	0.5	OC	−	−	+	−	−	+	−
P. citrinum		1.0	0.5		−	+	−	−	−	+	−
P. claviforme	3.0		OC	4.6	+	+	−	−	−	−	−
P. chrysogenum			0.5	OC	+	+	+	−	−	−	−
P. diversum			4.6	0.5	−	+	+	−	−	+	−
P. expansum			2.3	0.5	+	−	−	−	−	+	−
P. frequentens		0.5	0.5	OC	−	+	−	−	−	−	−
P. jantinellum		1.3		OC	+	+	+	−	−	−	−
P. genseni	2.2	5.6	1.0	OC	+	+	+	−	−	−	−
P. lilacinum	0.5	0.5	0.5	OC	+	+	−	+	−	−	−
P. notatum	2.8				−	+	−	+	−	−	−
P. rugulosum	2.8	OC	0.5	OC	+	+	−	+	−	−	−
P. stoloniferum	2.4	1.3	0.5	OC	+	+	+	−	−	−	−
P. thomii				1.3	−	+	+	+	−	−	−
Penicillium sp. (GLP 17)					+	+	−	+	−	−	−
Phialophora fastigiata (group)	1.5	2.3	1.5	0.5	+	+	+	+	+	−	−
P. cf. malorum	0.5	3.5		0.5	−	+	+	+	+	+	+

Table 1 (Continued)

Organisms	Standing dead in phyllosphere	Litter	Soil 0–10 cm	Soil 10–20 cm	Activity on:						
					Pectin	Starch	Xylan	Cellulose	Humic acid	Gallic acid	Wooden dowels (lignocellulose)
Phialophora sp. (GL 106)	0.5	OC				+	+	+	+	+	
Phoma glomerata		1.0	2.6		+	+	+	+	+	+	+
P. cf. *herbarum*	OC	0.5	OC		+	+	−	−	+	+	+
Rhizoctonia sp. (GL 111)			2.6	2.8	−	+	−	+	−	−	−
Rhizoctonia sp. (GL 112)			OC	0.5	+	+	−	+	−	−	−
Scopulariopsis sp.			1.0	2.6	−	−	−	+	+	+	+
Sporobolomyces roseus	4.6	OC			−	+	+	+	−	−	−
Stachybotrys sp. (GL 41)	OC		0.5		+	+	+	(+)	+	+	+
Torula sp. (GL 88)	OC	1.6			+	+	−	−	−	+	+
Trichoderma koningii	1.0	0.5	0.5	OC	+	+	+	+	−	+	−
T. viride	0.5	1.0	1.0	OC	+	+	−	(+)	−	−	−
Trichoderma sp. (GL 93)	1.7	OC		2.8	+	+	+	−	−	−	−
Wardomyces sp. (GL 81)		0.5	0.5		+	+	−	−	−	−	−

Mycelia Sterilia											
Hyaline											
GL 1 (with clamp connection [CC])	6.7										
GL 6	11.7	OC	2.8	+	+	+	+	+	−	−	+
GL 11 [CC]	2.3	2.3		+	+	+	+	+	−	−	−
GL 18	4.6	OC		+	+	+	−	+	−	+	−
GL 74		OC		+	+	+	+	−	−	−	−
GL 62	1.0	1.0		+	+	+	−	−	−	−	−
GL 102 [CC]				+	+	+	+	−	−	+	−
Dematiaceous											
GL 15	4.6	OC		−	+	+	+	+	+	+	−
GL 27 [CC]	1.0	OC	4.0	+	+	+	+	+	+	+	+
GL 28	2.3	OC	4.6	+	+	+	+	+	+	+	+
GL 96 [CC]		1.0		+	+	+	−	−	−	−	−
GL 116			4.6	+	+	+	+	+	+	+	+
GL 126			1.6	−	−	−	+	+	+	+	+
GL 127	2.3			+	+	+	+	−	−	−	−

^aKey: OC = occurs infrequently; + = positive at room temperature only; [+] = very weak + reaction; − = no activity; (+) = positive activity at room temperature and below 5°C.

^bActivity blank = no test done; source blank = no occurrence.

Table 2 Taxonomy, utilization potential, and frequency (%) of fungi in taiga birch forests[a]

Organism	Cellulose	Gallic acid	Humic acid	Pectin	Starch	Birch S	Birch L	Spruce S	Spruce L
			Activity				Frequency		
PHYCOMYCETES									
Mucorales									
Mortierella									
parvispora	-	-	-	-	+	2.4	4.7	1.7	2.6
M. ramanniana	+	-	-	+	+			1.3	0.6
M. ramanniana var. angulispora	-	-	-	(+)	(+)	6.7	16.1	4.1	4.6
M. verticillata	-	-	-	+	+			1.0	
M. vinacea	-	-	-	+	+	5.5	3.5	11.7	0.3
Mucor corticolus	-	-	-	+	+	0.4		0.5	
M. microsporus	-	-	-	+	+	2.1	8.0	1.3	5.0
M. mucedo	-	(+)	-	(+)	(+)	1.2	6.3	0.3	3.3
M. saximontensis	-	-	-	+	+	8.7	2.5	14.4	5.3
ASCOMYCETES									
Sphaeriales									
Eupenicillium sp. F-27	+	-	-	-	+	0.2		5.5	0.6
Mycosphaerella cf. tassiana	-	+	+	-	+	1.2		0.3	
Hypocreales									
Nectria viridescens	-	(+)	+	(+)	(+)	0.2		2.7	1.0
FUNGI IMPERFECTI									
Moniliaceae									
Acremonium sp. F-120	(+)	-	-	(+)	+		2.0	0.3	
Aureobasidium pullulans	-	-	-	(+)	+			1.7	
Calcarisporium arbuscula	-	-	-		(+)	0.2	0.7		

Taxon								
Chrysosporium								
cf. *merdarium*	(+)	−	−	(+)		0.6		
C. pannorum var.	−	−	−	(+)	0.2	0.3		
C. pruinosum	+	(+)	−	−	0.4	0.2		
Penicillium								
cinereo-atrum	−	−	+	(+)	9.4	4.5	1.0	2.0
P. claviforme	−	+	+	+		0.3	1.0	6.0
P. daleae	−	−	+	+		0.2		
P. duclauxii	+	−	+	+	0.2	0.2		1.0
P. funiculosum	−	−	+	+				2.0
P. implicatum	−	−	+	+		0.3		1.0
P. pinetorum	−	(+)	−	+		0.3		
P. raistrickii	−	−	+	+	1.2	5.5	1.0	13.0
P. stoloniferum	−	−	+	+		2.2	1.0	2.6
P. thomii	−	−	+	+		1.0	1.3	
P. variabile F-140	(+)	+	−	+			1.0	
Penicillium sp. F-90	+	+	+	+	0.4			0.6
Penicillium sp. F-144	−	−	+	+	0.2			
Tolypocladium								
inflatum	−	−	+	+	18.1	10.0	10.0	6.0
Trichoderma								
polysporum	−	−	+	+	9.4	11.8	1.0	16.3
T. viride var. F-34	+	−	+	+	2.4	3.2	3.1	9.7
T. viride var. F-78	+	−	+	+	7.0	5.2	4.8	6.3
T. viride var. F-132	+	−	+	+	2.4	4.7	3.7	6.6
Trichoderma sp. F-2	−	−	+	−	1.4	0.5		0.3
Dematiaceae								
Cladosporium								
herbarum	−	+	+	+	0.2	0.5	1.3	
Oidiodendron								
griseum	−	−	+	+		0.2		
Phialophora								
hoffmanni	+	+	+	+	5.0	0.2		
P. cf. *malorum*	+	(+)	−	−		0.2	0.6	
P. malorum	+	−	+	+		0.2		
Stachybotrys atra	+	−	+	+		0.3	1.0	
Torula sp. F-119	−	−	−	−	1.2	0.2		
Sphaerioidaceae								
Ceuthospora								
cf. *phacidibides*	+	−	+	+	8.4	0.5		
Phoma sp. PB-8	+	+	+	+		0.2	0.3	

Table 2 (Continued)

| Organism | Activity | | | | | Frequency | | | |
	Cellulose	Gallic acid	Humic acid	Pectin	Starch	Birch S	Birch L	Spruce S	Spruce L
Excipulaceae									
Pilidium									
acerinum	−	+	−	−	−	0.2	0.7		
Tuberculariaceae									
Fusarium									
roseum	−	−	−	+	−		0.2		
MYCELIA STERILIA									
Mycelia sterilia F-13	+	+	+	−	+			0.6	
Mycelia sterilia F-62	−	−	−	+	−			1.0	
Mycelia sterilia HC-12	−	+	+	−	−			0.3	0.6
Sclerotium sp. F-41	+	+	−	+	−	1.4	1.2	1.3	
Sclerotium sp. F-79	+	+	−	−	−		0.2	2.0	

[a]Activity blank = no test done; source blank = no occurrence.

gi in the aspen site were not significantly different from those reported here for
the birch forest. The isolates from the latter sites were identified by the gross
characteristics of their growth on agar plates as compared with known isolates, given
a culture number, and tested for enzyme activity as described earlier (Sec. II.B).

Fungal species diversity, as measured from fungi isolated on agar plates, decreased markedly between the temperate grassland sites and the high arctic grassland
sites, where, respectively, 131 and 38 species were isolated. The temperate peat bog
site contained approximately 95 species. The birch, spruce, and aspen forest sites
and the alpine tundra site at Eagle Summit contained 48, 42, 50, and 28 species, respectively. Microbial species diversity in all groups including bacteria, yeasts,
basidiomycetes, and fungi in general in the northern sites is very low relative to
that in other biomes (Bunnell et al., 1980). The results of this study support the
view that fungal species diversity decreases in the latitudinal gradient from temperate to arctic ecosystems and that species diversity increases with less variable climate.

Seasonal changes in fungal distribution patterns were not detected at any of the
sites studied. The lack of seasonal colonization patterns for fungi had been observed
earlier for tundra by Flanagan and Scarborough (1974) and for a forest podzol by Williams and Parkinson (1964). However, many significant within-site differences in
taxa were obvious between the mycofloras of the phyllosphere, litter, and soil horizons. This spatial segregation of fungal taxa probably reflects the timecourse of
successional changes which would be observed if standing dead leaves or litter were
examined over a period of several years. Other studies in which this was done have
provided clear evidence for fungal succession (Webster, 1956, 1957; Watson et al.,
1974).

Many obvious taxonomic differences are seen between the mycofloras from different
sites (see Tables 1 and 2; also Flanagan and Scarborough, 1974; Dowding and Widden,
1974; Bunnell et al., 1980). The major differences involve a scarcity or absence
of certain genera (*Alternaria*, *Aspergillus*, *Botrytis*, *Fusarium*, and *Rhizopus*) from
the higher latitude sites. Further, in northern forests and in tundra sites respectively *Trichoderma* and *Penicillium* spp. are not numerous, apparently replaced by
greatly increased numbers of *Mycelia sterilia* (Flanagan and Scarborough, 1974).

B. Fungal Enzyme Activity

Starch, pectin and xylan were degraded by more than 65% of fungi in all ecosystems
studied; no significant differences in fungal abilities to utilize these polysaccharides were noted for any site. Mycofloral utilization capacity decreased in the order
pectin, starch, xylan (equally), cellulose, gallic acid, humic acid, and lignocellulose (wood) in all study sites. Across the ecosystems studied the percentages of
fungi capable of utilizing pectin, starch, xylan, cellulose, humic and gallic acids,
and wood were 65-98%, 78-95%, 65-96%, 22-28%, 11-17%, and 4-10%, respectively.

In general, the variation in specific hydrolytic potentials at room temperatures expressed as a percentage of isolated mycofloras between different sites was no greater than the variation in the same potential measured within a site at different times. Statistical analysis showed no significant between-site differences at the 5% level.

Forest ecosystems were expected to, but did not have, proportionally more cellulose and phenolic acid-degrading fungi. Forest mycofloras did have slightly higher average percentages of lignocellulose decaying fungi. However, statistical treatment of variation in this capacity within a site in a given year and between sites showed no significance at the 5% level.

C. Spatiotemporal Mycofloral Utilization Capacities

During the summer season at each site the percentage of fungi isolated having a specific capacity to hydrolyze plant structural constituents did not vary significantly. Although sites showed no seasonal successional pattern for mycofloral decay capacities, pronounced differences in such capacities were noted at different depths in the soil profile and in the phylloplane (Figs. 2-4). In the four sites studied in detail (Pt. Barrow tundra, Eagle Summit tundra, and the temperate peat bog and grassland sites), the proportion of cellulolytic and lignocellulolytic fungi in the resident microbial populations increased with increasing depth in the litter and soil. Phylloplane microbial populations contained a lower proportion of these fungi than did any of the soil microbial populations. Similar distribution patterns were observed for fungi capable of hydrolyzing pectin, starch, and xylan (Figs. 2-4).

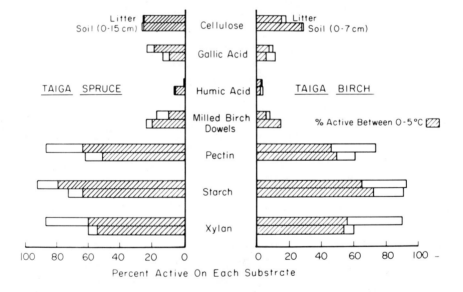

Fig. 2 Distribution by site and forest floor horizon of fungal enzyme capabilities at room temperatures and below 5°C.

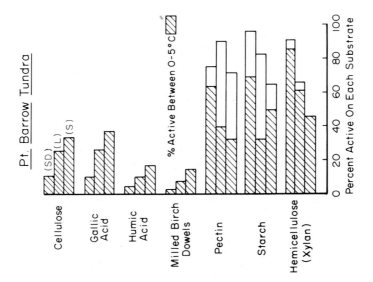

Fig. 3 Spatial distribution of fungal enzyme capability at room temperature and below 5°C in arctic tundra. Key: SD = standing dead vegetation in phyllosphere; L = litter; and S = soil 0-10 cm. (From Bunnell et al., 1980.)

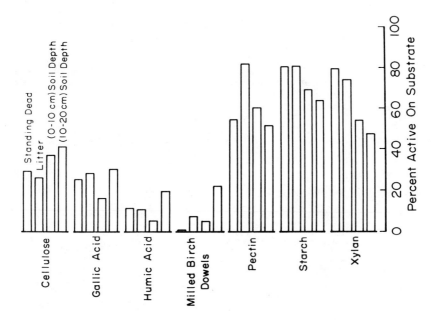

Fig. 4 Spatial distribution of fungal enzyme capability at room temperature in a temperate grassland site. Fungi capable of breaking down substrates at temperatures below 5°C were few, and results of tests were often dubious.

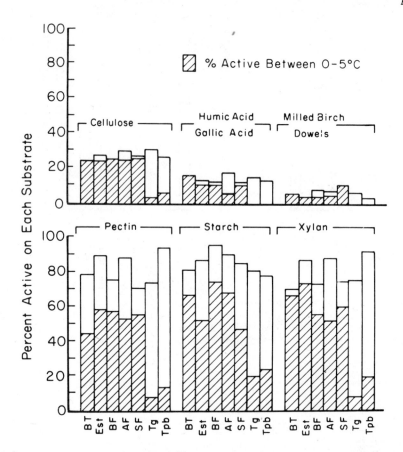

Fig. 5 Between site comparison of the percentage of fungi at each site capable of hydrolyzing the test substrates at room temperature and below 5°C. Data presented are based on tests performed on fungi isolated in midsummer season over a period of 6 years (BT, Barrow tundra; Est, Eagle Summit tundra), 5 years (BF, birch forest), 4 years (SF, spruce forest), 2 years (AF, aspen forest; TPb, temperate peat bog), and 1 year (Tg, temperate grassland). Within-site seasonal variation in a hydrolytic capacity and between-year variation within sites was greater than the between-site variation in hydrolytic capacity illustrated in the figure.

D. Temperature and Fungal Substrate Utilization Capacities

In marked contrast to the tundra and taiga mycofloras, the temperate ecosystem fungi showed very little capacity to decay any of the test substrates below 5°C. Low temperature (below 5°C in vitro) rendered the temperate site fungi inactive on humic acid, gallic acid, and birch wood and almost obliterated their potential to decay cellulose, pectin, starch, and xylan (Fig. 5). In the more northern sites fungi retain some capacity to utilize all test substrates below 5°C, albeit at reduced rates, but cold does not completely stop any of these activities until temperature drops below 0°C.

However, in all the northern sites a percentage of fungi do cease to be active on a given substrate at low temperature. This tendency is more pronounced with pec-

tin, xylan, and starch than it is with cellulose and lignin compounds. All of the
fungi from Pt. Barrow tundra retain some capacity to utilize cellulose and lignocellulose down to -2°C, but the percentage of fungi capable of utilizing pectin, starch,
and xylan decreases by 35%, 14%, and 3%, respectively, below 5°C. The tendency for
fungal populations to show decreased activity at low temperatures on pectin, starch,
and xylan to a significantly greater degree than on lignocellulose and cellulose
compounds was observed in all the northern sites (Figs. 2 and 3).

Another tendency related to fungal spatial position in the ecosystem is seen in
the northern sites and particularly the northern forest sites. In these sites the
proportion of fungi capable of utilizing pectin, starch, and xylan at low temperatures increased with increasing depth in the soil profile (Figs. 2 and 3). This
tendency is also seen, though not as clearly, with fungi capable of degrading cellulose or lignocellulose. In examining the temperate grassland mycoflora, no similar
relationships were observed (Fig. 4).

None of the fungi tested at any of the sites were specialists to the extent that
they used only one substrate.

At room temperature no significant differences in fungal substrate utilization
capabilities were seen, either within sites or between sites (Fig. 5). At low temperatures highly significant differences were seen between the substrate utilization
capacities of fungi from northern sites and those from temperate sites (Tables 1 and
2). The within-site variability in fungal capacity to decay a given substrate at
2-5°C was also highly significant. But between the northern sites there was no significant difference in mycofloral utilization potential at low temperatures (Fig. 5).
These data clearly indicate a low-temperature accomodation of the enzymes in fungi
from the northern sites which barely exists in fungi from temperate climes. Clearly
the temperature range wherein northern fungi may be active on a given substrate is
much broader, because the optimum temperature for activity of many of these northern
fungi is close to 40°C but the low temperature threshold is extended to 0°C or even
to subzero temperatures (Flanagan and Veum, 1974). The broader range for mycological
activity in northern ecosystems is consistent with the greater extremes of temperature
experienced in the northern ecosystem (Table 3).

The vast majority of fungi utilized starch, pectin, and xylan with more or less
the same level of efficiency. The *Phialophora* and *Phoma* isolates were the most versatile, being capable of vigorous growth on all substrates. Only a few sterile forms,
which because of clamp connections were probably basidiomycetes, shared this versatility.

E. Fungal Biomass

In this study fungal biomass as expressed in meters per gram DWS, almost always decreased with depth. Expressed in grams per square meter, fungal biomass varied widely
between sites (Table 4). When results were expressed in meters per gram of organic
matter (OM), the decrease in fungal biomass with depth was not marked; in fact it

Table 3 Climate and cellulose decomposition[a]

Site	Dates (years studied)	Percentage weight loss (±$s_{\bar{x}}$)	pH	Mean annual temp. (°C)	June 15 - Sept. 15		
					max.	min.	mean
BT	June 15-Sept. 15 (1970-75	3.8 ± 0.3	5.0	-12.5	12.7	-2.0	5.2
Est	June 10-Sept. 15 (1970-75)	3.4 ± 0.5	5.5	-21.0	22.5	-4.5	8.3
BF	June 30-Oct. 15 (1971-75	5.2 ± 1.3	6.0	-20.0	32.0	4.0	16.1
AF	June 5-Sept. 20 (1976-77)	5.0 ± 2.0	6.3	-16.0	31.6	5.3	17.2
SF	June 30-Sept. 30 (1972-75)	3.4 ± 1.0	4.4	-23.0	32.0	5.5	16.3
Tg	May 30-Sept. 10 (1973)	4.2 ± 1.1	6.7	10.5	24.3	8.2	14.7
TPb	May 25-Sept. 11 (1973-74)	3.8 ± 1.7	4.5	10.0	23.7	9.2	12.8

[a]Watman No. 1 cellulose strips were used. (See the text.)

often increased with depth, especially in the highly organic soils (Table 5).

By expressing fungal biomass in meters of live hyphae per gram OM, differences between most of the sites for similar layers (0-5 cm and 5-10 cm) disappeared (Table 4). The notable exception to this observation was the black spruce forest live fungal biomass. The reason for this exception will be discussed later. Apparent differences between sites in the 0-10 cm horizons when fungal biomass is expressed as meters per gram DWS, meters per gram OM, or grams per square meter all disappear when the fungal biomass is expressed as meters of live mycelium per gram OM. It seems that between-site differences in fungal biomass are due primarily to differences in amount and density of organic matter in the upper 10 cm of organic soil and to the fact that the proportion of live to dead mycelium varies significantly between sites and between horizons within sites (Table 4).

The regression of grams of live mycelium versus kilograms per square meter OM above mineral soil irrespective of organic layer depth (Fig. 6) shows a strong significant correlation between organic matter content and grams of live mycelium per square meter. If the taiga spruce forest and the temperate peat bog data are removed from the set, the fit of the regression line is extremely good and so we can venture to say that in sites other than these there is approximately 1.0 g of fungal tissue per kilogram OM. This relationship does not hold for the black spruce site nor for the temperate peat bog organic soil, possibly due to the poor quality of fungal substrates in both sites (Table 5). That both sites contain considerable amounts of sphagnum moss remains may also be significant in this regard since this moss decomposes very slowly and in general accommodates few fungi or bacteria (P. W. Flanagan, unpublished data). The high tannin, acidic, and humic content of the organic matter in these sites, together with resin, waxes, cutin, and gums in the black spruce site,

Table 4 Average live and dead fungal biomass in site horizons[a]

Site	Depth	Biomass (m/g DWS)	Biomass (m/g OM)	Meters of live mycelium per gram OM	Meters of dead mycelium per gram OM	Grams live mycelium per m² to mineral soil (cm)
Pt. Barrow	0-5	774	793	487	306	16.2(12)
	5-10	473	804	463	341	
Eagle Summit	0-5	660	676	427	249	7.5(8)
	5-10	312	614	477	137	
Birch forest	0-5	450	542	435	107	3.5(7)
	5-7	203	721	503	218	
Aspen forest	0-5	510	600	518	82	5.8(9)
	5-7	210	735	476	258	
Spruce forest	0-5	225	246	228	18	4.6(25)
	5-10	300	322	307	15	
	10-15	262	282	221	46	
Temperate grassland	0-5	536	687	544	143	6.5(10)
	5-10	202	663	558	105	
Temperate peat bog	0-5	494	537	438	99	16.5(35)[b]
	5-10	654	719	414	305	
	10-15	754	810	245	665	
	15-20	825	840	102	738	
	20-25	1177	1206	144	1062	
	25-30	696	717	87	630	

[a]Above the mineral soil from June to September.

[b]As far as measurements were made; mineral soil was not encountered in this site until the 1.5 m level.

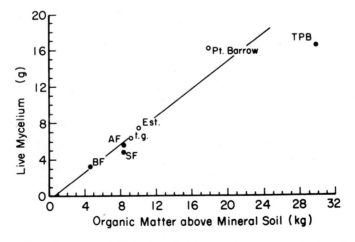

Fig. 6 Regression of grams of live fungal biomass by site against each site's organic matter content (n = 7, r = 0.90).

Table 5 Results at the various sites in terms of organic matter

Site	Kilograms per square meter of organic matter to mineral soil	Percentage of organic matter readily extractable[a]	$\bar{x}$ (mg/m^2) Soluble N	$\bar{x}$ (mg/m^2) Extractable P
Eagle Summit	10.5	14.5	118.5[b]	7.4[b]
Pt. Barrow	18.0	13.5	3600.0[c]	22.0[c]
Birch forest	4.3	14.8	422.0	188.0
Spruce forest	8.6	2.5	166.0	31.0
Aspen forest	8.2	12.8	215.0	165.1
Temperate grassland	9.0	11.7	566.0	160.5
Temperate peat bog	29.0[d]	3.3	192.0[e]	17.5[e]

[a] Warm water followed by 80% ethanol.

[b] From Anderson (1974).

[c] From Chapin et al. (1978).

[d] To 35 cm, mineral soil not encountered.

[e] Upper 10 cm.

all increase the recalcitrance to organic decay, paucity of fungal biomass, and low availability of essential nutrients (Table 5).

IV. SUMMARY AND DISCUSSION

There was very low substrate specificity among all the fungi (>400) tested, and at room temperature few differences existed among species for utilization of a given substrate, even among those from geographically widely separated sites. Between northern sites there was no significant difference in mycofloral substrate utilization potential at low temperatures, but between northern and temperate zone sites such differences proved extremely significant.

What is lacking here is a measure of the content of each test substrate between study sites. However, the measurement of "readily soluble" substrates between sites (Table 5) is positively related to the live fungal biomass measurements.

The residence of fungi in soils is the organic matter matrix. The data presented here indicates that this niche on average is occupied by 0.5-1.0 g dry weight of living fungal biomass per kilogram OM (Fig. 6) irrespective of site latitude. The biomass present is related positively to organic matter quantity, and the data indicate that there is a further positive relationship between substrate quality and fungal biomass.

Fluctuations in fungal biomass within a site are most readily explained in terms of input of organic matter in spring and fall (Bunnell et al., 1980; Söderström, 1977)

and to variations in site temperature and moisture (Flanagan and Van Cleve, 1977).
From the limited data available, the seasonal fluctuation (increases and decreases)
about the mean midseason fungal biomass seems to be by about two to three times.
Lower degrees of fluctuation, i.e., twofold or less, are associated with the lower-
quality organic remains (Flanagan and Van Cleve, 1977; Söderström, 1977) and the
higher fluctuations are associated with the higher quality substrates, e.g., tundra
litter and standing dead remains (Bunnell et al., 1980). Apparently, not only is the
average fungal biomass controlled by substrate quality, but also the amplitude of
seasonal fluctuations in biomass within a substrate is controlled by its quality;
hence seasonal fungal productivity in a site is controlled by substrate quality.

The component species of this biomass of enzyme-secreting fungal network, though
varying from site to site, do not seem to affect the process of decomposition in the
field, at least as far as cellulose is concerned (Table 3). The tundra sites do not
appear to accumulate organic matter (Flanagan and Bunnell, 1978) any more than the
temperate grassland or northern deciduous forest sites do. The peat bog site, basically a graminoid/sphagnum-dominated wet site, does accumulate organic debris, as do
the black spruce forests of interior Alaska.

Although only seven sites were studied, the data suggest that the main determinants of fungal biomass in these ecosystems are the quantity and quality of organic
matter. It seems that the temperature differences between sites do not influence
fungal biomass per gram OM and furthermore that, irrespective of site climatic conditions, if organic quality and availability of inorganic nutrients (Table 5) is low,
fungal biomass will be proportionally low. These observations indicate a temperature
adaptation of some members of the fungal community enabling them to fill the decomposer niche in northern climates (Flanagan, 1978) and an inability of fungi to exploit certain kinds of substrates irrespective of site climate differences.

The low fungal species diversity in northern ecosystems whether gramineous or
forest is to be expected and has been commented on previously (Dowding and Widden,
1974; Bunnell et al., 1980). The cause for this low fungal species diversity is unclear, but it may be related to an inability in most soil fungi to adapt to low temperatures and/or to the fact that tundra systems are geologically young and have not
as yet received their full complement of organisms. The observation that certain
cold-tolerant fungi represent a large portion of the colonists of organic matter,
e.g., *Tolypoclodium inflatum* (taiga forests), *Cladosporium herbarum*, and sterile
mycelium (tundra; see Flanagan and Veum, 1974) and that many northern fungi are
psychrophilic, or cold tolerant (Flanagan and Scarborough, 1974; Bunnell et al.,
1980; Flanagan, 1978), adds support to the view that the composite mycological niche
in tundra is occupied by cold-adapted fungi.

The strikingly high proportion (frequency sum) of *Trichoderma* sp., *Tolypoclodium
inflatum*, *Penicillium* spp., and *Mucor* spp. in the northern forest sites is noteworthy,
as is the absence therein of *Fusarium* sp. and other soil fungi such as *Botrytis*, *As-*

pergillus, and *Rhizopus*. The high proportion of sterile forms in tundra sites further accents generic and specific differences in the fungal flora between sites.

Despite these differences, however, the data from cellulose decay measurements, live biomass measurements, and analyses of decomposer potential between sites (Fig. 5) support the view that there is a decomposer system attuned to each ecosystem which ensures that decomposition goes on at a rate commensurate with the rate of primary production. Thus, these data suggest that the size of the composite niche occupied by fungal decomposers in ecosystems is determined directly by the quantity and quality of the primary production (see also Lockwood, Chap. 18). The diversity in fungal niche occupants seems to be related to climate and decreases with increasing latitudes. Fungal versatility in enzyme production and cold tolerance may compensate for lack of species diversity in northern climates. The obvious low-temperature tolerance of most northern fungi and the inability of most temperate fungi to function at near-zero temperatures helps explain how the fungal niche in northern ecosystems is filled and how the process of decomposition is sustained.

It appears that taxonomic differences in fungal communities between ecosystems do not significantly influence the decay rate of plant remains between sites. The biomass of fungi is related to quantity and quality of organic matter, and the decay rate of a given organic matter is controlled primarily by climate. In northern ecosystems with adverse climates the fungal decomposer niche comprises fungi with low-temperature adaptation and wide separation between minimum and maximum temperature thresholds.

ACKNOWLEDGMENTS

The time given and data willingly shared by my colleagues in the US/IBP, Tundra Biome study and the taiga forest biome study made it possible for this paper to be written. Special thanks are due to Dr. Keith Van Cleve, Mrs. Arla Scarborough, and Mrs. Judy Johnson. Research was financed jointly by National Science Foundation Grants 9V29342 and BMS7513998 to the University of Alaska to support the US/IBP Tundra Biome and taiga forest biome studies, respectively.

REFERENCES

Adametz, L. (1886). Untersuchungen über die mederen Pilze der Aekerkrume. Inaugural dissertation, University of Leipzig.

Anderson, J. H. (1974). Plants, soils, phytocoenology and primary productivity of the Eagle Summit Tundra Biome Site. U.S. Tundra Data Reports No. 74-42, NSF.

Anderson, J. P. E., and Domsch, K. H. (1973). Selective inhibition as a method for estimation of the relative activities of microbial populations in soils. *Bull. Ecol. Res. Comm.* (Stockholm) *17*: 281-282.

Babiuk, L. A., and Paul, E. A. (1970). The use of fluorescein isothiocyanole in the determination of the bacterial biomass of grassland soil. *Can. J. Microbiol. 16*: 57-62.

Bunnell, F. L., MacLean, S. F., Jr., and Brown, J. (1975). Barrow, Alaska, USA. In *Structure and Function of Tundra Ecosystems*, T. Roswall and O. W. Heal (Eds.). *Ecol. Bull.* (Stockholm) *20*: 73-124.

Bunnell, F. L., Miller, O. K., Flanagan, P. W., and Benoit, R. E. (1980). Tundra microflora: composition, biomass and environmental relations. In *An Arctic Ecosystem: The Coastal Tundra of Northern Alaska*, J. Brown, L. Tretzen, and F. L. Bunnell (Eds.). Dowden, Hutchinson, and Ross, Stroudsburg, Pa., in press.

Chapin, F. W., III, Barsdate, R. J., and Barel, D. (1978). Phosphorus cycling in Alaskan coastal tundra: A hypothesis for the regulation of nutrient cycling. *Oikos 31*: 189-199.

Clark, F. E. (1975). Viewing the invisible prairie. In *Prairie: A Multiple View*, M. K. Wali (Ed.). Univ. of North Dakota Press, Grand Forks, N.D.

Domsch, K. H., and Gams, W. (1969). Variability and potential of a soil fungus population to decompose pectin, xylan, and carboy methylcellulose. *Soil. Biol. Biochem. 1*: 29-36.

Dowding, P., and Widden, P. (1974). Some relationships between fungi and their environment in tundra regions. In *Soil Organisms and Decomposition in Tundra*, A. J. Holding, O. W. Heal, S. F. MacLean, and P. W. Flanagan (Eds.). Tundra Biome Steering Committee, Stockholm, pp. 123-149.

Flanagan, P. W. (1978). Microbial ecology and decomposition in arctic tundra and sub-arctic taiga ecosystems. In *Proceedings Life Sciences, Microbial Ecology*, M. Loutit (Ed.). Springer-Verlag, New York, pp. 168-179.

Flanagan, P. W., and Bunnell, F. L. (1976). Decomposition models based on climatic variables, substrate variables, microbial respiration and production. In *The Role of Aquatic and Terrestrial Organisms in Decomposition Processes*, J. M. Anderson and A. MacFadden (Eds.). Blackwell Sci. Publns., Oxford, England, pp. 439-457.

Flanagan, P. W., and Bunnell, F. L. (1980). Microfloral activities and decomposition. In *An Arctic Ecosystem: The Coastal Tundra of Northern Alaska*, J. Brown, L. Tretzen, and F. L. Bunnell (Eds.). Dowden, Hutchinson, and Ross, Stroudsburg, Pa., in press.

Flanagan, P. W., and Scarborough, A. M. (1974). Physiological groups of decomposer fungi on tundra plant remains. In *Soil Organisms and Decomposition in Tundra*, A. J. Holding, O. W. Heal, S. F. MacLean, and P. W. Flanagan (Eds.). Tundra Biome Steering Committee, Stockholm, pp. 159-181.

Flanagan, P. W., and Van Cleve, K. (1977). Microbial biomass respiration and nutrient cycling in a black spruce taiga ecosystem. *Ecol. Bull.* (Stockholm) *25*: 261-273.

Flanagan, P. W., and Veum, A. K. (1974). Relationships between weight loss, temperature and moisture in organic residues on tundra. In *Soil Organisms and Decomposition on Tundra*, A. J. Holding, O. W. Heal, S. F. MacLean, and P. W. Flanagan (Eds.). Tundra Biome Steering Committee, Stockholm, pp. 249-277.

Frankland, J. C. (1974). Importance of phase-contrast microscopy for estimation of total fungal biomass by the agar film technique. *Soil Biol. Biochem. 6*: 409-410.

Frankland, J. L. (1975). Estimation of live fungal biomass. *Soil Biol. Biochem. 7*: 339-340.

Gray, T. R. G., and Williams, S. T. (1971). Microbial productivity in soil. *Symp. Soc. Gen. Microbiol. 21*: 255-286.

Hanssen, J. F., Thingstad, T. F., and Goksøyr, J. (1974). Evaluation of hyphal lengths and fungal biomass in soil by a membrane filter technique. *Oikos 25*: 102-107.

Jensen, C. N. (1912). Fungus flora of the soil. *New York (Cornell) Agr. Exp. Sta. Bull. 315*: 415-501.

Jones, P. C. T., and Mollison, J. E. (1948). A technique for the quantitative estimation of microorganisms. *J. Gen. Microbiol. 2*: 54-69.

Nagel-de Boois, H. M., and Jansen, E. (1971). The growth of fungal mycelium in forest soil layers. *Rev. Ecol. Biol. Sol. 8*: 509-520.

Nikitinsky, J. (1902). Über die Zersetzung der Huminsaure durch physikalisch-chemische Agenten und durch Microorganismen. *Jahrb. Wiss. Bot. 37*: 365-420.

Odum, E. P. (1968). Energy flow in ecosystems: A historical review. *Amer. Zool. 8*: 11-18.

Parkinson, D., and Williams, S. T. (1961). A method for isolating fungi from soil microhabitats. *Plant and Soil 13*: 347-355.

Reinitzer, F. (1900). Über die Eignung der Huminsubstanzen zur Ernährung von Pilzen. *Bot. Ztg. 58*: 59-73.

Satchell, J. E. (1974). Introduction: Litter-interface of animate/inanimate matter. In *Biology of Plant Litter Decomposition*, C. H. Dickinson and G. J. E. Pugh (Eds.), Vol. 1. Academic Press, New York, pp. 13-64.

Söderström, B. E. (1977). Vital staining of fungi in pure cultures and in soil with fluorescein diacetate. *Soil Biol. Biochem. 9*: 59-63.

Thaysen, A. C., and Bunker, H. J. (1927). *The Microbiology of Cellulose, Hemicellulose, Pectin and Gums*. Oxford Univ. Press, New York.

Waksman, S. A. (1916a). Do fungi actually live in the soil and produce mycelium? *Science 44*: 320-322.

Waksman, S. A. (1916b). Soil fungi and their activities. *Soil Sci. 2*: 103-163.

Waksman, S. A. (1917). Is there any fungus flora of the soil? *Soil Sci. 3*: 565-589.

Watson, E. S., McCluskin, D. C., and Huneycutt, M. B. (1974). Fungal succession on loblolly pine and upland hardwood foliage and litter in North Mississippi. *Ecology 55*: 1128-1134.

Webster, J. (1956). Succession of fungi on decaying cocksfoot culms. I. *J. Ecol. 44*: 517-544.

Webster, J. (1957). Succession of fungi on decaying cocksfoot culms. II. *J. Ecol. 45*: 1-30.

Williams, S. T., and Parkinson, D. (1964). Studies of fungi in a podzol. I. Nature and fluctuation of the fungus flora of the mineral horizons. *J. Soil Sci. 15*: 331-341.

Part VII

FUNGI IN NUTRIENT CYCLING

Chapter 30

THE IMPACT OF FUNGI ON ENVIRONMENTAL BIOGEOCHEMISTRY

J. Remacle
Université de Liège
Sart-Tilman
Liège, Belgium

I. IDENTIFICATION OF ROLES

On earth, life is dependent upon the cycling of elements in the biosphere, cycles in which all living components play a role. For example, before industrial civilization, the budget of atmospheric carbon dioxide was governed fundamentally by photosynthesis and by biological oxidation; atmospheric CO_2 would have disappeared in a year or so owing to utilization by green plants if heterotrophic respiration and combustion had not produced CO_2 which recharged the atmosphere (Bormann and Likens, 1967). The decomposers, and among them the fungi, take a great part in CO_2 production. Indeed, in some forests 50% of the gross production disappears as CO_2 due to decomposer activities. The other elements are of no less importance to living organisms and can be described by biogeochemical or global cycles in which ecosystems function and interact in a complex system of inputs and outputs.

Alexander (1971) has summarized the biochemical processes catalized by microorganisms that serve to accelerate or restrain the cycles. These include:

1. Mineralization-immobilization
2. Oxidation-reduction
3. Volatilization-fixation
4. Precipitation-solubilization

It is now possible to assess the potential role played by the fungi in connection with the aforementioned processes.

A. Mineralization

Fungi have been widely implicated in the process of mineralization for a variety of important, widely distributed biopolymers, including protein (Remacle, 1975), cellulose (Gams, 1960), and lignin (Mangenot and Kiffer, 1972). Generally, during this process elements are liberated from organic material and released in an inorganic state so that they become available to primary producers. However, in systems supplied by large amounts of biodegradable materials, the exponential growth of heterotrophic microorganisms may result in a pronounced immobilization of inorganic nutrients, as might be expected from the total composition of the microorganisms (Gosz et al., 1976). Thus, competition between the soil and litter microflora and higher plants may develop (Dommergues and Mangenot, 1970; Garrett, 1974).

B. Oxidation and Reduction

Oxidation processes deserve mention because they are the main sources of energy with release of CO_2. Fungi are obviously involved in these processes. They may also be involved with other oxidations, for example, the oxidation of nitrogenous compounds to nitrate. In the acidic soils of Belgian spruce forests no autotrophic nitrifying bacteria are able to live, but some discrete nitrification has been detected that results from activities of *Aspergillus* sp. However, as shown by in situ techniques and by continuous cultures, this process is always very slow in comparison with autotrophic nitrification (Remacle, 1977). Significant biological oxidation of CO to CO_2 has been observed in acidic soils (Inman et al., 1971), hence it can be assumed that fungi are also implicated in this oxidation process. Conversely, several processes catalyzed by living organisms can contribute to the reduction of inorganic substances. These may be directly connected with cellular metabolism to produce energy or may occur as a side effect of other microbial activities, e.g., during a fall in E_h, a fall in pH, or the formation of inorganic or organic acids (Moureau, 1972). Fungi may be implicated in such processes. For example, Zalokar (1953) has observed that a *Neurospora* strain can reduce selenite to selenium.

C. Volatilization and Fixation

Volatilization-gasification and fixation can also be demonstrated qualitatively in the fungi. Fungi are able to liberate NH_3 to the atmosphere from urea-fertilized raw humus (Mahendrappa and Salonius, 1974), wherein fungal colonization is promoted. Conversely, fungi can fix some gaseous compounds such as SO_2 and CO_2. Craker and Manning (1974) have demonstrated SO_2 uptake by soil fungi; other fungi, including mycorrhizal mycelium, can fix atmospheric $^{14}CO_2$ (Reid and Wood, 1967). Various yeastlike fungi were thought to be involved in atmospheric nitrogen fixation, but this has now been discounted (Postgate, 1972).

D. Precipitation and Solubilization

The fourth of the important processes mentioned by Alexander has to do with the solubilization of rocks. Many studies have been devoted to the biological weathering and solubilization of rocky material with release of metallic ions and silicate. Some fungi excrete large amounts of organic acids including citric, oxalic, and gluconic acids, which bind ions; the ecological value of this process is obvious (Henderson and Duff, 1963; Bertelin, 1971; Krumbein, 1972). For example, Silverman and Munoz (1970) observed that fungi were able to extract 64% of the iron from various rocks. Symbiotic associations of fungi and algae, i.e., the lichens, also play an important role in the weathering of rocks (Togwell and Keller, 1970).

The reverse process, precipitation, can also be achieved by fungi and contribute to sediment formation. As early as 1913, Bejerinck isolated fungi that precipitated a manganese salt (see Mulder, 1972), work which was confirmed by microscopic studies of experimental soils (Jeanson, 1972).

II. QUANTIFICATION OF FUNGAL ACTIVITY

The interaction of all the aforementioned processes can best be visualized in a block diagram (Fig. 1). The complexity of these interactions requires that certain simplifying assumptions be made in attempting to deal with the quantification of fungal activities. For this purpose, it is useful to consider that the impact of fungi on nutrient cycles will be most significant in internal cycles within ecosystems where the most active exchange processes generally occur. For example, 95% of the nitrogen flow in the global terrestrial system takes place within the plant-microorganism-soil system (Rosswall, 1976). In a grassland Burns and Beever (1978) have shown that 1-10% of the total phosphorus was located in living microorganisms, an amount similar to that present in the vegetation. Satchell (1971), Visser and Parkinson (1975b), and Carter and Cragg (1976) have demonstrated that the fungal biomass is by far the largest of all the biotic components in forest soils. In arctic soils, Flanagan (1978) also observed that 90% of the microbial biomass was fungal. Harley (1971) has calculated that in most temperate forests 4% of the root biomass consists of fungal mycelium. However, it must be borne in mind that these figures are standing crop data which provide no information on the physiological status of the mycelium (i.e., whether it is living or dead).

A second simplifying assumption rests on the observation that the upper layers of soil and litter are most active in elemental cycling. Thus, Williams and Gray (1974) have demonstrated mineral contents for litter on the order of 2-6% for gymnosperms and 8-14% for angiosperms.

In considering the impact of fungi on nutrient cycling in upper layers of soil, several problems must be addressed: (1) the effects of physical and biological leaching must be separated; (2) the specific activities of the fungi must be distinguished from those of other microorganisms; and (3) relationships between fungi and other biotic components must be analyzed.

With regard to leaching, both the physical and biological processes are generally considered together, either by giving the total amounts of leached nutrients (Attiwill, 1966, 1967; Lemée and Bichaut, 1973; Parkinson and Lousier, 1975; Duvigneaud and Denaeyer-Desmet, 1969) or by calculating the nutrient resident time (Rochow, 1975). Bocock et al. (1960) and Gosz et al. (1976) have tried to partition physical from biological leaching by assuming that the first month's loss was largely due to physical processes. However, Beck et al. (1969) has shown that litter can still be toxic several months after the fall of the leaves, and Harrison (1971) has noted that oak litter tannins inhibit 70% of the fungal decomposer species. Thus, physical leaching may predominate for longer than 1 month.

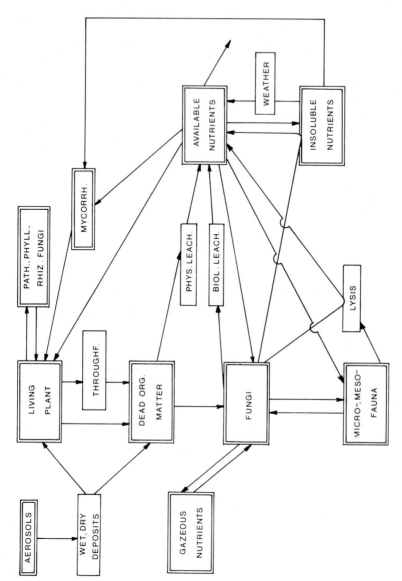

Fig. 1 Fungi and nutrient cycle. The large boxes represent biological and physical compartments, while the smaller rectangles are processes. Only some processes are mentioned in order to keep the diagram understandable and clear. The arrows show direction of fluxes. Key: path., phyll., rhiz. fungi—pathogenic, phylloplane, rhizosphere fungi; throughf.—throughfall; mycorrh.—mycorrhizae; dead org. matter—dead organic matter; phys. leach.—physical leaching; biol. leach.—biological leaching; weather.—weathering.

This problem can be attacked under more carefully controlled conditions in microcosm studies (Witkamp and Frank, 1970; Cowan, 1975; Rosswall et al., 1976; Witkamp and Ausmus, 1976). In our laboratory, Buldgen (1977) has tried to distinguish physical from biological leaching by incubating holorganic[*] layers of a beech forest in microcosms. Undisturbed holorganic layers were sampled (ca. 500 g) in polyvinylchloride (PVC) cylinders (20 cm diameter). These layers were incubated at 4.5° and 15°C and watered daily with distilled water for 31 days. At the end of the experiment, the salt contents were determined and the rates of leaching were calculated per 100 mg of dry layer. The respiration of the same samples was also recorded during 7 days at 4.5° and 15°C in order to determine the steady state rate of oxygen consumption. By assuming that the balance of biological leaching minus biological immobilization (i.e., the net biological leaching) is directly dependent on the respiration rate (Witkamp, 1971) and by evaluating a mean solubility coefficient for the salts of the considered element, we can write for each temperature the following equation:

$$L = L_P + (L_B - I_B)$$

where L = total leaching (mg/100 g),

$\quad L_P$ = physical leaching (mg/100 g),

$\quad L_B$ = biological leaching,

$\quad I_B$ = biological immobilization, and

$\quad L_B - I_B$ = net biological leaching (mg/100 g).

For example, in the case of the K^+ leaching from the holorganic layer of a beech forest we can write:

$\quad$ at 15°C: $1.2 L_P + 4(L_B - I_B) = 2.5$

$\quad$ at 4.5°C: $L_P + (L_B - I_B) = 1.87$

Here, 1.2 is the solubility coefficient between 4.5° and 15°C; 4 is the respiratory coefficient between 4.5° and 15°C. By solving these two equations, we have

$\quad$ at 4.5°C: $\dfrac{L_P}{L} = 0.94$

and

$\quad$ at 15°C: $\dfrac{L_P}{L} = 0.85$

The data shows that at 15°C, 15% of the leaching would depend on biological activities. This low percentage could be due to the fact that the mineralized K is quickly immobilized by incorporation into new cellular material. In the case of the Na salts, 56% of the leaching would be the result of biological activities. Thus, this type of experiment should provide information about the impact of decomposer activity upon the loss of different nutrients.

----

[*]Totally organic.

With regard to the second and third problems, many approaches can be conceived. As a first approximation, the role of fungi can be evaluated by examining the physiological data which give the nutritional requirements of the fungi and their chemical composition. Except for carbon concentrations which average between 40% and 63%, the concentration of elements in fungal cells is far from constant (Rennerfelt, 1934; Foster, 1949; Lilly, 1965). Indeed, fungi can accumulate mineral ions both in their mycelium and in spores (Rennerfelt, 1934). Physical processes such as membrane permeability and ionic equilibria must be involved in the absorption of mineral constituents from the media. Other phenomena such as biological substitution and ion antagonism are also of great interest as they affect the mineral content of cells (Brouwers, 1960). As a result the mineral content of fungal cell material is characterized by great fluctuations quantitatively as well as qualitatively. Moreover, the kinetics of adsorption and desorption of ions by the mycelium are not well understood. Thus, at present, physiological data about fungi are difficult to apply in situ because fungal cells have neither a typical mineral content nor characteristic nutritional requirements (Dowding, 1976).

Therefore other approaches must be investigated, and among them microcosm studies promise to be the most useful. Witkamp and Ausmus (1976) have shown the reduced losses of ^{137}Cs from leaves inoculated with *Penicillium* sp; they were able to distinguish between the net immobilization by the mycelium and the loss of ^{137}Cs from the old mycelium. Similarly, Cowan (1975) has demonstrated that the best retention of nitrate and phosphate occurs in soil colonized by fungi and that this extraction was greater when the leaching was slow. He suggests that this improved mineral retention may have resulted either from uptake by mycelium or from a change in the flow of elements through the soil columns due to the mycelial mat. The accumulation of fungal organic acids in the soil was proposed to explain the loss of nitrate and phsophate at low rates of leaching.

In situ observations are the most valid ecologically but also the most difficult methodologically. In some circumstances observations can become easier--for example, along fungal "fairy rings" (Grunda, 1976; Fischer, 1977) or in the basidiomycete rhizomorphs and sporocarps (Stark, 1972; Cromack et al., 1976). Grunda (1976) has compared the soil properties colonized by fungal fairy rings and control plots. The data revealed that the Ca content was lower in the fairy ring than in the other soils, but conversely the contents of soluble N, P, and K and of loosely bound humic fractions were higher. By chronological sampling of expanding fairy rings, Fischer (1977) showed that N and P were mobilized during active mycelial extension; he also noted the fast release of N and P into the soil by the growing mycelium.

Cromack and associates (1976) have analyzed nutrient accumulation in forest floor basidiocarps and in bulked rhizomorphs. Among the elements, K and Na were significantly more concentrated in higher fungi than in litter; but Ca was concentrated only by the rhizomorphs, where it occurred in concentrations 30 times higher than in the surrounding litter (up to 31,000 ppm of Ca). Using electron probe analysis, Todd and coworkers (1973) performed in situ observations and showed also that Ca was more concentrated in hyphae in comparison with the control. It must be noted that generalization in this matter is difficult: Stark (1972) observed that rhizomorphs of a pine temperate forest could accumulate Na and K in addition to Ca. Unfortunately, no data are available on the biomass of these rhizomorph-forming fungi, and therefore it is not possible to evaluate the impact of the fungi upon the immobilization of nutrients.

Another type of observation can be conducted with labeled substances buried in soils (Mayaudon and Simonart, 1965; McBrayer and Reichle, 1971); Mayaudon and Simonart (1965) have followed the fate of ^{14}C in soil colonized by *Aspergillus*. Such combinations of field and laboratory data allow the calculation of nutrient immobilization by the decomposers (Ausmus et al., 1976) and more specifically by the fungi (Visser and Parkinson, 1975a). During the early phase of decomposition where maximum immobilization occurs, a low percentage of the nutrients can be immobilized in the litter by mycelium. Obviously this immobilization depends on the percentage of living mycelium, which drops from some 50-80% in the freshly colonized litter layer to 15-20% in the fermentation layer in a deciduous forest (Nagel-de Boois and Jansen, 1971; Waid et al., 1973; Visser and Parkinson, 1975b). Further, Nagel-de Boois and Jansen (1967) have demonstrated that fungi are destroyed rapidly in a mull soil owing to the high lytic activities in such soils. Thus, after the first stages of decomposition, it can be supposed that most of the decomposer biomass is composed of inactive biomass with low concentrations of nutrients (Ausmus et al., 1976). More accurate information certainly could be obtained by techniques which allow discrimination between bacteria and fungi such as those devised by Faegri and colleagues (1977). By homogenization, fractionation, and centrifugation of soil samples, these authors obtained two fractions—one containing all the fungi; the other, 50-80% of the bacteria. This approach could be combined with Söderström's technique (1977) which permits discrimination between live and dead mycelia. However, in the fractionation process many living cells are killed, thus leading to overestimation of the amount of dead mycelium.

Up to this point this discussion has focused on the processes of mineralization and immobilization. Fungi are also involved with other processes that relate to elemental cycling. These include translocation of elements from soil to litter, a phenomenon studied mainly with mycorrhizae and wood-decomposing fungi (Reid and Wood, 1967; Kinden and Brown, 1976; Harley, 1971; Gianinazzi-Pearson and Gianinazzi, Chap. 33; Sollins et al., Chap. 31). Grazing of fungi by microarthropods constitutes another process of interest (Mignolet, 1972). In a temperate forest, McBrayer and colleagues (1974) observed that 86% of the net production of fungi was eaten by fungivores which assimilated mainly K and P and which could consume 24 mg of fungi per square meter per day (McBrayer and Reichle, 1971). In an aspen woodland soil, 2% of the fungal standing crop would be eaten by oribatid mites in a year (Mitchell and

Parkinson, 1976), whereas Coleman and McGinnis (1970) found that only 0.2% of fungal standing crops were consumed by animals.

Carter and Cragg (1976) have studied arthropod food chains in soil and litter in an aspen woodland and have concluded (1) that because of their high standing crops in relation to those of the microfauna, the fungi play a preeminent role in nutrient immobilization; and (2) that as grazers, the microfaunal populations are important in regulating fungal standing crops. Thus, the microfauna may exert an indirect influence on immobilization and release of minerals through grazing.

To this point only the direct influence of fungi have been examined. Other, indirect results of fungal activity may also be of interest, albeit difficult to quantify. For example, cellulolytic fungi release significant amounts of reducing sugars which are utilized by noncellulolytic fungi (Hudson, 1975). Humic substances which are directly concerned with the exchange capacity of the soil may be produced (Martin and Haider, 1971). Bacterial activities may be repressed through the synthesis and release of fungal antibiotics. Pectinolytic rhizosphere fungi may foster the growth of nitrogen-fixing species of *Azotobacter* in the rhizosphere (Remacle, 1972).

III. CONCLUSIONS

In conclusion, it can be stated that fungi play a dominant role in nutrient cycling in terrestrial ecosystems (Ausmus et al., 1976; Gosz et al., 1976; Witkamp and Ausmus, 1976). As decomposers they are principally responsible for substrate breakdown and conversion of the nutrient pools to a state available to roots of higher plants. In this role the fungi regulate the release rates of various elements and thus govern the stability and productivity of ecosystems. If we wish to understand the extent of chemical fluxes through fungal standing crops we must move beyond reliance on standing-crop data alone to studies of the kinetics of fungal communities in relation to environmental factors (Gray et al., 1974; Wagner, 1975; Bunnell et al., 1977). The construction of simulation models to predict seasonal fluxes in nutrients based on the effects of environmental parameters on decomposition processes will be a logical next step toward that goal.

ACKNOWLEDGMENTS

Thanks are due to P. Dowding for amending the manuscript. I wish to express my gratitude to G. C. Carroll for improving the English text and for his valuable suggestions.

REFERENCES

Alexander, M. (1971). *Microbial Ecology*. Wiley, New York, 511 pp.
Attiwill, P. M. (1966). The chemical composition of rain water in relation to cycling of nutrients in mature eucalyptus. *Plant and Soil 3*: 390-406.

Attiwill, P. M. (1967). The loss of elements from decomposing litter. *Ecology 49*: 142-145.

Ausmus, B. S., Edwards, N. T., and Witkamp, M. (1976). Microbial immobilization of carbon nitrogen, phosphorus and potassium. Implication for forest ecosystem processes. In *The Role of Terrestrial and Aquatic Organisms in Decomposition Processes*, J. M. Anderson and A. Macfayden (Eds.). Blackwell Sci. Publns., Oxford, England, pp. 397-416.

Beck, G., Dommergues, Y., and Van den Driessche, R. (1969). L'effet litière. II. Etude expérimentale du pouvoir inhibiteur des composés hydrosolubles des feuilles et des litières forestières vis-à-vis de la microflore tellurique. *Oecol. Plant. 4*: 237-266.

Bertelin, J. (1971). Altération microbiologique d'une arène granitique. *Sc. du Sol 1*: 11-29.

Bocock, K. L., Gilbert, O., Capstick, C. K., Twinn, D. C., Waid, J. S., and Woodman, M. J. (1960). Changes in leaf litter when placed on the surface of soils with contrasting humus types. I. Loss in dry weight of oak and ash leaf litter. *J. Soil Sci. 11*: 1-9.

Bormann, F. H., and Likens, G. E. (1967). Nutrient cycling. *Science 155*: 424-429.

Brouwers, L. (1960). Contribution à l'étude de la nutrition minérale des microfungi: Les équilibres ioniques optimum pour *Aspergillus niger*. Doctoral thesis, Inst. Agron., Gembloux, Belgium, 201 pp.

Buldgen, P. (1977). Les formations forestières de la Helle et la minéralisation de leurs horizons holorganiques en microcosmes. Mémoire de licence, Université de Liège, Belgium, 115 pp.

Bunnell, F. L., Tait, D. E. N., Flanagan, P. W., and Van Cleve, K. (1977). Microbial respiration and substrate weight loss. I. A general model of the influences of abiotic variables. *Soil Biol. Biochem. 9*: 33-40.

Burns, D. J. W., and Beever, R. E. (1978). Physiology of P release from fungi: Implication for the P cycle. In *Microbial Ecology*, Intern. Symp., Dunedin, M. W. Loutit and J. A. R. Miles (Eds.). Springer-Verlag, New York, pp. 156-160.

Carter, A., and Cragg, J. B. (1976). Concentration and standing crops of calcium, magnesium, potassium and sodium in soil and litter arthropods and their food in an aspen woodland ecosystem in the Rocky Mountains (Canada). *Pedobiologia 16*: 379-388.

Coleman, D., and McGinnis, J. (1970). Quantification of fungus small arthropod food chains in soils. *Oikos 21*: 134-137.

Cowan, M. C. (1975). Observations on the leaching of nitrate, phosphate and potassium in soil colonized by fungi. *Plant and Soil 42*: 711-715.

Craker, L. E., and Manning, W. J. (1974). SO_2 uptake by soil fungi. *Envir. Poll. 6*: 309-311.

Cromack, K., Todd, R. L., and Monk, C. D. (1976). Patterns of basidiomycete nutrient accumulation in conifer and deciduous forest litter. *Soil Biol. Biochem. 7*: 265-268.

Dommergues, Y., and Mangenot, F. (1970). *Ecologie microbienne du sol*. Masson, Paris, 796 pp.

Dowding, P. (1976). Allocation of resources. Nutrient uptake and release by decomposer organisms. In *The Role of Terrestrial and Aquatic Organisms in Decomposition Processes*, J. M. Anderson and A. Macfayden (Eds.). Blackwell Sci. Publns., Oxford, England, pp. 169-183.

Duvigneaud, P., and Denaeyer-Desmet, S. (1969). Biological cycling of minerals in temperate deciduous forest. In *Analysis of Temperate Forest Ecosystems*, D. Reichle (Ed.). Springer-Verlag, New York, pp. 199-225.

Faegri, A., Torsvik, V. L., and Goksøyr, J. (1977). Bacterial and fungal activities in soil: Separation of bacteria and fungi by a rapid fractionated centrifugation technique. *Soil Biol. Biochem. 9*: 105-112.

Fischer, R. F. (1977). Nitrogen and phosphorus mobilization by fairy ring fungus, *Marasmius oreades* (Bolt) Fr. *Soil Biol. Biochem. 9*: 239-241.

Flanagan, P. W. (1978). Microbial ecology and decomposition studies in arctic tundra and taïga ecosystems. In *Microbial Ecology, International Symposium, Dunedin*, M. W. Loutit and J. A. R. Miles (Eds.). Springer-Verlag, New York, pp. 161-168.

Foster, J. W. (1949). *Chemical Activities of Fungi*. Academic Press, New York, 648 pp.

Gams, W. (1960). Studium Zellulolytischer Bodenpilze mit Hilfe der Zellophanstreifen-Methode mit Carboxymethyl-Zellulose (C.M.C.). *Sydovia 14*: 295-307.

Garrett, S. D. (1974). Cellulose decomposition and saprophyte survival by *Phialophora radicicola*. *Trans. Brit. Mycol. Soc. 62*: 622-625.

Gosz, J. R., Likens, G. E., and Bormann, F. H. (1976). Organic matter and nutrient dynamics of the forest and forest floor in the Hubbard Brook forest. *Oecologia 22*: 305-320.

Gray, T. R. G., Hissett, R., and Duxbury, T. (1974). Bacterial populations of litter and soil in deciduous woodland. II. Numbers, biomass and growth rates. *Rev. Ecol. Biol. Sol. 11*: 15-26.

Grunda, B. (1976). Effects of fungal "fairy rings" on soil properties. *Ceska Mykol. 30*: 27-32.

Harley, J. L. (1971). Fungi in ecosystems. *J. Ecol. 59*: 653-668.

Harrison, A. F. (1971). The inhibitory effect of oak leaf litter tannins on the growth of fungi, in relation to litter decomposition. *Soil Biol. Biochem. 3*: 167-172.

Henderson, M. E. K., and Duff, R. B. (1963). The release of metallic and silicate ion from minerals, rocks and soil by fungal activity. *J. Soil Sci. 14*: 236-246.

Hudson, H. F. (1975). Secondary saprophytic sugar fungi. In *Biodégradation et Humidification*, G. Kilbertus, O. Reisinger, A. Mourey, and J. A. Cansela da Fonseca (Eds.). Pierron, Sarreguemines, France, pp. 15-18.

Inman, R. E., Ingersoll, R. B., and Levy, E. A. (1971). Soil: A natural sink for carbon monoxide. *Science 172*: 1229-1231.

Jeanson, C. (1972). Etude microscopique de dépôts de fer, manganèse et de calcium dans un sol expérimental. *Rev. Ecol. Biol. Sol. 9*: 479-490.

Kinden, D. A., and Brown, M. F. (1976). Electron microscopy of vesicular-arbuscular mycorrhizae of yellow poplar. IV. Host endophyte interactions during arbuscular deterioration. *Can. J. Microbiol. 22*: 64-75.

Krumbein, W. E. (1972). Rôle des microorganismes dans la genèse, la diagenèse et la dégradation des roches en place. *Rev. Ecol. Biol. Sol. 9*: 283-319.

Lemée, G., and Bichaut, N. (1973). Recherches sur les écosystèmes des réserves biologiques de la forêt de Fontainebleau. II. Décomposition de la litière de feuilles des arbres et libération des bioéléments. *Oecol. Plant 8*: 153-174.

Lilly, V. G. (1965). Chemical constituents of the fungal cell. I. In *The Fungi*, Vol. I, G. C. Ainsworth and A. S. Sussman (Eds.). Academic Press, New York, pp. 163-177.

McBrayer, J. F., and Reichle, D. E. (1971). Trophic structure and feeding rate of forest soil invertebrate populations. *Oikos 22*: 381-388.

McBrayer, J. F., Reichle, D. E., and Witkamp, M. (1974). Energy flow and nutrient cycling in a cryptozoan food-web. EDFB-IBP 73-8, Oak Ridge National Laboratory, Oak Ridge, Tenn., 77 pp.

Mahendrappa, M. R., and Salonius, P. O. (1974). Ammonia volatilization from black spruce raw humus created with normal and controlled release urea. *Soil Sci. 117*: 117-119.

Mangenot, F., and Kiffer, E. (1972). Pouvoir ligninolytique des sols de la R.C.P. 40. *Rev. Ecol. Biol. Sol 9*: 21-39.

Martin, J. P., and Haider, K. (1971). Microbial activity in relation to soil humus formation. *Soil Sci. 111*: 54-63.

Mayaudon, J., and Simonart, P. (1965). Métabolisme de la glycine C^{14} dans le sol. *Ann. Inst. Pasteur 3* (Suppl.): 224-234.

Mignolet, R. (1972). Etat actuel des connaissances sur les relations entre la microfaune et la microflore édaphiques. *Rev. Ecol. Biol. Sol 9*: 655-670.

Mitchell, M., and Parkinson, D. (1976). Fungal feeding of oribatid mites (Acari: Cryptostigmata) in an aspen woodland soil. *Ecology 57*: 302-312.

Moureau, C. (1972). Influence des facteurs microbiologiques sur la solubilisation d'éléments minéraux à partir de sol ferrallitique malgache et à partir de biotite en présence de litière tropicale. *Rev. Ecol. Biol. Sol 9*: 539-547.

Mulder, E. G. (1972). La cycle biologique tellurique et aquatique du fer et du manganèse. *Rev. Ecol. Biol. Sol 9*: 321-348.

Nagel-de Boois, H. M., and Jansen, E. (1967). Hyphal activity in mull and mor of an oak forest. In *Progress in Soil Biology*, Vieweg, Braunschweig (Brunswick), W. Germany, pp. 27-36.

Nagel-de Boois, H. M., and Jansen, E. (1971). The growth of fungal mycelium in forest soil layers. *Rev. Ecol. Biol. Sol 8*: 509-520.

Parkinson, D., and Lousier, J. D. (1975). General aspect of leaf litter decomposition in a cool temperate deciduous forest. In *Biodégradation et humification*, G. Kilbertus, O. Reisinger, A. Mourey, and J. A. Cancela de Fonseca (Eds.). Pierron, Sarreguemines, France. pp. 75-87.

Postgate, J. (1972). *Biological Nitrogen Fixation*. Merrow, England, 61 pp.

Reid, C. P. P., and Wood, F. W. (1967). Atmospheric transfer of C^{14}: A problem in fungus translocation studies. *Science 157*: 712-713.

Remacle, J. (1972). Mixed culture of micro-organisms in the rhizosphere of ivy (*Hedera helix* L.). *Plant and Soil. 36*: 199-203.

Remacle, J. (1975). La dégradation des protéines dans les forêts à Mirwart (Ardennes belges). In *Biodégradation et humification*, G. Kilbertus, O. Reisinger, A. Mourey, and J. A. Cancela da Fonseca (Eds.). Pierron, Sarreguemines, France, pp. 127-137.

Remacle, J. (1977). Microbial transformation of nitrogen in forests. *Oecol. Plant. 12*: 33-43.

Rennerfelt, E. (1934). Untersuchungen über die salzaufnahme bei *Aspergillus niger*. *Planta 22*: 221-237.

Rochow, J. J. (1975). Mineral nutrient pool and cycling in a Missouri forest. *J. Ecology 63*: 985-994.

Rosswall, T. (1976). The internal nitrogen cycle between microorganisms, vegetation and soil. *Ecol. Bull.* (Stockholm) *22*: 157-167.

Rosswall, T., Lohm, U., and Sohlenius, B. (1976). Développement d'un microcosme pour l'étude de la minéralisation et l'adsorption radiculaire de l'azote dans l'humus d'une forêt de conifères. *Lejeunia 84*: 1-24.

Satchell, J. E. (1971). Feasibility study of an energy budget for Meathop Wood. In *Productivity of Forest Ecosystems*, P. Duvigneaud (Ed.). UNESCO-PBI, Paris, pp. 619-630.

Silverman, M. P., and Munoz, E. (1970). Fungal attack on rock: Solubilization and altered infrared spectra. *Science 169*: 985-987.

Söderström, B. E. (1977). Vital staining of fungi in pure cultures and soil with fluorescein diacetate. *Soil Biol. Biochem. 9*: 59-64.

Stark, N. (1972). Nutrient cycling pathways and litter fungi. *Bioscience 22*: 355-360.

Todd, R. L., Cromack, K., and Stormer, J. C. (1973). Chemical exploration of the microhabitat by electron probe microanalysis of decomposer organisms. *Nature 243*: 544-546.

Togwell, A. J., and Keller, W. D. (1970). A comparative study of the role of lichens and "inorganic" processes in the chemical weathering of recent Hawaian lava flows. *Amer. J. Science 269*: 446-466.

Visser, S., and Parkinson, D. (1975a). Litter respiration and fungal growth under low temperature conditions. In *Biodégradation et humification*, G. Kilbertus, C. Reisinger, A. Mourey, and J. A. Cancela da Fonseca (Eds.). Pierron, Sarreguemines, France, pp. 88-97.

Visser, S., and Parkinson, D. (1975b). Fungal succession on aspen poplar leaf litter. *Can. J. Bot. 53*: 1640-1651.

Wagner, G. H. G. (1975). Microbial growth and carbon turnover. In *Soil Biochemistry*, E. A. Paul and A. P. McLaren (Eds.), Vol. 3. Dekker, New York, pp. 269-305.

Waid, J. S., Preston, K. J., and Harris, P. J. (1973). Autoradiographic techniques to detect active microbial cells in natural habitats. In *Modern Methods in the Study of Microbial Ecology*, Th. Rosswall (Ed.). *Bull. Ecol. Res. Comm.* (Stockholm) *17*: 317-322.

Williams, S. T., and Gray, T. R. G. (1974). Decomposition of litter on the soil surface. In *Biology of Plant Litter Decomposition*, C. H. Dickinson and G. J. F. Pugh (Eds.), Vol. 2. Academic Press, New York, pp. 611-632.

Witkamp, M. (1971). Forest soil microflora and mineral cycling. In *Productivity of Forest Ecosystems*, P. Duvigneaud (Ed.). UNESCO-PBI, Paris, pp. 413-424.

Witkamp, M., and Frank, M. (1970). Effects of temperature, rainfall and fauna on transfer of ^{137}Cs, K, Mg and mass in consumer decomposer microcosms. *Ecology 51*: 465-474.

Witkamp, M., and Ausmus, B. S. (1976). Processes in decomposition and nutrient transfer in forest systems. In *The Role of Terrestrial and Aquatic Organisms in Decomposition Processes*, J. M. Anderson and A. MacFadyen (Eds.). Blackwell Sci. Publns., Oxford, England, pp. 375-396.

Zalokar, M. (1953). Reduction of selenite by *Neurospora*. *Arch. Biochem. Biophys. 44*: 330-337.

Chapter 31

ROLE OF LOW-MOLECULAR-WEIGHT ORGANIC ACIDS IN THE
INORGANIC NUTRITION OF FUNGI AND HIGHER PLANTS

Phillip Sollins, Kermit Cromack, Jr.,
Robert Fogel[*]

Oregon State University
Corvallis, Oregon

Ching Yan Li

Pacific Northwest Forest and Range
  Experimental Station
U.S. Forest Service
Corvallis, Oregon

I. INTRODUCTION

Since the late 1800s, oxalate compounds have been known to be important constituents of higher plants and important by-products of fungal metabolism. Recently there has been considerable interest in the role of these compounds in elemental cycling processes. Papers by Bruckert (1970a,b), Boyle et al. (1974), and Graustein et al. (1977) have demonstrated the importance of the oxalate ligand in mineral silicate weathering, calcium immobilization, and iron and aluminum transport during soil profile development. In this chapter we give additional evidence for the ubiquity and massiveness of oxalate accumulation in natural systems and explore the role of these compounds in the inorganic nutrition of fungi and higher plants.

Some properties of oxalates are shown in Table 1. Oxalic acid is the strongest of the low-molecular-weight (LMW) organic acids containing only C, H, and O. Oxalic acid is also the most highly oxidized organic form of carbon and releases only about one-seventh as much energy as is released during combustion of an equal weight of sucrose (Table 1). Production of oxalic acid by an organism therefore constitutes a minor energy drain compared with production of other LMW acids such as citric acid. Also, oxalic acid, because of its low heat of combustion, does not encourage growth of potentially competing organisms. In fact, it cannot be used as a sole energy source by most organisms (Harder, 1973). Unlike other LMW organic acids, oxalic acid forms a sparingly soluble precipitate with Ca [pK_{sp} = 8.64 (Ringbom, 1963)]. The Mg salt is substantially more soluble [pK_{sp} = 4.07 (Ringbom, 1963)]. The oxalate ligand forms exceedingly stable complexes with the transition metals, which accounts for its ability to extract iron and aluminum from feldspars and clays and to transport the elements downward in solution through the soil profile (Graustein et al., 1977).

Oxalate concentrations of up to 23% dry weight have been reported in rhizomorphs of *Hysterangium crassum*, a probable ectomycosymbiont of Douglas-fir trees. This

---

[*]*Present address:* Herbarium, University of Michigan, Ann Arbor, Michigan

Table 1 Properties of short-chain organic acids and other compounds[a]

Compound	Structure	pK_1	pK_2	Heat of Combustion (kcal/gm)
Oxalic acid	HO_2C-CO_2H	1.23	4.19	0.67
Malonic acid	$HO_2C-CH_2-CO_2H$	1.83	6.07	2.00
Maleic acid	$HO_2C-CH=CH-CO_2H$	2.83	5.69	2.81
Formic acid	$H-CO_2H$	3.75	--	1.37
Carbonic acid	$HO-CO_2H$	6.37	10.25	0
Acetic acid	H_3C-CO_2H	4.75	--	3.48
Malic acid	$HO_2C-CH_2-CHOH-CO_2H$	3.40	5.11	2.44
Succinic acid	$HO_2C-(CH_2)_2-CO_2H$	4.16	5.61	3.02
Citric acid	$HOC-(CH_2-CO_2H)_2-CO_2H$	3.08	4.74	2.47
Sucrose	--	--	--	3.94

[a]Based on data in the *Handbook of Chemistry and Physics*, CRC Press, Cleveland, and the *International Critical Tables*, McGraw-Hill, New York.

fungus forms dense mats in the A_1 horizon, such that the average oxalate content of the A horizon is over 850 kg/ha (Cromack et al., 1977, 1979). Maxwell and Bateman (1968) found that *Sclerotium rolfsii*, a plant pathogen, produced more than 1 g of oxalic acid for each gram dry weight of hyphal growth.

The presence of large amounts of calcium oxalate explains the discrepancy between the findings by Stark (1972), Ausmus and Witkamp (1973), and Cromack et al. (1975) of high Ca concentrations in fungal tissues and the findings of Steinberg (1948) that solutions "containing only faint spectroscopic traces of calcium" permitted maximum growth of *Aspergillus niger*, *Sclerotium rolfsii*, and other fungi. It seems reasonable to assume that Ca is required in only trace quantities but is present in large amounts because it precipitates in and on fungal tissues as the oxalate salt. The question remains, however, why so many fungi produce prodigious amounts of oxalate rather than oxidizing metabolites to CO_2. What advantages do oxalate production and release offer which compensate for the energy expended in maintaining an additional enzyme system?

In attempting to explain this, we have uncovered two other phenomena which we believe are important in the inorganic nutrition of mycorrhizae and saprophytic fungi. One is cation translocation by fungi, which we suggest is for pH equilibration. The other is a generally occurring effect of bicarbonate concentration on cation uptake by mycorrhizae and nonmycorrhizal roots.

II. METHODS

Soils colonized by the closely related basidiomycetes *Gautieria* sp. and *Hysterangium crassum* were studied in a stand of 40- to 65-year-old second-growth Douglas-fir trees with sparse understory. The stand is located about 16 km west of Philomath, Oregon, at 460 m elevation. Soils and vegetation of this Woods Creek site are described more fully by Fogel (1976) and Cromack et al. (1979).

The litter layer was removed first to locate the mats. Intact mats consisting of rhizomorphs, mycorrhizae, and other Douglas-fir roots, sporocarps, and attached soil were excavated from the A horizon to a depth of approximately 15 cm. Soil within 10 cm of the perimeter of the colonized areas was collected to the same depth.

Bulk soil samples (25 g) were taken from the A horizon of three areas colonized by *Hysterangium* and three areas colonized by *Gautieria*. Adjacent uncolonized areas were also sampled. Samples were dried at 50°C and then sieved through a 0.5-mm screen to remove roots, rhizomorph fragments, and sporocarps.

Douglas-fir roots for decay experiments were collected in the 1974-1975 winter at the Woods Creek site, which was selected because the sparse understory ensured that roots were almost purely Douglas-fir. Roots were returned to the laboratory, washed free of soil with double-distilled water while still fresh, separated into two diameter classes (2-3 mm and 3-5 mm), dried at 50°C in a forced-air oven, and stored in airtight plastic bags.

For the decay study, approximately 5 g of small Douglas-fir roots (2-3 mm diameter) and 7 g of larger roots (3-5 mm diameter) were placed in 20- X 20-cm bags of 1-mm nylon mesh. The bags were set out in May 1975 at 1-m spaces in a row at the litter-mineral soil interface on Reference Stand No. 2 of the H. J. Andrews Experimental Forest (RS-2). Reference Stand 2 is an old-growth Douglas-fir stand previously used by Fogel and Cromack (1977) for litter-decay experiments. It contains a *Tsuga heterophylla*, *Rhododendron macrophyllum*, *Berberis nervosa* understory and is typical of the Rhma-Bene community type (Dyrness et al., 1974).

Root bags were collected after 190 and 350 days, returned to the laboratory, and dried at 50°C. Roots were then removed, weighed, and ground to pass a 40-mesh sieve. Samples of the orginal undecayed roots were processed similarly. Fungal rhizomorphs colonizing the roots were removed from two of the bags collected after 350 days, dried, weighed, and analyzed separately for nutrient elements and oxalate. Several rhizomorphs were also mounted on aluminum stubs, air-dried in a desiccator, coated with gold-palladium alloy, and examined under an AMR model 1200 scanning electron microscope (SEM).

In all analyses, N was determined by the micro-Kjeldahl method (Jackson, 1958), and total C by the Walkley-Black method (Jackson, 1958). After digestion with perchloric acid, P was determined by molybdate reduction (APHA, 1971) and cations by atomic absorption spectroscopy. Lanthanum oxide was added for Ca and Mg determinations to prevent interference.

Oxalate was determined by the method of Mee and Stanley (1973) with a Microtek 2000R gas chromatograph (GC). Soil samples (200 mg) were placed in airtight tubes containing 5-ml aliquots of 5% HCl-methanol solution, which both solubilizes Ca oxalate and forms dimethyl oxalate. Commercial dimethyl oxalate samples were added to 5-ml aliquots of the HCl-methanol reagent and run as controls. We used a stainless steel coil column (0.33-cm inside diameter) which contained 15% diethylene glycol succinate on 0.17-mesh Chromosorb W washed with acid. Column life was prolonged by sample decolorization with activated charcoal for 5 min.

III. RESULTS

Oxalate concentrations were much higher in soil colonized by both fungal species than in adjacent uncolonized soil (Table 2). Soil carbon (excluding roots) was high in all areas sampled [3-8% dry weight (DW)]. It was lowest in the soil colonized by *Gautieria*. In the three *Gautieria* mats which we sampled, oxalate accounted for about 6% DW of the total soil carbon.

Concentrations of N, P, K, Ca, and Mg in undecayed roots were similar to those reported elsewhere (Table 3). Results of the root decay experiments showed about 80% loss in dry weight within the first 180 days for both root size classes (Table 4). Amounts of Ca increased significantly ($p < 0.05$ for both size classes and dates), but N remained essentially constant in small roots and increased slightly ($p < 0.05$) in the larger roots. Other elements decreased in amount as decomposition proceeded. Potassium was unusual in that it decreased during the first 190 days, then increased significantly ($p < 0.01$) between 190 and 350 days. Concentrations in rhizomorphs picked from the decayed roots were generally higher than in the bulk samples (decayed roots plus rhizomorphs). These rhizomorphs were abundant (Fig. 1) and accounted for an average of 1.14% of the bulk dry weight, 5.1% of the N, 6.4% for both P and K, 6.9% of the Ca, and 4.0% of the Mg in two root bags. These percentages are conservative because they include only those rhizomorphs which could be removed easily.

Table 2 Oxalate and carbon content of mat occupied and adjacent soil[a]

	Carbon (% DW)	Oxalate (% DW)	Oxalate-C (% total C)
Hysterangium			
Colonized	8.78 ± 2.35	0.73 ± 0.23[b]	2.30 ± 0.30
Adjacent	3.90 ± 1.76	0.02 ± 0.006[b]	0.19 ± 0.09
Gautieria			
Colonized	3.02 ± 0.73	0.65 ± 0.15	6.20 ± 2.20
Adjacent	4.27 ± 1.63	0.03 ± 0.01	0.24 ± 0.21

[a]Values are ±1 SD.
[b]After Cromack et al. (1979).

Table 3 Mean elemental concentrations in live roots and decaying roots of Douglas-fir trees and in fungal rhizomorphs colonizing decayed roots

	Number of samples	N (% DW)	P (% DW)	K (% DW)	Ca (% DW)	Mg (% DW)
Live roots						
Woods Creek (3-5 mm)	1	0.28	0.07	0.19	0.28	0.09
Woods Creek (2-3 mm)	1	0.42	0.08	0.28	0.48	0.11
Dinner Creek[a] (<2 mm)	3	0.48	0.09	0.21	0.46	0.07
WS-10, H. J. Andrews[b] (<5 mm)	243	0.62	0.10	0.17	0.69	--
Decaying roots[c]						
RS-2, H. J. Andrews (3-5 mm)	14	0.42	0.04	0.13	0.52	0.03
RS-2, H. J. Andrews (2-3 mm)	5	0.55	0.06	0.14	0.64	0.05
Fungal rhizomorphs[d]						
RS-2, H. J. Andrews	2	2.48	0.31	0.63	3.80	0.13

[a]Site near Woods Creek (R. Fogel, unpublished data).

[b]Includes hardwood roots (Santantonio et al., 1977).

[c]After 350 days of decomposition.

[d]Rhizomorphs attached to 3-5 mm roots after 350 days of decomposition.

From the root-decay results, we infer that oxalate precipitation reduces loss of Ca during root decomposition but does not affect loss of other metallic cations:

Fig. 1 Rhizomorphs attached to decayed Douglas-fir roots (X2).

Table 4 Accumulation and loss of dry weight and elements from decaying roots of Douglas-fir trees (Pseudotsuga menziesii)[a]

Days elapsed	Weight (%)	N (%)	P (%)	K (%)	Ca (%)	Mg (%)
Small roots (2-3 mm)						
190 (n = 5)	21.7 ± 0.9	102.4 ± 13.0	72.5 ± 11.2	26.3 ± 8.8	126.9 ± 19.5	64.0 ± 5.9
350 (n = 5)	24.8 ± 1.1	101.6 ± 11.4	54.6 ± 9.7	37.6 ± 5.5	100.4 ± 11.0	31.5 ± 13.0
Large roots (3-5 mm)						
190 (n = 10)	20.6 ± 2.7	115.8 ± 13.0	61.4 ± 8.7	33.3 ± 6.2	169.7 ± 16.4	54.2 ± 10.0
350 (n = 14)	21.3 ± 1.5	119.2 ± 10.5	48.8 ± 5.8	54.2 ± 5.8	144.1 ± 14.6	26.0 ± 7.0

[a]Data are mean ± 95% confidence interval.

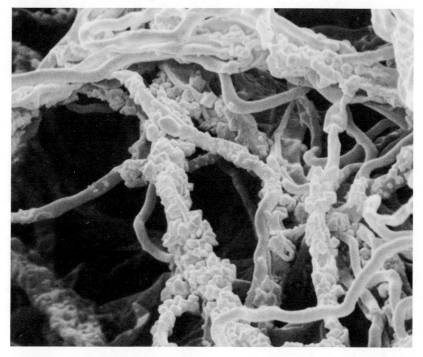

Fig. 2 Crystals (presumably Ca oxalate) adhering to hyphae of rhizomorphs shown in Fig. 1 (X2100). (SEM photograph by R. B. Addison, U.S. Forest Service.)

Ca, which forms a sparingly soluble oxalate salt [pK_{sp} = 8.64 (Ringbom, 1963)], accumulates during root decay; Mg and K, both of whose solution activity is essentially unaffected by the presence of oxalate anions, are leached rapidly from the dead roots. Although x-ray diffraction patterns were not run on the root specimens, SEM photographs (Fig. 2) show crystals which are similar in morphology to those identified definitively as Ca oxalate by Graustein et al. (1977). Gas chromatographic analysis of the rhizomorph samples showed presence of large amounts of oxalate, so large in fact that the columns became overloaded and no precise estimate could be obtained. There is little doubt that the Ca accumulated as the oxalate salt.

Also from the root decay results, we can infer that fungal colonization accounts for much of the accumulation of elements in the decaying roots. This does not mean necessarily that fungi translocated elements into the root bags. We believe that water percolating downward through the bags may have provided the input. The Ca accumulated because it precipitated as the oxalate salt; other elements, because they were efficiently absorbed and immobilized by the colonizing fungi. Translocation may in fact have occurred (see Sec. IV, below), but water percolation through the bags would also account for the increases.

IV. DISCUSSION

Massive accumulation of calcium oxalate in habitats colonized by fungi is now well documented, as are the effects of this on calcium, iron, and aluminum dynamics (Cromack et al., 1975, 1977, 1979; Graustein et al., 1977). What advantage the fungi derive by maintaining an extra enzyme system to excrete oxalate rather than bicarbonate or CO_2 remains unknown. Bacterial solubilization of "rock phosphate" has been clearly demonstrated in the laboratory (Sperber, 1957) and in the field (Azcon et al., 1976). Rose (1957) showed that both *Aspergillus niger* and *Sclerotium rolfsii* solubilized various "insoluble" phosphates by releasing oxalic acid. Various studies have shown that *Endogone* and several vesicular-arbuscular (V-A) mycorrhizal fungi are unable to solubilize rock phosphate (Sanders and Tinker, 1971; Hayman and Mosse, 1972), but there is no evidence that these species produce oxalic acid. Nevertheless, P release from so-called insoluble inorganic forms cannot be the entire explanation for oxalic acid production because P is not generally present in inorganic forms in the organic substrates frequented by wood-rotting fungi and other accomplished oxalate producers. Oxalic and other organic acids may help release P from resistant organic compounds. Bateman and Beer (1965) found that oxalic acid produced by *S. rolfsii* facilitated polygalacturonase digestion from cell walls of a host plant by removing Ca from the cell wall pectin and precipitating it as Ca oxalate. Malic acid has been observed to solubilize Ca phytate under laboratory conditions (K. Cromack, Jr., unpublished data). This P compound constitutes an important fraction of organic P in soils (Russell, 1973). However, a definite role for oxalic acid in P release from organic substrates remains to be demonstrated.

In vascular plants, factors regulating oxalate production are much better understood. Oxalate (or malate) is synthesized in response to a decrease in intracellular acidity caused, for example, by nitrate assimilation (Osmond, 1967; Raven and Smith, 1976; Kirkby and Knight, 1977). The H^+ used when NO_3^- is converted to organically bound NH_3 is supplied by conversion of neutral carbohydrate to organic acids (usually malic or oxalic acid). If malate is produced, then it must be excreted from the plant to prevent an endless increase in osmotic potential. Excretion of large amounts of malate by roots has been demonstrated by Smith (1976).

Excessive osmotic potentials will also be prevented if the organic anion precipitates in an inert form. Many plants in which nitrate is assimilated in the foliage maintain constant pH by producing oxalic rather than malic acid. The oxalate then precipitates as the inert Ca salt in the foliage (Raven and Smith, 1976). Calcium oxalate thus serves not to sequester excess Ca but rather to dispose of oxalate. In fact, it is easy to induce Ca deficiency in plants accumulating calcium oxalate (Brumagen and Hiatt, 1966; Lötsch and Kinzel, 1971)--hardly what one would expect if the oxalate served to dispose of excess Ca.

The majority of elements required for plant growth (N, P, K, Ca, and Mg) occur in the soil solution as cations. In meeting only nutritional requirements plants would absorb more cations than anions, but because electrical neutrality must be maintained either other cations must be excreted or other anions absorbed. This phenomenon, unfortunately termed "excess" cation absorption, was first studied by Ulrich (1941), who showed that either the plant releases H^+ in exchange for the "excess" absorbed cations or the plant absorbs HCO_3^- to balance the "excess" cation uptake. This causes not only an increase in cell pH but also a corresponding decrease in the pH of the external solution. To supply the H^+ which is either excreted or used to neutralize the absorbed HCO_3^-, malic acid (or oxalic acid in a few plants) is synthesized from carbohydrate. This mechanism operates in fungi also (Shere and Jacobson, 1970) but is not as well studied as in higher plants.

In fungi, the oxalate is apparently secreted through the cell membrane and precipitates externally as the Ca salt. The function, i.e., maintenance of electrical neutrality, could be the same as in vascular plants, though this apparently has not been studied. If the function is the same, then oxalate yields should be greater when N is supplied as NO_3^- than when it is supplied as NH_4^+. *Aspergillus niger* produces large amounts of oxalic acid when KNO_3 is the sole N source (Müller, 1965), but the effect of different N sources does not appear to have been compared directly.

Also interesting is the possibility that fungi release oxalate rather than bicarbonate as a respiratory end product to avoid reabsorbing an unwanted anion. The oxalate precipitates and cannot compete with other more useful anions for uptake sites on the hyphal surfaces.

Another curious phenomenon documented several times and perhaps relevant to this discussion is that decayed wood, particularly rotten portions of standing live trees, often shows high concentrations of certain cations, particularly K (Rennerfelt and Tamm, 1961; Johansson and Theander, 1974; Safford et al., 1974). The ammounts of K per unit volume appear to be much greater in the decay zones than in uninfected wood, which indicates that the fungi must have translocated K into the decay zone. Amounts of H^+ typically decrease during decay (increase in pH), which suggests that H^+ is translocated out of the zone by the fungus as K^+ is translocated in.

Large quantities of LMW organic acids are released from roots (Smith, 1976; one must consider the effects on the rhizosphere. Malic acid in particular is a high-energy intermediate of the tricarboxylic acid cycle and can act as the sole carbon and energy source for many microorganisms (Palmer and Hacskaylo, 1970). Recently it has been shown that malic acid is a necessary and sufficient carbon source for at least one strain of the N_2 fixer *Rhizobium* in soybean nodules (Tjepkema and Evans, 1975). Carbohydrate alone is not sufficient, although it can be utilized if LMW organic acids (such as malic or succinic acid) are also supplied (Postgate, 1975).

Malic acid released by roots is unquestionably used as a carbon and energy source by a wide range of rhizosphere bacteria and fungi. An interesting speculation is whether any rhizosphere organisms have evolved which can increase the rate at which roots release these high-energy compounds into the rhizosphere. This process would create half of the requirement for a symbiosis; the remaining requirement would be that the process be beneficial for the rooted plant.

Such a mechanism not only is easily hypothesized, but many of the links have been demonstrated in laboratory experiments. Several studies have shown that an increase in HCO_3^- concentration in the external solution causes increased cation absorption (Carrodus, 1966, 1967) and synthesis of organic acids (Lee and Woolhouse, 1971). Jacoby and Laties (1971) showed that increased absorption of "excess" cations results in higher HCO_3^- levels inside the root cells and that this elicits organic acid synthesis. The presence of large amounts of respiring organisms in the rhizosphere almost certainly increases the HCO_3^- concentration near the root surfaces. The only remaining evidence required, other than a demonstration of all of these processes for a single species, is that an increase in external HCO_3^- concentrations increases release of organic acids from the roots, an experiment which is easily performed.

The other half of the symbiosis, the microbially mediated process which benefits the rooted plant, is also easily demonstrated. Heterotrophic bacteria, able to release nutrients from detrital material, provide an obvious example. They produce cations in excess of needs, which are absorbed by the plant roots and cause synthesis of malic acid. The malic acid could be stored in vacuoles or released into the external solution. The latter would of course be more beneficial to the heterotrophs because it would provide them with a high-energy carbon source.

Fungi extending from root surfaces into the litter or soil matrix could also take advantage of this mechanism by absorbing cations in regions far from the roots and transporting them to the root surfaces where their release would cause a flow of high-energy carbon compound from the root to the fungus. The mechanism is intriguing because we have seen that some fungi translocate cations into regions of low pH, perhaps as a means of increasing the pH. The release of malic acid by the roots certainly results in a low pH, which would be a further link in an elegant symbiosis.

It may be unnecessary to add that we believe that this symbiosis exists and is called the mycorrhizal condition. Although many links have not been demonstrated, and although we can raise many objections, we feel that this scheme's simplicity and elegance call for serious thought and experimentation. Too many questions are unanswered by other hypotheses. Why do plant roots release high-energy substrate to their mycorrhizal associate? Pathogenic fungi obtain their energy by destroying root cell walls and actively removing the energy substrate, but how does a mycorrhizal fungus obtain its food? Why does a mycorrhizal fungus release its hard-won nutrients to the plant roots? Obviously the root efficiently removes any nutrient released from the fungus, perhaps even creating a concentration gradient greater than exists at other points along the hyphae. Possible effects of the fungus on the membrane structure of the host plant and effects of substrate availability on active transport of P across the host cell membrane have been considered (Woolhouse, 1975). But a mechanism whereby hyphae would actively release nutrients and thereby obtain some advantage over other nonmycorrhizal fungi seems at least as plausible an explanation.

We began this chapter by presenting data on cation and LMW organic anion accumulation in fungal-colonized habitats and argued from this for the existence of mechanisms whereby inorganic nutrition is intimately tied to problems of pH regulation. Other authors have noticed the importance of H^+ creation or absorption in mineral cycles (Likens et al., 1969; Sollins et al., 1974; Reuss, 1975, 1976) and in biochemical processes (Raven and Smith, 1976). Sollins et al. (1974, 1980) show how information on carbon, water, and nutrient cycles can be integrated, summarized, and better interpreted by constructing an overall H^+ budget for the system. Many, if not most, biological and physical processes either cause fluxes of H^+ or are affected by pH. We have shown that careful consideration of such pH interactions, usually viewed as a complicating nuisance, can provide insights into a wide range of biochemical and ecological questions. We hope that this paper prompts further examination of these questions as well as a careful testing of our hypotheses.

ACKNOWLEDGMENTS

Chemical analyses (other than gas chromatographic) were performed by the U.S. Forest Service Chemical Laboratory, Corvallis, Oregon. The SEM photograph was prepared by R. G. Addison, U.S. Forest Service. Research was supported by National Science Foundation Grants DEB 77-06075 and DEB 74-20744A06. This is contribution No. 319 from the Coniferous Forest Biome and paper 1254 from the Forestry Research Laboratory, Oregon State University, Corvallis.

REFERENCES

American Public Health Association (1971). *Standard Methods for the Examination of Water and Waste Water*, 13th ed. APHA, Washington, D.C.

Ausmus, B. S., and Witkamp, M. (1973). Litter and soil microbial dynamics in a forest stand. USAEC Rep. No. EDFB-IBP-73-10, Oak Ridge National Laboratory, Oak Ridge, Tenn.

Azcon, R., Barea, J. M., and Hayman, D. S. (1976). Utilization of rock phosphate in alkaline soils by plants inoculated with mycorrhizal fungi and phosphate-solubilizing bacteria. *Soil Biol. Biochem. 8:* 135-138.

Bateman, D. F., and Beer, S. V. (1965). Simultaneous production and synergistic action of oxalic acid and polygalacturonase during pathogenesis by *Sclerotium rolfsii*. *Phytopathology 55:* 204-211.

Boyle, J. R., Voight, G. K., and Sawhney, B. L. (1974). Chemical weathering of biotite by organic acids. *Soil Sci. 117:* 42-45.

Bruckert, S. (1970a). Influence des composés organiques solubles sur la pédogénese en mileu acide. I. Etudes en terrain. *Ann. Agron. 21:* 421-452.

Bruckert, S. (1970b). Influence des composés organiques solubles sur la pédogénese en mileu acide. II. Experiences de la laboratoire modalites d'action des agents complexants. *Ann. Agron. 21:* 725-757.

Brumagen, D. M., and Hiatt, A. J. (1966). The relationship of oxalic acid to the translocation and utilization of calcium in *Nicotiana tabacum*. *Plant Soil 24:* 239-249.

Carrodus, B. B. (1966). Absorption of nitrogen by mycorrhizal roots of beech. I. Factors affecting the assimilation of nitrogen. *New Phytol. 65:* 358-371.

Carrodus, B. B. (1967). Absorption of nitrogen by mycorrhizal roots of beech. II. Ammonium and nitrate as sources of nitrogen. *New Phytol. 66:* 1-4.

Cromack, K., Jr., Todd, R. L., and Monk, C. D. (1975). Patterns of basidiomycete nutrient accumulation in conifer and deciduous forest litter. *Soil Biol. Biochem. 7:* 265-268.

Cromack, K., Jr., Sollins, P., Todd, R. L., Fogel, R., Todd, A. W., Fender, W. M., Crossley, M. E., and Crossley, D. A., Jr. (1977). The role of oxalic acid and bicarbonate in calcium cycling by fungi and bacteria: Some possible implications for soil animals. *Ecol. Bull.* (Stockholm) *25*: 246-252.

Cromack, K., Jr., Sollins, P., Graustein, W. C., Speidel, K., Todd, A. W., Spycher, G., Li, C. Y., and Todd, R. L. (1979). Calcium oxalate accumulation and soil weathering in mats of the hypogeous fungus *Hysterangium crassum*. *Soil Biol. Biochem.* *11*: 463-468.

Dyrness, C. T., Franklin, J. F., and Moir, W. H. (1974). A preliminary classification of forest communities in the central portion of the western Cascades in Oregon. *Coniferous For. Biome Bull. No. 4.* University of Washington, Seattle.

Fogel, R. (1976). Ecological studies of hypogeous fungi. II. Sporocarp phenology in a western Oregon Douglas-fir stand. *Can. J. Bot. 54*: 1152-1162.

Fogel, R., and Cromack, K. (1977). Effect of habitat and substrate quality on Douglas-fir litter decomposition in Western Oregon. *Can. J. Bot. 55*: 1632-1640.

Graustein, W. C., Cromack, K., and Sollins, P. (1977). Calcium oxalate: Occurrence in soils and effect on nutrient and geochemical cycles. *Science 198*: 1252-1254.

Harder, W. (1973). Microbial metabolism of organic Cl and C2 compounds. *Antonie van Leeuwenhoek J. Microbiol. Serol. 39*: 650-652.

Hayman, D. S., and Mosse, B. (1972). Plant growth responses to vesicular-arbuscular mycorrhiza. III. Increased uptake of labile P from soil. *New Phytol. 71*: 41-47.

Jackson, M. L. (1958). *Soil Chemical Analysis*. Prentice-Hall, Englewood Cliffs, N.J.

Jacoby, B., and Laties, C. G. (1971). Bicarbonate fixation and malate compartmentation in relation to salt-induced stoichiometric synthesis of organic acid. *Plant Physiol. 47*: 525-531.

Johansson, M., and Theander, O. (1974). Changes in sapwood of roots of Norway spruce attacked by *Fomes annosus*. Part I. *Physiol. Plant. 30*: 218-225.

Kirkby, E. A., and Knight, A. H. (1977). Influence of the level of nitrate nutrition on ion uptake and assimilation, organic acid accumulation, and cation-anion balance in whole tomato plants. *Plant. Physiol. 60*: 349-353.

Lee, J. A., and Woolhouse, H. W. (1971). The relationship of compartmentation of organic acid metabolism to bicarbonate-ion sensitivity of root growth in calcicoles and calcifuges. *New Phytol. 70*: 103-111.

Likens, G. E., Bormann, F. H., and Johnson, N. M. (1969). Nitrification: Importance to nutrient losses from a cutover forested ecosystem. *Science 163*: 1205-1206.

Lötsch, B., and Kinzel, H. (1971). Zum Calciumbedarf von Oxalatpflanzen. *Biochem. Physiol. Pflanz. 162*: 209-219.

Maxwell, D. P., and Bateman, D. F. (1968). Oxalic acid biosynthesis by *Sclerotium rolfsii*. *Phytopathology 58*: 1635-1642.

Mee, J. M. L., and Stanley, R. W. (1973). A gas chromatographic method for determination of oxalic acid in biological material. *J. Chromatogr. 76*: 242-243.

Müller, H. (1965). Untersuchungen zum Sauerstoffwechsel von *Aspergillus niger*. *Arch. Mikrobiol. 52*: 251-265; *53*: 77-91, 277-287, 303-316.

Osmond, C. B. (1967). Acid metabolism in *Atriplex*. I. Regulation of oxalate synthesis by the apparent excess cation absorption in leaf tissue. *Aust. J. Biol. Sci. 20*: 575-587.

Palmer, J. G., and Hacskaylo, E. (1970). Ectomycorrhizal fungi in pure culture. I. Growth on single carbon sources. *Physiol. Plant. 23*: 1187-1197.

Postgate, J. (1975). *Rhizobium* as a free-living nitrogen fixer. *Nature 256*: 363.

Raven, J. A., and Smith, F. A. (1976). Nitrogen assimilation and transport in vascular land plants in relation to intracellular pH regulation. *New Phytol. 76*: 415-431.

Rennerfelt, E., and Tamm, C. O. (1961). The contents of major plant nutrients in spruce and pine attacked by *Fomes annosus* (Fr.) Cke. *Sond. Phytopathol. Z. 43*: 371-382.

Reuss, J. O. (1975). Chemical/biological relationships relevant to the ecological effects of acid rainfall. U.S. Environmental Protection Agency Rep. No. EPA-660/3-75-032, National Environmental Research Center, Corvallis, Ore.

Reuss, J. O. (1976). Chemical and biological relationships relevant to the effect of acid rainfall on the soil-plant system. In *Proceedings of the First International Symposium on Acid Precipitation and the Forest Ecosystem*, U.S. Dept. of Agriculture, For. Serv. Gen. Tech. Rep. NE-23, L. S. Dochinger and T. A. Seliga (Eds.), Northeast For. Exp. Sta., Upper Darby, Pa., pp. 791-813.

Ringbom, A. (1963). *Complexation in Analytical Chemistry*. Wiley (Interscience), New York.

Rose, R. E. (1957). Techniques for determining the effect of micro-organisms on insoluble organic phosphates. *N.Z. J. Sci. Technol. 38B*: 773-780.

Russell, E. W. (1973). *Soil Conditions and Plant Growth*, 10th ed. Longman, London.

Safford, L. O., Shigo, A. L., and Ashley, M. (1974). Gradients of cation concentration in discolored and decayed wood of red maple. *Can. J. For. Res. 4*: 435-440.

Sanders, F. E., and Tinker, P. B. (1971). Mechanism of absorption of phosphate from soil by *Endogone* mycorrhizas. *Nature 233*: 278-279.

Santantonio, D., Hermann, R. K., and Overton, W. S. (1977). Root biomass studies in forest ecosystems. *Pedobiologia 17*: 1-31.

Shere, S. M., and Jacobson, L. (1970). The influence of phosphate uptake on cation uptake in *Fusarium oxysporum* f. sp. *vasinfectum*. *Physiol. Plant. 23*: 294-303.

Smith, W. H. (1976). Character and significance of forest tree root exudates. *Ecology 57*: 324-331.

Sollins, P., Waring, R. H., and Cole, D. W. (1974). A systematic framework for modeling and studying the physiology of a coniferous forest ecosystem. In *Integrated Research in the Coniferous Forest Biome, Coniferous Forest Biome Bull. No. 5*, R. H. Waring and R. L. Edmonds (Eds.). University of Washington, Seattle, pp. 7-20.

Sollins, P., Grier, C. C., McCorison, F. M., Cromack, K., Jr., Fogel, R., and Fredriksen, R. L. (1980). The internal element cycles of an old-growth Douglas-fir ecosystem in western Oregon. *Ecol. Monogr. 50*: 261-285.

Sperber, J. I. (1957). Solution of mineral phosphates by soil bacteria. *Nature 180*: 994-995.

Stark, N. (1972). Nutrient cycling pathways and litter fungi. *BioScience 22*: 355-360.

Steinberg, R. A. (1948). Essentiality of calcium in the nutrition of fungi. *Science 107*: 423.

Tjepkema, J., and Evans, H. J. (1975). Nitrogen fixation by free-living *Rhizobium* in a defined liquid medium. *Biochem. Biophys. Res. Commun. 65*: 625-628.

Ulrich, A. (1941). Metabolism of non-volatile organic acids in excised barley roots as related to cation-anion balance during salt accumulation. *Amer. J. Bot. 28*: 526-537.

Woolhouse, H. W. (1975). Membrane structure and transport problems considered in relation to phosphorus and carbohydrate movements and in regulation of endotrophic mycorrhizal associations. In *Endomycorrhizas*, F. E. Sanders, B. Mosse, and P. B. Tinker (Eds.), Academic Press, New York, pp. 209-239.

Chapter 32

NUTRIENT UPTAKE AND ALLOCATION DURING SUBSTRATE
EXPLOITATION BY FUNGI

Paul Dowding

Trinity College
Dublin University
Dublin, Ireland

I. INTRODUCTION

Fungi are widely regarded as the primary agents of decomposition of terrestrial plant remains (Hudson, 1968; Swift, 1976). They effect the first sequestration of mineral nutrients as well as carbon and nitrogen from the substrate to their mycelia. The efficiency with which they do this is governed by the complex chemical nature of the substrate and by the ability of various fungi to produce appropriate coenzymes as well as to actually grow through the body of the substrate. During decomposition the chemical composition of the substrate changes (Garrett, 1956; Flanagan and Scarborough, 1974; Swift, 1976) and with it the species of fungi obviously involved (Hudson, 1968). The most readily utilized compounds are absorbed first and their carbon atoms either lost as CO_2 or incorporated into fungal tissue. The fate of N and mineral atoms is different; on all except the richest substrates all of the absorbed N is allocated to mycelial growth and to exoenzyme production (Dowding, 1976). Minerals in ionic form may be leached out in advance of microbial uptake, but the residue and all nonionic or insoluble mineral elements can be taken up and allocated to fungal protoplast.

Bulk analysis of the decaying sample (Dowding, 1974) fails to reveal the processes just outlined because fungus and substrate are inseparable. All that is noted in average litter types is an accelerated loss of C, an accumulation of N and P, and decelerating loss of K, over the first year or so. Consequently, at first the gravimetric and spatial concentration of both N and P increases as decay proceeds. In some instances even the total content increases due to nutrient import. In addition to spatial concentration changes, there are locational changes as well.

In the exploratory phase of colonization, concentration of nutrients may be observed along the advancing mycelial margin. In portions of the substrate which have been colonized, the availability of mineral nutrients to animals is increased by the different chemical compounds (e.g., amino acids in active protoplasts) into which the minerals have been incorporated.

Fungi are dependent on exoenzyme production for their carbon and nitrogen nutrition in complex substrates and must also incorporate a certain amount of N into new cell walls (Burnett, 1969) to exploit new space. Nitrogen supply and in particular the spatial concentration of available nitrogen sources must in part determine the acquisition of carbon by a fungus (Levi and Cowling, 1969). If nitrogen is in too low a concentration the fungus will be unable to absorb enough N to allocate sufficient N to growing tips, which are in turn required to derive more carbon and mineral nutrients. Nitrogen limitation of growth is reduced by the ability of the majority of fungi so far tested to recycle protoplasm (Merrill and Cowling, 1966; Dowding, 1976), but there must still be some ultimate point where an excess of an insoluble carbon source is rendered unavailable by too great a dilution of N (Park, 1976).

Similar arguments can be advanced for other metabolically important mineral elements. Phosphorus is required for nucleic acids, sugar phosphates, and membranes, and an adequate supply is therefore necessary for protein synthesis and for new protoplasm. Phosphorus appears to be even more mobile in mycelium than nitrogen, and this must be due in part to its being entirely protoplasmic in location. Phosphate limitation of nitrogen uptake and growth has already been demonstrated for higher plants (e.g., Pigott and Taylor, 1964) and is reported for *Botrytis cinerea* (Dowding and Royle, 1972), *Agaricus bisporus* (Watson, 1974), and *Chaetomium globosum* (a study by the present author, discussed later in this chapter).

Potassium nutrition has not been studied widely among fungi as there is only one source (ionic K). Potassium would appear to be the major ionic component of protoplast and vacuole, and as such could not be moved to the extent that N and P are moved out of old areas of mycelium to the growing tips. Similarly there is no exoenzymic mechanism involved in K uptake, and the only limitation to K uptake is spatial concentration (K uptake is discussed by Sollins et al. in Chap. 31).

Since most substrates for fungi eventually disappear, fungi must have some means to move away from old substrate to new--either in space or in time. Dispersal requires specialized structures in all but a few instances, as well as special allocation of nutrients since spores are richer in nutrients than any part of the vegetative mycelium. Even the growing tips of hyphae do not contain the stores of carbon compounds and ribonucleic acid that spores seem to have. Dispersal of spores in space also requires specialized fruiting structures, ranging from a simple conidiophore in *Cephalosporium* to an enormous sporocarp in *Calvatia gigantea*. Large sporocarps of polypores and related genera are often perennial and therefore require carbon compounds and mineral nutrients for their upkeep as well as for their initial formation. The larger the fruit body, the more extensive the mycelial network that needs to be maintained to provide nutrients and carbon compounds for spore formation. Since the size and shape of sexual spores, at least, are genetically fixed, it would seem logical to assume that their chemical composition is also largely predetermined and unaffected by the nutrient economy of the parent mycelium. As with higher plants, the fungus response to nutrient starvation is to economize on the number of reproductive structures and propagules formed, not on their size. Some fungi may be able to accumulate nutrients over a variable period of time and so adapt to near starvation conditions by delayed fruiting. The smaller the fruiting unit, the more adaptable the fungus should be, as much less nutrient capital has to be expended on sporocarp and associated transport networks. It would appear that *B. cinerea* is more flexible in the proportion of mycelial nutrients it can devote to fruiting than is *A. bisporus* (Dowding and Royle, 1972; Watson, 1974; Dowding, 1976).

The study described below represents an attempt to build a model situation where the allocation of nutrients to the various aforementioned compartments could be measured or estimated. Further, it was hoped to examine the effect of spatial concentration of nutrients independently of their gravimetric concentration.

II. EXPERIMENTAL APPROACH

The simplest experimental approach is to partition fungus from substrate and reproductive propagules or fruit bodies from vegetative mycelium and to analyze each separately. To date most of this type of work has been done on industrially used molds and yeasts in liquid culture (Prescott and Dunn, 1959; Smith and Berry, 1975). Work on the more important naturally occurring fungi has been hampered by the general inapplicability of liquid culture methods to these fungi and by their inseparability from the substrate. Where much work has been done, it has been concentrated on nitrogen (Merrill and Cowling, 1966; Nicholson, 1972; Fogel, 1976).

The model used in the present study employed a vigorous cellulolytic fungus, *Chaetomium globosum*, growing on pure cellulose filter papers supplemented with mineral nutrients. Ascocarps were produced on the surface of the filter paper and were easily scraped off for the partitioning of fruiting bodies from the mycelium. An approximate value for the weight of mycelium in the filter paper was obtained by weighing the filter papers, dissolving the mycelium out with 15 M KOH at 121°C for 1 hr, and weighing the residual cellulose after washing (Swift, 1976). Experiments were run for up to 6 weeks at 20°C, by which time less than 50% of the cellulose remained and fungal activity had slowed greatly. Uninoculated filter papers were used as controls for each harvest period and were subjected to exactly the same analytical procedures as the test papers. Three replicate papers were dried, weighed, and analyzed for each treatment at each harvest interval. Residual cellulose was determined after KOH digestion, and nitrogen and phosphorus after wet-ashing, by Kjeldahl and molybdate-blue methods, respectively.

The first set of experiments were designed to test for a nitrogen/phosphorus interaction on fungal growth, nutrient uptake, and cellulolytic activity. The concentrations of inorganic sources used are given in Table 1. Fungal growth and cellulose loss (Fig. 1) at high phosphorus concentration are greatly affected by nitrogen concentration, a 10-fold dilution reducing growth and gross decay by eight times during a 6-week period. At low phosphorus concentrations nitrogen level has a small, insignificant effect. A 10-fold reduction in phosphorus concentration has no effect at low nitrogen concentrations but halves growth and decay at high nitrogen levels.

Table 1 Concentration of nutrients used in experiments with *C. globosum*[a]

Nutrient	Level	Concentration		C/nutrient ratio
		mg/liter	mg/microcosm	
Nitrogen	High	500	10	16.5
	Low	50	1	165
Phosphorus	High	125	2.5	66
	Low	12.5	0.25	660

[a]Carbon = 0.165 g per microcosm.

Nitrogen uptake is most affected by nitrogen concentration (Fig. 2); however, at high nitrogen levels, low phosphorus concentration slows nitrogen uptake significantly during the first 4 weeks. At low phosphorus concentrations, there is some evidence of luxury N uptake, as mycelial concentration of N is 7% compared with 4% when phosphorus is not limiting. At low nitrogen concentrations, mycelial N concentration is reduced to 3%. The proportion of initial N taken up does not vary from about 40% after 6 weeks.

Phosphorus uptake (Fig. 3) is significantly affected by phosphorus concentration. At high phosphorus levels where N was not limiting, mycelial P concentration was 0.75%; but where N limited growth, mycelial P concentration was 1.8%, showing luxury uptake. Where high N levels allowed reasonable growth at low P levels, the mycelial P concentration fell to 0.5%. The proportion of initial P taken up was 3-4% at high P levels and 6-10% at low P levels.

If inorganic nutrients in the bathing medium were replaced by organic nutrients (see Sec. IV) so that the total amounts of N and P per disk were 500 and 80 µg, respectively (cf. Table 1), both growth and decay were affected.

Early gross weight loss was high compared with growth, presumably because the easily assimilated milk carbohydrates were being respired. By 4 weeks both growth and decay were greater than at the lowest inorganic nutrient concentration, even though the total N and P content in the organic supplement was less than half that in the lowest inorganic combination. This could be caused either by some extra and unidentified source of nutrient in the milk or by the less acid environment when milk was used as a nutrient source. A third possible explanation is that although the amounts of nutrients per filter paper were less with the milk supplement, their spatial concentration was higher (Table 2). This last hypothesis was subsequently investigated.

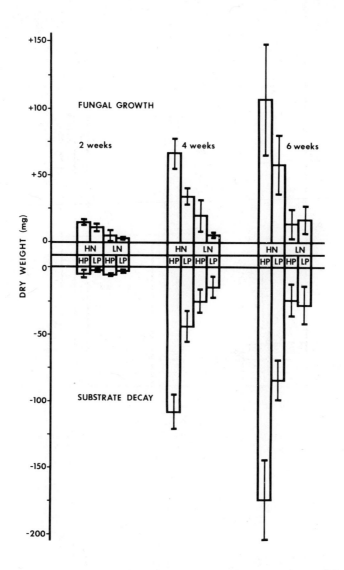

Fig. 1 Fungal dry weight gain and substrate decay shown by *Chaetomium globosum* inoculated on α-cellulose in soluble nutrient solutions. Key: HN = high nitrogen
(10 mg N/microcosm); LN = low nitrogen (1 mg N/microcosm); HP = high phosphorus
(2.5 mg P/microcosm); LP = low phosphorus (0.25 mg P/microcosm). Ninety-five percent
confidence intervals around the mean computed from the range of three replicates
according to Dean Dixon (1951).

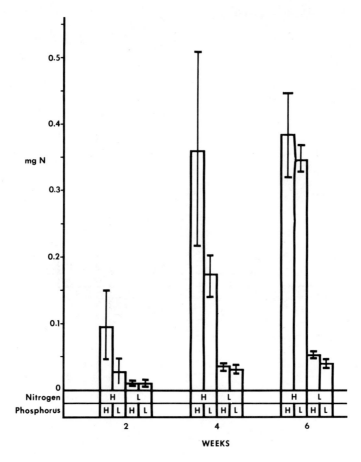

Fig. 2 Nitrogen uptake by *C. globosum* mycelia on α-cellulose in soluble nutrient solutions. Bars around mean values show 95% confidence intervals. Nutrient regimes as in Fig. 1.

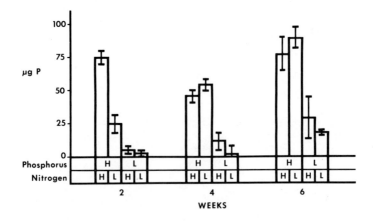

Fig. 3 Phosphorus uptake by *C. globosum* mycelia on α-cellulose in soluble nutrient solutions. Bars around mean values show 95% confidence intervals. Nutrient regimes as in Fig. 1.

Table 2 Gravimetric and spatial concentrations of nutrients in insoluble organic-amended cellulose, compared with soluble inorganic solution concentrations[a]

Nutrient	Level	µg/microcosm	C/nutrient ratio	mg/liter
Nitrogen	0.1% MS[b]	15	11,000	15
	1.0% MS	40	4,100	40
	5 ml high N	10,000	16.5	2,000
Phosphorus	0.1% MS	1	165,000	1
	1.0% MS	7.5	22,000	7.5
	5 ml high P	2,500	66	500

[a]Space bounded by 7-cm filter disk = 1 cm^3.

[b]MS = milk supplement.

III. SPATIAL NUTRIENT CONCENTRATION EFFECTS

Soluble nutrient regimes with 10 mg N (as urea) and 25 mg P (as K_2HPO_4) per filter paper were adopted. The solution volumes ranged from 1 to 20 ml (which had previously been standard). Net decay and nutrient uptake were determined at 9 and 16 days after inoculation. There was some weight increase at zero days (Fig. 4) due to nutrient absorption, and significant net weight increases had occurred by 9 days with nutrient uptake by the fungus exceeding CO_2 loss; the increase was significantly greater at 1 ml solution volume than at 2-20 ml volumes. By 16 days substantial net weight loss had occurred, with the 5 ml volume stimulating significantly greater weight loss than the 1, 2, and 20 ml volumes. The 1 ml volume showed significantly less net decay than all others at 16 days, but this was in part due to the greater net weight increase at 9 days.

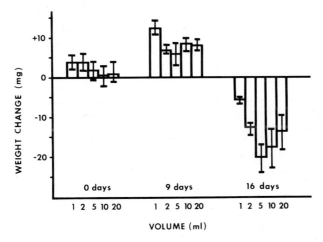

Fig. 4 Dry weight gain by *C. globosum* and cellulose weight loss on α-cellulose microcosms in different volumes of nutrient solution containing 10 mg N and 2.5 mg P. Bars around mean values show 95% confidence intervals.

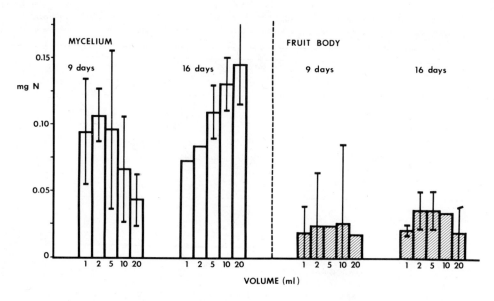

Fig. 5 Nitrogen uptake by *C. globosum* in variable volume microcosms. Initial nitrogen content = 10 mg N/microcosm. Ninety-five percent confidence intervals around the
mean computed from the range of three replicates according to Dean and Dixon (1951).

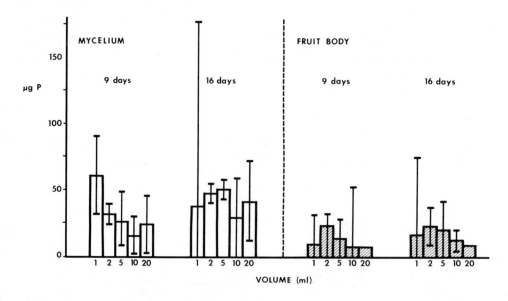

Fig. 6 Phosphorus uptake by *C. globosum* in variable volume microcosms. Initial
phosphorus content = 2.5 mg P/microcosm. Ninety-five percent confidence intervals
around the mean computed from the range of three replicates according to Dean and
Dixon (1951).

Nutrient uptake was significantly affected by spatial concentration, with mycelial N (Fig. 5) (as distinct from fruit body N) greatest in the most dilute situations after 16 days. Mycelial phosphorus (Fig. 6) was, however, significantly greater at the median concentration, which corresponded with the most active mycelium (Fig. 4). After 16 days 12-16% of initial N and 4-6% of initial P had been taken up by the fungus at the different spatial concentrations.

Fruit bodies were separated by scraping from the filter papers and analyzed for N and P in the spatial concentration experiment. There were differences in fruit body N and P between dilutions and between harvest times. Fruit body nutrient amounts were greatest in the median dilution and comprised 3% of total fungus N and 30-40% of total fungus P at 16 days. At extreme dilutions fruit body total N was decreased but fruit body total P was not altered.

The significant effects of spatial nutrient concentration on growth, exoenzyme production (as measured by gross decay), nutrient uptake, and allocation to fruit bodies called into question the whole basis of liquid and semisolid culture experiments. In nature solid substrates may have high C/N ratios, but the spatial concentration of nutrients within them is also relatively high (Park, 1976). The filter paper analog was felt to be a good enough experimental situation to warrant an attempt to modify it by incorporation of solid phase nutrients. As a first attempt, dried skim milk solutions of various dilutions (Table 2) were incorporated by soaking filter papers to runoff point and drying at 80°C. The papers were then reweighed to check the milk addition and were autoclaved to dry. *Chaetomium globosum* was inoculated in the normal manner onto filter papers floating in 5 ml sterile distilled water.

At the highest gravimetric concentrations of nutrients, radial growth and fruiting were both retarded. Growth and fruiting patterns like those on high soluble nutrient regimes were shown by 1% and 2% MS (milk supplement) levels (Table 2). Even at 0.1% MS there was reasonable growth and fruiting, but nutrient limitation was apparent (as in low N-low P levels in the soluble nutrient experiments). At concentrations of 10% MS, carbon/nutrient ratios are very high, but spatial concentrations are of the same order as high and low N and P levels, respectively (Tables 1 and 2). Since the mycelium does not have to explore the liquid to obtain nutrients, more of the nutrients taken up are diverted to fruiting. This is shown in Fig. 7, where the allocation to fruiting can be seen to be a much greater proportion of microcosm nutrients. (It was impossible to separate fungus mycelium from the substrate for N and P analysis.) Where P is limiting (1% MS) the proportion of microcosm P appearing in fruit bodies varied from 33% to 45%, while the proportion of N was ca. 25%. At 10% MS the proportion of microcosm N in fruit bodies increased to half, but the proportion of P was reduced to 10%. At 10% MS there was mobilization of N and P to the liquid phase, but none was detected at 1% MS. Where solid phase nutrients were used, the liquid phase from each set of replicates was wet digested and analyzed for N by Kjeldahl and for P by molybdate blue methods.

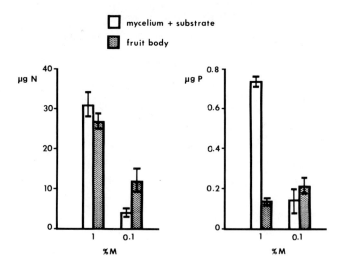

Fig. 7 Nitrogen and phosphorus partitioning in α-cellulose microcosms with incorporated nutrients. Concentrations: 10% MS = 40 μg N/microcosm and 7.5 μg P/microcosm; 1% MS = 15 μg N/microcosm and 1 μg P/microcosm. Bars around mean values show 95% confidence intervals.

IV. DISCUSSION OF METHODOLOGY

The implications of the work just reported and of Park's (1976) work are considerable for all studies involving measurement of nutrient uptake, reproduction, and growth on artificial media. The plate model with 20 ml or so of semisolid medium as well as stationary liquid cultures present the microorganism with carbon source and nutrients in unnatural forms, concentrations, and spatial arrangements. The widespread use of C/N ratios in soil and decomposition microbiology can be defended on analytical grounds (Gotz et al., 1973). The application of C/N ratios to artificial media where either one or both carbon and mineral sources are soluble is not analogous and is indefensible unless conclusions are drawn about the effects of gravimetric and spatial concentrations as well. Even in field situations conclusions based on C/N ratios should be drawn with care, and it would be better if concentrations were used instead (Park, 1976). The single great advantage of liquid culture--that of separation of microorganism from substrate for analysis--is not entirely shared by cultures on semisolid media. Ease of partitioning has been the chief reason for the preponderance of liquid culture studies in work on fungal growth and nutrition (e.g., Treschow, 1944; Bajaj et al., 1954). Where agar cultures have been used (e.g., Käärik, 1960; Buston and Basu, 1948; Leach and Moore, 1966), growth and sporulation have been estimated by indices rather than by absolute measures.

The use of pure solid substrate (e.g., α-cellulose) allows the estimation of fungal growth by chemical partitioning after growth. If soluble nutrients are used and their residues carefully washed from the cellulose-fungus complex, nutrient uptake by the fungus can be estimated. However, spatial concentration of nutrients affects fungal growth significantly, and a pure cellulose matrix floating in nutrient solution is manifestly unreal. The initial attempts at incorporation of an insoluble nutrient source within the cellulose matrix have been partially successful. *Chaetomium globosum* grew and reproduced at much lower gravimetric N and P concentrations than have been previously reported for any fungus. Spatial concentrations were in the normal range for natural substrates. Yet the relatively undefined nature of the nutrient amendment used prevented complete partitioning of fungus and substrate for analysis. The method has still to be devloped along more defined lines for rigorous partitioning.

V. ASSIMILATION EFFICIENCY

The volume of substrate which can be exploited for carbon source and for nutrients by a single hypha depends on diffusion gradients to the hypha, on the presence of commensals in the space between substrate and hypha, and on time. The nature of the nutrient sources and the enzymatic capability of the fungus will determine how much of a substrate is potentially available.

In order to exploit a given volume of substrate, a fungus has to expend capital on a network of hyphae and maintain the network with supplies of C for respiration and C and N for coenzymes. If the substrate is rich, carbon compounds and nutrients can be exported from the first volume for exploitation of fresh volumes and/or for reproduction. The poorer the substrate, the poorer the net return for the fungus. It would appear from the work reported here and from other references cited that an individual fungus cannot extract all the nutrients from a substrate, even if these are all initially available. The mechanisms underlying this are undoubtedly complex, but an important one seems to be the limitation on growth imposed by some essential nutrient in short supply. The availability of carbon is determined by the enzymatic spectrum of the fungus; once suitable enzymes for their digestion are produced, most insoluble carbon sources are present in such large amounts that they do not limit growth. From the studies above *nitrogen* appears to be the most critical nutrient at natural levels of concentration. Most fungi are equipped with proteases, so the availability of at least part of substrate N is not often in question unless the substrate protein is so tangled with polyphenols as to make it unavailable. Fungi, however, must allocate N irrevocably to cell walls and exoenzymes, and hence nitrogen uptake is always a net expenditure. In very dilute situations (such as liquid or semisolid media) large amounts of N relative to C have to be supplied for the fungus to form a network sufficiently large to explore the medium.

Phosphorus is much more mobile in mycelium than is nitrogen (Watkinson, 1971), and furthermore P is neither secreted nor laid down in cell walls in any great amount. The fungus can therefore recycle much of its phosphorus, whether in nucleic acids, phospholipids, or other metabolic cofactors. The phosphorus concentration of old mycelium can become very low indeed, as cell space is filled by large vacuoles and carbohydrate storage granules. Phosphorus can then be diverted to volumes of substrate currently being exploited, and eventually be redeployed to fruiting structures where P is much more mobile than C or N (which are inevitably expended on construction of the fruit body).

Nitrogen concentrations therefore can frequently limit exploitation of nutrients and of carbon: in experiments with *C. globosum* at 50 mg N per liter only 10%, 50%, and 4% of available C, N, and P, respectively, are taken up; at 500 mg N per liter, 50% C, 40% N, and 10% P are taken up. At the lower medium concentration, the mycelial N concentration was 1/3 of the mycelial N concentration at the higher medium N. A similar variation is observed in mycelial phosphorus concentrations with age and medium concentrations. On liquid media the efficiency of extraction of P by *C. globosum* is much lower (10% at 12.5 mg/liter) probably because this concentration was not limiting. *Botrytis cinerea* and *Agaricus bisporus* were both able to extract N more efficiently than P (Dowding, 1976) from pure cultures with mineral salts supplied.

VI. ALLOCATION TO FRUITING

Fungi which form large, complex fruit bodies must allocate C and N to their construction and C to their respiratory upkeep. Rapidly sporulating fungi, on the other hand, have simple sporophores and do not have to take up and store large amounts of carbohydrate and other nutrients for fruit body construction. In liquid culture *B. cinerea* and *C. globosum* allocate variable but rather similar amounts of total nutrients in fruiting; their comparative inefficiency in both uptake and allocation could be due to the formation of extensive mycelial networks necessary for exploitation of soluble nutrients or simply to the unsuitability of liquid media for fungal growth. Sporulation on natural substrates always appears to be much more luxuriant than on the media used in the present study. On more natural substrates, i.e., artificial composts of varying nutrient concentrations, *A. bisporus* did very little better than *B. cinerea* and *C. globosum* in allocating nutrients to fruiting bodies (Dowding, 1976). There, however, it faced microbial competition. When *C. globosum* was grown on cellulose with incorporated nutrients, the proportion of nutrients in the microcosm ending up in fruit bodies was much higher (Fig. 7), as only a small mycelial network was necessary for exploitation of the substrate.

Unfortunately the limitations of the method do not allow estimation of what proportion of substrate nutrients was taken up by the mycelium. If the conditions under which *C. globosum* was grown are considered as ideal--with no competition, no water stress, and, in theory at least, a completely available substrate--then the proportion of nutrients allocated to fruiting can be considered maximal. In that case *C. globosum* is unable to allocate more than half of the substrate N and 40% of the substrate P to fruiting. In a succession of organisms the remainder, whether or not it were incorporated into the mycelium left in the substrate, would become potentially available for later occupants of the niche when *C. globosum* had died.

Partial extraction of nutrients by each phase of a fungal succession is one feature which allows the later organisms to survive and to fruit. Another feature of succession is the gradual loss of carbon or, more tangibly, dry weight, so that the gravimetric concentration of nutrients does not fall as fast as might be expected from their extraction in fruit bodies and spores (Dowding, 1974). However, the substrate does not usually shrink in proportion to carbon loss, so that the spatial concentration of nutrients falls more rapidly than their gravimetric concentration.

Concurrent with the rather complex changes in nutrient concentration with decomposition are changes in location caused by shifts from original substrate to dead microbial cells (Swift, 1976), as well as changes in chemical composition. The N originally in the substrate as proteins and possibly lignins is converted into polysaccharides in microbial cell walls and into microbial protein. Substantial amounts of P are converted to microbial inositol polyphosphate. Both change in location and chemical composition must qualify the gross concentration changes measurable by analytical techniques and may or may not make the nutrient elements more available than before. It is now fairly certain that one group of organisms does benefit from this reassortment of nutrient pools and from the disappearance of structural carbohydrate-- the invertebrate animals that feed in and on decaying plant material. Their invasion and digestion of partially decayed plant material is facilitated either by removal of toxic or unpalatable compounds and/or by accumulation of more digestible compounds and essential vitamins or amino acids. The feces which result from detritivore digestion processes have a higher spatial concentration of nutrients than the partially decayed plant material from which they were derived, and so initiate a cyclical succession of microorganisms and invertebrates (Hargrave, 1976). Rapidly sporing actinomycetes and microfungi (e.g., penicillia) with poor translocating ability are better able to utilize fecal pellets than whole plant debris. Paradoxically, the relative permanence of fecal pellet complexes (if not of individual pellets) as they make up soil organic matter or humus also allows basidiomycetes with very long-lived and extensive mycelia to flourish and to produce large fruit bodies. The scale (of time as well as of space) over which long-lived fungi with large fruit bodies work forces us to consider the soil organic matter/nutrient system not as a series of batch cultures exhibiting successions (Hudson, 1968) but as a continuous culture system to which nutrients are constantly being added and from which plants, animals (e.g., tipulids), fungi, and rain water are constantly abstracting nutrients. In such a system partitioning of nutrient pools by size becomes less important than evaluation of fluxes between pools. That is another day's journey, to be considered by a variety of methods in the next three chapters.

REFERENCES

Bajaj, V., Damle, S. P., and Krishnan, P. S. (1954). Phosphate metabolism of mold spores. (1) Phosphate uptake by spores of *Aspergillus niger*. *Arch. Biochemistry* *50*: 451-460.

Burnett, J. (1969). *Fundamentals of Mycology*. Arnold, London.

Buston, H. W., and Basu, S. N. (1948). Some factors affecting the growth and sporu-lation of *Chaetomium globosum* and *Memnoniella echinata*. *J. Gen. Microbiol.* 2: 162-192.

Dean, R. B., and Dixon, W. J. (1951). Simplified statistics for small numbers of observations. *Anal. Chem.* 23: 636-638.

Dowding, P. (1974). Nutrient losses from litter on IBP Tundra sites. In *Soil Organisms and Decomposition in Tundra*, A. J. Holding, O. W. Heal, S. F. Maclean, Jr., and P. W. Flanagan (Eds.). Tundra Biome Steering Committee, Stockholm.

Dowding, P. (1976). Allocation of resources: Nutrient uptake and release by decom-poser organisms. In *The Role of Terrestrial and Aquatic Organisms in Decomposition Processes*, J. M. Anderson and A. MacFayden (Eds.). Blackwell Sci. Publns., Oxford, England.

Dowding, P., and Royle, M. C. I. (1972). Uptake and partitioning of nitrate and phos-phate by cultures of *Botrytis cinerea*. *Trans. Brit. Mycol. Soc.* 59: 193-203.

Flanagan, P. W., and Scarborough, A. (1974). Physiological groups of decomposer fun-gi in tundra plant remains. In *Soil Organisms and Decomposition in Tundra*, A. J. Holding, O. W. Heal, S. F. Maclean, Jr., and P. W. Flanagan (Eds.). Tundra Biome Steering Committee, Stockholm.

Fogel, R. (1976). Ecological studies of hypogeous fungi. II. *Can. J. Bot.* 54: 1152-1162.

Garrett, S. D. (1956). *The Biology of Root Infecting Fungi*. Cambridge Univ. Press, New York.

Gotz, J. R., Likens, G. E., and Bormann, F. H. (1973). Nutrient release from decom-posing leaf and branch litter in the Hubbard Brook Forest, New Hampshire. *Ecol. Monogr.* 43: 173-191.

Hargrave, B. T. (1976). The central role of invertebrate faeces in sediment decom-position. In *The Role of Terrestrial and Aquatic Organisms in Decomposition Process-es*, J. M. Anderson and A. Macfayden (Eds.). Blackwell Sci. Publns., Oxford, England.

Hudson, H. J. (1968). The ecology of fungi on plant remains above the soil. *New Phytol.* 67: 837-874.

Käärik, A. (1960). The growth and sporulation of *Ophiostoma* and some other blueing fungi on synthetic media. *Symbolae Bot. Upsal.* 16: 1-159.

Leach, R., and Moore, K. G. (1966). Sporulation of *Botrytis fabae* on agar cultures. *Trans. Brit. Mycol. Soc.* 49: 593-601.

Levi, M. P., and Cowling, E. B. (1969). Role of nitrogen in wood deterioration. VII. *Phytopathology* 59: 460-468.

Merrill, W., and Cowling, E. B. (1969). The role of nitrogen in wood deterioration: Amount and distribution of nitrogen in fungi. *Phytopathology* 56: 1083-1090.

Nicholson, D. L. (1972). The succession of organisms on pine logs and its effect on nitrogen and phosphorus. B.A. (Mod.) thesis, University of Dublin, Ireland.

Park, D. (1976). Carbon and nitrogen levels as factors influencing fungal decompos-ers. In *The Role of Terrestrial and Aquatic Organisms in Decomposition Processes*, J. M. Anderson and A. Macfayden (Eds.). Blackwell Sci. Publns., Oxford, England.

Pigott, C. D., and Taylor, K. (1964). The distribution of some woodland herbs in relation to the supply of nitrogen and phosphorus in the soil. *J. Ecol.* 52 (Jubilee Suppl.): 175-185.

Prescott, S. C., and Dunn, R. G. (1959). *Industrial Microbiology*, 3rd ed. McGraw-Hill, New York.

Smith, J. E., and Berry, D. R. (1975). *The Filamentous Fungi*, Vol. I: *Industrial Mycology*. Arnold, London.

Swift, M. J. (1976). Species diversity and the structure of microbial communities in terrestrial habitats. In *The Role of Terrestrial and Aquatic Organisms in Decomposition Processes*, J. A. Anderson and A. MacFayden (Eds.). Blackwell Sci. Publns., Oxford, England.

Treschow, C. (1944). Nutrition of the cultivated mushroom. *Dansk. Bot. Arch. 11*: 1-180.

Watkinson, J. C. (1971). Phosphorus translocation in stranded and non-stranded mycelium of *Serpula lachrymans*. *Trans. Brit. Mycol. Soc. 57*: 535-539.

Watson, J. M. (1974). The nitrogen and phosphorus nutrition of *Agaricus bisporus*. M.Sc. thesis, University of Dublin, Ireland.

Chapter 33

*ROLE OF ENDOMYCORRHIZAL FUNGI IN PHOSPHORUS CYCLING
IN THE ECOSYSTEM*

V. Gianinazzi-Pearson[*] and S. Gianinazzi[*]

*Station de Physiopathologie Végétale
Institut National de la Recherche Agronomique
Dijon, France*

I. INTRODUCTION

In natural ecosystems plants depend largely on the activity of soil microorganisms
for the supply of mineral nutrients essential to their growth. It is evident that
microorganisms that form symbiotic associations with plant roots, for example, nitrogen-fixing bacteria and mycorrhizal fungi, are particularly well placed to intervene
in plant nutrition.

A distinguishing feature of mycorrhizal fungi is that after root infection part
of the mycelium remains active in the soil. Plant roots provide them with an ecological niche with abundant substrate from which their hyphae extend outward through the
soil and effectively explore a much greater volume than nonmycorrhizal roots. The
two most important groups of mycorrhizal fungi consist of (1) those forming ectomycorrhizas which are characterized by mycelial sheaths around the roots and intercellular hyphal invasion of the root cortex and (2) those forming endomycorrhizas with
a loose external hyphal network in the soil and extensive intracellular hyphal growth
in the root cortex.

The beneficial effects of ectomycorrhizas on the growth and nutrition of tree
species in soils of low nutrient status have been recognized for a long time (Mitchell
et al., 1937; McComb, 1938), and the role of ectomycorrhizal fungi in supplying mineral nutrients to the host plant has been conclusively established (Melin and Nilsson,
1950, 1952, 1953a,b; Melin et al., 1958; Harley, 1969). It is only within the last
few years, however, that the potential importance of endomycorrhizas in the growth
and mineral nutrition of their host plants has been widely appreciated (Mosse, 1973;
Gianinazzi-Pearson, 1976). Two groups of endomycorrhizas[†] are able to benefit their
host plants in improving mineral nutrient uptake: the ericoid and the vesicular-arbuscular mycorrhizas. This chapter deals in particular with the role of the fungi
that form these endomycorrhizas in phosphorus uptake by plants and discusses the importance of this role in the cycling of phosphorus in the ecosystem.

---

[*]Station d'Amélioration des Plantes, Institut National de la Recherche Agronomique, Dijon, France.

[†]A third group, found in the Orchidaceae and concerned with the carbohydrate
nutrition of developing seedlings, is not considered here.

II. ENDOMYCORRHIZAL INFECTIONS: ECOLOGY, MORPHOLOGY, AND ENDOPHYTE TAXONOMY

Ericoid mycorrhizas are restricted to genera of the Ericaceae, of which *Calluna*, *Vaccinium*, and *Erica* are examples. These plants occur widely as dominant and codominant members of calcifuge plant communities and are normally associated with mor-humus soils of low nutrient status. The endomycorrhizal fungi can be isolated and cultured axenically (Pearson and Read, 1973a), and one true mycorrhizal isolate has been identified as *Pezizella erica* sp. nov. (Read, 1974). It is probable that all ericoid endophytes will ultimately be recognized as Ascomycetes of this or a related genus. Endomycorrhizas are formed annually with development of the lateral hair roots of the host plant. Endophytic hyphae present in the soil or originating from previously infected roots penetrate the host cells after formation of appressoria and by repeated branching develop compact intracellular mycelial complexes, or "hyphal coils," in the outer cortical cells (Fig. 1a). These intracellular hyphae appear to be separated from the host cytoplasm by the surrounding intact host plasmalemma (Fig. 1b). Nutrient exchange is thought to take place by lysis of these intracellular hyphae, but this has not yet been demonstrated. In young seedlings the extent of mycorrhizal infection can reach up to 70% of the total root system and the number of hyphal entry points well over 1000 per centimeter of root (Read and Stribley, 1975). The mycelium spreads around the root and into the soil, establishing frequent hyphal connections between infected host cells and soil particles around the root.

Vesicular-arbuscular (VA) mycorrhizas, unlike ericoid mycorrhizas, are not limited to any one plant family and have an exceptionally wide range of hosts and habitats. There are only a few families where they are not found, and these include some forming ectomycorrhizas or non-VA endomycorrhizas and those not forming mycorrhizas at all (Chenopodiaceae and Cruciferae). VA mycorrhizas are formed by phycomycetous fungi which cannot be cultured axenically but which, on the basis of their spore morphology, have been identified as members of the genera *Glomus*, *Gigaspora*, *Acaulospora*, and *Sclerocystis* of the Endogonaceae (Gerdemann and Trappe, 1974). There is a marked lack of host specificity among the different VA endophyte strains or species. Typical aspects of VA infections are shown in Figs. 1c, d, and e for soybean, onion, and clover roots, respectively. An infecting hypha, originating from a spore or previously infected root in the soil, enters the root without forming a well-defined appressorium, then ramifies rapidly and spreads intercellularly along the inner layers of the cortex (Fig. 1c). At intervals hyphae penetrate the cortical cells and form the highly branched haustoria-like structures known as arbuscules (Fig. 1d). These intracellular hyphae do not penetrate into the host cytoplasm but remain enveloped by the host plasmalemma (see Fig. 6a in Sec. IV), thus creating a large surface area of contact between the fungus and the host cell (Cox and Sanders, 1974; Cox and Tinker, 1976; Dexheimer et al., 1979). Although there is no direct proof, this is generally regarded as the site of transfer of material between the symbionts. Lipid-containing vesicles, which are probably storage organs, may form as the infection ages (Fig. 1e). The extent of VA infection varies considerably according to the host plant, the endo-

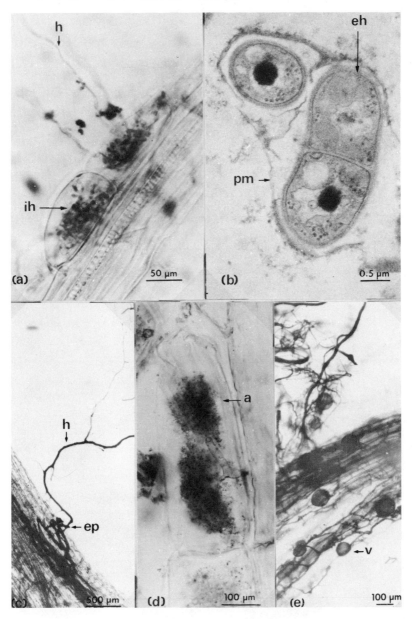

Fig. 1 Ericoid (a,b) and VA (c,d,e) endomycorrhizal infections. (a) Field infected hair root of *Calluna vulgaris* showing penetrating hyphae (h) and intracellular hyphal complexes (ih). (b) Electron micrograph of endophytic hyphae (eh) surrounded by host plasmalemma (pm) in a cortical root cell of *C. vulgaris* (by courtesy of P. Bonfante-Fasolo, CSMT-CNR, Turin). (c) Soyabean root infected with *Glomus mosseae* showing external hypha (h), entry point (ep) and intercellular mycelium. (d) Arbuscule (a) of *G. mosseae* in onion root. (e) Vesicles (v) and intercellular hyphae of *G. mosseae* in clover root. External vesicles and mycelium are also present.

phyte strain, and the habitat, but it may attain 95% of the root system (Khan, 1975). Hyphae spreading along the root surface make new entry points [2-20 per cm (Mosse,

1959; Sanders and Tinker, 1973)] and some extend away from the root surface so as to
establish hyphal continuity between the host cells and the surrounding soil.

III.   IMPROVED GROWTH AND NUTRITION OF HOST PLANTS

In the few studies that have been made on the response of ericaceous plants to mycorrhizal infection there is general agreement that ericoid mycorrhizas can improve the
growth of their host plants in rooting media or soils of low nutrient status (Table
1) (Freisleben, 1936; Brook, 1952; Burgeff, 1961; Read and Stribley, 1973; Stribley
and Read, 1974; Stribley et al., 1975). However, both the mycorrhizal effect and
infection are highly dependent on the nutrient status of the rooting medium, and both
are reduced by increases or imbalances in available phosphorus and nitrogen concentrations (Brook, 1952; Morrison, 1957; Pearson, 1971; Bannister and Norton, 1974;
Stribley et al., 1975; Stribley and Read, 1976).

On the other hand, there are many reports of large growth responses to VA infections in a wide range of plant species developing in soils or rooting media containing little available phosphorus (see examples cited in Table 1 and Mosse, 1973).
Growth responses can almost always be reproduced in nonmycorrhizal plants by the addition of large amounts of available phosphorus to the rooting medium (Fig. 2). This
has the effect of diminishing infection in mycorrhizal plants (Sanders and Tinker,
1973).

It is currently believed that the better nutrition of endomycorrhizal plants is
probably the sole cause of the growth responses to endophyte infection. In most cases
plants showing growth responses to ericoid or VA infections contain larger amounts of
nutrients. In ericoid mycorrhizal plants important increases in both phosphorus and
nitrogen concentrations have been observed (Pearson, 1971; Read and Stribley, 1973;

Fig. 2 Effect of VA mycorrhiza and additional available phosphorus on growth of
soyabeans in gamma-irradiated soil (0.8 Mrad). All plants inoculated with *Rhizobium*
(+ R). Key: +R -M, nonmycorrhizal; +R +M, mycorrhizal (*G. mosseae*); +R +P, nonmycorrhizal plus soluble phosphorus (1.0 g KH_2PO_4/kg soil). (From Asimi et al., 1978.)

Table 1 Responses to ericoid and vesicular-arbuscular (VA) endomycorrhizas

Plant and duration of experiment	Medium	Type of endomycorrhiza	Mycorrhizal status[a]	Yield (mg dry weight per plant)	Reference
Calluna vulgaris 12 weeks	Sterile sand	Ericoid	M NM	2.5 0.6	Pearson (1971)
Vaccinium macro-carpon 6 months	Sterile soil/agar	Ericoid	M NM	235 184	Read and Stribley (1973)
Lycopersi-cum escu-lentum (30) 9 weeks	Sterile sand	VA	M NM	535 198	Daft and Nicolson (1966)
Coprosma robusta 3 months	Irradiated soil	VA	M NM	160[b] 20	Hayman and Mosse (1971)
Allium cepa 10 weeks	Irradiated soil	VA	M NM	255[b] 52	Gianinazzi-Pearson and Gianinazzi (1978)

[a]M, mycorrhizal; NM, nonmycorrhizal.
[b]Shoots only.

Stribley and Read, 1974), and it is thought that the mycorrhizas aid in the uptake of both these elements by the host plants. In VA infections, however, the only consistently important differences between mycorrhizal and nonmycorrhizal plants is the higher phosphorus content of the former (Gerdemann, 1964; Holevas, 1966; Bowen and Theodorou, 1967; Gray and Gerdemann, 1967; Sanders and Tinker, 1971; Sanders et al., 1977). Although other elements such as nitrogen (Ross, 1971), zinc (Gilmore, 1971), and sulfur (Gray and Gerdemann, 1973) have been shown to be involved occasionally, phosphorus is regarded as by far the most important nutrient concerned in the growth responses.

Studies of phosphate uptake from isotopically labeled solutions have shown that this higher accumulation of phosphorus in endomycorrhizal plants is the result of an enhanced uptake by infected roots (Table 2). Roots of *Calluna* seedlings exposed to labeled phosphate solution have 3-4.5 times more activity when mycorrhizal; VA mycorrhizal clover and *Liriodendron* roots have about twice as much activity as nonmycorrhizal roots. The rate of transfer of this absorbed phosphorus to the shoot differs in the two types of endomycorrhiza. In mycorrhizal ericaceous plants there is some accumulation of phosphorus-32 in the roots with a relatively slow release to the shoots (Pearson, 1971; Pearson and Read, 1973b), whereas in VA-infected plants there is a rapid translocation of the element to the aerial portions of the plant (Gray and Gerdemann, 1969; Rhodes and Gerdemann, 1975). These results demonstrate clearly that

Table 2 ^{32}P Uptake by ericoid, VA mycorrhizal, and nonmycorrhizal roots from radio-labeled orthophosphate solutions

Plant and duration of experiment	Type of endomycorrhiza	Mycorrhizal status[a]	Mean radioactivity in roots (counts min^{-1} mg^{-1})	Reference
Calluna vulgaris Ericoid				
48 hr		M	1149[b]	Pearson (1971)
Intact plants		NM	249	
Detached roots		M	7365[c]	
		NM	2296	
Liriodendron				
tulipifera	VA	M	6416[c]	Gray and
Intact plants				Gerdemann (1967)
4 days		NM	3365	
Trifolium				
subterraneum	VA	M	26160[c]	Bowen and
Intact plants				Theodorou (1967)
30 min		NM	10620	

[a]M, mycorrhizal; NM, nonmycorrhizal.

[b]Fresh weight basis.

[c]Dry weight basis.

both ericoid and VA endomycorrhizas absorb more available phosphate than do nonmycorrhizal roots and that this is translocated to the shoot.

Plant-available or -exchangeable phosphate in soils, however, accounts for only about 1-5% of the total phosphate content. In low-nutrient heath soils, for example, the concentration can be as low as 5-8 ppm (Hayman and Mosse, 1972). It has therefore been suggested that endomycorrhizas, while improving the uptake of available phosphate, might also solubilize sources of phosphorus that would otherwise be only slightly or not at all available to uninfected roots.

Earlier reports that VA mycorrhizal plants respond greatly to additions of relative insoluble forms of phosphorus such as bonemeal, apatite, and rock phosphate (Daft and Nicolson, 1966; Murdoch et al., 1967; Jackson et al., 1972) seemed to indicate that VA mycorrhizas do indeed solubilize these less available forms. In order to test this possibility directly, an isotopic dilution method has been used which involves labeling the available or exchangeable phosphorus in a soil and comparing the ratio of ^{32}P to ^{31}P (specific activity) in plants grown on it with and without mycorrhizas. If mycorrhizal plants obtain phosphorus from insoluble sources, their specific activity should be lower than that in non-mycorrhizal plants. Studies on several different soils have shown that although VA mycorrhizal plants always absorb greater amounts of phosphorus, no difference exists between their specific activity and that of nonmycorrhizal plants in any one soil (Sanders and Tinker, 1971; Hayman

Table 3 Effect of ericoid and VA mycorrhizas on acid phosphatase activity of detached roots

Plant	Type of mycorrhiza	Mycorrhizal status[a]	Surface acid phosphatase activity (μmol p-nitrophenol mg^{-1} dry weight of root)
Calluna vulgaris	Ericoid	M	14.6[b]
		NM	8.0[b]
Allium cepa	VA	M	1.5
		NM	1.6

[a]M, mycorrhizal; NM, nonmycorrhizal

[b]Significantly different at 5% level.

and Mosse, 1972; Mosse et al., 1973). It has therefore been concluded that both VA mycorrhizal and nonmycorrhizal plants draw their phosphate from the same source and, from measurements of the specific activity of the soil solutions (J. C. Fardeau, personal communication, 1978), that this source is the soil solution or adsorbed phosphate in equilibrium with it. The fact that VA mycorrhiza formation does not modify root surface acid phosphatase activity (Table 3), which is believed to contribute to the mobilization of insoluble organophosphorus compounds by plants (Weissflog and Mengdehl, 1933; Rogers et al., 1940; Saxena, 1964; Wild and Oke, 1966), provides further evidence to support this conclusion. The VA mycorrhizal effect thus appears to be due to a more efficient absorption of available phosphorus and not solubilization. The growth responses of mycorrhizal plants in the presence of relatively insoluble inorganic phosphates could be due to a more efficient uptake of the chemically dissociated ions drawn into solution from solid phase phosphate as the solution phosphate is depleted. This would explain the decline in the relative advantage of VA mycorrhizal over nonmycorrhizal plants in the presence of large amounts of bonemeal (Daft and Nicolson, 1966, 1972).

There is evidence that ericoid mycorrhizas, on the contrary, are active in the mobilization of insoluble organophosphorus compounds in the soil; mycorrhizal roots have a much higher surface acid phosphatase activity than do nonmycorrhizal roots (Table 3). It seems possible therefore that mycorrhizal infection may enhance phosphorus nutrition in ericaceous plants by both a more efficient uptake of available phosphorus and an increased utilization of insoluble phosphorus complexes in the soil. Further work is clearly necessary to verify this point.

IV. ROLE OF ENDOMYCORRHIZAL FUNGI

Several hypotheses have been postulated to explain the increased phosphorus uptake of plants following endomycorrhizal infection (Harley, 1969; Sanders and Tinker,

1973; Tinker, 1975). Of these, modifications in root anatomy or increased longevity
of the roots' absorbing function seem improbable as there is no evidence that endomycorrhizal fungi induce such changes (Read and Stribley, 1975; Tinker, 1975). The
possibility that the endomycorrhizal fungi may be directly involved in the increased
phosphorus transfer from the soil to the plant has been closely studied following
earlier observations that in VA mycorrhizal roots most of the increased P[32]-labeled
phosphate uptake is localized in the fungal structures (Mosse and Bowen, 1967).

The frequent hyphal connections (discussed earlier) between infected cells and
soil particles around ericoid and VA mycorrhizal roots could provide a pathway for
such a transfer. Indirect evidence that this pathway may be functional was first
obtained from measurements of phosphate inflow rates of VA mycorrhizal and nonmycorrhizal roots in soil (Sanders and Tinker, 1971, 1973). Inflow rates increased some
3- to 16-fold when roots became mycorrhizal, and it was concluded that this additional
phosphate inflow into mycorrhizal roots must have entered through the external hyphae
of the endomycorrhizal fungus.

This hyphal pathway of phosphate transport has since been confirmed for both
ericoid and VA endomycorrhizal fungi, under sterile conditions, using the system depicted in Fig. 3 (Pearson and Read, 1973b; Pearson and Tinker, 1975). When radio-
actively labeled phosphate is fed to the external mycelium, it rapidly appears and
accumulates in the mycorrhizal seedlings (Fig. 4). Hyphae of both ericoid and VA
endomycorrhizal fungi clearly have the capacity to absorb and translocate phosphorus
from a solution source to the host plant. The distance of this hyphal transport can
reach what are probably ecologically important values: distances of up to 2 cm have

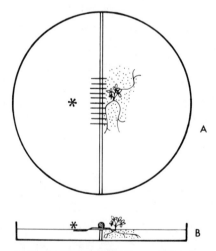

Fig. 3 Surface view (A) and plan view (B) of the split plate method developed to
study translocation by endomycorrhizal fungi. The lines crossing the dividing wall
represent external hyphae of the mycorrhizal fungus, the cross-hatched area shows
the position of the lanolin diffusion barrier, and the asterisk indicates where the
isotope is applied. The stippled area is the soil/agar mixture in which the seedling
is growing.

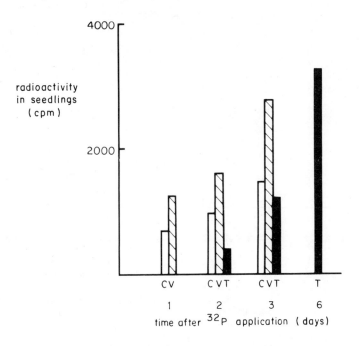

Fig. 4 Phosphorus-32 translocation to seedlings by ericoid (C, V) and VA (T) mycorrhizal fungi. Key: C, *Calluna vulgaris* (cpm/whole seedling X 10²) (Pearson and Read,
1973b); V, *Vaccinium oxycoccos* (cpm/mg fresh weight of shoot X 10) (Pearson and Read,
1973b); T, *Trifolium repens* (cpm/shoot) (V. Pearson and P. B. H. Tinker, unpublished
data).

been reported for ericoid fungi (Pearson and Read, 1973b) and up to 8 cm, in a different system, for VA fungi (Rhodes and Gerdemann, 1975). The ramifying external
mycelia of endomycorrhizal fungi in the soil can thus provide the host plant with a
means of absorbing available phosphorus from nondepleted sources in the soil at an
appreciable distance from the root.

Using the aforementioned system, Pearson and Tinker (1975) measured fluxes of
0.3 to 1.0 X 10^{-9} mol P cm^{-2} sec^{-1} in hyphae of a VA endophyte at some distance
from the host root, values which are not very different from that of 3.8 X 10^{-8} mol
P cm^{-2} sec^{-1} computed theoretically by Sanders and Tinker (1973) for the same fungus.
Since these values are too high to be explained by simple diffusion, an active transport mechanism must be involved and, with the finding of polyphosphate granules in
the vacuoles of VA fungi (Cox et al., 1975), Tinker (1975, p. 339) has proposed "cy-
closis, plus bulk flow, with loading and unloading of polyphosphate into vacuoles as
the method of varying the phosphorus concentration of the streaming protoplasm."
There is now evidence that the subsequent transfer of phosphorus from the fungus into
the host cell is also an active process, taking place across the living interface
(Cox and Tinker, 1976), and that it does not result from digestion of the fungus as
previously believed.

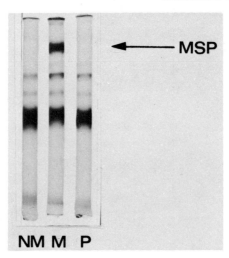

Fig. 5 Electrophoretic gels of soluble extracts of roots of 6-week-old onions stained for alkaline phosphatase activity. Key: NM, nonmycorrhizal; M, VA mycorrhizal (*G. mosseae*); P, nonmycorrhizal plus soluble phosphorus (1.0 g KH_2PO_4/kg soil); MSP, mycorrhiza-specific phosphatase activity.

Mycorrhiza-specific alkaline phosphatase (MSP) activity (EC 3.1.3.1.) (Fig. 5) has recently been reported in VA endomycorrhizas (Gianinazzi-Pearson and Gianinazzi, 1976; Bertheau, 1977). A close correlation exists between the activity maximum of this phosphatase and both the moment of mycorrhizal growth stimulation and the arbuscular phase of the infection (Gianinazzi-Pearson and Gianinazzi, 1978), which seems to indicate that it is somehow linked to the mechanisms involved in the assimilation of phosphorus by VA mycorrhiza. There is strong evidence that mycorrhiza-specific alkaline phosphatase is of fungal origin (Gianinazzi-Pearson et al., 1978), and an intense alkaline phosphatase activity has been localized within the vacuoles of the VA mycorrhizal fungus (Gianinazzi et al., 1979) (Figs. 6c,d). These findings, together with the fact that the vacuole is an active system where the tonoplast appears to play a very important role in ion transport mechanisms (Matile and Wiemken, 1976), suggest that alkaline phosphatase may be involved in the active phosphate transport and/or transfer mechanisms in VA fungi, as has been proposed for certain basidiomycetes (North and Lewis, 1971).

No such studies have been reported on the mechanisms of transport and release of nutrients by ericoid fungi to ericaceous mycorrhizal plants. The accumulation of phosphorus in ericoid mycorrhizal roots and its relatively slow release to the shoots could be a reflection of a passive transfer mechanism such as lysis of intracellular endophytic hyphae (Nieuwdorp, 1969). No evidence of fungal digestion by host cells, however, has been found in ultrastructural studies (P. Bonfante-Fasolo and V. Gianinazzi-Pearson, unpublished data). Pure culture studies have shown that the mycorrhizal fungus possesses a high acid phosphatase activity and readily utilizes sodium phytate (sodium inositol hexaphosphate) (Pearson and Read, 1975), providing further evidence that ericoid mycorrhizas may be active in the mobilization of organo-phosphorus compounds in the soil.

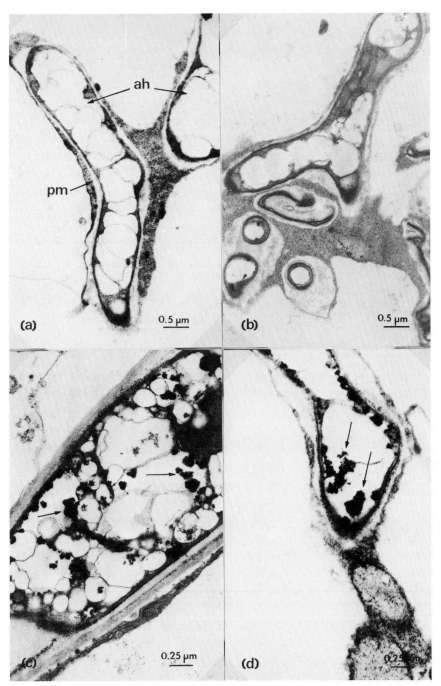

Fig. 6 Electron micrographs of localization of alkaline phosphatase activity within hyphae of the VA mycorrhizal fungus *G. mosseae* in onion roots. Black precipitate (arrows) indicates enzyme activity within fungal vacuoles. (a) Substrate omitted (ah, arbuscular hyphae; pm, host plasmalemma). (b) Substrate (α-naphthyl phosphate) plus KCN (alkaline phosphatase inhibitor). (c) α-Naphthyl phosphatase activity. (d) β-Glycerophosphatase activity.

V. CONCLUSIONS

Endomycorrhizal fungi are by no means the only soil microorganisms influencing the phosphorus nutrition of plants, but because of their symbiotic nature they are much better placed than free-living phosphate-dissolving fungi or bacteria to intervene in phosphorus uptake by roots. Their external hyphae can extend and ramify in the soil to a region of phosphorus supply well beyond the phosphorus-depletion zone that normally surrounds strongly absorbing roots (Nye, 1969). The apparent avidity of endomycorrhizal fungi for available phosphate must enable them to compete efficiently with other microorganisms in the soil. Owing to this efficiency and their capacity to translocate phosphorus over large distances (microbiologically speaking), they provide the plant with an alternative pathway for phosphorus transfer through the soil, parallel to that of normal diffusion, as well as a larger, better-distributed surface for its absorption.

Although the aforementioned hypotheses concerning the sources of soil phosphorus utilized by endomycorrhizal fungi need to be more widely tested, from studies already performed it seems improbable that VA endophytes release phosphates from insoluble complexes. At first sight, it may appear that in natural ecosystems these fungi simply exhaust the limited supplies of phosphate in the soil solution. However, as the soil solution becomes depleted, more phosphate ions will be drawn into solution from the soil reserves by desorption processes and thus more phosphate will come into circulation. There is also evidence that VA fungi may absorb phosphorus present in the soil solution at such low concentrations that it is practically unavailable to nonmycorrhizal roots of some plants (Beslow et al., 1970; Mosse et al., 1973). This could confer a big advantage on mycorrhizal plants growing in conditions where roots are so close together that depletion zones overlap, as in woodlands and pastures. Since VA fungi are not limited to any one plant family or habitat, their contribution to phosphorus cycling in any given ecosystem will depend mainly on the plant species forming endomycorrhizas, the competition for phosphorus between them (Fitter, 1977), and the efficiency of the VA endophyte strains or species involved to absorb and transfer phosphorus to the host plants.

Ericoid endophytes can, in addition to available phosphorus, utilize phytate as a phosphorus source. This should enormously benefit ericaceous plants and place them in a strongly advantageous competetive position in their ecosystem, where the mor-humus soils normally have very low free inorganic phosphorus levels and where the commonest sources of this element are phytates (Cosgrove and Tate, 1963).

It is now evident that since nearly all soils contain endomycorrhizal fungi they must play a vital role in the cycling of phosphorus. Intensive studies of their behavior in soil and of the mechanisms which determine their efficiency in phosphate utilization and transfer to the host plant should promote our understanding of this role in the ecosystem.

REFERENCES

Asimi, S., Gianinazzi-Pearson, V., Gianinazzi, S., Obaton, M., and Berthau, S. (1978). Interactions entre les endomycorrhizes à vésicules et arbuscules (VA) et le *Rhizobium* chez le soja. *103e Congr. Nat. Soc. Sav.*, Vol. 1, pp. 247-256.

Bannister, P., and Norton, W. M. (1974). The response of mycorrhizal rooted cuttings of heather (*Calluna vulgaris* (L) Hull) to variations in nutrient and water regimes. *New Phytol. 73*: 81-88.

Bertheau, Y. (1977). Etudes des phosphatases solubles des endomycorhizes à vésicules et arbuscules. D.E.A. thesis, University of Dijon, France.

Beslow, D. T., Hacskaylo, E., and Melhuish, J. H. (1970). Effects of environment on beaded root development in red maple. *Bull. Torrey Bot. Club 97*: 248-252.

Bowen, G. D., and Theodorou, C. (1967). Studies on phosphate uptake by mycorrhizas. *Proc. 14th IUFRO Congr. 24*: 116-138.

Brook, P. J. (1952). Mycorrhiza of *Pernettya macrostigma*. *New Phytol. 51*: 388-397.

Burgeff, H. (1961). *Mikrobiologie des Hochmoores*. Fischer, Stuttgart.

Cosgrove, D. J., and Tate, M. E. (1963). The occurrence of *neo*-inositol hexaphosphate in soil. *Nature 200*: 568-569.

Cox, G., and Sanders, F. (1974). Ultrastructure of the host-fungus interface in a vesicular-arbuscular mycorrhiza. *New Phytol. 73*: 901-912.

Cox, G., and Tinker, P. B. (1976). Translocation and transfer of nutrients in vesicular-arbuscular mycorrhiza. I. The arbuscule and phosphorus transfer: A quantitative ultrastructural study. *New Phytol. 77*: 371-378.

Cox, G. C., Sanders, F. E. T., Tinker, P. B., and Wild, J. (1975). Ultrastructural evidence relating to host-endophyte transfer in a vesicular-arbuscular mycorrhiza. In *Endomycorrhizas*, F. E. Sanders, B. Mosse, and P. B. Tinker (Eds.). Academic Press, New York, pp. 297-312.

Daft, M. J., and Nicolson, T. H. (1966). Effect of *Endogone* mycorrhiza on plant growth. *New Phytol. 65*: 343-350.

Daft, M. J., and Nicolson, T. H. (1972). Effect of *Endogone* mycorrhiza on plant growth. IV. Quantitative relationships between the growth of the host and the development of the endophyte in tomato and maize. *New Phytol. 71*: 287-295.

Dexheimer, J., Gianinazzi, S., and Gianinazzi-Pearson, V. (1979). Ultrastructural cytochemistry of the host-fungus interfaces in the endomycorrhizal association *Glomus mosseae/Allium cepa*. *Z. Pflanzenphysiol. 92*: 191-206.

Fitter, A. H. (1977). Influence of mycorrhizal infection on competition for phosphorus and potassium by two grasses. *New Phytol. 79*: 119-125.

Freisleben, R. (1936). Weitere Untersuchungen über die Mykotrophie der Ericaceen. *Jahrb. Wiss. Bot. 82*: 413-459.

Gerdemann, J. W. (1964). The effect of mycorrhiza on the growth of maize. *Mycologia 56*: 342-349.

Gerdemann, J. W., and Trappe, J. M. (1974). The Endogonaceae of the Pacific Northwest. *Mycol. Memoir 5*: 75 pp.

Gianinazzi, S., Gianinazzi-Pearson, V., and Dexheimer, J. (1979). Enzymatic studies on the metabolism of vesicular-arbuscular mycorrhiza. III. Ultrastructural localisation of acid and alkaline phosphatase in onion roots infected by *Glomus mosseae* Nichol. & Gerd. *New Phytol. 82*: 127-132.

Gianinazzi-Pearson, V. (1976). Les mycorhizes endotrophes: état actuel des connaissances et possibilités d'application dans la pratique culturale. *Ann. Phytopath. 8*: 249-256.

Gianinazzi-Pearson, V., and Gianinazzi, S. (1976). Enzymatic studies on the metabolism of vesicular-arbuscular mycorrhiza. I. Effect of mycorrhiza formation and phosphorus nutrition on soluble phosphatase activities in onion roots. *Physiol. Vég. 14*: 833-841.

Gianinazzi-Pearson, V., and Gianinazzi, S. (1978). Enzymatic studies on the metabolism of vesicular-arbuscular mycorrhiza. II. Soluble alkaline phosphatase specific to mycorrhizal infection in onion roots. *Physiol. Plant Pathol. 12*: 45-53.

Gianinazzi-Pearson, V., Gianinazzi, S., Dexheimer, J., Bertheau, Y., and Asimi, S. (1978). Les phosphatases alcalines solubles dans l'association endomycorhizienne à vésicules et arbuscules. *Physiol. Vég. 16*: 671-678.

Gilmore, A. E. (1971). The influence of endotrophic mycorrhizae on the growth of peach seedlings. *J. Amer. Soc. Hort. Sci. 96*: 35-38.

Gray, L. E., and Gerdemann, J. W. (1967). Influence of vesicular-arbuscular mycorrhiza on the uptake of phosphorus-32 by *Liriodendron tulipifera* and *Liquidambar styraciflua*. *Nature 213*: 106-107.

Gray, L. E., and Gerdemann, J. W. (1969). Uptake of phosphorus-32 by vesicular-arbuscular mycorrhizae. *Plant Soil 30*: 415-422.

Gray, L. E., and Gerdemann, J. W. (1973). Uptake of sulphur-35 by vesicular-arbuscular mycorrhizae. *Plant Soil 39*: 687-689.

Harley, J. L. (1969). *The Biology of Mycorrhiza*. Hill, London.

Hayman, D. S., and Mosse, B. (1971). Plant growth responses to vesicular-arbuscular mycorrhiza. I. Growth of *Endogone*-inoculated plants in phosphate-deficient soils. *New Phytol. 70*: 19-27.

Hayman, D. S., and Mosse, B. (1972). Plant growth responses to vesicular-arbuscular mycorrhiza. III. Increased uptake cf labile P from soil. *New Phytol. 71*: 41-47.

Holevas, C. D. (1966). The effect of a vesicular-arbuscular mycorrhiza on the uptake of soil phosphorus by strawberry (*Fragaria* sp. var. Cambridge Favourite). *J. Hort. Sci. 41*: 57-64.

Jackson, N. E., Franklin, R. E., and Miller, R. H. (1972). Effects of VA mycorrhizae on growth and phosphorus content of three agronomic crops. *Proc. Soil Sci. Amer. 36*: 64-67.

Khan, A. G. (1975). Growth effects of VA mycorrhiza on crops in the field. In *Endomycorrhizas*, F. E. Sanders, B. Mosse, and P. B. Tinker (Eds.). Academic Press, New York, pp. 419-435.

McComb, A. L. (1938). The relation between mycorrhizae and the development and nutrient absorption of pine seedlings in a prairie nursery. *J. For. 36*: 1148-1154.

Matile, Ph., and Wiemken, A. (1976). Interactions between cytoplasm and vacuole. In *Transport in Plants III, Encyclop. Plant. Physiol.* [*N.S.*] *3*: 255-287.

Melin, E., and Nilsson, H. (1950). Transfer of radioactive phosphorus to pine seedlings by mycorrhizal hyphae. *Physiol. Plant. 3*: 88-92.

Melin, E., and Nilsson, H. (1952). Transfer of labelled nitrogen from an ammonium source to pine seedlings through mycorrhizal mycelium. *Svensk. Bot. Tidskr. 46*: 281-285.

Melin, E., and Nilsson, H. (1953a). Transfer of labelled nitrogen from glutamic acid to pine seedlings through the mycelium of *Boletus variegatus* (Sw.) Fr. *Nature 171*: 134.

Melin, E., and Nilsson, H. (1953b). Transport of labelled phosphorus to pine seedlings through the mycelium of *Cortinarius glaucopus* (Schaeff. ex Fr.) Fr. *Svensk. Bot. Tidskr. 48*: 555-558.

Melin, E., Nilsson, H., and Hacskaylo, E. (1958). Translocation of cations to seedlings of *Pinus virginiana* through mycorrhizal mycelia. *Bot. Gaz. 119*: 241-246.

Mitchell, H. L., Finn, R. F., and Rosendhal, R. O. (1937). The relation between mycorrhizae and the growth and nutrient absorption of coniferous seedlings in nursery beds. *Black Rock For. Papers 1*: 58-73.

Morrison, T. M. (1957). Host-endophyte relationship in mycorrhiza of *Pernettya macrostigma. New Phytol. 56*: 247-257.

Mosse, B. (1959). The regular germination of resting spores and some observations on the growth requirements of *Endogone* species causing vesicular-arbuscular mycorrhizas. *Trans. Brit. Mycol. Soc. 42*: 274-286.

Mosse, B. (1973). Advances in the study of vesicular-arbuscular mycorrhiza. *Ann. Rev. Phytopath. 11*: 171-196.

Mosse, B., and Bowen, G. D., quoted by Bowen and Theodorou (1967), above.

Mosse, B., Hayman, D. S., and Arnold, D. (1973). Plant growth responses to vesicular-arbuscular mycorrhiza. V. Phosphate uptake by three plant species from P-deficient soils labelled with ^{32}P. *New Phytol. 72*: 809-815.

Murdoch, C. L., Jackobs, J. A., and Gerdemann, J. W. (1967). Utilization of phosphorus sources of different availability by mycorrhizal and non-mycorrhizal maize. *Plant Soil 27*: 329-334.

Nieuwdorp, P. J. (1969). Some investigations on the mycorrhiza of *Calluna, Erica,* and *Vaccinium. Acta Bot. Neerl. 18*: 180-196.

North, J., and Lewis, D. (1971). Phosphatases of *Coprinus lagopus*: The conditions for their production and the genetics of the alkaline phosphatase. *Genet. Res.* (Cambridge) *18*: 153-166.

Nye, P. H. (1969). The soil model and its application to plant nutrition. In *Ecological Aspects of the Mineral Nutrition of Plants, Symp. Brit. Ecol. Soc. 9*: 201-213.

Pearson, V. (1971). The biology of mycorrhiza in the Ericaceae. Ph.D. thesis, University of Sheffield, England.

Pearson, V., and Read, D. J. (1973a). The biology of mycorrhiza in the Ericaceae. I. The isolation of the endophyte and synthesis of mycorrhiza in aseptic culture. *New Phytol. 72*: 371-379.

Pearson, V., and Read, D. J. (1973b). The biology of mycorrhiza in Ericaceae. II. The transport of carbon and phosphorus by the endophyte and the mycorrhiza. *New Phytol. 72*: 1325-1331.

Pearson, V., and Read, D. J. (1975). The physiology of the mycorrhizal endophyte of *Calluna vulgaris. Trans. Brit. Mycol. Soc. 64*: 1-7.

Pearson, V., and Tinker, P. B. H. (1975). Measurement of phosphorus fluxes in the external hyphae of endomycorrhizas. In *Endomycorrhizas*, F. E. Sanders, B. Mosse, and P. B. Tinker (Eds.). Academic Press, New York, pp. 277-287.

Read, D. J. (1974). *Pezizella ericae* sp. nov. the perfect state of a typical mycorrhizal endophyte of Ericaceae. *Trans. Brit. Mycol. Soc. 63*: 381-383.

Read, D. J., and Stribley, D. P. (1973). Effect of mycorrhizal infection on nitrogen and phosphorus nutrition of ericaceous plants. *Nature 244*: 81-82.

Read, D. J., and Stribley, D. P. (1975). Some mycological aspects of the biology of mycorrhiza in the Ericaceae. In *Endomycorrhizas*, F. E. Sanders, B. Mosse, and P. B. Tinker (Eds.). Academic Press, New York, pp. 105-117.

Rhodes, L. H., and Gerdemann, J. W. (1975). Phosphate uptake zones of mycorrhizal and non-mycorrhizal onions. *New Phytol. 75*: 555-561.

Rogers, H. T., Pearson, R. W., and Pierre, N. H. (1940). Absorption of organophosphates by corn and tomato plants and the mineralizing action of exoenzyme systems of growing roots. *Soil. Sci. Soc. Amer. Proc. 5*: 285-291.

Ross, J. P. (1971). Effect of phosphate fertilization on yield of mycorrhizal and non-mycorrhizal soybeans. *Phytopathol. 61*: 1400-1403.

Sanders, F. E., and Tinker, P. B. (1971). Mechanism of absorption of phosphate by *Endogone* mycorrhizas. *Nature 233*: 278-279.

Sanders, F. E. T., and Tinker, P. B. H. (1973). Phosphate flow into mycorrhizal roots. *Pestic. Sci. 4*: 385-395.

Sanders, F. E. T., Tinker, P. B., Black, R. L. B., and Palmerley, S. M. (1977). The development of endomycorrhizal root systems. I. Spread of infection and growth-promoting effects with four species of vesicular-arbuscular endophyte. *New Phytol. 78*: 257-268.

Saxena, S. N. (1964). Phytase activity of plant roots. *J. Exp. Bot. 15*: 654-658.

Stribley, D. P., and Read, D. J. (1974). The biology of mycorrhiza in the Ericaceae. IV. The effect of mycorrhizal infection on uptake of ^{15}N from labelled soil by *Vaccinium macrocarpon* Ait. *New Phytol. 73*: 1149-1155.

Stribley, D. P., and Read, D. J. (1976). The biology of mycorrhiza in the Ericaceae. VI. The effects of mycorrhizal infection and concentration of ammonium nitrogen on growth of cranberry (*Vaccinium macrocarpon* Ait.) in sand culture. *New Phytol. 77*: 63-72.

Stribley, D. P., Read, D. J., and Hunt, R. (1975). The biology of mycorrhiza in the Ericaceae. V. The effects of mycorrhizal infection, soil type and partial soil sterilization (by gamma-irradiation) on growth of cranberry (*Vaccinium macrocarpon* Ait.). *New Phytol. 75*: 119-130.

Tinker, P. B. H. (1975). Effects of vesicular-arbuscular mycorrhizas on higher plants. In *Symbiosis, Symp. Soc. Exp. Biol. 29*: 325-349.

Weissflog, J., and Mengdehl, H. (1933). Studien zum Phosphorstoffwechsel in den höheren Pflanzen. III. Aufnahme und Verwertbarkeit organischer Phosphosäureverbindungen durch die Planze. *Planta 19*: 182-241.

Wild, A., and Oke, O. L. (1966). Organic phosphate compounds in calcium chloride extracts of soil: Identification and availability to plants. *J. Soil Sci. 17*: 356-371.

REFERENCES ADDED IN PROOF

Asimi, S., Gianinazzi-Pearson, V., and Gianinazzi, S. (1980). Influence of increasing soil phosphorus levels on interactions between vesicular-arbuscular mycorrhizae and *Rhizobium* in soybeans. *Can. J. Bot. 58*: 2200-2206.

Bonfante-Fasolo, P., and Gianinazzi-Pearson, V. (1979). Ultrastructural aspects of endomycorrhiza in the Ericaceae. I. Naturally infected hair roots of *Calluna vulgaris* L. Hull. *New Phytol. 83*: 739-744.

Callow, J. A., Capaccio, L. C. M., Parish, G., and Tinker, P. B. (1978). Detection and estimation of polyphosphate in vesicular-arbuscular mycorrhizas. *New Phytol. 80*: 125-134.

Cooper, K. M., and Tinker, P. B. (1978). Translocation and transfer of nutrients in vesicular-arbuscular mycorrhizas. II. Uptake and translocation of phosphorus, zinc and sulphur. *New Phytol. 81*: 43-52.

Cox, G., Moran, K. J., Sanders, F., Nockolds, C., and Tinker, P. B. (1980). Translocation and transfer of nutrients in vesicular-arbuscular mycorrhizas. III. Polyphosphate granules and phosphorus translocation. *New Phytol. 84*: 649-659.

Cress, W. A., Throneberry, G. O., and Lindsey, D. L. (1979). Kinetics of phosphorus absorption in mycorrhizal and nonmycorrhizal tomato roots. *Plant Physiol. 64*: 484-487.

Chapter 34

ROLES OF LITTER-DECOMPOSING AND ECTOMYCORRHIZAL FUNGI IN NITROGEN CYCLING IN THE SCANDINAVIAN CONIFEROUS FOREST ECOSYSTEM

Gösta Lindeberg
*Swedish University of Agricultural Sciences
Uppsala, Sweden*

I. INTRODUCTION

Among natural terrestrial ecosystems, the forests have the highest sustained productivity because of the effective recycling of nutrient elements (Witkamp and Ausmus, 1976). In this chapter the role of fungi in the recycling of nitrogen will be discussed, particularly with reference to Scandinavian coniferous forests.

One of the most obvious and interesting mycological problems in this ecosystem is the relation between the litter-decomposing fungi and the mycorrhizal fungi associated with the trees. Since this relationship is intimately involved in the transfer of nitrogen from litter to trees, a discussion of this special aspect will be included.

II. AMOUNTS AND COMPOSITION OF LITTER

The litter fall in the Scandinavian coniferous forests is of the order of 2000-4000 kg ha^{-1} year^{-1} (Bonnevie-Svendsen and Gjems, 1957). The humus layer is, in most cases, of the mor type, and the soil profile is a podzol.

The chemical composition of newly fallen pine needle litter is shown in Table 1. Characteristic of this type of litter is that the cell walls of the plant residues

Table 1 Composition (%) of dead needles of *Pinus sylvestris*

Cellulose	30
Hemicelluloses	19
Lignin	28
Water-soluble fraction	14.5
Acetone-soluble fraction	8.4
Total C content	51
Total N content	0.4

Source: Unpublished data of B. Berg (Swedish Coniferous Forest Project).

are strongly lignified, there are large amounts of resins and terpenes, and the C/N ratio is high (>100). The pH of the litter, as well as the F and H layers, is about 4.

III. NITROGEN DISTRIBUTION AND TURNOVER

The nitrogen distribution and turnover in a Scots pine (*Pinus sylvestris*) forest on a sand plain in central Sweden has been extensively studied within the Swedish Coniferous Forest Project, an interdisciplinary ecological research project. Some results are presented in Table 2 (Bringmark, 1978).

 The nitrogen content of the vegetation was found to be about twice that in the litter layer. By far the largest part of the nitrogen in the ecosystem is stored in the slowly decomposing humus which occurs not only in the F and H layers but also in considerable amounts in the mineral soil to a depth of at least 30 cm. The nitrogen taken up annually by the vegetation is about one-third of the amount found in the litter. This indicates a slow mineralization rate. In fact, it takes about 7 years for the needles to be completely decomposed. It has been shown (Bringmark, 1978) that about half of the nitrogen taken up by the vegetation goes into the trees and half into the ground cover. The nitrogen input into the system mainly consists of inorganic nitrogen compounds, deposited from the atmosphere. Microbial nitrogen fixation is very small. The total accumulation is about 4 kg of nitrogen per hectare per year. The system is limited by nitrogen (and incidentally also by phosphate). Any nitrogen mineralized is immediately immobilized by microbes, and none remains in solution, so that the drainage loss is practically zero. Thus the metabolism of the fungi and bacteria in the soil layer maintains the nitrogen content of the system.

Table 2 Nitrogen distribution and turnover in a mature Scots pine forest in central Sweden[a]

N content of vegetation (including roots)	184
N content of surface litter	69
N content of root litter	13
N content of humus (F + H) layer	162
N content of mineral soil 0-30 cm	780
N uptake by vegetation	33
N input	
Deposition from the atmosphere	3.6
Microbial fixation	0.4
Drainage loss at 20 cm depth	(0.13)
Accumulated	3.9

[a]Kilograms per hectare per year.

Source: Data from the Swedish Coniferous Forest Project; see Bringmark (1978).

Table 3 Amount of nitrogen in inorganic form and in biomass in humus incubated at 60% WHC[a] and 15°C[b]

	Weeks after start	
	0	28
Amount of nitrogen in:		
Fungi	290	550
Bacteria	156	144
Nematodes	0.03	5.0
Microarthropods	0.02	0.2
Biomass total	446	699
Inorganic N	8.4	13.0
Total movable N	454	712

Total N: 9.8 mg/g DW
pH 4.0

[a]WHC = water-holding capacity.

[b]Numbers represent µg N/g DW.

Source: Sohlenius et al. (1976).

An estimate of the amount of nitrogen in inorganic form and in the biomass of the various soil organisms was made in an experiment where samples of humus from the FH layer were incubated in plastic bags for 28 weeks at 15°C and 60% WHC (Sohlenius et al., 1976). Some results are presented in Table 3.

This experiment again shows that the largest part of the nitrogen was bound in the humus components. The "total movable N" (= N bound in organisms + inorganic N) constituted only a small part of the total nitrogen of the humus: 4.6% at the start and 7.3% at the end of the experiment. Some increase in mineralized nitrogen occurred during the experiment: inorganic nitrogen (ammonium nitrogen) rose from 8.4 to 13.0 µg/g dry weight (DW).

Of the nitrogen bound in living organisms, the largest part was found in the fungal mycelia, and the increase in biomass during the experiment was mainly due to fungal growth. The bacterial biomass remained practically unchanged. The amount of nitrogen bound in nematodes and microarthropods was very small.

Because of the apparent importance of the fungi, a critical question is to what extent the fungal mycelia, measured by the Jones and Mollison (1948) technique, are indeed living, i.e., contain living protoplasm, or to what extent they consist of empty cell walls. Söderström (1977), using a method for vital staining of fungal mycelia with fluorescein diacetate, found that during periods of high fungal activity, only 1-5% of the total mycelial mass was metabolically active. Consequently, the relative importance of the bacterial population might be greater than has been thought in forest soil of the mor type.

IV. LITTER-DECOMPOSING FUNGI

Fungi and bacteria compete in and on litter for all that is available to them. The
success of each species in a particular environment is dependent on enzyme production,
uptake efficiency, growth potential, spreading efficiency, and potential to inhibit
the growth of competitors (Goksøyr, 1975).

A. Microfungi

As the different constituents of the litter are gradually decomposed and utilized
by microorganisms, a succession of fruiting organs of microfungi, i.e., phycomycetes
and ascomycetes, as well as conidial stages of fungi imperfecti, generally occurs.
This succession may partly be explained in terms of the time taken for the various
species to reach the fruiting or conidial stage, as shown by Harper and Webster (1964)
with the fungi developing on rabbit dung. Another cause of this succession is the
gradual change in the substrate due to the activities of the decomposers.

     Detailed studies on the succession of microfungi on decomposing leaf litter of
Pinus sylvestris have been carried out by various workers, including Kendrick (1958;
summarized by Kendrick and Burges, 1962) and Hayes (1965). Söderström (1975) studied
the microfungi occurring on litter and in the different soil horizons in a plantation
of Norway spruce (*Picea abies*) in southern Sweden. Reviews of the work on the fungal
flora which colonizes coniferous litter were published by Hudson (1968) and Millar
(1974). Several authors, however, have emphasized that frequency of isolation is a
poor guide to the importance of a fungus species in decomposition (Burges, 1960; Hering, 1967). Therefore, to elucidate the possible role of soil microfungi in decomposition of the various chemical components of plant litter, a number of studies on
the enzyme production and decomposing capacity of the soil fungal flora have been
carried out (e.g., Domsch and Gams, 1969; Flanagan and Scarborough, 1974; Bååth and
Söderström, 1974). In most of these studies, the capacity of the fungi to decompose
cellulose, xylan (hemicellulose), pectin, chitin, and protein was tested. A large
number of fungi have been found to produce enzymes necessary for the decomposition
of these substances.

     It has not been possible to test directly the capability of the fungi to decom-
pose lignin in a corresponding way. Flanagan and Scarborough (1974) used an indirect
combined test (decolorization of humic acid, darkening of gallic acid, and cellulose
decomposition) as an indication of lignin-decomposing ability. So far, however, there
is no clear evidence that effective lignin breakdown is brought about by individual
microfungi. The present view is that in soil a complex flora of microfungi and bacteria brings about a partial degradation and conversion of lignin into humus.

B. Basidiomycetes

The fungi occurring in the litter and humus layers comprise basidiomycetes as well
as the microfungi already discussed. The basidiomycetes themselves comprise litter
decomposers and mycorrhizal fungi.

It is not easy to estimate the abundance of basidiomycetes. When the dilution plate technique is used, basidiomycetes are not recorded. When fungal biomass is estimated, the length and average diameter of the hyphae per gram dry weight is determined, but the methods used do not differentiate between different groups of fungi. Although clamped (basidiomycete) and unclamped hyphae may be separated (Miller and Laursen, 1974), several basidiomycetes, e.g., the genus *Suillus* (*Boletus*), lack clamp connections, which makes the discrimination between basidiomycetes and nonbasidiomycetes uncertain. Differentiation between mycelia of litter decomposers and mycorrhizal fungi by morphological methods is not possible. The difficulties are illustrated by a quotation from Flanagan and Scarborough (1974, p. 159), who, referring to the physiological groups of decomposer fungi in tundra, write that "the role of the basidiomycete fungi in the tundra has not been studied, but hyphae with clamp connections show that basidiomycetes account for at least one third of the total soil fungal biomass in the autumn." The fact that research on microfungi and on basidiomycetes mostly seems to proceed with little contact between the different mycologists working in each area exacerbates the difficulties inherent in these studies.

It is well known that certain basidiomycetes, the white rot fungi, can decompose both lignin and cellulose in wood. Falck (1923) suggested that various basidiomycetes growing in forest litter decompose the litter in the same way as the wood-rotting fungi decompose wood, later proving in experiments with pure cultures of a few species that this was the case (Falck, 1930). Lindeberg (1944, 1946) cultured 46 different species of basidiomycetes that occur in nature on decomposing litter (belonging to the genera *Clavaria*, *Clitocybe*, *Collybia*, *Flammula*, *Hypholoma*, *Lepiota*, *Marasmius*, *Mycena*, *Panus*, *Pholiota*, and *Tricholoma*) on sterilized leaf litter and determined the amounts of dry matter, cellulose, and lignin decomposed. All the species studied decomposed the lignin and, with one single exception, the cellulose of the substrate and caused a total dry weight loss of 13-54% in some 6-7 months. Maximum lignin and cellulose losses amounted to 75% and 85%, respectively. Hering (1967) obtained similar results with some other basidiomycetes.

V. IMPORTANCE OF THE C/N RATIO

The carbon/nitrogen ratio of the litter is often used as an indicator of potential decomposability but is, in fact, not a particularly simple concept (Park, 1976). Usually this ratio is expressed as total C/total N and is determined by chemical methods. The ratio thus obtained may be termed the provision ratio (Park, 1976). However, with respect to each species of microorganism that takes part in the decomposition process, only the carbon and nitrogen contents of those constituents of the litter are important which are available to the specific organism in the presence of other microorganisms.

This is illustrated in the "saprophytic sugar fungi" (Burges, 1939), mainly phycomycetes, which are able to utilize sugars and other carbon constituents of plant tissues simpler than cellulose or lignin. The cellulose and lignin of the litter do not represent carbon and energy sources for these organisms. Only after the cellulose has been hydrolyzed by cellulase-producing organisms and soluble decomposition products (glucose, cellobiose) have been produced will further development of the sugar fungi occur. Some of the nitrogen of the litter may be locked up in humus-like compounds at an early stage (Swaby and Ladd, 1962, 1966) and may therefore be unavailable to the sugar fungi.

The lignin- and cellulose-decomposing fungi are probably able to utilize a portion of the nitrogen bound in these compounds, which are to a high degree resistant to attacks by other microorganisms (cf. Sec. VII, below), and hence have a larger nitrogen supply than the sugar fungi. However, the effective C/N ratio of the substrate available to the aforementioned fungi will be much higher than the C/N ratio of the substrate available to the sugar fungi.

VI. NITROGEN NUTRITION OF LITTER-DECOMPOSING BASIDIOMYCETES

Nitrogen nutrition of the litter-decomposing basidiomycetes has been studied by several workers, but many problems are still unsolved.

As a rule, these fungi are able to use ammonium nitrogen as the sole nitrogen source, whereas only a few of the species investigated can utilize nitrate (Lindeberg, 1944; Lundeberg, 1970). The ability of the same fungi to utilize amino acids has not been extensively studied. In several species, however, glutamic and aspartic acids as well as asparagine could be used as nitrogen sources (Norkrans, 1950). Other amino acids were utilized only by certain species.

In *Marasmius perforans*, which decomposes spruce needles, an addition of pyridoxine was required for the utilization of casein hydrolyzate and individual amino acids (Lindeberg and Lindeberg, 1964). However, requirements for specific amino acids seem to be uncommon among these fungi.

Growth of the lignin- and cellulose-decomposing fungi will be limited by the available nitrogen. It has been shown by Cowling (1970) that the wood-destroying fungi economize on their limited nitrogen resources by recycling available nitrogen in a dynamic and continuous system of autolysis of less active cells and reutilization of nitrogenous constituents by more active mycelia without significant loss. To what extent the litter-decomposing basidiomycetes are able to conserve their nitrogen supply in a similar way is not known. A recycling of nitrogen within the mycelia of these fungi seems, however, most probable.

VII. ROLE OF FUNGI IN THE TRANSFER OF NITROGEN TO TREE ROOTS

Since the classical works of Frank (1885) and Melin (1917), the beneficial effect of ectomycorrhizae to trees has been well established. Using the tracer technique, Melin and Nilsson (1952, 1953) showed that ^{15}N, absorbed as ammonium ions or glutamic acid by the mycelium of the fungal symbiont, was translocated to the host plant, whereas ^{14}C-tagged products of photosynthesis were translocated to the fungal symbiont (Melin and Nilsson, 1957).

On the other hand, the hypothesis that the ectomycorrhizae of forest trees might function as nitrogen-fixing organs has not been supported by critical experimental work, although the possibility remains that nitrogen-fixing bacteria may to some extent be stimulated in the rhizosphere of mycorrhizae (Harley, 1969).

Since the findings of Melin (1925) that various mycorrhizal fungi could use essentially only glucose as a carbon and energy source, the restricted ability of these fungi to utilize organic compounds other than a few mono- and disaccharides and, sometimes, starch has been confirmed by several authors (Norkrans, 1950; Palmer and Hacskaylo, 1970). Björkman (1942) demonstrated a correlation between the frequency of mycorrhizae and the concentration of soluble carbohydrates in the roots of *Pinus sylvestris*.

Two factors have been suggested as causing a rapid flow of assimilates from the host tree to the mycorrhizal fungus: (1) Lewis and Harley (1965) found that the mycorrhizal fungus in beech mycorrhiza absorbed sugars from the host and converted them into forms (mannitol, trehalose, and glycogen) that were unavailable for reciprocal flow. This "sink" in the fungal layer effectively contributes to the rapid translocation of photosynthate to the fungal 'sheath. (2) Auxin secretion of the mycorrhizal fungi may possibly enhance translocation of sugars to the infected roots from the upper parts of the tree (Meyer, 1974).

The mycorrhizal fungi which have so far been studied in the laboratory are mostly unable to produce extracellular enzymes such as cellulase, pectinase, proteinase, and laccase (Lindeberg, 1948; Norkrans, 1950; Lyr, 1963; Lundeberg, 1970; Lindeberg and Lindeberg, 1977). Lyr (1963) concluded that in these fungi, compared with litter and wood decomposers, the biosynthetic power is more specifically directed toward mycelial than toward ectoenzyme production. It seems probable that this is facilitated by the ample supply of sugars that the symbionts obtain from the trees.

According to Goksøyr (1975, p. 232), the ecological importance of the mycorrhizal fungi probably lies on the primary production side, in relation to mineral nutrition: "These fungi should then, in an ecological sense, be considered as extensions of the plant root system, and their biomass and growth (including fructifications) as an energy expenditure by the plant, necessary to obtain a sufficient supply of inorganic nutrients."

An estimation of the production of fruit bodies of mycorrhizal fungi which may occur in a pine forest is given by Romell (1939). During one season, the fruit bodies of *Boletus* (*Suillus*) *bovinus* amounted to 180 kg DW (containing about 5 kg nitrogen) per hectare. The formation of this quantity of fruit bodies would require about 400 kg of carbohydrates, i.e., the equivalent of 1 m^3 of timber. Lower estimates for the production of fruit bodies by mycorrhizal fungi were made by Hering (1966) and Richardson (1970). Harley (1971) suggests that the values given by Romell may be accepted as representing the upper limit of the carbon drain due to mycorrhizal fungi.

Because mycorrhizal fungi do not depend on litter for their carbon and energy, they compete successfully with litter-decomposing fungi for available inorganic nutrients. The effectiveness of the nitrogen uptake of the mycorrhizal fungi probably contributes to the prevention of leakage of nitrogen from the forest ecosystem.

The question as to what extent actively growing mycelia of litter-decomposing and mycorrhizal basidiomycetes occur within the same layers of litter is as yet unresolved. It is generally assumed that the mycelia of mycorrhizal fungi extend from the mycorrhizal roots into the surrounding soil so that a larger volume of soil is penetrated and the absorption of water and nutrient salts is increased. According to Bowen (1968), those mycorrhizal fungi which form mycelial strands are especially able to exploit the soil. On the other hand, there is some evidence that exploitation of the litter layers by mycorrhizal fungi may be inhibited. Cold water extracts of leaf litter contain substances which inhibit the growth of various ectomycorrhizal fungi (Melin, 1946), whereas the growth of litter-decomposing fungi is strongly stimulated by the same extracts (Lindeberg, 1944). The main growth-inhibiting components of leaf litter of aspen (*Populus tremula*) were identified by Olsen et al. (1971) as benzoic acid and catechol. These compounds, when added alone to synthetic nutrient solutions, had a strong inhibiting effect on mycorrhizal fungi, e.g., *Boletus* (*Suillus*) *luteus* and *B. variegatus*, but only a weak inhibiting effect on litter-decomposing fungi, e.g., *Marasmius perforans* and *M. scorodonius* (Table 4). In the presence of aspen leaf litter extract, growth of the litter decomposers was strongly stimulated even after an extra addition of benzoic acid and catechol.

The sensitivity of mycorrhizal fungi to aromatic compounds in litter probably restricts their growth in fresh litter. However, the litter-decomposing basidiomycetes generally produce extracellular phenol oxidases (Lindeberg, 1948) and probably

Table 4 Effect on growth of fungi of catechol (Cat.), benzoic acid (BA), and extract of aspen leaf litter[a]

Addition	Mycorrhizal fungi		Litter decomposers	
	Boletus luteus	*Boletus variegatus*	*Marasmius perforans*	*Marasmius scorodonius*
None	100	100	100	100
Cat.	3	16	44	66
Cat. + BA	0.6	2	51	69
Extract	0.9	57	617	520
Extract + Cat.	1	7	200	456
Extract + Cat. + BA	0.9	2	295	457

[a]Concentration: Cat. and BA 0.5 mM, extract 20% v/v. Dry weight in percentage of control.

Source: Olsen et al. (1971).

cause an oxidative breakdown of aromatic compounds in the litter. This may suggest that the aromatic inhibitors are gradually eliminated so that the mycorrhizal fungi can penetrate somewhat further into the litter layers, where they could compete for mineralized nitrogen.

A suppressing effect of mycorrhizal roots of *Pinus radiata* on litter decomposition was reported by Gadgil and Gadgil (1971, 1975). These authors claim that the ectomycorrhizal fungi permeate the litter, inhibit the activity of the litter decomposers, and indirectly contribute to the formation of mor humus. Mycorrhizal control of litter decomposition would act as a conservation mechanism whereby the nutrient pool in the upper soil layers is protected against losses by leaching or volatilization. In view of the differences between mycorrhizal and litter-decomposing fungi with respect to the availability of their respective energy sources (Harley, 1971), it seems possible that the former group might be favored in the competition for nitrogen and other nutrients, which may explain the effects observed by Gadgil and Gadgil (1971). To what extent the conclusions of Gadgil and Gadgil are compatible with the aforementioned inhibiting effects of litter extracts is not quite clear.

Certainly, more research is needed on the relations between mycorrhizal and litter-decomposing basidiomycetes in the litter and humus layers.

In forest management it is important to know to what extent mobilization of the large pool of nitrogen bound in humus may be enhanced and which soil organisms are able to mineralize this type of nitrogen. It has been suggested that mycorrhizal fungi may contribute to the mineralization and that this activity might be one of the advantages of mycorrhizal formation as summarized by Lundeberg (1970).

This problem has been studied experimentally by Lundeberg (1970). Mor humus from a pine stand was labeled with ^{15}N by storage for 6 months at optimum temperature and humidity after the addition of [^{15}N]ammonium sulfate and glucose. To eliminate nonincorporated nitrogen, the humus was extracted with dilute sulfuric acid. The capability of different ectomycorrhizal and litter-decomposing basidiomycetes to utilize the organically bound nitrogen in the humus was tested in culture experiments. The litter-decomposing species but none of the mycorrhizal species were able to utilize this nitrogen.

Lundeberg (1970) also studied the capability of mycorrhizal and litter-decomposing fungi to stimulate the translocation of nitrogen bound in humus to pine seedlings grown in ^{15}N-containing humus. None of the mycorrhizal fungi tested had any significant effect. Only in the presence of one litter-decomposing nonmycorrhizal isolate of *Boletus subtomentosus* did the seedlings show considerably stimulated growth. With other isolates of litter decomposers, no positive results were obtained.

VIII. CONCLUSIONS

The quantitative roles of different groups of fungi in decomposition of various types of litter is not known. The enzymatic and other metabolic potentials of different microfungi have been studied, and enzymes catalyzing the breakdown of cellulose, hemicelluloses, pectin, chitin, and protein have been found. Lignified cellulose in forest litter is probably attacked to a large extent by basidiomycetes capable of lignin and cellulose decomposition. Nitrogen is a limiting factor in litter decomposition, and a large part of the available nitrogen will be utilized and immobilized in mycelia and fruit bodies of the fungi. These mycelia and fruit bodies will serve as sources of food for the soil fauna, resulting in a release of nutrients in feces. Lytic bacteria and actinomycetes will contribute to the decomposition of mycelia, and autolysis is most probably part of a mechanism for recycling of nitrogen within mycelia of higher fungi. The transfer of nitrogen from the decomposers to the roots of trees and ground cover vegetation is very effective and is probably due predominantly to uptake by mycorrhizal fungi. The effective conservation of nitrogen within the forest ecosystem is probably also due, in part, to the activity of bacteria in the humus layer. The humus contains a large pool of bound nitrogen. Preliminary experiments indicate that certain litter-decomposing basidiomycetes but not the mycorrhizal fungi may contribute to the mobilization of this nitrogen. The relationships between the litter-decomposing and the mycorrhizal fungi should be further studied.

REFERENCES

Bååth, E., and Söderström, B. (1974). Decomposition of cellulose, xylan, chitin, and protein by microfungi isolated from coniferous forest soils. *Swedish Coniferous Forest Project, Internal Rep. No. 19* (Uppsala).

Björkman, E. (1942). Über die Bedingungen der Mykorrhizabildung bei Kiefer und Fichte. *Symb. Bot. Upsal. 6*(2): 1-191.

Bonnevie-Svendsen, C., and Gjems, O. (1957). Amount and chemical composition of the litter from larch, beech, Norway spruce and Scots pine stands and its effect on the soil. *Meddl. Norske Skogforsöksvesen No. 48.*

Bowen, G. D. (1968). Phosphate uptake by mycorrhizas and uninfected roots of *Pinus radiata* in relation to root distribution. *Trans. 9th Int. Congr. Soil. Sci. 1968* Vol. 2, p. 219.

Bringmark, L. (1978). A bioelement budget of an old Scots pine forest in central Sweden. *Silva Fennica 11*(3): 201-209.

Burges, A. (1939). Soil fungi and root infection. *Broteria 8*: 64-81.

Burges, A. (1960). Dynamic equilibria in the soil. In *The Ecology of Soil Fungi*, D. Parkinson and J. S. Waid, (Eds.). Liverpool University Press, Liverpool, England, pp. 185-191.

Cowling, E. (1970). Nitrogen in forest trees and its role in wood deterioration. *Acta Univ. Upsal. 164.*

Domsch, K. H., and Gams, W. (1969). Variability and potential of a soil fungus population to decompose pectin, xylan, and carboxymethylcellulose. *Soil Biol. Biochem. 1*: 29-36.

Falck, R. (1923). Erweiterte Denkschrift über die Bedeutung der Fadenpilze für die Nutzbarmachung der Abfallstoffe zur Baumernährung im Walde. *Mykol. Untersuch. Ber. 2*: 38-72.

Falck, R. (1930). Nachweise der Humusbildung und Humuszehrung durch bestimmte Arten höheren Fadenpilze im Waldboden. *Forstarch. 6*: 366-377.

Flanagan, P. W., and Scarborough, A. (1974). Physiological groups of decomposer fungi on tundra plant remains. In *Soil Organisms and Decomposition in Tundra*. IBP Tundra Biome Steering Committee, Stockholm, pp. 159-181.

Frank, B. (1885). Über die auf Wurzelsymbiose beruhende Ernährung gewisser Bäume durch unterirdische Pilze. *Ber. Deut. Bot. Ges. 3*: 128-145.

Gadgil, R. L., and Gadgil, P. D. (1971). Mycorrhiza and litter decomposition. *Nature 233*: 133.

Gadgil, R. L., and Gadgil, P. D. (1975). Suppression of litter decomposition by mycorrhizal roots of *Pinus radiata*. *N.Z. J. For. Sci. 5*: 31-41.

Goksøyr, J. (1975). Decomposition, microbiology, and ecosystem analysis. In *Fennoscandian Tundra Ecosystems*. *Ecol. Stud. No. 16*, F. E. Wielgolaski (Ed.). Springer-Verlag, New York, pp. 230-238.

Harley, J. L. (1969). *The Biology of Mycorrhiza*, 2nd ed. Plant Science Monographs, N. Polunin (Ed.). Hill, London.

Harley, J. L. (1971). Fungi in ecosystems. *J. Ecol. 59*: 653-668.

Harper, J. E., and Webster, J. (1964). An experimental analysis of the coprophilous fungus succession. *Trans. Brit. Mycol. Soc. 47*: 511-530.

Hayes, A. J. (1965). Some microfungi from Scots pine litter. *Trans. Brit. Mycol. Soc. 48*: 179-185.

Hering, T. F. (1966). The terricolous higher fungi of four Lake District woodlands. *Trans. Brit. Mycol. Soc. 49*: 369-383.

Hering, T. F. (1967). Fungal decomposition of oak leaf litter. *Trans. Brit. Mycol. Soc. 50*: 267-273.

Hudson, H. J. (1968). The ecology of fungi on plant remains above the soil. *New Phytol. 67*: 837-874.

Jones, P. C. T., and Mollison, P. E. (1948). A technique for the quantitative estimation of soil microorganisms. *J. Gen. Microbiol. 2*: 54-69.

Kendrick, W. B. (1958). Ph.D. thesis, University of Liverpool, England.

Kendrick, W. B., and Burges, A. (1962). Biological aspects of the decay of *Pinus sylvestris* leaf litter. *Nova Hedwigia 4*: 313-344.

Lewis, D. H., and Harley, J. L. (1965). Carbohydrate physiology of mycorrhizal roots of beech. III. Movement of sugars between host and fungus. *New Phytol. 64*: 256-269.

Lindeberg, G. (1944). Über die Physiologie ligninabbauender Bodenhymenomyzeten. *Symbolae Bot. Upsal. 8*(2): 1-183.

Lindeberg, G. (1946). On the decomposition of lignin and cellulose in litter caused by soil-inhabiting hymenomycetes. *Ark. Bot. 33A*: No. 10.

Lindeberg, G. (1948). On the occurrence of polyphenol oxidases in soil-inhabiting basidiomycetes. *Physiol. Plant. 1*: 196-205.

Lindeberg, G., and Lindeberg, M. (1964). The effect of pyridoxine and amino acids on the growth of *Marasmius perforans* Fr. *Arch. Microbiol. 49*: 86-95.

Lindeberg, G., and Lindeberg, M. (1977). Pectinolytic ability of some mycorrhizal and saprophytic hymenomycetes. *Arch. Microbiol. 115*: 9-12.

Lundeberg, G. (1970). Utilization of various nitrogen sources, in particular bound soil nitrogen, by mycorrhizal fungi. *Stud. For. Suec. 79*: 1-95.

Lyr, H. (1963). Zur Frage des Streuabbaues durch ektotrophe Mykorrhizapilze. In *Mykorrhiza: Internationales Mykorrhizasymposium, Weimar, 1960*. Fischer, Jena, pp. 123-145.

Melin, E. (1917). Studier över de norrländska myrmarkernas vegetation. *Norrl. Handbibl.* (Uppsala) *No. 7*.

Melin, E. (1925). *Untersuchungen über die Bedeutung der Baummykorrhiza.* Fischer, Jena.

Melin, E. (1946). Der Einfluss von Waldstreuextrakten auf das Wachstum von Bodenpilzen, mit besonderer Berücksichtigung der Wurzelpilze von Bäumen. *Symbolae Bot. Upsal. 8*(3): 1-116.

Melin, E., and Nilsson, H. (1952). Transfer of labelled nitrogen from an ammonium source to pine seedlings through mycorrhizal mycelium. *Svensk Bot. Tidskr. 46:* 281-285.

Melin, E., and Nilsson, H. (1953). Transfer of labelled nitrogen from glutamic acid to pine seedlings through the mycelium of *Boletus variegatus* (Sw.) Fr. *Nature 171:* 134.

Melin, E., and Nilsson, H. (1957). Transport of C^{14} labelled photosynthate to the fungal associate of pine mycorrhiza. *Svensk Bot. Tidskr. 51:* 166-186.

Meyer, F. H. (1974). Physiology of mycorrhiza. *Ann. Rev. Plant Physiol. 25:* 567-586.

Millar, C. S. (1974). Decomposition of coniferous leaf litter. In *Biology of Plant Litter Decomposition*, C. H. Dickinson and G. J. F. Pugh (Eds.). Academic Press, New York, pp. 105-128.

Miller, O. K., Jr., and Laursen, G. A. (1974). Belowground fungal biomass on US tundra biome sites at Barrow, Alaska. In *Soil Organisms and Decomposition in Tundra.* IBP Tundra Biome Steering Committee, Stockholm, pp. 151-158.

Norkrans, B. (1950). Studies in growth and cellulolytic enzymes of *Tricholoma.* *Symbolae Bot. Upsal. 11*(1): 1-126.

Olsen, R., Odham, G., and Lindeberg, G. (1971). Aromatic substances in leaves of *Populus tremula* as inhibitors of mycorrhizal fungi. *Physiol. Plant. 25:* 122-129.

Palmer, J. G., and Hacskaylo, E. (1970). Ectomycorrhizal fungi in pure culture. I. Growth on single carbon sources. *Physiol. Plant. 23:* 1187-1197.

Park, D. (1976). Carbon and nitrogen levels as factors influencing fungal decomposers. In *The Role of Terrestrial and Aquatic Organisms in Decomposition Processes*, 17th Symp. Brit. Ecol. Soc. Blackwell Sci. Publns., Oxford, England, pp. 41-59.

Richardson, M. J. (1970). Studies on *Russula emetica* and other agarics in a Scots pine plantation. *Trans. Brit. Mycol. Soc. 55:* 217-229.

Romell, L. G. (1939). Barrskogens marksvampar och deras roll i skogens liv. *Svenska SkogsvFör. Tidskr.* pp. 348-375.

Söderström, B. E. (1975). Vertical distribution of microfungi in a spruce forest soil in the south of Sweden. *Trans. Brit. Mycol. Soc. 65:* 419-425.

Söderström, B. E. (1977). Vital staining of fungi in pure cultures and in soil with fluorescein diacetate. *Soil Biol. Biochem. 9:* 59-63.

Sohlenius, B., Berg, B., Clarholm, M., Lundkvist, H., Popović, B., Rosswall, T., Staaf, H., Söderström, B., and Wirén, A. (1976). Mineralization and soil organism activity in a coniferous humus: A model building experiment. *Swedish Coniferous Forest Project, Internal Rep. No. 40*, 51 pp.

Swaby, R. J., and Ladd, J. N. (1962). Chemical nature, microbial resistance, and origin of soil humus. *Trans. Joint Meeting Comm. IV and V, Int. Soc. Soil Sci.* pp. 197-202.

Swaby, R. J., and Ladd, J. N. (1966). Stability and origin of soil humus. In *The Use of Isotopes in Soil Organic Matter Studies.* Spec. suppl. to *J. Appl. Rad. Isotopes*, pp. 153-159.

Witkamp, M., and Ausmus, B. S. (1976). Processes in decomposition and nutrient transfer in forest systems. In *The Role of Terrestrial and Aquatic Organisms in Decomposition Processes*, 17th Symp. Brit. Ecol. Soc. Blackwell Sci. Publns., Oxford, England, pp. 375-396.

Chapter 35

*ROLE OF FUNGI IN CARBON FLOW AND NITROGEN IMMOBILIZATION
IN COASTAL MARINE PLANT LITTER SYSTEMS*[*]

Jack W. Fell and Steven Y. Newell[†]

*University of Miami
Miami, Florida*

I. INTRODUCTION

In discussing the role of fungi in marine environments, Hughes (1975, p. 144) notes
that "Ecological studies of these organisms and their roles in marine and marine-
dominated systems have hardly progressed beyond the descriptive phase, with a strong
emphasis on distributional ecology." Hughes draws a conclusion that "In all areas
dealing with ecology our information is so fragmentary that meaningful conclusions
regarding the relationship of fungi to either substrate or environmental parameters
can rarely be made" (p. 169). This lack of information has apparently led some ob-
servers to conclude that fungi are unimportant in marine systems. For example,
Fenchel (1972, p. 14) states, "It is probable that the only significant primary decomposers of plant material in the sea are bacteria," while more pointedly Hanson
and Wiebe (1977, p. 326) claim that fungi "do not appear to be quantitatively important in most marine environments."

The mycological literature indicates that fungi have a variety of roles in marine habitats. They are possibly best known for their association with wood decay
(Jones, 1976) and as pathogens of marine invertebrates and fishes (Alderman, 1976).
One of their most important potential functions is in the decomposition of plant
litter. As discussed in several other chapters in this volume, the capacity of fungi to decompose plant litter is adequately documented in nonmarine systems, particularly soils (Jackson, 1975; Parkinson, 1975); wood (Kirk, 1973; Swift, 1977); forest litter (Stark, 1972; Witkamp, 1974; Visser and Parkinson, 1975; de Boois, 1976)
and freshwater lakes and streams (Bärlocher and Kendrick, 1974; Kaushik, 1975; Suberkropp and Klug, 1976; Bärlocher et al., 1978).

In marine coastal systems macrophytes (algae, seagrasses, marsh grasses, and
mangroves) form an important base for the food web leading to the animals of commer-

---

[*]This is a contribution from the Rosenstiel School of Marine and Atmospheric
Science of the University of Miami.

[†]*Present affiliation*: Marine Institute, University of Georgia, Sapelo Island,
Georgia

cial and sport fisheries (Mann, 1976). Production of these macrophytes ranges from
500-1000 g cm^{-2} yr^{-1} in contrast to open ocean phytoplankton annual production of
50 g cm^{-2} yr^{-1} and coastal zone phytoplankton at 100 g cm^{-2} yr^{-1} (Mann, 1972). Only
about 5% of the macrophyte material is directly consumed by herbivores (Fenchel, 19-
72; Odum et al., 1973); the remaining material must be converted to microbial biomass
prior to utilization by primary consumers (Hargrave, 1976; Yingst, 1976; Heinle et
al., 1977; Tenore, 1977). The purpose of this review is to present some of the direct and indirect evidence that implicates fungi in the decomposition of coastal
plant litter with the hope of stimulating further research in an area that is vital
to understanding of marine food webs.

II. CARBON AND NITROGEN CHANGES IN COASTAL MARINE LITTER

From a marine mycological ecologist's viewpoint, two basic questions to be resolved
are (1) what chemical changes take place in decomposing coastal marine litter and
(2) to what extent are the fungi functional in bringing about these changes? As
discussed by Dowding (Chap. 32), the content of carbon is reduced and that of nitrogen increased in the combination of decomposing cellulosic substrate and infiltrating
fungal mycelia. Thus, in terrestrial systems, where fungi have proven to be prominent among the microbes involved in litter decomposition, the fungi are essential
conservers of nitrogen and other nutrients (Stark, 1972; Jackson, 1975; O'Neill,
1976; also see Lindeberg, Chap 34 of this volume). The general terrestrial pattern
of decreasing carbon-to-nitrogen ratio which results from fungal and other microbial
decomposing activity (Parnas, 1975, 1976) appears also to apply to coastal marine
plant litter (de la Cruz and Gabriel, 1974; Hunter, 1976; Fell and Master, 1980; Newell
et al., in preparation). De la Cruz and Gabriel reported a drop in C/N ratio of black-
rush (*Juncus roemerianus*) litter from 74 to 42 during the first year of decomposition.
Fell and Master (1980) reported decreases in C/N ratio of red mangrove (*Rhizophora
mangle*) leaves of approximately 60% (100 to 40) as leaves passed from the yellow,
senescent stage to the black, nearly disintegrated stage. For black-rush leaves, decreases of approximately 30% were recorded by Newell and his colleagues (75 to 51) in
the first year of residence on the marsh floor, with a further 40% decrease after 2
years (final C/N ratio ≃ 30). Hunter (1976) found that fronds of a decomposing brown
alga (rockweed, *Fucus vesiculosus*) which were predried at 105°C, changed in C/N ratio
from 28 to about 12 during a 60-day period in litterbags in both salt-marsh and rocky-
shore environments.

    Leaves of sea grasses, although characterized by a low C/N ratio, appear to
undergo first relatively small increases and then decreases in C/N during decomposition. For example, Thayer et al. (1977) found the range of mean C/N ratios in
living, dead, and detrital seagrass (*Zostera marina*) leaves to be from 16 (living
leaves) to 19 (dead leaves); there was a small decrease (from 19 to 17) in C/N between the dead attached and detrital (dead detached) stages. We (Newell and Fell,

unpublished data) have recorded changes in the mean C/N ratio of turtlegrass (*Thalassia testudinum*) leaves from 18 for senescent leaves, to as high as 23 for leaves
at 1 month after detachment, and then to as low as 17 for leaves some 2-5 months
after detachment.

III. EVIDENCE FOR FUNGAL INVOLVEMENT

These changes in carbon and nitrogen content, in addition to indicating increased
nutritive value of litter for detritivorous animals (Russell-Hunter, 1970; Tenore,
1977), may be indirect indicators of fungal activity in decomposing coastal marine
litter. Their detection, of course, does not specifically signify fungal activity
as opposed to other microbial activity; bacterial and algal biomass is also high in
nitrogen. In addition to this evidence for general microbial activity, there is some
evidence that the fungi in particular may be responsible for a substantial portion
of the changes in carbon and nitrogen content of decomposing coastal marine litter.
This evidence is mostly inferential and comes from three types of findings: (1) demonstration that successions in fungal communities take place as decomposition proceeds; (2) direct detection of fungal presence as hyphal networks by scanning electron microscopy (SEM); and (3) detection of changes in biochemical content which may
imply quantitative changes in fungal biomass.

A. Regularly Occurring Mycoseres

The literature is replete with articles demonstrating that fungi virtually always
occur on autochthonous and allochthonous plant litter in marine systems. This literature has been thoroughly reviewed by Johnson and Sparrow (1961) and updated by
Hughes (1975) and Jones (1976). Kohlmeyer (e.g., 1977; see also Kohlmeyer and Kohlmeyer (1979), relying on direct observation rather than cultural techniques, has
provided an abundance of evidence for in situ fungal reproduction on coastal marine
plant litter. Fell and Master (1973, 1975; also Fell et al., 1975), Gessner (1977;
also Gessner and Kohlmeyer, 1976), Newell (1976), and Fell and Hunter (1979) have
described successions which take place in the mycoflorae of decomposing mangrove
leaves and seedlings, leaves and stems of cordgrass, and leaves of black rush. In
all of these cases, measures were taken to ensure that the fungal species recorded
were actively involved in the decay process. Gessner recorded only those fungi
which had produced reproductive structures in the field, and the other aforementioned
investigators applied surface sterilants to samples prior to their enculturation, so
that only or primarily internally produced structures could grow. On the whole, the
successions observed were similar to the general fungal successional scheme proposed
by Hudson (1968) for terrestrial litter. Green and senescent plant structures were
inhabited by a combination of substrate-specific fungi, some of which appeared to
be weak parasites (Hudson's *restricted primary saprophytes*), and members of a ubiquitous living phylloplane mycoflora (Hudson's *common primary saprophytes*). Examples

of fungi of the weak-parasite category are *Pestalotia* sp. in red mangrove leaves and seedlings, *Leptosphaeria juncina* on black-rush leaves, and *Buergenerula spartinae* on cordgrass. Among the common primary saprophytes were *Cladosporium cladosporioides* (three of the four substrates) and *Alternaria alternata* (all four substrates). These were then joined or replaced by fungi characteristic of Hudson's *ascomycetous and deuteromycetous secondary saprophytes* (e.g., *Lulworthia* spp. on all except black-rush leaves, *Halosphaeria hamata* on black-rush and cordgrass, and *Zalerion varium* on red mangrove seedlings; note that these are marine-restricted fungi). Finally, in the latest stages of the coastal marine fungal successions, as the litter substrates reached the point of breakdown to the fine particulate stage, fungi characteristic of Hudson's *"soil-inhabitant" secondary saprophyte* stage appeared. The foremost example here was *Trichoderma viride*; although this and other "soil" fungi were not recorded from cordgrass, upon which only direct observation was used, it should be noted that Lee and Baker (1973) did record *T. viride* on red mangrove roots by direct observation. Major elements of Hudson's successional scheme which have not been recorded as prevalent parts of the mycoseres of coastal marine litter are the *basidiomycetous secondary saprophytes* and the *zygomycetous "soil-inhabitant" secondary saprophytes*.

Oomycetes have been reported as early colonizers of leaf litter [*Dictyuchus monosporus* (Apinis et al., 1972; Taligoola et al., 1972)] and cellulose [*Pythium fluminum* (Park and McKee, 1978)] in freshwater environments. Similarly, marine oomycetes are unique members of the mycoseres of red mangrove leaves and seedlings and other coastal marine leaf litter [e.g., *Phytophthora vesicula*, *Pythium grandisporangium* (Anastasiou and Churchland, 1969; Fell and Master, 1975)]. On red mangrove leaves, these oomycetous fungi are the first to appear after leaves enter the water, apparently colonizing as early as the first 24-hr period of submergence (Fell and Master, 1975). In contrast, on seedlings, which are not senescent when they fall into the water, the oomycetes appeared to fall into Hudson's "soil-inhabitant" secondary saprophyte group, colonizing at the same late stage as *Trichoderma viride*. Since senescent leaves release substantial quantities of easily decomposable carbohydrates soon after submergence (e.g., Suberkropp and Klug, 1976), it may be that the marine oomycetes occupy niches analogous to those of the terrestrial zygomycetes, both in their capacity as early colonizers of dung and compost (substrates containing high levels of easily decomposable carbon compounds) and as late colonizers ("soil fungi") of decayed plant tissue (Hudson, 1968).

Recently, the authors (Newell and Fell, 1980) investigated the mycoflora of living and decaying turtlegrass leaves. As opposed to the marsh grasses and mangroves, this source of coastal marine litter is rarely if ever emersed during production. Methods of surface sterilization used were validated in pilot experimentation with control systems (sterilized, reinoculated, and surface-contaminated leaf disks). Incubation of field samples in seawater with no additional nutrients was also used as a means of discouraging the growth of inactive fungi (Bärlocher and Kendrick, 1974; Suberkropp and Klug, 1976). It was found that a megachytridiaceous species (probably *Nowakowskiella* sp.) was virtually the only colonizer of green turtlegrass leaves (averaging 85% frequency), and its frequency of occurrence declined in senescent and decaying leaves. It also appeared that the colonization pattern on leaves which are detached and float into the littoral zone (where they undergo daily emersion) was distinct from the pattern on leaves which decay while continuously submerged. Leaves in unshaded sandy littoral zones were colonized by *Lindra thalassiae*, *Dendryphiella salina*, and *Acremonium* sp., which replaced the megachytrid. Leaves which remained submerged showed decreasing frequencies of the megachytrid; but, after surface sterilization, frequencies of *L. thalassiae*, *D. salina*, and *Acremonium* sp. were much lower than they had been on littoral-zone leaves (0-40%, as opposed to 70-100%, depending on time and site). *Cladosporium*, *Aspergillus*, and *Penicillium* species occurred very rarely or not at all, on either submerged or littoral leaves--a result which differed from the findings of Meyers et al. (1965), who used leaf-washing rather than surface-sterilization techniques.

B. Electron Microscopy

The occurrence of regular successions with similarities to fungal successions on terrestrial litter suggests that (with the possible exception of submerged seagrass litter) the fungi are as active in decomposition of coastal marine litter as they have proven to be in terrestrial litter. A more direct form of proof of fungal activity associated with decomposing coastal marine litter derives from the SEM analyses of cordgrass litter provided by Gessner et al. (1972), Sieburth et al. (1974), and May (1974). Their electron micrographs revealed fungal hyphae being produced upon and within the tissues of decaying cordgrass. In contrast, bacteria were relatively rare, and May concluded that fungi are the primary agents of cordgrass litter decomposition.

C. Biochemical Indicators

Inferential biochemical evidence for fungal involvement in decomposition has been developed for four types of coastal marine litter. Schultz and Quinn (1973) found that the concentrations of total fatty acids and branched-chain, 15-carbon acids were greater in detrital cordgrass than in stems and leaves of the living plants. These increases appeared to be the result of microbial activities. Therefore, they examined three fungal species (*Buergenerula spartinae*, *Dendryphiella salina*, and *Mucor* sp.) and found that *B. spartinae* contained 15-carbon acids, but at an insufficient level to account for the entire increase in these acids. They concluded that a combination of fungi, bacteria, and other microbes contributed to the fatty acid composition of decaying cordgrass.

Casagrande and Given (1974) measured amino acid content of decomposing red mangrove leaves and found a reduction in the C/N ratio associated with increased amounts of total amino acids--particularly an increased proportion of nonprotein acids, including those found in bacterial cell walls. This suggested that the nitrogenous compounds of the mangrove leaves had been microbially reworked. While Casagrande and Given did not report quantities of individual amino acids, they found that the nonprotein amino acids were in greater abundance in yellow leaves than in brown leaves. They postulated a greater importance of bacterial growth in early stages of decay and of fungal attack later.

White et al. (1977) and Morrison et al. (1977) suggested that a similar sequence occurs on decomposing live-oak leaves and slash pine needles when these are placed in estuarine water. They postulated an initial bacterial colonization followed by fungi and other microbes. The results of Morrison et al. were based on data that indicated an initial colonization by populations with a high content of muramic acid relative to the ATP content. These populations were apparently succeeded by microbes with lower muramic acid/ATP ratios. White and colleagues examined the rate of incorporation of $[^{32}P]H_3PO_4$ and $[^{14}C]$sodium acetate into lipids. After 4 weeks they recorded a rapid increase in ^{32}P incorporation, followed by a rapid decrease. This suggested the initial presence of bacteria, which are characterized by relatively high percentages of phospholipids, followed by fungi, with low percentages of phospholipids. These studies were terminated after 6 weeks of leaf submergence in the estuary, when the oak leaves had lost less than 20% of initial dry weight.

These apparent successions reported by White and his coworkers and by Casagrande and Given may be the result of rapid bacterial colonization of leaf surfaces in the initial stage of leaf decay, during which the bacteria utilize the organic carbon leachates. While bacteria are often restricted to substrate surfaces, fungi are particularly well suited for litter breakdown due not only to the production of extracellular enzymes lytic to plant structural compounds but also to the formation of hyphae; the combination permits effective penetration of plant cells (Harley, 1971).

Thayer et al. (1977) determined total amino acid, plus glucosamine concentrations for living, dead attached, and dead detached (detrital) eelgrass leaves. The most pronounced increase in the sequence from living to dead and especially dead to detrital was in glucosamine content, which increased by an order of magnitude. A similar increase (sixfold) was found for cordgrass. Since taurine was not detected in their samples, and the authors had invariably found this amino acid in hydrolysates of estuarine arthropods, it was concluded that the glucosamine represented a component of hydrolyzed bacterial cell wall material (murein). However, glucosamine is also a breakdown product of fungal chitin, so it is possible that the glucosamine increase at least in part indicated an increase in fungal biomass within the leaves. It is notable that alanine content of detrital leaves was lower than that of glucosamine; L-alanine, in addition to being a common protein amino acid, is present in murein in a 1/1 ratio to glucosamine (Stanier et al., 1970).

D. Experimentation with Microecosystems

One means of resolving questions regarding activity of components of complex ecosystems is the construction and manipulation of physical models. A variety of models have been developed to delve into the function of detrital decomposition systems. These have included some thought-provoking simulation models of terrestrial and aquatic systems (Boling et al., 1975; Parnas, 1975; O'Neill, 1976; Hunt, 1977; Reuss and Innis, 1977). In addition, several laboratory model ecosystems have been constructed, including some of marine systems.

An acknowledged problem with laboratory models is their oversimplification of field conditions. Problems of design include inadequate supply or unintentional alteration of critical nutrients, as well as exclusion of natural decomposer organisms. The basic experiment consists of placing litter in containers with seawater and adding various test substances. In many cases the leaves have been dried prior to testing; however, drying and rewetting increases the rate of decomposition (e.g., see Jager and Bruins, 1975). Heat has been applied for drying, which could structurally alter the carbon compounds and destroy the associated microbial leaf community. Maintenance or restoration of the normal flora is often ignored; addition of ambient seawater inoculum does not necessarily reestablish the same microbes that inhabit the decomposing litter in the field. Grinding of the litter has been employed in an attempt to simulate particle production, but this structurally alters the leaves at an unnaturally early stage, breaking cell walls and creating an increased surface area that may be more suitable for bacteria than fungi. Grinding also disrupts fungal hyphae. Antibiotics have been employed to differentiate bacterial and fungal activities; this is a technique which must be used with caution. Yetka and Wiebe (1974) found that bacterial respiration was not satisfactorily inhibited by some antibacterial agents, and I. M. Master and J. W. Fell (unpublished data, 1977) observed that fungi can take up nitrogen from streptomycin and/or penicillin, resulting in recording of an artificially high level of nitrogen immobilization. Nitrogen and phosphorus are often limiting factors in litter decomposition. Due to the low levels at which these nutrients occur in seawater, one must ascertain if the batch or continuous-flow system has environment-simulating quantities of both elements. Batch cultures can be particularly troublesome, as there may be a buildup of leachates such as tannins that are inhibitory to some microbes.

Despite the difficulties associated with laboratory models, a great deal can be learned from them about the normal detrital production process as well as deleterious effects of perturbants. Laboratory models of decomposition of marine plants have been limited in number and have centered on mineralization/immobilization of carbon and nitrogen, utilization of leachates, and synthesis of fatty acids. Two of the limiting factors in litter decomposition, nitrogen and phosphorus, have been tested in freshwater stream models (Kaushik and Hynes, 1971). These investigators added 5, 10, and 100 mg N/liter to suspensions of leaves of forest trees in the presence

and absence of phosphorus (5 mg/liter). They found little or no immobilization in
the absence of both or either factor and greatest increases with the combination of
phosphorus and 100 mg N/liter. Similarly Howarth and Fisher (1976), in a freshwater
model system, found that leaves enriched with nitrogen (10 mg N/liter) and phosphorus
(10 mg P/liter) immobilized nitrogen, in contrast to leaves lacking enrichment, which
released nitrogen. Nitrate/phosphorus enrichment also increased the rates of weight
loss and respiration associated with the leaves. However, Triska and Sedell (1976),
using quantities of added nitrogen (NO_3^-) much lower (100 times) than those of Kaushik
and Hynes or Howarth and Fisher, found that in their model freshwater streams there
was no effect of ambient nitrogen concentration upon weight loss or nitrogen immobilization in leaf litter.

The Kaushik and Hynes (1971) model demonstrated by use of antibacterial and
antifungal agents that fungi were more important than bacteria in decomposition, particularly in nitrogen immobilization. Fallon and Pfaender (1976) also used antibiotics, in a salt-marsh model system, and investigated the uptake of ^{14}C-labeled
Spartina leachates. They found that uptake of the label was more rapid in systems
with bacteria alone and with a mixed microbial community than with fungi alone. They
reported that the label was extensively mineralized in the mixed and bacterial systems, in contrast to a more efficient immobilization to microbial biomass in the fungal system.

Studies in our laboratory (Fell and Master, 1980; Fell et al., 1980) centered on
the potential role of fungi in nitrogen immobilization in the red mangrove (*Rhizophora
mangle*) leaf litter system. The laboratory model consisted of partially sterilized
leaves in a semibatch culture, i.e., a flask system in which the water was changed
routinely (every 24 hr) to maintain nutrient levels and to reduce buildup of toxic
compounds. The effect of varied concentration of ambient nitrogen was tested in the
presence of marine *Phytophthora* spp. Nitrogen levels ranged from 0.025 to 17.8 mg
at N/l [$(NH_4)_2SO_4$], with phosphorus constant at 161 µg at P/l (KH_2PO_4). Fungi were
excluded in sterile controls. The results demonstrated that these rapid-colonizing
fungi are potential agents of nitrogen immobilization, i.e., nitrogen immobilization
similar to that previously observed in the field occurred in the presence of fungi
and inorganic combined nitrogen, whereas leaves in sterile controls did not immobilize
nitrogen, nor did control leaves in the presence of fungi but lacking additions of
nitrogen. Additionally, both weight loss and change in carbon content were more
pronounced in the presence of fungi.

IV. FUTURE DIRECTIONS

Further experimentation with physical models of coastal litter systems of known microbial construct should continue to provide a clearer picture of the capabilities
of fungi in the decomposition of this marine litter material. As a complement to
this type of work, the methods which have been developed for estimation of the mag-

nitude of fungal roles in decomposition of terrestrial litter must be applied in
field study of coastal marine litter decomposition to obtain firm answers to the
question of extent of fungal participation. Particularly valuable among these in-
vestigative approaches are those designed to measure the changes in fungal biomass
within decomposing litter and sublittoral soil; if fungi are active converters of
litter material into microbial material, then changes in fungal biomass should be
detectable as decomposition proceeds.[*] Among these techniques are the direct-count
methods of Jones and Mollison (1948) as modified by Thomas et al. (1965; also Parkinson et al., 1971: an agar-film technique) and Hanssen et al. (1974; also Sundman
and Sivelä, 1978: a membrane filter technique), and the indirect chemical-estimation
methods of Ride and Drysdale (1972; also Swift, 1973a,b: determination of glucosamine content). Although direct-count methods would ideally be more sound, they
suffer from problems of subjectivity, incomplete extraction of hyphae intimately
associated with internal plant tissue, and recognizability of fungal structure disrupted during homogenization (Frankland et al., 1978; S. Y. Newell and J. W. Fell,
unpublished data). The estimation of biomass from glucosamine is subject to inaccuracies due to variability of the glucosamine/biomass ratio (e.g., see Sharma et al.,
1977), negligible content of glucosamine in most oomycete cell walls (Rosenberger,
1976), and contribution of bacterial and animal glucosamine to the measured total
(e.g., see Thayer et al., 1977). The indirect chemical estimation of fungal biomass
is probably best suited for use in laboratory microecosystems with known microbial
members, a situation analogous to plant-fungal symbioses and unifungal decomposition
systems with which this sort of estimation has been successfully used (e.g., by Hepper, 1977; Swift, 1978). The paper by Frankland et al. (1978) should be consulted
for an experimental comparison of the efficiencies and major sources of imprecision
and inaccuracy of the glucosamine and Jones-Mollison methods. With both methods,
the determination of the percentage of measured biomass which is alive presents problems (Swift, 1973a). Frankland (1974, 1975), Anderson and Slinger (1975a,b), Paul
and Johnson (1977), and Söderström (1977) have developed phase-contrast and fluores-
cent-staining techniques designed to maximize recognition of disrupted fungal tissue,
provide a degree of differentiation among fungal and bacterial structures, and permit
estimation of the living fraction of measured biomass.

   The Jones-Mollison method for estimation of fungal biomass has been tested by
us with one type of coastal marine litter, viz., turtlegrass litter (Newell and Fell,
unpublished data). Preliminary results indicate that efficiency of hyphal extraction
(3-min. homogenization, setting 6, Virtis Model 23 Microhomogenizer) may average as
low as 20%, based on the portion of the substrate reduced below 250 μm in largest

---

[*]In setting forth this hypothesis, one must assume that mycophagous detritivores
do not selectively remove mycelial biomass as rapidly as it is produced (see Wood,
1974; Padgett, 1976; also Bärlocher and Kendrick, Chap. 40).

dimension (the size of particle above which it was judged that hyphae were mostly
retained). Longer homogenizing might improve the percentage of extraction, but it
is also likely to decrease recognizability of hyphal fragments, resulting in sharp
reductions in biomass estimates (Swift, 1973b). Also, comparison with cleared and
stained preparations prepared from some of the same turtlegrass leaf samples revealed
that ascocarps and zoosporangia should have been but were not seen (or recognized)
in the agar films. Clearly, corrections for hyphal extraction inefficiency and exclusion of mass of reproductive structures should be made in direct-count estimates
of fungal biomass. Additionally, perhaps as a consequence of the microbial communities peculiar to submerged seagrass litter, the average hyphal diameter was approximately 1 μm. (Did this represent fine chytrid rhizomycelium, filamentous bacteria,
or actinoplanacean filaments?) Application of the recent improvements developed
for agar-film and membrane-filter fungal biomass estimations should provide answers
to this type of question, for seagrass and for other types of coastal marine litter.
Then, if reliable measures can be made of the dynamics of fungal biomass, in parallel
with measures of carbon and nitrogen flux, the clarification of the role of fungi in
these fluxes should be possible.

ACKNOWLEDGMENTS

The portions of the work herein discussed which were conducted at the University of
Miami were supported by grants from Florida Power and Light Company, the U.S. Department of Energy, and the National Science Foundation.

REFERENCES

Alderman, D. J. (1976). Fungal diseases of marine animals. In *Recent Advances in
Aquatic Mycology*, E. B. G. Jones (Ed.). Elek, London, pp. 223-260.

Anastasiou, C. J., and Churchland, L. M. (1969). Fungi on decaying leaves in marine
habitats. *Can. J. Bot. 47*: 251-257.

Anderson, J. R., and Slinger, J. M. (1975a). Europium chelate and fluorescent bright-
ener staining of soil propagules and their photomicrographic counting. I. Methods.
Soil Biol. Biochem. 7: 206-209.

Anderson, J. R., and Slinger, J. M. (1975b). Europium chelate and fluorescent bright-
ener staining of soil propagules and their photomicrographic counting. II. Efficiency. *Soil Biol. Biochem. 7*: 211-215.

Apinis, A. E., Chesters, C. G. C., and Taligoola, H. K. (1972). Microfungi colonizing
submerged standing culms of *Phragmites communis* Trin. *Nova Hedwigia 23*: 473-480.

Bärlocher, F., and Kendrick, B. (1974). Dynamics of the fungal population on leaves
in a stream. *J. Ecol. 62*: 761-791.

Bärlocher, F., Kendrick, B., and Michaelides, J. (1978). Colonization and conditioning of *Pinus resinosa* needles by aquatic hyphomycetes. *Arch. Hydrobiol. 81*: 462-474.

Boling, R. H., Goodman, E. D., Van Sickle, J. A., Zimmer, J. O., Cummins, K. W.,
Petersen, R. C., and Reice, S. R. (1975). Toward a model of detritus processing in
a woodland stream. *Ecology 56*: 141-151.

Casagrande, D. J., and Given, P. H. (1974). Geochemistry of amino acids in some Florida peat accumulations. I. Analytical approach and total amino acid concentrations.
Geochim. Cosmochim. Acta 38: 419-434.

de Boois, H. M. (1976). Fungal development on oak leaf litter and decomposition potentialities of some fungal species. *Rev. Ecol. Biol. Sol. 13:* 437-448.

de la Cruz, A. A., and Gabriel, B. C. (1974). Caloric, elemental, and nutritive changes in decomposing *Juncus roemerianus* leaves. *Ecology 55:* 882-886.

Fallon, R. D., and Pfaender, F. K. (1976). Carbon metabolism in model microbial systems for a temperate salt marsh. *Appl. Environ. Microbiol. 31:* 959-968.

Fell, J. W., and Hunter, I. L. (1979). Fungi associated with the decomposition of the black rush, *Juncus roemerianus*, in South Florida. *Mycologia 71:* 322-342.

Fell, J. W., and Master, I. M. (1973). Fungi associated with the decay of mangrove (*Rhizophora mangle* L.) leaves in South Florida. In *Estuarine Microbial Ecology*, L. H. Stevenson and R. R. Colwell (Eds.). Univ. of South Carolina Press, Columbia, pp. 455-466.

Fell, J. W., and Master, I. M. (1975). Phycomycetes (*Phytophthora* spp. nov. and *Pythium* sp. nov.) associated with degrading mangrove (*Rhizophora mangle*) leaves. *Can. J. Bot. 53:* 2908-2922.

Fell, J. W., and Master, I. M. (1980). The association and potential role of fungi in mangrove detrital systems. *Bot. Mar. 23:* 257-263.

Fell, J. W., Cefalu, R. C., Master, I. M., and Tallman, A. S. (1975). Microbial activities in the mangrove (*Rhizophora mangle* L.) leaf detrital system. In *Proceedings of the International Symposium on Biology and Management of Mangroves*, C. E. Walsh, S. Snedaker, and H. Teas (Eds.). Institute of Food and Agricultural Science, University of Florida, Gainesville, pp. 661-679.

Fell, J. W., Master, I. M., and Newell, S. Y. (1980). Laboratory model of the potential role of fungi in the decomposition of red mangrove (*Rhizophora mangle*) leaf litter. In *Marine Benthic Dynamics*, K. R. Tenore and B. C. Coull (Eds.). Univ. of South Carolina Press, Columbia, pp. 359-372.

Fenchel, T. (1972). Aspects of decomposer food chains in marine benthos. *Verhandl. Deut. Zool. Ges. 65:* 14-23.

Frankland, J. C. (1974). Importance of phase-contrast microscopy for estimation of total fungal biomass by the agar-film technique. *Soil Biol. Biochem. 6:* 409-410.

Frankland, J. C. (1975). Estimation of live fungal biomass. *Soil Biol. Biochem. 7:* 339-340.

Frankland, J. C., Lindley, D. K., and Swift, M. J. (1978). A comparison of two methods for the estimation of biomass in leaf litter. *Soil Biol. Biochem. 10:* 323-333.

Gessner, R. V. (1977). Seasonal occurrence and distribution of fungi associated with *Spartina alterniflora* from a Rhode Island estuary. *Mycologia 69:* 477-491.

Gessner, R. V., and Kohlmeyer, J. (1976). Geographical distribution and taxonomy of fungi from salt marsh *Spartina*. *Can. J. Bot. 54:* 2023-2027.

Gessner, R. V., Goos, R. D., and Sieburth, J. McN. (1972). The fungal microcosm of the internodes of *Spartina alterniflora*. *Mar. Biol. 16:* 269-273.

Hanssen, J. F., Thingstad, T. F., and Goksøyr, J. (1974). Evaluation of hyphal lengths and fungal biomass in soil by a membrane filter technique. *Oikos 25:* 102-107.

Hanson, R. B., and Wiebe, W. J. (1977). Heterotrophic activity associated with particulate size fractions in a *Spartina alterniflora* salt-marsh estuary, Sapelo Island, Georgia, USA, and the continental shelf waters. *Mar. Biol. 42:* 321-330.

Hargrave, B. T. (1976). The central role of invertebrate faeces in sediment decomposition. In *The Role of Terrestrial and Aquatic Organisms in Decomposition Processes*, J. M. Anderson and A. Macfayden (Eds.). Blackwell Sci. Publns., Oxford, England, pp. 301-321.

Harley, J. L. (1971). Fungi in ecosystems. *J. Appl. Ecol. 8:* 627-642.

Heinle, D. R., Harris, R. P., Ustach, J. F., and Flemer, D. A. (1977). Detritus as food for estuarine copepods. *Mar. Biol. 40:* 341-353.

Hepper, C. M. (1977). A colorimetric method for estimating vesicular-arbuscular mycorrhizal infection in roots. *Soil Biol. Biochem. 9*: 15-18.

Howarth, R. W., and Fisher, S. G. (1976). Carbon, nitrogen, and phosphorus dynamics during leaf decay in nutrient-enriched stream ecosystems. *Freshwater Biol. 6*: 221-228.

Hudson, H. J. (1968). The ecology of fungi on plant remains above the soil. *New Phytol. 67*: 837-874.

Hughes, G. C. (1975). Studies of Fungi in oceans and estuaries since 1961. I. Lignicolous, caulicolous, and foliicolous species. *Oceanogr. Mar. Biol. Ann. Rev. 13*: 69-180.

Hunt, H. W. (1977). A simulation model for decomposition in grasslands. *Ecology 58*: 469-484.

Hunter, R. D. (1976). Changes in carbon and nitrogen content during decomposition of three macrophytes in freshwater and marine environments. *Hydrobiologia 51*: 119-128.

Jackson, R. M. (1975). Soil fungi. In *Soil Microbiology*, N. Walker (Ed.). Halsted Press, New York, pp. 165-180.

Jager, G., and Bruins, E. H. (1975). Effects of repeated drying at different temperatures on soil organic matter decomposition and characteristics, and on the soil microflora. *Soil Biol. Biochem. 7*: 153-159.

Johnson, T. W., and Sparrow, F. K. (1961). *Fungi in Oceans and Estuaries*. Cramer, Weinheim, W. Germany, 668 pp. [Reprint: 1970, Macmillan (Hafner Press), New York.]

Jones, E. B. G. (1976). Lignicolous and algicolous fungi. In *Recent Advances in Aquatic Mycology*, E. B. G. Jones (Ed.). Elek, London, pp. 1-50.

Jones, E. B. G. (Ed.) (1976). *Recent Advances in Aquatic Mycology*. Elek, London, 749 pp.

Jones, P. C. T., and Mollison, J. E. (1948). A technique for the quantitative estimation of soil microorganisms. *J. Gen. Microbiol. 2*: 54-69.

Kaushik, N. K. (1975). Decomposition of allochthonous organic matter and secondary production in stream ecosystems. In *Productivity of World Ecosystems*. National Academy of Sciences, Washington, D.C., pp. 90-95.

Kaushik, N. K., and Hynes, H. B. N. (1971). The fate of dead leaves that fall into streams. *Arch. Hydrobiol. 68*: 465-515.

Kirk, T. K. (1973). The chemistry and biochemistry of decay. In *Wood Deterioration and Its Prevention by Preservative Treatments*, D. Nicholas (Ed.). Syracuse Univ. Press, Syracuse, N.Y., pp. 149-181.

Kohlmeyer, J. (1977). New genera and species of higher fungi from the deep sea (1615-5315 m). *Rev. Mycol. 41*: 189-206.

Kohlmeyer, J., and Kohlmeyer, E. (1979). *Marine Mycology. The Higher Fungi*. Academic Press, New York, 690 pp.

Lee, B. K. H., and Baker, G. E. (1973). Fungi associated with the roots of red mangrove, *Rhizophora mangle*. *Mycologia 65*: 894-906.

Mann, K. H. (1972). Macrophyte production and detritus food chains in coastal waters. *Mem. Ist Ital. Idrobiol. 29* (Suppl.): 353-383.

Mann, K. H. (1976). Decomposition of marine macrophytes. In *The Role of Terrestrial and Aquatic Organisms in Decomposition Processes*, J. M. Anderson and A. Macfayden (Eds.). Academic Press, New York, pp. 429-440.

May, M. S. (1974). Probable agents for the formation of detritus from the halophyte, *Spartina alterniflora*. In *Ecology of Halophytes*, R. J. Reimold and W. H. Queen (Eds.). Academic Press, New York, pp. 429-440.

Meyers, S. P., Orpurt, P. A., Simms, J., and Boral, L. L. (1965). Thalassiomycetes. VII. Observations on fungal infestation of turtlegrass, *Thalassia testudinum* König. *Bull. Mar. Sci. 15*: 548-564.

Morrison, S. J., King, J. D., Bobbie, R. J., Bechtold, R. E., and White, D. C. (1977). Evidence for microfloral succession on allochthonous plant litter in Apalachicola Bay, Florida, USA. *Mar. Biol. 41*: 229-240.

Newell, S. Y. (1976). Mangrove fungi: The succession in the mycoflora of red mangrove seedlings (*Rhizophora mangle* L.). In *Recent Advances in Aquatic Mycology*, E. B. G. Jones (Ed.). Elek, London, pp. 51-91.

Newell, S. Y., and Fell, J. W. (1980). Mycoflora of turtlegrass (*Thalassia testudinum* König) as recorded after seawater incubation. *Bot. Mar. 23*: 265-275.

Newell, S. Y., Fell, J. W., and Tallman, A. S. (in preparation). Dry matter, carbon, and nitrogen content of decomposing rush (*Juncus roemerianus*) and mangrove (*Rhizophora mangle*) leaves in South Florida.

Odum, W. E., Zieman, J. C., and Heald, E. J. (1973). Importance of vascular plant detritus to estuaries. In *Proceedings of the Coastal Marsh and Estuary Management Symposium*, R. H. Chabreck (Ed.). Louisiana State University, Division of Continuing Education, Baton Rouge, La., pp. 91-114.

O'Neill, R. V. (1976). Ecosystem persistence and heterotrophic regulation. *Ecology 57*: 1244-1253.

Padgett, D. E. (1976). Leaf decomposition in a tropical rainforest stream. *Biotropica 8*: 166-178.

Park, D., and McKee, W. (1978). Cellulolytic *Pythium* as a component of the river mycoflora. *Trans. Brit. Mycol. Soc. 71*: 251-259.

Parkinson, D. (1975). Terrestrial decomposition. In *Productivity of World Ecosystems*. National Academy of Sciences, Washington, D.C., pp. 55-60.

Parkinson, D., Gray, T. R. G., and Williams, S. T. (1971). *Methods for Studying the Ecology of Soil Microorganisms*, IBP Handbook No. 19. Blackwell Sci. Publns., Oxford, England, 116 pp.

Parnas, H. (1975). Model for decomposition of organic material by microorganisms. *Soil Biol. Biochem. 7*: 161-169.

Parnas, H. (1976). A theoretical explanation of the priming effect based on microbial growth with two limiting substrates. *Soil Biol. Biochem. 8*: 139-144.

Paul, E. A., and Johnson, R. L. (1977). Microscopic counting and adenosine 5'-triphosphate measurement in determining microbial growth in soils. *Appl. Environ. Microbiol. 34*: 263-269.

Reuss, J. O., and Innis, G. S. (1977). A grassland nitrogen flow simulation model. *Ecology 58*: 379-388.

Ride, J. P., and Drysdale, R. B. (1972). A rapid method for the chemical estimation of filamentous fungi in plant tissue. *Physiol. Plant Pathol. 2*: 7-15.

Rosenberger, R. F. (1976). The cell wall. In *The Filamentous Fungi*, Vol. 2: *Biosynthesis and Metabolism*, J. E. Smith and D. R. Berry (Eds.). Wiley, New York, pp. 328-344.

Russell-Hunter, D. (1970). *Aquatic Productivity*. Macmillan, New York, 306 pp.

Schultz, D. M., and Quinn, J. G. (1973). Fatty acid composition of organic detritus from *Spartina alterniflora*. *Est. Coastal Mar. Sci. 1*: 177-190.

Sharma, P. D., Fisher, P. J., and Webster, J. (1977). Critique of the chitin assay technique for estimation of fungal biomass. *Trans. Brit. Mycol. Soc. 69*: 479-483.

Sieburth, J. McN., Brooks, R. D., Gessner, R. V., Thomas, C. D., and Tootle, J. L. (1974). Microbial colonization of marine plant surfaces as observed by scanning electron microscopy. In *Effect of the Ocean Environment on Microbial Activities*, R. R. Colwell and R. Y. Morita (Eds.). University Park Press, Baltimore, pp. 418-432.

Söderström, B. E. (1977). Vital staining of fungi in pure cultures and in soil with fluorescein diacetate. *Soil Biol. Biochem. 9*: 59-63.

Stanier, R. Y., Doudoroff, M., and Adelberg, E. A. (1970). *The Microbial World*, 3rd ed. Prentice-Hall, Englewood Cliffs, N.J., 873 pp.

Stark, N. (1972). Nutrient cycling pathways and litter fungi. *Bioscience 22*: 355-360.

Suberkropp, K., and Klug, M. J. (1976). Fungi and bacteria associated with leaves during processing in a woodland stream. *Ecology 57*: 707-719.

Sundman, V., and Sivelä, S. (1978). A comment on the membrane filter technique for estimation of length of fungal hyphae in soil. *Soil Biol. Biochem. 10*: 399-401.

Swift, M. J. (1973a). The estimation of mycelial biomass by determination of the hexosamine content of wood tissue decayed by fungi. *Soil Biol. Biochem. 5*: 321-332.

Swift, M. J. (1973b). Estimation of mycelial growth during decomposition of plant litter. *Bull. Ecol. Res. Comm.* (Stockholm) *17*: 323-328.

Swift, M. J. (1977). The ecology of wood decomposition. *Sci. Progr.* (Oxford) *64*: 175-199.

Swift, M. J. (1978). Growth of *Stereum hirsutum* during the long-term decomposition of oak branch-wood. *Soil Biol. Biochem. 10*: 335-337.

Taligoola, H. K., Apinis, A. E., and Chesters, C. G. C. (1972). Microfungi colonizing collapsed aerial parts of *Phragmites communis* Trin. in water. *Nova Hedwigia 23*: 465-472.

Tenore, K. (1977). Food chain pathways in detrital-feeding benthic communities: A review, with new observations on sediment resuspension and detrital recycling. In *Ecology of Marine Benthos*, B. C. Coull (Ed.). Univ. of South Carolina Press, Columbia, pp. 37-53.

Thayer, G. W., Engel, D. W., and Lacroix, M. W. (1977). Seasonal distribution and changes in the nutritive quality of living, dead and detrital fractions of *Zostera marina* L. *J. Exp. Mar. Biol. Ecol. 30*: 109-127.

Thomas, A., Nicholas, D. P., and Parkinson, D. (1965). Modifications of the agar film technique for assaying lengths of mycelium in soil. *Nature 205*: 105.

Triska, F. J., and Sedell, J. R. (1976). Decomposition of four species of leaf litter in response to nitrate manipulation. *Ecology 57*: 783-792.

Visser, S., and Parkinson, D. (1975). Fungal succession on aspen poplar leaf litter. *Can. J. Bot. 53*: 1640-1651.

White, D. C., Bobbie, R. J., Morrison, S. J., Oosterhof, D. K., Taylor, C. W., and Meeter, D. A. (1977). Determination of microbial activity of estuarine detritus by relative rates of lipid biosynthesis. *Limnol. Oceanogr. 22*: 1089-1099.

Witkamp, M. (1974). Direct and indirect counts of fungi and bacteria as indexes of microbial mass and productivity. *Soil Sci. 118*: 150-155.

Wood, T. G. (1974). Field investigations on the decomposition of leaves of *Eucalyptus delegatensis* in relation to environmental factors. *Pedobiologia 14*: 343-371.

Yetka, J. E., and Wiebe, W. J. (1974). Ecological application of antibiotics as respiratory inhibitors of bacterial populations. *Appl. Microbiol. 28*: 1033-1039.

Yingst, J. Y. (1976). The utilization of organic matter in shallow marine sediments by an epibenthic deposit-feeding holothurian. *J. Exp. Mar. Biol. Ecol. 23*: 55-69.

Part   VIII

THE AQUATIC HYPHOMYCETE COMMUNITY--AN EXAMPLE

Chapter 36

BIOLOGY AND ECOLOGY OF AQUATIC HYPHOMYCETES

John Webster

University of Exeter
Exeter, Devon, England

I. AQUATIC HYPHOMYCETES

Although a few aquatic hyphomycetes had been described earlier, it was the pioneer
studies of Ingold (1942) which drew the attention of mycologists to this fascinating
group of organisms. In the 35 years which have followed, a large number of descriptive papers have been published, so that more than 150 species are now recognized.
Much of this taxonomic literature has been summarized by Ingold in his admirable illustrated guide (Ingold, 1975). The term *aquatic hyphomycetes* has come to have a
rather restricted meaning for "Ingoldian fungi"--that is, conidial aquatic fungi to
be found in abundance on decaying leaves and twigs of deciduous trees in rapidly flowing, unpolluted streams. Their spores are large and are often branched or sigmoid.
Representative form genera with branched conidia are *Actinospora*, *Alatospora*, *Articulospora*, *Clavariopsis*, *Dendrospora*, *Heliscus*, *Lemonniera*, *Tetrachaetum*, *Tetracladium*,
Tricladium, and *Varicosporium*, while form genera with sigmoid conidia include *Anguillospora*, *Flagellospora*, and *Lunulospora*. Ingold's studies of spore development have
shown that both kinds of spore shape may be the result of fundamentally different
developmental processes. Among the forms with branched, tetraradiate spores, the
form genus *Lemonniera* is an excellent example of phialidic development (Descals et
al., 1977). The form genera *Actinospora* and *Brachiosphaera* include species which
have spores so similar superficially that they have frequently been confused, yet
studies of their conidial development show that they are quite dissimilar (Webster
et al., 1975; Descals et al., 1976). In *Actinospora*, the conidia develop from the
tips of dichotomous conidiophores, and a second conidium is formed from the cell immediately behind the first conidium. In *Brachiosphaera*, a succession of conidia is
formed by sympodial proliferation. Many other examples can be found among aquatic
fungi of forms with superficially similar spore shape but dissimilar development.
This suggests that the branched or tetraradiate propagule may be the result of convergent evolution. This idea receives support from the known relationships of aquatic organisms with such propagules (see Table 1). Tetraradiate or branched propagules
are known from three major groups of "perfect" fungi and, of course, from many deuteromycetes, some of which may yet prove to have ascomycete or basidiomycete perfect

Table 1 Relationships of aquatic organisms with branched or tetraradiate
          propagules

Species	Group	Reference
Entomophthora spp.	Zygomycotina, Entomophthorales	Webster et al. (1978)
Actinospora megalospora (conidial *Miladina lechithina*)	Ascomycotina, Pezizales	Descals and Webster (1978)
Varicosporium sp. (conidial *Hymenoscyphus varicosporoides*)	Ascomycotina, Helotiales	Tubaki (1966)
Heliscus lugdunensis (conidial *Nectria lugdunensis*)	Ascomycotina, Sphaeriales	Webster (1959b)
Clavariopsis bulbosa (conidial *Corollospora pulchella*)	Ascomycotina, Sphaeriales	Shearer and Crane (1971)
Clavariopsis aquatica (conidial *Massarina* sp.)	Ascomycotina, Pseudosphaeriales	Webster and Descals (1979)
Taeniospora gracilis (conidial *Leptosporomyces galzinii*)	Basidiomycotina	Nawawi et al. (1977a)
Dendrosporomyces prolifer	Basidiomycotina	Nawawi et al. (1977b)
Ingoldiella hamata (conidial *Sistotrema* sp.)	Basidiomycotina	Nawawi (unpublished data, 1977)
Digitatospora marina (tetraradiate basidiospores)	Basidiomycotina	Doguet (1962)
Nia vibrissa (tetra- or pentaradiate basidiospores)	Basidiomycotina, Gasteromycetes	Moore and Meyers (1959); Doguet (1967)
Sphacelaria spp.	Phaeophyta, Sphacelariales	--
Unidentified bryophyte	Probably Musci	Nawawi (unpublished data, 1977)

states. It is of interest that, within the Ascomycotina, tetraradiate or branched
conidia are known from several orders which are only distantly related. Within the
Basidiomycotina, tetraradiate propagules may be found as basidiospores (*Digitatospora*,
Nia) or conidia (*Ingoldiella*, *Leptosporomyces*, *Dendrosporomyces*). *Ingoldiella hamata*
is unusual in possessing monokaryotic unclamped and dikaryotic clamped conidia (A.
Nawawi, unpublished data). Tetraradiate propagules are well known in the brown seaweed *Sphacelaria*. Tetraradiate propagules possibly belonging to a moss have been
found in tropical streams in Malaysia (A. Nawawi, unpublished data). A previously
well-known but unidentified tetraradiate propagule has recently been identified as a
secondary conidium of a member of the *Entomophthora curvispora* group. In this case,
the primary cornute conidia may develop balloon-shaped, violently projected secondary
conidia in air or tetraradiate propagules under water. It is tempting to speculate
that the two different secondary propagules are adaptations to infection of an aquat-

Table 2 Relationships of aquatic organisms with sigmoid propagules

Species	Group	Reference
Anguillospora crassa (conidial *Mollisia* sp.)	Ascomycotina, Helotiales	Webster (1961)
Anguillospora furtiva ined (conidial *Rutsroemia* sp.)	Ascomycotina, Helotiales	Webster and Descals (1979)
Anguillospora longissima (conidial *Massarina* sp.)	Ascomycotina, Pseudosphaeriales	Willoughby and Archer (1973)
Anguillospora sp. 1 (conidial *Orbilia* sp.)	Ascomycotina, Helotiales	Webster and Descals (1979)
Flagellospora penicillioides (conidial *Nectria penicillioides*)	Ascomycotina, Sphaeriales	Ranzoni (1956)

ic insect with a submerged larval stage and a free-flying adult stage (Webster et al., 1978).

The significance of most tetraradiate propagules in aquatic environments has probably to do with the fact that, at any rate for aquatic hyphomycetes, this type of propagule is more effectively trapped on underwater surfaces than are spores of more conventional shape (Webster, 1959a). One possible explanation of this is that, when tetraradiate spores are impacted on an underwater surface or are allowed to settle onto a surface, three of the spore arms come into contact with the surface, forming a stable "tripod" attachment. There appears to be a contact stimulus to germination, because the three arms in contact with a surface develop germ tubes and appressoria whereas the arm not in contact with a surface does not. However, Bandoni (1974, 1975) has suggested that the tetraradiate spore shape, which is certainly not confined to aquatic organisms or aquatic habitats, may be adapted to dispersal in surface films of water such as might be formed on or between leaves in terrestrial habitats.

The sigmoid spore form is also the result of convergent evolution. Phialidic conidial development takes place in *Flagellospora*; holoblastic development occurs in *Lunulospora* and *Anguillospora*. Table 2 shows the known relationship of some forms with sigmoid spores. The form genus *Anguillospora* is a particularly interesting example of convergent evolution of the form of aquatic conidia. This form genus contains species which have mature detached conidia only critically distinguishable from each other but with "perfect" states from widely different groups of ascomycetes (e.g., *A. longissima* = *Massarina* sp. conidia, and *A. furtiva* = *Rutstroemia* sp. conidia). The significance of the sigmoid spore form in aquatic organisms is not understood, although it is probably related to dispersal. It is of interest that the pollen of some aquatic angiosperms which are pollinated in water also has this shape (Proctor and Yeo, 1972).

   Both sigmoid and branched spore forms are readily trapped in foam which accumulates in suitable places in turbulent streams. Foam thus affords the investigator an easy way of sampling a stream to gain an indication of its complement of aquatic hyphomycetes. It is also possible to make isolations from spores in foam. A large number of species are known so far only as spores in foam or in cultures derived from such spores. Care should be exercised in interpreting such occurrences. The mere presence of spores in river foam does not, in itself, indicate that an organism is aquatic, in the sense that it can grow vegetatively and sporulate under water. Indeed, it is likely that many terrestrial fungi have spores which are washed into streams and which may accumulate in foam. At the time of collection in stream foam, the majority of spores held in it are ungerminated; but within a few hours of collection, especially if the temperature is allowed to rise, germination will occur. Experimental studies by Iqbal and Webster (1973a) have shown that the tetraradiate type of spore is removed preferentially by air bubbles from mixed spore suspensions containing spores of tetraradiate or sigmoid shape, i.e., an air bubble of given size has a higher trapping efficiency for tetraradiate propagules than for propagules of other shape. Highly branched propagules of *Varicosporium elodeae* were more readily removed from suspension by air bubbles than less branched spores. Comparison of the spore content of a stream, as indicated by Millipore filtration, with the spores caught in foam showed a higher proportion of tetraradiate spores in foam than in the filtered water sample. This supports the view that the tetraradiate spore form is preferentially trapped by air bubbles, i. e., that the spore content of foam samples is likely to be biased in favor of branched spores. It is not known whether the presence of spores in foam is of any significance in nature or is merely fortuitous. It has been suggested that one way in which aquatic hyphomycetes may be transported from one body of water to another unconnected one may be by waterfowl, but there is no firm evidence for this. This problem will be discussed later in relation to the dispersal of "perfect" spores.

   Because the spores of many aquatic hyphomycetes are large and readily identifiable (although care may be needed to discriminate between some forms), quantitative estimations of spore concentrations in stream water are easy to make if water samples are filtered through Millipore filters which are then rendered transparent and stained. Iqbal and Webster (1973b, 1977), using 8-μm filters, followed the changing spore concentration in several streams in Devon, England, during a 2 – year period. As might be expected, the highest numbers of conidia were recorded shortly after the peak season for deciduous tree leaf fall, and total spore numbers approaching 10^4 per liter were estimated in the river Creedy, a small lowland stream, in November and December. Values even higher than this, between 2-3 X 10^4 per liter, have been obtained in the nearby River Teign in November. In moorland streams on Dartmoor that drain areas largely devoid of trees, spore numbers were much lower, and the aquatic hyphomycete flora contained a number of less frequently encountered species which

could be traced to the debris of grasses and rushes that form the dominant monocotyl-
edonous vegetation of the moorland. In the lowland streams bordered by deciduous
trees, the spore concentrations of most species mirrored the availability of leaf
litter, and the spore concentrations sometimes fell below detectable levels in the
months of May through July. With the availability of a fresh supply of leaf litter
in August, spore numbers began to rise rapidly. Some species showed distinctive
patterns of seasonal abundance of spores not explained in terms of substrate availability. The conidia of *Lunulospora curvula* were only detected in the River Creedy
from August through Novemeber. In contrast, the conidia of *Tricladium gracile* were
only detected from December through April. The most likely explanation of these differences is that the temperature optima for growth and sporulation of *Lunulospora*
are higher than for *T. gracile*. Variations in stream temperature during the period
of study ranged between 18°C in August to 3°C in January (Webster et al., 1976). The
high temperature optimum for *L. curvula* is in keeping with its known world distribution, because it appears to be particularly abundant in the tropics. An alternative
explanation for its sudden decline in October and November is that it may have a preference for leaves of alder (*Alnus glutinosa*), which are soft in texture and palatable to aquatic animals. These leaves therefore quickly disappear from streams.

The effect of physical factors on growth and sporulation of a number of aquatic
hyphomycetes has been tested. Thornton (1963) showed that many of the common British
species have temperature optima for growth and sporulation near 15°C. Although many
species of aquatic hyphomycetes can be grown on a wide range of culture media, most
do not fruit on dry agar. Submersion of culture strips in water will often induce
sporulation. Sporulation is enhanced if culture pieces are placed in water and forcibly aerated with a stream of compressed air (Webster and Towfik, 1972; Webster,
1975). This effect appears to be associated with increasing turbulence of the water,
rather than with any effects of the amount of dissolved oxygen or of removal of volatile sporulation inhibitors. Turbulence affects sporulation in two ways: (1) by
causing an increase in the branching of conidiophores, and (2) by hastening the development and detachment of conidia. Another technique for inducing sporulation is
to irrigate culture strips with sterile water. When this is done in a specially constructed flow chamber designed for microscopic observation, the details of spore development under a range of environmental conditions can be followed visually (Descals
et al., 1976). Using similar techniques and collecting the spores from the outflow
of the culture chambers, Sanders and Webster (1980) have shown that certain species
(e.g., *Alatospora acuminata*, *Anguillospora crassa*, *Articulospora tetracladia*,
Clavariopsis aquatica, *Lemonniera aquatica*, *Tetrachaetum elegans*, *Tetracladium setigerum*, and *Tricladium chaetocladium*) respond to increasing linear flow rates by increased sporulation, whereas others (e.g., *Anguillospora longissima*, *Heliscus lugdunensis*, *Lemonniera cornuta*, *L. terrestris*, *Lunulospora curvula*, *Pyricularia aquatica*,
Tetracladium marchalianum, *Tricladium splendens*, and *Varicosporium elodeae*) appear

to be little affected by flow above a critical threshold value. One reason for the
increased ssporulation from culture strips subjected to higher irrigation rates may
be the removal of soluble materials released from the culture strips. The introduction of small amounts of solid culture media upstream of the sporulating culture can
result in depression of sporulation, probably associated with the fact that many of
the hyphae growing out from the culture strip into the water then continue to grow
vegetatively instead of being converted into conidiogenous cells, which happens when
pure water is used to irrigate the cultures. These two sets of observations may provide an explanation of why aquatic hyphomycetes appear to sporulate better in streams
which are clear and turbulent than in stagnant or polluted waters.

The aquatic branched yeast *Candida aquatica* has propagules whose shape is greatly
affected by their nutrition. The fungus was first isolated from foam in a mountain
tarn, but it was later discovered in the hollow internodes of dead *Equisetum* shoots.
If supplied with available nutrients in the form of dilute malt extract, the fungus
buds readily, forming short, fat cells with dense cytoplasmic contents. Transfer of
such well-grown cells to distilled water results in the development of long, tapering
arms. Colonies with short, fat cells settle more rapidly in water than do starved
colonies, whereas the trapping efficiency of starved colonies is higher. It has been
suggested that when nutrients are available, e.g., when the organism is growing in
contact with decaying tissues of aquatic plants, it is sedentary, but when nutrients
are scarce the organism becomes planktonic and is dispersed by water currents (Webster
and Davey, 1975).

The foregoing discussion and the use of the term *aquatic* has implied that we are
dealing with fungi that live exclusively in water. Evidence is accumulating that,
although rapidly flowing streams and some types of lake or pond represent habitats
where conidia of many of these fungi can be found in abundance and with a high degree
of certainty, conidia can also be found in terrestrial situations, e.g., in leaf litter well away from watercourses. Bandoni (Chap. 37) has reviewed the relevant literature and has described useful techniques for the demonstration of the presence
of conidia in such habitats. Our own experience (Webster, 1977), based on a 1-year
survey of the occurrence of conidia on *Quercus* leaves on a steeply sloping wooded
hillside above the River Teign in Devonshire, is that the conidia of "aquatic" hyphomycetes were very rarely found except close to the flood level of the river, although
some species were certainly present as conidia, and less frequently as conidiophores
bearing conidia, at higher levels. Sanders and Webster (1978) have also studied the
survival of "aquatic" hyphomycetes in terrestrial situations using several techniques.
In one method, autoclaved oak leaf disks were inoculated with aquatic hyphomycetes
and left for 1 month to become fully colonized. The leaf disks were then buried in
leaf litter, either in nylon mesh litterbags or sandwiched directly between sterile
oak leaves, in a woodland site at Woodbury, Devon. The leaf disks were recovered
at intervals and tested for the presence of aquatic hyphomycetes; they were

incubated in water and then searched for evidence of conidium formation. After 8-10 weeks of exposure to a leaf litter environment, conidium formation was restricted to the vein ends. Survival of fungi was as follows: *Clavariopsis aquatica* and *Articulospora tetracladia*, 52 weeks; *Anguillospora crassa*, 47 weeks; *Lemonniera aquatica*, 33 weeks; *Varicosporium elodeae* and *Lunulospora curvula*, 29 weeks; *Tetracladium marchalianum* and *Tricladium splendens*, 15 weeks; *Heliscus lugdunensis*, 10 weeks; *Dendrospora erecta*, 8 weeks. Ten species were tested for their ability to colonize sterilized oak leaves in contact with an infected leaf disk. Only two species were shown to be capable on invasive growth: *Articulospora tetracladia* and *Varicosporium elodeae*.

In a second method we followed the survival of aquatic hyphomycetes in leaves brought out from a river at a time of flood and stranded in the branches of adjacent trees. After 13 weeks of exposure, five species could be detected: *Anguillospora longissima*, *Articulospora tetracladia*, *Clavariopsis aquatica*, *Flagellospora curvula*, and *Lemonniera aquatica*. However, after 17 weeks no aquatic hyphomycetes could be detected. Similar results were found by using artificially inoculated leaf disks enclosed in litter bags suspended 1 m above the ground, where the longest survival period was 15 weeks for *Anguillospora crassa*. We have no evidence that aquatic hyphomycetes produce resting structures such as chlamydospores in leaf tissues, although sclerotia are common in *Lemonniera* (Descals et al., 1977).

The fact that these fungi may survive in stranded leaves may suggest a limited method of dispersal in windblown litter which was once in a stream or river, but the range of dispersal is unlikely to be great.

A possible explanation of our relative lack of success in demonstrating terrestrial activity of the hyphomycetes which we studied may be that the water content of the litter, which obviously reflects not only the incident rainfall but also soil drainage and topography, may be substantially lower than in other areas of the world where terrestrial occurrence has been reported.

In Tables 1 and 2, the known perfect states of "aquatic" conidial fungi have been listed. For the most part, these occur on twigs and logs half-submerged in water and are of rare occurrence on leaf litter. Willoughby and Archer (1973) have shown that prolonged incubation of twigs collected from streams and placed in moist chambers may result in the development of ascocarps. We have confirmed this and have shown that, in culture, apothecia, perithecia, and pseudothecia may develop after several months in petri dishes taped to minimize evaporation, when these are incubated at fairly low temperatures (10-12°C) and illuminated with near-ultraviolet light (Webster and Descals, 1979). The occurrence of perfect states capable of discharging ascospores and basidiospores into the air may help to explain how such fungi are distributed from one watercourse to an unconnected one. It is also possible that the perfect spores, falling onto moist litter on the forest floor, may result in the colonization of leaves, given favorable conditions, and this may be followed by the development of conidia.

The examples of the perfect-imperfect connections listed in Tables 1 and 2 suggest that more aquatic conidial fungi may have perfect states, and a more thorough investigation of this problem, by systematic collection of ascomycetes and basidiomycetes from twigs and branches in or near water, would be rewarding. In this respect, the tropics are virtually unexplored.

It is tempting to speculate that some aquatic hyphomycetes have evolved from terrestrial ancestors and have adapted the biology of their imperfect states to life in the water while retaining terrestrial habitats for their perfect states. If this idea is correct, it might be more appropriate to term them amphibious fungi.

II. AERO-AQUATIC HYPHOMYCETES

Contrasting with the clean water and turbulent flow which characterize the typical habitats for Ingoldian fungi, there are other aquatic habitats such as water-filled ditches and stagnant pools in which leaf and twig material is frequently colonized by a distinctive group of aero-aquatic fungi. The term *aero-aquatic* implies that these fungi have an aquatic vegetative state, i. e., mycelium growing on submerged leaves and twigs, but that their reproduction only takes place when their substrates are brought into the air. This commonly happens when a pond or ditch dries up and the carpet of litter which has accumulated on the bottom of the pond is exposed to air. Conidia of these fungi then develop freely on the exposed substrates. These fungi are simple to collect by dredging litter from the mud surface of stagnant ponds. If the litter is then rinsed and incubated for a few days at low temperature (10-15°C), conidia will develop readily and, from such conidia, cultures can be prepared. Representative form genera are *Aegerita*, *Beverwijkella*, *Candelabrum*, *Clathrosphaerina*, *Helicodendron*, *Helicoon*, *Helicosporium*, and *Spirosphaera*. The characteristic feature which all these fungi share is the production of a buoyant propagule, usually by entrapping air as the propagule develops. It is fascinating to study the range of methods by which this is achieved. In the helicosporous form genera such as *Helicoon* and *Helicodendron*, the conidia are usually barrel shaped and formed by the helical coiling of a single hypha, coiled either clockwise or anticlockwise. As the coil forms at the water-air interface at the surface of the moist substratum, air is enclosed within the coil. Spore formation does not occur under water. The spores are extremely difficult to wet or to sink, although following detergent treatment it is possible to suspend them in water. In nature it is presumed that, following a refilling of the pond within which they grow, the conidia float off and are held at the water meniscus and are available for dispersal. Such spores occasionally accumulate in foam, but much less frequently than those of the Ingoldian type. In *Clathrospherina*, the mature propagule is a clathrate sphere, somewhat resembling a hollow practice golf ball. It is formed by the repeated dichotomy of branches which make up the conidium. The tips of the branches are incurved and grow towards each other, so that an air-containing network results. Irregular, incurved branches entrap air

in *Spirosphaera*, whereas in *Aegerita* and *Beverwijkella* clusters of subglobose cells
are separated by intercellular air spaces. The taxonomic literature of this group
is listed by Fisher and Webster in Chap. 38. Moore (1955) has provided keys to the
literature. About 40 species are known.

   Although the habitats within which these fungi can be collected with certainty
are of the type described earlier, i.e., stagnant ponds and ditches, they, like the
Ingoldian fungi, have also been reported from other types of habitat. For example,
they are quite common on accumulations of wood in rivers and streams, for example,
piles of flotsam half-submerged in the water. They have occasionally been isolated
from soil and also from terrestrial leaf litter (see Bandoni, Chap. 37). Less is
known about their perfect states than for those of the Ingoldian fungi. *Clathrosphaerina zalewskii* has a *Hyaloscypha* perfect state (Descals and Webster, 1976).
Helicosporium spp. are known to have perfect states within the related genera of bi-
tunicate ascomycetes *Ophionectria* and *Tubeufia*. *Aegerita candida*, which has a clamp
connection at the base of each cell making up the propagule, is the conidial state
of a *Bulbillomyces* which grows on moist wood.

   Ecological studies on this group of fungi are now being carried out (see Fisher
and Webster, Chap. 38). It is difficult to summarize or draw brief conclusions on
the biology of what is certainly a taxonomically diverse group of organisms which
appear to have adopted similar strategies to enable themselves to take advantage of
their special niche. It seems likely that many of them rely for underwater dispersal on leaf-to-leaf contact or on fragmentation of hyphae. Relatively few of them
have thick-walled chlamydospores or sclerotia, and their ability to survive in dry
leaf litter for long periods is presumably associated with the possession of thick-
walled hyphae. The mud surface on which the leaves accumulate may at times be very
deficient in oxygen and may also contain other substances such as sulfides. There
is no evidence as yet that these fungi can grow (i. e., increase in dry weight) in
an anaerobic environment, but it is certainly true that they can survive for several
months, both in the field and in the laboratory, at O_2 tensions close to zero. Possibly their ability to survive under these conditions enables them to compete effectively with other organisms for colonization of the substrate. Consider a deciduous
tree leaf about to drop into a pond. The leaf would already contain a resident population of fungi and bacteria adapted to a terrestrial existence. The fungi present
would include such common primary colonizers as *Aureobasidium*, *Sporobolomyces*, *Cladosporium*, *Alternaria*, *Epicoccum*, and *Botrytis*. The survival of these fungi in submerged leaves has not been studied in sufficient detail, but preliminary observations
show that they are replaced in a matter of months by some of the aero-aquatic fungi.
It is not yet known whether the "phylloplane" flora are unsuited physiologically to
growth and survival under water or whether a more active form of competition, e.g.,
antibiotic production, is involved in their displacement, but experiments to test
this are in progress.

One of the exciting discoveries of the role of Ingoldian fungi in nature concerns the way in which they improve the palatability and nutrient content of leaf
litter for invertebrates. Experiments in progress here at Exeter (S. J. Smith, unpublished data) have shown that colonization of oak leaf litter by *Helicoon* and *Helicodendron* spp. makes them much more attractive as food for *Gamarus* and *Asellus*, so it
is likely that in their own special habitat the aero-aquatic fungi play an analogous
role.

REFERENCES

Bandoni, R. J. (1974). Mycological observations on the aqueous films covering decaying leaves and other litter. *Trans. Mycol. Soc. Japan 15*: 309-315.

Bandoni, R. J. (1975). Significance of the tetraradiate form in dispersal of terrestrial fungi. *Rep. Tottori Mycol. Inst.* (Japan) *12*: 105-113.

Descals, E., and Webster, J. (1976). *Hyaloscypha*: Perfect state of *Clathrosphaerina
zalewskii*. *Trans. Brit. Mycol. Soc. 67*: 525-528.

Descals, E., and Webster, J. (1978). *Miladina lecithina* (Pezizales), the ascigerous
state of *Actinospora megalospora*. *Trans. Brit. Mycol. Soc. 70*: 466-472.

Descals, E., Nawawi, A., and Webster, J. (1976). Developmental studies in *Actinospora* and three similar aquatic hyphomycetes. *Trans. Brit. Mycol. Soc. 67*: 207-222.

Descals, E., Webster, J., and Dyko, B. S. (1977). Taxonomic studies on aquatic hyphomycetes. I. *Lemonniera* de Wildeman. *Trans. Brit. Mycol. Soc. 69*: 89-109.

Doguet, G. (1962). *Digitatospora marina* n.g. n.sp.: Basidiomycète marin. *C. R.
Acad. Sci. Paris 254*: 4336-4338.

Doguet, G. (1967). *Nia vibrissa* Moore & Meyers, remarquable basidiomycète marin.
C. R. Acad. Sci. Paris 265: 1780-1783.

Ingold, C. T. (1942). Aquatic hyphomycetes of decaying alder leaves. *Trans. Brit.
Mycol. Soc. 25*: 339-417.

Ingold, C. T. (1975). *An Illustrated Guide to Aquatic and Water-Borne Hyphomycetes
(Fungi Imperfecti) with Notes on Their Biology*. Freshwater Biological Association,
Sci. Publ. No. 30. Ferry House, Ambleside, Cumbria, England, 96 pp.

Iqbal, S. H., and Webster, J. (1973a). The trapping of aquatic hyphomycete spores by
air bubbles. *Trans. Brit. Mycol. Soc. 60*: 37-48.

Iqbal, S. H., and Webster, J. (1973b). Aquatic hyphomycete spora of the river Exe
and its tributaries. *Trans. Brit. Mycol. Soc. 61*: 331-346.

Iqbal, S. H., and Webster, J. (1977). Aquatic hyphomycete spora of some Dartmoor
streams. *Trans. Brit. Mycol. Soc. 69*: 233-241.

Moore, R. T. (1955). Index to the Helicosporae. *Mycologia 47*: 90-103.

Moore, R. T., and Meyers, S. P. (1959). Thallasiomycetes. I. Principles of delimitation of the marine mycota with the description of a new aquatically adapted deutero-
mycete genus. *Mycologia 51*: 871-876.

Nawawi, A., Descals, E., and Webster, J. (1977a). *Leptosporomyces galzinii*, the
basidial state of a clamped branched conidium from fresh water. *Trans. Brit. Mycol.
Soc. 68*: 31-36.

Nawawi, A., Webster, J., and Davey, R. A. (1977b). *Dendrosporomyces prolifer* gen.
et sp.nov., a basidiomycete with branched conidia. *Trans. Brit. Mycol. Soc. 68*:
59-63.

Proctor, M. C. F., and Yeo, P. (1972). *The Pollination of Flowers*. Taplinger, New
York.

Ranzoni, F. V. (1956). The perfect state of *Flagellospora penicillioides*. *Amer. J. Bot. 43*: 13-17.

Sanders, P. F., and Webster, J. (1978). Survival of aquatic hyphomycetes in terrestrial situations. *Trans. Brit. Mycol. Soc. 71*: 231-237.

Sanders, P. F., and Webster, J. (1980). Sporulation responses of some 'aquatic Hyphomycetes' in flowing water. *Trans. Brit. Mycol. Soc. 74*: 601-605.

Shearer, C. A., and Crane, J. L. (1971). Fungi of the Chesapeake Bay and its tributaries. *Mycologia 63*: 237-260.

Thornton, D. R. (1963). The physiology and nutrition of some aquatic hyphomycetes. *J. Gen. Microbiol. 33*: 23-31.

Tubaki, K. (1966). An undescribed species of *Hymenoscyphus*, a perfect stage of *Varicosporium*. *Trans. Brit. Mycol. Soc. 49*: 345-349.

Webster, J. (1959a). Experiments with spores of aquatic hyphomycetes. I. Sedimentation and impaction on smooth surfaces. *Ann. Bot.* (London) *23*: 595-611.

Webster, J. (1959b). *Nectria lugdunensis* sp.nov., the perfect stage of *Heliscus lugdunensis*. *Trans. Brit. Mycol. Soc. 42*: 322-327.

Webster, J. (1961). The *Mollisia* perfect stage of *Anguillospora crassa*. *Trans. Brit. Mycol. Soc. 44*: 559-564.

Webster, J. (1975). Further studies of sporulation of aquatic hyphomycetes in relation to aeration. *Trans. Brit. Mycol. Soc. 64*: 119-127.

Webster, J. (1977). Seasonal observations on "aquatic" hyphomycetes on oak leaves on the ground. *Trans. Brit. Mycol. Soc. 68*: 108-111.

Webster, J., and Davey, R. A. (1975). Sedimentation rates and trapping efficiency of cells of *Candida aquatica*. *Trans. Brit. Mycol. Soc. 64*: 437-440.

Webster, J., and Descals, E. (1979). The teleomorphs of water-borne Hyphomycetes from fresh water. In *The Whole Fungus*, Vol. 2, W. B. Kendrick (Ed.). National Museum of Natural Sciences, National Museums of Canada and the Kananaskis Foundation, Ottawa, Canada, pp. 419-451.

Webster, J., and Towfik, F. H. (1972). Sporulation of aquatic hyphomycetes in relation to aeration. *Trans. Brit. Mycol. Soc. 59*: 353-364.

Webster, J., Nawawi, A., and Descals, E. (1975). *Spore Development in Aquatic Hyphomycetes*, 16mm cine-film. University of Exeter, Devon, England.

Webster, J., Moran, S. T., and Davey, R. A. (1976). Growth and sporulation of *Tricladium chaetocladium* and *Lunulospora curvula* in relation to temperature. *Trans. Brit. Mycol. Soc. 67*: 491-549.

Webster, J., Sanders, P. F., and Descals, E. (1978). Tetraradiate aquatic propagules in two species of *Entomophthora*. *Trans. Brit. Mycol. Soc. 70*: 472-479.

Willoughby, L. G., and Archer, J. F. (1973). The fungal spora of a freshwater stream and its colonization pattern on wood. *Freshwater Biol. 3*: 219-239.

Chapter 37

AQUATIC HYPHOMYCETES FROM TERRESTRIAL LITTER

R. J. Bandoni

The University of British Columbia
Vancouver, British Columbia, Canada

I.   INTRODUCTION

In discussing the occurrence of aquatic hyphomycetes on terrestrial litter, no attempt will be made to define or delimit the group of fungi concerned. The term
aquatic hyphomycetes might be expected to include most of the Moniliales found growing in any aquatic environment. In practice, however, the term has come to be applied
to a restricted group of fungi found in freshwater streams.

There have been relatively few reports of aquatic hyphomycetes in nonaquatic
environments. Furthermore, not all of these reports are concerned with the occurrence
of such fungi on materials which could be classified as litter. Consequently, little
is known about either the frequency of occurrence or the role of these fungi in terrestrial habitats. For this reason, my discussion is largely concerned with techniques that have been applied, or could be applied, in ecological studies of the
group.

II.   SURVEY OF SPECIES REPORTED

Most of the early reports of freshwater hyphomycetes from nonaquatic habitats have
been discussed elsewhere (Mäkelä, 1972; Bandoni, 1972; Park, 1974b; Dyko, 1976), and
I shall not discuss these reports in detail. All of the species thus far reported
are listed, together with references and substrates, in Table 1.

Recent reports--the species from which are also listed in Table 1--are by Park
(1974b), Dyko (1976), Webster (1977), and Singh and Musa (1977). Park found *Gyoerffyella craginiformis*, *Tetracladium setigerum*, *Candelabrum spinulosum*, and *Spirosphaera
floriformis* on litter in a well-drained shrubbery. He also observed sporulation of
either aquatic hyphomycetes or fungi that closely resemble them on slides buried in
beakers of soil in the laboratory. These species could not be isolated, and their
identity is therefore unknown.

Dyko (1976) did a comparative study of aquatic hyphomycetes in forest litter as
contrasted with those in nearby streams in several southeastern states in the United
States. She observed *Alatospora acuminata*, *Clavariopsis aquatica*, *Lemonniera aquatica*,

Table 1 Aquatic and aero-aquatic hyphomycetes reported from nonaquatic habitats
(including reports in which only conidia were observed)

Species	Substrates	References
Alatospora acuminata Ingold	Litter	Hudson and Sutton (1964); Bandoni (1972); Dyko (1976)
Anguillospora longissima (Sacc. & Syd.) Ingold	Roots	Nemec (1969)
Articulospora tetracladia Ingold	Litter	Tubaki (1960); Nilsson (1964)
Camposporium pellucidum (Grove) Hughes	Litter	Gönczöl (1976); Webster (1977)
Candelabrum spinulosum van Beverwijk	Litter	Park (1974b)
Centrospora acerina (Hartig) Newhall	Living plants	See Newhall (1946) for several references.
Clathrosphaerina zalewskii van Beverwijk	Roots	Waid (1954)
Clavariopsis aquatica de Wild.	Litter	Dyko (1976)
Dendrospora erecta Ingold	Litter; rabbit dung	Webster (1961, 1977)
Dendrospora sp.	Litter	Webster (1977)
Flagellospora curvula Ingold	Litter	Webster (1977)
Gyoerffyella gemellipara Marvanová	Litter	Bandoni (1972, as *I. cragini-formis*; see also Marvanová (1975)
Gyoerffyella rotula (Hohn.) Marvanová	Litter	von Höhnel (1904); Nilsson (1964); Bandoni (1972); Park (1974b); W. Gams (cited in Marvanová, 1975)
Isthmotricladia sp.	Litter	Webster (1977)
Lateriramulosa uni-inflata Matsushima		Matsushima (1971)
Lemonniera aquatica de Wild.	Litter	Tubaki (1958); Webster (1977)
Lemonniera terrestris Tubaki	Litter	Tubaki (1958); Dyko (1976)
Lunulospora curvula Ingold	Litter	Dyko (1976)
Pleuropedium tricladioides Marvanová & Iqbal	Litter	Webster (1977)
Speiropsis pedatospora Tubaki	Litter	Tubaki (1958)
Spirosphaera floriformis van Beverwijk	Litter	Park (1974b)
Tetrachaetum elegans Ingold	Litter	Tubaki (1960); Dyko (1976)
Tetracrium amphibium Price & Talbot	Decaying wood	Price and Talbot (1966)
Tetracladium marchalianum de Wild	Soil, roots, litter	Gams et al. (1969); Nemec (1969); Dyko (1976)
Tetracladium maxilliforme (Rostrup) Ingold	Litter	Rostrup (1894); Mäkelä (1972)

Table 1 (cont.)

Species	Substrates	References
Tetracladium setigerum (Grove) Ingold	Litter	Scourfield (1940); Hudson and Sutton (1964); Nilsson (1964); Bandoni (1972); Park (1974b); Dyko (1976)
Tetraploa aristata Berk. & Br.	Litter, air spora	Gregory (1961); Webster (1977)
Tetraploa ellisii Cooke	Roots	Nemec (1969)
Tricellula aquatica Webster	Litter	Hudson and Sutton (1964); Mäkelä (1972)
Tricladium gracile Ingold	Litter	Dyko (1976)
Triscelophorus monosporus Ingold	Litter	Singh and Musa (1977)
Trisulcosporium acerinum Hudson & Sutton	Litter	Hudson and Sutton (1964)
Varicosporium elodeae Kegel	Roots, soil, litter, etc.	Bessey (1939); Waid (1954); etc.
Volucrispora graminea Ing. McD. & Dann	Litter	Webster (1954); Mäkelä (1972)

Lunulospora curvula, *Tetrachaetum elegans*, *Tetracladium marchalianum*, *T. setigerum*, *Tricladium gracile*, and *Varicosporium elodeae* in forest litter.

Webster (1977) carried out an extensive study of aquatic hyphomycetes on leaf samples collected adjacent to the River Teign in Devon, England (see Chap. 36). The samples, examined both for the presence of conidia and for conidiophores, were collected along a transect leading from the water's edge up a steep bank. Conidia of 17 species were observed, but conidiophores were found on only 11 of 560 leaves examined. In addition, some species were found only at the water's edge or at levels subjected to occasional flooding. I have omitted species observed only at the water-line, but have included the other species in Table 1.

Singh and Musa (1977) observed abundant sporulation of *Triscelophorus monosporus* and the basidiomycete *Ingoldiella hamata* Shaw on leaf skeletons from high, well-drained sites in Sierra Leone. Goos (1978) has also collected *T. monosporus* on terrestrial litter samples in Hawaii. These samples came from areas subjected to periodic drying as well as from areas of high rainfall.

Table 1 includes some species that are not definitely known to occur in streams, e.g., *Tetracrium amphibium*, conidia of which were observed in foam. The description of this species (Price and Talbot, 1966) is based upon isolates from decaying terrestrial wood samples. No attempt has been made to evaluate individual studies, but it should be noted that some reports are based upon observations of conidia only, i.e., not growth, in terrestrial habitats.

III. METHODS

A. Spore Sampling

As has been pointed out elsewhere (Park, 1974b; Bandoni, 1972; Webster, 1977) the
presence of aquatic hyphomycete conidia in terrestrial sites is not in itself proof
of activity in those sites. However, if such conidia are present in large numbers,
are there consistently, or show distinctive patterns of distribution (specificity),
the probability of activity within the site increases greatly. In general, spore densities can be expected to be greatest near points of production and to decline with
increasing distances from such sources. In streams, the foam spora is unusual in
that the number of propagules is very great and they are not necessarily close to
their points of production. To a certain extent, films of water on terrestrial litter resemble foams in this respect.

     Observations of aquatic hyphomycete *conidia* on terrestrial litter samples have
mostly involved washing techniques (Scourfield, 1940; Hudson and Sutton, 1964) or
simple flotation and interface examination (Bandoni, 1972; Webster, 1977). In several instances (Scourfield, 1940; Hudson and Sutton, 1964; Bandoni, 1972) detection
of conidia was, at least initially, fortuitous. Both of the aforementioned methods
are satisfactory for the detection of conidia and, to a certain extent, for estimating relative abundance. However, washing can lead to heavy losses of conidia if
special precautions are not taken. That is, conidia of aquatic hyphomycetes typically float free of litter materials when these are placed in water; the conidia remain at the air-water interface. If the vessel used is agitated violently, submersion of the surface conidia will occur, but many of the conidia will be returned to
the surface by rising air bubbles. If the wash water is poured from one container
to another, surface conidia can be carried away from the direction of pour by surface
tension movements and these conidia will remain in the first vessel. In addition,
the washing technique should be followed by centrifugation to concentrate the conidia
present.

     A simpler technique can be used for both qualitative and quantitative studies
of conidia in litter. A narrow graduated cylinder (25 ml) is partially filled with
distilled water, and the volume is recorded. Small pieces of leaves or other litter,
selected at random from a litter sample, are then submerged in the water one at a
time. Sufficient litter is used to give a measureable displacement. Without removing
the litter fragments, the film present at the air-water interface can then be picked
up on clean coverslips. A thoroughly cleaned coverslip is held at one edge with
tweezers, passed vertically through the interface, then withdrawn. The coverslip is
blotted to remove materials from one surface and is placed on a slide with the remaining wet side down. A number of coverslips must be used in order to remove most of
the conidia at the interface. This can be seen clearly in counts from an original
sample, as shown in Table 2.

Table 2 Conidia on decaying leaves of *Prunus* (Japanese Cherry): May 30, 1977

Type	1	2	3	4	5	6	7	8	9	T[a]	C/L[b]
					Slide No.						
Alatospora	132	105	115	30	10	3	7	15	4	421	210,500
Articulospora	17	2	2	1	1	1	--	--	--	24	12,000
Camposporium	5	5	6	1	3	--	--	--	--	20	10,000
Gyoerffyella	129	56	154	26	9	2	9	7	1	393	196,500
Helicosporeae	33	20	24	16	17	13	3	1	2	129	32,250
Tetracladium	4	1	3	--	--	1	--	1	--	10	5,000
Tricladium	5	7	1	1	--	--	3	2	--	19	9,500

[a] T = total count for 4-ml sample (displacement).

[b] C/L = estimated conidia per liter of litter.

Theoretically, both sides of the coverslip should carry roughly equal loads of conidia when retrieved from the water. Since half the load is destroyed by cleaning one side of the coverslip, the observed numbers must be doubled. If the submerged portions of the litter sample have been chosen randomly, numbers of conidia per liter of litter (displacement) can be estimated. Alternatively, numbers per unit of surface area can be calculated. Whichever calculation is made, is must be remembered that the conidia observed need not have been produced on the litter surface examined.

The use of a narrow, deep vessel is important, for it results in the concentration of conidia from a great surface area--that of the litter fragments--onto a very small surface area. The deposition of conidia on coverslips is not very uniform, and consequently it appears to be necessary to scan each slide completely in order to obtain accurate counts. Checks of the water below the surface indicate that virtually all tetraradiate and helicoid conidia remain at the interface when fragments are submerged. However, it is possible that some conidia are still trapped on the surface of the fragments.

The kinds of hyphomycetes listed in Table 2, and some additional ones, are common in litter of deciduous trees in southwestern British Columbia. To show the kinds of qualitative differences that can be found in different types of litter, conidia from a single sample of *Scirpus microcarpus* Presl are illustrated in Fig. 1. This sample, although collected near a stream, was taken from a plant located about 1.5 m above the water level and some 2 m from the stream bank.

B. Plating Methods

In their studies of fungi associated with living roots, Harley and Waid (1955) developed a serial washing technique to remove all surface propagules on root samples.

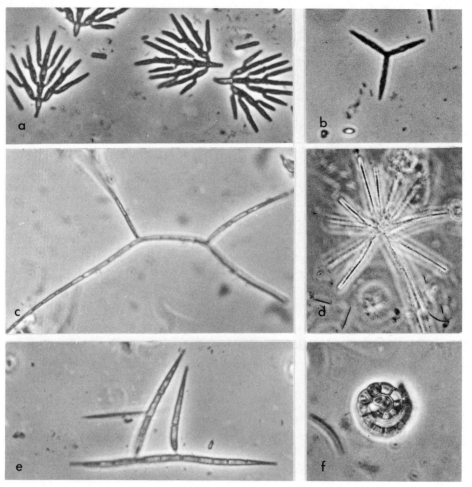

Fig. 1 Representative conidia from litter spora of *Scirpus microcarpus*: (a) unidentified genus abundant in spora and fungus isolated from senescing leaves; (b) *Scolecobasidium* sp.; (c,e) *Tricladium* spp.; (d) *Dendrospora* sp.; (f) *Helicoma* sp. (All figures except d, ca. X1000; d, ca. X250.)

The samples were washed with repeated changes of sterile water, agitated with each
change, then placed on nutrient media. This procedure permitted isolation of the
slow-growing root-surface fungi without interference from soil fungi. Waid (1954),
utilizing this technique, isolated *Varicosporium elodeae* and *Clathrosphaerina zalewskii* from living beech roots. A similar technique was employed by Taylor and Parkinson (1965) in demonstrating that *V. elodeae* was a common colonizer of living *Phaseolus roots*.

I have used Harley and Waid's (1955) washing technique essentially as originally
described to isolate aquatic hyphomycetes from nonaquatic litter samples. Parts of
freshly collected litter samples, such as leaf disks, segments of petioles, and twigs,
were washed in 30 successive changes of sterile distilled water. Each aliquot of
water was contained in a washed, sterilized jar. The samples were violently agitated
on a rotary shaker for a period of 2 min. in each jar. One modification of the technique as originally described was the introduction of a 50-sec. rinse in 50% ethanol-
distilled water solution between washes 26 and 27. As a check of the efficiency of

the washing procedure, cultures were prepared from each of the wash-water samples. The results were essentially as observed by Harley and Waid with roots.

The washed litter materials were either plated on an isolation medium or were placed in sterile distilled water for incubation. Initially, media containing tannic acid, gallic acid, vanillic acid, and similar lignin degradation products were used. A weak medium containing 0.5 g each of yeast extract, tannic acid, and gallic acid (the last two filter-sterilized and added to the autoclaved medium) was reasonably successful. However, the medium most used consisted of yeast extract (Difco), 0.5 g; sorbose, 4 g; agar, 15 g or equivalent; distilled water, 1 liter. After autoclaving, 100 mg of water-soluble tetracycline was added to each liter of medium. The use of this antibiotic to discourage bacterial growth was first suggested to me by G. C. Hughes, and I have used it successfully in isolation of numerous types of fungi.

The medium contains a relatively low concentration of nutrients; this, together with the restrictive action of sorbose and incubation at low temperatures, helps to discourage rapid spread of weed fungi. Sorbose was first used by *Neurospora* geneticists (de Terra and Tatum, 1961) to induce colonial growth, and it inhibits rapid spread of many fungi. However, such spreading did occur on many plates of litter fragments.

Washed litter fragments were drained dry on sterile absorbent paper, placed on the surface of the isolation medium, then incubated at either 5° or 10°C in chambers with 12-hr-light, 12-hr-dark cycles. After 24 hr, each piece of litter was inverted and was placed on a fresh spot on the agar surface. All fragments were so inverted and moved each day until all of the agar surface had been utilized; the fragments were left on the last spot.

Initial observations were generally made about 5 days after plating of the washed samples. Species of *Mucor*, *Botrytis*, *Cladosporium*, *Penicillium*, and *Cylindrocarpon*-like fungi sometimes developed quickly and overran the plates. However, this was not a major problem. Perhaps more troublesome was the heavy development of yeasts in the initial incubation spots. The brief rinse in 50% ethanol during the washing procedure was inserted to inhibit yeast development, but this was not entirely successful.

Isolated colonies of interest were removed from the initial plates and were grown on a malt-peptone-yeast extract medium for examination and identification. This medium had previously been found to be suitable for growth and sporulation of terrestrial isolates of aquatic hyphomycetes (Bandoni, 1972).

Apart from the frequency in general terms, no quantitative data can be given for aquatic hyphomycetes in washed, plated litter samples. That is, only a limited number of the total colonies on any plate was isolated and examined. The primary objective in these studies was to demonstrate that aquatic hyphomycetes do occur in non-aquatic litter samples. From this standpoint, it was deemed unnecessary to obtain quantitative results in the initial studies.

The overall effect of daily inversion of litter fragments, low-temperature incubation, and restrictive medium is to provide a partial separation of species inhabiting litter. Presumably, actively growing species, or those which initiate growth from resistant propagules very quickly, will appear on the first several spots. Subsequently, those which either grow slowly or are present only as resistant structures will commence growth and colonize the agar surface. The time required for recovery from the washing process, the ethanol rinse, and tetracycline and sorbose inhibition may also be involved. Initial spots were most frequently colonized by yeasts, especially heterobasidiomycetous types, and *Aureobasidium*. The chytrid *Entophlyctis confervae-glomeratae* (Cienowski) Sparrow, and, frequently, species of aquatic hyphomycetes could be isolated from the early spots. The later spots were generally overgrown by species of *Cladosporium*, *Penicillium*, *Botrytis*, *Trichoderma*, *Alternaria*, and similar fungi. With respect to filamentous fungi that were present, it could be seen that most colonies extended through the entire leaf thickness. That is, patterns of colonies of adjacent spots formed reverse images of one another.

Plates from which colonies had been transferred were not discarded but were incubated for 2-3 weeks and then examined again. On litter fragments, helicosporous and sometimes aquatic hyphomycetes could be detected. Where these could not be seen, flooding of the plate and examination of the floating spores often revealed their presence.

The leaf sample for which spore counts are listed in Table 2 was washed and plated using the foregoing methods. From this sample, *Articulospora tetracladia*, *Camposporium* sp., *Gyoerffyella gemellipara*, and a species of *Tricladium* were obtained on the plates. In addition, *Alatospora acuminata*, which did not develop on the plates, sporulated on washed, submerged leaf disks within 48 hr of submersion.

Table 3 lists all aquatic or aero-aquatic hyphomycetes isolated from washed litter samples, together with substrates from which they were obtained. Although the number of species isolated is not large, relatively few litter samples were examined by this method. Plating of additional litter types from this and other localities would undoubtedly yield additional species.

In general, the numbers of colonies of these hyphomycetes in leaves was small. The use of relatively intact leaves or other litter materials, necessary because of the agitation in washing, might account in part for the low numbers. Aquatic hyphomycetes in streams have generally been reported to be most abundant in well-decayed leaves.

C. Stem Flow and Street Spora

In 1976, I was shown photographs of some unusual conidia by G. C. Carroll. The conidia had been collected in canopy throughfall in a conifer forest in Oregon. Subsequently, in the winter of 1976-1977, I monitored stem flow of several standing trees in a forest area adjacent to the University of British Columbia campus. The collec-

Table 3 Species isolated from washed, plated, or submerged litter

Species	Substrates
Alatospora acuminata	Submerged *Prunus* leaves
Articulospora tetracladia	Plated leaves, *Prunus*, *Acer*
Camposporium antennatum	Plated leaves, *Gaultheria*, *Prunus*
Gyoerffyella biappendiculata	Plated *Acer* leaves
Gyoerffyella gemellipara	Plated leaves, *Acer* and *Prunus*
Tetracladium maxilliforme	Plated *Alnus* leaves
Tetracladium sp.	*Sambucus* twigs, *Salix* and *Plantago* leaves, plated
Tricladium (?) attenuatum	*Alnus* twig and leaves, *Acer* leaf, plated
Tricladium sp.	*Alnus*, *Prunus*, *Acer* leaves, plated
Varicosporium elodeae	Plated leaves, *Alnus*, *Acer*, *Scirpus*, etc.
Varicosporium (?) delicatum	Plated leaves, *Scirpus*

ting device consisted of a length of split plastic tubing, a plastic bag, and a sling. The tubing was attached to the tree trunk in a steep downward spiral, split side uppermost, by tacking the inner edge to the tree. The plastic bag was attached to the lower, unsplit end of the hose, and the sling was used to hold the bag in position. To prevent germination of conidia, 100 ml of ethanol was added to each collecting bag. Retrieval of the water samples was at irregular intervals, depending upon rainfall. The water collected from the bags was first centrifuged, then examined microscopically for the presence of conidia. Some of the kinds of conidia observed are illustrated in Fig. 2.

Stem flow of five trees was examined; these included a single tree each of *Thuja plicata* Donn, *Pseudotsuga menziesii* (Mirb.) Franco, and *Acer macrophyllum* Pursh, and two trees of *Alnus rubra* Bong. The types of conidia present varied to a certain extent with tree species, but some types appeared to be common to all of the trees. The ethanol, necessary to prevent germination, also prevented isolation and identification. Some of the conidia were recognizable as to genus or as forms that have been illustrated in studies of the foam spora.

As for aquatic hyphomycetes or forms considered as such, only conidia of *Gyoerffyella biappendiculata* (Fig. 2a) appeared regularly; they were found only in stem flow of the *Alnus* trees. One *Alnus* tree bore a number of dense fascicles of small branches along the straight trunk, and many decaying leaves were trapped in the fascicles. However, the second alder had a relatively clean trunk with no fascicles of branches and few trapped leaves. *G. gemellipara* (Fig. 2b) conidia were observed in the stem flow of the *Acer*. This tree supported a large community of epiphytic ferns,

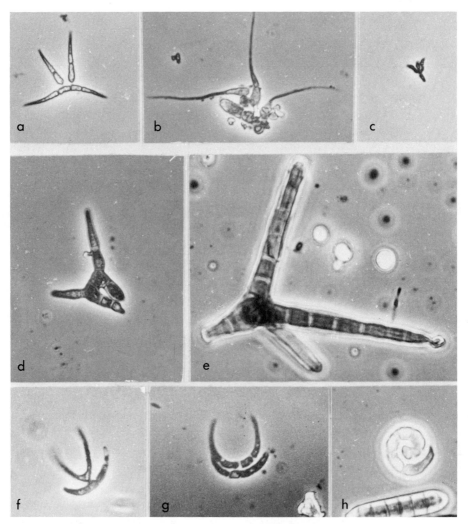

Fig. 2 Conidia from stem flow of living trees: (a) *Gyoerffyella biappendiculata*;
(b) *G. gemellipara*; (c) *Volucrispora* sp.; (d) *Tripospermum* sp.; (e) *Ceratosporella*
sp.; (f,g) unidentified arcuate-tetraradiate conidia; (h) unidentified helicoid,
tapered conidium. (All figures, ca. X1000.)

bryophytes, and lichens along its trunk. A single small water sample was obtained
from the tree, but this was the richest sample in terms of the variety of conidia
present. A species of *Volucrispora* (Fig. 2c) also was common in the stem flow of the
two alders.

The commonest branched conidia in the stem flow were of species of *Tripospermum*
spp. (Fig. 2d), *Ceratosporella* sp. (fig. 2e), *Ceratosporium* and the unidentified,
arcuate-tetraradiate conidium shown in Fig. 2f,g. Conidia of species of *Titaea* and
Cornutispora, and elongate conidia, especially the helicoid, tapered form shown in
Fig. 2h, were common in most samples.

Gönczöl (1976), in Hungary, found *Alatospora acuminata*, *Articulospora tetra-*
cladia, *Camposporium pellucidum*, and a *Tricladium* sp. growing on leaves submerged in
water in hollows in beech trees. He later found *Dactylella submersa*, *Helicodendron*

westerdijkae and *H. paradoxum* on such leaves, the last two present only as conidia. The trees of which I monitored stem flow did not have any visible hollows in which water might have collected. This possibly accounts for the small number of aquatic hyphomycete conidia found.

In the abstract of his paper, Gönczöl (1976) mentions unusual spores in "gutters," but no details are given. During the winter of 1976-1977, I periodically examined rain drainage in a residential district of Vancouver. The samples, collected from the surface of a paved lane, were from drainage of a portion of a single residential block. The samples were collected within a few minutes of the start of a rainstorm; falling rain was also caught and examined. Both kinds of samples were centrifuged to concentrate spores before microscopic examination.

The falling rain yielded only a few nondescript spores and cells, but the surface runoff contained numerous unusual conidia. Relatively few samples were examined and no attempt was made to count numbers of conidia. Those which appeared to be most abundant were of species of *Tripospermum*, but *Alatospora*, *Articulospora*, *Tricladium*, and *Varicosporium* conidia were also common.

D. Submerged Incubation

In examining terrestrial litter, Park (1974a), Dyko (1976), Singh and Musa (1977), and Webster (1977) all have submerged litter for incubation and examination of aquatic hyphomycetes. This technique was first used by Ingold (1942) in his studies of leaves from streams. The problems connected with the use of this technique were discussed by Park (1974b) and earlier by Nilsson (1964). One aspect discussed by Park, the matter of time required for appearance of conidia, is an important feature that should be noted in studies of both aquatic and land litter samples. That is, if conidia do not develop within a few hours of the start of incubation, uncertainty arises as to whether the conidia were produced by mycelia actively growing at the time of submersion. Contaminating surface conidia could germinate and produce a mycelium and conidia in the changed environment of the culture dish. Fouling, discussed by Nilsson (1964), could prevent development of conidia by mycelia active at the time of collection. Nilsson also pointed out that sporulation on submerged materials, such as agar blocks and mycelium or litter, was not in itself satisfactory evidence of adaptation to the aquatic environment. He noted that some nonaquatic fungi will sporulate under these conditions and, alternatively, that aquatic hyphomycetes would sporulate without the necessity of submerging blocks of mycelium. He used weak media for this purpose and observed that sporulation often occurred in plates on cut surfaces where blocks had been removed.

The submersion technique proved useful in my studies of litter in isolating *Alatospora acuminata* as mentioned earlier. In fact, if used with the precautions concerning time in mind, this would certainly be a simpler method than plating to obtain aquatic hyphomycetes from terrestrial samples. Since I was interested in other fungi in addition to this group, plating was preferable.

E. Other Methods

Park's use of the slide burial technique (1974b) is of interest because it demonstrates the ability of certain fungi, either aquatic hyphomycetes or forms with similar conidia, to grow and sporulate in soil. The limitation of this method is that
the species could not be isolated from the soil samples and identification could not
be made from the slides.

Mäkelä (1972) examined grass samples directly for the presence of conidia and
conidiophores. Because of the low frequency of observed occurrences and hence the
large number of samples that must be examined, this is not a very attractive method.
It should also be noted that one species found, *Tetracladium maxilliforme*, was present
on *Dactylis* leaves that had been incubated at 5-10°C for almost 4 months.

IV. DISCUSSION

In my opinion, there can be little criticism of the methods used by Waid (1954) and
subsequently by Taylor and Parkinson (1965) in demonstrating that *Varicosporium elodeae* was associated with living roots of beech and bean. Nemec's (1969) isolation
methods were not described in detail, and the validity of his reports could therefore be questioned. However, he did indicate that *Tetracladium marchalianum* might
play a role in disease development, as it was primarily isolated from plants with
early root-rot symptoms. *Centrospora acerina*, listed in a number of reports on aquatic hyphomycetes, is known to attack living plants of several types and to be of
broad distribution in terrestrial habitats (Newhall, 1946). My studies indicate that
an undescribed genus and species (Fig. 1a), the conidia of which have been found in
foam, can be isolated from senescing leaves of *Scirpus microcarpus*. Thus, there is
evidence of early colonization or actual parasitism of land plants by some species
of aquatic hyphomycetes.

Apart from the above, *Heliscus lugdunensis* was originally described from pine
bark (Saccardo, 1880) and *Volucrispora graminea* was first found on orchard grass
(Webster, 1954), both from nonaquatic habitats. Although the habitat of *Gyoerffyella
rotula* was not specified in the original decription, it appears probable that this
species also was first found on land (see Marvanová, 1975). Judging by the litter
spora, *Volucrispora graminea* is one of the commonest fungi on grass litter in the
area that I have studied. There also can be little doubt that species of *Alatospora,
Articulospora, Tetracladium, Tricladium, Gyoerffyella, Triscelophorus, Speiropsis,
Lemonniera,* and *Varicosporium* inhabit terrestrial habitats. Finally, Singh and Musa
(1977) have found the basidiomycete conidial fungus *Ingoldiella hamata* in upland
litter; Nawawi et al. (1977) have shown that a second type of branched conidium from
water is that of *Leptosporomyces galzinii*, a common terrestrial basidiomycete.

Aero-aquatic fungi have been isolated from roots by Waid (1954) and from litter
in a shrubbery bed by Park (1974b). I have isolated *Helicodendron triglitziense*
(Jaap) Linder from washed, plated leaves obtained near the margins of (but not sub-

merged in) temporary pools. Also, conidia of species of *Clathrosphaerina* and *Candelabrum* have frequently been observed in the spora of upland litter.

Species of aquatic hyphomycetes isolated from soil include *Tetracladium marchalianum* (Gams et al., 1969) and *Varicosporium elodeae* (Bessey, 1939). The isolations of *Varicosporium* and *Tetracladium* from roots, as well as observations of Park (1974b) on buried slides, all indicate that some aquatic hyphomycetes are active in soils.

Although washing and plating techniques have been used in studies of litter decay (Hering, 1965; Hogg and Hudson 1966; Visser and Parkinson 1975), I know of no reports of aquatic hyphomycetes in such studies. It is probable that the type of medium used and the daily inversion of litter parts were responsible for my success in isolating these fungi. In addition to recognizable species described from aquatic habitats (Table 3), the types isolated include some undescribed species with branched spores. One, a species of *Scolecobasidium* (Fig. 1b), has conidia similar to triradiate forms depicted by Ingold (1975) from foam. A second form, the conidia of which have a number of branches (Fig. 1a) does not appear to belong to any described genus. Conidia of both of these types can be found in the foam sporas of local streams.

The climate of the area in which the isolations were made could, in part, be responsible for the presence of aquatic hyphomycetes in upland litter samples. Maximum angiosperm leaf decay occurs during the months of November through May. During these months, rainstorms are frequent and the total average rainfall for the period is about 88 cm. The temperatures are cool but commonly remain above freezing. The precipitation and an abundance of overcast days contribute to maintenance of saturation of litter for prolonged periods. However, widespread occurrences of aquatic hyphomycetes (Table 1) in decaying plant material on land suggest that saturation of the litter layer is not necessary for their development.

Conidia of aquatic hyphomycetes other than those isolated have been observed in terrestrial litter samples using the flotation technique described earlier. Although the presence of such conidia should not be taken as proof of activity in a given sample, distribution patterns also rule out the possibility that such conidia are transported from streams and deposited on the litter. That is, the "litter spora" of adjacent plants of different species can show marked differences as to types of spores present. For example, abundant conidia of a species of *Tetracladium* can be found on the dead lower leaves of the rosette plants of *Plantago* and *Taraxacum* in my area of study. Adjacent grass plant litter commonly yields conidia of *Volucrispora graminea* and those of a *Tetraploa*, but not the *Tetracladium*. The flotation method is simple, rapid, and helpful in indicating potential substrates of a variety of hyphomycetes. (This is also an excellent exercise for classes in mycology.)

The conidia found in the stem flows of living trees are an indication that some colonization of trapped decayed leaves can occur by aquatic hyphomycetes on materials not in the litter layer. I have also isolated *Varicosporium elodeae*, an undescribed hyphomycete with rhomboidal conidia resembling those of *Margaritispora*, and in one

instance *Gyoerffyella gemellipara* from naturally trapped leaves. It is highly probable that conidia and other propagules are carried by rain, either directly or indirectly from trees to streams. Gönczöl (1976) observed a large number of aquatic hyphomycetes in a foam that developed at the base of a tree. The presence of aquatic
hyphomycete conidia in drainage during rainstorms also is an indication that transport
of aquatic hyphomycete conidia can occur from land to water.

V. CONCLUSIONS

The following conclusions can be drawn:

1. Some aquatic hyphomycetes actively grow and sporulate in terrestrial litter
    and in soil.

2. Many branched conidia of hyphomycetes and other fungi are produced on land
    and are carried to streams by overland drainage or are splashed by rain from
    overhanging plants directly into streams.

3. Further study is needed before the roles of aquatic hyphomycetes in terres-
    trial habitats, as well as their abundance there, can be ascertained. These
    fungi are not dominant species in the litter samples studied, but they might
    be more common in later stages of decay.

ACKNOWLEDGMENTS

I wish to thank S. Brezden and D. Mozell for assistance during the initial phase of
my studies, J. DeLange for her help and for preparation of the plates, and G. C.
Hughes for reading the manuscript. Financial assistance of the National Research
Council of Canada (Grant No. NRC A 801) is gratefully acknowledged.

REFERENCES

Bandoni, R. J. (1972). Terrestrial occurrence of some aquatic hyphomycetes. *Can.
J. Bot. 50*: 2283-2288.

Bessey, E. A. (1939). *Varicosporium elodeae* Kegel, an uncommon soil fungus. *Pap.
Mich. Acad. Sci. 35*: 15-37.

de Terra, N., and E. L. Tatum (1961). Colonial growth of *Neurospora*. *Science 134*:
1066-1068.

Dyko, B. S. (1976). A preliminary study of aquatic hyphomycetes on leaves in forest
and stream litter. *J. Tennessee Acad. Sci. 51*: 7-8.

Gams, W., Domsch, K. H., and Weber, E. (1969). Nechweis signifikant verschiedener
Pilzpopulationen bei gleichenden Bodennutzung. *Plant Soil 31*: 439-450.

Gönczöl, J. (1976). Ecological observations on the aquatic hyphomycetes of Hungary.
II. Observations on biotypes of aquatic hyphomycetes in S.W. Hungary. *Acta Bot.
Acad. Sci. Hung. 22*: 51-60.

Goos, R. D. (1978). Occurrence of *Triscelophorus monosporus* in upland sites in Oahu,
Hawaii. *Mycologia 70*: 188-189.

Gregory, P. H. (1961). *The Microbiology of the Atmosphere*. Hill, London.

Harley, J. L., and Waid, J. S. (1955). A method of studying active mycelia on living
roots and other surfaces in the soil. *Trans. Brit. Mycol. Soc. 38*: 104-108.

Hering, T. F. (1965). Succession of fungi in the litter of a Lake District oakwood. *Trans. Brit. Mycol. Soc. 48*: 391-408.

Hogg, B. M., and Hudson, H. J. (1966). Micro-fungi on leaves of *Fagus sylvatica*. *Trans. Brit. Mycol. Soc. 49*: 185-192.

Hudson, H. J., and Sutton, B. C. (1964). *Trisulcosporium* and *Tetranacrium*, two new genera of Fungi Imperfecti. *Trans. Brit. Mycol. Soc. 47*: 197-203.

Ingold, C. T. (1942). Aquatic hyphomycetes of decaying alder leaves. *Trans. Brit. Mycol. Soc. 25*: 339-417.

Ingold, C. T. (1975). *An Illustrated Guide to Aquatic and Water-Borne Hyphomycetes (Fungi Imperfecti) with Notes on Their Biology*. Freshwater Biological Association, Sci. Publn. No. 30. Ferry House, Ambleside, Cumbria, England, 96 pp.

Mäkelä, K. (1972). Some aquatic hyphomycetes on grasses in Finland. *Karstenia 13*: 16-22.

Marvanová, L. (1975). Concerning *Gyoerffyella* Kol. *Trans. Brit. Mycol. Soc. 65*: 555-565.

Matsushima, T. (1971). *Microfungi of the Solomon Islands and Papua-New Guinea*. Published by the author, Kobe, Japan, 78 pp.

Nawawi, A., Descals, E., and Webster, J. (1977). *Leptosporomyces galzinii*, the basidial state of a branched conidium from water. *Trans. Brit. Mycol. Soc. 68*: 31-36.

Nemec, S. (1969). Sporulation and identification of fungi isolated from root-rot-diseased strawberry plants. *Phytopathology 59*: 1552-1553.

Newhall, A. G. (1946). More on the name *Ansatospora acerina*. *Phytopathology 36*: 893-895.

Nilsson, S. (1964). Freshwater hyphomycetes. Taxonomy, morphology and ecology. *Symbolae Bot. Upsal. 18*: 1-130.

Park, D. (1974a). *Tricladium terrestre* sp. nov. *Trans. Brit. Mycol. Soc. 63*: 179-181.

Park, D. (1974b). Aquatic hyphomycetes in non-aquatic habitats. *Trans. Brit. Mycol. Soc. 63*: 183-187.

Price, I. P., and Talbot, P. H. B. (1966). An aquatic hyphomycete in a lignicolous habitat. *Aust. J. Bot. 14*: 19-23.

Rostrup, O. (1894). Mykologiske Meddelser. IV. *Bot. Tidskr. 19*: 36-47.

Saccardo, P. A. (1880). Conspectus generum fungorum Italiae inferiorum. *Michelia 2*: 1-38.

Scourfield, D. J. (1940). The microscopic life of the "leafcarpet" of wood and forests. *Essex Natur. 26*: 231-246.

Singh, N., and Musa, T. M. (1977). Terrestrial occurrence and the effect of temperature on growth, sporulation and spore germination, of some tropical aquatic hyphomycetes. *Trans. Brit. Mycol. Soc. 68*: 103-106.

Taylor, G. S., and Parkinson, D. (1965). Studies in the root region. IV. Fungi associated with the roots of *Phaseolus vulgaris* L. *Plant Soil 22*: 1-20.

Tubaki, K. (1958). Studies on the Japanese hyphomycetes. V. Leaf and stem group with a discussion of hyphomycetes and their perfect stages. *J. Hattori Bot. Lab. 20*: 142-144.

Tubaki, K. (1960). On the Japanese hyphomycetes. Scum & foam group, referring to the preliminary survey of the snow group. *Nagaoa 7*: 15-29.

Visser, S., and Parkinson, D. (1975). Fungal succession on aspen and poplar leaf litter. *Can. J. Bot. 53*: 1640-1651.

von Höhnel, F. X. R. (1904). Mycologische Fragmente (Forsetzung). *Ann. Mycol. 2*: 38-60.

Waid, J. S. (1954). Occurrence of aquatic hyphomycetes upon the root surfaces of beech grown in woodland soils. *Trans. Brit. Mycol. Soc. 37*: 420-421.

Webster, J. (1954). The micro-fungi of *Dactylis glomerata* with particular reference to their occurrence and succession on decaying culms. Ph.D. thesis, London University. [Original reference not seen; cited in Park (1974b), above.]

Webster, J. (1961). The *Mollisia* perfect state of *Anguillospora crassa*. *Trans. Brit. Mycol. Soc. 44*: 559-564.

Webster, J. (1977). Seasonal observations on "aquatic" hyphomycetes on oak leaves on the ground. *Trans. Brit. Mycol. Soc. 68*: 108-111.

Chapter 38

ECOLOGICAL STUDIES ON AERO-AQUATIC HYPHOMYCETES

Peter Jack Fisher and John Webster
University of Exeter
Exeter, Devon, England

I. INTRODUCTION

When leaves fall into lakes, ponds, and ditches they are frequently colonized by a
group of fungi known as aero-aquatic hyphomycetes (van Beverwijk, 1953). The aero-
aquatic fungi were characterized by the development of a purely vegetative mycelium
under water, sporulation occurring only when the substrate on which the fungus was
growing was exposed to a moist atmosphere. The term has been criticized on account
of the fact that many terrestrial fungi imperfecti can form vegetative mycelium under
water and will sporulate on being brought to the surface. Park (1972) has outlined
a scheme for the ecological classification of heterotrophic microorganisms, based on
the extent to which they can maintain themselves in an aquatic environment. He clas-
sifies as indwelling an organism fully adapted to an aquatic life, i.e., able to
maintain its biomass at a more or less constant level from year to year utilizing
substrata and nutrients that become available there. Fisher (1977) has redefined
the term *aero-aquatic* to include Park's classification as follows: Aero-aquatic fungi are indwelling organisms characterized by the production of a purely vegetative
mycelium in substrata under water and by the formation of conidia with a special flotation device, formed only when the substrate on which the fungus is growing is exposed to a moist atmosphere (Figs. 1-6).

Comparatively few references have been made to this group. Linder's publications on helicosporous fungi imperfecti (1925, 1929, 1931) include two genera, *Helicoon* and *Helicodendron*, which can be described as aero-aquatic. Linder collected
his material chiefly from weathered bark and decaying wood at the borders of a variety of aquatic habitats. It is not clear in the majority of cases whether the material had originated from water.

More recently several species have been described (Glen-Bott, 1951, 1955; van
Beverwijk, 1951a,b, 1953, 1954). Van Beverwijk included in her investigations a number of genera that are not helicosporous: *Candelabrum, Clathrosphaerina, Papulospora,*
and *Spirosphaera.*

It has been assumed that aero-aquatic hyphomycetes may be adapted to life under
anaerobic or microaerobic conditions in leaf substrates at the bottom of static freshwater habitats (Glen-Bott, 1951; Tabak and Cooke, 1968).

Fig. 1

Fig. 2

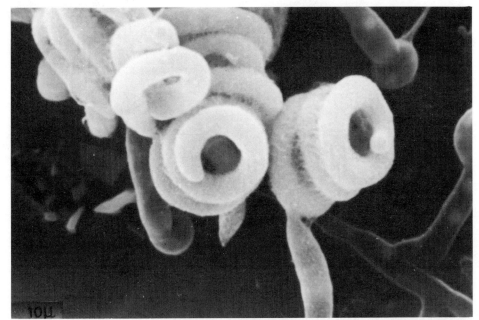

Fig. 3

Fig. 4

Flotation devices of the spores of aero-aquatic hyphomycetes:

Fig. 1 *Helicodendron tubulosum.*

Fig. 2 *Helicodendron giganteum.*

Fig. 3 *Helicodendron triglitziense.*

Fig. 4 *Helicoon plurispetatum.*

Air is trapped inside the hollow coils of the spores.

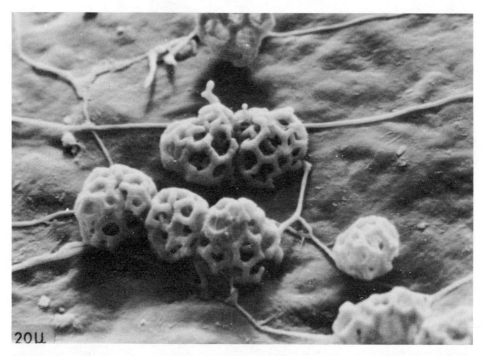

Fig. 5 Clathrosphaerina zalewskii. Air is trapped inside the hollow network of the spores.

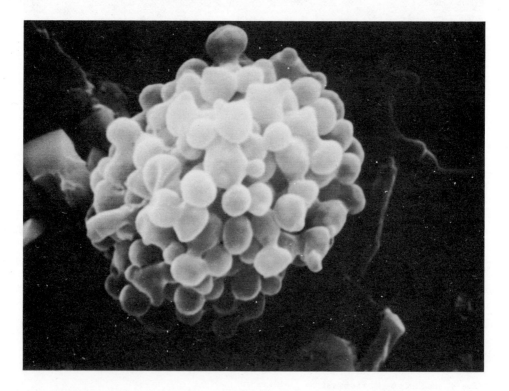

Fig. 6 Papulospora sp. Air is trapped between the projections of the spores.

The aim of these investigations was to provide information about the general
ecology of the aero-aquatic hyphomycetes in two habitats. Information was sought
regarding the general pattern of distribution in relation to the physical and chemical characteristics of the water. Various factors affecting colonization of leaves
were studied, such as colonization of leaves under anaerobic or microaerobic conditions in the field. Parallel aerobic laboratory experiments were designed to show
whether an ample supply of oxygen would affect colonization rate. Further, field
studies were undertaken to compare colonization of freshy abscissed and decayed beech
(*Fagus sylvatica*) leaves. Survival on land, possibly important with respect to spread
and distribution, was investigated in the field and by laboratory experiments.

Nearly all field work was carried out within the county of Devon, England.
A eutrophic and an oligotrophic habitat were investigated. The eutrophic site referred to as Stoke Woods (Map Ref. SX 915958) was a water-filled channel 600 m long
by 2-6 m wide bordered on one side by a steep wooded bank with *Fagus sylvatica* and
Acer pseudoplatanus and on the other by pastures which separated it from the River
Exe. The channel was 60-80 cm deep in the center. Leaf deposits in the bottom of
the channel varied from 2-15 cm in depth. Experiments were carried out about 50 cm
from the bank at a depth of 15-20 cm. Prevailing weather conditions had a marked
effect on the physical and chemical characteristics of the water. During periods of
low rainfall and high summer temperatures the oxygen dissolved in the water decreased
to zero and solute concentrations remained high. In contrast, periods of high rainfall led to a sharp rise in oxygen concentrations and a fall in solute concentrations.

The oligotrophic site referred to as Bystock (Map Ref. SY 034843) was a freshwater reservoir draining a heath on Bunter sandstone rock bordered by woodland with
Pinus sylvestris, *Pinus nigra*, *Betula pendula*, *Fagus sylvatica*, *Quercus robur*, and
Salix capraea. The physical and chemical characteristics of the water remained stable
throughout the year and were little affected by rainfall and temperature. Table 1
contrasts some of the characteristics of the waters at these two sites during periods
of drought and flooding. Leaves from deciduous trees shed into the water at Bystock
were rapidly skeletonized mainly due to the feeding activities of aquatic invertebrates, the so-called shredders (Cummins, 1973), so that about 6 months after leaf
fall, deposits of pine needles varying from 2 to 30 cm in depth formed the principal
substrate in the littoral zone around the lake shore with very little deciduous litter remaining. Fisher and colleagues (1977) demonstrated cellulolytic activity for
a number of aero-aquatic hyphomycetes. This gives the fungi a nutrient supply not
available to some aquatic arthropods because of their inability to utilize cellulose.
Kaushik and Hynes (1971) suggest that the occurrence of certain crustaceans and insects may be limited by the availability of decomposing leaves. Similarly, the absence of certain species of aero-aquatic hyphomycetes at Bystock may be due to a
shortage of deciduous leaf detritus. Hence in certain habitats aero-aquatic hypho-

Table 1 Water characteristics[a] Stoke Woods (SW) and Bystock (B) during periods of drought and flooding

Period	Temp. (°C)		pH		O$_2$ (% saturation)		BOD[b]		Nitrogen as N	
	SW	B	SW	B	SW	B	SW	B	SW	B
1976--Drought										
April	10	11	7.2	7.5	0.9	82	15.0	1.6	1.12	1.4
May	11	13	7.3	7.7	0.0	87	30.0	1.4	9.50	1.6
June	16	17	7.6	7.4	0.0	80	59.0	1.2	36.05	1.2
July	20	22	7.3	7.2	0.0	82	28.0	1.0	56.00	1.7
1977--Flooding										
March	8	7	7.3	7.2	84	99	1.3	1.1	2.38	1.03
April	8	7	7.3	7.2	83	97	1.6	1.2	2.02	1.03

[a]Milligrams per liter.

[b]BOD = biological oxygen demand.

mycetes may be in direct competition for available leaf detritus with aquatic arthropods. It would be of interest to know whether insect feces containing undigested cellulose are used as a nutrient source by these fungi.

In contrast, shredders were absent at the eutrophic site at Stoke Woods so that beech leaves remained largely intact for periods of up to 2 years.

II. COLONIZATION OF BEECH LEAF DISKS UNDER ANAEROBIC OR
MICROAEROBIC CONDITIONS IN THE FIELD COMPARED
WITH AEROBIC LABORATORY CONDITIONS

A. Introduction and Methods

Fisher (1977) used freshly abscissed unsterilized beech (*Fagus sylvatica*) leaf disks as baits to investigate colonization by aero-aquatic fungi under conditions of oxygen depletion in the eutrophic habitat at Stoke Woods during a period of drought and high summer temperatures. Several of the techniques employed in these experiments were adaptations of the Cambridge method (see Garrett, 1950; Butler, 1953) for investigating colonization of dead plant tissue by root infecting fungi.

The fungus *Helicodendron giganteum* was introduced into the habitat in the form of an inoculum/substrate dilution series. The inoculum consisted of a pure culture of the fungus grown in sterile beech leaf mash, subsequently mixed with quartz sand. For the preparation of the substrate, decaying leaves were removed from a selected quadrat 15-20 cm below the water surface of the Stoke Woods site. These leaves were broken up in a vegetable blender and mixed with quartz sand in the same proportion as the inoculum.

Table 2 Colonization of leaf disks by aero-aquatic fungi[a,b]

	Inoculum control 100:0			Composition of dilution series 90:10			50:50			10:90			Substrate control 0:100		
	L	F	FI	L	F	FI	L	F	FI	L	F	FI	L	F	FI
Helicodendron giganteum Glen-Bott[c]	100	0	54	60	0	32	38	0	25	54	0	10	0	0	0
Helicodendron tubulosum (Riess) Linder	0	0	0	8	0	0	40	0	0	29	0	0	37	0	0
Helicodendron conglomeratum Glen-Bott	0	0	0	4	0	0	13	0	0	17	0	0	31	0	0
Helicodendron fuscum (Berk. & Curt.) Linder	0	0	0	0	0	0	0	0	0	3	0	4	12	0	0
Helicodendron luteo-album Glen-Bott	0	0	0	0	0	0	22	0	0	15	0	0	28	0	0
Helicodendron triglitziense (Jaap) Linder	0	0	0	0	0	0	0	0	33	0	0	5	5	0	10
Helicoon ellipticum (Pk) Morgan	0	0	0	0	0	0	0	0	0	0	0	0	8	0	0
Helicodendron westerdijkiae Beverwijk	0	0	0	0	0	0	0	0	0	3	0	0	12	0	0
Aegerita sp. 1	0	0	0	0	0	0	22	0	0	26	0	0	46	0	0

[a]Number of disks colonized by indicated species is expressed as percentage of disks colonized in each petri dish.

[b]Key: L, aerated laboratory setup; F, anaerobic or microaerophilic field setup; FI, nonsporulating disks of field setup after 7-day incubation in aerated distilled water.

[c]Introduced species.

The experiments were set up in open 5-cm diameter petri dishes placed 10-15 cm below the water surface, to show whether an introduced species could colonize 60 unsterilized freshly abscissed beech leaf disks in each petri dish under prevailing anaerobic field conditions in competition with the naturally occurring flora of decaying leaves.

A similar experimental series was set up in the laboratory to test whether vigorous aeration would affect the rate of colonization of beech leaves by the introduced and indigenous species. Aeration was supplied by placing the experimental setups individually into 200-ml glass jars filled with water from the experimental site and then bubbling air through them. The oxygen concentration in the water above the dishes remained 9.1 ± 1 ppm (saturation) at 16°C. Sporulation on the leaf disks after moist chamber incubation was taken as the criterion for colonization in all experiments.

B. Results

The results of these experiments have been summarized in Table 2. Eight aero-aquatic fungi could be identified in the aerated laboratory setup (Table 2 substrate control, column L), whereas only four species were identified by incubating leaves taken directly from the habitat (*H. tubulosum*, *H. conglomeratum*, *H. triglitziense*, and *Papulospora* sp.).

The well-aerated substrate control thus provided an improved assay technique over previously adopted methods of incubating leaves in moist chambers. By this substrate control method, the eutrophic and oligotrophic habitats at Stoke Woods and Bystock were surveyed for indigenous species of aero-aquatic fungi in decaying leaf litter. The species in the accompanying tabulation were found to be indigenous in the two habitats.

Stoke Woods	Bystock
Helicodendron tubulosum (Riess) Linder	*Helicodendron fractum* Fisher
Helicodendron conglomeratum Glen-Bott	*Helicodendron giganteum* Glen-Bott
Helicodendron fuscum (Berk. & Curt.) Linder	*Helicodendron hyalinum* Linder
Helicodendron luteo-album Glen-Bott	*Helicoon pluriseptatum* Beverwijk
Helicodendron triglitziense (Jaap) Linder	*Clathrosphaerina zalewskii* Beverwijk
Helicodendron westerdijkiae Beverwijk	*Spirosphaera floriformis* Beverwijk
Helicoon ellipticum (Pk.) Morgan	*Papulospora* sp.[a]
Helicoon fuscosporum Linder	*Candelabrum spinulosum* Beverwijk
Papulospora sp.[a]	
Helicosporium sp.	

[a]*Papulospora* sp. was the only species common to both habitats.

The survey was conducted during a period of drought when surface water supplies to the two habitats had dried up, so that the organisms identified as indigenous may have been indwellers (Park, 1972). Later surveys conducted during periods of heavy rainfall when surface water flooded into the habitat, carrying with it deposits of leaf detritus from surrounding static water sources, showed an increase in the number of aero-aquatic species identified. These so-called immigrants (Park, 1972) included *Helicoon richonis* Boudier, *Helicodendron paradoxum* Peyronel, *Beverwykella pulmonaria* (Beverwijk) Tubaki and *Helicodendron* sp.n. Their presence may only be transient in these habitats, depending on a particular set of environmental conditions which favor their growth.

C. Conclusions

The results of these experiments indicate that many aero-aquatic fungi are adapted to exist in anaerobic or micro-aerobic situations (Table 1) but that colonization of leaf detritus is slow under these conditions. On the other hand, a well-oxygenated environment speeds the colonization of leaf detritus and enables the organisms to sporulate when the leaves in which they are growing are exposed to a moist atmosphere.

The inability of these fungi to sporulate after subjection to a period of oxygen depletion is highlighted in Table 2, column F. This shows that all the leaf disks of the field setup gave an initial zero sporulation after moist chamber incubation for 20 days. Only after the disks were aerated for 7 days in distilled water before incubation could it be demonstrated that *H. triglitziense*, *H. fuscum*, and the inoculum fungus *H. giganteum* had colonized the disks under anaerobic or micro-aerobic conditions (Table 2, column FI). None of the other naturally occurring species could be detected, and it is therefore possible that *H. triglitziense* and *H. fuscum* are more competitive under anaerobic or micro-aerobic conditions in the field than the other indigenous species. The results from the laboratory experiments indicate that under well aerated conditions neither may have this advantage (Table 2, substrate control, column L).

The artificial introduction of a species into a habitat poses a number of problems. In the experiment outlined above *H. giganteum* was introduced in a dilution series and was able to colonize unsterilized beech leaves in the field setup in the presence of the natural substrate (Table 2, column FI).

This result cannot, however, be interpreted to mean that *H. giganteum* is more competitive than those species not detected in the field setup. This is because the inoculum of the introduced species was grown under well-aerated conditions in a nutrient-rich substrate (sterile leaf mash) to produce a vigorously growing mycelium which readily sporulates when brought to the surface, whereas the fungi present in the natural substrate had survived for many weeks in an oxygen depleted environment and may therefore have had their competitive ability seriously impaired.

III. FIELD STUDIES WITH AERO-AQUATIC FUNGI COMPARING COLONIZATION
      OF FRESHLY ABSCISSED LEAVES WITH DECAYING LEAVES

A. Introduction

Glenn-Bott (1951) noted that aero-aquatic fungi are found most frequently on black
decaying leaves in standing water. During our own study similar observations were
made.

    The experiments were designed to test whether freshy abscissed or black decaying
beech leaves are more readily colonized by a selected number of species taken from
this group.

    Hogg and Hudson (1966) have studied the succession of microfungi on leaves of
beech from the time of unfolding up to 18 months after leaf fall. They found that
at leaf fall beech leaves are frequently colonized by *Discula quercina* (the conidial
state of *Gnomonia errabunda*), *Cladosporium herbarum* (the conidial state of *Mycosphaerella tassiana*), *Aureobasidium pullulans* (the conidial state of *Guignardia fagi*),
Alternaria tenuis, and *Botrytis cinerea*.

    During the present study beech leaves which had reached the abscission stage
were stripped from the trees and incubated in moist chambers to confirm the absence
of aero-aquatic species on such leaves.   Three experiments (referred to as a, b, and
c) were conducted simultaneously at the eutrophic site at Stoke Woods and the oligotrophic site at Bystock during March-April 1977.

    The following species were investigated:

Experiment a at Stoke Woods

   Helicodendron triglitziense (indigenous)

   Papulospora sp. (indigenous)

   Helicodendron giganteum (introduced)

   Clathrosphaerina zalewskii (introduced)

Experiment b at Bystock

   Helicodendron triglitziense (introduced)

   Papulospora sp. (indigenous)

   Helicodendron giganteum (indigenous)

   Clathrosphaerina zalewskii (indigenous)

Experiment c at both sites

   Helicodendron fractum (indigenous at Bystock; introduced at Stoke Woods)

B. Method

    1. Preparation of inocula

    The Cambridge method for investigating saprophytic colonization of dead plant
tissue (Garrett, 1950; Butler, 1953) has been criticized because the inocula were
prepared by growing the fungi to be investigated in maize meal instead of noting

their mode of survival in the soil and using inocula as similar as possible to those found in nature (Dhingra et al., 1976). This investigation required the use of inocula as similar as possible to those found in nature.

There is no evidence to suggest that aero-aquatic fungi colonize leaf detritus under water solely by means of their aerially produced floating spores. Neither is there evidence that they produce chlamydospores or other resting structures under water except for *Helicodendron westerdijkiae.*

Both laboratory and field observations suggest that underwater spread is often brought about by leaf to leaf contact and mycelial fragments. Pure mycelium mixed with quartz sand was therefore used as inoculum for all underwater experiments.

2. *Standard units of dead plant tissue*

Freshly abscissed beech leaves (*Fagus sylvatica*) will be referred to as "brown leaves" and the decaying beech leaves collected from the mud surface as "black leaves." The brown leaves were stripped from the trees the previous autumn and stored at -10°C. The black leaves were recovered from below the water surface at Stoke Woods prior to the start of the experiments. Black leaves used for the experiments at Stoke Woods and Bystock during March and April were sorted by holding them in front of a 60-W electric light bulb. Leaves from the previous autumn usually transmit some light and retain traces of autumn coloration if viewed by transmitted light, whereas older leaves are opaque. Only opaque leaves were chosen for the experiments. All beech leaves which had become slimy and showed signs of disintegration were excluded from the experiments. Black and brown unsterilized beech leaf disks, 3.6 mm in diameter, were used throughout the experiments for assay purposes. About 6000 disks each served as the stock from which all those for the experiments were taken. The stock was deemed to be necessary to ensure an even spread of leaf variation throughout the experiment.

For experiment d with *Helicodendron fractum* beech leaf disks of both kinds as well as pine needles (*Pinus sylvestris*) (6 mm lengths) were used for the assays. The pine needles were of two types. Freshly abscissed brown needles were stripped from the trees and will be called "fresh" needles. The second type consisted of identical needles placed into 1-mm mesh nylon bags which were leached at Stoke Woods for 18 months prior to the experiment. These needles will be referred to as "leached" needles.

Assays to establish the presence of aero-aquatic fungi in the black disks and the leached pine needles were conducted on 300 black disks randomly chosen from the stock of 6000 and on 50 leached pine needle segments chosen similarly from a stock of 1000.

3. *Preparation of the experimental setups*

Twenty cubic centimeters of the respective inocula were placed into separate petri dishes and the appropriate leaf baits were buried evenly in each dish. The dishes were then submerged some 15-20 cm below the water surface at the experimental field sites. After 4 weeks all leaf disks were harvested, washed well in distilled water, and placed on moist filter paper in closed petri dishes to be incubated for 12 days at 10° ± 2°C. Sporulation was used as the criterion for colonization of the disks and pine needles by the inoculum species.

C. Results

The results are expressed as number of leaf units colonized out of 60 (Table 3).

Table 3 Comparison between colonization of black and brown disks and fresh and leached pine[a,b]

Species	Brown leaves	Black leaves
Experiment a: Stoke Woods (eutrophic)		
Helicodendron triglitziense (indigenous)	35.3 ± 4.2****	7.1 ± 2.5
Helicodendron giganteum (introduced)	56.3 ± 3.8****	23.7 ± 3.1
Clathrosphaerina zalewskii (introduced)	47.3 ± 4.7****	2.8 ± 3.4
Papulospora sp. (indigenous)	41.7 ± 5.0****	8.9 ± 1.5
Experiment b: Bystock (oligotrophic)		
Helicodendron triglitziense (introduced)	60.0 ± 0 ***	30.4 ± 7.5
Helicodendron giganteum (indigenous)	59.7 ± 0.6 [NS]	60.0 ± 0
Clathrosphaerina zalewskii (indigenous)	60.0 ± 0 [NS]	60.0 ± 0
Papulospora sp. (indigenous)	25.3 ± 8.1*	7.5 ± 4.2

Experiment c: with *Helicodendron fractum* introduced at Stoke Woods, indigenous at Bystock

Species	Brown leaves	Black leaves	Fresh pine	Leached pine
Bystock				
Helicodendron fractum	9.7 ± 4.5	0 ± 0	0 ± 0	20.0 ± 3
Stoke Woods				

No colonization could be recorded on any of the four substrates.

[a]Mean deviations and significance levels for individual species. Key to significance levels: ****, $P = 0.001$; ***, $P = 0.01$; **, $P = 0.02$; *, $P = 0.05$; [NS], P = not significant.

[b]Of 300 black disks taken from the stock, 7% and 3% were colonized by *H. triglitziense* and *Papulospora* sp., respectively, before the start of the experiment. These have been deducted from the results. Sixty leaf baits were used in each petri dish. There were three replicates.

D. Conclusions

The results of experiment a at the eutrophic Stoke Woods site indicate that brown leaves are more readily colonized than black leaves by all the fungi. A similar experiment, b, at the oligotrophic site Bystock showed that two of the inoculum fungi colonized brown and black leaves equally effectively. Two principal reasons are suggested for these results. First, the inoculum species surrounded by quartz sand faces little competition from other aquatic microorganisms under the experimental conditions encountered in the field. Typical terrestrial species present in brown leaves such as *Cladosporium herbarum* probably become relatively inactive under water (P. J. Fisher and J. Webster, unpublished results). Black leaves, however, that have been in the water for 6 months or longer are colonized by numerous aquatic microorganisms which would provide active competition to the inoculum species, reducing colonization by the latter. Second, an improved oxygen supply in the littoral zone of the oligotrophic habitat may favor certain of the aero-aquatic fungi, allowing them to colonize black leaves equally effectively. The importance of an adequate supply of oxygen in the colonization of leaf litter by these fungi has already been stressed in Sec. 1.

The results of experiment c are of particular interest. It was possible to demonstrate that *Helicodendron fractum* colonized leached pine needles more effectively than it colonized beech leaves. This may explain why this fungus is abundant at Bystock, where pine needles form the greater part of the leaf litter. The absence of this species from Stoke Woods may be explained by its apparent inability to colonize black decaying leaves in competition with other microorganisms and the absence of suitable pine needle substrates there. This species has been recovered on numerous occasions during these studies from oligotrophic sites in Great Britain but appears to be absent from eutrophic habitats.

While an adequate supply of oxygen is probably the most important single factor which determines the frequency and distribution of these organisms in static water, the presence of growth inhibitors in polluted waters may also be responsible for their absence from leaf litter in certain habitats, but as yet we have no evidence of this.

IV. SURVIVAL OF AERO-AQUATIC HYPHOMYCETES ON LAND

A. Introduction

Existing records indicate that aero-aquatic hyphomycetes are adapted for life under water and in localities with high humidities (Linder, 1929; Glen-Bott, 1951, 1955; van Beverwijk, 1951a,b, 1953, 1954; Gönczöl, 1976). Our own records of recovery indicate that the preferred habitat of the group is static water. A number of species, notably *Helicoon pluriseptatum* and *Candelabrum spinulosum*, may be adapted for life in terrestrial and aquatic habitats. Both species have been recovered from moist leaf litter on land as well as from underwater sources. Experiments were designed to test survival of a selected number of species on garden soil and in a desiccator. Pure mycelia and spores were investigated for resistance to drying.

B. Method

1. *Survival on garden soil*

Eight species were cultured separately in sterile beech leaf mash. The colonized mash was spread into cages with wooden sides and nylon mesh (1 mm^2) as a base and top. These cages were placed outside onto garden soil. Unsterilized freshly abscissed beech leaves were loosely placed on top of each lot of mash.

At chosen intervals 3- to 5-g samples were removed from the cages and examined at 25X magnification for the presence of spores of the inoculum species. The samples were then used to test for the presence of living mycelium by incubating the leaf litter in sterile distilled water at 20°C for 7 days. Subsequent sporulation on the litter surface was noted according to the moist chamber incubation technique.

2. *Survival in a desiccator*

Petri dishes, 5 cm in diameter, were filled with air-dried colonized leaf mash and placed over blue, coarse self-indicating silica gel in a desiccator at 20°C. The relative humidity within the desiccator remained at 5% throughout the experiment.

3. *Survival of mycelium*

All species were grown in beech leaf decoction. After 4 weeks of incubation at 20°C the mycelia were harvested on a 0.5-mm plastic sieve and washed with sterile distilled water. Each mycelium was placed on dry filter paper in 9-cm-diameter petri dishes and air-dried at 20°C. At the time of harvest, small samples of the wet mycelia were incubated in moist chambers to confirm that each species was viable at the beginning of the experiment.

At chosen intervals, sections of the dried mycelia were cut out together with the filter paper on which they were resting, submerged under sterile distilled water in petri dishes, and incubated for 2 weeks at 20°C.

The water was then poured away, and the now soaked mycelia were incubated in moist chambers as previously described. Mycelia that showed active growth but no spores were transferred onto 0.1% malt extract agar for a further incubation period of 10 days.

In all cases sporulation was taken as the criterion for survival.

Mycelia were grown separately for determination of moisture content, then dried to constant weight in aluminum foil boats at 95°C.

4. *Survival of conidia after desiccation*

The fungi were grown in sterile beech leaf disks. After a period of 4 weeks of growth all disks were incubated in moist chambers for sporulation. Ten disks from each species bearing spores were now placed over blue self-indicating silica gel in a

Table 4 Survival in leaf litter in the field

| | 7-day drying period[a] (25-30°C) | Dates of sampling[b,c] | | | | | | | | |
| Species | | 1976 | | | 1977 | | | | | |
		8/8 M L	9/12 M L	10/10 M L	1/9 M L	2/13 M L	3/13 M L	4/10 M L	5/15 M L	6/12 M L
Helicodendron triglitziense	+	+ -	+ -	+ +	+ +	+ +	+ +	- -	- -	- -
Helicodendron tubulosum	+	+ -	+ -	+ +	+ +	+ +	+ +	+ +	+ +	+ +
Helicodendron conglomeratum	+	+ -	+ -	+ +	+ +	+ +	+ +	+ +	+ +	+ +
Helicodendron giganteum[d]	+	+ -	+ -	+ +	+ +	+ +	+ +	+ +	+ +	+ +
Helicodendron fuscum	+	+ -	+ -	+ -	+ -	- -	- -	- -	- -	- -
Papulospora sp.	+	+ -	+ -	+ +	+ +	- -	- -	- -	- -	- -
Helicodendron hyalinum	+	+ -	+ -	- -	- -	- -	- -	- -	- -	- -
Helicoon ellipticum	+	+ -	+ -	- -	- -	- -	- -	- -	- -	- -
Water content of leaf litter (% of dry weight)[e]	12.1	11.3	15.6	49.2	53.1	35.4	21.9	20.5	15.0	16.5

[a] July 1-8, 1976.

[b] From Aug. 8, 1976, through June 12, 1977.

[c] Key: M, colonized leaf mash; L, uncolonized beech leaf; + or -, survival or extinction in leaf mash; L+, colonization of leaf.

[d] *H. giganteum* was the only species that produced spores in the field.

[e] Mean of four determinations.

desiccator at 20°C. After 10 days of desiccation, approximately 50 spores from each species were streaked onto 0.1% malt extract agar and incubated at 20°C. Mycelial growth followed by sporulation was taken as the criterion for spore survival.

The moisture content for spores of *H. giganteum* was determined after the period of desiccation by drying the spores to constant weight at 95°C.

C. Results

The results are presented in Tables 4 through 7.

Table 5 Survival in leaf litter in a desiccator[a]

Species	7-day drying period (25–30°C)	Number of days desiccation (15–20°C)									
		31	66	94	122	157	185	220	248	276	311
Helicodendron triglitziense	+	+	+	+	+	+	+	+	+	+	−
Helicodendron tubulosum	+	+	+	+	−	−	−	−	−	−	−
Helicodendron conglomeratum	+	+	−	−	−	−	−	−	−	−	−
Helicodendron giganteum	+	+	−	−	−	−	−	−	−	−	−
Helicodendron fuscum	+	+	−	−	−	−	−	−	−	−	−
Papulospora sp.	+	+	−	−	−	−	−	−	−	−	−
Helicodendron hyalinum	+	−	−	−	−	−	−	−	−	−	−
Helicoon ellipticum	+	−	−	−	−	−	−	−	−	−	−
Water content of leaf litter (% of dry weight)[b]	11.0	3.5	3.2	3.2	3.2	3.4	3.1	2.9	3.2	3.0	3.1

[a] Key: +, survival; −, extinction.
[b] Mean of four determinations.

Table 6 Survival of dried mycelia under laboratory conditions[a]

Species	Number of days dried (15-20°C)						Water content of mycelia (% of dry weight)
	0	35	62	94	124	155	
Helicosporium sp.	+	+	+	-	-	-	--
Helicodendron sp.n. 1	+	+	+	-	-	-	--
Papulospora sp.	+	+	+	+	+	-	5.8
Helicodendron tubulosum	+	+	+	+	+	+	6.4
Helicodendron westerdijkiae	+	+	+	+	+	+M	7.0
Helicodendron conglomeratum	+	+M	+M	+M	+M	+M	6.9
Helicodendron triglitziense	+	-	-	-	-	-	--
Helicodendron giganteum	+	-	-	-	-	-	--
Helicoon fuscosporum	+	-	-	-	-	-	6.5
Helicodendron hyalinum	+	-	-	-	-	-	--
Helicodendron luteo-album	+	+	-	-	-	-	--
Clathrosphaerina zalewskii	+	+	-	-	-	-	5.1

[a]Key: M, sporulation after transfer to 0.1% malt agar extract; +, survival; -, extinction.

Table 7 Survival of spores in a desiccator[a]

Species	Period of desiccation (days) at 15-20°C		Water content of spores (% of dry weight)[b]	
	10	20	Before desiccation	After desiccation
Helicodendron giganteum	+	-	125.4	6.7
Helicodendron westerdijkiae	+	-	--	--
Helicodendron conglomeratum	+	-	--	--
Helicoon fuscosporum	+	-	--	--
Clathosphaerina zalewskii	-	-	--	--
Helicodendron tubulosum	-	-	--	--
Papulospora sp.	+	-	--	--
Helicodendron hyalinum	+	-	--	--

[a]Key: +, survival; -, extinction.

[b]Mean of three determinations. Moisture content was determined after 10 days.

D. Conclusions

Survival of aero-aquatic hyphomycetes on land is of ecological interest in relation
to their distribution and spread in water.

All species grown in leaf litter survived an initial drying period of 7 days at
prevailing laboratory temperatures (Tables 4 and 5). *Helicodendron tubulosum, H.
conglomeratum,* and *H. giganteum* have survived in leaf litter after 11 months of contact with garden soil (Table 4). Only *H. giganteum* produced spores on the litter
surface during this period, but none of these germinated when transferred onto 0.1%
malt extract agar, indicating that they were probably dead since spores of this species readily germinate on this medium.

Detection of these fungi by an examination of the litter surface was therefore
not possible. In all cases incubation of the litter under water followed by moist
chamber incubation was necessary before viable spores were produced; this, we suggest, provides evidence that water is the preferred habitat of this group.

All species except *H. ellipticum, H. hyalinum,* and *H. fuscum* spread to some of
the non-sterile beech leaves placed in contact with the leaf litter.

An inspection of the litter at 25X magnification showed it to be well colonized
with *Cladosporium herbarum, Botrytis cinerea,* and numerous unidentified discomycetes.
Mites and other soil animals able to pass through the nylon mesh were numerous. The
inoculum species therefore survived in competition with other organisms.

The early part of the field experiment fell into the hot and dry summer of 1976.
Later heavy rainfall alternating with drier spells subjected the fungi to changeable
conditions.

The result of the experiment in the desiccator (Table 5) suggests that some species can withstand prolonged periods of drought. Particularly impressive in this
respect is the resistance to desiccation in leaf litter of *H. triglitziense* (Table
5).

The survival of pure mycelia under dry conditions is of interest because it offers an opportunity to observe possible resting structures and the origin of new
growth when the mycelium is returned to water (Table 6). Of the species studied,
only *H. westerdijkiae* forms chlamydospores both within and on the surface of beech
leaves. None of these germinated during the various incubation procedures, nor were
they present in pure mycelium. Resting structures were also absent in the other mycelia studied except for the presence of thick-walled hyphae. These were prominent
in *H. giganteum, H. tubulosum, H. conglomeratum, H. hyalinum,* and *H. westerdijkiae.*
Such hyphae were notably absent in *H. triglitziense,* which failed to survive as pure
mycelium. Its remarkable resistance to desiccation in leaf litter can at present not
be explained and requires further investigation.

Spore survival on land has only briefly been investigated here. Conditions of
severe desiccation allowed the spores of six species to survive for 10 days.

The following additional information is of interest. Leaves bearing spores of
H. conglomeratum, H. tubulosum, H. giganteum, H. hyalinum, and *Clathrosphaerina zalewskii* were air-dried at 20°C and kept for 1 year in a refrigerator at 5°C. None
of the spores germinated on 0.1% malt extract agar after this period.

The spores formed by *H. giganteum* sometime during the field experiment were,
as previously mentioned, unable to germinate.

This evidence suggests that spore survival on land may be very limited and that
the floating propagules of many of the aero-aquatic fungi serve to spread the species
from one habitat to another via rivers and streams.

The survival of many of these fungi in leaf litter on land and their considerable
resistance to drying would enable them to survive during periods of drought and assist
their spread via windblown leaf fragments.

V. DISCUSSION

Throughout this study aero-aquatic hyphomycetes have been found abundantly colonizing
leaf detritus sampled from Bystock Reservoir and Stoke Woods. It seems, therefore,
that this group of fungi may have an ecological role as agents of leaf litter decomposition in static freshwater habitats ranging from well oxygenated to microaerobic.
It is, however, important to note that neither of the habitats studied so far have
remained permanently anaerobic or microaerobic.

The water at Stoke Woods showed long periods of oxygen depletion only during
periods of low rainfall and high temperatures, and it remains to be seen whether members of the group can survive in permanently anaerobic or microaerobic habitats. The
results of experiments described in Secs. II and III, however, argue against such a
generalized view. It could be shown that colonization of freshly abscissed leaf
disks by indigenous species at Stoke Woods was extremely slow under conditions of
oxygen depletion. Further, the results of experiments described in Sec. III showed
conclusively that freshly abscissed leaves are preferentially colonized to black decaying leaves. It can, therefore, be assumed that colonization of natural leaf substrates at Stoke Woods during summer periods of deoxygenation, when only black decaying leaves remain from the previous autumn, must be even slower than that demonstrated
for the freshly abscissed leaf disks during the experiments described in Sec. II.
Thus, it appears likely that the aero-aquatic fungi differ in their physiological
adaptation to life under water in two principal ways from terrestrial species present
on the leaves at leaf fall: (1) they can probably survive prolonged periods at low
or zero oxygen levels as live mycelia in leaves in a dormant condition without incur-
ring damage, whereas typical terrestrial species probably cannot (P. J. Fisher and
J. Webster, unpublished work); (2) their rate of spread through leaf detritus during
periods of oxygenation may well be considerably faster than that of terrestrial species. Methods developed during these studies can be adapted to compare the role of
terrestrial leaf litter colonizers with that of aero-aquatic fungi in situations

under water. The dependence of aero-aquatic hyphomycetes on oxygen for rapid spread through leaf detritus may confine their role to a relatively shallow area of the litter layer in the littoral zone of static habitats.

This view can be supported if the physical and gaseous conditions are considered which leaves and twigs will encounter at the mud-water interface. The topmost few millmeters of mud have a high diffusion rate due to the stirring action of the overlying turbulent water. Below that level, however, the diffusion rate is virtually molecular, within the actual sediment (Willoughby, 1974).

Terrestrial survival may be the explanation of the widespread distribution of aero-aquatic fungi in many isolated static waters. Land may represent a marginal habitat for the majority of the group, but it is not thought that any of the species investigated for land survival form a regular component of the leaf litter flora on land.

With the exception of *Clathrosphaerina zalewskii*, which has the perfect state *Hyaloscypha zalewskii* (Descals and Webster, 1976), none of the fungi included in these studies is known to have a perfect state. Should such perfect states exist, spore dispersal from them may contribute to their spread on land and water.

There remains the question as to why the aero-aquatic flora in the Bystock Reservoir differs from that at the Stoke Woods site, with only one fungus, *Papulospora* sp., common to both. The following reasons are suggested. All the fungi found at Bystock can colonize *Pinus sylvestris* and *Pinus nigra*, with *Helicodendron fractum* and *Helicodendron hyalinum* abundant on these substrates. Yet pine needles submerged for 18 months in nylon litter bags with 1-mm mesh remained uncolonized at Stoke Woods, whereas freshly abscissed beech leaves submerged in similar bags at Stoke Woods became colonized by *H. triglitziense*, *H. conglomeratum*, and *H. tubulosum* within 6-8 months after submersion. This suggests that pine needles may be an unsuitable substrate for certain fungi occurring at Stoke Woods. However, the results of experiment b described in Sec. III indicate that *H. triglitziense*, indigenous at Stoke Woods, can also colonize leaf detritus at Bystock providing suitable substrates are available there. Its apparent absence from Bystock Reservoir may be due to a scarcity of suitable deciduous leaves, which are rapidly eaten by animals, coupled with an inability to compete with other microorganisms on pine needle substrates. This may also be true of other species from the Stoke Woods site, such as *H. tubulosum* and *H. conglomeratum*, which have not been found in the Bystock Reservoir.

In this connection it is of considerable interest that a collection of conifer leaf litter made at the Environmental Sciences Centre, University of Calgary, Kananaskis, Seebe, Alberta, Canada, during September, 1977 yielded *Helicodendron triglitziense* on species of *Pinus*. The needles were only sparsely colonized. The total absence of *Helicodendron fractum* and *H. hyalinum*, so common in pine leaf litter at Bystock Reservoir, is noteworthy. It proves that certain species of *Pinus* can be colonized in the field by *H. triglitziense*, and it strongly suggests that its absence in pine litter in Bystock Reservoir may be due to interspecific competition with *Helicoden-*

dron hyalinum and *Helicodendron fractum*, which grew abundantly in this substrate. This idea, which would be worth testing, receives further support by the fact that sterile freshly abscissed needles of *P. sylvestris* and *P. nigra* can be successfully colonized by *H. triglitziense* in the laboratory.

Experiment c, Sec. III, showed that *Helicodendron* sp.n. 1 was unable to colonize beech or pine leaves at Stoke Woods but did so at Bystock. Possibly this fungus requires an environment free of pollutants not provided by the Stoke Woods site. In this connection it is of interest that in Scotland this fungus was recovered on a number of occasions from pine needles taken from freshwater lochs but never from stagnant pools.

Two species, *H. giganteum* and *C. zalewskii*, introduced to the Stoke Woods site from Bystock Reservoir colonized both brown freshly abscissed and black decaying beech leaves under the prevailing experimental field conditions (Sec. III, experiment a). Their absence from the indigenous aero-aquatic flora at Stoke Woods may be explained by the finely balanced nature of most ecological situations. This means that certain of these fungi may well be able to colonize leaf detritus at Stoke Woods for brief periods when conditions favor their growth. Yet, in the longer term, unfavorable periods for their growth may be more prevalent there, preventing species such as *H. giganteum* and *C. zalewskii* from effectively competing with the indigenous microflora.

The absence of species from a particular habitat may be explained by several other factors--including those involving chance. Thus, even a suitable habitat may remain uncolonized by certain species if their inocula have failed to arrive there. In this respect the age of a habitat and the methods of species dispersal are of importance. Dispersal by airborne spores is likely to be fast and efficient. This method, however, is probably not available to the aero-aquatic fungi investigated, since their spores often remain in clumps and are difficult to separate. Most of their spores are also relatively large when compared with typically airborne spores such as those of *Cladosporium herbarum*. The methods of spread available to the species investigated include distribution by windborne leaf fragments and spore and mycelium dispersal by flowing water. Neither method is likely to be as efficient as dispersal by airborne spores. This may only occur where perfect states exist.

Thus, the time required for a newly formed water body such as a pond to become colonized will depend on the presence of suitable substrates and on climatic factors such as rainfall and wind, as well as on the origins and quantity of the water entering via streams and surface runoff. It is not thought likely that the distinct aero-aquatic floras of the Stoke Woods site and the Bystock Reservoir are due to a lack of interchange of inocula. The sites are only 18 miles (ca. 29 km) apart, and both appear to have been established more than 50 years ago. Therefore, other factors previously discussed are likely to be responsible for the maintenance of different species in each habitat.

REFERENCES

Butler, F. C. (1953). Saprophytic behaviour of some cereal root rot fungi. I. Saprophytic colonization of wheat straw. *Ann. Appl. Biol. 40*: 284-297.

Cummins, K. W. (1973). Trophic relations of aquatic insects. *Ann. Rev. Entomol. 18*: 183-206.

Descals, E. C., and Webster, J. (1976). *Hyaloscypha*: Perfect state of *Clathrosphaerina zalewskii*. *Trans. Brit. Mycol. Soc. 67*: 525-528.

Dhingra, O. D., Tenne, F. D., and Sinclair, J. B. (1976). Method for the determination of competitive saprophytic colonization of soil fungi. *Trans. Brit. Mycol. Soc. 66*: 447-456.

Fisher. P. J. (1977). New methods of detecting and studying the saprophytic behaviour of aero-aquatic hyphomycetes from stagnant water. *Trans. Brit. Mycol. Soc. 68*: 407-411.

Fisher, P. J., Sharma, P. D., and Webster, J. (1977). Cellulolytic ability of aero-aquatic hyphomycetes. *Trans. Brit. Mycol. Soc. 69*: 495-520.

Garrett, S. D. (1950). Ecology of root-inhabiting fungi. *Biol. Rev. Cambridge Phil. Soc. 25*: 220-254.

Glen-Bott, J. I. (1951). *Helicodendron giganteum* n.sp. and other aerial sporing hyphomycetes of submerged dead leaves. *Trans. Brit. Mycol. Soc. 34*: 17-30.

Glen-Bott, J. I. (1955). On *Helicodendron tubulosum* and some similar species. *Trans. Brit. Mycol. Soc. 38*: 17-30.

Gönczöl, J. (1976). Ecological observations on the aquatic hyphomycetes of Hungary. II. *Acta Bot. Hung. 22*: 51-60.

Hogg, B. M., and Hudson, H. J. (1966). Micro-fungi on the leaves of *Fagus sylvatica*. *Trans. Brit. Mycol. Soc. 49*: 185-192.

Kaushik, N. K., and Hynes, H. B. N. (1971). The fate of the dead leaves that fall into streams. *Arch. Hydrobiol. 68*: 465-515.

Linder, D. H. (1925). Observations on the life history of *Helicodesmus*. *Amer. J. Bot. 12*: 259-269.

Linder, D. H. (1929). A monograph of the helicosporous fungi imperfecti. *Ann. Mo. Bot. Garden 16*: 227-388.

Linder, D. H. (1931). Brief notes on the Helicosporae with descriptions of four new species. *Ann. Mo. Bot. Garden 18*: 9-16.

Park, D. (1972). On the ecology of heterotrophic micro-organisms in fresh-water. *Trans. Brit. Mycol. Soc. 68*: 407-411.

Tabak, H. H., and Cooke, W. B. (1968). The effects of gaseous environments on the growth and metabolism of fungi. *Bot. Rev. 34*: 126-251.

van Beverwijk, A. L. (1951a). Zalewski's *Clathrosphaerina spirifera*. *Trans. Brit. Mycol. Soc. 34*: 280-290.

van Beverwijk, A. L. (1951b). *Candelabrum spinulosum* a new fungus species. *Antonie van Leeuwenhoek J. Microbiol. Serol. 17*: 278-284.

van Beverwijk, A. L. (1953). Helicosporous hyphomycetes. I. *Trans. Brit. Mycol. Soc. 36*: 111-124.

van Beverwijk, A. L. (1954). Three new fungi: *Helicoon pluriseptatum* n.sp., *Papulospora pulmonaria* n.sp. and *Tricellula inaequalis* n.gen. n.sp. *Antonie van Leeuwenhoek J. Microbiol. Serol. 20*: 1-16.

Willoughby, L. G. (1974). Decomposition of litter in fresh water. In *Biology of Plant Litter Decomposition*, C. H. Dickinson and C. J. F. Pugh (Eds.), Vol. 2. Academic Press, New York, pp. 659-681.

Chapter 39

OBSERVATIONS ON THE ECOLOGY OF LIGNICOLOUS AQUATIC HYPHOMYCETES

E. B. Gareth Jones

Portsmouth Polytechnic
Portsmouth, England

I. INTRODUCTION

Decaying twigs, branches, and logs are frequently observed in streams and rivers
(Fig. 1), yet little attention has been devoted to the organisms involved in the decay of these substrates in freshwater habitats. Mycologists interested in freshwater
hyphomycetes have concentrated their efforts on the fungi colonizing leaves in freshwater habitats (Ingold, 1959, 1965). In this section I shall present a review of the
fungi known to occur on timber in freshwater habitats and their role in the decay of
wood.

A. Lignicolous Freshwater Fungi

Species of Phycomycetes and Basidiomycetes have been observed on wood in freshwater
habitats (Shaw, 1972; Leightley and Eaton, 1977; E. B. G. Jones, unpublished data),
but the most common freshwater fungi belong to the Ascomycetes and Fungi Imperfecti
(Jones, 1972; Eaton, 1976). Eaton and Jones (1971) recorded 42 species from *Fagus*
sylvatica and *Pinus sylvestris* test blocks submerged in freshwater at the Chester
pumping station on the River Dee in Cheshire, England. Only a few were frequently
present, namely, *Monodictys putredinis*, *Trematosphaeria pertusa*, *Clasterosporium*
caricinum, *Helicoon sessile*, *Tricladium splendens*, *Ceratosphaeria lampadophora*, and
Sterigmatobotrys macrocarp. Eaton (1972) lists 71 microfungi from timber in water
cooling towers (most of the towers with fresh water circulating) and the most frequently encountered species, in order of frequency, were *Chaetomium globosum*, *Savory-*
ella lignicola, *Trematosphaeria pertusa*, *Ceratosphaeria lampadophora*, ascomycete I
(subsequently identified as *Ophioceras dolichostomum*), and *Melogramma* sp. (Ascomycetes);
Monodictys putredinis, *Asteromella* sp., *Graphium* sp., *Doratomyces microsporus*, *Pyreno-*
chaeta sp., *Fusarium* sp., *Septonema* sp., *Cephalosporium* sp., and *Torula herbarum*.

Aquatic hyphomycetes, described so frequently from decaying leaves (Ingold, 19-
59), are also found on wood in freshwater habitats (Perrott, 1960; Webster, 1961;
Sládečková, 1963; Jones and Oliver, 1964; Irvine, 1974; Tubaki, 1966; Price and Talbot, 1966; Kirk, 1969; Archer and Willoughby, 1969; Fisher et al., 1976). Willoughby
and Archer (1973) reported 130 fungi from timber in freshwater habitats, the most fre-

Fig. 1 Branches, twigs, and leaves undergoing decomposition in the River Meon. Foam rich in the conidia of aquatic hyphomycetes.

quent being *Fusarium* spp., *Heliscus lugdunensis*, *Anguillospora longissima*, and *Clavariopsis aquatica*. Twenty-three species of aquatic hyphomycetes were recorded on twigs, while 41 were present in the foam samples (Willoughby and Archer, 1973).

Kane (1980) reported 50 fungi from beech and pine test panels submerged in the River Severn at Tewkesbury, Gloucestershire. Aquatic hyphomycetes were present in significant numbers, and a total of 11 were observed. On *Fagus sylvatica* the most common species were *Camposporium pellucidum* (95% frequency of occurrence) with *Anguillospora crassa* (25%), *A. longissima*, *Dactylella aquatica*, *Helicomyces scandens*, and *Heliscus lugdunensis* (all 15%). Kane (1980) reported fewer fungi present on *Pinus sylvestris* test panels. The most common species reported was *Cremasteria* sp. (subsequently identified as *Trichocladium lignincola*) (95% frequency of occurrence) with the freshwater hyphomycetes *Heliscus lugdunensis* (35%), *Alatospora acuminata* (20%), and *Camposporium pellucidum* and *Campylospora chaetocladia* (both 15%). Kane (1980) demonstrated that it is important to examine incubated wood at regular intervals for up to 3 months. *Alatospora acuminata* and *Heliscus lugdunensis* were observed on the wood on removal from the river and were present for up to 1 week on incubation but were not observed on the panel at 1 and 3 months. Other fungi were not found sporulating on the wood initially but subsequently appeared, e.g., *Camposporium pellucidum*, *Peyronelina glomerulata*, and *Melanospora* sp. appeared 48 hr, 1 week, and 3 months, respectively, on incubation.

B. Decay of Wood in Freshwater Habitats

Considerable information is available on the decay of wood in marine habitats, but little published information is available for timber in freshwater habitats.

Eaton and Jones (1971) have shown that beech and Scots pine test blocks placed in Ince water cooling tower (freshwater circulating) had lost 18 and 20% of their weight, respectively, at 18 weeks whereas at 108 weeks weight losses of 56.2 and 56.6%, respectively, were obtained. Irvine and associates (1972) investigated the weight loss of beech and Scots pine test blocks at nine test sites. Considerable variation was noted (Table 1) with low weight losses at Elland water cooling tower (21.6 and 3.1%). In all cases beech was decayed at a more rapid rate than Scots pine. Irvine (1974) in a study on the breakdown of timber in cooling towers and rivers obtained low weight losses at Elland (34.8 and 21.6%), with high weight losses for beech at Maerdy Mill on the River Ceirw (60.7%) (Table 2).

When timber submerged in freshwater is examined typical soft rot decay attack is observed (Eaton, 1976), although bacterial erosion cavities have also been recorded (Irvine et al., 1972; Furtado, 1978; Holt and Jones, 1978).

Eaton and Jones (1971) and Eaton (1976) have shown that a number of the fungi present on timbers in rivers and water cooling towers have the ability to degrade wood under laboratory conditions. The highest weight loss reported by Eaton and Jones (1971) was for *Chaetomium globosum* on beech test blocks (33.5%). The decay pattern observed, in all the fungi tested, was of the soft rot type.

Table 1 Percentage weight loss of beech and Scots pine test blocks[a] after 40 weeks of exposure in cooling towers and in the sea

Test site	Weight loss, beech (%)	Weight loss, Scots pine (%)
Elland (fresh water)	21.6	3.1
Agecroft (fresh water)	27.1	23.8
High Marnham (fresh water)	31.8	26.4
Bold (domestic water and sewage effluent)	25.4	14.2
Chadderton (fresh water and sewage water)	27.5	15.1
Stella North (brackish water)	28.9	30.8
Fleetwood (seawater)	22.8	19.0
Burnham-on-Crouch (seawater)	27.6	19.8
Langstone (seawater)	19.7	13.2

[a]Dimensions: 5 X 2.5 X 0.5 cm.

Source: After Irvine et al. (1972).

Table 2 Weight loss of beech (*Fagus sylvatica*) and Scots pine (*Pinus sylvestris*) sapwood test blocks submerged in different water types

| Site | Weight loss at 98 weeks (%) | | Water type | Temp. (°C) | No. of fungi | Habitat | PO_4^{3-} (ppm) | NO_3^- (ppm) | NO_2^- (ppm) |
	Beech	Scots pine							
Connahs Quay	48.3 ± 6.2	53.2 ± 5.2	Fresh water, sometimes brackish	25-33	29	Cooling tower	0.26	1.8	0.1
Agecroft	45.9 ± 5.6	49.2 ± 3.4	Fresh water	22-30	27	Cooling tower	0.72	4.4	0.1
Ince	40.9 ± 7.8	41.4 ± 11.1	Fresh water	17-26	11	Cooling tower	0.32	0.7	0.1
Bold	38.6 ± 2.3	35.9 ± 9.2	Sewage effluent	21-32	26	Cooling tower	38.30	48.4	0.5
Maerdy Mill	60.7 ± 3.5	29.6 ± 5.4	Fresh water	0-20	21	River Ceirw	0.85	0.8	0.1
Elland	34.8 ± 4.3	21.6 ± 2.0	Fresh water	20-30	20	Cooling tower	1.60	1.2	0.1
Budds Farm	12.5 ± 1.4	3.2 ± 0	Sewage effluent	--	21	Sewage farm	6.4	14.6	0.1

Source: After Irvine (1974).

Leightley and Eaton (1977) have examined in greater detail the ability of *Heliscus lugdunensis* to degrade wood and have demonstrated the production of cellulase, xylanase, and mannase under laboratory conditions. Although *H. lugdunensis* was able to form typical soft rot cavities in beech, none was observed in Scots pine. This fungus combined cavity (Type I) and erosion (Type II) decay patterns in its mode of attack, the latter being the most apparent.

Leightley and Eaton (1977) also examined the ability of an unnamed basidiomycete, isolated from test blocks placed in the River Severn at Tewkesbury, to decay wood under laboratory conditions. This fungus caused wall lysis in the form of erosion troughs with connecting strands present between the hypha and the edge of the trough (white rot type of attack).

It has been demonstrated that timber submerged in freshwater habitat soon becomes degraded (soft rot attack), and a range of fungi are implicated (Jones, 1974, 1975; Leightley and Eaton, 1977) in this decay process. Aquatic hyphomycetes can colonize such wood (Kane, 1980), and in the following sections we set out to investigate further their ability to degrade wood under laboratory conditions.

II. METHODS AND MATERIALS

The fungi employed are listed in Table 3 and were obtained from the Culture Collection of the Department of Biological sciences, University of Exeter.

A. Weight Losses on Cellulose

The source of cellulose was Solka floc. Each experimental flask contained 50 ml Petri medium [$(NH_4)_2 SO_4$, 0.5 g; KH_2PO_4, 1 g; KCl, 0.5 g; $MgSO_4 \cdot 7H_2O$, 0.2 g; $CaCl_2$, 0.1 g; 1 liter distilled water], to which was added 1 g Solka floc and the flasks then autoclaved at 15 psi for 15 min. Each flask was inoculated with a 0.5 mm agar disk of an actively growing culture of the fungus to be tested. Flasks were incubated at 20°C on a rotary shaker and harvested at the times indicated. Fungal infested Solka floc

Table 3 Percentage weight loss of cellulose (solka floc) in Petri medium after 3 weeks and at 20°C

Species	Weight loss (%)
Anguillospora crassa	18.0
Anguillospora longissima	13.2
Articulospora tetracladia	25.0
Heliscus lugdunensis	22.0
Lemonniera aquatica	7.5
Tricladium splendens	13.9
Varicosporium elodeae	14.6

medium from each flask was filtered through individual weighed sintered glass crucibles containing glass wool. The crucibles were washed through with distilled water,
dried for 18 hr at 104°C, cooled, and weighed. Weight loss was expressed as a percentage without a correction for the mycelium. Three flasks per fungus were analyzed
for weight loss of cellulose.

B. Weight Losses on Wood

The decay capacity of the isolates was determined as the weight loss of small sapwood
blocks (10 X 10 X 10 mm or 10 X 6 X 2 mm) of Scots pine (*Pinus sylvestris*), beech
(*Fagus sylvatica*), birch (*Betula pubescens*), and balsa (*Ochroma lagopus*) by using a
modification of the technique described by Kaune (1968). Petri medium (80 ml) was
added to 15 g vermiculite in 250-ml Erlenmeyer flasks and autoclaved at 15 psi for
15 min. Six blocks were present in the flasks, and eack block was inoculated with
an 0.5-mm disk from an actively growing culture and incubated for 13 weeks at 20°C
(Fig. 2). After the test period, one block was sectioned for soft rot decay and the
remainder were dried down and weighed after superficial mycelium had been removed
from the wood surface.

Fig. 2 Scots pine test blocks (10 X 10 X 10 mm) inoculated with *Lunulospora curvula*
on vermiculite medium in an Erlenmeyer flask.

III. RESULTS

A. Weight Losses on Cellulose

All the fungi tested caused weight loss of the cellulose at 3 weeks (Table 3), with the highest weight loss recorded for *Articulospora tetracladia* (25.0%). With yeast extract medium replacing the Petri medium, lower weight losses were obtained, e.g., *A. crassa,* 8.0%; *H. lugdunensis,* 10.5%; and *T. splendens,* 7.0%.

B. Weight Losses on Wood

The highest weight loss was obtained with *Anguillospora longissima* (24.0% on balsa), whereas no significant weight losses occurred with *H. lugdunensis* and *A. tetracladia* (Table 4). If a fungus caused a weight loss, then all the timbers tested were de-

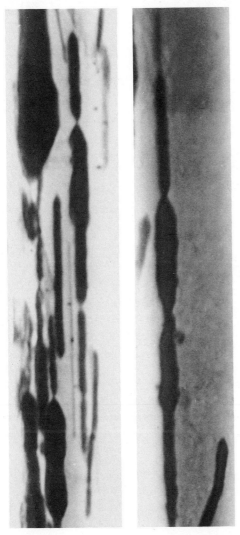
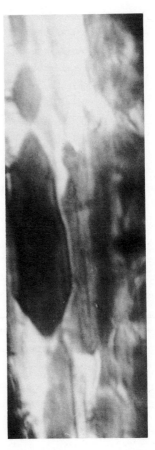

Figs. 3 and 4 Longitudinal sections of balsa wood, viewed under polarized light, showing soft rot cavities caused by the aquatic hyphomycete *Lunulospora curvula.*

Fig. 5 Soft rot cavities formed by *Lunulospora curvula* in birch wood. Longitudinal section viewed under polarized light.

Table 4 Percentage weight loss of wood exposed on vermiculite culture medium for
13 weeks at 20°C[a]

Species	Birch	Beech	Balsa	Soft rot	Pine[b]	Poplar[b]
Anguillospora crassa	7.3	7.6	16.6	+	NT	NT
Anguillospora longissima	11.2	12.5	24.0	+	NT	NT
Articulospora tetracladia	2.4	0.8	--	-	NT	NT
Heliscus lugdunensis	0.2	0.7	1.3	-	NT	NT
Tricladium splendens	7.6	8.9	6.2	+	NT	NT
Lunulospora curvula[c]	10.8	10.8	23.4	+	9.7	8.3

[a]Block size: 1 X 0.6 X 0.2 cm.

[b]NT = not tested.

[c]Grown on wood in shake culture with petri medium for 15 weeks at 20°C.

graded. Four of the fungi tested caused soft rot attack (Figs. 3-5), but no soft rot
cavities were observed in wood colonized by *H. lugdunensis* and *A. tetracladia*.

IV. DISCUSSION

From Table 2 it can be seen that timber in freshwater habitats is decayed, the rate
of decay varying from site to site. Wood decomposition is typically slower than that
of other plant materials, as illustrated in Table 5. To decay wood to 50% of its
original weight takes 56 weeks, whereas oak leaves take half that time when submerged
in the River Thames but take 70 weeks when in ground contact. The nutrient content
of woody materials is generally lower than that of other plant tissues (Swift, 1977),
and this may account for the slower rate of decay of wood. Sharp and Millbank (1973)
and Sharp and Levy (1974) have shown that bacteria can fix nitrogen in wood blocks in
ground contact and suggest that they are a prerequisite to colonization by cellulolytic fungi.

The freshwater hyphomycetes have previously been regarded as noncellulolytic
sugar fungi by Hudson (1972) and Willoughby (1974). The work presented here shows
that all the aquatic hyphomycetes tested were able to utilize cellulose as Solka floc
and cause appreciable weight losses. Fisher and coworkers (1977) have also shown
that a number of aero-aquatic hyphomycetes have the ability to utilize cellulose (as
filter paper), with *Helicoon fuscosporum* and *Aegerita* sp. giving the highest weight
losses (19.9 and 19.3%, respectively).

Only four of the aquatic hyphomycetes tested (*Anguillospora crassa, A. longissima,
Lunulospora curvula, T. splendens*) produced significant weight loss of wood under
laboratory conditions. Soft rot cavities were also formed by these fungi on all the
timbers tested.

The weight losses recorded here are higher than those reported by Leightley and
Eaton (1977) for a number of aquatic fungi but lower than those reported by Eaton
and Jones (1971) for some cooling tower fungi, e.g., *Chaetomium globosum* (35.5% on
beech), *Asteromella* sp. (24.5% on beech), and *Graphium* sp. (20.0% on beech).

Table 5 The rate of decomposition of materials in various habitats

Substrate	Species	Time (weeks)	Weight loss (%)	Habitat	Reference[a]
Elm leaves	*Ulmus* sp.	6	24.0	Stream	Hudson (1972)
Phragmites litter	*Phragmites communis*	5	53.0	Laboratory test	Mason (1976)
Birch leaves	*Betula verrucosa*	16	50.0	Mull soil	Mason (1976)
Oak leaves	*Quercus robur*	30	50.0	River Thames	Mason (1976)
Spartina leaves	*Spartina alterniflora*	31	50.0	Salt marsh	Mason (1976)
Beech sapwood	*Fagus sylvatica*	56	50.9	Cooling tower	Irvine (1974)
Pine sapwood	*Pinus sylvestris*	56	48.8	Cooling tower	Irvine (1974)
Beech sapwood	*Fagus sylvatica*	56	40.1	River Ceirw	Irvine (1974)
Pine sapwood	*Pinus sylvestris*	56	14.6	River Ceirw	Irvine (1974)
Oak leaves	*Quercus petraea*	70	50.0	Mull soil	Mason (1976)
Pine leaves	*Pinus sylvestris*	52	50.0	Mor soil	Mason (1976)
Posts		[6[b]]	Advanced decay	Ground contact	Käärik (1974)
Conifer needles	*Abies amabilis,* *Tsuga mertensiana*	52	80–90	Mountain lake	Rau (1978)
Leaves	*Alnus rubra*	26.5 & 19.5	90.0	Two streams	Sedell et al. (1975)
	Acer circinatum	48.5 & 16.5	90.0	Two streams	Sedell et al. (1975)
	Acer macrophyllum	137.0 & 30.5	90.0	Two streams	Sedell et al. (1975)
	Pseudotsuga menziesii, *Tsuga heterophylla*	131.0 & 25.0	90.0	Two streams	Sedell et al. (1975)

[a]For a complete review of literature see Singh and Gupta (1977).

[b]Years.

The decomposition of materials in freshwater is brought about by a range of organisms: bacteria, fungi, nematodes, and oligochaete worms. Efford (1969) showed that in Marion Lake, British Columbia, Canada, some 86% of the energy budget is derived from forest debris, while Kaushik and Hynes (1968, 1971) concluded that fungi are far more successful than bacteria in the colonization and degradation of leaves, at least in the early stages, and bring about substantial protein increment. The observations presented here confirm the view that aquatic hyphomycetes have the ability to utilize cellulose and may therefore decay a wide range of materials in freshwater streams and rivers, thus contributing significantly to the recycling of materials in these habitats.

ACKNOWLEDGMENTS

I am grateful to Professor J. Webster and colleagues for supplying the cultures of the aquatic hyphomycetes, to Mr. K. Purdy for photographic assistance, and to Mr. A. A. Dizer for completing the experimental work.

REFERENCES

Archer, J. F., and Willoughby, L. G. (1969). Wood as the growth substratum for a freshwater foam spore. *Trans. Brit. Mycol. Soc. 53*: 484-486.

Eaton, R. A. (1972). Fungi growing on wood in water cooling towers. *Int. Biodet. Bull. 8*: 39-48.

Eaton, R. A. (1976). Cooling tower fungi. In *Recent Advances in Aquatic Mycology*, E. B. G. Jones (Ed.). Elek, London, pp. 359-387.

Eaton, R. A., and Jones, E. B. G. (1971). The biodeterioration of timber in water cooling towers. I. Fungal ecology and the decay of wood at Connah's Quay and Ince. *Material und Organismen 6*: 51-80.

Efford, I. E. (1969). Energy transfer in Marion Lake, British Columbia, with particular reference to fish feeding. *Verhandl. Int. Verein. Limnol. 17*: 104-108.

Fisher, P. J., Kane, D., and Webster, J. (1976). *Peyronelina glomerulata* from submerged substrate in Britain. *Trans. Brit. Mycol. Soc. 67*: 351-354.

Fisher, P. J., Sharma, P. D., and Webster, J. (1977). Cellulolytic ability of aero-aquatic hyphomycetes. *Trans. Brit. Mycol. Soc. 69*: 495-520.

Furtado, S. E. J. (1978). The interactions of organisms in the decay of timber in aquatic habitats. Ph.D. thesis, C.N.A.A., Portsmouth Polytechnic, Portsmouth, England.

Holt, D., and Jones, E. B. G. (1978). Bacterial cavity formation in delignified wood. *Material und Organismen 13*: 15-30.

Hudson, H. J. (1972). *Fungal Saprophytism*, Studies in Biology No. 32. Arnold, London.

Ingold, C. T. (1959). Submerged aquatic hyphomycetes. I. *Quekett. Microsc. Club., Ser. 4*: 115-130.

Ingold, C. T. (1965). Hyphomycete spores from mountain torrents. *Trans. Brit. Mycol. Soc. 48*: 453-458.

Irvine, J. (1974). An investigation of the factors affecting the biodeterioration of treated timber in aquatic habitats. Ph.D. thesis, C.N.A.A., Portsmouth Polytechnic, Portsmouth, England.

Irvine, J., Eaton, R. A., and Jones, E. B. G. (1972). The effect of water of different ionic composition on the leaching of a waterborne preservative from timber placed in cooling towers and in the sea. *Material und Organismen 7*: 45-71.

Jones, E. B. G. (1972). The decay of timber in aquatic environments. *British Wood Preserving Association, Annual Convention* pp. 1-18.

Jones, E. B. G. (1974). Aquatic fungi: Freshwater and marine. In *Biology of Plant Litter Decomposition*, C. H. Dickinson and G. J. F. Pugh (Eds.). Academic Press, New York, pp. 337-383.

Jones, E. B. G. (1975). Observations on lignicolous aquatic fungi. In *Proc. 1st Int. Congr. IAMS*, T. Hasegawa (Ed.), Vol. 2, pp. 374-382.

Jones, E. B. G., and Oliver, A. C. (1964). Occurrence of aquatic hyphomycetes on wood submerged in fresh and brackish water. *Trans. Brit. Mycol. Soc. 47*: 45-48.

Käärik, A. A. (1974). Decomposition of wood. In *Biology of Plant Litter Decomposition*, C. H. Dickinson and G. J. F. Pugh (Eds.). Academic Press, New York, pp. 129-174.

Kane, D. (1980). The effect of sewage effluent on the growth of micro-organisms in the marine environment. Ph.D. thesis, C.N.A.A., Portsmouth Polytechnic, Portsmouth, England.

Kaune, P. (1968). Vermiculite as a carrier for culture media. Organization for Economic Cooperation and Development, Working Doc. No. 27, DAS/CSI/M/185, 6 pp.

Kaushik, N. K., and Hynes, H. B. N. (1968). Experimental study on the role of autumn-shed leaves in aquatic environments. *J. Ecol. 56*: 229-243.

Kaushik, N. K., and Hynes, H. B. N. (1971). The fate of the dead leaves that fall into streams. *Arch. Hydrobiol. 68*: 465-515.

Kirk, P. W. (1969). Aquatic hyphomycetes on wood in an estuary. *Mycologia 61*: 177-181.

Leightley, L. E., and Eaton, R. A. (1977). Mechanisms of decay of timber by aquatic micro-organisms. *British Wood Preserving Association, Annual Convention* pp. 1-26.

Mason, C. F. (1976). *Decomposition*, Studies in Biology No. 74. Arnold, London, 58 pp.

Perrott, E. (1960). *Ankistrocladium fuscum* gen nov., sp. nov., an aquatic hyphomycete. *Trans. Brit. Mycol. Soc. 43*: 556-558.

Price, I. P., and Talbot, P. H. B. (1966). An aquatic hyphomycete in a lignicolous habitat. *Aust. J. Bot. 14*: 19-23.

Rau, G. H. (1978). Conifer needle processing in a subalpine lake. *Limnol. Oceanogr. 23*: 356-358.

Sedell, J. R., Triska, F. J., and Triska, N. S. (1975). The processing of conifer and hardwood leaves in two coniferous forest streams. I. Weight loss and associated invertebrates. *Verh. Int. Verein. Limnol. 19*: 1617-1627.

Sharp, R. F., and Levy, J. F. (1974). Colonization and decay of limewood. Part 1. Untreated wood. *Material und Organismen 9*: 51-72.

Sharp, R. F., and Millbank, J. W. (1973). Nitrogen fixation in deteriorating wood. *Experientia 29*: 895-896.

Shaw, D. E. (1972). *Ingoldiella hamata* gen. et sp. nov., a fungus with clamp connections from a stream in North Queensland. *Trans. Brit. Mycol. Soc. 59*: 255-259.

Singh, J. S., and S. R. Gupta (1977). Plant decomposition and soil respiration in terrestrial ecosystems. *Bot. Rev. 43*: 449-527.

Sládečková, A. (1963). Aquatic deuteromycetes as indicators of starch campaign pollution. *Int. Rev. Ges. Hydrobiol. 48*: 47-64.

Swift, M. J. (1977). The ecology of wood decomposition. *Sci. Progr.* (Oxford) *64*: 175-199.

Tubaki, K. (1966). An undescribed species of *Hymenscyphus*, a perfect stage of *Varicosporium*. *Trans. Brit. Mycol. Soc. 44*: 559-564.

Webster, J. (1961). The *Mollisia* perfect state of *Anguillospora crassa*. *Trans. Brit. Mycol. Soc. 44*: 559-564.

Willoughby, L. G. (1974). Decomposition of litter in freshwater. In *Biology of Plant Litter Decomposition*, C. H. Dickinson and G. J. F. Pugh (Eds.). Academic Press, New York, pp. 659-681.

Willoughby, L. G., and Archer, J. F. (1973). The fungal spora of a freshwater stream and its colonization pattern on wood. *Freshwater Biol. 3*: 219-239.

Chapter 40

ROLE OF AQUATIC HYPHOMYCETES IN THE TROPHIC STRUCTURE OF STREAMS

Felix Bärlocher

Universität Basel
Basel, Switzerland

Bryce Kendrick

University of Waterloo
Waterloo, Ontario, Canada

I. INTRODUCTORY REMARKS

Plants or plant parts that are eaten while still alive are said to enter the *grazing food chain*; if they die before being eaten they fuel the *detritus food chain* (Odum, 1971). The proportion of primary production eaten alive is usually small in terrestrial habitats (e.g., 1-20% in forests). It is often higher in aquatic environments, but even there detritus (dead organic material, including any attached microorganisms) supplies a substantial portion of the food web with energy. Valleys often support dense forests whose shade effectively curtails primary productivity in the streams that flow through them. However, the riparian vegetation sheds leaves, branches, and twigs which are trapped by the "sticky" water surface and thus become available to stream organisms (Hynes, 1963, 1970). The contribution made by this allochthonous detritus to the total energy budget of stream communities has been estimated to vary between 50 and 99% (Cummins et al., 1966; Fisher and Likens, 1973; Nelson and Scott, 1962; Teal, 1957).

The direct nutritional value of higher plant remains is usually poor. To begin with, much of the energy they contain is locked up in recalcitrant substances like cellulose and lignin, which together account for between 35 and 95% of leaves and wood (Daubenmire and Prusso, 1963; Hering, 1967; Käärik, 1974; Saitô, 1957). These substances are further protected by tannins, polyphenols, or even outright poisons (Levin, 1976; Swain, 1977) that actively depress the palatability and digestibility of the substrate. Undoubtedly, this simply means that the plant's chemical defenses against herbivores and pathogens persist beyond its death.

Very few animals indisputably produce enzymes active on native cellulose [the silverfish *Ctenolepisma*, the edible snail *Helix pomatia*, the marine isopod *Limnoria* (Dickinson and Pugh, 1974; Nielsen, 1962; Whitaker, 1971)], and none is known to decompose lignin. By contrast, a large number of bacteria and fungi can use cellulose, although many of them are less effective when substantial amounts of lignin are also present. Heavily lignified cellulose, such as occurs in wood, is usually attacked most successfully by fungi, particularly by basidiomycetes (Käärik, 1974). Therefore, if animals and microorganisms were to attack plant remains independently, one would expect the microorganisms as a group to get much the larger share of the available energy. This is probably true in any case (for reviews see Dickinson and Pugh, 1974; Anderson and Macfadyen, 1976). But the natural decomposition of plant materials actually involves a complex series of interactions among many groups of organisms. Some of these interactions clearly benefit animals by giving them access to a larger proportion of the total plant material than their own enzymatic equipment would permit. An obvious way for the detritus feeder to accomplish this would be to enrich its diet with microorganisms that have used cellulose and lignin as food sources. An elegant solution has been evolved by some fungus-growing termites: they ingest plant material and presumably assimilate its less resistant fractions. The fecal leftovers are then deposited in a moist chamber where a fungus, tended by the termites, breaks them down further and uses them to increase its own biomass, which is subsequently harvested and eaten by the termites (Batra and Batra, 1967). Attine ants have gone a step further and no longer ingest the plant material they collect; instead, they use it as a substrate for a fungus which is their only source of food (Batra and Batra, 1967).

It was suggested at least as early as 1891 (H. Simroth; quoted by Baier, 1935) that a similar mechanism operates in the diet of less conspicuously specialized detritus-feeders in terrestrial environments; its significance for food webs in pond sediments was recognized by Baier (1935). Both authors assumed that detritus feeders primarily or exclusively assimilate microbial cells regardless of whether these are ingested with or without their substrate. This hypothesis has recently received new support from students of many detritus-dominated aquatic habitats, ranging from mangrove swamps (Heald and Odum, 1970), salt marshes (de la Cruz, 1975; de la Cruz and Gabriel, 1974; Fenchel, 1970, 1972; Harrison and Mann, 1975; Odum and de la Cruz, 1967), seaweed beds (Mann, 1972), reed swamps (Mason and Bryant, 1975), beaver ponds (Hodkinson, 1975a,b,c), caves (Dickson, 1975), temporary pools (Bärlocher et al., 1978b), to freshwater streams (Bärlocher and Kendrick, 1973a,b, 1974, 1976; Cummins, 1973, 1974; Cummins et al., 1973; Iversen, 1973; Kaushik and Hynes, 1971; Kostalos and Seymour, 1976; Mackay and Kalff, 1973; Madsen, 1974; Petersen and Cummins, 1974; Sedell et al., 1975; Suberkropp and Klug, 1976; Suberkropp et al., 1976; Triska, 1970; Triska et al., 1975).

It can often be observed that decomposing detritus initially undergoes increases in nitrogen or protein levels (de la Cruz, 1975; de la Cruz and Gabriel, 1974; Heald and Odum, 1970; Harrison and Mann, 1975; Hodkinson, 1975b; Mason and Bryant, 1975; Odum and de la Cruz, 1967), or in palatability to detritus feeders (Bärlocher et al., 1978a; Kostalos and Seymour, 1976; Mackay and Kalff, 1973; Madsen, 1974; Triska, 1970), or both (Bärlocher and Kendrick, 1974; Bärlocher et al., 1978b; Iversen, 1973; Kaushik and Hynes, 1971). These observations have been interpreted as meaning that microbial cells, whose growth may cause a protein increase, constitute the actual food of the animals. Cummins (1973) compared microbially conditioned detritus to peanut butter crackers in which the protein-rich peanut butter (microbial cells) is embedded in a nutritionally poor cracker (plant remains). Obviously, only very small animals could ingest microbial cells exclusively and exclude the substrate; larger invertebrates ("shredders") are thought to digest the bacteria and fungi selectively while the remainder passes through their system virtually unchanged. It has indeed been observed that bacteria disappear during the passage of detritus through the intestines of invertebrates, while the substrate shows no visible changes (Fenchel, 1970, 1972). The fact that most detritus feeders pass food very rapidly through their guts (Baker and Bradnam, 1976; Berrie, 1976; Hynes, 1970) also makes extensive exploitation of recalcitrant substances unlikely.

Nevertheless, the problem of whether microbial colonization is in fact *necessary* for the growth and survival of detritus-feeding invertebrates was generally ignored for many years.

II. ECOLOGICAL STUDIES OF AQUATIC HYPHOMYCETES
     IN STREAM SYSTEMS

To return to stream systems: Kaushik and Hynes (1971) investigated the decay of autumn-shed leaves falling into streams and observed the rise in protein levels and palatability just described. The fresh leaves most preferred by several detritus feeders were also those which showed the highest protein increase and whose rate of weight loss was highest when invertebrates were excluded. By using bactericides and fungicides in laboratory experiments, Kaushik and Hynes showed that the fungi are mainly responsible for these phenomena, at least in the initial stages of decay. Triska (1970) reached similar conclusions by measuring respiration rates of leaf disks recovered from a stream and incubated with or without bactericides. Mackay and Kalff (1973) found that leaves inoculated with fungi were more palatable to caddis fly larvae (Trichoptera) than those inoculated with bacteria.

In these and many other cases fungi seem to be more important than bacteria in the early colonization of leaves. Their presence is therefore crucial to the unlocking of a major energy source in streams. Which are the dominant fungal forms? An obvious first step is to distinguish between the terrestrial fungi colonizing the phyllosphere of senescent leaves (*Aureobasidium*, *Cladosporium*, *Alternaria*, etc.), and the group of the aquatic hyphomycetes [or freshwater hyphomycetes (Nilsson, 1964) or amphibious hyphomycetes (Michaelides and Kendrick, 1978)] discovered by Ingold on dead leaves decaying in streams (see Ingold, 1942, 1975). But attempts at careful evaluation of the relative impact of the two groups are rare. From the outset, Triska (1970) restricted his study of the fungal flora on leaves to aquatic hyphomycetes, on the grounds that preliminary studies (which he did not specify) had shown their predominance. The same is true for Padgett's study (1976) of leaf decomposition in a tropical rainforest stream. The results of both investigators are based on the production of conidiophores on leaf disks kept under water, a procedure clearly favoring aquatic hyphomycetes. By contrast, Kaushik and Hynes (1971) plated leaf particles on a rich nutrient medium incubated at room temperature, a method just as

blatantly biased toward the fast-growing terrestrial genera they recovered. We repeated their study, but used both nutrient-rich (malt extract agar) and nutrient-
poor (water agar, leaf agar) media, and incubated the plates either at room or at
stream temperature (Bärlocher and Kendrick, 1974). Aquatic hyphomycetes proved to
be consistently more common than terrestrial fungi on nutrient-poor agar incubated
at the lower stream temperature (under these conditions aquatic forms were four to
nine times more numerous than terrestrial species). In decomposition experiments
with sterile leaves at low temperatures, significant weight losses could usually be
achieved only when inoculum of an aquatic hyphomycete was used. When the leaves were
sterilized before being introduced into the stream, terrestrial fungi could not be
found in significant numbers until 4 to 5 months later, even on malt agar plates incubated at room temperature; nevertheless, such leaves lost weight and gained protein
at the same rate as control leaves that had not been sterilized, and both groups had
the same community of aquatic hyphomycetes. These results suggest that terrestrial
fungi which are on the leaves before these fall into the stream have little influence
on the subsequent direction or rate of leaf decay, at least during the colder seasons. These conclusions are corroborated by Suberkropp and Klug (1976). In their
thorough study, direct examination and environment-simulating incubations showed
that aquatic hyphomycetes were the dominant members of the mycoflora, whereas particle plating on nutrient medium at room temperature revealed the presence of terrestrial fungi. Willoughby and Archer (1973) showed that aquatic hyphomycetes are also
the most common fungi colonizing presterilized twigs.

    We have obtained additional evidence for the ecological importance of aquatic
hyphomycetes from a recent study of conifer needles decaying in a stream (Bärlocher
et al., 1978a). Needles are known to make poor food for invertebrates. Of four
streams within a square mile, that running through a white pine plantation had the
lowest standing crop of individuals and biomass of invertebrates (Woodall and Wallace,
1972). When areas originally covered with deciduous forests or heath were reaffor-
ested with spruce, there was a drastic reduction of the aquatic fauna in nearby
streams (Huet, 1951). Sedell et al. (1975) found that needle packs of *Pseudotsuga
menziesii* and *Tsuga heterophylla* had to be exposed to the stream environment for 140
days before any appreciable invasion by invertebrates took place. In addition Ingold
(1966, 1975), who examined a large variety of substrates for the growth of these fungi, never observed them on conifer needles. A simple hypothesis linking these four
observations is as follows: Aquatic hyphomycetes are the fungal group best adapted
to the stream environment. If conifer needles, just like deciduous leaves, have to
be colonized by fungi before they are accepted by invertebrates, their apparent resistance to attack by aquatic fungi will automatically protect them against animal
consumers as well. Conversely, if it is possible to remove or destroy the factor
inhibiting fungal growth, the needles should also become palatable to detritus feeders in a much shorter period of time. We found that the inhibition could be removed
to some extent by treating needles with hot water. Hot alcohol was much more effec-

tive, but hot NaOH solution produced the most spectacular results of all. After 1 month of stream exposure, control needles had an average of less than 0.5 conidiophores of aquatic hyphomycetes per centimeter. Needles treated with NaOH carried almost 100 conidiophores per centimeter. As we expected, feeding started much earlier on the needles with heavy fungal colonization. After only 1 month, NaOH- or alcohol-treated needles showed unmistakable traces of animal feeding. After 4 months, about 80% of the dry weight of NaOH-treated needles had disappeared. By contrast, the control needles had lost only about 30% of their dry weight in the same period, and they did not show any traces of animal feeding until they had been in the stream for 5 months. During the last 4 months, the concentration of conidiophores on control needles had risen slowly from 0.5 to about 15 per centimeter. The stream in which we conducted this experiment flows through a mixed forest, so that detritus feeders presumably had plenty of alternative, more palatable food and could afford to ignore the unpalatable needles until they became fully "conditioned."

When the amphipod *Gammarus pseudolimnaeus* was given a choice between control needles and similar needles freshly treated with hot NaOH (needles never exposed to the stream and therefore carrying no aquatic hyphomycetes), the animals preferred the NaOH-treated needles over control needles at a ratio of about 2:1. But after the two needle types had been exposed to the stream for a month, NaOH-treated needles were preferred over control needles at a ratio of about 20:1. This is the same phenomenon we have mentioned before: those substrates which are preferred by *Gammarus* also allow best growth by fungi, and their conditioning proceeds fastest.

As a first approximation, we can assume that the food of leaf-eating stream invertebrates consists of two main components--the actual leaf material and fungal mycelium. In a series of experiments we compared their relative merits in the nutrition of *Gammarus pseudolimnaeus* (Bärlocher and Kendrick, 1973a,b, 1975b). When feeding on elm or maple leaves with negligible microbial populations, *Gammarus* assimilated 10-20% of dry weight, protein, or caloric content (assimilation = amount ingested minus amount defecated). This agrees well with data from other aquatic invertebrates, whose assimilation of detritus has been estimated to range between 7 and 35% (for review, see Berrie, 1976). By contrast, when *Gammarus* was feeding on fungal mycelium of one of 10 fungi, it could assimilate between 42.6 and 75.6% of the dry weight and 73.3 to 96.4% of the fungal protein. Hargrave (1970) estimated that *Hyalella azteca* (Amphipoda) assimilated 81-97% of its food when feeding on bacteria.

We also measured body weight increases and daily consumption for different sets of *Gammarus* which had as their sole food supply either maple or elm leaves or mycelium of one of 10 fungi. While the actual amounts consumed in all-leaf diets were roughly 10 times greater than in all-fungus diets, the highest weight increases were still found in those animals feeding on four of the fungi. Two other fungi not only failed to support growth, but appeared lethal, and the remaining four fell between these extremes. Cummins et al. (1973) could not observe any growth of *Stenonema* sp. (Ephemeroptera) on mycelium of *Lunulospora curvula*, whereas Willoughby and Sutcliffe

(1976) observed growth of *Gammarus pulex* on one of two fungi examined. An interesting practical application of such studies was tried by Newell and Fell (1975). Their
approach was to use agricultural waste products (e.g., straw) as a substrate for
fungi and then to feed the harvested mycelium to peneid shrimps. Growth of these
commercially important animals was described as ranging "from very poor to encourag-
ingly high." The variability of these results is undoubtedly due in large measure
to the natural variability of fungi. In an investigation of the suitability of many
fungi as food for larvae of Cecidomyidae (gall midges), Nikolei (1961) showed that
such seemingly small changes in the cultivation of the fungus as a temperature increase of 4°C, an increased nutrient concentration in the medium, or the use of different strains of the same species could decide whether a fungus was nutritionally
acceptable or toxic.

Results are more consistent when sterile leaves are compared with microbially
conditioned leaves. Survival of *Gammarus minus* was highest on fungus-enriched leaves
and lowest on sterile leaves (Kostalos and Seymour, 1976). Nilsson (1974) found
that *Gammarus pulex* assimilated leaves microbially conditioned for 10 days in artificial stream water better than untreated leaves; also, chironomid larvae continued
growth and hatched as adult winged insects when kept on a diet of decomposing cellulose (Park, 1974).

As one might expect, invertebrates have generally been shown to select the food
which they use most efficiently. For example, in short-term experiments *Gammarus
pseudolimnaeus* prefers fungal mycelium to unconditioned leaves (Bärlocher and Kendrick,
1973b) and *Gammarus minus* prefers fungus-enriched leaves to sterile leaves (Kostalos
and Seymour, 1976). From these observations it appears that although these *Gammarus*
species can survive and grow on freshly shed and perhaps even on sterile leaves, their
food will be much improved, both in palatability and nutritional value, by prior microbial conditioning. The frequently observed preference for microbially conditioned
leaves among other invertebrate taxa suggests that this is generally true. We know
of only one study in which microbially conditioned leaves appeared to be less palatable than nonconditioned leaves. Otto (1974) found that green leaves of *Fagus* were
preferred over leaves which had been in a stream for about 7 months. But beech leaves
are known to become increasingly unpalatable during their development (Feeny, 1970),
so dead leaves would be both less acceptable and less nutritious than living ones; in
addition, Otto may have collected the stream-exposed leaves after their period of
maximum palatability was over (cf. Iversen, 1973).

The aforementioned food selection experiments were done in the laboratory, where
the invertebrates had ready access to good as well as poor food. This need not be
the case in streams, where animals may be forced to search for conditioned food. The
question arises as to whether the profit gained from higher food quality exceeds the
extra cost involved in hunting for it (this includes the loss incurred by not using
available, less palatable food). The available evidence indicates that the search
generally pays off. The number of shredders colonizing leaf packets were distinctly

higher after a rise in microbial respiration, suggesting discrimination against unconditioned leaves (Anderson and Grafius, 1975; Sedell et al., 1975). A rise in invertebrate numbers with time was also found by Hart and Howmiller (1975). This lag
period is generally longer and more distinct in nutritionally poor substrates (Petersen and Cummins, 1974). Selective feeding is also the basis for the staggered appearance of telltale signs of invertebrate feeding when the natural decay of several substrates is compared (Bärlocher et al., 1978a; Kaushik and Hynes, 1971). But food
selection may take place on a much smaller scale: we found evidence (Bärlocher and
Kendrick, 1974) that leaf areas with highest fungal biomass disappear fastest, again
indicating preferential feeding by invertebrates. This possibility has also been
suggested by Padgett (1976).

But some invertebrates may more profitably accept less-than-optimally conditioned
food. This may be the case with relatively slow-moving shredders (such as some caddis
fly larvae), whose chances of finding fully conditioned food are low when they are
sharing their habitat with highly mobile competitors and who therefore cannot afford
to turn down food that is only partly conditioned. *Gammarus* undoubtedly belongs to
the more mobile class of shredders. Similar conditions may prevail for the whole
community if the food supply is dominated by a strongly seasonal, nutritionally poor
type of leaf whose conditioning requires several months. If no alternative food is
available (e.g., algae), and unless they can fine-tune their life cycle to the point
where their demand for food is highest a few months after leaf-fall, shredders may
simply have to eat inadequately conditioned food. A possible consequence is that
conditioning is of less consequence in those habitats where its effect would be most
beneficial. In streams with a great variety of leaves, shredders can better afford
to ignore poor substrates until they are considerably improved by microbial colonization, thus reaping the full benefits of conditioning. If this is the case, lag
periods in the invasion of nutritionally poor leaf packets should be most distinct
in streams running through mixed forests, or where leaf fall is spaced out evenly
throughout the whole year. Some relevant observations are reported by Haeckel et al.
(1973). In stream regions with alder, poplar, and beech trees, autumn-shed beech
leaves were still largely untouched the following March, while most alder leaves were
completely skeletonized. Poplar leaves were moderately decayed. But in a stream
reach dominated by beech trees, the few poplar leaves were heavily damaged and the
breakdown of beech leaves had progressed much farther.

How much of the food intake of detritus feeders in their natural habitat consists of microbial cells? This is a difficult question to answer, since the leaves
we find in streams are often those which have been refused by the animals, presumably
in favor of better-conditioned food. Microbial populations on such leaves may therefore be lower than those on the actual food. In addition, the biomass of microbial
populations is notoriously difficult to estimate. A traditional approach is to measure mycelial length per unit of bleached leaves and, with the help of empirical ratios,
convert it into fungal biomass per unit weight of decaying substrate. Visser and

Parkinson (1975) estimated that the highest densities of hyphae present in aspen
poplar leaves decaying in a terrestrial habitat correspond to about 8% of the total
dry weight. When the same ratios are applied to our data on maple leaves in a stream
(Bärlocher and Kendrick, 1974), the highest average density would correspond to about
2%, but the maximum density found in small heavily colonized areas would be almost
twice as high. Suberkropp et al. (1976) measured ATP concentrations in decaying hickory and oak leaves. ATP occurs only in living cells, and the ratio of ATP to total
cell carbon is thought to be fairly constant (Ausmus, 1973; note, however, Greaves
et al., 1973). When Suberkropp's values are converted into microbial biomass (using
the average of bacterial and fungal values as given by Ausmus), the living microbial
cells would generally vary between 5.9 and 11.8% of the dry weight of hickory leaves
and between 1.6 and 3.3% of oak leaves.

Leaves unpalatable to invertebrates are apparently also more resistant to fungal
colonization. Highest ATP levels coincided with highest activity of fungi observed
by other methods (direct observations or plating out). Extremely low values were
found by Iversen (1973) on decomposing beech leaves: fungal biomass (estimated by
measuring hyphal length) would correspond to only 0.004% of the total substrate
weight; bacteria were more common and contributed about 0.3%. Similar conclusions
were reached by Davis and Winterbourn (1977) in a study on the breakdown of *Nothofagus*
leaves: hyphae were seldom observed until about 4-5 months after immersion in the
stream. These observations suggest that bacteria may be more active in the early
colonization of very recalcitrant leaves; however, their biomass was very low (0.3%
on beech leaves), and they may be simply existing on organic substances dissolved in
the stream water and concentrated at the solid-liquid interfaces of submerged leaves.
More detailed studies are certainly needed to examine the degradation of sterile
leaves by selected bacteria and fungi. But despite the very low microbial biomass
on beech or *Nothofagus* leaves, the palatability of the substrate still increased, as
did nitrogen levels. Iversen estimates that only between 4 and 14% of the new nitrogen is actually bound to microbial cells, and he assumes that the remaining nitrogen
is present in organic substances released by bacteria. A possible mechanism is suggested by Suberkropp et al. (1976). They found that a substantial fraction of the
nitrogen is fixed in highly refractory complexes, presumably formed by leaf polyphenols and the microbial organic nitrogen compounds (possibly exoenzymes), which in
conventional analytical methods show up as "lignin." The relative amount of total
leaf nitrogen immobilized in such complexes was higher in the slow-decomposing oak
(26-36%)--with higher initial levels of soluble polyphenols--than in the faster decomposing hickory (16-22%). Such complexes have been shown to be very resistant to
enzymes, so that an invertebrate may derive little benefit from nitrogen levels increased in such ways. However, their stability is greatly influenced by the pH, and
invertebrates with a high intestinal pH may be able to metabolize them at least in
part (Davies et al., 1964; Feeny, 1970). Of equal importance to the detritus feeder
may be the fact that the polyphenols in such leaves are saturated and no longer inter-

fere with the animal's digestive enzymes. Protein may be precipitated directly from stream water, but at least in laboratory experiments protein increments have been strictly dependent on the presence of actively growing fungi.

If we accept the foregoing estimates of microbial biomass, their share of the total substrate weight would usually amount to some 2-10% (but below 1% in very recalcitrant leaves). If we further assume that microbial cells contain proportionally three times as much digestible material as the leaf, the average proportion of nutrients provided by microbial cells in the diet of leaf-eating invertebrates would vary between 6 and 25%, and be around 1% in the most refractory substrates. For comparison, Baker and Bradnam (1976) estimate that the filter-feeding blackfly *Simulium* derives about 2.1% of its food from ingested bacteria. These figures are quite substantial, and (taking into account the potential sources of error in their estimation) they may fully explain the observed improvement in the leaves' palatability and digestibility. But at least in some cases, additional mechanisms seem to be called for. For example, Nilsson (1974) reports that *Gammarus* assimilates virtually nothing of fresh beech leaves but can use 35% of fully conditioned leaves. If microbial biomass is indeed as low as that reported by Iversen (Nilsson offers no statement about it), the rise in digestibility cannot be due exclusively to enrichment with microbial cells.

It is possible that the leaf substrate itself becomes more digestible under the influence of microbial activity. We showed that, at least with maple leaves and with cellulose, palatability can be improved when the substrate is exposed to fungal exoenzymes or to hot HCl, a hydrolyzing agent (Bärlocher and Kendrick, 1975a). Nielsen (1962) and Bjarnov (1972) found that all detritus feeders they examined could use cellobiose, although most of them had little or no ability to degrade even modified cellulose. The presence of cellobiases in animals lacking cellulases has also been reported by Luxton (1972) and Kristensen (1972). Free cellobiose is very rare in nature (Pazur, 1970). Its most important occurrence is as a subunit of cellulose, and it is an intermediate product of cellulose degradation, a process which typically takes place in a series of steps (Whitaker, 1971). In aquatic detritus feeders cellulases, if present at all, are generally so weak that extensive degradation of native cellulose during its rapid passage through the gut appears unlikely (Monk, 1976). Kristensen (1972) suggested that, in noncellulolytic animals, enzymes able to degrade cellobiose were in fact nonspecific and acting on quite different substances in the animals' actual diet. This may well be the case. But the occurrence of enzymes active on cellobiose, and possibly on modified cellulose, would be an obvious advantage to detritus feeders if they had access to partly degraded cellulose. This would require that some microorganisms did not completely degrade and metabolize all the cellulose they attack. Some fungi behave like this, and it has already been established that other microorganisms can take advantage of it. For example, Tribe (1966) showed that a noncellulolytic oomycete, *Pythium oligandrum*, was able to grow on cellulose in association with the strongly cellulolytic hyphomycete *Botryotrichum piluliferum*, but not without it. Similarly, Frankland (1966, 1969) observed many "sugar fungi" (fungi dependent on simple carbohydrates) late in the decay of *Pteridium aquilinum* petioles, even though soluble carbohydrates were less than 1%. Additional experiments with pure cultures demonstrated that the sugar fungi were dependent on the release of sugars from the substrate brought about by the activities of basidiomycete mycelium. The significance of such processes may well remain hidden when we use conventional analytical methods, which do not distinguish between stages in the stepwise degradation of cellulose and which rely on average values of large samples.

Decomposing leaves are mosaics made up of patches that range from virtually intact areas to heavily colonized ones. Even if areas with a high level of easily digestible substrate are constantly becoming available, the average overall value will not change perceptibly since these areas will quickly be removed by preferential feeding. Unless potential consumers are barred, the composition of the leaf at any particular time resembles the leftovers of a meal rather than the actual food of detritus feeders. This point is well illustrated by Saitô's (1965) experiments with two nutritionally different fungi. When sterile leaves were inoculated with a strongly cellulolytic basidiomycete, the content of soluble carbohydrates increased rapidly. But when a second fungus was introduced simultaneously, the increase was slower and only temporary. The soluble carbohydrates were now removed and used by the second species.

One factor which clearly regulates the amount of soluble substances produced during decomposition is oxygen. Under anaerobic conditions, cellulose is still degraded by various microorganisms, but only a small percentage is built into microbial cells. Most is converted into volatile fatty acids. Ruminants and some insects exploit this process by harboring in their own body microorganisms which break down inaccessible food sources such as cellulose into digestible subunits which are siphoned off by the animal (Blaxter, 1967; Buchner, 1965; Hungate, 1966). In this process, only a small percentage of the energy is lost through microbial respiration, and the advantage to the animal lies not in the production of microbial cells, as in the fungus-growing attine ants, but rather in microbial catalysis.

But most detritus feeders lack such far-reaching adaptations and have little control over the direction of microbial activity on plant remains. Of necessity, they must be more opportunistic and flexible in their feeding strategy. Besides directly digesting as much plant material as they can, they may enrich their diet with microbial cells and possibly profit from increased substrate digestibility brought about by previous microbial colonization. This argument may even be extended to some carnivorous stream insects. Cummins (1973) states that predators feeding on early instars take in food with little fat or protein, surrounded by an indigestible chitinous shell. The most nutritious portion of such prey may be the digestive tract, packed with food particles. At least to some extent, predators might be considered to feed on prey digestive tracts (Cummins, 1973). It is therefore not very surprising that cellobiases, or even weak cellulases, have been found in some carnivorous invertebrates (Kristensen, 1972; Monk, 1976). It remains to be seen whether or not microbial enzymes remain active, and for how long that activity might persist, when the substrate is chewed and bathed in the invertebrate's digestive fluids.

III. CONCLUSION

Much of the work on detritus processing has been done or inspired by zoologists. Not surprisingly, leaf conditioning has been described from an animal's point of view. But to an unbiased observer it is obvious that this process, so advantageous to invertebrates, has serious drawbacks for leaf-colonizing fungi. Their resources are depleted, and their biomass directly reduced, by animal consumption. The often-heard statement that shredding increases surface area and thereby stimulates microbial activity may be true, but this benefits primarily microorganisms specialized in growing on fine particles and there is no evidence to suggest that aquatic hyphomycetes belong to this group. To anybody who has ever observed the luxuriant growth of fungi on leaves from which all potential consumers are excluded and then contrasted that with the scarcity of visible mycelium on leaves just recovered from a stream, it seems obvious that predation on fungi in natural streams must be severe. One would therefore expect the fungi to have evolved some countermeasures, analogous to the plants' defenses against herbivores. The most obvious one would be for the fungus to produce unpalatable substances or poisons deterring leaf-eating animals. As we have already noted, some fungi appear to have adopted this mechanism. Alternatively, the fungus might spread its mycelium so thinly that the resulting increase in substrate palatability was unnoticeable. Predation would then be more or less accidental, and its effects on the fungus minimal. A third possibility and, to judge from the scant information available, the most likely is that the fungus also colonizes plant parts which are safe from consumers for a considerable length of time and thereby provide a refuge. This may be within wood (Willoughby and Archer, 1973), conifer needles (Bärlocher et al., 1978a), or the veins and petioles of deciduous leaves. Skeletonized leaves are often the most profitable source of aquatic hyphomycetes (Ingold, 1942, 1975). We found that, in early samples taken from sets of maple leaves exposed to a stream, *Tricladium angulatum* was much more common than *Heliscus lugdunensis* on lamina whereas *Heliscus* was slightly more common than *Tricladium* on veins and stalks. But after about 3 months in the stream, the frequency of occurrence of *Tricladium* declined drastically, while *Heliscus* continued to rise. At the same time skeletonization became more and more noticeable, so that the relative amount of veins increased. A simple explanation is that *Tricladium* grew much better on the softer leaf parts between the veins. Its rapid growth made these areas palatable, and they were removed by preferential feeding. As a result veins, which carried more *Heliscus* than *Tricladium*, became relatively more abundant. The seemingly paradoxical conclusion is that the fastest-growing fungus may also decline most rapidly because of its very success. *Tricladium* appeared to be better adapted to a good but dangerous substrate, and *Heliscus* to a poorer but safer habitat. In the first case, fast growth

and reproduction are crucial; in the second, efficient use of the substrate is probably more important. To translate this (under protest) into the jargon of evolutionary ecologists: *Tricladium* is probably more r selected, and *Heliscus* more K selected
(MacArthur and Wilson, 1967).

Preferential removal of areas with high hyphal densities may have a quite different effect on the number of species coexisting on a leaf: density-dependent predation can allow inferior competitors of the prey population to exist in habitats
where they would otherwise be outcompeted (MacArthur, 1972; Menge and Sutherland,
1976; Paine, 1966). The issue of whether such effects play any role at all or are
overridden by the seasonality of leaf fall, or by the numbers and types of leaves
falling into a stream, or other factors, cannot at present be resolved on firmly
empirical grounds.

As we have already emphasized, the leaves which even before microbial colonization are most palatable to invertebrates are also those which lose weight fastest
when attacked solely by microorganisms (Bärlocher and Kendrick, 1973b; Bärlocher et
al., 1978a,b; Kaushik and Hynes, 1971). In other words, a source of food more nutritious to the detritus feeder is also more profitably exploited by the fungus (more
nutritious probably means less cellulose, lignin, tannins, etc., and more proteins).
Despite its ability to subsist on structural carbohydrates and dissolved inorganic
nutrients, the average aquatic hyphomycete apparently grows better and faster when
more easily digested substances are available. But the higher the original food
quality, the greater the probability that it will reach a level where it is acceptable to animals without prior conditioning. For example, several *Gammarus* species
are also carnivorous, and some ingest algae (Hynes, 1970; Moore, 1975). It is highly
unlikely that this protein-rich food could be improved by fungal growth. But even
if we restrict the argument to leaves, some of these are already so palatable that
the improvement due to conditioning is slight. Nilsson (1974) estimated that *Gammarus pulex* assimilates 30% of freshly fallen alder leaves and 40% of microbially
conditioned leaves. The gain might have been even lower or nonexistent had he used
young, green alder leaves (cf. Otto, 1974). Not surprisingly, alder leaves are also
a particularly good substrate for aquatic hyphomycetes (Nilsson, 1964): indeed, it
was on decaying alder leaves that these organisms were originally observed by Ingold
(1942). If certain leaves are indeed acceptable to animals without prior conditioning, the fungi and invertebrates will be in direct competition. Considering the
ability of animal communities to shred and ingest vast amounts of materials in a
short time, the outcome is hardly in doubt: unless there is a superabundance of
this high-quality food, the share of the fungi would be limited to leftovers too
small to be efficiently captured by the animals. But these particles might also be
too small to permit an aquatic hyphomycete to develop from colonizing conidium to
newly formed conidium. Thus the superior enzymatic equipment of the fungi may have
evolved partly as compensation for their inability to compete with animals for the
more easily exploitable substrates.

The relationships between fungi and invertebrates can be summarized as follows: In general an initially very unpalatable substrate will be improved only a little per unit time by fungal colonization, at least in the early stages of conditioning. As palatability increases, conditioning will also proceed more rapidly, until a maximum is reached, after which the speed of conditioning will start to decline, eventually reaching zero. Beyond that point, any further fungal colonization is likely to decrease the nutritive value, simply by depleting the substrate. The profit to the invertebrate of full leaf conditioning may actually be highest in the case of substrates which are initially so poor as to be virtually useless to the animal. This requires that the animal have alternative sources of food while the conditioning is going on and that the losses due to downstream transport during the long conditioning period are not too high (or are compensated for by gains from upstream). Depending on the position of a particular substrate in this spectrum, the relationships between fungus and invertebrate can be very different. On less palatable leaves the animal will prey on the fungus, while on palatable substrates they will be competitors. In the first case, the animal will try to ingest as much fungal biomass as possible; in the second case, to avoid it altogether by eating the leaf substrate as soon as it falls into the stream. This illustrates the inadequacy of certain terms common in ecological literature, such as *first* and *second trophic level*, and the distinction between *consumer* and *decomposer*. All heterotrophs consume organic substances; all of them have to decompose it at least partly to gain energy and assimilate or rearrange subunits. Microorganisms are simply able to use a wider range of natural substances, and they are more thorough in breaking down their substrates (e.g., they may produce inorganic nitrogen rather than the organic nitrogen excreted by terrestrial animals). In the context of detritus food chains, the significance of microorganisms may lie less in their ability to break down complex structures into organic substances than in the production of new microbial cells.

In assessing or rather predicting the significance of aquatic hyphomycetes in the trophic structure of natural streams, it is convenient to start with the trivial assumption that their impact is greatest where their activity and production are highest. With given chemical and physical characteristics of stream water, this seems to be the case when a wide range of substrates of different quality are available. This avoids a direct clash between fungi and invertebrates, thus reducing the likelihood of the substrate being reduced to fine particles before the fungi have had a chance to colonize it. It allows the fungi to stay one step ahead of their predators--colonizing substrates not yet attacked by invertebrates, building up their population, and producing propagules which will colonize the next substrate. A probable consequence is that a stream receiving leaves belonging to several species will support more shredders per unit weight of leaf than the same stream receiving the same total input of leaves of only one species. Nor is this likely to be restricted to shredders: the better the food, the more nutritious the feces derived from it, and the better the supply for fine particle feeders. This amplifying effect is largely due to aquatic hyphomycetes.

REFERENCES

Anderson, J. M., and Macfadyen, A. (1976). *The Role of Terrestrial and Aquatic Organisms in Decomposition Processes*. Blackwell Sci. Publns., Oxford, England.

Anderson, N. H., and Grafius, E. (1975). Utilization and processing of allochthonous material by stream Trichoptera. *Verhandl. Int. Verein. Limnol. 19*: 3083-3088.

Ausmus, B. S. (1973). The use of ATP assay in terrestrial decomposition studies. *Bull. Ecol. Res. Comm.* (Stockholm) *17*: 223-234.

Bärlocher, F., and Kendrick, B. (1973a). Fungi in the diet of *Gammarus pseudolimnaeus* (Amphipoda). *Oikos 24*: 295-300.

Bärlocher, F., and Kendrick, B. (1973b). Fungi and food preferences of *Gammarus pseudolimnaeus*. *Arch. Hydrobiol. 72*: 501-516.

Bärlocher, F., and Kendrick, B. (1974). Dynamics of the fungal population on leaves in a stream. *J. Ecol. 62*: 761-791.

Bärlocher, F., and Kendrick, B. (1975a). Leaf-conditioning by microorganisms. *Oecologia 20*: 359-362.

Bärlocher, F., and Kendrick, B. (1975b). Assimilation efficiency of *Gammarus pseudolimnaeus* (Amphipoda) feeding on fungal mycelium or autumn-shed leaves. *Oikos 26*: 55-59.

Bärlocher, F., and Kendrick, B. (1976). Hyphomycetes as intermediaries of energy flow in streams. In *Recent Advances in Aquatic Mycology*, E. B. G. Jones (Ed.). Elek, London, pp. 435-446.

Bärlocher, F., Kendrick, B., and Michaelides, J. (1978a). Colonization and conditioning of *Pinus resinosa* needles by aquatic hyphomycetes. *Arch. Hydrobiol. 81*: 462-474.

Bärlocher, F., Mackay, R. J., and Wiggins, G. P. (1978b). Detritus processing in a temporary vernal pool in southern Ontario. *Arch. Hydrobiol. 81*: 269-295.

Baier, R. G. (1935). Studien zur Hydrobakteriologie stehender Binnengewässer. *Arch. Hydrobiol. 29*: 183-264.

Baker, J. H., and Bradnam, L. A. (1976). The role of bacteria in the nutrition of aquatic detritivores. *Oecologia 24*: 95-104.

Batra, S. W. T., and Batra, L. R. (1967). The fungus gardens of insects. *Sci. Amer. 217*: 112-120.

Berrie, A. D. (1976). Detritus, micro-organisms and animals in fresh water. In *The Role of Terrestrial and Aquatic Organisms in Decomposition Processes*, J. M. Anderson and A. Madfadyen (Eds.), 17th Symposium of the British Ecological Society. Blackwell Sci. Publns., Oxford, England, pp. 323-338.

Bjarnov, N. (1972). Carbohydrases in *Chironomus*, *Gammarus* and some Trichoptera larvae. *Oikos 23*: 261-263.

Blaxter, K. L. (1967). *The Energy Metabolism of Ruminants*. Hutchinson, London.

Buchner, P. (1965). *Endosymbiosis of Animals with Plant Microorganisms*. Wiley, New York.

Cummins, K. W. (1973). Trophic relations of aquatic insects. *Ann. Rev. Entomology 18*: 183-206.

Cummins, K. W. (1974). Structure and function of stream ecosystem. *BioScience 24*: 631-641.

Cummins, K. W., Coffman, W. P., and Rolf, P. A. (1966). Trophic relations in a small woodland stream. *Verhandl. Int. Verein. Limnol. 16*: 627-638.

Cummins, K. W., Petersen, R. C., Howard, F. O., Wuycheck, J. C., and Holt, V. I. (1973). The utilization of leaf litter by stream detritivores. *Ecology 54*: 336-345.

Daubenmire, R., and Prusso, D. C. (1963). Studies of the decomposition rates of tree litter. *Ecology 44*: 589-592.

Davies, R. I., Coulson, C. B., and Lewis, D. A. (1964). Polyphenols in plant, humus, and soil. III. Stabilization of gelatin by polyphenol tanning. *J. Soil Sci. 15*: 299-309.

Davis, S. F., and Winterbourn, M. J. (1977). Breakdown and colonization of *Nothofagus* leaves in a New Zealand stream. *Oikos 28*: 250-255.

de la Cruz, A. A. (1975). Proximate nutritive value changes during decomposition of salt marsh plants. *Hydrobiologia 47*: 475-480.

de la Cruz, A. A., and Gabriel, B. C. (1974). Caloric, elemental, and nutritive changes in decomposing *Juncus roemerianus* leaves. *Ecology 55*: 882-886.

Dickinson, C. H., and Pugh, G. J. F. (1974). *Biology of Plant Litter Decomposition*, 2 vols. Academic Press, New York.

Dickson, G. W. (1975). A preliminary study of heterotrophic microorganisms as factors in substrate selection of troglobitic invertebrates. *Nat. Speol. Soc. Bull. 37*: 89-93.

Feeny, P. (1970). Seasonal changes in oak leaf tannins and nutrients as a cause of spring feeding by winter moth caterpillars. *Ecology 51*: 565-581.

Fenchel, T. (1970). Studies on the decomposition of organic detritus derived from the turtle grass *Thalassia testudinum*. *Limnol. Oceanogr. 15*: 14-20.

Fenchel, T. (1972). Aspects of decomposer food chains in marine benthos. *Verhandl. Deut. Zool. Ges. 65*: 14-22.

Fisher, S. G., and Likens, G. E. (1973). Energy flow in Bear Brook, New Hampshire: An integrative approach to stream ecosystem metabolism. *Ecol. Monogr. 43*: 421-439.

Frankland, J. C. (1966). Succession of fungi on decaying petioles of *Pteridium aquilinum*. *J. Ecol. 54*: 41-63.

Frankland, J. C. (1969). Fungal decomposition of bracken petioles. *J. Ecol. 57*: 25-36.

Greaves, M. P., Wheatley, R. E., Shepherd, H., and Knight, A. H. (1973). Relationship between microbial populations and adenosine triphosphate in a basin peat. *Soil Biol. Biochem. 5*: 685-687.

Haeckel, J. W., Meijering, M. P. D., and Rusetzki, H. (1973). *Gammarus fossarum* Koch als Fallaubzersetzer in Waldbächen. *Freshwater Biol. 3*: 241-249.

Hargrave, B. T. (1970). The utilization of benthic microflora by *Hyalella azteca* (Amphipoda). *J. Anim. Ecol. 39*: 427-437.

Harrison, P. G., and Mann, K. H. (1975). Detritus formation from eelgrass (*Zostera marina* L.): The relative effects of fragmentation, leaching, and decay. *Limnol. Oceanogr. 6*: 924-934.

Hart, S. D., and Howmiller, R. P. (1975). Studies on the decomposition of allochthonous detritus in two southern California streams. *Verhandl. Int. Verein. Limnol. 19*: 1665-1674.

Heald, E. J., and Odum, W. E. (1970). The contribution of mangrove swamps to Florida fisheries. *Proc. Gulf and Carib. Fish. Inst. 22*: 130-135.

Hering, H. T. (1967). Fungal decomposition of oak leaf litter. *Trans. Brit. Mycol. Soc. 50*: 267-273.

Hodkinson, I. D. (1975a). A community analysis of the benthic fauna of an abandoned beaver pond. *J. Anim. Ecol. 44*: 533-551.

Hodkinson, I. D. (1975b). Dry weight loss and chemical changes in vascular plant litter of terrestrial origin, occurring in a beaver pond ecosystem. *J. Ecol. 63*: 131-142.

Hodkinson, I. D. (1975c). Energy flow and organic matter decomposition in an abandoned beaver pond ecosystem. *Oecologia 21*: 131-319.

Huet, M. (1951). Novicité des boisements en Epicéas (*Picea excelsa* Link.). *Verhandl. Int. Verein. Limnol. 11*: 189-200.

Hungate, R. E. (1966). *The Rumen and Its Microbes*. Academic Press, New York.

Hynes, H. B. N. (1963). Imported organic matter and secondary productivity in streams. *Proc. 16th Int. Congr. Zool., Washington* Vol. 4, pp. 324-329.

Hynes, H. B. N. (1970). *The Ecology of Running Waters*. Univ. of Toronto Press, Toronto.

Ingold, C. T. (1942). Aquatic hyphomycetes of decaying alder leaves. *Trans. Brit. Mycol. Soc. 25*: 339-417.

Ingold, C. T. (1966). The tetraradiate aquatic fungal spore. *Mycologia 58*: 43-56.

Ingold, C. T. (1975). *Guide to Aquatic Hyphomycetes*. Freshwater Biological Association, Sci. Publn. No. 30. Ferry House, Ambleside, Cumbria, England, 96 pp.

Iversen, T. M. (1973). Decomposition of autumn-shed leaves in a springbrook and its significance for the fauna. *Arch. Hydrobiol. 72*: 305-312.

Käärik, A. A. (1974). Decomposition of wood. In *Biology of Plant Litter Decomposition*, C. H. Dickinson and G. J. F. Pugh (Eds.), Vol. I. Academic Press, New York, pp. 129-174.

Kaushik, N. K., and Hynes, H. B. N. (1971). The fate of the dead leaves that fall into streams. *Arch. Hydrobiol. 68*: 465-515.

Kostalos, M., and Seymour, R. L. (1976). Role of microbial enriched detritus in the nutrition of *Gammarus minus* (Amphipoda). *Oikos 27*: 512-516.

Kristensen, J. H. (1972). Carbohydrases of some marine invertebrates with notes on their food and on the natural occurrence of the carbohydrates studied. *Mar. Biol. 14*: 130-142.

Levin, D. A. (1976). The chemical defenses of plants to pathogens and herbivores. *Ann. Rev. Ecol. Syst. 7*: 121-159.

Luxton, M. (1972). Studies on the oribatid mites of a Danish beech wood soil. I. Nutritional biology. *Pedobiologia 12*: 434-463.

MacArthur, R. H. (1972). *Geographical Ecology: Patterns in the Distribution of Species*. Harper & Row, New York.

MacArthur, R. M., and Wilson, E. O. (1967). *The Theory of Island Biogeography*. Princeton Univ. Press, Princeton, N.J.

Mackay, R. J., and Kalff, J. (1973). Ecology of two related species of caddis fly larvae in the organic substrates of a woodland stream. *Ecology 54*: 499-511.

Madsen, B. L. (1974). A note on the food of *Amphinemoura sulcicollis* (Plecoptera). *Hydrobiologia 45*: 169-175.

Mann, K. H. (1972). Macrophyte production and detritus food chains in coastal waters. *Mem. Ist. Ital. Idrobiol. 29* (Suppl.): 353-383.

Mason, C. F., and Bryant, R. J. (1975). Production, nutrient content and decomposition of *Phragmites communis* Trin. and *Typha angustifolia* L. *J. Ecol. 63*: 71-95.

Menge, B. A., and Sutherland, J. P. (1976). Species diversity gradients: Synthesis of the roles of predation, competition, and temporal heterogeneity. *Amer. Natur. 110*: 351-369.

Michaelides, J., and Kendrick, B. (1978). An investigation of factors retarding colonization of conifer needles by amphibious hyphomycetes in streams. *Mycologia 70*: 419-430.

Monk, C. D. (1976). The distribution of cellulase in freshwater invertebrates of different feeding habits. *Freshwater Biol. 6*: 471-475.

Moore, J. W. (1975). The role of algae in the diet of *Asellus aquaticus* L. and *Gammarus pulex* L. *J. Anim. Ecol. 44*: 719-730.

Nelson, D. J., and Scott, D. C. (1962). Role of detritus in the productivity of a rock-outcrop community in a Piedmont stream. *Limnol. Oceanogr. 7*: 396-413.

Newell, S. Y., and Fell, J. W. (1975). Preliminary experimentation in the development of natural food analogues for culture of detritivorous shrimps. Univ. of Miami, Sea Grant, Tech. Bull. No. 30.

Nielsen, C. O. (1962). Carbohydrases in soil and litter invertebrates. *Oikos 13*: 200-215.

Nikolei, E. (1961). Vergleichende Untersuchungen zur Fortpflanzung heterogoner Gall-mücken unter experimentellen Bedingungen. *Z. Morphol. Ökol. Tiere 50*: 281-329.

Nilsson, L. M. (1974). Energy budget of a laboratory population of *Gammarus pulex* (Amphipoda). *Oikos 25*: 35-42.

Nilsson, S. (1964). Freshwater hyphomycetes. *Symb. Bot. Upsal. 18*: 1-130.

Odum, E. P. (1971). *Fundamentals of Ecology.* Saunders, Philadelphia.

Odum, E. P., and de la Cruz, A. A. (1967). Particulate organic detritus in a Georgia salt marsh-estuarine ecosystem. In *Estuaries*, G. Lauff (Ed.). *Amer. Assoc. Adv. Sci. Publ. No. 83*: 383-388.

Otto, C. (1974). Growth and energetics in a larval population of *Potamophylax cingulatus* (Steph.) (Trichoptera) in a South Swedish stream. *J. Anim. Ecol. 43*: 339-361.

Padgett, D. E. (1976). Leaf decomposition by fungi in a tropical rainforest stream. *Biotropica 8*: 166-178.

Paine, R. T. (1966). Food web complexity and species diversity. *Amer. Natur. 100*: 65-75.

Park, D. (1974). On the use of the litter bag method for studying degradation in aquatic habitats. *Int. Biodet. Bull. 10*: 45-48.

Pazur, J. H. (1970). Oligosaccharides. In *The Carbohydrates: Chemistry and Biochemistry*, W. Pigman and D. Horton (Eds.), Vol. II, Pt. A. Academic Press, New York, pp. 69-137.

Petersen, R. C., and Cummins, K. W. (1974). Leaf processing in a woodland stream. *Freshwater Biol. 4*: 343-368.

Saitô, T. (1957). Chemical changes in beech litter under microbiological decomposition. *Ecol. Rev. 14*: 209-216.

Saitô, T. (1965). Coacting between litter-decomposing hymenomycetes and their associated microorganisms during decomposition of beech litter. *Sci. Rep. Tohoku Univ. 31*: 255-273.

Sedell, J. R., Triska, F. J., and Triska, N. S. (1975). The processing of conifer and hardwood leaves in two coniferous forest streams. I. Weight loss and associated invertebrates. *Verhandl. Int. Verein. Limnol. 19*: 1617-1627.

Suberkropp, K., and Klug, M. J. (1976). Fungi and bacteria associated with leaves during processing in a woodland stream. *Ecology 57*: 707-719.

Suberkropp, K., Godshalk, G. L., and Klug, M. J. (1976). Changes in the chemical composition of leaves during processing in a woodland stream. *Ecology 57*: 720-727.

Swain, T. (1977). Secondary compounds as protective agents. *Ann. Rev. Plant Physiol. 28*: 479-501.

Teal, J. M. (1957). Community metabolism in a temperate cold spring. *Ecol. Monogr. 27*: 283-302.

Tribe, H. T. (1966). Interactions of soil fungi on cellulose film. *Trans. Brit. Mycol. Soc. 49*: 457-466.

Triska, F. J. (1970). Seasonal distribution of aquatic hyphomycetes in relation to the disappearance of leaf litter from a woodland stream. Ph.D. thesis, University of Pittsburgh.

Triska, F. J., Sedell, J. R., and Buckley, B. (1975). The processing of conifer and hardwood leaves in two coniferous forest streams. II. Biochemical and nutrient changes. *Verhandl. Int. Verein. Limnol. 19*: 1628-1639.

Visser, S., and Parkinson, D. (1975). Fungal succession on aspen poplar leaf litter. *Can. J. Bot. 53*: 1640-1651.

Whitaker, D. R. (1971). Cellulases. In *The Enzymes*, P. D. Boyer (Ed.), Vol. V. Academic Press, New York, pp. 273-290.

Willoughby, L. G., and Archer, J. F. (1973). The fungal spora of a freshwater stream and its colonization pattern on wood. *Freshwater Biol. 3*: 219-239.

Willoughby, L. G., and Sutcliffe, D. W. (1976). Experiments on feeding and growth of the amphipod *Gammarus pulex* (L.) related to its distribution in the River Duddon. *Freshwater Biol. 6*: 577-586.

Woodall, W. R., and Wallace, J. B. (1972). The benthic fauna in four small southern Appalachian streams. *Amer. Midl. Natur. 88*: 393-407.

Chapter 41

DEGRADATION OF LEAF LITTER BY AQUATIC HYPHOMYCETES

Keller Suberkropp*   M. J. Klug

Indiana-Purdue Universities   *Kellogg Biological Station*
at Fort Wayne   *Hickory Corners, and*
Fort Wayne, Indiana   *Michigan State University*
                                      East Lansing, Michigan

I. INTRODUCTION

The maintenance of community structure and function in small woodland streams depends on inputs of organic matter derived from the terrestrial system through which they flow (Fisher and Likens, 1973; Sedell et al., 1973; Cummins, 1974). In temperate climates, a significant portion of this organic matter is in the form of autumn shed leaf litter (Fisher and Likens, 1973; Sedell et al., 1973). These litter inputs are decomposed and utilized as energy sources by a diverse group of aquatic organisms.

As a component of the stream community, microorganisms play a central role in the decomposition of litter inputs through mineralization and enhancement of invertebrate feeding on decomposing litter (Nelson and Scott, 1962; Egglishaw, 1964; Minshall, 1967; Kaushik and Hynes, 1971; Cummins, 1974; Petersen and Cummins, 1974). Many invertebrates only feed on litter previously colonized by microorganisms and appear to derive the majority of their nutrition from the microorganisms associated with the ingested litter (Cummins, 1974; MacKay and Kalff, 1973; Bärlocher and Kendrick, 1973a,b, 1975, 1976).

Although the role of individual groups of microorganisms in the decomposition of litter is just beginning to emerge, fungal activity appears to be most significant during the initial stages of decomposition (Triska, 1970; Kaushik and Hynes, 1971; Suberkropp and Klug, 1976; Bärlocher and Kendrick, 1974, 1976). Bacterial activity increases during the latter stages of leaf decomposition when the surface area of the litter has increased through fungal and invertebrate activity (Triska, 1970; Suberkropp and Klug, 1976). Comparative ecological studies on the types of fungi associated with leaf litter in streams indicate that aquatic hyphomycetes dominate the fungal species associated with decomposing litter in streams (Bärlocher and Kendrick, 1974; Triska, 1970; Suberkropp and Klug, 1976). The ubiquity of these fungi

_____

*
Present affiliation: New Mexico State University, Las Cruces, New Mexico

in stream environments (Ingold, 1942, 1976; Nilsson, 1964; Padgett, 1976; Willoughby and Archer, 1973) suggests that they possess certain characteristics which allow them to favorably compete with other microorganisms in the stream habitat. Previously the common spore morphologies within this group were shown to be favorable adaptations for their dispersal and colonization in flowing water (Webster, 1959; Ingold, 1966, 1976).

The purpose of this chapter is to examine the physiological adaptations of these fungi which allow them to grow on leaf litter in temperate streams. It is also hoped that it will help clarify the role that these microorganisms play in the decomposition of leaf litter and show how their activities affect those of other microorganisms and invertebrates in the stream ecosystem.

II. PHYSIOLOGICAL ADAPTATIONS

A. Temperature

Seasonal extremes in temperatures of small woodland streams in temperate climates are typically found to range between 0° and 25°C (Hynes, 1972; Fisher and Likens, 1973; Iqbal and Webster, 1973; Iversen, 1975; Suberkropp et al., 1975; Bärlocher and Kendrick, 1974, 1976). Although the majority of species commonly found in temperate and cold regions of the world (Nilsson, 1964) have temperature optima for growth in the laboratory of 15-25°C (Table 1), e.g., *Alatospora acuminata*, *Anguillospora longissima*, *Articulospora tetracladia*, *Flagellospora curvula*, *Lemonniera aquatica*, *Tetracladium marchalianum*, *T. setigerum*, *Tricladium gracile*, and *T. splendens*, they also exhibit growth below 5°C. However, species characterized as common in tropical regions of the world (Nilsson, 1964) or found only during the summer or early autumn in temperate climates (Triska, 1970; Iqbal and Webster, 1973; Suberkropp and Klug, 1974), exhibit higher temperature optima and no growth at or below 5°C (e.g., *Clavatospora tentacula*, *Flagellospora penicillioides*, *Lunulospora curvula*). These laboratory studies indicate that aquatic hyphomycetes should be extremely well adapted to the temperature regimes of the habitats in which they commonly occur.

In deciduous woodland streams, particularly in temperate climates, increased numbers of aquatic hyphomycetes species occur at the time of large autumn pulses of leaf litter into the streams. Interestingly, these pulsed inputs and increased number of species are accompanied by decreasing or yearly minimum temperatures (Nilsson, 1964; Conway, 1970; Triska, 1970). In a series of English streams increases in both species and total spore concentration occurred from August to December followed by a gradual decline to periods of very low productivity during May to July (Iqbal and Webster, 1973). The onset of spore production occurred at the time of maximum deciduous leaf litter inputs. Such studies suggest that overall productivity of aquatic hyphomycetes is dependent on the seasonal availability of suitable substrates, such as leaf litter, for growth and their physiological adaptation to degrade these substrates at low temperatures.

Table 1 Temperature characteristics of aquatic hyphomycetes

Species	Minimum	Growth at lowest temp. examined	Optimum	Maximum
Alatospora acuminata Ingold		1^e	$20^{b,e}$	$25\text{-}30^e$
Anguillospora longissima (Sacc et Syd) Ingold		1^e	$15\text{-}25^e$	$25\text{-}30^e$
Articulospora tetracladia Ingold		5^b	$20\text{-}25^a,20^b$	
Clavariopsis aquatica DeWild.		1^e	$25\text{-}30^e$	$30\text{-}35^e$
Clavatospora tentacula (Umph.) Nilsson	$5\text{-}10^e$		25^e	$30\text{-}35^e$
Flagellospora curvula Ingold		1^e	$15^a,20^e$	$25\text{-}30^e$
Flagellospora penicillioides Ingold	$5\text{-}10^e$		$25^a,$ $25\text{-}30^e$	$35\text{-}40^e$
Gyoerfyella craginiformis (Peters.) Maranova		5^b	20^b	
Heliscus lugdunensis Sacc & Therry			$20\text{-}25^a$	
Lemonniera aquatica DeWild.		1^e	$20^{a,e}$	$25\text{-}30^e$
Lunulospora curvula Ingold	$5\text{-}10^{d,e}$		$20^d,25^e$	$30\text{-}35^{d,e}$
Margaritispora aquatica Ingold		5^b	$20^c,25^b$	
Tetracladium marchalianum DeWild.		1^e	$20^{c,e}$	$25\text{-}30^e$
Tetracladium setigerum (Grove) Ingold		5^b	20^b	
Tricellula aquatica Webster		5^b	$10^a,20^b$	
Tricellula aurantiaca (Haskins) von Arx		5^b	20^b	
Tricellula curvata Haskins		5^b	20^b	
Tricellula inaequalis van Bevern.		5^b	20^b	
Tricladium chaetocladium Ingold		5^b	15^d	$25\text{-}30^d$
Tricladium gracile Ingold			20^a	
Tricladium splendens Ingold		5^b	$15^a,20^b$	
Varicosporium elodeae Kegel		5^b	$20\text{-}25^b$	
Volucrispora graminea Ingold, McDougall and Dann		5^b	20^b	

[a] Thornton (1963).

[b] Koske and Duncan (1974).

[c] Nilsson (1964).

[d] Webster et al. (1976).

[e] K. Suberkropp (unpublished data).

Direct experimental evidence is lacking for a selection of these fungi over other members of the microbial community during periods of low temperatures. However, when activities of terrestrial and aquatic hyphomycetes isolated from leaf litter were compared in pure culture, species of aquatic hyphomycetes consistently caused greater weight loss and protein increases associated with the leaves (Bär-locher and Kendrick, 1974). This was particularly true when cultures were incubated at low temperatures (5-10°C) typical of streams during winter months. The ability of these fungi to compete favorably with other fungi at low temperatures is also suggested by comparing frequencies of occurrence of fungi on leaf litter colonized in streams and incubated in the laboratory at different temperatures (Bärlocher and Kendrick, 1974; Suberkropp and Klug, 1976). In general, higher frequencies were found for aquatic hyphomycetes when incubations were carried out at stream tempera-tures than when room temperature incubations were employed. These results were re-versed for fungi commonly associated with terrestrial litter. The latter results are consistent with the fact that the bulk of leaf litter is decomposed in a terres-trial ecosystem during the spring, summer, and early fall at higher temperatures. Thus it would appear that species of aquatic hyphomycetes are adapted to grow at the ambient temperature of the stream when the major litter inputs occur, and have a selective advantage over the other members of the fungal community.

B. Nutrient Utilization

While relatively little is known about the nutrition of aquatic hyphomycetes, avail-able information indicates that they have simple carbon and nitrogen requirements which would be readily available in streams. Of seven species studied by Thornton (1965), six (*Articulospora tetracladia*, *Flagellospora penicillioides*, *Heliscus lug-dunensis*, *Tricladium angulatum*, *T. splendens*, and *Varicosporium elodeae*) used most of the amino acids found in leaf litter as both carbon and nitrogen sources. The same species were found to be able to utilize both ammonium and nitrate ions as sole nitrogen source (Thornton, 1963). The observed increases in nitrogen content of de-composing leaves in streams (Kaushik and Hynes, 1968; Iversen, 1973; Triska et al., 1975; Suberkropp et al., 1976) suggests that these fungi not only can utilize nitro-gen containing compounds present in leaves but also immobilize nitrogen from stream water.

The species examined by Thornton (1963) were equally ubiquitous with respect to carbon source and utilized all carbohydrates tested with the exception of methyl cellulose. These included monosaccharides (glucose, fructose, and xylose), disac-charides (sucrose, maltose, and cellobiose) as well as starch. Two species of *An-guillospora* were also found to utilize a variety of mono- and disaccharides in addi-tion to starch, cellulose, and pectin (Ranzoni, 1951).

Since soluble carbohydrates are lost from leaf litter very rapidly after immer-sion in streams (Krumholz, 1972; Suberkropp et al., 1976), aquatic hyphomycetes ought to be able to degrade insoluble polymers if they are to obtain carbon and energy from litter over the period of time they are assoicated with it. Recently Jones (see Chap. 39) has demonstrated that a number of aquatic hyphomycetes can utilize native cellulose in the form of Solka floc and that they carry out a soft rot action on wood submerged in aquatic environments. In our laboratory the majority of isolates exam-

Table 2 Cellulose degradation by aquatic hyphomycetes

Species	Number of isolates tested	Degree of clearing of cellulose agar[a]
Tetracladium marchalianum	14	++
Lunulospora curvula	7	++
Flagellospora curvula	6	++
Lemonniera aquatica	8	+
Clavatospora tentacula	3	+
Claviaropsis aquatica	1	+
Alatospora acuminata	1	+
Flagellospora penicillioides	3	Trace

[a]Ability to utilize cellulose determined relative to time and amount of cleared zone on agar containing 12.5 mM KNO_3, 2 mM $MgSO_4$, 5 mM KH_2PO_4, pH 7.0, 6.7 mM NaCl, 1.5% agar, and 0.05% ball-milled Whatman No. 1 filter paper.

ined grew and caused clearing of cellulose containing agar (Table 2). In addition, when growing in pure culture on leaf litter, *Alatospora acuminata*, *Claviaropsis aquatica*, *Flagellospora curvula*, *Lemonniera aquatica*, and *Tetracladium marchalianum* released enzymes which liberated reducing sugars from the pure substrates carboxymethyl cellulose, xylan, and polygalacturonic acid. Clearly additional information is needed on the nutrition of these organisms in order to understand their metabolism of complex polysaccharides. However, their relatively simple nutritional requirements and the present realization of their more complete array of hydrolytic enzymes parallels the field observations of their ability to compete successfully in the woodland stream environment.

III. ACTIVITY ON LEAF LITTER

Laboratory nutritional studies using pure substrates give an indication of the physiological capabilities of microorganisms and provide indications of their activity in natural habitats. In complex organic matter such as leaf litter, these same constituents may be complexed with one another, making enzymatic attack more difficult or impossible. In lignocellulose complexes, for example, the cellulose is not as readily available for enzymatic attack as in extracted cellulose preparations (Bailey et al., 1968; Greaves, 1971). Other types of complexes may also occur during decomposition, e.g., the complexing of humic acids with organic constituents such as protein, rendering them unavailable for microbial utilization (Haider et al., 1965; Schnitzer et al., 1974).

With these limitations of laboratory studies in mind we felt that native leaf material must be used as sole carbon source in order to assess the potential role of

aquatic hyphomycetes in the decomposition of leaf litter. Activities of aquatic
hyphomycetes on leaf material in pure culture were compared with those found in association with leaf litter colonized by the microbial community in the stream. Similarities between activities in the field and the laboratory point toward the activity of these fungi in the natural stream environment, and the results of these
studies are summarized below.

A. Pure Culture Studies

When isolates of *T. marchalianum*, *F. curvula*, *L. aquatica*, *C. aquatica*, and *A. acuminata* were grown in pure culture on ethylene oxide sterilized hickory (*Carya glabra*)
leaf disks as the sole carbon source, they demonstrated the capacity to soften and
cause the release of the leaf matrix. This macerating activity can cause the loss
of most of the leaf epidermal and parenchyma cells, leaving only skeletonized leaf
disks (Fig. 1). The leaf cells released constitute a fine particulate organic matter (FPOM) which differs in both its physical and chemical properties from either
the original uncolonized or remaining skeletonized leaf material [i.e., the coarse
particulate organic matter (CPOM)]. The FPOM released in this manner has a high
surface area and contains a lower percentage of the structural constituents cellulose
and hemicellulose (Suberkropp and Klug, 1980).

   In a closed culture system, FPOM can be separated from CPOM, dried, and the
amount released determined. An example of the kinetics of the weight loss of CPOM
and corresponding increases of FPOM caused by *T. marchalianum* in shake culture at
18°C is given in Fig. 2. Since production of FPOM in an open environment is much
more difficult to collect and determine, another measurement of the macerating capa-

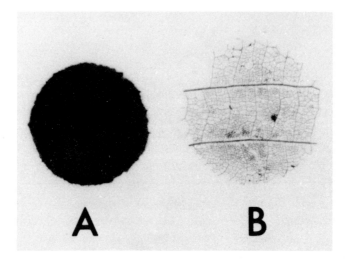

Fig. 1 Skeletonization of hickory leaf disks by *Tetracladium marchalianum*. (A) Uninoculated controls. (B) Incubated with *T. marchalianum*.

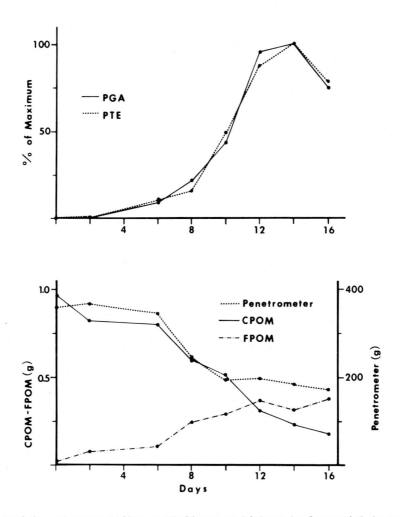

Fig. 2 Activity of *Tetracladium marchalianum* on hickory leaf material in pure culture. Ethylene oxide-sterilized hickory leaf disks were added to a mineral salts
medium containing 2.0 g KNO_3, 2.0 g KH_2PO_4, 0.5 g NaCl, 0.2 g $MgSO_4 \cdot 7H_2O$, 0.1 g $CaCl_2$,
0.05 M bicine buffer (Sigma), 1 liter distilled water, adjusted to pH 7.5, inoculated
with a washed hyphal fragment suspension of *T. marchalianum* and incubated on a re-
ciprocating water bath shaker at 18°C. Lower graph: CPOM > 1 mm, FPOM < 1 mm, and
sedimented after centrifugation at 1000 X *g* for 10 minutes. Both particulate fractions were dried at 50°C and weighed. Penetrometer measurements indicate the mass
necessary to cause a rod, 0.5 cm in diameter, to break leaf disks held between two
plexiglass blocks. Upper graph: Enzyme activities of culture supernatants. Enzyme
reaction mixtures consisted of 1 ml culture supernatant or boiled culture supernatant,
1 ml 0.2 M buffer (either sodium acetate at pH 5.0 or Tris-HCl at pH 8.0), and 1 ml
of 1% polygalacturonic acid (Sigma, grade III). Reactions were carried out at 30°C
for 2 hr. Polygalacturonase (PGA) activity was measured as the release of reducing
sugars (Nelson, 1944; Somogyi, 1952) at pH 5 during the incubation. Polygalacturonate transeliminase activity (PTE) was measured as the increase in absorbance at 550
nm after reaction with thiobarbituric acid (Ayers et al., 1966). The activities of
both enzymes are expressed as the percentage of the maximum activity observed in each
of the assay procedures.

Table 3 Partitioning of leaf organic matter by *Tetracladium marchalianum* as percentage of control dry weight

Day of incubation	CPOM[a]	FPOM[a]	DOM[a]	Difference lost as mineralization
6	75.9	17.8	4.9	1.4
12	26.5	55.3	10.1	8.1
26	14.7	53.6	13.3	18.4

[a]Organic matter fractions separated CPOM > 1 mm, FPOM < 1 mm. DOM not pelleted after centrifugation at 1000 X *g* for 10 min. CPOM and FPOM dried at 50°C; DOM freeze-dried.

bility of these fungi was also used. The relative toughness of leaf material, measured by determining the amount of weight necessary to cause a rod, 0.5 cm in diameter, to penetrate individual leaf disks (Feeney, 1970), decreased in a manner similar to FPOM production (penetrometer curve, Fig. 2). Decreases in the weight required for penetration began at the same time (6 days) as the maximum rate of FPOM release in cultures and continued to decrease until leaf disks were approaching complete skeletonization (10-12 days).

The most significant result of fungal activity was found to be the relase of FPOM. Portions were also released as dissolved organic matter (DOM) and carbon dioxide. Table 3 summarizes the percentage of organic matter recovered in these three fractions after incubation of *T. marchalianum* on hickory leaves. Although over 85% of the weight of the leaf litter was lost, most of this was converted to another pool of organic matter (FPOM, DOM). Less than one-fifth of the original organic matter was unaccounted for and presumably lost through respiration.

The mechanism responsible for FPOM production by these fungi remains under investigation in our laboratory. However, initial findings with *T. marchalianum* suggests the involvement of pectin-degrading enzymes in the observed softening of the leaf matrix. This class of enzymes have also been implicated in the softening of plant tissue by certain plant pathogens (Bateman and Millar, 1966). Two such enzymes differing in both mode of action and pH characteristics were detected in leaf culture supernatants (Fig. 2). Both polygalacturonase (PGA), a hydrolytic enzyme with maximum activity near pH 5, and polygalacturonate transeliminase (PTE), a lyase with maximum activity near pH 8, increased in activity during the period of FPOM release and reached their maximum activity shortly after FPOM release had ceased.

B. Stream Studies

The observed activities of aquatic hyphomycetes on leaf material in culture were compared with the action of the microbial community on hickory leaflets incubated in a third-order section of August Creek, Michigan (Suberkropp and Klug, 1976). Leaflets

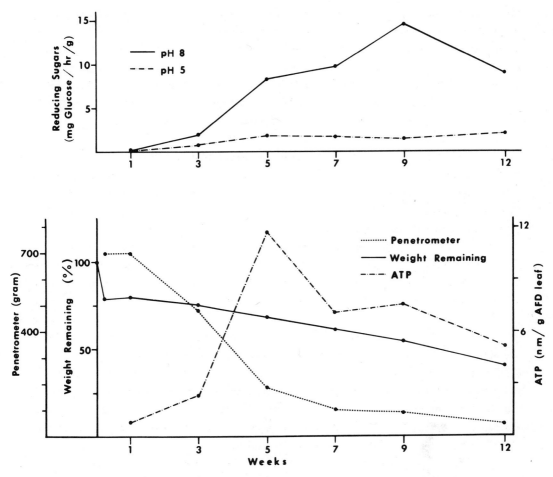

Fig. 3 Activity of the natural stream community on hickory leaf material. Individual hickory leaflets incubated in stream from Nov. 29 up to Feb. 16; see text for details of colonization. Lower graph: Weight remaining of individual leaflets removed at indicated time intervals as percentage of the original weight of those leaflets. Penetrometer measurements as in Fig. 2. Adenosine triphosphate was determined (Suberkropp and Klug, 1976) and expressed as nanomoles ATP per gram ash-free dry weight of leaf material. Upper graph: Enzyme activities associated with naturally colonized leaflets. Enzyme reaction mixtures as in Fig. 2 except that 1 ml of a leaf homogenate used in place of culture supernatant. Enzyme activities at both pH 5.0 and 8.0 were measured as the release of reducing sugars from PGA and expressed as milligram glucose equivalents released per hour at 30°C per gram of leaf material.

were attached individually near their basal end between two aluminum strips placed perpendicular to the current in a covered, flow-through box designed to minimize algal and invertebrate colonization. Leaflets were removed at intervals and subjected to analysis.

After an initial leaching loss, dry weight of the leaf material declined at a relatively constant rate over the 12-week study (Fig. 3). In the stream, there was no period of rapid weight loss corresponding to that observed in culture during the

period of maximum FPOM release. Major differences in the physical conditions of the
two systems are thought to be responsible for this difference. In culture, temperatures remained constant at 18°C, whereas during the period of the field study (Nov.
29 to Feb. 16) the mean weekly temperature was less than 4°C and for most of this
period was below 2°C. In addition, cultures were agitated on a mechanical shaker
to provide gaseous exchange, which may have caused a more rapid loss of the softened
leaf matrix than occurred in the stream. Shaking, however, had little effect on uninoculated leaf disks (see Fig. 1).

 Penetrometer determinations of stream-incubated leaf material (Fig. 3) were
similar to the pattern observed in culture (Fig. 2). Shortly after immersion in the
stream, the leaf material began softening, i.e., it was more easily penetrated, and
this trend continued until the amount of weight necessary to break through the leaflets stabilized between 5 and 7 weeks.

 Activities of pectin-degrading enzymes (Fig. 3), particularly when assayed at
pH 8 (presumably PTE activity), were detected in leaf litter incubated in the stream.
Activity measured at pH 8 (the pH of stream water was 7.2-8.0) increased to a maximum after the penetrometer readings had leveled off (at 9 weeks) in a manner similar
to that observed in culture.

 During the period of maximum decreases in penetrometer measurements, both ATP
content (Fig. 3) and frequencies of aquatic hyphomycetes sporulating on the leaf litter (Table 4) increased to maximum values. The corresponding maxima in ATP content
and frequency of aquatic hyphomycetes suggest that they are dominant members of the
microbial community. These data are supported by our more detailed study (Suberkropp
and Klug, 1976) on the decomposition of hickory leaflets in this stream that showed
similar patterns of ATP and hyphomycete frequency.

 The activities of pure cultures of aquatic hyphomycetes (see Sec. III.A, Fig.
2) and the microbial community associated with decomposing leaves in the stream
(Fig. 3) are quite similar. These similarities and the observed dominance of these
organisms on decomposing leaves suggest that they are responsible for the softening
of leaf tissue in the stream.

Table 4 Occurrence of aquatic hyphomycetes on leaf litter[a]

Species	Frequencies (%) on leaf litter, data taken up to 12 wks					
	1	3	5	7	9	12
Flagellospora curvula	13	47	60	7		
Lemonniera aquatica		20	47	10	7	3
Alatospora acuminata	10	3	47	36	23	23
Tetracladium marchalianum			20	3		7

 [a]Frequencies determined as percentage of 30 leaf disks, 0.5 cm in diameter, on
which each species was found sporulating upon direct microscopic examination during
weeks 1, 3, 5, 7, 9, and 12.

   The common observation of aquatic hyphomycetes together with skeletonized leaves
(Nilsson, 1964; Ingold, 1966), as well as the masses of hyphae (Padgett, 1976) associated with losses of mesophyll tissue, further suggests their macerating activity.

IV. SYNTHESIS

Disappearance of organic matter during decomposition normally implies total mineralization. The stage of leaf litter decomposition in streams which leads to the loss
of the original integrity of the leaf (i.e., skeletonization or complete disappearance of macroscopically recognizable leaf parts) includes both mineralization and
the production of a series of transformation intermediates (DOM, FPOM). We view
this stage as a continuum of dynamic interactions between microorganisms and macroinvertebrates which lead to the various observed pools of organic matter. The latter
are further metabolized by other groups of the stream community which are incapable
of metabolizing the original leaf material (Fig. 4). Although the interactions of
each group of organisms in this continuum are not known, we consider the aquatic
hyphomycetes as having a pivotal role since other members of the community are dependent on their activity.

   Their morphological adaptation to life in flowing waters (Webster, 1959; Ingold,
1966) and their physiological adaptation to growth on leaf material at low tempera-

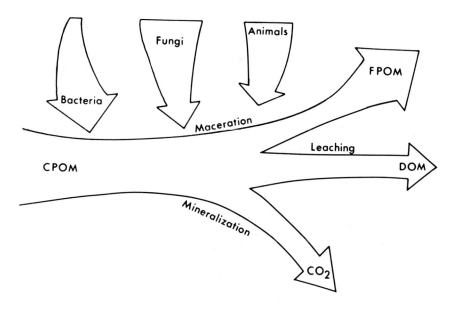

Fig. 4 Conceptualization of leaf processing in streams. Leaf litter (CPOM) is converted by the various populations of the stream community not only to carbon dioxide
but also to other pools of organic matter (FPOM and DOM), which are further processed
at other sites. See text for discussion of roles of biological community.

tures (Sec. II.A) are bases to support the observed increased frequencies of these organisms in streams following inputs of leaf material. In addition all available evidence indicates that fungal populations are metabolically more active than other members of the microbial community in the early stages of leaf decomposition in streams (Triska, 1970; Kaushik and Hynes, 1971; Suberkropp and Klug, 1976).

This initial microbial colonization of the leaf material leads to stimulation of the feeding by detritivores, primarily aquatic insect larvae (Triska, 1970; Mac-Kay and Kalff, 1973; Kaushik and Hynes, 1971; Bärlocher and Kendrick, 1973a,b, 1976; Petersen and Cummins, 1974). While the nature of this stimulation has not been completely clarified, both physical and biological mechanisms are suggested. The physical softening of the leaf matrix through microbial enzymatic activity (Sec. III.B) results in a more palatable substrate for macroinvertebrates to feed upon. In our laboratory fourth instar larvae of *Tipula abdominalis* were found to have the greatest consumption of decomposing hickory leaves when the microbial biomass (expressed as ATP) associated with the leaves was the highest. Since the leaf material in this latter study was greatly softened at the time that the microbial biomass (ATP content) was the greatest, it is difficult to separate the physical from the biological factors as a cause for the feeding stimulation. Evidence is also available to indicate that certain invertebrates can select leaf material colonized by certain fungal species (Bärlocher and Kendrick, 1973a). This suggests a chemotactic response or differential softening of detritus colonized by certain microorganisms. The former is similar to the feeding responses observed in terrestrial invertebrates (Allen et al., 1964; Smythe et al., 1967).

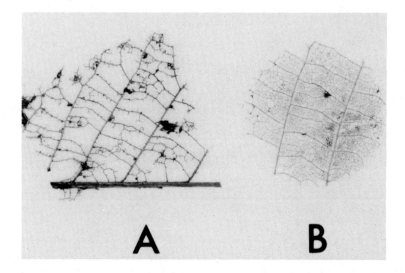

Fig. 5 Comparison of skeletonization of leaf material by invertebrate feeding and fungal activity. (A) Hickory leaf fragment used as food source by larvae of *Tipula abdominalis*. (B) Hickory leaf disk grown in pure culture with *Tetracladium marchalianum*.

The ingestion and subsequent egestion by the animals results in a pool of FPOM similar to that produced in culture through fungal maceration of leaf material. The chemical nature of the FPOM resulting from the two sources appears to be different. Figure 5 illustrates the reamining leaf skeleton after the feeding by a large-particle-feeding detritivore, *T. abdominalis*, and that derived through the enzymatic activity of *T. marchalianum*. The animal ingests considerably more of the structural components of the leaf than is released through fungal activity. Further it is not known at this time what changes in structural carbohydrates occur as the ingested leaf material passes through the gut tract of large-particle-feeding detritivores.

The contribution of fungi and macroinvertebrates in the conversion of leaf material to other pools of organics is difficult to determine since their activities are intimately related. However differences in the rate of loss of CPOM in the presence and absence of macroinvertebrates has been noted. The stimulation of the conversion of leaf material to other pools of organic matter was found to be 20% (Cummins et al., 1973) by coarse-particle-feeding detritivores and 13% by *Gammarus pulex* (Mathews, 1967). Thus the pool of FPOM derived from the two sources would result in differences in physical (particle size) and chemical properties leading to a diverse substrate for filter- and sediment-feeding invertebrates and other microorganisms.

Activity and biomass of bacteria during leaf litter decomposition has been shown to increase with time of decomposition (Triska, 1970; Suberkropp and Klug, 1976). Initially leaf material presents large planar surfaces which increase in surface area significantly through the macerating activity of fungi and the feeding of large-particle detritivores. Increase in surface area results in a competitive advantage for bacteria during the latter stages of leaf litter decomposition (Harley, 1971; Bärlocher and Kendrick, 1976; Suberkropp and Klug, 1976).

Although visually less dramatic than FPOM release, DOM is constantly being lost from leaf litter after it is immersed in water. The loss of DOM is through both abiological and biological mechanisms. Abiological losses of soluble components (sugars, amino acids, and phenolic compounds) are through leaching, which occurs immediately after immersion in water and biological losses are through activity of the stream community. Depending on the species of leaf, the former can result in losses of some 5-20% of the dry weight of the leaf Nykvist, 1963; McConnell, 1968; Bretthauer, 1971; Krumholz, 1972; Lush and Hynes, 1973; Petersen and Cummins, 1974; Suberkropp et al., 1976). While values for biological losses are not known in situ, pure cultures (Table 3) of *T. marchalianum* were shown to release 13.6% of the dry weight of hickory leaves as DOM. In mixed cultures the latter value would undoubtedly be lower due to metabolism of the released organics by associated microorganisms.

Clearly additional studies are necessary to quantify the contribution of all members of the stream community in the described sequence of events. The pivotal role of aquatic hyphomycetes in this sequence is becoming increasingly clear, however, and is a result of their morphological and physiological adaptation to life and growth on leaf substrates in flowing water.

ACKNOWLEDGMENTS

The technical assistance of K. Hogg and R. Greening during various aspects of this study is greatly appreciated. The authors thank Dr. K. W. Cumins for reading and criticizing the manuscript in its preparation. Research was supported in part by the National Science Foundation (Grants 6B-3600-68X, BWS-75-0733) and the U.S. Department of Energy (EY-76-S-02-2002-A001). Contribution No. 359 of the W. K. Kellogg Biological Station and Journal Article No. 8565 of the Michigan Agricultural Experiment Station.

REFERENCES

Allen, T. C., Smythe, R. V., and Cappell, H. C. (1964). Responses of 21 termite species to aqueous extracts of wood invaded by the fungus *Lenzitestrabea* Pers. ex Fr. *J. Econ. Ent. 57*: 1009-1011.

Ayers, W. A., Papavizas, G. C., and Diem, A. F. (1966). Polygalacturonate transeliminase and polygalacturonase production by *Rhizoctonia solani*. *Phytopathology 56*: 1006-1011

Bärlocher, F., and Kendrick, B. (1973a). Fungi and food preferences of *Gammarus pseudolimnaeus*. *Arch. Hydrobiol. 72*: 501-516.

Bärlocher, F., and Kendrick, B. (1973b). Fungi in the diet of *Gammarus pseudolimnaeus* (Amphipoda). *Oikos 24*: 295-300.

Bärlocher, F., and Kendrick, B. (1974). Dynamics of the fungal population on leaves in a stream. *J. Ecol. 62*: 761-791.

Bärlocher, F., and Kendrick, B. (1975). Assimilation efficiency of *Gammarus pseudolimnaeus* (Amphipoda) feeding on fungal mycelium or autumn-shed leaves. *Oikos 26*: 55-59.

Bärlocher, F., and Kendrick, B. (1976). Hyphomycetes as intermediaries of energy flow in streams. In *Recent Advances in Aquatic Mycology*, E. B. G. Jones (Ed.). Elek, London, pp. 435-446.

Bailey, P. J., Liese, W., and Rösch, R. (1968). Some aspects of cellulose degradation in lignified cell walls. In *Biodeterioration of Materials*, A. H. Walters and J. J. Elphinck (Eds.). Elsevier, New York, pp. 546-557.

Bateman, D. F., and Millar, R. L. (1966). Pectic enzymes in tissue degradation. *Ann. Rev. Phytopathol. 4*: 119-146.

Brettnauer, R. (1971). Quantitative estimation of low molecular ninhydrin-positive matter in waters rich in autumn shed leaves. *Int. Rev. Ges. Hydrobiol. 56*: 123-128.

Conway, K. E. (1970). The aquatic hyphomycetes of central New York. *Mycologia 62*: 516-530.

Cummins, K. W. (1974). Structure and function of stream ecosystems. *Bioscience 24*: 631-641.

Cummins, K. W., Petersen, R. C., Howard, F. O., Wuycheck, J. C., and Holt, V. I. (1973). The utilization of leaf litter by stream detritivores. *Ecology 54*: 336-345.

Egglishaw, H. J. (1964). The distributional relationship between the bottom fauna and plant detritus in streams. *J. Anim. Ecol. 33*: 436-476.

Feeny, P. (1970). Seasonal changes in oak leaf tannins and nutrients as a cause of spring feeding by winter moth caterpillars. *Ecology 51*: 565-581.

Fisher, S. G., and Likens, G. E. (1973). Energy flow in Bear Brook, New Hampshire: An integrative approach to stream ecosystem metabolism. *Ecol. Monogr. 43*: 421-439.

Greaves, H. (1971). The effect of substrate availability on cellulolytic enzyme production by selected wood-rotting microorganisms. *Aust. J. Biol. Sci. 24*: 1169-1180.

Haider, K., Frederick, L. R., and Flaig, W. (1965). Reaction between amino acid compounds and phenols during oxidation. *Plant Soil 22*: 49-64.

Harley, J. L. (1971). Fungi in ecosystems. *J. Ecology 59*: 653-668.

Hynes, H. B. N. (1972). *The Ecology of Running Waters*. Univ. of Toronto Press, Toronto.

Ingold, C. T. (1942). Aquatic hyphomycetes of decaying alder leaves. *Trans. Brit. Mycol. Soc. 25*: 339-417.

Ingold, C. T. (1966). The tetraradiate aquatic fungal spore. *Mycologia 58*: 43-56.

Ingold, C. T. (1976). The morphology and biology of freshwater fungi excluding Phycomycetes. In *Recent Advances in Aquatic Mycology*, E. B. G. Jones (Ed.). Elek, London, pp. 335-357.

Iqbal, S. H., and Webster, J. (1973). Aquatic hyphomycete spora of the River Exe and its tributaries. *Trans. Brit. Mycol. Soc. 61*: 331-346.

Iversen, T. M. (1973). Decomposition of autumn-shed beech leaves in a spring brook and its significance for the fauna. *Arch. Hydrobiol. 72*: 305-312.

Iversen, T. M. (1975). Disappearance of autumn-shed beech leaves placed in bags in small streams. *Verhandl. Int. Verein. Limnol. 19*: 1687-1692.

Kaushik, N. K., and Hynes, H. B. N. (1968). Experimental study on the role of autumn-shed leaves in aquatic environments. *J. Ecol. 56*: 229-243.

Kaushik, N. K., and Hynes, H. B. N. (1971). The fate of dead leaves that fall into streams. *Arch. Hydrobiol. 68*: 465-515.

Koske, R. E., and Duncan, I. W. (1974). Temperature effects on growth, sporulation and germination of some "aquatic" hyphomycetes. *Can. J. Bot. 52*: 1387-1391.

Krumholz, L. A. (1972). Degradation of riparian leaves and the recycling of nutrients in a stream ecosystem. *Univ. of Kentucky Water Resources Inst. Res. Rep. No. 57*, 36 pp.

Lush, D. L., and Hynes, H. B. N. (1973). The formation of particles in freshwater leachates of dead leaves. *Limnol. Oceanogr. 18*: 968-977.

McConnell, W. J. (1968). Limnological effects of organic extracts of litter in a southwestern impoundment. *Limnol. Oceanogr. 13*: 343-347.

MacKay, R. J., and Kalff, J. (1973). Ecology of two related species of caddis fly larvae in the organic substrates of a woodland stream. *Ecology 54*: 499-511.

Mathews, C. P. (1967). The energy budget and nitrogen turnover of a population of *Gammarus pulex* in a small woodland stream. *J. Anim. Ecol. 36*: 1-62.

Minshall, G. W. (1967). Role of allochthonous detritus in the trophic structure of a woodland spring brook community. *Ecology 48*: 139-149.

Nelson, D. J., and Scott, D. C. (1962). Role of detritus in the productivity of a rock-outcrop community in a Piedmont stream. *Limnol. Oceanogr. 7*: 396-413.

Nelson, N. (1944). A photometric adaptation of the Somogyi method for the determination of glucose. *J. Biol. Chem. 153*: 375-380.

Nilsson, S. (1964). Freshwater hyphomycetes. *Symbolae Bot. Upsal. 18*(2): 1-130.

Nykvist, N. (1963). Leaching and decomposition of water-soluble organic substance from different types of leaf and needle litter. *Stud. For. Suec. 3*: 3-29.

Padgett, D. E. (1976). Leaf decomposition by fungi in a tropical rainforest stream. *Biotropica 8*: 166-178.

Petersen, R. C., and Cummins, K. W. (1974). Leaf processing in a woodland stream. *Freshwater Biol. 4*: 343-368.

Ranzoni, F. V. (1951). Nutrient requirements for two species of aquatic hyphomycetes. *Mycologia 43*: 130-141.

Schnitzer, M. F., Sowden, F. J., and Iverson, K. C. (1974). Humic acid reactions with amino acids. *Soil Biol. Biochem. 6*: 401-407.

Sedell, J. R., Triska, F. J., Hall, J. D., Anderson, W. H., and Lyford, J. H. (1973). Sources and fates of organic inputs in coniferous forest streams. Contribution No. 66, Coniferous Forest Biome, U.S. International Biological Program, Oregon State University, Corvallis, 23 pp.

Smythe, R. V., Cappell, H. C., and Allen, T. C. (1967). The responses of *Recticulitermes* spp. and *Zootermopsis augusticollis* (Isoptera) to extracts from woods decayed by various fungi. *Ann. Entomol. Soc. Amer. 60*: 8-9.

Somogyi, M. (1952). Notes on sugar determination. *J. Biol. Chem. 195*: 19-23.

Suberkropp, K. F., and Klug, M. J. (1974). Decomposition of deciduous leaf litter in a woodland stream. I. A scanning electron microscopic study. *Microbial Ecology 1*: 96-103.

Suberkropp, K., and Klug, M. J. (1976). Fungi and bacteria associated with leaves during processing in a woodland stream. *Ecology 57*: 707-719.

Suberkropp, K., and Klug, M. J. (1980). Maceration of deciduous leaf litter by aquatic hyphomycetes. *Can. J. Bot. 58*: 1025-1031.

Suberkropp, K., Klug, M. J., and Cummins, K. W. (1975). Community processing of leaf litter in woodland streams. *Verhandl. Int. Verein. Limnol. 19*: 1653-1658.

Suberkropp, K., Godshalk, G. L., and Klug, M. J. (1976). Changes in the chemical composition of leaves during processing in a woodland stream. *Ecology 57*: 720-727.

Thornton, D. R. (1963). The physiology and nutrition of some aquatic hyphomycetes. *J. Gen. Microbiol. 33*: 23-31.

Thornton, D. R. (1965). Amino acid analysis of fresh leaf litter and the nitrogen nutrition of some aquatic hyphomycetes. *Can. J. Microbiol. 11*: 657-662.

Triska, F. J. (1970). Seasonal distribution of aquatic hyphomycetes in relation to the disappearance of leaf litter from a woodland stream. Ph.D. thesis, University of Pittsburgh, 189 pp.

Triska, F. J., Sedell, J. R., and Buckley, B. (1975). The processing of conifer and hardwood leaves in two coniferous forest streams. II. Biochemical and nutrient changes. *Verhandl. Int. Verein. Limnol. 19*: 1625-1639.

Webster, J. (1959). Experiments with spores of aquatic hyphomycetes. I. Sedimentation, and impaction on smooth surfaces. *Ann. Bot.* (London) [*N.S.*] *23*: 595-611.

Webster, J., Moran, S. T., and Davey, R. A. (1976). Growth and sporulation of *Tricladium chaetocladium* and *Lunulospora curvula* in relation to temperature. *Trans. Brit. Mycol. Soc. 67*: 491-495.

Willoughby, L. G., and Archer, J. F. (1973). The fungal spora of a freshwater stream and its colonization pattern on wood. *Freshwater Biol. 3*: 219-239.